The Printing Ink Manual

The Printing Ink Manual

FOURTH EDITION

Edited by

Dr R. H. Leach, Dr C. Armstrong, J. F. Brown,
M. J. Mackenzie, L. Randall and Dr H. G. Smith

International

in association with

SBPIM

The Society of
British Printing Ink
Manufacturers

Published by Van Nostrand Reinhold (International) Co. Ltd.
11 New Fetter Lane, London EC4P 4EE

First published 1961
Revised impression 1963
Second edition 1969
Third edition 1979
Reprinted 1984
Fourth edition 1988
Reprinted 1988, 1989

Typeset in Meridien by Best-set Typesetter Ltd, Hong Kong.
Printed in Great Britain by T. J. Press (Padstow) Ltd,
Padstow, Cornwall.

British Library Cataloguing in Publication Data

The Printing ink manual. — 4th ed.
  1. Printing-ink
  I. Leach, R.H.
  667'.5 Z247

ISBN 0-7476-0000-7

Contents

Preface

The Printing Ink Manual was first published in 1961 under the auspices of the Society of British Printing Ink Manufacturers with the object of providing an authoritative work on printing ink technology. This, the fourth edition, continues that purpose and presents a comprehensive study of the current 'state of the art' in the ink industry. For those starting in the printing ink industry it is a textbook dealing with all aspects of the formulation and manufacture of printing ink. For the ink technician it is a practical manual and useful source of reference. For printers and users of printed material the manual supplies helpful information on the nature and behaviour of ink both on the printing press and as the finished print. Readers with a little scientific knowledge will have no difficulty in using the manual, but as in previous editions, sufficient chemistry and physics have been introduced to assist the advanced technician and research scientist.

In the eight years since the last edition very substantial changes have taken place in ink technology, and accordingly the book has been entirely rewritten by a new panel of authors all of whom are currently engaged in the printing ink or associated industries. While retaining the general structure of previous editions, the chapters have been arranged in a more logical order. Those dealing with ink formulation reflect the considerable changes that have taken place in the printing industry; letterpress inks have largely been superseded by lithographic inks for both packaging printing and newspaper production, and gravure and flexographic inks are now applied to a much greater variety of substrates for packaging. Developments in manufacturing methods have also been substantial. Both milling and mixing techniques have benefited by new designs in machinery and the introduction of electronic and computerised controls.

Two important chapters have been added to this edition of the manual, one dealing with Radiation Curable Systems, the other with Health, Safety and Environment. Radiation curable systems have now become widely used for drying printing inks and are based on quite different ink technology. This technology is explained in depth together with details of the special raw materials required. Matters of health and safety at work and environmental pollution have assumed great importance since the last edition of the manual. Therefore a chapter has been included which offers a survey of the current legislation both in the United Kingdom and other countries concerning the requirements for the safe manufacture, handling and transport of printing inks and also considers the limitations in the use of raw materials.

It is hoped that with its new edition, this highly respected reference book will continue to serve as a guide for all involved in the printing ink industry worldwide.

R. H. Leach
Editor-in-Chief

ACKNOWLEDGEMENTS

The editors wish to thank Barry Hermiston of Coates Brothers Inks Ltd, who compiled the index. They also wish to thank those who supplied diagrams and illustrations not acknowledged elsewhere and those members of the Society of British Printing Ink Manufacturers who have kindly checked the ink formulations and chemical structural formulae at the proofing stage.

The Editors and Authors

Editors

Dr R. H. Leach, Editor-in-Chief
Dr C. Armstrong, Technical Director, Coates Brothers Inks Ltd.
Mr J. F. Brown, Scientific Services Manager, Mander-Kidd (UK) Ltd.
Mr M. J. Mackenzie, Deputy Technical Director, Usher-Walker plc.
Mr L. Randall, Deputy Managing Director, Usher-Walker plc.
Dr H. G. Smith, Technical Director, Fishburn Division, BASF Coatings +
  Inks Ltd

Authors

Chapter			
	1	Dr C. Armstrong	Coates Brothers Inks Ltd.
	2	Mr B. W. Blunden	PIRA
		Mr J. W. Birkenshaw	PIRA
	3	Mr J. F. Brown	Mander-Kidd (UK) Ltd.
	4	Mr M. J. Clayton	Sun General Printing Ink
	5	Mr G. B. Burdall	Usher-Walker plc
	6	Mr P. I. Ford	US Printing Ink Corporation
		Mr R. J. Tuft	Coates Brothers Inks Ltd.
	7	Dr E. Cawkill	BASF Coatings + Inks Ltd.
		Mr B. A. Ellison	BASF Coatings + Inks Ltd.
	8	Mr F. C. Wyatt	Lorilleux & Bolton Ltd.
	9	Mr G. Joannou	Coates Brothers Inks Ltd.
	10	Dr R. J. Holman	BASF Coatings + Inks Ltd.
	11	Mr R. Marsh	Coates Brothers Inks Ltd.
	12	Mr M. J. Heath	Lorilleux & Bolton Ltd.
		Mr A. J. Wild	Usher-Walker plc.
	13	Mr A. Tabbernor	Mander-Kidd (UK) Ltd.
	14	Dr E. Cawkill	BASF Coatings + Inks Ltd.
		Mr D. Easterby	BASF Coatings + Inks Ltd.
	15	Mr G. J. May	Coates Brothers Inks Ltd.
		Mr J. P. Orpwood	Coates Brothers Inks Ltd.
Index		Mr B. N. Hermiston	Coates Brothers Inks Ltd.

CHAPTER 1
The Nature of Printing Inks

Inks are coloured, usually liquid, materials for writing or printing. They are used for many reasons, primarily as a tool to convey a message and give instant decoration. Inks are extremely versatile and can be applied to a wide variety of different surfaces, whatever their texture, size or shape. Paper, plastic, metal, glass and textiles will all accept ink. Printing inks are applied to such surfaces by printing presses of various designs and are conveniently divided into five classifications according to the type of press. Ink films, unlike paint films, are very thin varying from 2 to 30 μm depending on the print process. Inks consist essentially of two components, the colourant, an insoluble solid or a dye, and the vehicle, a liquid in which the colourant is suspended or dissolved. The combined components form a fluid capable of printing from a printing press. The secondary function of the vehicle is to 'dry' and bind the colourant firmly to the substrate.

Although the primary functions of an ink are to convey a message or decorate, it is of no value if it will not transfer to the substrate nor adhere satisfactorily after printing. In addition to (1) its visual characteristics, therefore, an ink is formulated to:

(2) print by a given process;
(3) dry under specified conditions;
(4) adhere to a given material;
(5) have specific resistance properties, dictated by the intermediate processing of, and the final end use of, the printed matter.

This chapter will discuss each of these properties of an ink in turn and outline how each of the fundamental building blocks of an ink — colourants, resins, solvents and additives — affect each property. Individual chapters in this manual will deal with each of these characteristics in more detail.

1.1 VISUAL CHARACTERISTICS OF INKS

The visual characteristics of an ink are defined in terms of its colour, its transparency or opacity and its gloss and are largely determined by the quantity and nature of colourant (pigment or dye) used.

The colour of inks

Colour is a complex concept and needs to be described in three interrelated ways. Firstly, the 'hue' of a colour indicates what kind of colour it is, i.e. red, blue, green, etc. Secondly, the 'strength' of a colour is a measure of its intensity or saturation. The third property is its 'purity' which indicates how bright or dark it is.

There are many factors that affect each of these parameters to one degree or another. Clearly, the chemical structure of the colourant has a fundamental effect on its hue, but so do particle size, surface characteristics and the amount of colourant contained in a given ink film. In general, the higher the concentration of a particular colourant, the greater the strength of the colour, although in many cases an optimum concentration occurs beyond which there is no further increase in strength.

The types of resins, oils or solvents used in the ink vehicle can alter the hue or purity of a particular colourant. This may be due to the colour of the vehicle itself or because different ink vehicles develop different colours when dispersing the same colourant. Sometimes additives selected to aid dispersion can cause alterations in hue and strength.

The choice of ink vehicle can be very important in affecting the practical strength of an ink. The ink vehicle must be capable of keeping the colourant particles dispersed otherwise they may settle out and there will be a tendency towards loss of strength. Also, an ink vehicle that penetrates the substrate will take some of the pigment with it, effectively lowering its strength. Similarly, a slow drying solvent can cause increased penetration into absorbent stocks with the same end result. Generally, therefore, higher amounts of resin in the vehicle and faster drying solvents give better hold-out and greater colour strength.

The transparency and opacity of printing inks

To achieve the desired visual appearance, inks need to have varying degrees of opacity and transparency. One of the major influences on the transparency of an ink is the choice of colourant and the degree to which it is dispersed.

Different colourants behave differently towards light. The more opaque colourants have a greater tendency to reflect and refract light, and this is influenced by particle size and refractive index.

Thus, the high refractive index and the particle size of titanium dioxide ensure that it effectively reflects, refracts and, therefore, scatters light of all visible wavelengths, making it one of the most opaque pigments in use in printing inks. Soluble dyes, on the other hand, have a relatively low refractive index and small particle size and give highly transparent ink films.

The ink vehicle can influence the opacity of an ink by its pigment dispersing properties and in some cases by its intrinsic refractive index which can influence the amount and degree of light scattering. Its interraction with the substrate can also influence the transparency/opacity of the ink film.

The gloss of printing inks

The gloss of an ink is a measure of its ability to reflect incident light and depends to a large extent on whether or not the ink forms a smooth film on the surface of the substrate and masks any irregularities. When an ink penetrates a substrate it tends to lose gloss. The degree of gloss depends on the nature of the colourant, its particle size, shape and surface characteristics and the amount of resin and its ability to form a continuous film. In general, the more resin that is present in relation to the colourant, the higher the gloss will be. The choice of resin can also be critical to gloss through its ability to disperse pigment — the better the pigment dispersion, the better the gloss. Also, thicker ink films will be more likely to hide substrate surface defects.

Solvent choice is also important in controlling gloss, as it can affect pigment dispersion, hold out, the percentage of resin in the ink at its printing viscosity and the ink's flow characteristics. If the solvent is not a true solvent, a poor flowing ink can result which can affect gloss. Careful control of drying may also be needed to ensure optimum gloss development, as solvents that evaporate too fast may cause poor flow out and loss of gloss. Alternatively, too slow drying on absorbent stocks can lead to excessive penetration with a resultant loss of gloss.

Additives such as plasticisers can often improve gloss by aiding ink flow and hence giving a smoother film. Waxes, widely used in printing inks, tend to reduce gloss because they move to the surface of the ink film forming an uneven profile which scatters the incident light, creating a matt finish.

1.2 THE NATURE OF PRINTING INKS AS DETERMINED BY THE PRINTING PROCESS

The formulation of a printing ink and its physical appearance are determined greatly by the method by which the image is produced and the transfer of the inked image on to the substrate to be printed.

Flexographic and gravure inks

Inks which are printed by the flexographic and gravure printing processes are characterised by their extremely fluid nature. They are generally termed liquid inks.

The fluidity of gravure inks allows the recessed cells of the gravure cylinder to be filled rapidly. Similar fluidity is required for the flexographic process as the inking systems depend upon an engraved cylinder metering the ink on to the raised rubber image. Both inking systems involve only a short time and distance between the bulk volume supply of the ink and the accurately metered film of ink on the printing image. Therefore, highly volatile solvents can be used and most flexographic and gravure inks are fast drying and the ink is transferred to the substrate as quickly as possible.

Liquid inks are printed on a wide range of substrates from absorbent stocks, where they dry by absorption and evaporation, to non-absorbent ones, where they dry through their film forming ability and evaporation drying to which they are particularly suited.

Neither print process puts any significant constraints on pigment choice, but flexography requires careful selection of the solvent which in turn limits the choice of resin systems. Resin choice, together with specific additive selection, can influence the printability, with certain systems giving better inking and transference.

Lithographic and letterpress inks

When compared to the fluid nature of flexographic and gravure inks, lithographic and letterpress inks are much more viscous and paste-like. Although the mechanisms of image generation of the latter processes are very different, the thick nature of these inks is determined primarily by the method by which the image is inked and then transferred to the substrate.

Both processes use a series of inking rollers which can meter accurately a very thin, uniform film of this viscous ink to give the characteristic high image definition. Because of the time this takes and the large area of ink film exposed on the rollers, lithographic and letterpress inks have to be formulated on relatively non-volatile solvents so that they do not evaporate during the long transfer sequence through the roller chain on the press.

In the relevant chapters of this manual, the reader will appreciate how few demands the letterpress process makes on the raw materials used compared to lithography despite their outward similarities. The choice of raw materials for lithographic inks, including pigments, vehicles and additives, is greatly influenced by the printing process mainly because of the presence of water and the planographic imaging technique used.

Screen inks

In screen printing, the ink is forced through the open areas of a stencil supported on a mesh of synthetic fabric, stretched across a frame. The ink is forced through the mesh on to the substrate underneath by drawing a squeegee across the stencil. The inks have to remain fluid and coherent on the mesh before printing, transfer through the mesh without sticking it to the substrate, flow to make the ink film continuous and so avoid the pattern of the mesh but maintain the image, and then dry rapidly once applied to the substrate. Inks which are able to print by this method are characteristically of household paint consistency. They are intermediate in viscosity between the fluid, liquid, flexographic and gravure inks and the paste-like lithographic inks.

Screen inks can be formulated to dry by a variety of different methods depending upon the type of substrate, which can range from absorbent papers and boards to non-absorbent plastics and metals. The most widely used technique is evaporation, although there are limitations on the volatility of the solvent which can be used. They tend to be intermediate between liquid and paste ink in order to ensure adequate print definition and stability on the screen.

1.3   THE DRYING CHARACTERISTICS

An ink in its supply form is a liquid but after application it is required to change to a solid. There are many mechanisms by which the liquid ink is transformed into a solid permanent mark. This change of state is referred to as ink drying and can be physical, chemical or a combination of both processes.

Absorption drying

An ink dries by absorption when it penetrates by capillary action between the fibres of a substrate and also when it is absorbed into the substrate itself. The ink remains liquid but because of the degree of penetration, is effectively 'dry'.

Oxidation drying

An ink dries by oxidation when the oxygen in the atmosphere chemically combines with the resin system converting it from a liquid to a solid. The systems used generally depend upon catalytically induced autoxidation. The process is relatively slow, taking several hours, and the ink film will remain tacky for a long time. To overcome this problem on absorbent substrates, the technique of quicksetting has been developed which combines absorption drying with that of oxidation. The inks contain a fluid component which can separate from the rest of the inks and be preferentially absorbed into the substrate leaving a film of ink on the surface which is dry to the touch but still soft and moveable. However, as autoxidation proceeds, the ink film becomes hard and tough.

The oxidation process can be affected by a number of factors, including pigments and additives, as well as the printing process and the substrates. The inkmaker's task is to produce carefully balanced formulations which ensure that high quality printing and high production speeds can be maintained whilst the drying process proceeds at a satisfactory rate.

Evaporation drying

A wide range of inks is formulated to dry by the physical removal of volatile solvents from the ink formulation, leaving the resin material behind to bind the pigment to the substrate. The rate of drying depends upon the evaporation rate of the solvents selected and also the affinity of the resin system for the solvents. Generally, the greater the affinity of the resin for the solvents the slower the rate of solvent release. This can greatly affect the properties of the ink especially with respect to print-ability, drying speed and retention of solvents in the ink film. Each type of ink requires a very careful balance of properties if the characteristics of the ink are to meet the requirements of the printing process and the end use of the printed article.

Chemical drying

Oxidation drying has already been explained, but there are other chemical reactions used by the inkmaker to convert a liquid into a solid. Some systems comprise a polymerisable compound which requires the presence of a catalyst, and sometimes heat, to set off a chemical cross-linking reaction. Other rely on the reaction of at least two different chemical types which when mixed together combine to produce a solid that binds the colourant to the substrate. Each of the various systems used has its own specific conditions under which the chemical reactions occur and give the dry ink films required. They can be two-part inks which have to be mixed prior to use and once mixed they have limited usable life. Alternatively, systems are available which are stable under normal ambient conditions. When in use, the chemical reaction is initiated and the process turning the liquid to a solid proceeds.

Radiation induced drying

There are several forms of radiation used to dry inks. These include ultra-violet radiation, infra-red, electron beams, microwave and radio frequency. Each drying method determines the chemistry of the ink vehicle and the nature of the printing ink used.

Ultraviolet, or UV, drying employs a process known as photopolymerisation; UV inks contain photoinitiators which absorb UV radiation and in so doing produce highly active chemical compounds known as free radicals. These free radicals initiate a chain reaction with the ink vehicle, which rapidly polymerises and changes from a fluid to a highly cross-linked solid, the dried ink film. While there are other photochemical reactions which can be used as the basis of UV inks, the majority dry by this mechanism. It is the rapid cure of UV that is its major benefit and this has been exploited principally in the lithographic printing of cartons to eliminate the extended drying of the autoxidation process.

Short and medium wave infrared radiation is used as a means of efficiently providing heat and is widely used to bring about the evaporation of volatile solvents. It can also be applied in lithographic printing to assist penetration and to speed up the autoxidation of conventional quicksetting inks.

Electron beam is similar to UV but uses a beam of high energy electrons to generate the free radicals needed to bring about rapid polymerisation of the ink vehicle. Inks cured by this technique are very similar to UV inks but do not include photoinitiators.

Microwave and radio frequency radiation are used to dry inks that contain a high proportion of very polar molecules such as water. Radiation is absorbed by the polar molecules and the ink heats up very rapidly. The water evaporates and leaves a solid ink film behind.

1.4   THE ADHESIVE NATURE OF PRINTING INKS

One of the fundamental functions of an ink is that it adheres and binds the colourant to the substrate on which it is printed, and keeps it there during the lifespan of the printed product.

Colourants have little or no effect on the adhesive nature of the inks, although if they are not adequately bound by the vehicle, maybe because of poor dispersion, the ink will not exhibit good adhesion.

It is the type of resin used in the ink that will, to a large extent, determine its adhesive characteristics. On absorbent substrates, adhesion is influenced by the degree of ink penetration. Pigments do not tend to be absorbed and if too much of the resin is drawn into the substrate, this can leave the pigment underbound. Under these circumstances, the pigment can powder off.

On non-absorbent substrates, adhesion is primarily controlled by the film forming ability of the resin and its molecular affinity for the substrate. The choice of resin for a specific substrate is therefore extremely important, although its performance can be modified by applying other techniques. The correct blend of solvents can influence adhesion by creating better wetting of the substrate and improved flow-out. Some solvents can actually be used to soften the surface of the substrate enabling the ink film to key more readily.

Adhesive properties can also be influenced by using surface active and chemical additives. Such materials modify the surface of the substrate either through chemical reaction or by physical interaction.

1.5 THE RESISTANCE PROPERTIES OF PRINTING INKS

All inks have to resist certain forms of chemical and physical attack during their 'life span'. Firstly, the raw materials must withstand the manufacturing process used to make the ink in the first place. Having manufactured the ink, it must be able to withstand the rigours of the printing process. For example, it may have to perform under conditions of high shear and in the presence of fount solutions in lithographic printing.

After printing, the ink may then have to withstand cutting, creasing gluing and forming before the printed article is finally completed and ready for use. The ink film may then be subjected to new influences such as filling with abrasive products and heat sealing with heated jaws in direct contact with the ink film. The ink may also have to withstand the chemical nature of the packaged product or the environmental conditions in which it is placed. By their very nature, inks need to have specific chemical and physical resistance properties.

Lightfastness

The degree of resistance to light will depend on the exposure, application and life expectancy of the ink. The action of light can cause a colour to become weaker, dirtier and/or change its shade.

The light resistance of an ink is primarily determined by the lightfastness of the colourant used. Colourants can be placed in broad classes of lightfastness and before attempting to use a colourant in any application, its lightfastness against a shaded scale, i.e. blue wool scale, should be checked. A few pigments, including carbon black, ultramarine and some iron oxides, can be considered permanent, but the remainder are all to some extent fugitive.

In general, the lightfasteness of a pigment decreases with its degree of dilution by other pigments, especially white. Highly concentrated coloured inks have better resistance to light because of the self-masking action of the pigments. Hence, inks of pastel shades are generally not as lightfast as the stronger colours.

The lightfastness of a colourant will vary according to the type of resin into which it is dispersed. The amount of resin used can also have an effect on lightfastness. In general, the greater the amount of binding material present the better is the resistance to light. Thus screen printing, which can print a thick film of ink with a high ratio of resin to colourant, produces the most lightfast image of all the printing processes.

Heat resistance

Printed packages sometimes have to resist excessive heat in their processing or application and the choice of resin and pigment are both critical to the heat resistance properties of the ink.

Pigments vary in their ability to resist heat. Heat can dirty a colour, cause strength loss and alter shade. Pigments must be evaluated and chosen so that they will withstand the temperatures encountered both during manufacture and end use of the printed article.

Abrasion resistance

Most, if not all, printed articles are subject to different forms of rub and abrasion during their life cycle. All types of printing inks are formulated with this in mind.

The degree to which a dried ink film can resist abrasive forces depends greatly on the degree to which the pigment is bound. The more resin that is present the greater is the ink's resistance to abrasion, and the hardness and flexibility of the ink film will affect its ability to withstand abrasion and mechanical wear. Additives such as waxes, which promote surface slip, are used extensively to increase the scuff resistance of inks. By their very nature, most printing inks contain some wax of some kind.

Product resistance

Ink in its many applications in the packaging field has to be able to withstand contact with the substances packaged. Colourants vary widely in their resistance to acids, alkalis, oils, fats, detergents and other substances and must, therefore, be selected with the end use of the ink in mind. Resins, also, have differing resistance properties and must be evaluated in conjunction with the colourants to determine their suitability for a particular application. Specific additives can be used to impart specific product resistance properties. For food packaging work, ink components must be selected to avoid odour and taint problems.

Weathering

An ink will have to withstand attack from physical and chemical forces when it is exposed to the weather, or chemical elements. Pigments must be selected for resistance to the environment and in the case where specific chemical contaminants are present, it is essential to choose pigments which will resist these. Some pigments become more fugitive when exposed to moist or acid conditions. Attempts to correlate accelerated weathering tests with actual conditions are continuously being made, but, in general, they can only be used as a guideline. Reliable ink formulations are arrived at by careful selection of both pigments and binders, in combination with detailed formulation experience.

CHAPTER 2
The Printing Processes

Worldwide the printing industry, together with its associated supplier industries, represent an important sector in all economies. In the developed economies printing and its related industries are likely to rank in the first 10 industry sectors by size. When all branches of printing are combined with those of publishing and the related industries of packaging, paper and board making, ink making and machinery manufacturing, they represent a sector equal to such industries as aerospace and automobile manufacturing.

In the United Kingdom the structure of the printing industry consists of several employers' trade associations, trades unions, the trade associations of the supplier industries and industry bodies such as PIRA, The Institute of Printing, the Worshipful Company of Stationers and the NEDO Printing Industries Economic Development Committee. In addition there is a network of educational institutions serving the printing sector and emergent bodies such as CICI — The Confederation of Information Communication Industries.

In the sixteenth and seventeenth centuries one could talk of the printing industry in precisely defined terms. Business was highly regulated and there existed only one process of printing. Today the picture is very different. The role of the printer is changing and it is not only in terms of technology. The relationship between the printer and his customer, particularly in publishing, is evolving under the impact of new communications technology. The newspaper sector has undergone a revolution in the last 25 years and this continues today. The packaging printer is now part of a sophisticated multi-disciplined distribution business. This fast changing environment causes the printer to make heavy demands on his supplier industries, and not least the manufacturers of printing ink.

There are in the region of 6000 firms engaged in general printing, manufactured stationery and the printing of packaging in the United Kingdom. In addition to this there are several hundred newspaper companies and a similar number of publishing houses. These companies employ something in excess of 300 000 persons. In 1983 the paper and paper products, printing and publishing sectors spent well over £20m. on research and development.

The main printing processes are:

(1) Letterpress, which is a relief printing process. The image area is raised above the non-printing areas, is inked by rollers and pressed into contact with the substrate.

(2) Lithography, which uses a flat printing plate. The image area is ink receptive, while non-printing areas are wetted by water and repel ink.

(3) Gravure, where the image is sunk into the plate surface. The entire plate surface is flooded with ink and the excess removed with a doctor blade. Ink-filled cells remain and transfer the ink to the substrate.

(4) Flexography, which is a relief printing process. This has some similarities to letterpress but is a rotary process using rubber or photopolymer plates and an inking system suited to low-viscosity solvent-based inks.

(5) Screen printing, which employs a stencil principle. A rubber squeegee is used to push ink through the stencil on to the substrate.

There are other minor processes such as intaglio or recess printing and collotype which are rarely used except for special purposes.

Of rapidly increasing significance are the so-called non-impact printing processes such as electrophotographic printing (commonly used in photocopiers and laser printers) and ink jet printing. The common element in all these systems is that the image to be printed is computer generated during the printing process.

2.1 THE LETTERPRESS PROCESS

The letterpress process requires a relief printing image carrier. That is, the areas to be printed are raised above the blank parts i.e. non-printing areas. A film of ink, having sufficient tack and body to adhere to the raised image surface, is applied to it using rollers.

The substrate to be printed is brought into contact with the inked surface and sufficient pressure is applied to transfer the ink from the face of the image carrier to the surface of the substrate. In order to achieve a good and even transference of ink the pressure must be uniform, in proportion to the area, over the whole image surface (see Fig. 2.1).

Press configurations

There are three main types of letterpress press:

(1) platen;
(2) flat-bed cylinder;
(3) rotary.

The platen press

A platen press achieves impression by bringing two flat surfaces into contact. The forme and impression surface may be either in a horizontal

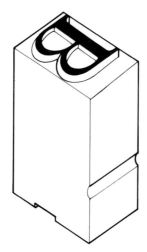

Fig. 2.1 Letterpress type character

or vertical plane. There are two types of platen press: the clamshell in which the platen rocks up against the forme; and the sliding platen in which the platen is first positioned parallel with the bed and then drawn up against it (Fig. 2.2).

   With this method the whole of the forme is under impression at the same time. As it requires a considerable force to transfer ink from a forme to paper clearly and smoothly, it naturally follows that machines of this class are strictly limited in size.

Flat-bed cylinder press
Flat-bed cylinder machines have the forme placed on a flat bed, usually but not necessarily in a horizontal position and the paper is fed round a

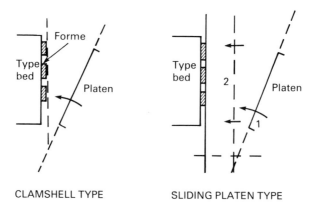

CLAMSHELL TYPE   SLIDING PLATEN TYPE

**Fig. 2.2 Platen presses: (a) clamshell type; (b) sliding platen type: (1)
platen moves to vertical position; (2) parallel approach to forme**

cylinder. The rolling action results in only a small portion of the forme being under impression at one time, so much larger machines can be successfully used. As, however, the type-bed travels to and fro under the cylinder, and as its direction must be changed at each end of its travel, there is a definite limit to the speed, measured in output of printed sheets, at which such machines can run. The impression cylinder usually makes two revolutions for each impression. It prints while making the first revolution and is raised during the second to clear the forme and permit inking on the return stroke. The type bed is driven back and forth by what is basically a crank mechanism. This would cause the bed to vary in speed during the impression stroke (giving changes in ink transfer) and consequently ingenious gear mechanisms are used to avoid this. Some presses only make one revolution for each impression. This is achieved by using an impression cylinder with a circumference larger than the print length. The type bed moves at high speed back to the start position during the passage of the non-printing part of the impression cylinder. This is cut away to clear the type bed, so it does not have to be continually raised and lowered (see Fig. 2.3).

There are variations in design to print two colours or print both sides of the sheet (perfecting). A two-colour press (Fig. 2.4) consists of two printing units, each with an impression cylinder and inking system. The bed carries two formes, one for each colour. After the first colour is printed, the sheet is taken from the first cylinder by a transfer cylinder and is conveyed to the second impression cylinder.

A perfecting press is similar to a two colour press in general construction except that there is no transfer cylinder. The sheets are taken directly from the first cylinder to the second, being turned over in the process.

Rotary presses

Rotary presses use two cylinders, one carrying a curved plate specially made to fit the particular cylinder and the other having the same circumference as the surface of the plate (Fig. 2.5). As both these

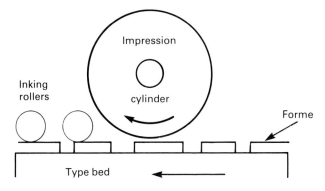

Fig. 2.3 Flat-bed cylinder press arrangement

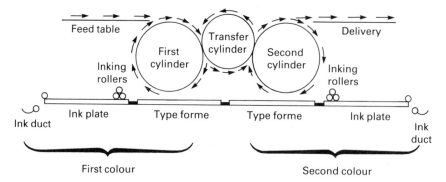

Fig. 2.4 Two-colour press arrangement

cylinders are continuously revolving, with the paper between them, very high speeds can be attained.

Rotary presses may be sheet fed or web fed. Sheet fed presses operate on either the unit principle or the common impression principle.

In the unit system, for each impression cylinder there is a plate cylinder, and the number of colours printed is governed by the number of units on the press. Two units would print two colours, five units would print five colours, and so on. Presses of this type can be assembled in any desired number of units.

The common impression system has only one impression cylinder on which one to five colours may be printed at one time. Up to five inking

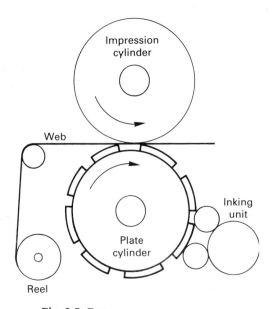

Fig. 2.5 Rotary press arrangement

systems are arranged with the same number of plate cylinders around a common impression cylinder.

Web fed presses print on a continuous roll of paper on both sides, as the web passes from one printing cylinder to another.

For single-colour printing the web is perfected, or printed both sides, by passing it through a pair of these units, generally known as a printing couple (Fig. 2.6). In newspaper printing, multiple printing couples can be arranged together, each perfecting a single web, with the multiple webs being assembled in a common folder.

In most high speed multi-colour rotaries, two or more plate cylinders are grouped around a single common impression cylinder, with successive colours being printed almost simultaneously. Special inks and dryers are used. These presses are operated at speeds of more than 1500 feet per minute. Paper is fed with automatic on-the-run splicing of one roll to another. The delivery mechanism can be of three types — sheet, roll or folded — but in most cases the printing is delivered folded. Attachments to perform other finishing operations have been designed.

Letterpress forme production

Type matter for letterpress printing can be produced by two main methods:

(1) Line casting, in which lines of characters are cast in a lead alloy. A single line is referred to as a slug.

(2) Single characters, which may be assembled by hand or cast using Monotype machines.

The galleys of type from the casting machines are assembled into a chase (a metal frame) and locked into position. The result is the printing forme.

The casting of type by either method is a precision operation. Type

Fig. 2.6 Perfecting web press

metal is an alloy made from lead, antimony and tin, different proportions being used depending on the application. A slug is cast from 11−12% antimony, 3−4% tin, the balance being lead. The result is soft and not very hard wearing but perfectly adequate for use in making stereos for newspaper printing. Monotype characters are cast from 15−16% antimony, 6−7% tin, the rest being lead. The result is harder and more wear resistant. Still harder alloys can be made by raising the antimony and tin levels higher.

The composition of the lead alloy, the temperature and rate of cooling of the mould must all be well controlled in order to cast good type. Also the dimensions of a character must be closely controlled otherwise characters may move about when printing, leading to uneven impression.

Pictures are printed from plates produced by some form of etching process, and duplicate plates are often also made.

Original plates

Original plates are produced by process engraving. This is a process of photographically producing an acid resist on a metal plate and then etching away unprotected parts by means of an acid. The etched parts become lower than the protected parts which remain in relief and so become the printing part of the plate.

Original plates can be classified in two groups — line and halftone.

Line plates

These are used for the reproduction of black-and-white line orginals. The plate is made of zinc, etched in nitric acid and then mounted so that it is type high.

Line plates are used for advertisements in newspapers, label work, letter headings, diagrams and work involving clear lines.

Colour line plates are also used — particularly in childrens' books.

Halftone plates

These are used to reproduce continuous-tone illustrations. The copy can be a photograph, wash drawing, water-colour or oil painting.

Halftone plates are made of copper, zinc or magnesium, and mounted on blocks to bring them to type height. The method differs from the line plate by the use of the halftone screen. Screens vary from 26 to 40 lines/cm for coarse paper, 40−50 lines/cm for calendered papers, and 50 lines/cm upwards for art papers. Electronic engraving methods are also available for these plates.

Plates are generally etched by what is known as the powderless etching process. A blank plate is coated with a light sensitive resist which is exposed to a negative of the required image. The exposed image areas become insoluble. The plate is developed and baked to harden the resist. The plate is then mounted in the etching machine which uses an arrangement of sprays or paddles to throw the etching solution forcefully against the plate. Additives are present in the etchant which protect the side walls of the image as etching progresses. The rate of etch downwards is therefore greater than the rate sideways (see Fig. 2.7).

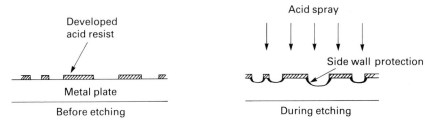

Fig. 2.7 Etching a letterpress plate

The concentration of etchant, additive, spray force, temperature and time of etch must all be accurately controlled in order to achieve repeatable results. When etching line plates, the additive used is an oily substance which acts as a physical barrier on the side walls. The spray force and heat of reaction all have an effect. Copper-powderless etching, however, depends on a chemical reaction between the additive (thiourea) and cuprous chloride — a gelatinous precipitate is formed which protects the side walls. The cuprous chloride is a product of reaction between the ferric chloride etchant and the copper plate.

Plates from original copy can also be made from plastic materials by photo-chemical methods, e.g. Dycril and Nyloprint plates. These are particularly useful as they can be used flat on a platen or flat-bed cylinder machine, or curved and used on a rotary press. One-piece 'photopolymer' plates for rotary printing are called 'wrap-around' plates.

Duplicate plates

Letterpress would be very restricted if it could only print from original type and photo-engravings. In package printing, multiples of the same subject need to be printed on the same press. National advertisements have to be printed in different publications at different times. This is made possible by the use of duplicate plates. There are several kinds of duplicate plates, but they are all produced from existing relief plates or complete formes by making an intermediate, called a matrix.

Duplicate plates have many advantages: they can be made to run for 500 000 impressions (as against 50 000 for type); some are lighter and less bulky than original materials; duplicate plates make it possible to preserve original materials; they can be made curved as well as flat.

Duplicate plates can be classified into four main kinds: stereotypes; electrotypes; plastics and rubber plates, but other types have been tried (Fig. 2.8).

To produce a metal *stereotype*, a mould is produced from the original plate. This is then placed in a casting box, surrounded by bearers. The box is closed and a molten mixture of lead, tin and antimony poured in. The casting takes only a few minutes, and the stereotype can be removed as soon as the metal has had time to cool. It can be nickel- or chromium-plated to improve its wearing qualities.

Flat stereotypes can be used on platen or flat-bed presses, but the main use of stereotypes is in newspaper printing; here, curved stereotypes are made to fit on the rotary cylinders.

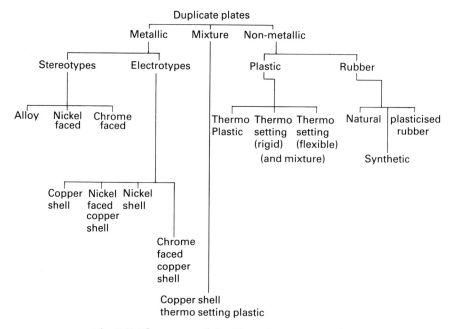

Fig. 2.8 The range of duplicate letterpress plates

Another kind of duplicate plate, an *electrotype*, is made by electro-plating. The original plate is coated with graphite to aid separation, and then a thermoplastic mould made from it under heat and pressure. This is coated with silver to make it electrically conductive, and then electro-lytically plated with nickel or copper. The thin plated shell is then strengthened by adding backing metal. These plates have high quality and long press life — 250 000 impressions from a nickel electrotype and up to 500 000 from a chromium-plated one.

Rubber and plastic plates are made from moulds in hydraulic presses under heat and pressure. They have the advantage of light weight, good wearing qualities, low cost and high speed of manufacture, but have limitations in the rendering of fine halftone images.

Rubber and plastic plates are used mainly for bag printing, business forms, books, tags, foil wraps, wrapping papers, continuous stationery and paperbacks.

Make-ready

Letterpress printing requires 'make-ready' to distribute pressure evenly over the printing plate, and to compensate for small variations in type height or plate mountings. Solids require greater printing pressure than highlights to ensure proper ink transfer and a number of 'pre-make-ready' systems have been developed to achieve this fairly easily and more quickly than the old fashioned technique of sticking tissue paper in various thicknesses on the impression cylinder or platen. These all involve taking a print from the forme and converting this into a slight relief image by some means. One method uses a resin powder which is dusted over the freshly printed sheet and therefore sticks to the ink film. It is then heated in an oven causing the resin to fuse, swell and harden. More powder adheres to solids than highlights so the end result is that the resin is thickest in the shadow areas. This sheet is then placed on the impression cylinder or platen, in register with the forme, and is known as an overlay. Further make-ready by hand is usually necessary. Underlays are also possible in which case the additional sheet is placed under the printing plate. A good make-ready is essential for high-quality letterpress printing.

Substrates

A wide variety of substrates but not plastic films can be printed by letterpress provided the press and inking levels are suitably chosen. For halftone work, the screen ruling must suit the paper roughness. Thus 22 lines/cm (55 lines/inch) is used with newsprint and up to 60 lines/cm (150 lines/inch) for high-quality colour work. See Chapter 5 for details of the types of inks used.

Applications

Each type of machine has its particular uses.

Platen and flat-bed cylinder machines are in much more extensive use in the jobbing or commercial sections of the industry than rotary presses, owing to their greater flexibility.

Automatic platen machines, such as the Heidelberg, are used mainly for simple jobbing work such as letter headings and business cards. The feeders will deal with almost any kind of paper with equal ease. Automatic platens can also be obtained for embossing, cutting and creasing, carbon printing and numbering. Specially adapted machines are obtainable for gold and metallic blocking.

Flat-bed cylinder machines, are capable of producing a wide range of work, at speeds of about 5000 impressions/hour in the case of small presses, while large machines are slower because of the mechanical inertia involved.

Sheet-fed rotary machines give tremendous mechanical advantages over the traditional flat-bed machines and permit a greater volume of work to be produced in a given time — about three times more sheets per hour. Perfectors — that is presses which are able to print both sides of a sheet at one pass through the press — have proved to be popular for bookwork.

Web-fed rotary presses have been widely used in newspaper production owing to their high speed output. Any number of units can be coupled together and the several webs brought into the same folder so that complete copies cut, stitched and counted, if required, can run out of the delivery end of the press.

The rotary type of machine is also extensively used for the production of office stationery and continuous stationery. As the size of the page or copy must of necessity be a subdivision of the circumference of the cylinder, these machines cannot be adapted to such a wide range of jobs as other methods of printing.

Narrow-width rotary presses are widely used for label printing, currently a rapidly growing area.

2.2   THE OFFSET LITHOGRAPHIC PROCESS

Lithography is a *planographic* process. The image and non-image areas are in the same plane on the printing image carrier.

While it is fairly simple to define an image area by raising it above the background, and inking it with a roller, as in letterpress, when all areas are on the same level, chemical treatments have to be applied to ensure that ink adheres to some areas and not others.

Lithographic printing ink will adhere to a dry surface, but not to a wet one. Lithographic plates are treated so that the image areas are water repellent and ink accepting, while the non-image areas are water accepting.

Thus the plate must be damped *before* it is inked (although there is one newspaper press where this is reversed). Water will form a film on the water-accepting areas, but will contract into tiny droplets on the water-repellent areas. When an inked roller is passed over the damped plate, it will be unable to ink the areas covered by a water film — the non-image areas — but it will push aside the droplets on the water-repellent areas — the image areas — and these will ink up.

The process is called *offset* lithography because the inked image on the plate does not print directly on to paper, but is first 'offset' on to a rubber blanket, and thence transferred to paper. This process was first used for printing on tinplate, to overcome the difficulty of printing directly from metal on to metal. Originally, lithography was direct, but the image on the plate was rapidly worn away by contact with paper. The direct process is, however, used for the printing of some newspapers.

The printing unit

Litho presses may be sheet or web fed. Several different designs exist for sheet-fed presses, the basic one being shown in Fig. 2.9.

This shows a single-colour press consisting of a plate, blanket and impression cylinder. These three printing cylinders, together with the inking and damping systems, make up the printing unit.

The plate is clamped round the plate cylinder. At each revolution it is contacted first by the damping rollers, then by the inkers. The blanket is clamped round the blanket cylinder, and the paper passes between the blanket and impression cylinders, but does not contact the plate.

The damping system

A damping system supplies water, or fountain solution to the plate. It is necessary for the damping system to meter a carefully controlled water

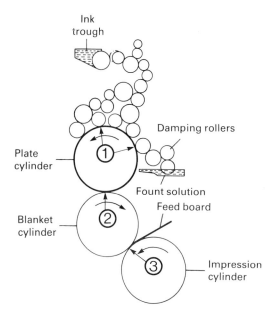

Fig. 2.9 Single-colour rotary offset machine: (1) damping and inking; (2) plate/blanket nip; (3) blanket/paper nip

film to the plate. There are various forms of damping units, for example a fountain reservoir and rollers, rotary brushes and canvas flaps.

One of the most common systems consists of a reservoir for the fountain solution, and five rollers. The fountain roller is partly immersed in the reservoir (called the fountain). As this rotates it carries a film of water up to the second roller and the distributor. The distributor then applies the fountain solution to two plate-damping rollers, and has a reciprocating action as well as rotational motion. The amount of water fed to the plate dampers is governed by how long the feed roller stays in contact with the fountain roller.

The two main functions of the fountain solution are:
(1) to keep non-image areas clean and free from ink;
(2) to help minimise the amount of water necessary to obtain a clean print and so assist in the maintenance of ink/water balance.

When a plate leaves the plate-making department, it should have on its surface a fine film of gum arabic or similar desensitiser. This makes the plate surface extremely hydrophilic (water attracting). If the plate remained in this condition, it would be possible to run with water only in the fountain. However, the gum film is very fragile and is easily worn away by the contact of roller and the blanket. When this occurs the non-image areas tend to accept very minute amounts of ink which print as 'scum', a light tint over non-image areas.

The desensitised layer must then be continually replaced if scumming is to be avoided: this is achieved by using a desensitiser. It is the desensitiser which is the main active component of any fountain solution.

The fountain solution may be made up from a concentrate, supplied ready to mix with water. Some press operators still however, mix their own, usually consisting of gum arabic, phosphoric acid, and water, which produces an acid solution. Some of the fountain solution concentrates are formulated to give an alkaline solution, whereas others may be slightly acidic or neutral.

In addition to the desensitiser which all fountain solutions contain, other additives may be included to improve the performance such as:

(1) a pH buffer, to maintain the solution at the desired pH under adverse conditions;

(2) preservatives to prevent the growth of fungi and bacteria.

In addition to these components, other additives may be used to fulfil a particular task. One example is designed to prevent dry-oxidation scum, which occurs when a thin film of oily material from the ink is allowed to spread over the surface of the water film on the non-image areas. When the press is stopped and the moisture, on which the oily film floats, is allowed to evaporate, the film is deposited on the non-image areas of the plate, and this has the effect of forming small image areas which print as a scum when printing is resumed.

The usual way of preventing this fault is to remove the oily film and replace it with a film of gum arabic by gumming up at each press stoppage. However, by altering the interfacial surface tension characteristics between the moisture and oily film, the additive prevents the spreading of the oil film from the ink, making it unnecessary to gum the plate at each press stoppage.

A further form of scum may be encountered when printing some coated papers which contain casein; if this is not fully hardened, it may remain soluble in water and can therefore be removed by the fountain solution. The casein in the fountain solution causes the non-image areas to become sensitised and print as an overall background scum. On a multi-colour machine the scum is generally most severe on the later units. The solubility of the casein in the fount solution can be reduced by adding Epsom salts.

The addition of alcohol to the fountain solution has become very popular during recent years, particularly with the advent of vibratorless dampening systems. It is claimed that a higher rate of evaporation of moisture from the plate surface between dampening and inking gives less emulsification, improved drying of inks, quicker ink/water balance and easier maintenance of this balance. Other claims made are that less fluff and debris is picked up from the plates and better trapping of ink on the paper is obtained. However, in addition to the extra cost of using alcohol in the fountain solution, there are number of drawbacks. Refrigerated circulation tanks, measuring devices and mixers are required to maintain the alcohol content.

Many printing problems are associated with the pH of the fountain solution. High pH values can cause excessive ink/water emulsification problems, while low pH values can cause ink drying difficulties, loss or sharpening of image on the plate and poor definition of dots. The ideal running pH is generally regarded as being between 5 and 7.

The offset blanket

In offset printing the image is printed from a rubber blanket on to paper. The resiliency of the blanket makes it possible to transfer fine halftone images to rough-surfaced papers, giving much greater detail than could be obtained by a direct printing process, and widening the scope of lithography.

Blankets are made in various thicknesses and constructions, and the top layer must be suitable for the kind of ink used.

Press configurations

Sheet-fed presses can be made to print from one to six colours. The common types are one-, two- and four-colour, with three-, five- and six-colour available for special requirements. Perfecting presses are also available.

Figures 2.10 and 2.11 show how two-colour and perfecting presses are built up. A four-colour press can either consist of two, two-colour units linked together, or a number of single colour units placed in line (see Fig. 2.12). By assembling basic printing units together any number of colours can be printed. A combination of single-, two- and four-colour presses, plus a prefector, can be used to print efficiently any number of colours (by using different machines to print successive colours).

All sheet-fed machines have a sheet-feeding mechanism attached to the back of the press and a pile delivery unit attached to the front. To complete the picture, a sheet-fed printing press must have a method of transporting the sheet from feeder to printing unit, around the printing cylinders and finally, to the delivery pile (Fig. 2.12, p. 25).

The small offset press

In some ways, these machines are the big brothers of duplicating machines. Their size is generally in the A4−A3 range. Because they are intended for use in in-plant printing works and other small companies, they are simple machines to operate, and yet, with care, can handle a wide range of work. Some machines are highly automated to optimise their use for short runs. These machines incorporate features for automatic plate change, wash up and start up and can be programmed to print from several plates in succession without any manual intervention.

Larger sheet-fed presses

Heavy offset machines take the larger sheet sizes up to a maximum of 1372 × 1956 mm (54 × 77 inches). They have a much heavier construction, are more highly specialised, and are usually bought with a particular type of work in mind. Whereas small offset machines can be operated by relatively unskilled staff, these machines require much greater knowledge and training.

Web offset presses

Three types of web offset press are in use, the most common being blanket-to-blanket perfectors with the web going either vertically or

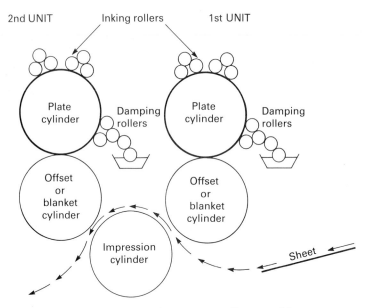

Fig. 2.10 Two-colour rotary offset machine

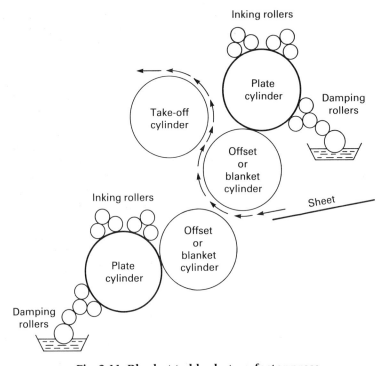

Fig. 2.11 Blanket to blanket perfector press

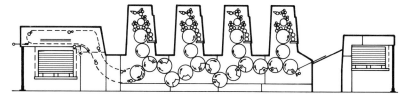

Fig. 2.12 A typical complete four-colour press

horizontally through the units. Secondly, there are common-impression units, and finally the three-cylinder principle press.

Blanket-to-blanket press
In these the blanket cylinder acts as the impression surface for the printing on the opposite side of the web. Any number of identical units may be mounted in-line to form a single printing press, although five is normally the maximum (see Fig. 2.13).

A five-unit press would enable two webs to run, with black both sides on one, and four colours both sides on the other. Other combinations can be used if needed.

The vertical web-fed printing unit is mostly found in newspaper plants. On the other hand, commercial web-offset printers find the horizontal-web blanket-to-blanket press the most suitable.

Common-impression drum presses
In these, up to five plate/blanket units are arranged circumferentially around a single-impression drum of large diameter and consequently only one side of the web is printed in one operation. (see Fig. 2.14).

Three-cylinder presses
This type of press is the least popular but is used for packaging, continuous stationery or small jobbing work. (see Fig. 2.15).

Lithographic platemaking

A wide variety of plates are used for offset-litho printing. The choice of plate is governed primarily by the length of run required, and also by cost and technical factors such as time and complexity of preparation.

A light-sensitive coating is the basis of most lithographic platemaking, enabling the image to be formed on the plate photographically. In some

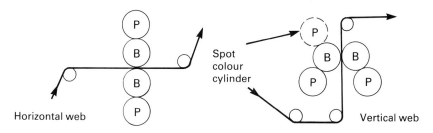

Fig. 2.13 Blanket to blanket perfector press units

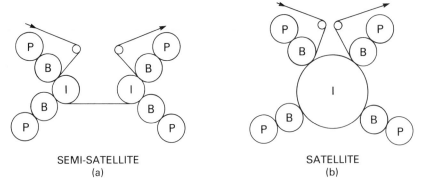

SEMI-SATELLITE   SATELLITE
(a)   (b)

Fig. 2.14 Web printing units with common impression cylinders: (a) semi-satellite; (b) satellite

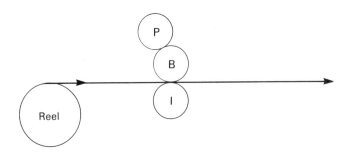

Fig. 2.15 Three-cylinder principle

cases the coating is put on at the printing works by whirling in a machine, or by wiping on with a sponge. In others, the plates are supplied ready-coated by the manufacturer — these are called pre-sensitised.

Presensitised surface plates

These plates are the most common now in use. They are produced using a number of different base materials including plastic, paper and aluminium. Various surface treatments can be applied to aluminium to modify its performance as a non-image metal, the most common being that of graining. The detailed nature of the grain depends on the method of producing it. Many plates are now available which have a hard porous layer formed in the aluminium by anodising. These plates give increased wear resistance with easier handling on the press.

A light-sensitive coating is applied to the base material by the plate manufacturers. Several types of pre-sensitised coating are now used, the two main ones being diazo and photopolymer. These can be either positive or negative working.

For negative working plates the coating becomes insoluble during exposure to light and forms a bond with the plate base, unexposed (non-image) areas are removed with a developing solvent. With positive working plates the coating becomes soluble upon exposure and is removed during subsequent development. In both cases the coating which remains forms the ink-receptive image area.

Production of the plate is fairly simple but some special equipment is required. The film negative or positive is positioned on the plate and held in contact in a vacuum frame during its exposure to UV light. After exposure the plate is developed by applying developer with a cotton wool swab until non-image areas are clear. The plate is then rinsed under running water, excess water being removed before gumming. Alternatively the plate may be fed into an automatic plate processor which will carry out all these operations in one pass.

Print runs of up to 250 000 impressions are possible from certain of these plates. Many of the newer plates can have their press life extended by baking the plate image after processing, making runs of up to 1 million impressions possible.

Wipe-on plates

Production of a wipe-on plate is basically the same as that required for the pre-sensitised plate. However wipe-on plates are supplied uncoated, the light sensitive resin coating is applied by the platemaker prior to use.

The coating is poured on to the centre of the plate and then smoothed evenly over the whole surface using a lint free pad. When dry it is exposed and processed in much the same way as the pre-sensitised plate. An alternative to the hand application of wipe-on coating is the use of the roller coating machine. The machine is normally uncomplicated and easy to use. It produces a smoother and more consistent coating in a shorter time than coating by hand. The main advantage to be gained from the use of wipe-on plates compared with pre-sensitised is their low material cost. The press life is not normally as good, 100 000 impressions being the maximum obtainable.

Deep-etch plates

Deep etch plates are so called because the image forming layer is very slightly below the surface of the non-image base metal. They are positive working and require long processing times. However, where these factors are not of paramount importance they do at present fill a gap between positive working pre-sensitised plates and multi-metal plates in terms of run length and cost. However, the continued improvement in the press life of pre-sensitised plates is resulting in a decline in the use of deep-etch plates.

Multi-metal plates

These have both the image and non-image areas composed of metal. A grease-receptive metal (e.g. copper or brass) is used for image areas, while metals which have an affinity for water (e.g. aluminium, stainless steel and chrome) are used for non-image areas. Only two metals are used for the lithographic surface, but a third metal may be used as a base support.

Multi-metal plates give high quality and a life of up to 2 million impressions. They are expensive but are suited to long runs or repeat work.

Electrostatic imaged plates

The electrostatic process employs photo-conductive materials to produce the image. Photo-conductors have the ability to hold an electrostatic charge until exposed to light.

When the charged photo-conductive coating is exposed to an illuminated copy the charge remains in image areas (black) but is dissipated in non-image areas (white). The coating is then developed with toner (powder or liquid) which has an electrostatic charge of the opposite polarity.

Print runs up to 100 000 impressions are possible on aluminium backed plates and makes them suitable for some newspaper printing. The quality of reproduction will depend to a great extent on the equipment which is used. The process is not normally suitable for large areas of solid or halftone screen rulings greater than 34 lines/cm (85 lines/inch).

Chemical diffusion transfer plates

The chemical or diffusion transfer process is so called because the silver salts which form the image are transferred from one surface to another (paper negative to plate) under the pressure of rollers in the presence of developer into which the salts dissolve. Diffusion transfer negatives are available in two speeds, slow for contacting and fast for projection.

The process is primarily intended for text and line copy although screen rulings up to about 40 lines/cm can be reproduced. Print runs of up to 40 000 are possible from metal plates.

Photodirect plates

The majority of these plates employ a dimensionally stable paper base in roll or sheet form, impregnated with plastic on which is coated an emulsion composed of three layers:
   (1) a pre-fogged top layer with moisture attracting characteristics but also possessing an affinity for litho inks when developed;
   (2) a middle layer of light-sensitive silver halides in gelatine;
   (3) a bottom layer containing a developer which will bring out the latent ink affinity of the pre-fogged top layer.

During exposure, light reflected from the white areas of the original passes through the top layer without affecting it producing a latent image in the light-sensitive layer. No light is reflected from the image, therefore the emulsion is unaffected.

Using this system it is possible to produce a large number of plates quickly at very low cost. They are only suitable for print runs up to 10 000 therefore their ideal application is work where only short print runs are required from a large number of different masters.

Laser exposed plates

The first systems for the production of these plates consisted of two separate laser units (reader and writer) working in unison. The reader scanned the copy to digitise the information. The digital data is used to modulate an exposing laser in the writer unit. The units can be sited together, or separate but in the same plant, or used in a facsimile mode between separate sites. More than one writer can be supplied by one reader, which makes this system ideally suited to newspaper installations where a large number of plates are required in a short time.

The logical development is to drive laser exposure systems directly from the digital data in a composition system capable of full page make-up. Such systems exist but are not yet widely used.

Direct image plates

The direct image plate is the simplest of all plates to produce. The image is applied by direct contact to the plate base, which may be aluminium, plastic or paper, using special crayons, inks, pencils or typewriter ribbons. Printing by letterpress or litho with a suitable ink may also be used to apply the image to the plate base.

The plates are cheap and quick to produce but are only suitable for short runs up to about 5000 impressions, where good print quality is not of prime importance.

The driographic plate.

The driographic plate is not lithography in the true sense of the word. It is, however, a planographic plate.

The driographic plate is different from all other plates previously mentioned, in that no water is required to prevent the non-image areas inking-up. The non-image areas of the plate are composed of a silicone rubber coating which has the ability to prevent the special inks used sticking to its surface. Plates are currently available, pre-sensitised and either negative or positive working.

In addition to the advantage of being able to print without water, the process is capable of printing exceptionally sharp dots with very good colour strength. The silicone rubber surface is rather fragile and care is required to avoid scratching.

Process control

Lithographic printing has many variables and only works well when carefully controlled. In order to reduce make-ready times and maintain consistent quality on presses which are gradually becoming faster, it has become necessary to develop automatic systems firstly for making the basic settings of the machine (such as the ink duct settings) and secondly to maintain settings during the run (such as register). These control systems are widely considered an essential part of a modern press and are the area of greatest innovation in press design at present and for the foreseeable future. Control of the platemaking process is also vital.

Platemaking control

For consistent results it is necessary to control *image transfer* from film positive or negative to plate. Some light spread always occurs when platemaking, resulting in undercutting or sharpening of the image. Thus, when working with positive film images, the effect will be to reduce slightly the size of the dots or line widths. In practice this effect is used to reduce the problems of dust spots and film or tape edges. Negative film images suffer from opposite effects.

Control can be achieved by making use of special test elements which are printed down on every plate along with the rest of the working image.

Currently available control strips make use of three types of elements that are suitable for controlling image transfer. These are:

(1) *Micro lines* — (Brunner, Fogra and Ugra). These are very sensitive. The resolving power of the plate needs to be established for correct interpretation.

(2) *Small dot patches* — (Brunner, Fogra, Ugra and Gretag). These are less sensitive and less dependent on plate resolution.

(3) *Continuous tone wedge* — (Gretag, Ugra and Stouffer). These provide a good indication of the level of or change in exposure but do not really indicate differences in image transfer.

Control in colour printing

In lithographic printing in particular, various aids have been developed to assist the press operator to maintain a given press condition so that consistent colour work is produced. In general terms, similar or equivalent techniques are applicable to other printing processes but have not been so well developed, and so the following description refers mainly to lithographic printing.

In litho printing in order to achieve consistent results when colour printing it is necessary to monitor and control the following:

(1) dot gain;

(2) solid density;

(3) trapping of one colour on another;

(4) grey balance (which should be correct if (1), (2) and (3) are correct);

(5) register.

Dot gain is the term used to describe the change in dot size which occurs between the image on the film positive or negative and that on the final print. There are a number of contributory factors to this change in size:

(1) the image transfer from film to printing plate (hence the need for control of platemaking procedures, use of light-integrating meters to control exposure, and use of control elements such as the UGRA wedge);

(2) the image transfer from plate to rubber blanket;

(3) the image transfer from blanket to paper.

The last two occur on the printing press and are in turn influenced by characteristics of the ink (especially viscosity), the blanket and paper. Dot gain is an inherent characteristic of lithographic printing (as is squash in letterpress and flexo) and not a fault. However, excessive and especially uncontrolled dot gain is bad practice.

Dot gain is normally measured using a densitometer to measure the density of tint patches of known original dot area. Density can be approximately converted to percentage dot area using the Murray–Davies equation:

$$\text{Area} = \frac{1 - 10^{-D_t}}{1 - 10^{-D_s}}$$

where D_t = density of tint;

$\quad\;\;\; D_s$ = density of corresponding solid.

Dot gain is then the measured area on the print minus the original area on the film.

If this is done for a whole range of original dot sizes and the measurements plotted graphically the type of diagram shown in Fig. 2.16 results. The curve is intuitively of this general shape since a dot of 0% cannot grow and neither can a dot of 100% grow any larger. Clearly dots in the 50% region to start with have the greatest potential for growth, and on coated paper figures of 18% are quite typical. Thus a 50% dot on film can be expected to grow to a 68% dot on paper.

The curve is not generally smooth or symmetrical. Figure 2.17 shows a more realistic typical case, the kinks being caused by the halftone screen dot join up.

It is important to measure and control dot gain. If it is known that dots are going to grow by a consistent amount between film and print, then a corresponding allowance can be made on the films, i.e. dot sizes are reduced on the films by an amount proportional to the growth which will occur when printing.

In the last few years a number of 'standards for lithographic printing' have been published, the single most significant component of which is the specification of dot gain. These specifications give the dot which should be induced in a proof so that it will be a reasonable representation of the average result from the production run. Making proofs with these amounts of dot gain can be difficult but special inks are now available in the UK which will give the desired results on standard proofing presses.

Solid density is clearly of considerable importance in colour reproduction since it has impact on the overall range of tones and colours which can be reproduced. However, variations in solid density do not influence the appearance of the printed result to the same degree as variations in

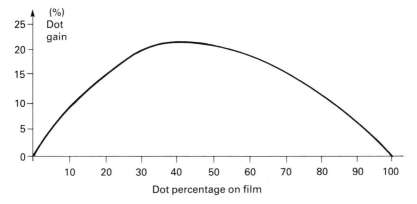

Fig. 2.16 Dot gain characteristic curve

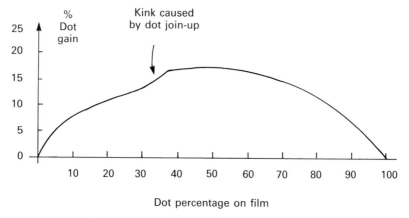

Fig. 2.17 **Real dot gain characteristic curves**

dot gain. There are various techniques for determining the optimum solid density for a given press/ink/paper combination (based, for example on maximising print contrast, i.e. the optimum solid density is taken to be that which gives the maximum difference between a tint and the solid. Such methods taken alone, however, can be misleading). But the preferred approach now is simply to follow the recommendations of one of the recognised standards.

The trapping or transfer of one ink on to another, whether it be wet on dry or wet on wet, is seldom 100%. That is to say that for a given ink film thickness, ink transfer to unprinted paper is greater than to a previously printed ink film. This in itself does not present too great a problem. Difficulty is encountered, however, if the trapping characteristics do not remain consistent or when trying to match prints on presses where the trapping characteristics are very different. The most common example of this occurs when trying to match on a multi-colour production press (wet on wet) prints produced on a single-colour proof press (wet on dry).

Some of the factors affecting trapping are:
— ink film thickness;
— area coverage;
— ink rheology;
— properties of the paper, e.g. absorbency;
— time interval between printings;
— wet on wet or wet on dry.

Greys in a colour reproduction are made up from the correct proportions of yellow, magenta and cyan. Clearly, if one of these changes significantly relative to the other two a colour shift will occur — greys will no longer be grey, and in fact most colours in the picture will be incorrect. So maintaining this correct balance is a key task for the press operator.

Various faults can also occur, the most common being slur and doubling, which superficially appear as excessive dot gain. Slur is caused by a movement of the sheet or web relative to the blanket in the printing nip and distorts the shape of dots in one direction. It is generally caused by slightly incorrect packings under the rubber blanket. Doubling occurs in wet on wet printing when, in a subsequent printing unit, dots of the previous ink colour become transferred back on to the blanket, then back to a sheet slightly out of register. Slur and doubling must be eliminated before reliable dot gain measurements can be made.

To assist the press operator in controlling these various aspects of litho colour printing, colour bars have been developed (see Fig. 2.18). These contain elements which in the main can be checked visually, and show up the above faults particularly well. They are independent of the job being printed and so provide a common reference point from one job to the next. It is now common practice to include such colour bars on all proofs, where they are also checked rigorously using a densitometer, and where possible on the production run. This is not always possible in the latter case but usually some reduced version can be fitted in somewhere. This is better than no control element at all.

Successful colour printing depends on good transfer of information about the printing process to those involved in making colour separations. There are various well-established techniques for measuring the printing characteristics of a printing press in such a way that the information is useful to a colour scanner operator.

In the halftone process the colour which is visually perceived from a combination of process colours can be influenced by the register of these colours to one another. In practice, however, these differences are only slight and do not generally have a significant influence on the final reproduction. What is of greater relevance is the effect which register has on the resolution and detail of the reproduction. The tolerance will vary depending on the class of work but will normally be in the region of ± 0.05 mm between colours.

All of the press settings which control the passage of the sheet/web through the press will affect register, as will the condition and properties of the stock on which printing is taking place.

Ambient conditions can also contribute to the variations in register. Many web presses are fitted with automatic register controls which maintain register to close tolerances.

Ink duct pre-setting and control
Maintaining solid density within suitable tolerances on the run is not too difficult, particularly if a densitometer is used. However, in colour reproduction, control of mid tones is vital and more difficult. Again, a densitometer can be used to make measurements of the colour control bar but the time involved becomes excessive in relation to the press output. Consequently efforts have been made to automate some of this measurement and control task.

A typical litho ink duct consists of a roller with a blade pushed against it to meter the ink (Fig. 2.19). The pressure of the blade is controlled by an

Fig. 2.18 Monochrome illustration of a typical colour bar
(courtesy of FOGRA)

array of ink duct keys spaced at intervals of a few centimetres along its length.

Control is achieved in two ways. Overall ink level is controlled by the angle that the ink duct roller moves through every time the oscillating transfer roller comes into contact with it. Control across the width of the press is achieved by using the ink duct keys. The objective in setting these is to match the ink transferred in each strip corresponding to each duct key with the demands of the image in that strip. This clearly varies from one job to the next and accurate setting is a quite difficult task and almost always involves an element of trial and error at the start of a run.

A fairly recent development has therefore been that of the plate scanner. This scans a plate and measures the printing image area in each strip of the plate corresponding to each ink duct key. The ink demand can then be calculated. Ink duct keys have been motorised with stepping motors so that their position can be controlled, and the information from the plate scanner is now automatically transferred to them. This results in a substantial reduction in make-ready time and start-up waste.

Attempts have also been made to mount densitometers on presses to measure the printed density of each colour across the width of the sheet and automatically adjust the ink duct. This has proved more difficult, partly at least because the printed density is also altered by the level of damping. Independent damp control systems have been developed based either on gloss measurement of the wet plate surface or infra-red absorption but these have not been widely adopted. At least one on-press densitometer system is now available.

Substrates and inks

The majority of sheet-fed litho inks are similar to letterpress inks, drying by oxidation, aided by absorption on porous materials.

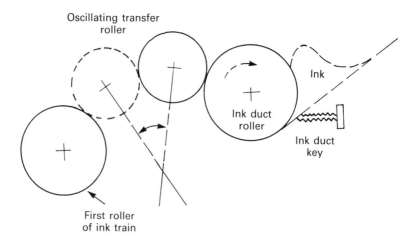

Fig. 2.19 Cross-section through simple ink duct on a sheet-fed press

When printing on non-absorbent substrates or where very rapid drying is required, it is necessary to use special equipment and inks which enable the drying to be accelerated.

Heat-set inks are used on a large number of web-offset presses. In these inks the vehicle consists of resin dissolved in a solvent, and drying takes place principally by evaporation.

In printing with heat-set inks, the printed web is passed through driers which originally used gas flames playing directly on the web. Now, however, forced hot air driers are more frequently used. These must raise the temperature of the web enough to cause evaporation of the solvent, leaving only the resin to bind the pigment into a film and to the paper. From the dryer the web passes on over chill rollers to cool it, after which it can be cut, folded and delivered. The cooling serves to set the ink and is necessary to prevent set-off or marking in the delivery.

Ultra-violet (UV) curing inks
The UV ink drying system involves specially formulated inks which contain a photo-initiator. After printing, the sheet is exposed to UV radiation from lamps housed within the press, this causes a chemical reaction to take place, producing a completely dry ink film virtually instantaneously.

This method of drying is used on some web-offset presses, but is most commonly found on sheet-fed presses printing cartons, where it permits cutting and creasing operations to be carried out immediately after printing. Some offset litho presses printing tin-plate and plastics also use this method for drying.

Infra-red radiation
This is also used to dry inks. Again specialised drying units are required but infra-red drying differs from UV curing in that specialised ink systems are not required. Infra-red drying is best considered simply as a machine improvement that generates heat in the print and accelerates the normal drying process.

2.3 THE GRAVURE PROCESS

In the gravure process the printing image is engraved into a cylinder in the form of cells which become filled with ink. Printing is achieved by passing the substrate between the gravure cylinder and an impression roller under pressure.

The printing unit

The printing unit of a gravure press consists of an ink duct in which the etched cylinder rotates in a fluid solvent-based ink. A metal doctor blade, which reciprocates from side to side scrapes excess ink from the cylinder surface. The substrate is fed from reels into a nip between the etched cylinder and a rubber-covered impression roller which supplies the pressure needed to transfer ink from the cells to the substrate (Fig. 2.20).

The printed web runs upwards through a heated drying system where the solvents are evaporated and extracted, and the ink is thus dried. In gravure printing each colour must be nominally dry before the succeeding colour is printed over it, therefore each printing unit has its own integral drying equipment. The ink which is usually stored underneath each unit, is pumped up to the ink trough and continuously circulated, and usually viscosity control is incorporated in this system. Because each printing unit has an integral drying system and impression roller, most presses consist of units arranged in line, where the web travels between units in a horizontal plane. As the impression cylinder is not gear driven, but obtains its drive through contact with the gravure cylinder, cylinders of

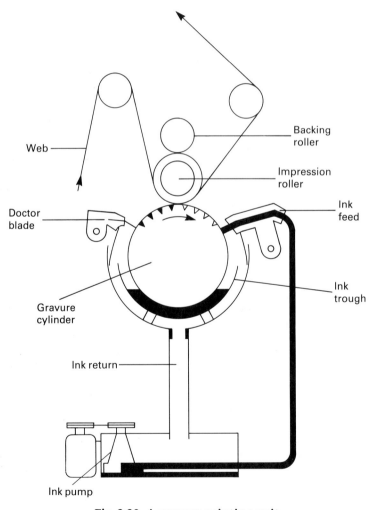

Fig. 2.20 A gravure printing unit

different size can be used to provide variable print repeat dimensions within certain limits.

The inking system

The oldest and simplest method which is still widely used is to place the cylinder in the ink trough. This has its deficiencies, however, since the ink is not well agitated, lots of fumes are given off and it is unsuitable for high speeds since the cylinder tends to carve a trough in the ink.

An alternative uses transfer rollers. The roller does not actually touch the cylinder — there is a gap of about 0.006 inches (0.15 mm). This method was used on some gravure machines printing from plates rather than cylinders.

Enclosed duct designs were introduced in 1924 and meant that more volatile solvents could be used which contributed to a rise in press speeds. Further refinement using the weir system provides the ability to run up to about 15 000 revs/hour. A circulation pump provides continuous agitation, and viscosity control is normally also a part of such a system. The whole arrangement is enclosed to avoid solvent losses. The pump takes the ink from a large tank and delivers it to a small trough in which the cylinder actually sits. Excess ink overflows through a sieve back into the large tank.

A further development of this is the spray system which can be used at even higher speeds. It is basically the same as the weir system except that the pump delivers the ink to nozzles pointing at the cylinder. It has the advantage that the cylinder surface never has the chance to dry out. Again, the system is totally enclosed.

Another system, instead of, or in addition to, spraying the cylinder, pumps the ink upwards through an open-mouthed jet directly under the doctor blade or a pre-metering blade. This is used at speeds of 25 000 revs/hour.

Several variations on the above systems have also been used.

Doctor blades

The function of the doctor blade is to remove surplus ink from the surface of the cylinder leaving the ink in the cells. There are many possible configurations for the doctor blade and they have an effect on the printed result. The thickness of the blade is generally 0.006–0.01 inches (0.15–0.25 mm) and it is made from high-carbon flexible steel. Doctor blades are usually supported by a backing blade to give extra support (say 0.03 inches or 0.76 mm thick).

Blades can be sharpened by hand but it is more usual, especially with wide presses, to use a special grinding machine. Blades are ground with a bevel edge and the angle of bevel is one of the factors influencing the printing results from the blade. A blunt blade will leave more ink in the cells than a sharp one. Similarly, the angle which the doctor blade makes with the cylinder affects the ink left in the cells. A steep angle gives a cleaner wipe (see Fig. 2.21).

Pressure must be exerted on the doctor blade or it will be forced up by the ink being pushed up underneath it. This can be done in several ways:

(1) Using a screw or air pressure. This has the effect of locking the blade rigidly in position. Cylinders must be perfectly cylindrical for this to work well but it is ideal for high-speed presses.

(2) Using weights. The advantage of this approach is that the blade will follow an uneven surface. It is used especially on plate machines.

(3) Using springs. This has similar advantages to weight systems.

Doctor blades are normally made to reciprocate by up to 6 cm. This gives a better wipe and disperses paper fibres which may get trapped under the blade. Blade mountings must have adjustments to cope with different sizes of cylinder and also movement for making the blade exactly parallel with the cylinder axis.

High speed presses may be equipped with a pre-doctoring system. This has one doctor blade which allows an ink film of 0.02 inches (0.5 mm) to remain, then another which performs the final doctoring. This has the advantages that the pressure on the second blade can be substantially reduced so that cylinder wear is less, and printed results are less affected by speed.

The impression roll

This has a steel core with a rubber covering 12−20 mm thick. It is a relatively hard rubber and the pressure applied between it and the printing cylinder is high in relation to other processes. This is one reason why the side frames of gravure presses are so substantial.

Gravure printing frequently suffers from speckle, caused by individual cells not printing on 'rough' papers. In this context it is the smoothness of the substrate under pressure which matters and consequently an uncoated, but compressible paper may well print better than a coated one which has a less compressible surface. In an effort to overcome this, electrostatically assisted ink transfer was introduced. A special impression roller made with an electrically conductive rubber is used, and in the common version of this equipment made by Crosfield Electronics Ltd, the inside of the roller contains an electrical generator. The turning action of the roller causes a high voltage to be generated (e.g. 1000 V). The

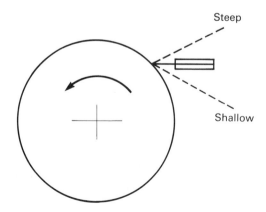

Fig. 2.21 Doctor blade angles

resulting electrical field encourages the ink to leave the cells and transfer to the paper even where contact is imperfect. Very marked improvements in the printed result are possible.

Drying system
Slow-speed sheet-fed machines may not have any drying system other than an extended delivery transport but some form of cold air blower is more normal. High-speed web machines demand more sophisticated driers between each unit. A web tends to carry a solvent layer with it, and consequently high-velocity hot air is used to break up this layer and remove the solvent. Steam-heated drums may be used to heat the web, and cold chill rollers are used on exit from the drier to cool the web. Some driers are split into sections so that highly saturated air is taken away first (to solvent recovery) while other sections recirculate the air thus conserving heat.

Press configurations

Most gravure printing is done on web-fed presses, which provide facilities for supporting and controlling the supply reel during unwinding. A variety of equipment is used for both manual and automatic splicing. An input feed control system is used to provide stability of web movement to the first printing unit.

Suitable folding equipment on the end of the machine is essential to produce a folded product such as magazines. For package printing where the printed product is required in reel form, rewinding is necessary, which employs either a centre or surface winding principle. Sometimes cartons are gravure printed on web-fed presses and are 'flat' die cut in line with the printing operation. Web-fed presses will also produce a sheeted product that registers with the printed image. There are sheet fed presses that print from solid cylinders, or plates fastened around a cylinder. Most fast running web-fed presses producing items such as magazines, cartons and packaging products on film are printed from hollow cylinders. With thin gauge film, delicate control of tension is essential, and often the tension in winding is graduated. With centre winding a lay-on roller is used to expel air between layers and most of the rollers in the machine are chevron engraved to smooth the film in the cross direction. Electronic control of register between colours is usually employed, and scanning equipment is provided to enable inspection of the web during the printing operation.

Gravure cylinder preparation

A gravure cylinder is generally made from a steel tube (although solid ones are also used) which is copper plated by electrolytic deposition from an acid copper sulphate bath. Additions to the plating bath are used to produce a surface finish which is almost ready for subsequent processing. Since it is not possible to electroplate steel directly from such a bath, a new steel cylinder is first lightly plated either with copper from a cyanide bath or with nickel. The cylinder is then transferred to the copper sulphate bath and grown to the appropriate size.

The base cylinder is obviously costly and so is reused. This can be done by turning off the image on a lathe and then growing the cylinder back to size by electroplating more copper. A more convenient and cheaper process is to use the Ballard skin technique. In this method, the base copper cylinder is grown to a few thousandths of a inch under the required diameter, and a layer of silver is then deposited followed by the rest of the copper. After use the silver layer allows easy separation of the copper skin on top, which can be pulled off by hand. The cylinder is cleaned, a new silver layer deposited and the process repeated for the next job.

There are several different ways of producing the image on a gravure cylinder.

Conventional etching

The image to be printed is transferred to the cylinder by means of a material called carbon tissue or pigment paper. Carbon tissue is a gelatine coating containing a light yellow/brown pigment on a paper base. The coating is made sensitive to light (peak sensitivity in the UV) by immersion in a solution of potassium dichromate (2–3% by weight in water) for 2–3 minutes. It is dried at room temperature in contact with glass or chrome glazing plates. When dry it can be lifted off and is generally stored in a freezer for 7–10 days before use. For repeatable results the temperature, concentration and replenishment of the dichromate bath must be closely controlled.

When exposed under a positive of the subject to be printed, the gelatine coating 'hardens' (i.e. becomes insoluble) to a depth depending on the amount of light received. Thus in shadow areas very little hardening occurs but it becomes progressively deeper as the tones become lighter. A gravure positive generally has a density range of approximately 0.35–1.65. The exact densities, once chosen, are very closely adhered to.

A second exposure is made under a gravure screen (Fig. 2.22). This screen is *not* used to produce a halftone dot pattern, but only to split the image up into tiny cells to produce a support for the doctor blade. It consists of a regular pattern of black squares, the width of the square being 2.5–3 times the space between them (i.e. the cell : wall ratio is said to be 3 : 1).

The exposed tissue is transferred to the copper-plated cylinder by rolling it into contact while feeding water into the nip. The gelatine layer adheres to the copper. The base paper is drenched in industrial methylated spirits, and then soaked in hot water at 45 °C until it can be lifted off.

Development of the remaining gelatine layer continues, washing away the unhardened coating, leaving a relief image of hardened gelatine corresponding to the tones of the photographic original. The cylinder is finally drenched in industrial methylated spirits again, wiped dry with a squeegee and possibly blown dry. It is allowed to adjust to room temperature (see Fig. 2.23).

Etching is carried out with ferric chloride solution through this gelatine

Fig. 2.22 A gravure screen

layer. The etch penetrates most rapidly through thin gelatine, giving deep cells in the shadows, and shallow cells in the highlights (Fig. 2.24).

Etching is a skilled operation. It has to begin with a strong solution of etch (conventionally measured in °Beame. 43–45 °Be would be the starting concentration) which causes the etching to start only in the shadows. The etcher will judge when to change to weaker solutions (down to perhaps 35 °Be) to start the etching of the mid-tones and finally the highlights. Considerable skill is needed to ensure that dark tones will have the correct depth of etch, just as etching starts in the highlights. At least five solutions of varying strength are used. The composition and temperatures of these solutions is critical. Usually they will have a certain amount of copper dissolved in them before use to give a smoother etch. Other additives may also be used.

The end result is a cylinder in which different tones are made up of square cells that are roughly the same area, varying only in depth. Deep cells carry more ink, and thus will print darker tones than shallow ones.

Single bath etching
In order to obtain more repeatable results than manual etching produces,

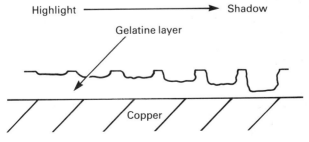

Fig. 2.23 Gelatine layer on copper cylinder after development

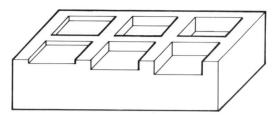

Fig. 2.24 The etched surface

single bath etching was introduced, and because only one strength of ferric chloride solution is used, is obviously easier to carry out with a machine. To be successful, times, temperatures and concentrations at all stages must be very closely controlled. Normally, a single strength solution of ferric chloride will not penetrate all the range of tones properly. However, the addition of ferric ammonium sulphate to the ferric chloride makes a single bath etch possible using a concentration of 38–40 °Be. In practice, completely repeatable results are almost impossible to achieve.

Cells vary in diameter with the screen size used; with a 60 line/cm screen the cells will be about 125 μm in diameter, the depth varying from about 2 μm in the highlights to 38 μm in the shadows. During etching, etching proceeds sideways as well as downwards. (There is no banking agent as in letterpress plate etching.) Consequently, the original cell : wall ratio of say 3 : 1 on the gravure screen changes to around 10 : 1 on the etched cylinder. Higher ratios are undesirable since the cell walls may collapse.

The finished cylinder is usually chromium-plated for longer life.

Halftone process

Double positive system — halftone gravure

This is similar to the conventional process except that instead of exposing the carbon tissue to a continuous tone positive and then a gravure screen, it is exposed to a continuous tone positive and a screened positive. The screened positive is made from the continuous tone positive by contact using a special contact screen and duplicating film (to yield a positive directly). The screened positive has dots of varying area, but ones which never join up completely because a cell wall pattern is always present.

Etching is similar to before, using baths of ferric chloride of different strengths or a single bath etching technique. The result is a cylinder in which different tones are made up of cells which vary in both depth and area.

The advantage of this technique is that etching is easier to control and the cylinder has greater tolerance to wear on long runs. This is because the highlight cells have smaller area but much greater depth so will not be affected by wear. Speckle can be a more prominent problem with this type of etch.

Halftone gravure

Another technique will produce cells which vary only in area and are all the same depth. Carbon tissue is not used and only a single halftone positive. A light sensitive coating is applied directly to the cylinder, exposed to the positive and developed. Etching can be done in a one bath etch. The Acigraf process is the most widely used process of this type although other methods based on powderless etching (as in letterpress platemaking) and electrolytic etching have been tried.

Mechanical engraving

This is a technique developed by the Hell organisation in Germany, who have produced a range of Helio-Klischograph machines. These use a diamond stylus to engrave cells shaped like inverted pyramids. For deeper cells the stylus penetrates deeper into the copper so that the area at the top of the cell becomes greater as well as the depth. Electron beam engraving systems are under development.

Lasergravure

This method has been developed by Crosfield Electronics Ltd in the UK. It uses a laser to vaporise a plastic filling to varying depths out of a pre-formed groove pattern in the cylinder surface. At the time of writing this is being used in one printing plant on a trial basis.

Press control systems

In contrast to the difficulty of controlling the cylinder making processes, the printing operation is more straightforward, and particularly on web presses, automatic control systems for register, web tension and viscosity control have been available for years.

Viscosity control of the ink is more or less essential on high-speed presses. There is a definite relationship between the speed at which a press is running and the viscosity which must be used in order to achieve good printing results. If the viscosity is too high then the cell wall pattern will still be evident even in shadow tones and the print will lack smoothness. If the viscosity is too low, then bleeding will occur (spreading of the image from shadow areas).

Many modern presses are equipped with solvent recovery systems. The hydrocarbon solvent laden air from the driers is passed through large tanks of activated charcoal. The solvent is adsorbed on the charcoal. When saturated, another tank of charcoal is brought into line. Steam is passed through the first tank and this drives the solvent off. When condensed, the solvent and water separate and can be pumped off separately. Recovery of 75–90% is possible. In the packaging industry, the use of alcohol/ester inks precludes the use of activated charcoal and other forms of solvent recovery are used.

Web tension is now generally automatically controlled but on less sophisticated presses this is not always so. On some multi-unit presses it is necessary to graduate cylinder sizes through the length of a machine so that each one is just slightly larger than the one in the previous unit. This ensures that each unit is very slightly pulling the paper from the previous unit, so satisfactory web tension is maintained.

Substrates and inks

Very fluid, solvent-based inks are used to print gravure, which dry by evaporation, leaving a dry film of resin and pigment on the substrate. (For details of the raw materials used see Chapters 4 and 7).

Evaporation is accelerated by moving air and by heat, so drying units are built on to each unit to speed up the drying.

The advantages of solvent-based inks are as follows:

(1) Rapid drying can be obtained on non-absorbent materials such as plastics films and metal foils.
(2) Converting operations can be done in-line.
(3) Solvents can be chosen to key on to a wide variety of different surfaces, and to dissolve film-forming resins that will impart specific properties to the inks.

Applications

Gravure printing is used primarily for long runs. Many magazines, at least in the past, were printed by gravure and in Europe and the USA they still are. The quality of the colour printing and the low waste of the printing operation are attractive features.

Much package printing of cartons, plastic films and metal foils is printed gravure. It is attractive since the rapid ink drying makes in line finishing processes possible.

Gravure is also used for a number of special purposes. Many postage stamps, reproductions of oil paintings and even some telephone directories are or have been printed by gravure.

2.4   THE FLEXOGRAPHIC PROCESS

Flexography is a process in which the printing images stands up in relief, like that in the letterpress process. A liquid ink is used which is mostly solvent-based, and dries mainly by solvent evaporation. Water-based inks can also be used.

A low printing pressure is essential to the process because of the combination of very fluid inks and soft and flexible printing plates that is used.

The distinctive features of the process are as follows:

(1) The use of liquid inks that dry rapidly by solvent evaporation, thus enabling fast printing speeds to be achieved on non-absorbent materials such as films and foils.
(2) The 'soft' and flexible relief printing plates that can be mounted and registered on a plate cylinder away from the printing press. Proofs can also be obtained. Individual plates can easily be changed or repaired, and a portion of a plate can be removed to enable items such as price or expiry date to be changed.
(3) The application of ink to the surface of the printing plate by means of a screened (Anilox) roller. The result is a simple ink feed system that consists of not more than two rollers.

(4) Although most flexographic printing is reel to reel, the machines
     enable changes in the print repeat length to be made simply.

The printing unit

The printing unit consists of three basic parts (Fig. 2.25):
  (1) the inking unit;
  (2) the plate cylinder;
  (3) the impression cylinder.
The function of the inking system is to meter out a fine and controlled
film of liquid ink, and apply this to the surface of the printing plate.

The inking system consists basically of an ink trough, a rubber covered
fountain roller, and a screened (Anilox) inking roller into which cells of
uniform size and depth are engraved. The fountain roller lifts ink to the
nip position, where it is squeezed into the cells in the screened inking
roller and by a shearing action is removed from the roller surface. The ink
in the cells is then transferred to the surface of the printing plates. To
regulate ink film thickness in printing, screened ink rollers are available
which have screens of from 40 to 200 cells/cm. These may be engraved or
etched metal or ceramic. The engraved cells are generally square in shape
with sloping side walls. When printing halftones the cells per centimetre
of the anilox roller needs to be about 3.5 times the halftone screen ruling.
The number of cells and their size regulate the volume of ink transferred.
Further regulation of the ink is achieved by varying the surface speed of
the fountain roller, altering the pressure between the fountain roller and
screened roller, and also altering the hardness of the rubber covering on
the fountain roller.

For high-quality flexographic printing, reverse angle doctoring of the
anilox roller has been introduced (Fig. 2.26). This is not speed dependent
as are the other methods, which is a distinct advantage.

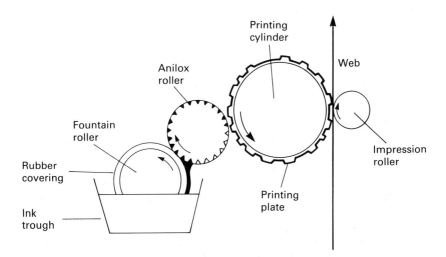

Fig. 2.25 Flexographic printing unit — ink doctoring by roller

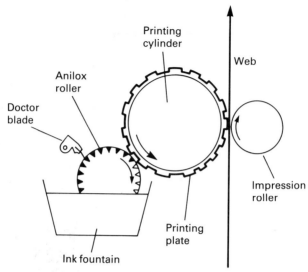

Fig. 2.26 Ink doctoring using a reverse angle doctor blade

The plate cylinder is usually made from steel. The printing plates, which have a thickness of approximately 2.9 mm are secured to the cylinder with two-sided self-adhesive material.

The impression cylinder is also made from steel. The substrate passes between the plate and impression cylinders, which generate printing pressure. The ink is transferred from the cells in the screened ink roller to the plate surface, and then to the substrate, during which it reaches virtually a uniform film.

For high-quality flexographic printing the components of the printing unit must be engineered to very tight tolerances (measured in tenths of thousandths of an inch). The ability to manufacture to these standards is one of the factors which is contributing to the current growth in flexographic printing, and its use for higher-quality products than was previously possible.

Press configurations

There are three basic types of configuration of flexographic presses:
  (1) stack;
  (2) common impression;
  (3) in-line.
The first two are the most common.

The stack press consists usually of two or three integral printing units arranged in vertical formation, and this is repeated to provide the number of printing units required. The stack press is mostly used for printing on paper of all grades, and is less suitable for printing on film. This machine enables reverse side printing on the web. Each printing unit has its own integral impression cylinder. The web enters the first upper unit and passes downwards through the units, where its direction is reversed to travel upwards towards the drying equipment. Where there is a second bank of printing units the web will travel through those to the drying equipment. Figure 2.27 shows that the web travels directly from one printing unit to the next being supported only from the non printed side, thus allowing six or more colours to be printed. Electronic register controls can be added to this type of press (and in-line) to enable higher quality colour work to be produced.

The common impression machine (Fig. 2.28) consists of a large cylinder around which are arranged either four or six printing units. The cylinder is very accurately made from steel. Usually the web enters the top unit on one side of the cylinder, travels to each unit with the cylinder, and emerges from the top unit on the opposite side of the cylinder. With the web wrapped around most of the impression cylinder well controlled register between colours, even with thin gauge extensible films, can be achieved. Most multi-colour work on extensible film that requires precise register is printed on common impression machines.

The in-line machine which is a less common configuration, consists of printing units arranged in horizontal formation, with the impression cylinder situated below the web, thus providing easy access to the plate cylinder. The web passes through each printing unit in a horizontal path. This configuration is used mostly for printing on lightweight board and less flexible materials that cannot be wrapped around rollers.

Many products printed by flexography are required in reel form for subsequent processing, and so machines provide suitably versatile winding equipment. Where necessary the printed web can be chopped into sheets in register.

The machine also provides facilities for supporting and controlling the supply reel during unwinding. A variety of equipment is available for both manual and automatic splicing, and also infeed control.

An ink drying system, which usually blows hot air on to the web is situated on top of the machine. This is to ensure that the inks dry rapidly

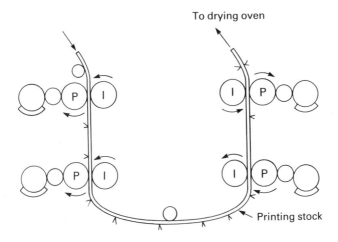

Fig. 2.27 Stack press

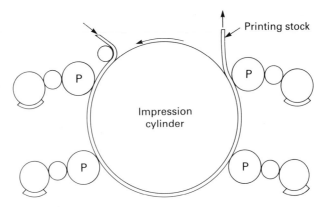

Fig. 2.28 Common impression cylinder

enough to enable printing and rewinding at a suitable speed without set-off and sticking.

As with gravure, most machines feature rewinding equipment which employs either a centre or surface winding principle.

Much development work is currently taking place in connection with flexographic printing. The process is being adapted to newspaper printing, electronic pre-setting and control systems are being added, and new types of ink and inking system are being introduced. New press configurations are sure to result. A recent novel development by Cobden Chadwick Ltd is a press which uses an electronic control system to vary the web length being drawn past the print roller, while ensuring that web and print rollers have the same relative speeds during impression.

Flexographic platemaking

There are several kinds of image carrier in flexography:
(1) the traditional rubber plate;
(2) photopolymer plates;
(3) laser engraved rubber plates or rubber rollers.

Rubber plates
The rubber plate is essentially a duplicate plate obtained from an original. The original may consist of various type elements such as, hand-set foundry type, Monotype or Linotype to provide the type display, and zinc plates to provide the line and halftone image. These are assembled and locked up in a steel chase for moulding. Alternatively, the type matter can be set on phototypesetting equipment to provide a negative and line image negatives combined with this to produce a one-piece zinc plate of the entire printing image. A one-piece plate can also be electronically engraved, which can be originated by hand-drawn artwork or photographic print. This equipment will engrave line and halftone images.

A mould is required to produce a rubber plate. The mould is made by pressing a heat-setting matrix material against the original under heat and pressure in a hydraulic moulding press. The thermosetting resin in the matrix material is initially softened by the heat to take the impression and then the material becomes hard and rigid in approximately 8 minutes at a temperature of 150 °C.

The rubber stereo is made by placing a layer of uncured rubber over the mould and pressing these together under heat and pressure. The grades of rubber used are usually cured in approximately 8 minutes at a temperature of about 150 °C. The raised areas of the stereo, which are the image areas, correspond to the image areas of the original. The back of the stereo has to be ground to achieve a suitable thickness.

Photopolymer plates

There are various photopolymer plates suitable for flexographic printing. These plates are made directly from photographic negatives.

The plates consist basically of a light-sensitive photopolymer, which when exposed to UV light through a photographic negative polymerizes and becomes resistant to the washing out medium. The non-image areas of the plate which have not been exposed to UV light and are therefore not polymerised or resistant, are removed in a plate processing machine. On completion of the plate processing, the plate is usually ready for printing. The washing-out or removal of the non-image area of the plate is done by the use of either solvent, an aqueous solution, or air. These plates have suitable thickness accuracy, and do not require back grinding.

Rubber suitable for flexographic printing can be engraved by laser techniques. The equipment will handle black and white positive copy for line work, and screened negatives or positives for halftone work. Engraving by this method can be done on either separate pieces of rubber, or rubber rollers. The ability to engrave rollers is unique, and an advantage in the printing of continuous designs.

Because flexographic printing is done from an image in relief it is essential that the shank of the image has a steep angle and is smooth. A suitable depth in the non-image area is also essential.

Plate mounting

Flexographic printing plates are usually secured to the printing cylinder by means of two-sided self-adhesive material. The plates are mounted on the plate cylinder and pre-registered in position on special equipment designed for this purpose. Several plate cylinders are normally available for one machine to enable pre-mounting of plates. This reduces the unproductive time on the machine to a minimum.

A 'cushion-back' adhesive layer behind the stereo is sometimes used to compensate for any inaccuracies in the plate or press. However in halftone printing this can lead to greater enlargement of highlight dots than would otherwise have been the case.

Applications

Flexographic printing is widely used for package printing, particularly of plastic films. However, paper-based products such as corrugated cardboard, for example, are also printed by this method.

Although still rather new, it is also now being used for some newspaper printing. The quality of flexographic printing has risen substantially in the last few years, enabling it to produce work that would otherwise have been printed gravure or litho.

2.5   THE SCREEN PRINTING PROCESS

Screen printing (formerly called silk-screen printing) is a stencil process whereby ink is transferred to the substrate *through* a stencil supported by a fine fabric mesh of silk, synthetic fibres or metal threads stretched tightly on a frame. The pores of the mesh are 'blocked-up' in the non-image areas and left open in the image area. This image carrier is called the screen.

During printing the frame is supplied with ink which is flooded over the screen. A squeegee is then drawn across it, thereby forcing the ink through the open pores of the screen. At the same time the substrate is held in contact with the screen and the ink is thereby transferred to it.

The principle is shown in Figure 2.29.

Because of their simplicity, screens can be produced cheaply and this makes it an attactive process for short-run work. Furthermore, since the image is produced *through* a screen rather than *from* a surface the impression pressure is very low. This makes it ideal for printing on fragile boxes or awkward shapes.

Irrespective of the type of machine the printing procedure is generally the same. A working supply of ink is placed at one end of the screen and the screen is then raised so that the stock may be fed to register guides or grippers on a base. The screen is then lowered and a rubber or plastic squeegee drawn across the stencil to produce the print. Ink replenishment is undertaken as necessary.

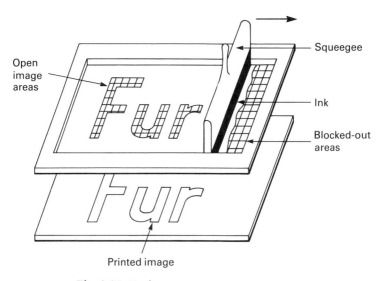

Fig. 2.29 Basic screen press arrangement

On most flat-bed machines the base to which the substrate is applied is of a vacuum type. This prevents the stock sticking to the screen and being lifted by tacky inks.

To a certain extent the thickness of the ink film printed can be controlled by the pressure, sharpness and angle of the squeegee blade. The more upright the blade the thinner the deposit of ink. Thus, in general, fine work requires a more upright blade. However, the type of ink, stock and machine govern the blade setting also.

Press configurations

Many printing machines still consist of a simple hand-operated unit. With these the substrate is fed in by hand, ink is placed in the frame and flooded over the screen and the squeegee is then hand drawn across the screen. These can be particularly useful where very thick or thin materials, which cannot be automatically fed, are printed or where a test run of a new package is required.

Semi-automatic machines also have the stock fed and taken off by hand but utilise a mechanised squeegee blade stroke.

Fully automatic flat-bed presses are also available where the substrate is fed in and taken off by means of automatic feed and delivery systems. After printing the sheets are taken through an air dryer to evaporate the solvent. Prints emerge dry and can be stacked ready for further processing. Such fully automatic machines have only really become feasible since the advent of thin ink films which dry more rapidly.

In order to speed up the process further, cylinder presses have been developed (Fig. 2.30). With these presses the squeegee remains stationary and the screen, cylinder and substrate all move in unison. This permits faster operation since the paper does not have to be brought to a halt and

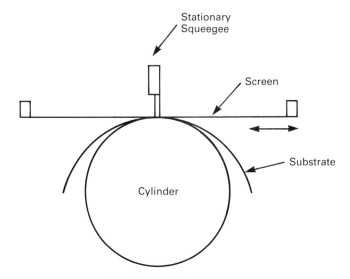

Fig. 2.30 Cylinder press

fed into a vacuum base as with flat-bed presses. Speeds of 5000 impressions per hour or more are not uncommon with such machines, but they are not suitable for rigid materials.

One of the major advantages of the screen process is the ability to obtain prints on non-flat objects. For example, printing on bottles or other cylindrical objects is achieved by using a press of the cylinder type described above but the object to be printed is placed in the machine where the impression cylinder is shown. After each impression the bottle is removed and another unprinted one substituted. There are few limitations on size or shape. Special screens and jigs are produced for printing on shaped objects such as cups with handles or tapering cylinders, and screens with high elasticity combined with shaped squeegees are used for conforming to irregular objects. Print heads can also be bolted to automatic production lines, so that printing becomes a part of the total production process of such objects as filled polythene bottles.

Screen stencil manufacture

The screen consists of a mesh of threads woven together. Early screens were made of silk and each element of the mesh consisted of a number of stranded threads woven together. These 'multi-filament' screens had a tendency to distort and swell, as ink was deposited between the fibres, and were difficult to clean. Because of these disadvantages, and the shortage of silk, alternative materials were sought after the war and today the materials used are almost wholly synthetic. Monofilament screens have become widely used, where each element of the mesh consists of a single strand of material. This minimises the disadvantages described above. Typical monofilament materials used are nylon and polyester, the latter being used where dimensional stability is important. Metal meshes are also used where a long life and complete dimensional stability is required but these are considerably more expensive. Multi-filament screens, usually polyesters, are occasionally used where a heavy ink film is required.

Every screen has a characteristic mesh opening and thickness. Which particular opening and thickness is chosen for a given application depends on the thickness of ink film required, the pigment particle size of the ink and the fineness of detail it is necessary to reproduce.

Mounting the screen

Prior to printing the screen has to be stretched and mounted on a frame. The correct tension is important to avoid misregister, premature wear of the stencil or splitting of the fabric. Although hobbyists and small printers may do this by hand, on small wooden frames, it is normally achieved in production by use of a pneumatic stretching frame. A number of clamps grip the screen and can be individually adjusted to achieve tension. The screen is then placed over a frame and stuck to it after which it is degreased before the stencil is applied.

Application of the stencil

For simple work a stencil may be hand-cut from a laminated film material and stuck down to the screen. Obviously this is only useful for large blocks of colour or simple geometric designs.

A *direct* photographic stencil is produced by applying a light sensitive coating such as a dichromated resin to the screen and drying it. (Usually a number of separate coatings are required.) Prior to exposure the coating is water soluble, but upon exposure to light it hardens. Thus the stencil is produced by coating the screen and then exposing it to a photographic positive. This hardens the coating in the non-image areas and leaves the image areas water soluble. By washing-out the screen with water the image areas will be cleared and the mesh left open.

Indirect stencils are produced using the same principle but during exposure and development the stencil is held on a temporary backing film. It is only after exposure that it is transferred to the screen, while in a moist state, and dried. The backing sheet can then be carefully removed leaving the stencil adhered to the screen.

Direct stencils have a longer life than indirect because the coating is bonded into the mesh but are much more difficult to expose (since the screen is attached to a frame during exposure). Furthermore because the washing out action tends to follow the mesh threads a serrated edge to the image may sometimes result. Indirect screens avoid these problems and therefore tend to produce finer detail. In addition they can be obtained pre-sensitised, so that they may be used direct from the packet without any of the mixing of chemicals and application to a base which was previously necessary.

After treatment

Following stencil production any remaining areas of the screen around the stencil have to be sealed to avoid ink penetration. A number of materials, such as lacquered tape or synthetic varnish, may be used which must be resistant to ink and yet remain flexible so that they do not crack under pressure of the squeegee.

Finally the stencil is inspected against a light box. Any areas not properly filled are then filled with an opaque filler.

Substrates and inks

Typically screen printing inks used to be very thick preparations based on oil-bound paints. They printed an ink film that literally stood out in relief on the surface. Such inks were highly suited for printing on almost any type of substrate, rigid or flexible, and it became a popular process for producing posters, display advertising, metal signs, glass and china decorating and textile printing. However, these oil based inks dried by aerial oxidation and being thick, drying times were long. For many years work was placed on drying racks stacked on top of each other and gradually these were replaced by mechanised racking systems. Even so the production speeds achievable were very low.

However, considerable developments are taking place with UV cured and other ink systems (refer to Chapters 9 and 10).

Applications

Since screen printing inks can be formulated to adhere to almost any surface, and the printing process itself can handle almost any substrate in a wide variety of shapes, screen printing is a very versatile process. It is used for package printing, especially of plastic containers, posters, point of sale materials and many other products. It has even been used for carpet tiles!

2.6   NON-IMPACT PRINTING PROCESSES

The two main non-impact processes of most importance to printing are:
(1) ink-jet printing;
(2) electrophotographic printing.
Many other processes exist but are not described here since at their present stage of development they are not capable of the production volumes normally anticipated in the printing industry.

Ink jet printing

Ink-jet printing is a non-impact means of generating images by directing small droplets or particles in rapid succession on to the surface of a substrate. There are various possible ways of generating and projecting droplets but only the following appear suited to real production applications.

Continuous jet
The primary ink-jet system is the continuous jet, formed by forcing ink under pressure out of a small nozzle (Fig. 2.31). In these circumstances, a liquid jet tends to break up into a stream of droplets of a size and frequency determined mainly by the surface tension of the liquid, the pressure applied to it and the size of the nozzle.

If regular pressure pulses are applied continuously to the ink within the reservoir at a suitable frequency, then the size and spacing of droplets become much more regular. This can be done by applying a high frequency alternating voltage to a piezo-electric crystal that is in contact with the ink.

A stream of uniformly sized and spaced droplets is generated, which if allowed to fall on to a moving web of paper produces a continuous line of dots, each having a diameter about three times that of the droplets.

For any practical message to be generated by a single nozzle, the droplets must be individually controlled and deflected and this is done electrostatically by surrounding the ink jet in the region of droplet formation by a charged electrode. The jet becomes charged by induction and each droplet carries a charge depending upon the voltage applied to the electrode at the instant the droplet separates from the jet. The droplets

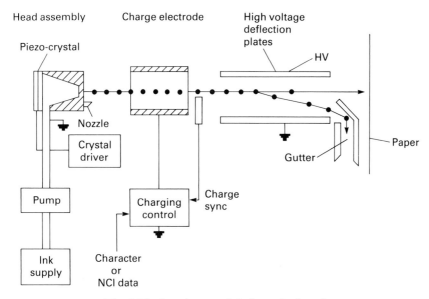

Fig. 2.31 Continuous ink-jet printhead

then pass between oppositely charged deflector plates and are deflected by the electric field to an extent proportional to the size of the charge it carried. Characters can be generated by depositing dots within a matrix: the minimum matrix size for reasonable legibility is 7 × 5 dots.

Thus by repeatedly increasing the voltage applied to the deflection electrodes in seven steps and then returning it to zero, a row of seven dots is printed on to paper. By moving the paper sideways, a parallel row of seven dots can follow. In this way, a rectangle of dots is printed. The other dots must be prevented from reaching the paper, and this is achieved by leaving them uncharged so that they are not deflected on to the paper but proceed straight to a trap where the unwanted ink is collected and can be filtered and returned to the reservoir.

Hence the stepped voltage pulses applied to the charging electrode are gated by signals that define the character matrix and allow dots to be printed as needed.

The physical constraints of useful droplet size, nozzle size and surface tension of the ink mean that in practice droplet frequencies will be somewhere in the region of 100 kHz and this determines, for a single nozzle, the limitations on writing speed and image definition. Thus, for 7 × 5 matrix characters, with some space between them, up to about 1500 characters per second (ch/s) may be printed, whereas for a 40 × 30 matrix, the speed may be under 100 ch/s.

For in-line numbering, coding and addressing applications 7 × 5 characters from a single continuous jet are widely used and, typically, messages of 100 or 200 characters can be printed at speeds of up to 1500 ch/s. According to the application, this corresponds to a web speed of 5 m/s or 90 000 articles/hour.

An alternative form of continuous jet system uses an array of closely spaced nozzles, each of which serves only one line of dots. Unwanted droplets only are charged and deflected into a gutter. Since the writing droplets are not charged, electrostatic interaction between adjacent droplets is avoided and flight distances are equal. The system is therefore in some ways simpler than the deflected droplet type. The complexity comes in assembling and controlling a multiplicity of nozzles. The Mead Dijit 2800, for example, has an array of nozzles spaced at 120/inch in a double row that may be up to 10.65 inches long and print up to 512 characters.

Impulse or drop on demand
With this method, the pressure on the ink reservoir is not maintained continuously but is applied by piezo-electric crystal when a droplet is needed to form part of a character (Fig. 2.32).

In a widely used system, an array of 12 nozzles, each actuated by its own piezo-electric crystal, is used to generate a matrix column, so that no deflection of droplets is needed to form the image. Since, also, the recovery and recirculation of non-printing droplets does not arise, the equipment can be both electrically and mechanically simpler than the other classes of ink-jet printer considered.

A variation on this type is the Canon bubble-jet (Fig. 2.33) which, instead of piezo-electric crystals, has very small heating elements behind each nozzle so that ink droplets can be ejected by rapidly- formed solvent vapour bubbles.

Drop on demand ink-jet systems are being used mainly in matrix printers but their application is not, in principle, limited to that. Examples of four-colour reproductions have been produced using the bubble-jet technique at resolutions of 16 points/mm suggesting that this may be a viable method for short runs in the future, and also for colour proofing.

Electrophotography

Electrophotographic printing is commonly used in photocopiers and laser printers. Some of the higher quality machines in both categories are

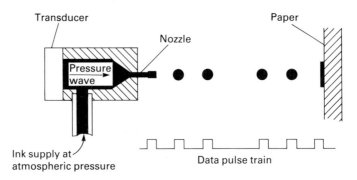

Fig. 2.32 Impulse or drop on demand printhead

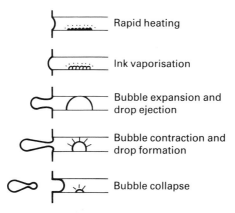

Fig. 2.33 Canon bubble jet

currently used for print runs where only up to a few hundred copies are required.

Electrophotography is the technique of producing an image by causing light to affect the charge distribution on a photoconductive drum or belt. The charge pattern is subsequently used to attract toner, which is then transferred to paper. The toner is then fixed by some means and forms the final image.

Figure 2.34 shows a schematic layout for an electrophotographic printer. The unit is based on a rotating drum or belt, coated with photoconductive material. The photoconductor is electrically insulating in the dark, but becomes more conductive when illuminated. Various materials are used, ranging from amorphous selenium to complex organic compounds. Each kind of material has a sensitivity which depends on wavelength, hence light sources have to be matched to the photo-conductor (Fig. 2.35).

The first step in the process leading to the printed image is to generate a uniform electric charge on the dark photoconductor surface. This charge is then dissipated by exposure to light in the non image areas. In a conventional copier, light reflected from the original is projected on to the drum. In a printer, the image is generated point by point using one of several types of imaging device.

Types of imaging devices include:
(1) laser;
(2) light-emitting diode (LED) array;
(3) liquid crystal array (LCD) (plus light source);
(4) magneto-optic array (plus light source);
(5) cathode ray tube with fibre optics.

The laser is the most common device. This is a suitable light source since it produces light which can be collimated and focused into small, high intensity spots. Three types are in use: HeNe gas, HeCd gas and diode. A complete laser scanning sub-system includes several essential components — the laser, a scanner to sweep the beam across the drum, a modulator to turn the beam 'on' and 'off' and several lenses and mirrors.

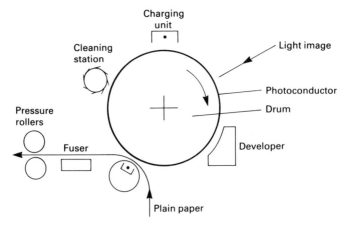

Fig. 2.34 Schematic layout of electrophotographic printer. Process steps: (1) charging; (2) imaging; (3) developing; (4) transfer; (5) fixing; (6) cleaning

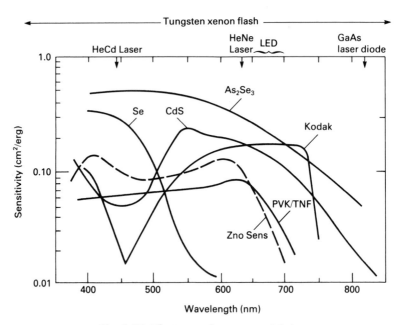

Fig. 2.35 Photoconductor sensitivity

Rather than have a single source, it is possible to have an array of light-emitting diodes. The total width of the drum is covered by several monolithic devices, each with a line of LEDS. The other types of light source, while they have been tried, are not widely used.

The next step in the electrophotographic process is to develop the latent electrostatic image. The ink in this process is known as toner, of which there are various types (Fig. 2.36).

(1) Liquid toner. The finer toner particles are dispersed in a non-conductive carrier liquid.

(2) Two-component toner. This is a dry powder consisting of fine particles (say 10 μm diameter) of toner pigment, and coarse beads (say, 200 μm diameter) known as the carrier. The toner and carrier become oppositely charged, and so the carrier beads become coated with toner.

(3) Mono-component toner. The toner particles are both magnetic and conductive, and are of intermediate size (say 20 μm diameter). The primary advantage is the simplicity of the system but the larger particle size limits resolution.

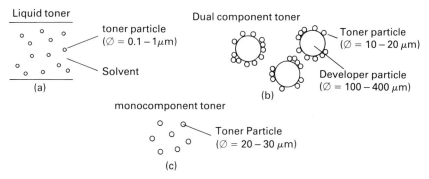

Fig. 2.36 Toner types: (a) liquid toner; (b) dual component toner; (c) monocomponent toner

After development, the image exists on the photoconductive drum as a distribution of toner particles. This is transferred to paper, which is moved past the drum so that the paper speed is the same as the drum surface speed. Behind the paper, at the point of contact between drum and paper, is a corona which charges the back of the paper so that toner is attracted from the drum to the paper. This process is not 100% efficient, and some toner (up to 20%) can remain on the drum and needs to be removed subsequently.

After transfer the toner is still only held on the paper electrostatically, and so a final fixing process is required. This may be done by heat, or heat and pressure, which melts the toner and fuses it to the paper. Heat can be applied by a radiant heater, a heated roller or some combination.

2.7   OTHER PRINTING PROCESSES

There are a few other printing processes which are rare but used in special applications.

Intaglio or recess printing (commonly used on bank notes) is the process of printing from an engraved steel, copper or brass plate or cylinder. The printing of fine tapering lines and a **very thick ink film** are characteristics of this process (Fig. 2.37).

Various press configurations are possible, but all require a very strong press frame indeed. Impression pressure is typically achieved with hydraulic rams.

The inking roller is typically like a flexo printing cylinder, that is, it has raised parts which transfer the ink on to the image areas only. A very thick ink film is transferred so that the engraved image is completely flooded with ink. The excess ink is then wiped off by a counter rotating roller which in turn is cleaned in a bath of solvent or by some other method.

Ink drying is typically by infra-red heaters.

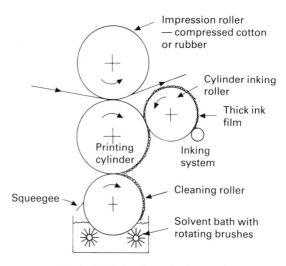

Fig. 2.37 Intaglio printing unit

Cylinder manufacture can be by a variety of methods including etching, but each print works tends to develop its own in-house methods.

Collotype printing reproduces illustrations in continuous tone — it is a screenless process which involves printing from a plate coated with gelatine. It is a difficult process to control and only relatively short runs are possible. It is still used very rarely for fine art reproductions and until a few years ago was used for some postcards.

2.8 PRINT RECOGNITION

It is not always possible to make a positive identification of the printing process used since the appearance characteristics of some processes are rather similar. However, in most cases the following features, examined under a magnifying glass, will provide a good guide:

(1) the edges of type characters;
(2) the appearance of halftone dots and solids;

(3) presence of embossment on either the top or reverse side of a printed sheet;

(4) the ink film thickness.

The kind of stock on which the print is made can modify its appearance considerably but may also be a clue to the printing process used. Also some processes have particular areas of application and are more likely to be used for certain kinds of work.

Letterpress

Letterpress prints are identified by reverse side embossment and 'ink-squeeze' or 'squash'. The absence of reverse side embossment, it should be noted, does not mean a print is not letterpress. However, the edges of letters are particularly characteristic showing a heavy rim of ink, slightly separated from the main body of the character. The amount of squash is influenced by a number of factors. On a rough, uncoated paper, the rim of ink is less clearly defined than it is on the smoother coated paper. The printing pressure, amount of ink applied and the material of the printing plate will also affect the degree of squash. Letterpress halftone dots are essentially circular; in the mid tones they form a checkerboard pattern and in dark tones they join together. They exhibit squash in the same way as the edges of letters. A highlight dot often appears as a rim of ink with a

Fig. 2.38 Letterpress print: (a) uncoated paper; (b) coated paper

hollow centre. Solid areas, particularly on uncoated papers, often show white spots on the print where the ink was unable to reach the bottom of the pits in the paper surface. This is called 'non-bottoming'. This can be minimised by increasing pressure and ink film thickness but this also increases ink squash on halftones, thereby changing tonal value. This degradation of quality is more pronounced the finer the screen, so coarse screens have to be used for halftones printed by letterpress on uncoated papers (see Fig. 2.38).

Flexography

Flexography is a relief process like letterpress, but of course uses rubber plates and very fluid inks. Ink squash is therefore a significant characteristic but is kept to a minimum by using very light pressures. Other characteristics of letterpress printing are, however, absent so there should be no possibility of confusing the two processes. There is no embossment of the sheet and because of the ability of the rubber plates to conform to the surface contours of the printing stock, non-bottoming should not arise. Flexography is widely used for packaging and especially where this involves plastic films, metal foils and waxed papers. The substrate and application are therefore also good guides to flexo printing (see Fig. 2.39).

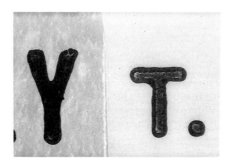

Fig. 2.39 Flexographic print: (a) on film; (b) on paper

Lithography

The main characteristics of a lithographic print are the uniform ink coverage and the absence of squash. There is no rim of ink around a type character, and even on rough paper, non-bottoming does not occur since the rubber blanket conforms well to the paper surface. Halftone dots are of the same shape and pattern as letterpresss dots, but the ink coverage is uniform and the smallest dots are smaller than is possible with letterpress. The edges of dots are fairly well defined on smooth coated stocks, but have rather fuzzy edges on uncoated stocks, so the dot pattern is less distinct. There are two defects specific to offset lithography which assist in identification if they happen to be present on a print. The first is scum which occurs when the non-image areas of a plate start to accept ink for some reason. Scum appears under magnification as randomly distributed fine spots of ink. A superficially similar fault is that of tinting which occurs when the ink becomes emulsified in the water. When this occurs, a thin wash of colour appears over the whole print which under *high* magnification can be seen to be made up of minute droplets of ink. (See Fig. 2.40).

Offset letterpress

A process which should be mentioned here is offset letterpress (also sometimes called dry offset or letterset) which combines features of both letterpress and lithographic printing. In this process, the image is raised — as in letterpress — but is offset on to a rubber blanket before printing on to paper. Prints made by this process can easily be confused with lithographic prints, and there is no certain means of identification. It is much

Fig. 2.40 Offset lithographic print

less common than lithographic printing and is mainly used for printing simple line, tone and solid designs, often for business forms and continuous stationery.

Gravure

Gravure printing exhibits the characteristic cell pattern, which gives a saw-tooth edge to lines and especially type characters. However, care is needed since type is sometimes screened at a finer screen ruling than the illustrations and this reduces the sawtooth effect. Also, special techniques have been used on mechanical engraving systems which also reduce the effect, making it difficult to be sure from this aspect alone. To produce a fully covered solid, ink has to flow out of each cell and join up to obliterate the cell wall pattern. If this action does not go to completion, white spots can be seen on the solids corresponding to the diamond-shaped areas between groups of four cells, and sometimes white lines corresponding to cell walls. The cell wall formation can, in any case, generally be seen in the darker mid-tones where the cell join up is not complete. If the ink is too fluid, a fully covered solid is produced, but with the mottle pattern.

The different gravure processes also have different appearances. Conventional gravure produces cells of approximately the same area, but of different depth to reproduce a range of tones. Cells only print as individual squares transferred from the two back edges of the cells giving a V-shaped pattern and as the cells increase in depth, ink first transfers from all four edges to give a cell outline, this then gradually fills in until a fully- covered square is produced (see Figs 2.41 and 2.42).

Invert or halftone gravure uses cells which vary in area to produce the tone range. Light tones are printed from small round cells, darker tones from larger cells similar to conventional gravure. The light to mid-tones can appear very similar to lithography, or even letterpress since the ink tends to transfer from the edges of cells, producing a rim of ink with a hollow centre — sometimes referred to as a 'doughnut'. This can be confused with 'squash'.

Two defects are particularly characteristics of gravure printing. Sometimes, thin streaks of ink will be seen running through the print. These

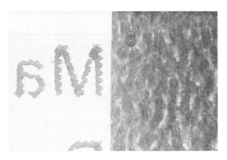

Fig. 2.41 Gravure print

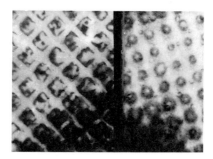

Fig. 2.42 Gravure prints: (a) conventional gravure; (b) halftone gravure

are called 'doctor streaks' and are caused by either a nick in the doctor blade, a piece of grit lodged under the doctor blade or a score in the cylinder surface. The other defect is called 'skipping' or 'speckle'. It usually appears in the light to mid tones and is caused by white spots where ink has failed to transfer from one or more cells. It is caused by rough paper and will occur wherever there is a pit in the paper surface larger than a gravure cell. In this case, the cell cannot make contact with the paper and no ink transfers.

Screen printing

The main characteristic of screen printing is the thickness of the ink film, which may be as much as ten times that deposited by letterpress. It can often be felt by running the fingertips lightly over the printed surface. Even with 'thin film' inks the deposit is still much thicker than that obtained from other processes. The print will also be patterned with the mesh structure of the screen fabric. This is most apparent at the edges of a print and gives an irregular zig-zag outline. Type characters also show this but generally, much care is taken to minimise it so the effect may not be obvious.

Non-impact printing

It is difficult to give guidelines on the recognition of non-impact processes since they are so varied. Ink jet processes usually use a fairly coarse matrix for type characters, and this is commonly found on packaging materials. The highest resolution ink jet system currently works at about 8 dots/mm so resolution will always be low compared with other printing processes. Ink jet images also tend to have a flat and matt look about them, and are low in density.

Electrographic prints, especially if of type only, can appear very similar to lithographic printing. However, the resolution of about 12 dots/mm will only allow the reproduction of very coarse halftones (say 30 lines/inch screen). Such prints can often be easily scratched off the surface and are in monochrome only, usually on A4 paper. Colour printers do exist but are relatively rare.

2.9 SUBSTRATE SELECTION

Paper requirements for printing can be classified into three main areas:
(1) general paper properties;
(2) runnability;
(3) printability.

General paper properties

Paper has a number of physical and chemical properties which influence printing characteristics. The main ones are:
(1) smoothness and softness;
(2) absorbency;
(3) surface strength;
(4) hygrostability;
(5) acidity or alkalinity;
(6) opacity;
(7) gloss;
(8) colour.

Optical properties that govern the visual appearance of paper — colour, opacity and gloss — will be of equal importance whatever printing process is used, as will absorbency. The level required of these properties will depend on the nature of the work being done.

The requirements of other properties — such as smoothness, surface strength, hygrostability and acidity — vary with the printing process being used.

Runnability

Most research on the suitability of paper for printing has been concentrated on its ability to produce a good quality print, but there is little point in buying paper of high printability if it is incapable of running through the press from feeder to delivery. This runnability requirement becomes of greater importance as press speeds increase. A paper that may run well can be ruined by bad handling and storage. For sheet-fed presses, good trimming, flatness and even moisture content are vital. Web printing imposes many additional requirements since web breaks are very undesirable.

Printability

The printing processes have certain requirements of the substrate. For letterpress printing, smoothness is the most important property. If a rough paper is used, solids will not 'bottom' properly. This can be overcome by increasing the ink film thickness but then dot squash will increase and coarser screens must be used for halftones.

For offset litho printing, the paper must have a strong surface. If fibres come away from the surface during printing, they will build up on the offset blanket, decreasing ink transfer. To return the print to normal, the press has to be stopped to wash the blanket. It is impossible for the printer to maintain constant colour throughout the run.

The deposits on the blanket accept water rather than ink, produce white fibrous markings on solids, and interfere with halftone printing, causing loss of contrast and definition. The fluff deposit on the blanket will wear the plate, especially if it contains abrasive loading materials from the paper.

Coated papers for offset must be strong enough to resist the pull of the tacky ink, otherwise areas of coating may be pulled away from the surface, giving a defect known as picking.

Paper for all processes, but especially litho printing must be treated properly or problems will occur. Creasing down the centre of the sheet, for example, happens on tight-edged paper. If a stack of paper is left standing in air that is too dry for it, the edges will lose moisture and shrink. The sheets of paper become saucer-shaped, with tight edges and a baggy middle. When this paper passes through the nip on the press, it is mangled flat and the baggy middle forms a crease.

Stacks of paper left standing in an atmosphere that is too damp will develop wavy edges, and these sheets will also crease when printed, but the creases are usually nearer the edges. Even if the distortions are not big enough to form creases, misregister will occur.

Creases are more likely in offset litho than in other methods of printing, due to the overall contact of the rubber blanket, both in printing and non-printing areas.

Paper must be reasonably dimensionally stable, i.e. change as little as possible with changes in humidity. It must also have the right moisture content for the printing conditions it is to encounter. Care must be taken to protect the edges of stacks from atmospheric variations, by using moistureproof wrappers and stack covers.

The problems of casein scumming have been mentioned above, although such coated papers can be printed without difficulty, provided the casein has been hardened so that it is insoluble in water and does not transfer to the plate. The problem is not often seen now since it is well known and therefore avoided.

This is an example illustrating an important property of all offset lithographic papers. They should not contain water-soluble materials that would adversely affect the plate if they transferred to it in the damping water. Paper coatings should be insoluble and unaffected by contact with water. If a coating swells in water, it will be weakened, and picking may occur on the second colour which is being printed on the weakened surface, even though the surface strength may be adequate when dry.

Gravure can be used to print a wide range of substrates including plastic films. Smoothness, under pressure, is perhaps the most important substrate property otherwise speckle occurs. Thus rough papers which are highly compressible may print satisfactorily while some 'higher quality' coated papers may not. The problem may be reduced by using electrostatic assist.

Flexographic printing also prefers a smooth substrate (for similar reasons to letterpress printing) and is extensively used on plastic films and foil. The rubber or polymer plates used do, however, provide some latitude and rough materials such as paper sacks and corrugated cardboard can be printed.

Screen printing is perhaps the most tolerant of all processes and can be made to print on almost any substrate, and on many pre-formed containers, providing the ink can be suitably formulated.

Electrographic printing generally has quite stringent requirements, and the faster the particular printer runs, the more exacting these requirements are. Often the properties are more to do with paper handling than printing — many of these printers heat the paper strongly to fuse the toner (e.g. 200 °C) and excessive curl and brittleness must be avoided. The paper surface is, however, important. Non-coated papers tend to give better image adhesion since the toner can key on. Difficulties can be experienced in trying to print over pre-printed areas (e.g. tinted areas on forms) — image adhesion is poor and tends to produce a form of set-off.

Ink-jet printing can be used on a wide range of substrates but since a very fluid ink is used, the final dot size printed is affected by the wetting characteristics of the substrate's surface. Quite objectionable dot spread can occur in bad cases.

Substrates should be selected not just to suit the characteristics or requirements of the printing process but also those of any subsequent processing stages and the end use of the product. Surprisingly, this is frequently overlooked. Thus, if a job is to be laminated after printing, the substrate and inks must be appropriately chosen (and an excess of anti-set off spray powder avoided). Prints on matt coated papers tend to have poor rub resistance, and the resulting product will be disappointing if the finishing processes involve rough handling. Much packaging work involves not only coating or laminating after printing, but cutting and creasing. Cracking of folds is a frequent problem, partly caused by wrong choice of materials. Subsequent performance on the packaging line is also important and depends very much on substrate properties.

2.10 THE NEED FOR COMMUNICATION

Almost all printing requires communication between the parties involved. These may be:

— the customer, e.g. a publisher or packaging company;
— the designer;

— a typesetting trade house;
— a colour reproduction trade house;
— an advertising agency;
— the printing company;
— a finishing processes trade house;
— the paper supplier;
— the ink manufacturer.

In order to end up with a satisfactory final product it is important that each organisation involved provides necessary information to the others. For example:

(1) It is helpful if the customer discusses the required substrate with the printer and finishing trade house. This should take into account the use to which the printed product is to be put.

(2) The printer should inform the colour trade house of the printing specification required so that separations can be made with correct tonal gradation, colour correction and dot gain allowance.

(3) The designer will hopefully choose typefaces appropriate to the printing process and substrate to be used. The reasons for a choice should be given so that arbitrary changes are not made at a later stage.

It is really quite difficult to supply too much information in relation to a print production.

CHAPTER 3
Colour and Colour Matching

Colour surrounds and affects us all. This is not new. Primitive man, we may assume, was depressed by a grey sky, cheered by a blue one, and awed by the sight of a full rainbow. Certainly the colours which could be obtained from the earth and plants intrigued him to the extent that he used them for body painting and to decorate his cave dwelling. Colour in commerce followed swiftly. Cloth, glass, stones and pottery, goods that were aesthetically pleasing, could be traded for food. This commercial aspect has continued to grow through the years until almost everything in our lives is coloured. Books, films, television, packages, and now more and more newspapers, appear in colour.

In recent years people involved in marketing have realised that colour sells. Brand names are frequently associated with house colours and manufacturers have become extremely demanding in their specifications. Ink manufacturers and printers are therefore under increasing pressure. Much of the basic physics and chemistry of colour has been understood for over 100 years. Even some of the complex mathematics involved in colour measurement and match prediction was explained 50 years ago [1]. We could not at that time measure colour with sufficient accuracy nor process the enormous quantity of data quickly enough to make effective use of these developments.

Colour measuring instruments which are much more sensitive than the human eye are now available. These can be directly connected or interfaced with relatively cheap computers capable of processing data at incredible speeds.

3.1   THE PHYSICAL NATURE OF COLOUR

When as children we learn to recognise the colour of everyday things, we are shown examples of objects which we are told are yellow, red or blue. Objects that are black, grey or white are not described as coloured, but the colour scientist must include them. The difference between the black, grey or white objects and the yellow, red and blue objects can be described in terms of hue. The former do not possess a hue, but they can be described in terms of their lightness. If we take a matt black object, like a piece of black velvet, we can say that it has no lightness at all, but if on the other hand we look at a piece of white chalk we can say that it has maximum lightness. The colour physicist in fact uses a block of pure barium sulphate as his 100% white object. A series of neutral grey objects can be arranged in order of their lightness, between the black and the white. If we call the black 0% light and the pure white 100% light, we can give values to our greys.

Now let us consider a similar arrangement of objects that have what we generally term colour, or more correctly hue. Newton [2] demonstrated that what we call white light can be split up into a full range of hues, namely red, orange, yellow, green, blue and violet, using a glass prism. This he called the visible spectrum. This natural sequence of hues can be arranged in a circle since after blue we have a red tone blue, we call violet, which can lead us through purple back to red.

This circle can be divided quantitatively like the lightness scale into say 100 divisions.

There is a third dimension to colour which is called intensity, strength, purity, vividness, or better still, saturation. If we take a pure white paint and add more and more pure blue paint to it we can make a range of specimens ranging from pure white through a series of increasingly strong blues to pure blue. This could be repeated for all the different hues, and again using all the greys to produce an almost infinitive range of colours. Each step can, as with the lightness and hue, be quantified.

We therefore have a description of colour in terms of three independent variables, lightness, hue and saturation. This is why it is impossible to arrange logically a full range of shades on an ink maker's shade card, which has only two dimensions. The answer is to arrange them in three-dimensional colour space, a concept we shall return to in considerable detail later in the chapter.

With a very few exceptions, objects do not emit coloured light, but only look coloured when they are illuminated. An object that we describe as white in normal daylight, looks as it does because it reflects most of the light that falls on it. If the object absorbs all of the light it appears black, but if it absorbs some portions of the spectrum more than others it will appear coloured. For example an object that absorbs only blue light will appear yellow, and one that absorbs red light will appear green–blue, or what we call cyan.

The term reflectance needs more careful definition. A good mirror reflects most of the light falling on it, but we do not call a mirror white. The mirror reflects all the light but also the images in front of it. This type of reflectance is called specular. On the other hand a wall painted with a matt white paint again reflects most of the light, but not images. It looks like this because the white pigment particles in the paint scatter the light in all directions. This type of reflectance is called diffuse. A wall painted with a glossy white paint will exhibit both specular and diffuse reflectance.

So far only white light has been considered. If the light used to illuminate the object is deficient in some parts of the spectrum, it too will be coloured. Light from a sodium vapour street lamp is pure yellow, containing no blue. The blue vehicles therefore look almost black under these conditions.

As Newton observed, however, the rays themselves are not coloured, it is only when they react with an object or the eye that the sensation we call colour is perceived.

The colour we see therefore depends not only on the nature of the object but also the light used to illuminate it.

Light sources

If a perfectly black object like a carbon rod becomes hot, the molecules begin to vibrate and emit energy, some as light. The more intense the heat, the brighter the emitted light, but not only does the light become brighter it changes colour. With increasing energy the colour of the light shifts from red towards the blue end of the spectrum.

It is convenient therefore to specify a light source as being equivalent to the so called black-body radiator at various temperatures. This is known as its colour temperature. The Kelvin scale is used (°Kelvin = °Celcius + 273). A tungsten filament lamp has a colour temperature of about 2800 K and north daylight about 6700 K [3].

The actual light sources, however, differ from this simple idea, and some standardisation is necessary. An international committee, the Commission Internationale de l'Éclairage (CIE) set up standards for illumination in 1931. They were called simply A, B and C. Illuminant A represents a tungsten filament lamp of colour temperature 2800 K, B sunlight at 4900 K and C north daylight at 6700 K. Most colour matching is carried out using north daylight, and it was found that the CIE illuminant C contains less UV than true north daylight. When a fluorescent component, for example an optical brightening agent is present the matchings made under illuminant C are unsatisfactory. A new illuminant of colour temperature 6500 K, with a UV content near to north daylight, has been introduced. This is known as D65. The more efficient fluorescent tube lamps are now more frequently used in shops, factories and laboratories than the tungsten filament lamp.

The primary source of light in the fluorescent tube lamp originates in the collision of mercury atoms during the passage of an electric current. The light emitted from this source is made up of the yellow, green and violet line spectrum of mercury. The inside of the tube is coated with a substance which fluoresces in the mercury discharge. These substances are known as phosphors and most of the light emitted by these lamps is produced as a result of this fluorescence. A number of phosphors are blended to produce lights of different colour temperature. Typically a cool, that is blue tone, lamp, will have a colour temperature of about 4000 K and a warm, redder tone lamp about 3000 K. A new range of high efficiency lamps using rare earth phosphors, similar to those employed in colour television tubes, have recently been introduced by Philips as their Colour 80 series. The light from these, although showing a good overall colour balance, is emitted in three relatively sharp bands, rather than the more normal wide band of continuous radiation.

Problems can arise when matching prints and fabrics coloured with different pigments and dyes. Two specimens may match under daylight, but not under artificial light. This phenomenon is known as metamerism. The problem can become more acute with the high efficiency lamps. Special fluorescent lamps are available for colour-matching areas of the laboratory and special colour-matching booths. These lamps emit light evenly over the whole of the visible spectrum [4]. Figure 3.1a shows a range of illuminants and Figures 3.1b, c and d are spectra from various fluorescent lamps.

3.2 THE PERCEPTION OF COLOUR

The eye

A cross-section of the eye is shown in Fig. 3.2. Light enters the eye through the cornea and is brought to focus on the light-sensitive retina by the lens. The light-sensitive cells in the retina can be divided into two types, rods and cones. The rods are the most sensitive cells, but cannot detect colour. The cones are concentrated in the central area of the retina, around the fovea and are sensitive to colour. It is now generally accepted that there are three types of cone, sensitive to red, green and blue light. At low levels of illumination, when only the rods are activated, the vision is said to be scotopic. At higher levels, when colour can be perceived through the cone cells the vision is photopic.

The sensitivity of the eye varies greatly with wavelength. The rods are most sensitive at 507 nm and the cones at 555 nm (Fig. 3.3) [5].

Defective colour vision

The spectral sensitivity of the eye varies from one person to another Wright studied the response of 10 observers so that a standard observer could be defined.

The majority of people have vision which corresponds quite closely with this standard, but about 5–8% of all men, and a very much smaller number of women, differ markedly from this standard and are quite often termed colour blind. True colour blindness, the inability to perceive any colour, is extremely rare. The most common defect is the observer's inability to distinguish red and green at low light levels. This red–green blindness varies in degree, but it is important to be able to detect it, since such people will generally be unable to match and assess the colour of inks and prints. The most satisfactory test was devised by Ishihara [6]. His test consists of a book of 38 coloured circles made up of a large number of coloured dots. The person under test is asked to read a number or trace a line on each plate. A reference book is provided for assessment, and if the test is carried out correctly it gives very reliable results. Even people with normal colour vision differ in their ability to match colours. Attempts have been made to assess colour matching aptitude, by selection of plastic chips made deliberately close, but not exact matches, to standards. This

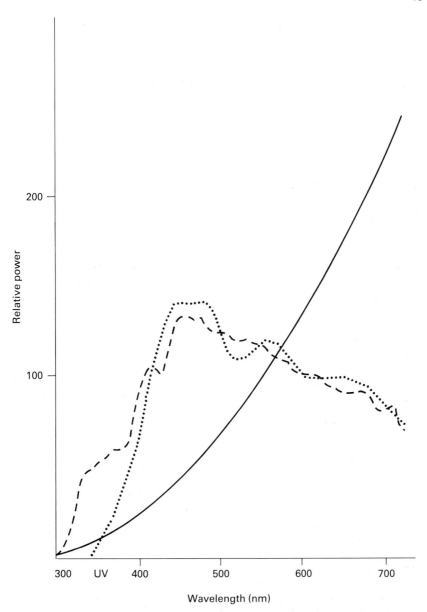

Fig. 3.1a A range of illuminants; (———) a tungsten filament lamp; (— — —) D65 north daylight, new CIE standard; (⋯⋯) C north daylight, original standard (courtesy of Philips Electronics)

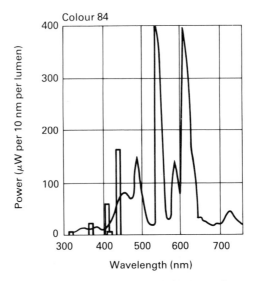

Fig. 3.1b Colour 84 (courtesy of Philips Electronics)

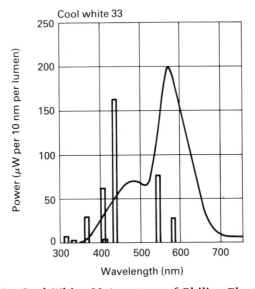

Fig. 3.1c Cool White 33 (courtesy of Philips Electronics)

type of test is not generally very satisfactory, and the best colour matchers are selected by practical ink-matching tests after a period of training.

Both the cornea and the lens become yellower with increasing age and this leads to a decrease in the blue sensitivity. This is not generally a big problem, but it should be borne in mind when the inevitable arguments on the acceptability of a match arise. 'Yellowness of vision' test equipment is available. [7].

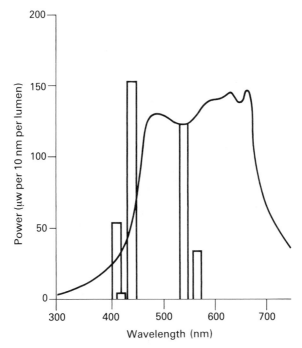

Fig. 3.1d Spectral power distribution—Graphica (courtesy of Philips Electronics)

Chromatic adaptation and colour constancy

If a print is viewed in a light cabinet under simulated daylight and the illumination is quickly changed to tungsten light, the colours will change. The most obvious change will be that the whites will turn yellow. After a very short time, however, the eye adapts to this change and again sees the white as white. This phenomenon is called chromatic adaptation.

When the eye is adapted to the illuminant change, the colour of most common objects will be accepted as perfectly normal, although the spectral composition of the light reaching the eye will have changed appreciably. The blues for example will appear much redder, but will still be described as blue rather than purple. The objects viewed under these changing conditions are said to possess colour constancy. Some objects do not possess colour constancy and the change in illuminant will result in a change in description. Some light sources, for example certain fluorescent lamps, also change the perceived colour in such a manner that the colour description is changed.

Metamerism

If the colourants in two objects differ appreciably in their spectral distribution, they may match in one illuminant but not in another. This is known as metamerism. It becomes a problem when an ink matcher uses a series of standard bases which differ from those used in the original

Fig 3.2a

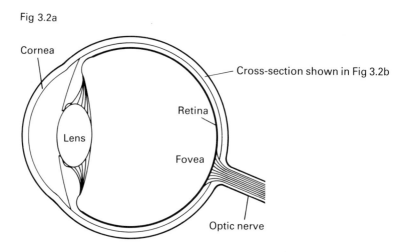

Fig 3.2b

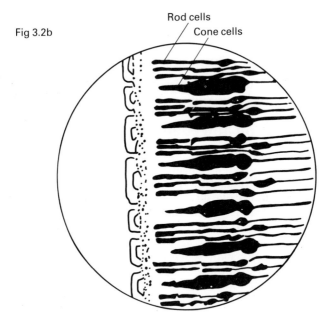

Fig. 3.2 (a) Section of the eye; (b) section of the surface of the retina

pattern. It is even more troublesome when an ink for a wallpaper has to match a fabric dye. The ink maker cannot use the same colourant and some degree of metamerism is inevitable.

The spectral reflectance curves of metameric matches always cross one another, sometimes several times (Fig. 3.4).

The eye sees them as a match under say D65 illuminant since the differences are averaged out in each of the red, green and blue regions. If

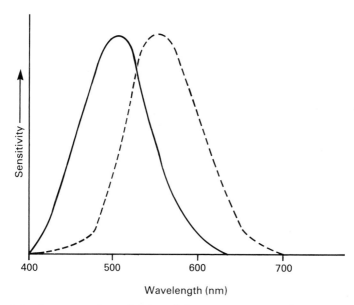

Fig. 3.3 The spectral sensitivity of the eye: (———) rods; (– – –) cones

the spectral profile of the illuminant changes, this may not then be the case and the shades do not match. The effect is particularly noticeable with fluorescent lamps (Fig. 3.1). If differences in the spectral curves of the specimen correspond with one or more of the sharp emission bands present in this type of illuminant marked colour differences will be seen in that region.

Matching specimens which differ in gloss is always troublesome. If the illumination is diffuse it may be possible to assess the colour differences but if the illumination is highly directional the top surface reflection, the specular component, dilutes the colour with white light, in other words, desaturates the glossy specimen. The specimens then may match when viewed at one angle but not at another. This phenomenon is termed geometric metamerism.

Dichroism

When assessing the colour of an ink, it is important to standardise the application to minimise differences due to many factors, a very important one being film thickness.

Apart from the obvious variation in apparent strength, film thickness also affects the hue. This change in hue with film thickness is called dichroism [8].

Illumination quality and levels

From the foregoing discussion it will be evident that if satisfactory matching of colours is to be made the quality and the level of the illumination used must be carefully standardised. All too often disagree-

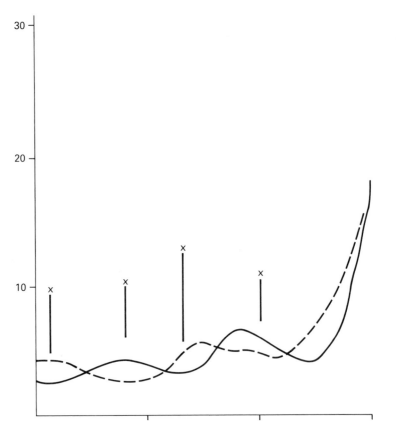

Fig. 3.4 Spectral curves of metameric browns: (———) brown 1; (— — —) brown 2; x = position of main absorption bands of a high efficiency fluorescent lamp

ments arise because of the use of non-standard lamps in the matching cabinets. The cabinets should be carefully maintained and cleaned. Objects not under test should be removed from the cabinet as these can change the illuminant effect and distract the observer.

The size of the specimens should be controlled. A very small specimen is difficult to assess against a very large standard. A minimum of 50 mm square should be used if possible.

3.3   ADDITIVE AND SUBTRACTIVE COLOUR MIXING

The additive primaries

The visible spectrum can be divided into three equal portions, the region from 700 to 600 nm being red, 600 to 500 green and 500 to 400 blue. It is impossible to produce light filters that will transmit 100% of each portion with a cut off exactly at the boundaries, but filters can be produced to give curves as shown in Fig. 3.5.

If red, green and blue filters are placed in front of three separate projectors and the colour spots allowed to superimpose, the red and green together will produce yellow, the green and blue, cyan and the blue and red, magenta. All three together theoretically produce white, but since in practice they differ from the ideal primaries the result will be somewhat grey. Since we are adding these lights to produce the colours, they are known as additive primaries.

The subtractive primaries

The printing of colour relies on the selective absorption of certain regions of the spectrum by dyes or pigments in the inks. Again the spectrum can be divided into three equal bands. If the blue third 400 to 500 nm is absorbed, the colour will be seen as yellow, likewise the removal by absorption of the green third 500 to 600 nm will produce magenta and removal of the red third 600 to 700 nm will give cyan.

It is theoretically possible therefore using yellow, magenta and cyan inks to produce all the required colours. Superimposition of all three producing black. This is a subtractive colour system (Fig. 3.6).

In fact the actual inks differ from the ideal subtractive primaries just as the filters differed from the ideal additive primaries. In process printing, therefore, it is necessary to correct this imperfection by the addition of a black printer and modifications of the primaries using colour masks. This procedure is dealt with under the graphic reproduction section.

The CIE system

In 1861 Clerk Maxwell [9] studied ways of representing the additive primaries, red, green and blue, graphically so that a quantitative description of colour could be obtained. He arranged the primaries at the three points of an equilateral triangle.

If we refer to the colour diagram of the additive primaries (Fig. 3.5), it can be seen that equal parts of blue and green produce a blue–green colour called cyan. This can be represented on Maxwell's triangle at a point midway between the blue and green points. Similarly red and green produce yellow; red and blue produce magenta. Addition of all three primaries gives white which Maxwell put in the centre of the triangle (Fig. 3.7). If we start at the middle and move along a line to the blue point, we pass from white through increasingly saturated blues to the completely saturated blue primary. Similar lines can be drawn from the white point to the green and red primaries.

If now we carry out a similar exercise with the intermediate colours, cyan, yellow and magenta, we find that point C on the triangle, while being correct for hue, is not a completely saturated cyan. It is possible to produce a more saturated cyan light than that made by adding green and blue lights, since these are not ideal primaries. The saturated cyan will be at a point C_1 outside the triangle. This presents problems if we wish to

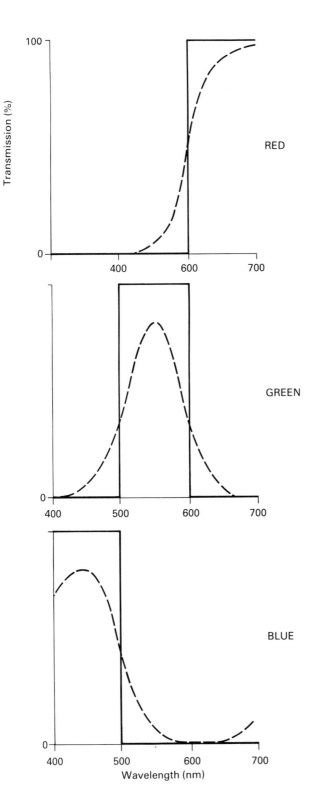

Fig. 3.5 The additive primaries: (———) spectral profile of an ideal filter, passing exactly ⅓ of the total spectrum; (— — —) actual profile of a filter

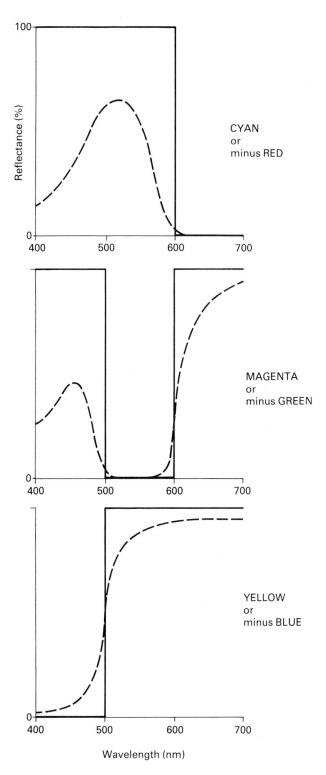

Fig. 3.6 The subtractive primaries: (———) spectral profile of ideal inks absorbing exactly ⅓ of the spectrum; (— — —) actual inks

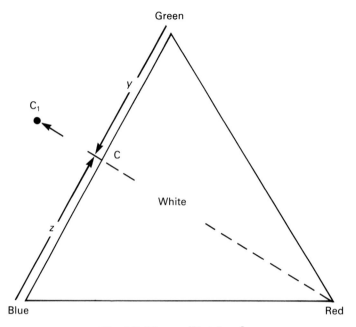

Fig. 3.7 Maxwell's triangle

quantify the colour. The desaturated cyan at point C can by described in terms of positive amounts of blue and green, say y parts of green and z parts of blue, but since the saturated cyan C_1 is outside the triangle we can only describe it by introducing a third quantity. This lies on the line from the red point through the centre, but since it is beyond the centre white point the values will be negative. We then have to describe the point C_1 on this diagram as being made up of y parts of green, z parts of blue and minus x parts of red, i.e.

$$C_1 = yG + zB - xR$$

If now all the saturated colours are plotted, they will be found to lie outside the triangle, but touching the red, green and blue points. This line is known as the spectrum locus (Fig. 3.8).

To avoid negative quantities in our description of a colour, the locus can be enclosed in a triangle X, Y, Z (Fig. 3.9).

In such a diagram the points X, Y, and Z represent colours which are more saturated than actual colours, they are therefore unobtainable with any light source and are described as imaginary primaries [10]. X is a reddish purple of a higher saturation than a real colour, Y has a saturation higher than any green and Z higher than any blue. Using this system we can define a colour by measuring it with an instrument and obtaining values for X, Y and Z.

These values can be represented as ratios:

$$x = \frac{X}{X + Y + Z} \qquad y = \frac{Y}{X + Y + Z} \quad \text{and} \quad z = \frac{Z}{X + Y + Z}$$

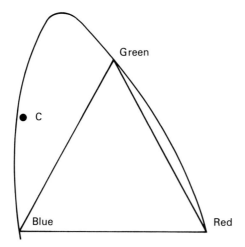

Fig. 3.8 The spectrum locus

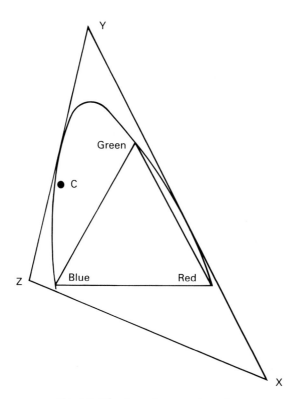

Fig. 3.9 The imaginary primaries

then $x + y + z = 1$.

It is therefore only necessary to use two of these values, known as chromaticity co-ordinates, to define the hue. The values can then be plotted on ordinary graph paper.

This description of a colour does not take luminance or brightness into account. Two colours can have the same x and y values, therefore the same hue, but differ in luminance. To quantify luminance it was decided to define the X and Z stimuli mathematically in such a way that they have zero luminance, while the value of Y, deliberately chosen to equal the response curve of the eye, gives the luminance information.

This system was adopted by the Commission International de l'Eclairage (CIE) in 1931. A full explanation is given by McLaren in his book *The Colour Science of Dyes and Pigments* [11].

3.4 ORIGINS OF COLOUR IN PRINTED MATERIAL

Nearly all printing inks contain three basic ingredients, a colourant which may be a pigment or a dye, a resin system and a thinning solvent.

Pigments

Pigments are finely divided particles relatively insoluble in the medium in which they are employed. These particles when adequately dispersed in the medium, absorb and scatter light. If the absorption is selective, the pigment will be coloured. Pigments may be entirely inorganic or organic; they may also be metallic (usually calcium or barium) salts of complex acids, or consist of organic dyes laked on to inorganic substances.

Dyes

If the colouring agent is soluble in the medium in which it is being used, it is called a dye. Dyes are normally pure organic or metallo-organic complexes. The colour of a dye is produced by selective absorption, but since no discrete particles are present, no scatter occurs and the system is transparent.

Origins of colour [12]

The configuration of atoms in the organic pigment molecules which give rise to selective absorption of light are called chromophores. Examples of these are:

$$-C=C-, \quad -C=O, \quad -C=N, \quad -N=N-, \quad -N=O$$

Certain other groups, while not themselves responsible for selective absorption, enhance the absorption of the chromophore and are known as auxochromes. Common auxochromes are:

$$-NH_2, \quad -OH, \quad -NO_2, \quad -CH_3, \quad -Cl, \quad -Br$$

The colour of inorganic pigments is also a function of chemical composition, but is strongly influenced by the crystal form of the substance.

Transparency and opacity
Since the average particle size of pigments is close to the wavelenghts of visible light, i.e. 0.4–0.7μ, the ordinary laws of optics do not apply, and radiation is scattered by these small particles. Scatter is a complex phenomenon, primarily dependent upon the relative refractive indices of the pigment and the medium in which it is dispersed, and the particle size. Particle shape, degree of aggregation and the wavelength of the light also affect scatter.

The opacity of a pigment depends normally upon its ability to scatter incident light, but absorption also plays a part, and the overall opacity is due to a combination of both properties. In the case of a white pigment, for example titanium dioxide, little or no absorption occurs and the opacity is due almost entirely to scatter. As the refractive index of a pigment approaches that of the medium in which it is dispersed, the amount of light scattered decreases and the pigment becomes more transparent.

Transparency of a pigment therefore is partially dependent upon the medium used to disperse it. The transparency of an ink is usually expressed in terms of the ratio of reflectance of the film printed over a black background to that over a white background. A 1–5 scale is employed, 5 indicating high transparency.

Colour strength
The colour or tinting strength of a pigment or dye is a measure of its ability to impart colour to a system. Although the colour strength of a pigment is primarily related to chemical composition, particle size and distribution, the actual value in the case of a printing ink will depend also upon the film thickness and the concentration of pigment in the ink. Colour strength can be measured either by reduction with a standard white pigment in a suitable medium or by measuring the pigmentation level required to give a standard-depth print at a specified film thickness.

Substrate effects
Since most printing inks are more or less transparent, the substrate on to which the ink is printed influences the final colour of the print. The colour, absorbency and surface roughness of the substrate, together with the ink film thickness will all affect the final colour of the print.

Substrate effect is the source of most of the problems encountered when using computer-assisted match prediction systems.

Colour Index classification
The most useful classification of pigments and dyes is the Colour Index, prepared by the Society of Dyers and Colourists. The index is prepared in five volumes; the first three list the colourants according to type and usage and give technical information under CI generic names. For example, Lake Red C is listed as CI Pigment Red 53. Volume 4, sometimes

referred to as Part 2 reference, lists the colourants according to their chemical constitution. The above example is listed as CI reference 15585. The final volume lists commercial names and manufacturers' code letters.

3.5 GRAPHIC REPRODUCTION [13]

It was shown in Section 3.4 that it is theoretically possible to produce colour prints having all the possible colours from three inks, cyan, magenta and yellow, each absorbing one-third of the visible spectrum (Fig. 3.6).

The first step in graphic reproduction is to analyse the original object in terms of the absorption of light in each of these wavebands, then print the amount of each ink necessary to produce that effect.

Three-colour printing

Originally the object was always photographed separately through three primary colour filters to produce negatives from which the printing plates were made. This colour separation process and many subsequent stages in graphic reproduction is now often done by very sophisticated electronic scanning equipment. The basic principles, however, are better explained by reference to the camera procedures. When the object is photographed through a red filter, the negative will have density in the red areas only. This negative is then used to produce a positive printing plate, on which the image area corresponds to the red, absorbing, that is the blue-green, region of the spectrum. This will be the cyan printing plate. Similarly the green filter will be used to produce the green absorbing, that is magenta printing plate and the blue filter the blue absorbing, that is the yellow printing plate (Fig. 3.10).

Four-colour printing

Since neither the separation filters nor the printing inks match the ideal primaries, the combination of all three inks will not produce a completely satisfactory black. It will appear as a very dark brown. To remedy this, a fourth impression, the black printer is introduced to strengthen shadows and ensure that the darkest tones are neutral black.

Under colour removal

Rather than just adding black in this way, it is possible to analyse the separation and substitute black for the proportion of the colours producing a neutral tone. This portion is sometimes referred to as the achromatic component; and the process is known as under colour removal.

Figure 3.11 shows an example of the relative concentrations of the three process colours required to produce a particular colour. It can be seen from this diagram that much of the colour is being used to produce

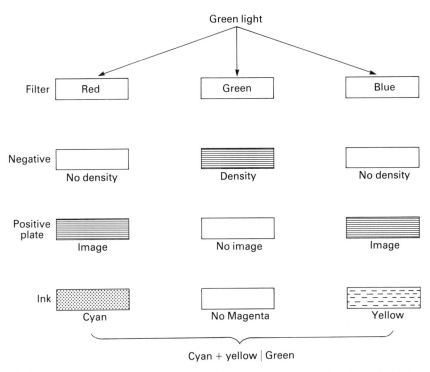

Fig. 3.10 The mechanism of three-colour reproduction, illustrated by the reproduction of a pure green subject

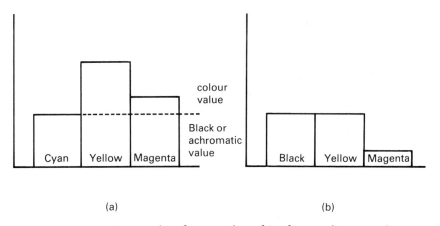

Fig. 3.11 (a) Conventional separation; (b) achromatic separation

black, sometimes known as the achromatic value. All or part of this portion can be substituted with black thereby reducing ink cost, and sometimes improving drying.

Masking

Sometimes a picture will have a bluish, greenish or reddish cast, which is due to an imbalance in the colours. The grey balance may be corrected by masking. The magenta ink absorbs too much blue light and the cyan too much green light. Only the yellow is satisfactory. It is necessary therefore to compensate for these deficiencies. The magenta ink cannot be made to reflect more blue light, therefore the yellow, which absorbs blue, is reduced. Similarly, the cyan cannot be made to reflect more green light therefore the magenta is reduced. Originally these alterations were made by hand retouching, but subsequently these reductions were made by masking the original negatives with low density positives produced from the other negatives, and more recently electronically by the scanner.

Half-tone dots

To produce a full range of shades it is necessary to vary the relative proportions of the three inks in any one area.

In the case of conventional gravure printing, it is possible to vary the thickness of the ink film by changing the cell depth but in the lithographic process a uniform ink film is applied. The variation in colour density in this case is achieved by breaking the image up into a screened dot pattern and altering the size of the dot. Superimposition of the dots varying in size and colour would not produce the required colour effect, they are therefore juxtaposed. Since they are so minute the eye cannot discern the individual dots and a single composite colour is seen.

Dot gain

Throughout the process the size of the dot will change to some extent. The greatest effect is the increase in dot size which occurs during printing. This is partly physical as the fluid ink dots are squashed, and partly optical due to the light scattering effect of the paper. Control of dot gain is important since differences from ink to ink will change the colour balance. Dot size can be measured indirectly with a densitometer using the Murray–Davies equation; the calculation being carried out in some cases with a built-in computer.

Dot generation

The dots are produced in the photographic process by placing a ruled screen in front of the negative. It is now possible to generate these dots electronically. Very accurate control of dot size and shape is possible with this type of equipment.

3.6   THE MEASUREMENT OF COLOUR [14]

Colour is a subjective sensation, therefore in strictly physical terms we cannot measure it. The amount of light reflected from the surface of an opaque object, or transmitted by transparent material can be measured, but the relationship between this instrumental value and some quantitative description of colour is complex. This section will deal with the measurement of the colour of opaque materials, in particular ink applied to a substrate, by instruments known as reflectance colorimeters and spectrophotometers. Those designed for the measurement of transparent materials, are known as transmittance instruments.

All colour-measuring instruments have three main components. A light source to illuminate the specimen, a system to select a specific region of the spectrum either by dispersing or filtering the light and a photo-detector.

Colorimeters

The light source in these instruments is normally a tungsten filament lamp, sometimes halogen filled. The electrical supply is stabilised to ensure a constant light output.

Selection of the region of the spectrum required is by use of three coloured filters designed to correspond as nearly as possible to the three visual stimuli of the standard observer as defined by Wright (Fig. 3.12). The X stimulus has both a blue and a red component, the latter design-ated $\bar{X}$. It is more convenient to use separate X and $\bar{X}$ filters. The so-called tristimulus colorimeter therefore quite often has four filters, X, Y, Z and $\bar{X}$ corresponding to Red, Green, Blue (Z) and Blue ($\bar{X}$). Alternatively, a proportion of the Z reading may be added to the X value to give a true X value.

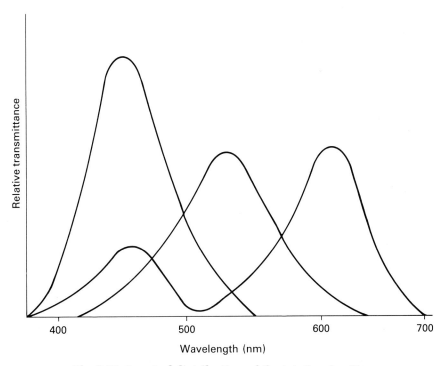

Fig. 3.12 Spectral distribution of the tristimulus filter

Reflected light is measured using a photodetector, usually a photodiode and the value displayed by a galvanometer or printed out (Fig. 3.13). The read-out is usually given as a percentage of reflectance, the zero being set with a matt black substance, for example black velvet and the 100% by a pure matt white block of barium sulphate. In practice these standards are difficult to maintain and secondary standards, usually black and white ceramic tiles are used.

Densitometers

The reflectance densitometer is a special form of colorimeter, used to measure the four process ink colours, black, cyan, magenta and yellow on colour prints. The filters are designed to select light of the wavelength mostly absorbed by these primary colours. A red filter is used to measure the cyan print, a green filter for the magenta print and a blue one for the yellow print. The black is measured using light filtered to correspond to the average response of the eye across the whole spectrum.

High-quality filters are required for densitometers. Typical spectral profiles of the three colour filters are shown in Fig. 3.14. Our assessment of the strength of a colour is not related to the reflectance in a linear manner, but is very nearly a logarithmic relationship. The value used is called density and is related to reflectance as follows:

$$\text{Density } (D) = \log_{10} \frac{1}{R}$$

where
$$R = \frac{\text{Reflected light intensity } (R_1)}{\text{Intensity of light reflected by the white paper } (R_w)}$$

$$\therefore \quad D = \log_{10} \frac{R_w}{R_1}$$

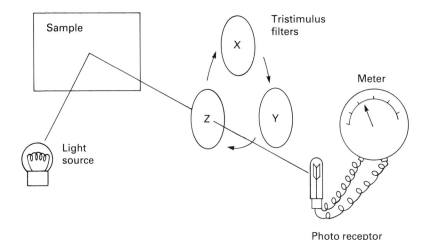

Fig. 3.13 Layout of a simple colorimeter

Colour Plates

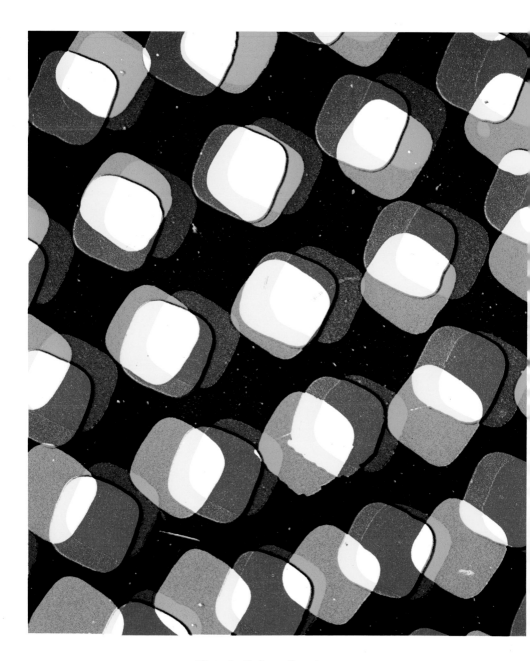

Plate 1. Colour Constancy.

View this reproduction under two quite different illuminants, such as tungsten and daylight, but ensure that the eye is properly adapted to each. You will see that, while the reproductions do not appear identical, they are very similar despite the significant difference in the colour of the illuminant.

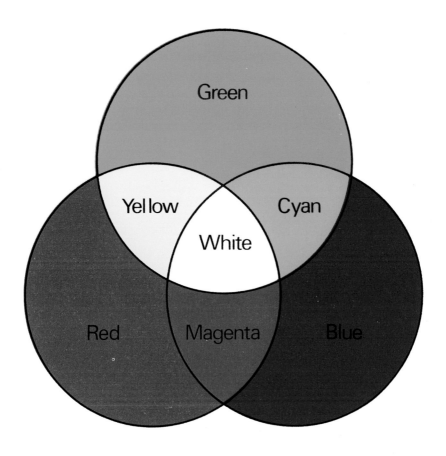

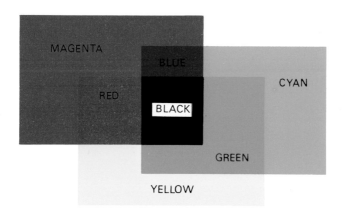

Plate 2. Colour Mixtures.

The result of colour mixture by the three primary colours for (above) additive and (below) subtractive colour mixture. Additive represents the mixing of three rays of coloured light; whereas subtractive represents the mixing of three coloured inks, each of which subtracts one third of the visible spectrum from the paper.

Plate 3. Four Colour Printing.

This shows the results of printing yellow, magenta, cyan, and black printing primaries in combination (above) and singly (below); using lithographic inks to BS 4666:1971.

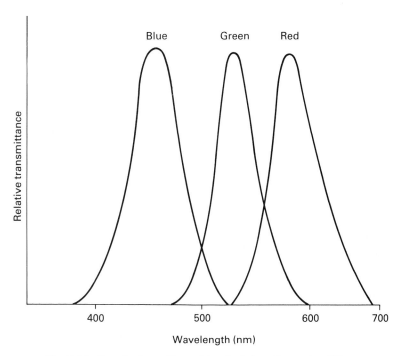

Fig. 3.14 Spectral profiles of typical densitometer filters

The sample is illuminated by a light source. The light beam first penetrates the ink layer and is thereby reduced in its intensity. The remaining light is then partly absorbed and partly scattered by the substrate. The scattered portion passes back through the ink film being further reduced in intensity before reaching the detector (Fig. 3.15).

In order to reduce directional effects, particularly with non-homogeneous samples, the light can be collected by a ring mirror before reaching the detector (Fig. 3.16).

Spectrophotometers

The light source in a spectrophotometer may be a tungsten filament lamp or a pulsed xenon discharge tube, often filtered to match D65 (diffuse north daylight).

Selection of the required wavelength of light is usually by dispersion with a prism or diffraction grating. By scanning the spectrum it is possible to construct a continuous plot of per cent reflected light versus wavelength.

Some instruments used for colour match prediction work employ an array of 16 photodiodes which measure the output from the diffraction grating at 20 nm intervals simultaneously. A complete spectral reflectance measurement can be made in 4 seconds with this equipment (Fig. 3.17). Instruments using a series of narrow band pass filters, typically 16 with peak transmittance values at 20 nm intervals, are known as abridged spectrophotometers.

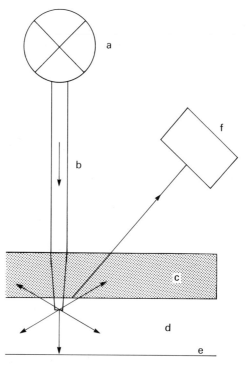

Fig. 3.15 Reflection densitometer: (a) light source; (b) illuminating beam; (c) ink layer; (d) substrate (paper); (e) substrate (film); (f) receiver

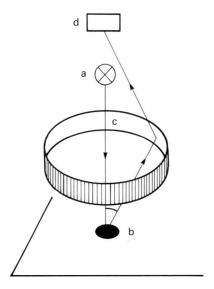

Fig. 3.16 Reflection densitometer: (a) light source; (b) measuring spot; (c) ring mirror; (d) receiver (courtesy of Gretag AG)

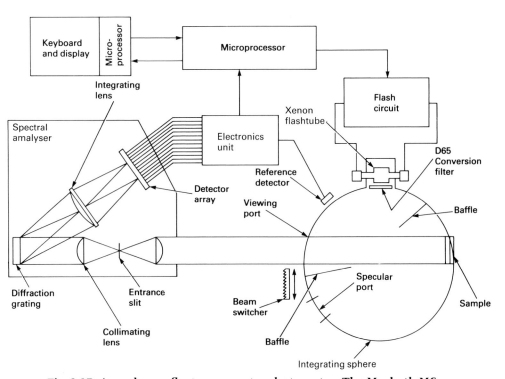

Fig. 3.17 A modern reflectance spectrophotometer: The Macbeth MS 2020. *Operation:* An operator-initiated command is issued at the keyboard which directs the command to the spectrophotometer microprocessor This triggers the flash circuit which causes the xenon flashtube to flash. The light from xenon flashtube passes through the D65 conversion filter which filters the light spectrum to approximate CIE standard illuminant D65. The light then illuminants the interior of the integrating sphere and illuminates the sample. Baffles prevent the light from directly striking the sample or the specular port. The specular port allows the inclusion or exclusion of the specular component. The beam switcher is a prism which automatically moves into the viewing port causing reference measurements of the sphere wall to be made. Light entering the spectral analyser via the fibre optics bundle is directed at the entrance slit which determines the wavelength resolution of the system. The beam is then transmitted through a collimating lens which directs the beam to the diffraction grating. The diffraction grating disperses the light into a spectrum and reflects the spectrum through an integrating lens, which directs the spectrum to the detector array. The detector array is a solid-state silicon photodiode array of 16 channels spaced at 20 nm intervals. The light is sensed by the the electronics unit for processing. A reference detector is used to monitor the intensity of the flashtube, and permits compensation for pulse-to-pulse variations in flashtube output (courtesy of Macbeth Division, Kollmorgen (UK) Ltd)

Optical geometry

The value required for colour measurement is diffuse reflectance but some of the specular component will also be included even if we use the special light traps designed to absorb this. The optical geometry of the instruments vary, and values obtained on one instrument can rarely be compared directly with those obtained on another of different design. Typical light paths are shown in Fig. 3.18.

Colour-measuring instruments are commonly interfaced, that is directly connected with computers, so that the often complex calculations required for calibration, colour difference and match prediction can be carried out more quickly and accurately.

3.7   THE RECORDING OF COLOUR DATA AND THE SPECIFICATION OF COLOUR

All printing ink manufacturers keep a large library of prints and standard ink samples against which production materials are compared. Prints must be carefully stored and be accompanied by data showing how the specimen was prepared, the weight of ink applied, the type of substrate employed and other details relating to special properties.

In addition to the manufacturer's own library, a number of national and international standard colour charts exist. One of the best known is

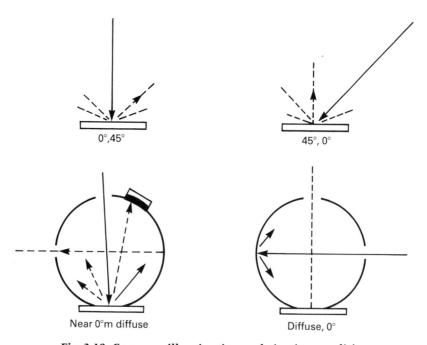

Fig. 3.18 Common illuminating and viewing conditions

the Pantone system in which eight basic colours, plus black and white are used to create a range of nearly 500 shades.

Formulations and prints on both coated and uncoated paper are included in the colour guide. Process colours are also included. The British Standards Institute issued a specification *'Inks for Offset Three — or Four — Colour Printing*, BS 4666:1971, which conforms to the European Standard CEI 13−67. Colour, colour strength, transparency, lightfastness and solvent resistance are specified. The master standard inks are held by PIRA.

PIRA have also published a *Guide to Standardised Lithographic Colour Printing* which includes a densitometer calibration procedure.

One of the earliest colour atlases was produced by Munsell [15] and is particularly valuable since the colours are arranged in such a way that there are equal visual differences in three dimensions which he called hue, value and chroma (Fig. 3.19).

The hue circle which follows the colour circle principle is divided into ten regions: red (R), yellow−red (YR), yellow (Y), green−yellow (GY), green (G), blue−green (BG), blue (B), purple−blue (PB), purple (P) and red−purple (RP), each region again being divided into ten sections. Value, which we now generally call lightness, is again divided into ten sections and chroma, or saturation, similarly divided on another axis, thereby producing a three-dimensional colour plot.

To define a colour by the Munsell system the hue, value and chroma are specified in that order. For example 3BG6/7 signifies a hue of 3 BG, a value of 6 and a chroma of 7.

Munsell therefore introduced the concept of colour space, a three-dimensional system in which all colours can be defined. Munsell's equal visual intervals, however, were approximate and could not form the basis of a quantitative description of colour.

If we now return to the X, Y and Z primaries introduced in Section 3.4, we can construct a colour space using Y as the vertical 'Lightness' axis and the x and y chromaticity co-ordinates as the other axes. This has the same configuration as the Munsell diagram, with a central vertical achromatic axis. While this is a quantitative description of colour the intervals in this colour space are not equal visual intervals (Fig. 3.20).

Colour difference [16]

The difference in colour between two specimens, can be represented by their distance apart in colour space. Mathematically this can be expressed

$$\Delta E = \sqrt{(\Delta H^2 + \Delta S^2 + \Delta L^2)}$$

where Δ in this case signifies 'difference in', E, total colour difference, from the German *Empfindung*, meaning sensation; H, hue; S, saturation; and L, lightness.

However, since the values derived from the colour-measuring instruments do not give an equal visual Munsell type space it is necessary to attempt to convert these values so that ΔE is the same visually for all parts of colour space. The equations used are numerous, but in spite of many years of research, no one system satisfies all applications.

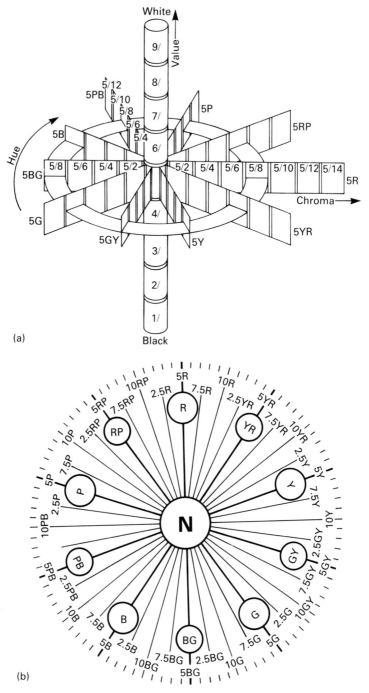

Fig. 3.19 (a) Munsell hue, value and chroma scales in colour space; (b) Munsell hue circle (reference [15]. Reprinted by permission of John Wiley & Sons Ltd)

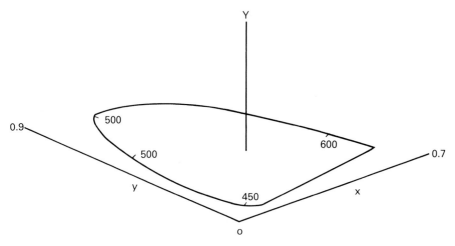

Fig. 3.20 Y, x y colour space

Ideally we should aim at a system where $\Delta E = 1$ is a good commercial match. A 'good commerical match' is difficult to define since it will depend upon the product and a customer's particular requirements. An acceptable match is still usually a matter of personal judgement and it is difficult to know where to draw the pass/fail boundary. In erring on the safe side we probably spend a lot of time getting matches closer than they need to be. Very frequently a fairly wide variation in strength is tolerated. On the other hand the hue is usually quite critical. A customer will often submit a standard pattern, sometimes including specimens illustrating colour tolerance boundaries. These, however, are again mainly to illustrate variations in strength.

The CIE have produced a number of equations to calculate ΔE the latest 'CIELab' equation was introduced in 1976. The L, a and b functions are derived from X, Y and Z. The chroma (C) is given by

$$C = (a^2 + b^2)^{0.5}$$

and hue (H), expressed as an angle from

$$H = \arctan b/a$$

with the $a+$ axis being at 0° and the $b+$ axis at 90°.

Adams and Nickerson's equation, ANLab, is also extensively used, but in 1976 McDonald used a different approach. Instead of attempting to derive a complex equation for all shades he constructed a range of ellipsoids in colour space defining the pass/fail boundaries of 55 colours, using a total of 640 matches. The axes of these microspaces are made linear with regard to visual response. Over a two-year period, Messrs J & P Coats, sewing thread manufactures, assessed 8454 matchings of 599 colours using 8 observers and these data were used to modify the original equations derived by McDonald. The equation is known as JPC 79. A similar approach was adopted by Taylor and Smart [17] in 1977 at

Instrumental Colour Systems working with Marks and Spencer. The latest version of this equation, M & S 83A, is included in the standard software supplied by Instrumental Colour Systems, but is as yet unpublished. This restricts its application to those using ICS equipment.

A similar approach is recommended by the Colour Measurement Committee of the Society of Dyers and Colourists using their equation CMC ($l:c$) [18]. The ellipsoid semi-axes are calculated from the CIELab, L, C and H values, but weighting factors for lightness, l and chroma, c have been introduced into the equations. If these factors are set at unity the value will correspond to the perceptible colour difference. However if wider tolerances for l and c are acceptable for a particular purpose, other values can be used. So far optimum values of l and c, for quantifying the acceptability of a colour match have only been determined for textiles. These are $l = 2$, $c = 1$. CMC ($l:c$) is the equation being considered by the British Standards Institution for a BS method for calculating colour difference for quantifying the perceptability and acceptability of small colour differences.

Total colour difference, ΔE, is normally calculated for illuminant D65. If the value changes with a change of illuminant the specimens are metameric.

Examples of a typical set of results produced on an ICS instrument are shown in Fig. 3.21.

3.8 COLOUR MATCHING

Selection of raw materials

The exact procedure used in colour matching of printing inks will depend upon the nature of the work and upon the skill and experience of the colour matcher. The pattern supplied may consist of a print, not necessarily produced by the process to be used or on the required substrate. It may be in the form of an artists colour wash, in which case the match will almost inevitably be metameric, since the pigmentation of artists materials differs from that acceptable for ink formulation.

A 'wet' sample is sometimes supplied, but although this can be very helpful it should be accompanied by a print. Before attempting a colour match, the exact requirements of the printer must be ascertained. This is probably best achieved by preparing a formal check list.

The general type of ink, the process, initial and anticipated order quantities are the vital pieces of information required by the laboratory manager before handing the project to the colour matcher. This should be followed by details of the press and the substrate. A sample of the substrate should be obtained. General ink specification details should include finish required, setting speed and colour sequence. Fastness properties must then be considered, special attention being paid to light and product fastness.

If the ink is required for food packages, pigments meeting the required restrictions must be selected and properties such as odour, rub resistance

```
COLOUR DIFFERENCE PROGRAM        DATE 08/09/1986
Standard SF345 NEW STD R Code  3324
Batch      SF345 PH08328 1ST SUB U/F R Code  5000
       Equation  CMC(2:1)     10 Degree Observer
Standard
(CIELAB)L              A        B        C        H
D65        82.35   -3.47   -10.27   10.84   251.32
Delta   DE     DL      DC       DH
D65    0.67   0.33    0.15     0.56
              DARKER   MORE     BLUER
                       CHROMA   (REDDER)
A      0.52   0.33    0.09     0.39
              DARKER   MORE     BLUER
                       CHROMA   (REDDER)
TL84   0.66   0.34    0.16     0.54
              DARKER   MORE     BLUER
                       CHROMA   (REDDER)
CWF    0.58   0.33    0.20     0.42
              DARKER   MORE     BLUER
                       CHROMA   (REDDER)


COLOUR DIFFERENCE PROGRAM        DATE 08/09/1986
Standard SF345 NEW STD R Code  3324
Batch      SF345 PH08328 1ST SUB U/F R Code  5000
M&S 83A   10 Degree Observer  Pass/Fail Tolerance  0.9
PASS   D65    DE   0.7   0.6 Bluer   0.3      0.2
                         (Redder)   Darker   Stronger
PASS   A      DE   0.7   0.6 Bluer   0.3      0.1
                         (Redder)   Darker   Stronger
PASS   TL84   DE   0.8   0.8 Bluer   0.3      0.2
                         (Redder)   Darker   Stronger
PASS   CWF    DE   0.7   0.6 Bluer   0.3      0.3
                         (Redder)   Darker   Stronger
```

Fig. 3.21 Two examples of colour differences of pale-blue specimens calculated with CMC (l : c) and M&S 83A for four illuminants

and suitability for deep freezing have to be taken into consideration. If after printing, further processes are involved, their effect will have to be taken into account. Over-printing, waxing, heat sealing, laminating and embossing can have a deleterious effect if unsuitable materials are chosen in formulation.

Matching techniques

When all these details have been assembled, the colour matcher can then start to examine the colour and decide the pigments most likely required. The very experienced matcher will probably be able to determine this fairly readily without reference to laboratory records and suppliers literature. In many cases, however, it will be necessary to examine full strength and reduced prints of single pigments against the pattern under various illuminants.

The colour circle

It is useful to consider the base colours in terms of a colour circle on which the purest and brightest colours lie. The arrangement is similar to that outlined in the discussion of the Munsell system. All the intermediate colours lie inside the circle with black at the centre (Fig. 3.22).

All the greens obtainable with a mixture of say a green tone blue and a green tone yellow will lie on a line joining these on the circle. Now consider the use of the same blue with a redder tone yellow. In this case the line passes closer to the centre black position indicating that the result will be 'dirtier'. In general therefore mixings represented by longer chords on the colour circle will be dirtier and those by short chords, cleaner. The colour circle representation does not take the third colour dimension, lightness, into account. Addition of white will increase this dimension but may also result in a hue shift.

Procedures

Inks are normally matched by blending pigments already dispersed in a suitable varnish in the form of finished inks, or pastes with high pigment loadings. The ink factory will normally manufacture pastes with the commonly used pigments dispersed in a range of vehicles. Quantities of these should be obtained in the laboratory and where possible matchings made with those available. This will avoid unnecessarily increasing the range of pre-dispersed pigments. Pigments with special properties, for example very high light and weather-fast materials for poster work, may be too expensive to keep in paste form. The laboratory may, however, have small quantities dispersed in suitable varnishes for matching purposes.

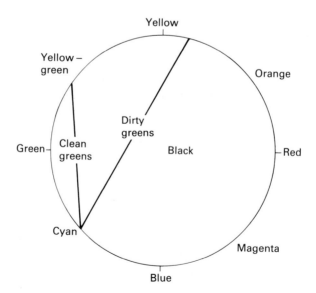

Fig. 3.22 Colour circle

Very small quantities of black and violet may be used in final shading. In these cases special pastes containing 5 or 10% of the concentrated paste may be used to avoid weighing errors. The use of dry colour in the matching process should be avoided if at all possible, because differences in degree of dispersion will add a further variable to the system.

It is sometimes convenient to carry out the initial matching in a conventional varnish even when a more unconventional system is dictated by the printer's requirements. This must then be translated into dry colour weights and checked using an automatic muller or small triple roll mill.

Oil inks

Between 10 and 15g of the selected pigment pastes or inks is put on the greaseproof paper circles (sometimes called 'skins') and weighed to three decimal places. These are then arranged around a clean glass plate. Initially small quantities of pastes are blended with a palette knife until the bulk, which should be about 5 g, closely matches the pattern. Using the index finger a little is 'tapped out' on to the chosen substrate. This process requires a lot of experience, but with practice a weight similar to that of each main printing process can be 'tapped out' and the masstone, undertone and strength assessed. Corrections for both shade and strength are now made, again followed by tapping out and close examination under various illuminants.

In some cases a 'drawdown' as described later is done at this stage. The bulk should now be increased to about 20–25 g, shade and strength being adjusted as before. It is advisable to work on the strong side, as it is comparatively easy to reduce strength with a clear 'tinting medium' or varnish. A weak result on the other hand will necessitate a complete rematch in order to be sure that the tint balance is maintained. The ink must now be test printed, again on the substrate to be used by the printer, using for example an IGT or Prüfbau machine set to print the desired weight. The first print will usually be some way out, more frequently for strength rather than hue. Corrections are again made.

The pastes on the greaseproof 'skins' are reweighed and the quantity of each one used is noted. A provisional formula can now be calculated and adjusted to a suitable bulk, ideally 100 units. At this stage the formula should be carefully examined and any apparently unusual features checked. The experienced matcher will have a good idea of the sort of combination likely to be required for a given shade. If it varies significantly form this, a rematch should be made.

If a very large bulk is to be manufactured, a small trial batch is made which may be printed on a suitable proofing press. Adjustments for properties such as tack can be made at this stage. When the material is put into production, it is often more convenient to use a 'drawdown' for quality control purposes. Since the 'standard' and 'batch' are of similar composition, this method yields an adequate control procedure. If the materials differ in composition in any way, the results of a 'drawdown' can be misleading and tapping out or test printing should be used.

The drawdowns are carried out by placing about a 0.25 ml of the inks to be compared about 2 cm apart on a suitable paper clipped on to a smooth board, or preferably a block of hard rubber. The inks are then drawn down the paper with a seven to eight centimeter wide knife blade. Towards the end of the drawdown, the pressure is relaxed to give a thicker film convenient for assessing masstone. It is advisable to use a paper with a black band printed across it. The drawdowns which should just touch, can then be examined over black and white substrates to assess opacity.

Liquid inks

The basic ideas discussed under oil ink procedures still apply, but the handling of these very rapid drying products presents special problems. The base inks used for the matching should be near to print viscosity, and a bulk of about 100 g should be used. It is most convenient to work in paper cups. Solvent loss should be kept to a minimum by storing the bases in tins with tight-fitting lids. Solvent lost while working must be replaced so that the viscosity of both standard and sample are the same and close to printing viscosity before application to a substrate.

The drawdown procedure using the broad knife is still useful, but on very absorbant substrates it is better to use a hand anilox roller. The anilox roller will ensure that a controlled quantity of ink is applied in a manner closer to the printing process than the drawdown.

A development of the drawdown technique involves the use of wire wound bars, generally known as K bars. The gauge of the wire used to wind the bar determines the quantity of ink carried. The bars are numbered 1 to 8, No. 1 applying a wet film approximately 6 μ thick up to No. 8 which applies 100 μ. No. 1, colour coded yellow, is normally used for flexo inks and No. 2, colour coded red for gravure. A rubber impression bed of controlled hardness is used to support the substrate. The inks being compared must be of the same viscosity and the drawdown carried out smoothly and quickly without allowing the bar to roll.

In order to standardise the weight applied to the bar and the speed of the drawdown an electrically or pneumatically driven unit has been developed.

Proofing of liquid inks is carried out using small laboratory units of various designs, but a very convenient machine is a small flat-bed gravure type.

3.9 INSTRUMENTAL COLOUR MATCH PREDICTION

The selection of colourants and the tinting to shade of a printing ink is time consuming. The delays in printing ink manufacture and the holding of printing machines while corrections for colour are made is extremely costly. It is not surprising therefore that newer computer-assisted methods of colour matching, extensively used in the paint industry are being developed for printing inks. In addition to time savings, more economical use of colourants and redundant inks can be made.

Basic theory

In Section 3.2 it was shown that light falling on a print is partly absorbed (K) and partly scattered (S). The scattered light can be measured as diffuse reflectance (R). The relationship between these values was worked out by Kubelka and Munk in 1931.

$$\frac{K}{S} = \frac{(1 - R)^2}{2R}$$

The model used by Kubelka and Munk [1] was an ideal surface coating. The film being of even thickness, opaque, with a pigment of very small particle size compared to the film thickness, evenly dispersed throughout (Fig. 3.23). While paints are often quite close to this model, printing inks are not. The films are relatively thin and transparent, the pigment particles therefore are large in relation to the film thickness. Perhaps of more significance is the often uneven thickness of the ink film (Fig. 3.24).

At present the original Kubelka and Munk equation forms the basis of most programmes used for colour match prediction of printing ink, but in spite of the numerous correction factors applied, results are frequently unsatisfactory. Nevertheless the procedures have proved valuable in some areas and research is continuing into more satisfactory relationships between measured reflectance and colourant composition.

In the 1940s Duncan, working at the Paint Research Station, showed that the K/S function is additive. This means that at any wavelength the K/S calculated from the reflectance is the sum of the K/S values of the constituents. Additionally the values are linear, that is for example, by doubling the concentration of a constituent the K/S contribution of that constituent is also doubled.

The reflectance curves and the derived K/S curves for a magenta and yellow ink are shown in Fig. 3.25. It can be seen that the resulting red made from a mixture of these primaries has a reflectance curve which cannot readily be seen as a combination of the magenta and the yellow.

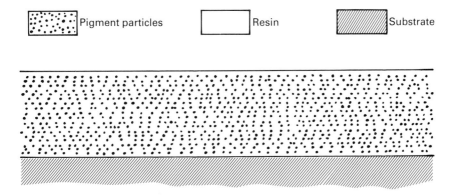

Fig. 3.23 **Ideal surface coating used in Kubelka–Munk analysis**

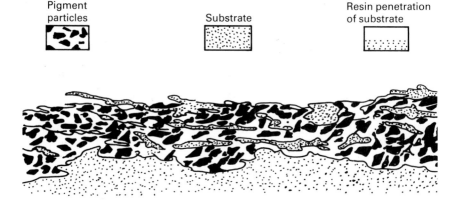

Pigment
particles

Substrate

Resin penetration
of substrate

Fig. 3.24 Cross-section of an ink on an absorbent substrate

On the other hand, the K/S curve of the red is clearly the K/S curves of the magenta and yellow added together. (Fig. 3.25).

The linearity of the K/S function is illustrated by a family of K/S curves for a rubine red ink, with increasing concentration of pigment (Fig. 3.26). Note that at maximum absorption ~560 nm, the K/S values are more nearly linear with respect to concentration, while this is not so in the case of the reflectance curves.

It is now possible to write an equation relating the reflectance at each wavelength to the concentrations of the constituent coloured pigments:

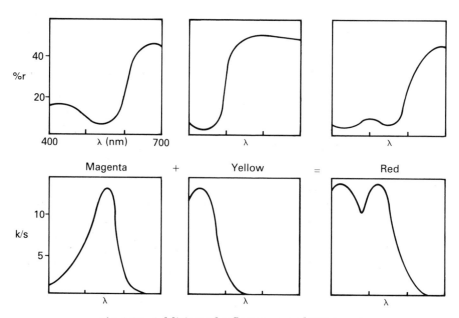

Fig. 3.25 Addition of reflectance and K/S curves

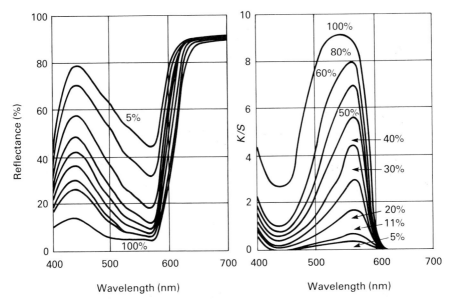

Fig. 3.26 Reflectance and derived K/S curves of a range of concentrations of a red ink

$$\frac{(1 - R)^2}{2R} = \frac{K}{S} = a\frac{(K)}{(S)_A} + b\frac{(K)}{(S)_B} + c\frac{(K)}{(S)_C}$$

where a, b and c are the concentrations of pigments A, B and C. In the case of printed materials, the K/S of the substrate must also be considered, since the film of ink is relatively transparent. However deducting the K/S values of the substrate is frequently unsatisfactory, since the observed colour is not simply the colour of the ink film plus the colour of the substrate.

To determine the concentrations of the pigments the data is weighted by the way we see the colour through the CIE $\bar{x}$, $\bar{y}$ and $\bar{z}$ coefficients, combined with a specific illuminant. It is then possible to write three simultaneous equations and solve them for the concentrations of the colourants.

Match prediction procedures [17]
The K/S values for all the pigments at a range of concentrations must first be prepared. This is normally done by taking a base ink and reducing with clear medium to give a range from 5 to 100%. Usually the intervals are as shown in Fig. 3.26. Prints of these must be very carefully prepared, usually on a standard substrate using an accurate proofing press. For paste inks a Prüfbau or Duncan Lynch press is frequently used and in the case of liquid inks an R–K coater. In the later case, care must be taken to ensure that the viscosity of the diluted samples is kept constant so that accurate film weights are applied. It is possible to make allowances for

slight changes in substrate colour and texture, but if any significant differences occur a new set of standards must be prepared.

The reflectance curves are measured, usually by taking 16 readings at 20 nm intervals; the data are then stored on a computer disc. If significant variations occur across or down a print then preparation procedures must be modified; slight variations can be dealt with by averaging a number of curves.

The K/S values of all the pigments, at usually nine concentrations plus the clear medium, are then calculated and the values again stored in the computer memory, to form what is generally known as a data base. Modern integrated match prediction systems are available. The ICS system is illustrated (Fig. 3.27). In this type of system the spectrophotometer is directly connected, that is interfaced, with a computer equipped with visual display unit (VDU) and a keyboard. A set of programmes, the software, is available from the suppliers so that the storage of data and all the necessary calculations can be more easily carried out.

Equipment suppliers also offer special training facilities for staff and a technical advisory service. The total cost of such equipment is about £50 000 at the time of writing.

The equipment should be installed close to the working area rather than in a remote laboratory. It should be kept in an environment, however, which is relatively clean, and this will often necessitate the construction of a small air conditioned colour laboratory on the shop floor.

This system should be made the responsibility of a trained colour

Fig. 3.27 A modern integrated Match Prediction System

technician, familiar with the raw material and finished product specifications and with the production procedures.

It is possible to calculate a colour match from all possible combinations of pigments, but the ink technologist will normally wish to select a range with special properties, e.g. light or product fastness, and take into consideration cost. It is also possible to store data on redundant inks which are to be 'worked-off' in current production. A match can therefore be made from base inks plus work-off or work-off alone.

Once a shade has been matched, data can be stored on disc in a shade library. It is possible therefore to search such a library for the nearest match, before attempting a new matching. Many other programmes to determine, for example quantitative colour differences, degree of metamerism and ink strength are available. It is now possible to interface this match prediction equipment with dispensing and weighing systems, so that automated production of ink from bases and waste ink is possible. Equipment of the type described in Chapter 12 has been employed in this way. It is further possible to incorporate computer stock control procedures and other business information systems with the installation. There is a danger with this type of equipment that it will be treated as a magic box. There are many limitations and results should be carefully checked and analysed.

One of the limitations is the basic theory used. While the main attraction of the Kubelka–Munk theory is its simplicity, it is at best an approximation. A critical review of Kubelka–Munk by Dr James Nobbs of the Department of Colour Chemistry at Leeds University was published in the *Review of Progress in Colouration* in 1985. In his conclusions, Dr Nobbs points out that in the case of semi-transparent layers the inherent limitations of the theory produce errors considerably greater than those from other causes.

The software used for printing ink colour work is very similar to that employed in textiles. The light scatter is assumed to come mainly from the substrate, and it is probably errors in quantifying the scatter values that leads to some of the poor results. Plots of K/S vs concentration are never linear as predicted by the theory, hence the need for a range of concentratons. If white ink is employed in the formulation some of the scatter comes from the ink and some from the substrate. Complex calculations are necessary therefore if white inks are included in the data bases.

It is this error in the scatter value that leads to poor results in the measurement of ink strengths. In this case it is much better to reduce the ink with a standard base white and apply a film of sufficient thickness to hide the substrate. The scatter then is almost entirely contributed by the white ink, and is predictable.

REFERENCES

1. Kubelka, P. and Munk, F. *Z. Tech. Physik* **12**, 593 (1931).
2. Newton, Isaac. *Opticks*. London Innys (1730), pp. 134–7.

3. Wright, W. D. *The Measurement of Colour.* (1969), pp. 4–12.
4. Philips Lighting, Fluorescent Lamp Data. Handbooks 6.2.1. and 6.2.10.
5. Wright, op. cit., Chapter 2, The perception of light and colour.
6. Ishihara, S. *Tests for Colourblindness* 38 plate edn. (1974).
7. McLaren, K. *The Colour Science of Dyes and Pigments.* (1983), pp. 85–8.
8. Chamberlin, G. J. and D. G. *Colour, its Measurement, Computation and Application.* (1979) pp. 9–12.
9. *The Scientific Papers of James Clerk Maxwell,* Cambridge University Press (1890).
10. McLaren, op. cit., pp. 96–102.
11. Ibid., pp. 102–7.
12. Patterson, D. *Pigments, An Introduction to their Physical Chemistry.* (1967).
13. Chambers, E. *Manual of Graphic Reproduction for Lithography.* Chapter 9–11.
14. Wright, op. cit.
15. Munsell, A. M. *Munsell Book of Color* (1905); Nickerson, D. History of the Munsell color system. *Color Eng.* **7**(5), (1969); Chamberlin and Chamberlin, op. cit., pp. 22–5.
16. McLaren, op. cit., Chapter 10, The quantification of colour differences.
17. *Handbook of Industrial Colour Technology.* Instrumental Colour Systems Ltd.
18. Clarke, F. J. J., McDonald, R. and Rigg, P. Modification to the JPC 79 colour difference formula. *J. Soc. Dyers and Colourists* **100**, 128–32 (April 1984).

CHAPTER 4
Raw Materials

The raw materials used throughout the printing ink industry have, like those of other industries, shown a constantly changing pattern in the last 35 years.

Prior to 1950 a considerable number of raw materials used in inks were pigments, resins and solvents produced mainly for the paint and other surface coating industries. Some were poorly adapted to meet the specific needs of what was at that time regarded by many manufacturers as a small and not very important industry.

Interest shown by raw material suppliers since then was generated by the knowledge that requirements of the industry could best be satisfied by the development of organic pigments, synthetic resins and solvents specifically for printing inks. New materials were designed to meet the stringent requirements of the more advanced inks that were being made for the newer methods of printing.

More recently some suppliers have installed laboratories containing both ink-making plant and printing machines. In this way they have kept abreast of ink development and are now able to advise ink manufacturers on formulation.

Following the expansion in the number of materials available during the last decade, use of some of what were very common materials has dwindled, and they have become uneconomic to produce. They have either disappeared from the market or are manufactured in small costly quantities. Other factors, such as the health and safety legislation have also contributed to changes which will continue to take place in the variety of raw materials. Shortages, some due to climatic, political, financial or transportation difficulties, have occurred from time to time. Joining the Common Market has made it necessary for all printing inks to be labelled with codings that give information about certain materials in the formulation.

For these reasons it is important that technologists are well informed about the properties of all the items they handle. The dissemination of literature supplied by manufacturers and journals is important, especially when new or different materials are being introduced. Unfortunately it is becoming increasing prohibitive to introduce new materials because of cost and health and safety pressures.

Printing ink formulation is a matter of precision combined with foresight and adaptability. It is imperative that the ink-maker has a clear understanding of the nature of the raw materials he uses. Few items are ever used in isolation so one must know not only their fundamental physical properties, but also the influence these properties exert on mixtures.

In this chapter the properties and uses of the raw materials used in the formulation of printing inks are described. Their composition, or source, is dealt with to a limited extent. Colour Index numbers are included for pigments and dyes. The Colour Index, produced by the Society of Dyers and Colourists should be consulted. It gives excellent references to the properties, formulations and tests for each individual colourant, together with patent information where applicable and the manufacturers' own nomenclature.

Materials have been grouped under the following headings:
   (1) pigments, inorganic and organic;
   (2) dyestuffs;
   (3) oils;
   (4) resins;
   (5) solvents;
   (6) plasticisers;
   (7) waxes;
   (8) driers;
   (9) miscellaneous additives;
 (10) raw materials for radiation curing systems;
 (11) Health and safety at work Act.

Section I: Pigments

Pigments, together with dyestuffs, are probably the most important items in printing ink formulation as they convey the visual identity of an ink and invariably contribute the major cost.

Both pigments and soluble dyes give colour to a substrate by altering its reflective characteristics. In the case of pigments, which have multi-molecular crystalline structures, produced to an optimum particle size distribution, insolubility is a key property. When applied in a vehicle to a substrate they either remain on the surface or have a tendency to fill the voids in paper or other irregular surfaces.

There are hundreds of different types of pigment. Some are formed by nature in mineral or vegetable forms, but most of them are synthetic materials produced from petroleum chemicals. A simple but not ideal classification is that of organic and inorganic pigments. Of the large number of types produced, few find their way into ink manufacture with its special demands. Many are uneconomical to produce, even in large quantities. Others do not meet the demands of tinctorial strength, particle size, cost and specific combinations of properties required for successful ink formulation.

This section is devoted to a detailed description of coloured pigments and extenders that are extensively used in all types of ink manufacture.

The ones selected are those most commonly used throughout the industry. Some pigments which are used for one specific purpose only, have been omitted, together with pigments that are no longer in general use.

Each pigment is identified by names in common use, followed by the part 1 Colour Index number (see Chapter 3, p. 85). Pigments have been grouped according to their chemical composition with the purpose of indicating their common structure.

Important physical and chemical properties have been tabulated, together with bulk value (100 divided by relative density) which indicates the volume relationship between equal weights of different pigments.

Owing to differences in flow characteristics, handleable concentrated pastes of pigment and varnish do not always follow exactly their bulk value relationships. The percentage of pigment in 28 poise linseed stand oil required to give such a base (concentrated paste) is also listed. It is felt that the combined information gives useful knowledge of the relative consistency characteristics which may be encountered in manufacture of paste inks.

White pigments and extenders, as well as carbon blacks, have been described separately from other pigments as special description of types and other properties is essential.

The raw materials from which pigments are prepared are stated (where known) together with the formula or structure.

4.1   YELLOW PIGMENTS

Mono azo or arylide yellows

Many arylide yellows are currently in production. Although they are often commercially known as Hansa, this in fact is a trade name. These yellow pigments are correctly referred to as aceto-acetarylamides. Shades range from very greenish yellow to a bright warm shade as shown in Table 4.1.

Table 4.1

Colour Index reference			
Part 1	Part 2		
Pigment			
Yellow	No.	Name	Shade
1	11 680	Arylide G	Warm yellow
3	11 710	Arylide 10G	Greenest yellow
4	11 665	Arylide 13G	Slightly greener than 5G
65	11 740	Arylide RN	Reddish yellow
73	11 738	Arylide GX	Bright yellow
74	11 741	Arylide 5GX	Strong lemon yellow
97	11 767	Arylide FGL	Bright Yellow

Properties
They are normally fairly opaque pigments, but there are some grades that are sufficiently transparent to be suitable for superimposition, arylamide yellows are widely used for their fastness to light and tinctorial strength as well as resistance to alkali and soap. They are unsuitable for inks that require stoving because they will sublime and bleed strongly in some non-polar organic solvents, paraffin wax and plasticizers. Tinting strength is inferior to diarylides.

Arylide yellows are not normally as strong as the diarylide yellows which have replaced them in many ink applications. The dominance of diarylides is helped by the fact that arylamide yellows are bulky and have poor flow properties in certain ink vehicles.

Uses
Arylide yellows are used in both paste and liquid ink formulations. In the former they print well on letterpress and lithographic machines. They are used on soap and detergent packages. Care is necessary in photogravure inks owing to their substantial solubility in toluene.

Structural formulae
CI Yellow 1, 3 and 4

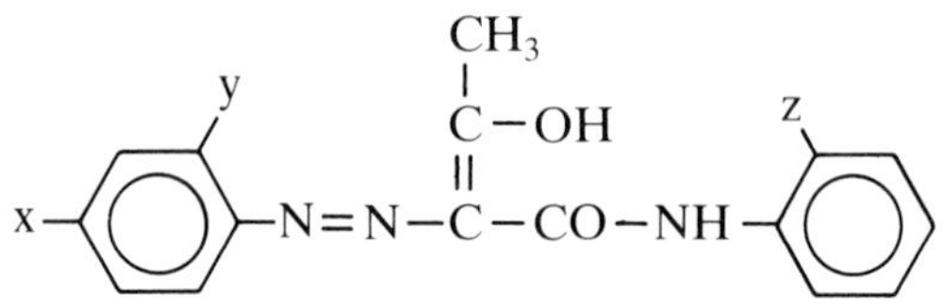

CI Yellow 65, 73, 74, 97 and 98

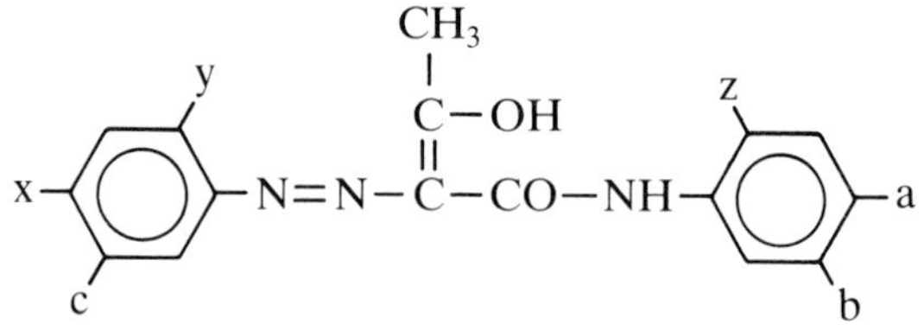

These are azo pigments formed by coupling the components given in Table 4.2 to give a, b, c, x, y or z substitutents.

Diarylide yellows

The diarylide yellows consist of a range of pigments based on dichlorobenzidine varying from lemon to golden hue and from opaque to transparent. Diarylide yellows are more widely used than arylide yellow. The most popular pigments in the range are given in Table 4.3.

Table 4.2

CI Pigment Yellow	Description	First Component	Second component	Structural variants					
				x	y	z	a	b	c
1	G	2-nitro-p-toluidine	acetoacetanilide	CH$_3$	NO$_2$				
3	10G	4-chloro-2-nitroaniline	acetoacet-o-chloroanilide	Cl	NO$_2$	Cl			
4	13G	p-nitroaniline	acetoacetanilide	NO$_2$					
65	RN	2-nitro-p-anisidine	o-acetoacetanisidide	OCH$_3$	NO$_2$	OCH$_3$			
73	GX	4-chloro-2-nitroaniline	acetoacet-o-anisidide	Cl	NO$_2$	OCH$_3$			
74	5GX	4-nitro-o-anisidine	acetoacet-o-anisidide	NO$_2$	OCH$_3$	OCH$_3$			
97	FGL	4-amino-2, 5-dimethoxy-sulfonanilide	4-chloro-2'5' dimethoxy-acetoacetanilide	So$_2$NNPh	OCH$_3$	OCH$_3$	Cl	OCH$_3$	OCH

Table 4.3

Colour Index reference

Part 1 Pigment Yellow	Part 2 No.	Name	Shade			Lightfastness
			Masstone	Undertone	Transparency	
12	21 090	Diarylide AAA	Warm-medium	Reddish	Transparent	Fair
13	21 100	Diarylide AAMX	Medium	Greenish	Transparent to very transparent	Very good
14	21 095	Diarylide AAOT	Medium	Greenish	Opaque	Good
17	21 105	Diarylide AAOA	Lemon	Greenish	Very transparent	Very good
55	21 096	Diarylide AAPT	Medium reddish	Warm	Opaque	Very good
83	21 108	Diarylide AADMC	Red shade	Warm	Transparent	Excellent

Properties

The tinctorial strength of this important range of pigments is considerably greater than the chrome and mono arylide yellows. Lightfastness ranges from fair to excellent, resistance to heat and soap is excellent. They are readily dispersed but as their particle size is very small and oil absorption high, care is necessary in formulations to obtain satisfactory ink flow. Improved flow has been obtained in the easily dispersible (ED) pigments, recently developed. The range of properties available in diarylide yellows is sufficient to satisfy the majority of the requirements of inks for most printing processes.

Uses

They are universally used in the three- and four-colour process inks for printing in any sequence, but if last down need good transparency. Diarylide yellows are based on diazotising 3,3'dichlorobenzidine and coupling with two molecules of various acetoacetarylides.

Structural formula:

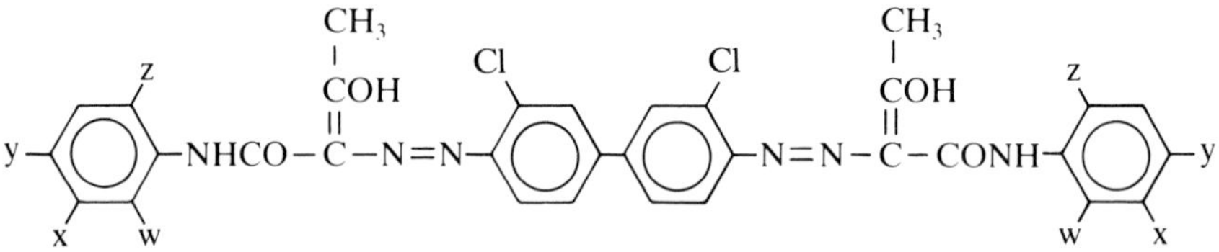

Substituent groups at w, x, y, z, yield the six diarylamide yellow variants shown in Table 4.4.

Pigment Yellow 60 CI No. 12705

CI Pigment	Description	1st Component	2nd Component
Yellow 60	Arylide Yellow 4R	*o*-chloroaniline	3-methyl-1-phenyl-5-pyrazolone

Shade — reddest and most lightfast arylide yellow.

Table 4.4

Colour Index Part 1 reference			Structural variants			
Pigment Yellow	Type	Coupling component	w	x	y	z
12	AAA	Acetoacetanilide	—	—	—	—
13	AAMX	2,4-acetoacetoxylidide	CH_3	—	CH_3	—
14	AAOP	O-aceto-aceto toluidide	CH_3	—	—	—
17	AAOA	*o*-acetoacetanisidide	OCH_3	—	—	—
55	AAPT	*p*-acetoacet-*o*-toluidide	—	—	CH_3	—
83	AADMC	4-chloro-2'5'dimethoxy-acetoacetanilide	—	OCH_3	Cl	OCH

Properties
Acid and alkali resistant and lightfast.

Uses
As arylide yellows with particular emphasis on the higher lightfastness.

Structural formula

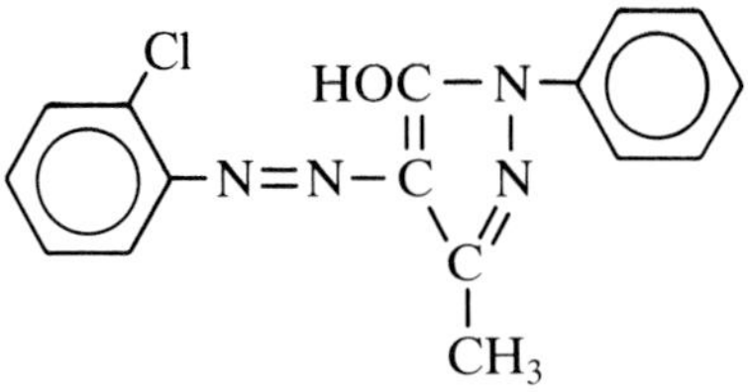

Iron oxide yellows

Pigment Yellow 42 CI No. 77492
Synthetic iron oxide yellow is hydrated ferrous oxide produced from ferrous sulphate.

Properties
Outstanding chemical and weather resistance. Many grades are very hard to disperse. The ink-maker is well advised to use micronised grade. They are economical and opaque.

Uses
Exceptional heat resistance makes them suitable for use in printed furniture surfaces other high light and heatfast specifications at low cost, and where low toxicity is required, e.g. 'cork tip' inks.

Tartrazine yellow lake

(*Pigment Yellow 100 CI No. 19140:1; Food and Dye and colour Yellow No. 5. EEC additive E102*)

Properties
FDA (Food and Drugs Act Regulations USA) certified and typical of aluminium salts of this kind. Useful where FDA approval and few other properties are paramount.

Uses
Metal decorating inks and nitrocellulose lacquers for food packaging.

Chrome yellows

(*CI Pigment Yellow 34 Nos 77600 and 77603*)

Properties
Lead chromates form a range of opaque yellow pigments varying from light primrose hue to golden orange. They have a high specific gravity;

this means that the weight ratio in inks is very high, giving a low pigment volume and coverage. It is sometimes desirable to incorporate extenders to control the rheological characteristics, but as the greener shades, being more bulky, are frequently extended with alumina hydroxide, they are sometimes prone to livering. When heated, chromes redden in colour and return to their original hue on cooling. The lightfastness is fair; they darken on exposure, especially in industrial areas; they are resistant to acids, wax, solvents, heat and soap, but decompose in the presence of strong alkali. They catalyse the oxidation of drying oils, so ink formulae must be adjusted accordingly.

Uses
The use of lead chromes had declined due to environmental factors such as effluent control and their unsuitability in inks for food wrappers or toys. They can be incorporated in all types of printing ink, especially where high hiding power is required. They were used as the first down colour in early process inks. The advent of transparent benzidines which can be printed in any sequence has superseded this. Future use will be very limited since an extremely low lead content is desirable in paper which is liable to be recycled as a secondary source of fibre.

Cadmium yellow

(CI Pigment Yellow 37 No. 77199)

Properties
Colours range from greenish to golden yellow. They are opaque, very resistant to light, heat, solvents and alkali. Cadmium sulphide pigments decompose in acids and, being sulphides, blacken when used in conjunction with lead pigments.

Uses
They are now seldom used in inks, except in applications needing very high resistance properties. Besides being expensive, they lack tinctorial strength and do not print well. Cadmium pigments are not permitted in inks on articles conforming to Food and Toy Regulations. They are not recommended for outdoor poster inks where they may decompose or darken under acidic atmospheric conditions. It is doubtful whether cadmium pigments will be permitted to be used in any form, due to toxic hazards in their manufacture and use.

Fluorescent yellow
Brilliant yellow 8GF

(CI Pigment Yellow 101 No. 48052)

Properties
A fluorescent green shade yellow, not very fast to light, heat or solvents, but it can be varnished or lacquered if care is taken to avoid bleeding. It is resistant to soap and detergents.

Uses
Suitable for lithographic, letterpress, gravure and screen process inks. Its main application is on cartons and display cards and it is commonly used in conjunction with other fluorescent colours on greetings cards. Brilliant Yellow 8GF has also been used as a first down process yellow to add lightness to prints often used as a gold on foil applications. Its price and poor lightfastness limited it use in the past. Recent developments, however, have improved brightness, and are faster to light and weathering.

Formula 2-hydroxy-1-naphthaldehyde, azine.

Structure

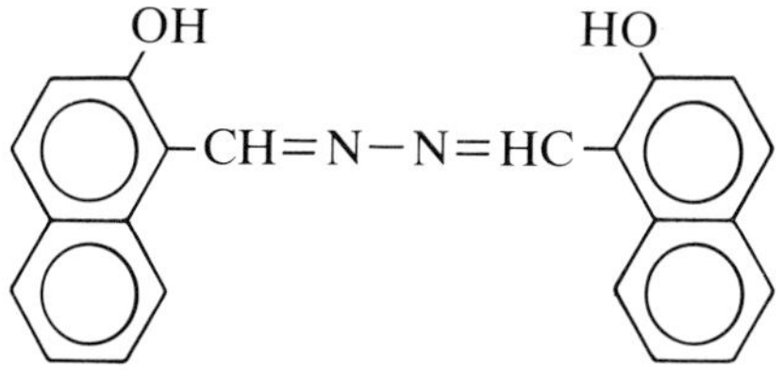

4.2   ORANGE PIGMENTS

DNA Orange

Orange 2G
Dinitroaniline Red (BS 3599/4)
Permanent Red 2G
(*CI Pigment Orange 5 No. 12075*)

Properties
Reddish orange with good lightfastness and fairly good heat resistance, with fair fastness to wax and butter, but excellent resistance to acid, alkali and water. It has high opacity, but lacks solvent resistance.

Uses
Orange 2G has only poor flow properties in inks and finds use mainly in cheap systems for printing on kraft paper and polythene where a bright reddish shade can be combined with good covering power. There is some doubt about its toxicity at the moment and its use is under investigation.

Structure

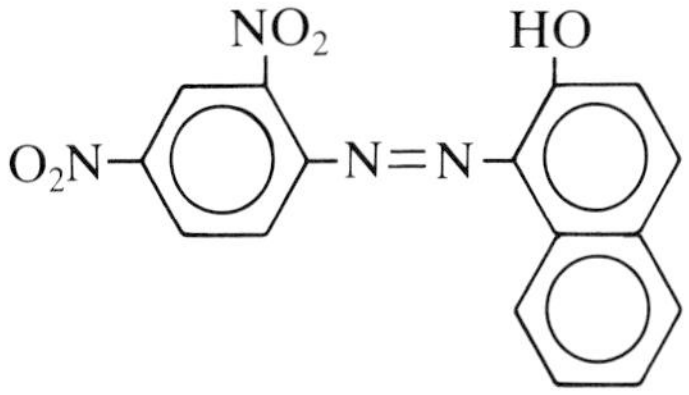

Pyrazolone orange

Permanent Orange G
(*CI Pigment Orange 13 No. 21110*)

Properties
A yellower shade orange which combines good fastness properties with high tinctorial strength. It is fast to acid, alkali, water, soap and wax. It also has good resistance to heat and to steam.

Uses
Orange G is semi-opaque and is a good working pigment which produces inks with a high finish for letterpress, litho and flexography. Although it does bleed in some gravure solvents, it is frequently used in liquid inks for packaging. It suffers from poor flow and printability problems.

Structure

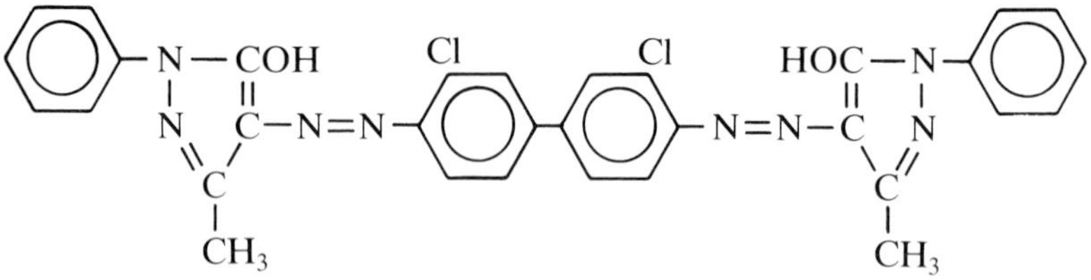

Diarylide orange

Dianisidine Orange
(*CI Pigment Orange 16 No. 21160*)

Properties
Typical of diarylide yellows having very high tinting strength. The hue is bright reddish orange, one of the few diarylides without a chloro group.

Uses
Often used to replace DNA Orange in kraft printing because of its superior properties.

Structure

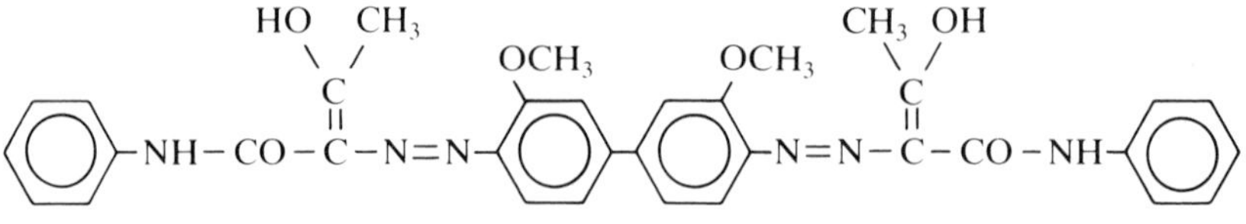

Fast orange F2G

(*CI Pigment Orange 34 No. 21115*)

Properties
The pigment is noted for its bright, reddish shade combined with high transparency. It has excellent lightfastness, resistance to alkali, acid soap and wax. It is fairly good for heat resistance, and steam sterilisation. It is also readily lacquered or varnished. Now available for liquid ink in good flow versions.

Uses
Suitable for all systems but is recommended for oil inks where its transparency and strength enable inks with a very high gloss to be obtained. Excess transparency can cause problems in liquid inks formulae.

Structure

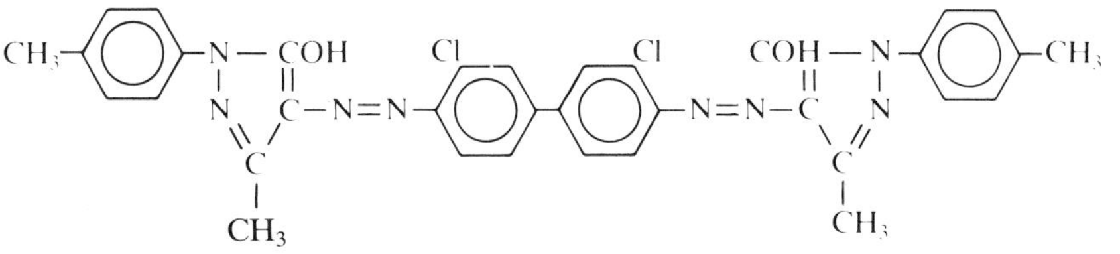

Benzimidazolone orange HL

(*CI Pigment Orange 36 No. 11780*)

Properties
Excellent light and heat-fastness and solvent stability. Slightly soluble in some organic solvents.

Uses
Expensive but used in products requiring these superior properties. Hue is reddish orange but dull in tint.

Structure

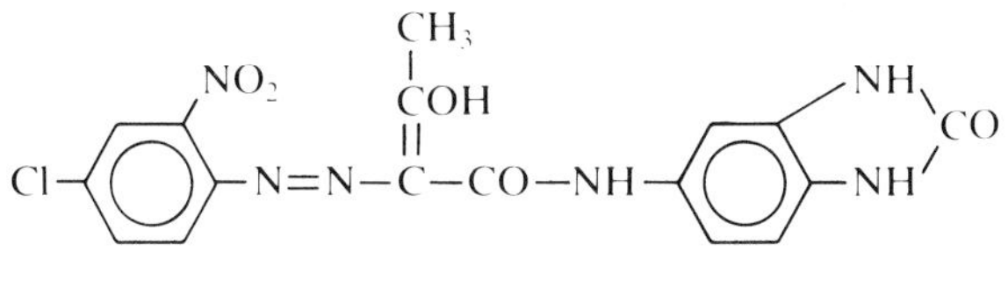

Ethyl lake red C

(*CI Pigment Orange 46 No. 15602*)

Properties
A yellow, clean version of Lake Red C (Pigment Red 53:1). Economical pigment (like 53:1) excellent brightness and tinting strength. Poor alkali resistance, and lightfastness and can bleed (like 53:1) in strong solvents.

Uses
Useful in water-based brown kraft work.

Structure

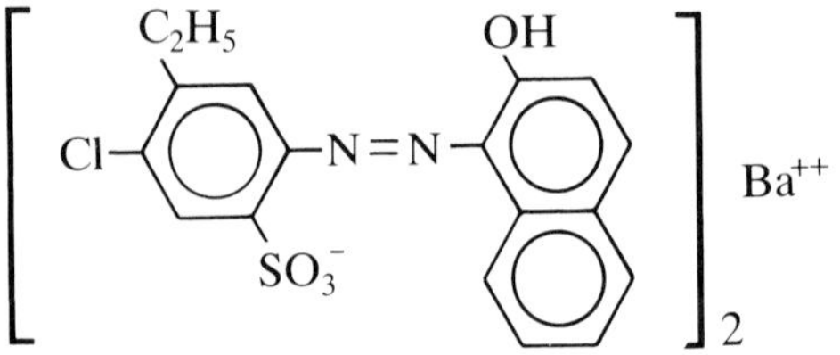

4.3   RED PIGMENTS

Para Red

(*CI Pigment Red 1 No. 12070*)

Properties
Yellow to blue shade reds of fairly good colour strength, but dull undertone. Para reds possess good lightfastness at full strength, dropping rapidly on reduction. They exhibit good resistance to acid and alkali, but tend to bleed badly in soap, oils and solvents. Dispersions of Para Red lack flow.

Uses
Inks based on Para Red were commonly used for poster printing but the bleeding tendency has led to a decline in use. Almost obsolete.

Structure

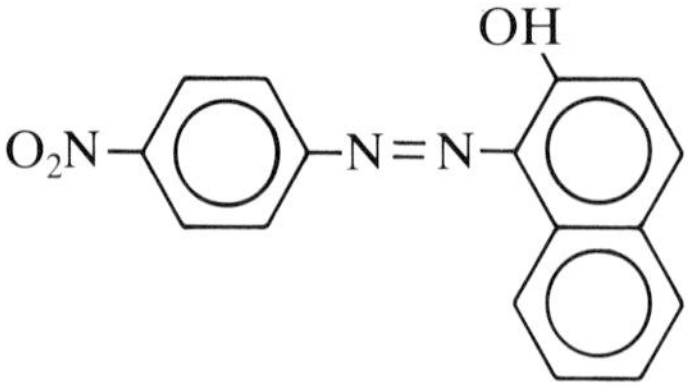

Naphthol Red (or Permanent Red FRR)

(*CI Pigment Red 2 No 12310*)

Properties
This is a strong, bright, clean, yellow shade red, with excellent resistance to acids, alkalis, soap and detergents. It is fairly lightfast.

Uses
It is used extensively in letterpress and lithographic carton inks, especially where soap resistance is important. It is also used for fertilizer sacks.

Structure

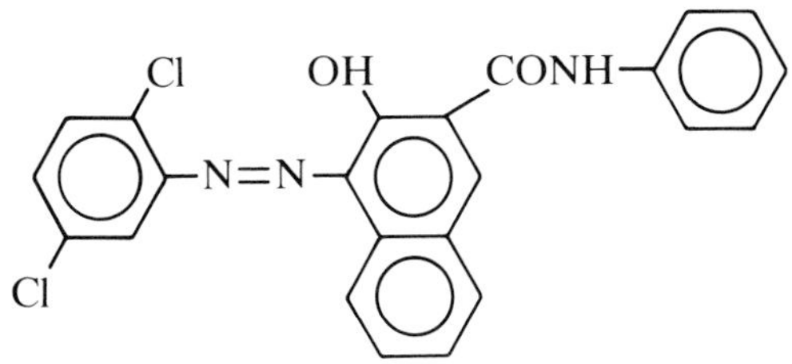

Toluidine Red

(*CI Pigment Red 3 No. 12120*)

Properties
A scarlet red of medium strength but dull in undertone. Fastness to light at full strength is excellent and resistance to alkali is good, but solvent resistance is poor and it discolours on stoving.

Uses
Used as a relatively low cost pigment, where lightfastness is of value. It is not recommended for use at low concentration in tints. Is very useful in waterbased inks.

Structure

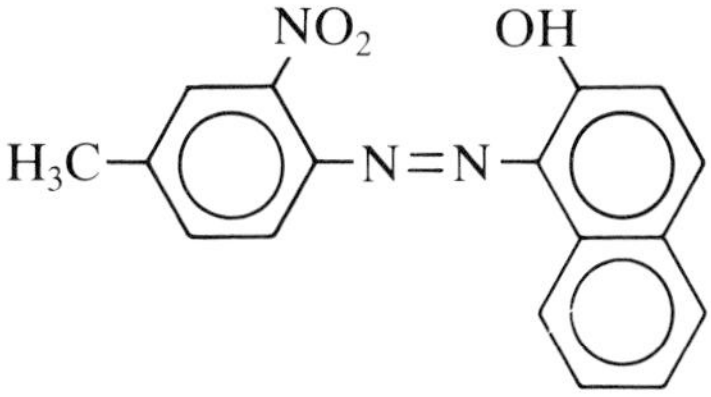

Permanent Red 'R' (Chlorinated Para Red)

(*CI Pigment Red 4 No. 12085*)

Properties
A bright orange red pigment of excellent lightfastness at full strength with good resistance to acid and alkali. Solvent resistance is poor and it is not suitable for stoving. Texture is soft but dispersions lack flow.

Uses
This red pigment is used in many applications but care in formulation is

required to produce inks of good printability. Used in large quantities due to its low price. It will sublime in heat transfer inks.

Structure

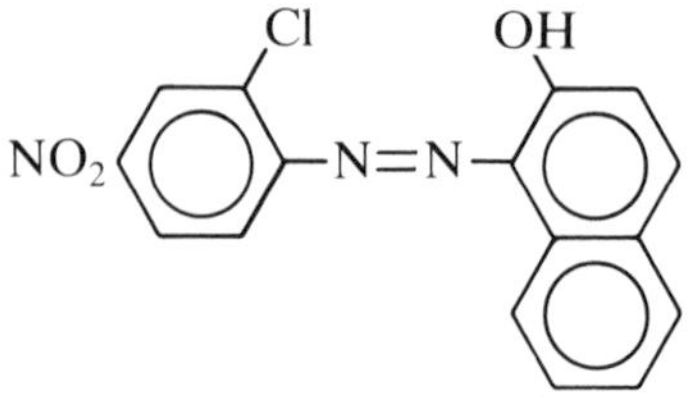

Carmine F.B.

(*CI Pigment Red 5 No. 12490*)

Properties
A strong blue-shade red of excellent lightfastness and resistance to water, acids, alkalis, soap and detergents possessing good stability to heat and steam sterilization. It disperses well, but lacks flow in full-strength inks.

Uses
High cost limits the use of this pigment to inks where high resistance properties are demanded.

Structure

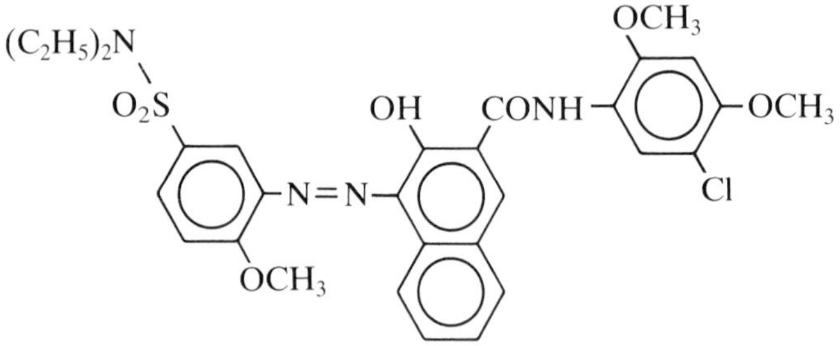

Naphthol Red F4R

(*CI Pigment Red 8 No. 12335*)

Properties
A blue shade red with a clean bluish undertone, having good heat and soap resistance. Suitable for autoclave sterilization. Lightfastness is fair.

Uses
Used in process inks for soap cartons and for tinprinting inks. Its cost and poor lightfastness considerably limit its use.

Structure

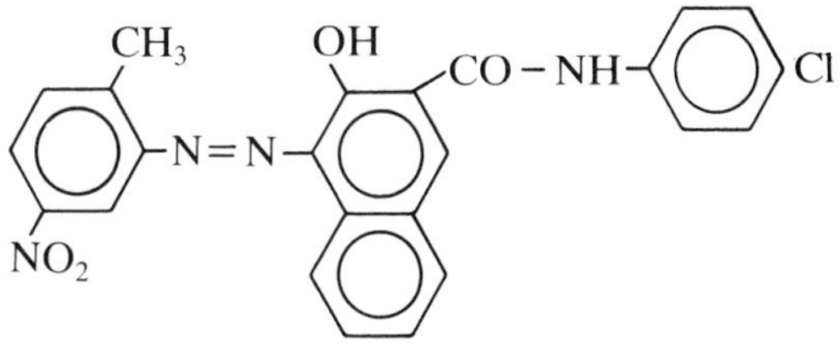

Naphthol Red LF

(*CI Pigment Red 9 No. 12460*)

Properties
A bright, transparent, yellow shade red with clean undertone, excellent lightfastness and heat stability plus steam sterilization, alkali and detergent resistance. It lacks fastness to strong solvents.

Uses
Used in tin printing applications, and in tints where fairly good lightfastness is needed.

Structure

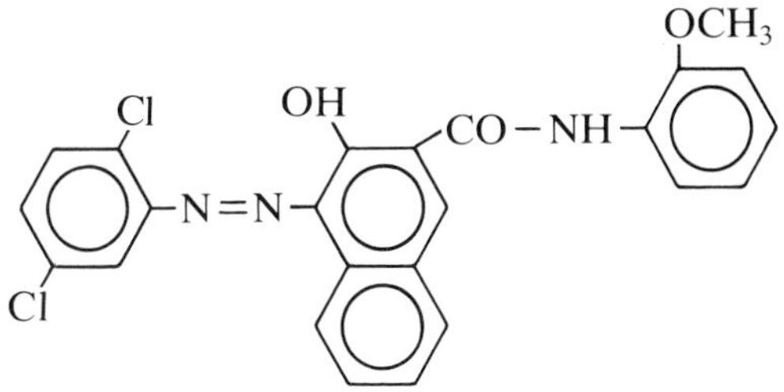

Permanent Red FRL

(*CI Pigment Red 10 No. 12440*)

Properties
A bright red with orange masstone and blue undertone, similar in shade to Lake Red C but with good chemical and heat resistance as well as improved lightfastness.

Uses
Used in all types of application which warrant improved resistance properties.

Structure

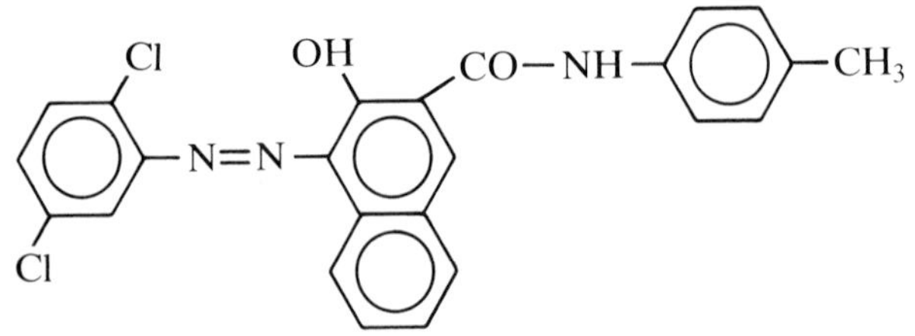

Bordeaux FRR (F4R)

(*CI Pigment Red 12 No. 12385*)

Properties
Extreme blue shade red toner with high tinctorial value, the undertone and tint-tone are slightly dull but extremely blue. It has good general fastness properties including high soapfastness and detergent resistance.

Uses
Used where good fastness properties are required at full and reduced strengths, and where that particular unique shade is required.

Structure

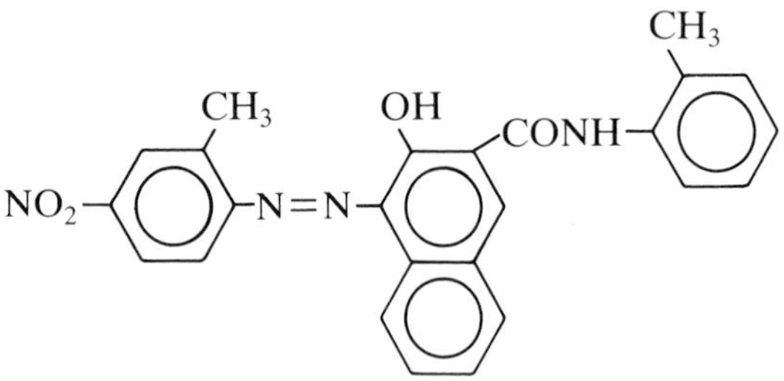

Naphthol Red

(*CI Pigment Red 17 No. 12390*)

Properties
A medium shade red with good soap and heat resistance, and fair lightfastness.

Uses
Used in packaging inks. Its cost and moderate lightfastness limit its use.

Structure

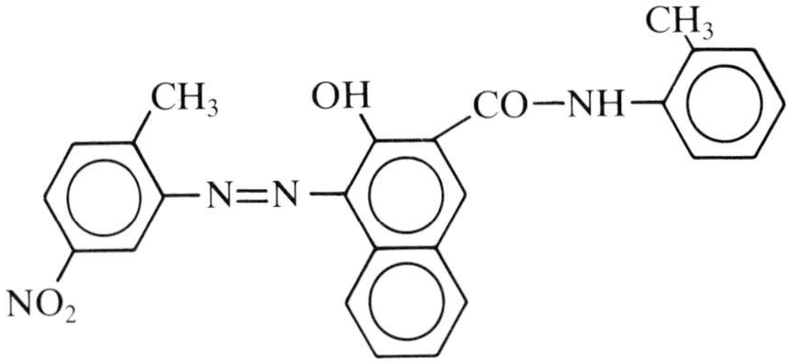

Naphthol Red Light

(*CI Pigment Red 22 No. 12315*)

Properties
A yellow shade red with good soap and heat resistance, and fair lightfastness.

Uses
Used in packaging inks. Its cost and moderate lightfastness limit its use.

Structure

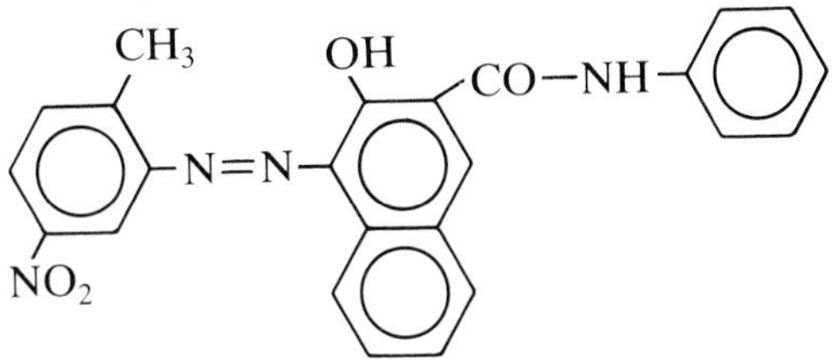

Naphthol Red Dark

(*CI Pigment Red 23 No. 12355*)

Properties
A dark blue shade red with good soap and heat resistance, and fair lightfastness.

Uses
Used in packaging inks for shade, chemical resistance. Particularly useful in nitrocellulose chips.

Structure

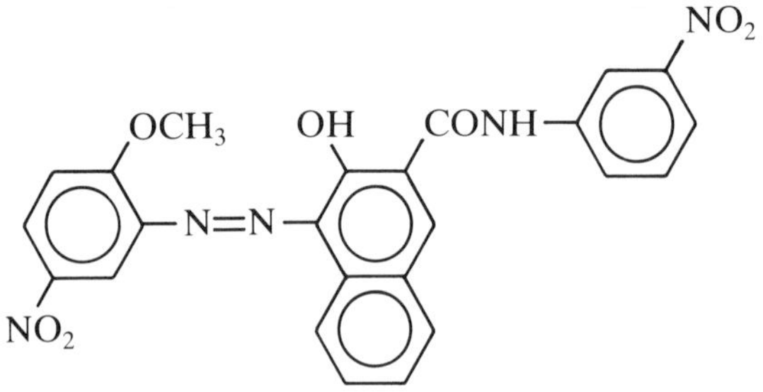

Rubine 2B, Permanent Red 2B, BON reds

(CI Pigment Red 48 No. 15865)
CI Pigment Red 48 (sodium salt), water soluble, seldom used.
CI Pigment Red 48:1 (barium salt), bright yellowish red.
CI Pigment Red 48:2 (calcium salt), very blue shade red.
CI Pigment Red 48:3 (strontium salt), less blue, but clean.
CI Pigment Red 48:4 (manganese salt), medium scarlet.

Properties
These range from a bright yellow shade red (barium) to a deep maroon
(calcium). They are primarily noted for their good value for money and
purity of hue. They possess adequate lightfastness, except when used as
tints. They are reasonably resistant to acids and solvents, but not to alkali.
They have good fastness to heat, oils and fats, but only fair fastness to heat
sterilization in autoclaves.

Uses
Very widely used. The barium and calcium salts are used extensively in all
types of liquid and paste ink, although the barium salt often has to be
excluded due to its toxicity. Their clean shades and working properties
make them invaluable for high quality inks. Low to medium cost.

Structure

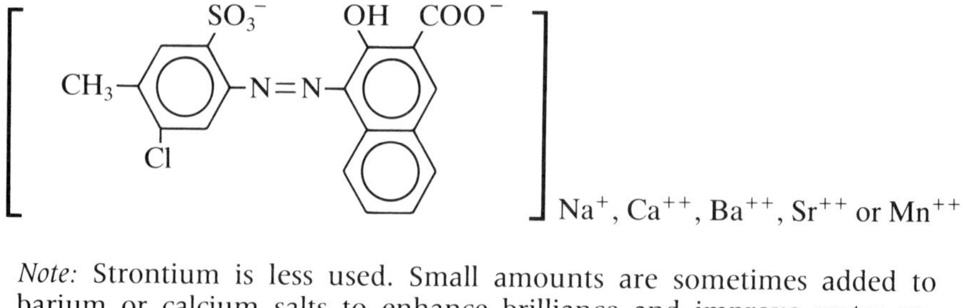

Note: Strontium is less used. Small amounts are sometimes added to
barium or calcium salts to enhance brilliance and improve water resistance. But all four (not the Na) have their uses for cost-effective mid-
red application.

Lithol reds

(CI Pigment Red 49 No. 15630)
CI Pigment Red 49 (sodium salt), yellow shade red.
CI Pigment Red 49:1 (barium salt), bright blue shade red.
CI Pigment Red 49:2 (calcium salt), strong blue shade red.

Properties

The lithols range from vivid yellow to blue shades of red dependent on the metal used. Their lightfastness and solvent fastness are poor, as is their resistance to acid and alkali. They do not have good heat resistance but are fairly resistant to oils and fats and are sometimes resinated to give transparency.

Uses

USA Publication Gravure Standard.

They are not used extensively in paste inks but are sometimes used for cheap jobbing colours. The barium salt gives a bright red and the calcium a deep red. These are invaluable for inexpensive liquid inks. Their poor lightfastness and solvent resistance restricts their use slightly. The very blue calcium salt gives a fine result as a liquid nitrocellulose or water-based ink.

Structure

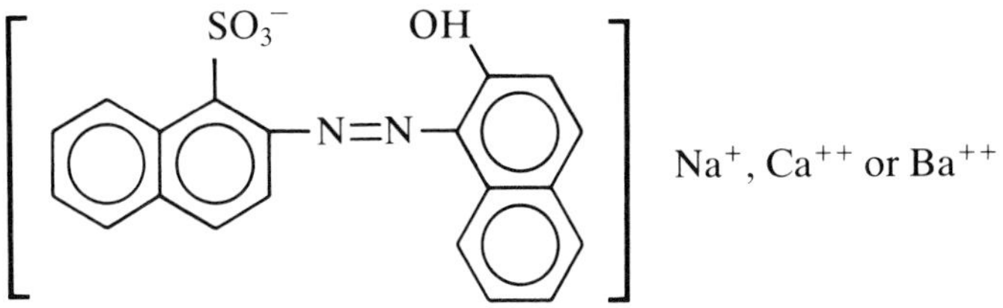

Na^+, Ca^{++} or Ba^{++}

BON Red (Lake Red C BON)

(CI Pigment Red 52 No. 15860)
CI Pigment Red 52 (sodium salt).
CI Pigment Red 52:1 (calcium salt).
CI Pigment Red 52:2 (manganese salt).

Properties

A red similar to Lake Red C but based upon a different coupling compound with C acid. It gives better solvent resistance, heat stability and lightfastness. It has good wax resistance. The calcium salt (the main one used in inks) is very close in shade to Rubine 4B.

Uses

A considerable amount is used in the USA in nitrocellulose and polyamide systems where good working qualities combined with bright shades are required. However, its popularity is lower in Europe. In solvent-based liquid inks, it gives good dispersion characteristics as well as a clean bright shade. It is particularly well suited for nitrocellolose chips.

Structure

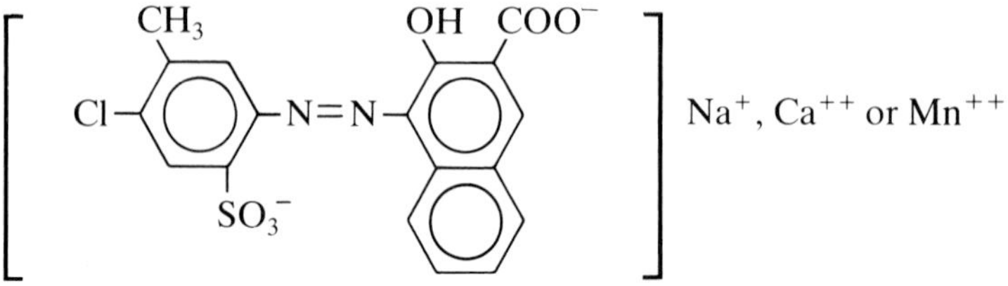

Na^+, Ca^{++} or Mn^{++}

Lake Red C

(CI Pigment Red 53 No. 15585)
CI Pigment Red 53 (sodium salt).
CI Pigment Red 53:1 (barium salt).
CI Pigment Red 53:2 (calcium salt).

Properties
Lake Red C is the name given to the barium salt. The other two salts are seldom used. It is the most widely used of all the organic reds in printing inks. It gives a strong, bright yellow masstone combined with a blue undertone. It is a fairly transparent pigment, and this is sometimes increased in the resinated forms, which also have good dispersibility. It has good heat, fat and oil resistance as well as good acid fastness, and waxability. It is unaffected by most hydrocarbons and solvents but has only fair lightfastness.

Uses
Used in all types of paste and liquid inks, where brightness and strength are required to be combined with an economic product. It is also the basic red pigment used in matching systems. Being a barium salt, like all other barium compounds it is toxic.

Structure

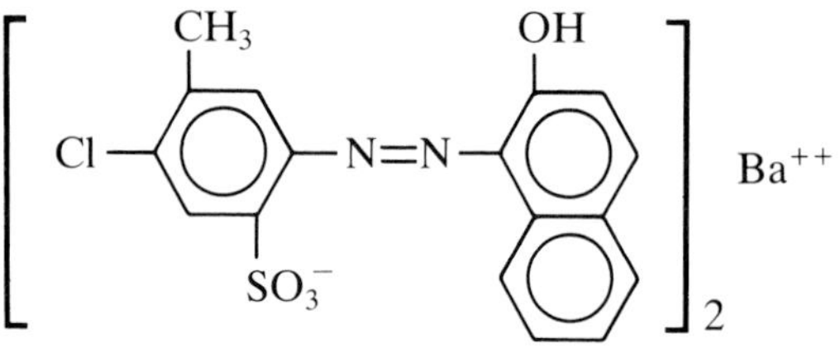

Ba^{++}

Lithol Rubine 4B

(*CI Pigment Red 57 No. 15850*)
CI Pigment Red 57 (sodium salt).
CI Pigment Red 57:1 (calcium salt).
CI Pigment Red 57:2 (barium salt).
CI Pigment Red 57:3 (strontium salt).

Properties
The Rubine 4B pigment most used by the printing ink industry is the calcium salt which is characterised by its brilliant deep masstone combined with a pronounced blue undertone. It has good resistance to most solvents and to oils, fats and waxes. it is fairly lightfast and also has medium resistance to heat and steam sterilisation. It has poor acid, alkali and soap resistance and only fair weatherproofness due to its sensitivity to rainwater. It disperses easily to make a good working ink. Much work has been done in recent years to achieve better water stability.

Uses
It is used mainly as a process magenta in all types of ink. Earlier versions had PMTA Rhodamine additions to obtain sufficient blueness for BS 3020 whereas the BS 4666 process sets rely upon the bluer brighter shades obtained by heat treatment and other modifications. Coated versions of this pigment are available with improved dispersibility and transparency. It is the blue shade red used in the base inks of colour-blending systems.

Structure

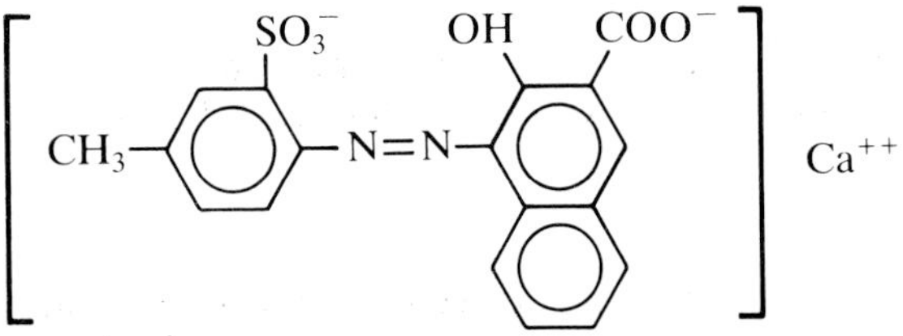

BON Maroon

(*CI Pigment 63 No. 15880*)
CI Pigment 63 (sodium salt), No. 15880.
CI Pigment 63:1 (calcium salt), No. 15880:1.
CI Pigment 63:2 (manganese salt), No. 15880:2

Properties
Deep blue shade red. Resistant to acids, moderately resistant to alkali and solvents. Resistant to fats with reasonable full tone lightfastness. Good in nitrocellulose chips.

Uses
Economical for these particular shades of Bordeaux and maroon.

Structure

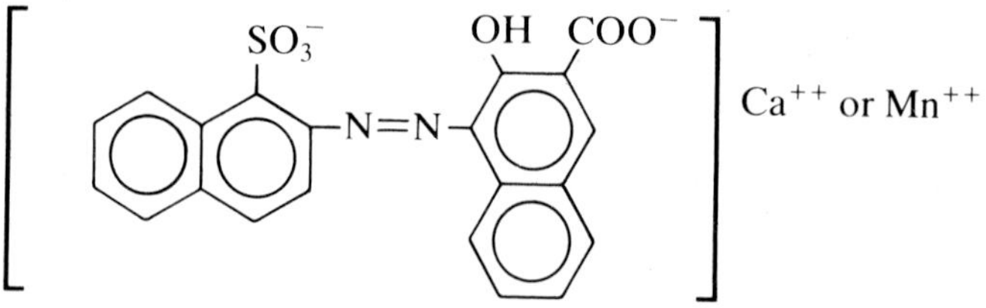

PMTA Pink, Rhodamine 6G

(*CI Pigment Red 81 No. 45160:1*)

Properties
Strong brilliant rose pink pigments produced by combining the basic dyestuff with phosphoric, molybdic and tungstic acids known as triple salts (PMTA). It is possible to omit the molybdic or tungstic acid (PTA, PMA) or to combine with a silicomolybdic acid alone (SM). These are generally known as double salts. The original trade name of Fanal Pink is sometimes given to the whole group.

The double salts are normally stronger and cheaper but have poor lightfastness, darkening rather than fading. None is resistant to alkali or soap and they all bleed readily in alcohol. Sometimes a slight oil bleed is apparent.

Note: Another pigment using the same basic dye with a copper ferrocyanide complex has been separately classified as CI Pigment Red 169.

Uses
It is used as the magenta in four-colour process inks, and in all types of ink where brilliance and tinting strength are the important factors. They print well and withstand paraffin wax but not spirit varnishing or lamination. Stabilised grades are available for liquid ink systems — unequalled in some applications both liquid and paste ink.

Formula of CI Basic Red 1 (No. 45160)

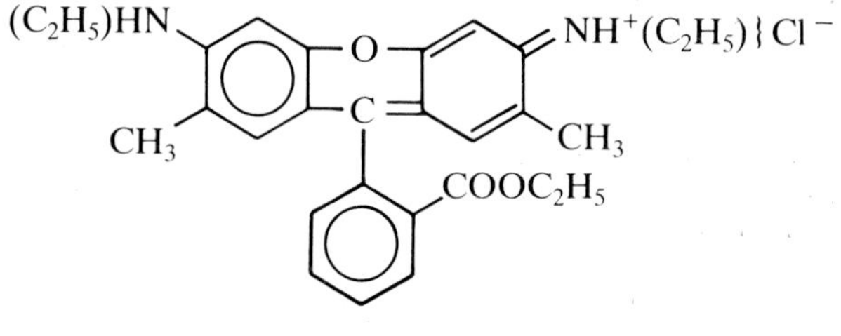

Note:
CI No. 45160:1 is the phosphomolybdotungstic salt.
CI No. 45160:2 is the copper ferrocyanide complex (see p. 133).

Molybdate Orange, Chrome Scarlet, Orange Chrome

(*CI Pigment Red 104 No. 77605*)

Properties
Bright, clean, dense orange shade reds with notable opacity. They have good lightfastness but tend to darken on exposure to atmosphere. They withstand heat and can be steam processed, but are affected by strong acids and alkalis. Prints are soap resistant and can be waxed.

Uses
A good working pigment of high opacity, suitable for use in a wide range of ink media. It has a low oil absorption which helps overcome its inherent lack of tinctorial strength, by allowing high pigmentation. Severe restrictions on its use are increasing, due to its toxicity.

Cadmium Red

(*CI Pigment Red 108 Nos 77196, 77202*)

Properties
A range of fairly opaque pigments from orange to deep maroon. They give excellent resistance to heat, light, alkali, very dilute acids, most solvents, soap and wax. They cannot be used for printing foodstuffs or toy wrappers, because of their toxicity, neither must they be used in conjunction with lead pigments, as the black lead sulphide may be formed.

Uses
Due to their toxicity the use of cadmium pigments is restricted to special applications in inks and coatings. Their use in inks that have to withstand exceedingly high temperatures or to have alkali resistance makes them worthy of note. They have a low bulk value with poor working qualities. They are opaque and have little tinctorial strength.

BON Arylamide Red, Naphthol Red FGR

(*CI Pigment Red 112 No. 12370*)

Properties
This is a bright, clean, medium shade red, with bluish undertones. It has high tinctorial strength and is fairly transparent. It has excellent lightfastness and is resistant to acid, alkali, water and heat but bleeds in most solvents.

Uses
Used in all types of paste inks, especially where cheaper pigments do not

have the resistance properties required. Also used for flesh-coloured tints and for making chip dispersions. Very useful soap fast scarlet.

Structure

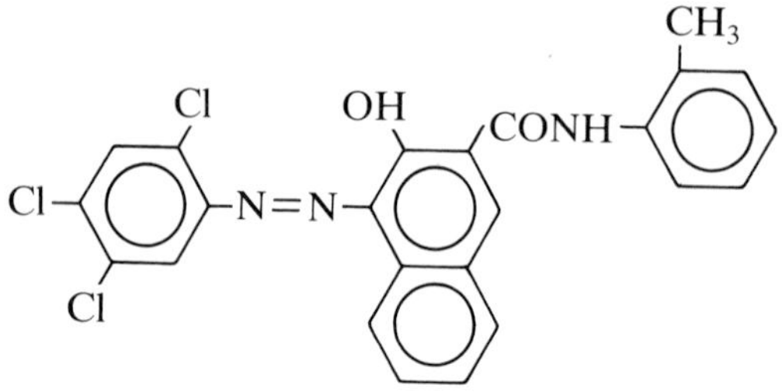

Quinacridone Magenta Y

(*CI Pigment Red 122 No. 73915*)

Properties
A bright blue shade red which is tinctorially weak and difficult to disperse. It is insoluble in organic solvents, unaffected by heat, light, acid, alkali, soap and wax.

Uses
Owing to its tinctorial weakness and high price the use of this pigment is limited to the applications which cannot be met in other ways. It works well in most ink systems.

Structure

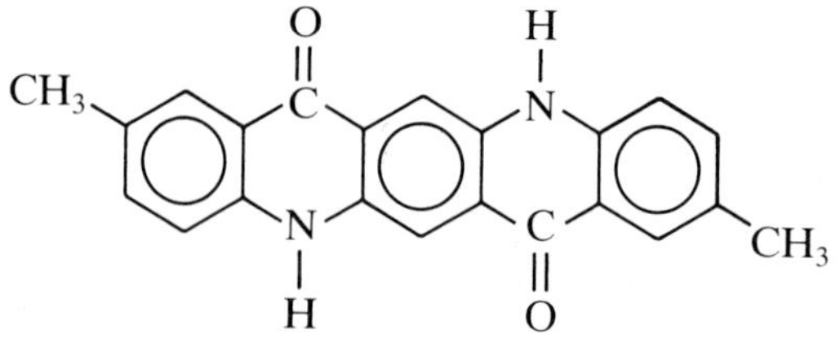

Naphthol Carmine FBB

Permanent Carmine FBB
(*CI Pigment Red 146 No. 12485*)

Properties
Multipurpose liquid, paste, and metal decorating, high property blue shade red pigment. Like many other Naphthols it needs good formulating to avoid poor flow. Hue is deep carmine. It has excellent fastness properties except to some organic solvents.

Uses
Fairly expensive, but is used where its superior properties are required.

Structure

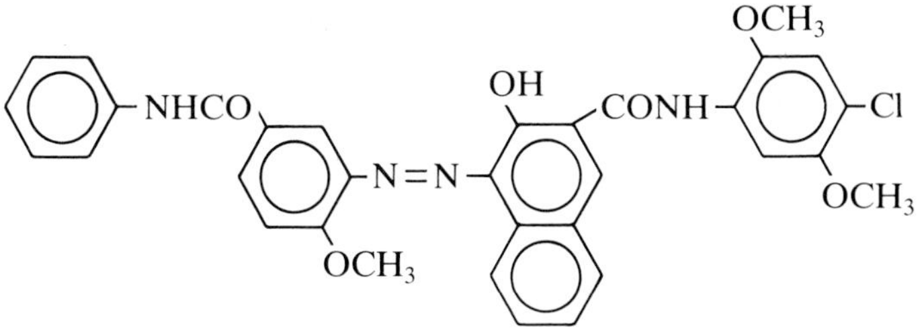

Copper Ferrocyanide Pink

(*CI Pigment Red 169 No. 45160:2*)

Properties
The colour is similar to the double and triple salts (CI Pigment Red 81). Its resistance to light is lower, with less resistance to heat and some solvents. It is stronger than the PMTA salts and has a higher oil absorption. Copper Ferrocyanide Pink is not soap fast, and has poor acid and alkali resistance.

Uses
Used as a substitute for the more expensive PMTA salts in publication and water-based inks. It normally has a brighter hue and excellent working properties. Stabilised grades exist for use in some solvent combinations, but care must be taken in formulation. They flow very well and find applications where cheap bright shades are required.

Structure

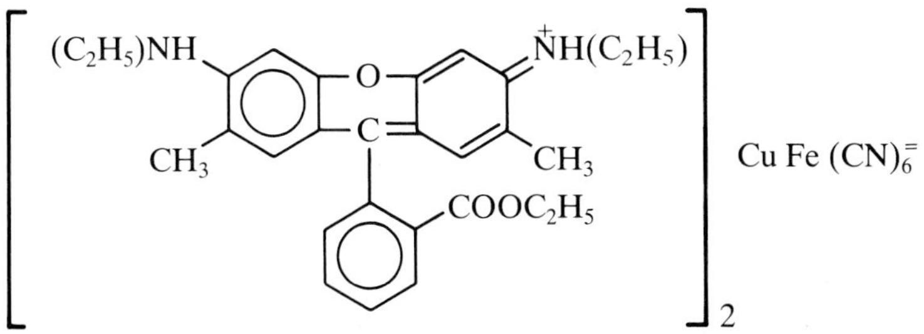

Naphthol Red F5RK

(*CI Pigment Red 170 No. 12475*)

Properties
It is very brilliant colour with a strong bright bluish red hue. Excellent light fastness, resistant to soap, wax, grease, acid and alkali.

Uses
It is suitable for all types of printing inks in virtually any system. High cost restricts its use.

Structure

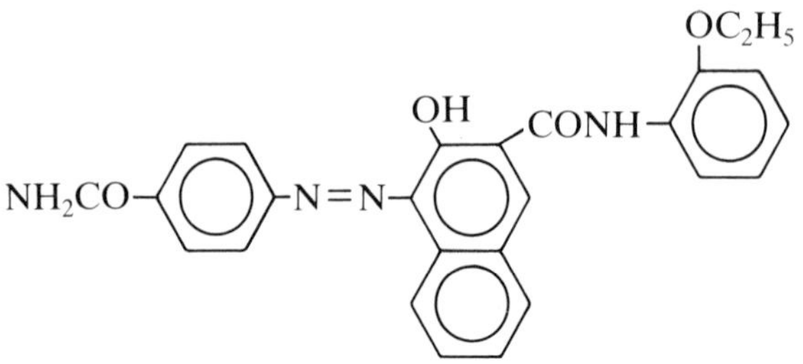

Benzimidazolone Carmine HF 3C

(CI Pigment Red 176 No. 12515)

Properties
The very bright and transparent hue is bluish red. It is lightfast, heat stable, acid, alkali, grease and soap resistant. It is widely used for packaging and metal decorating applications. Expensive.

Uses
PVC inks, alkyd resin enamels, lacquers and many liquid packaging uses where high specification is necessary.

Structure

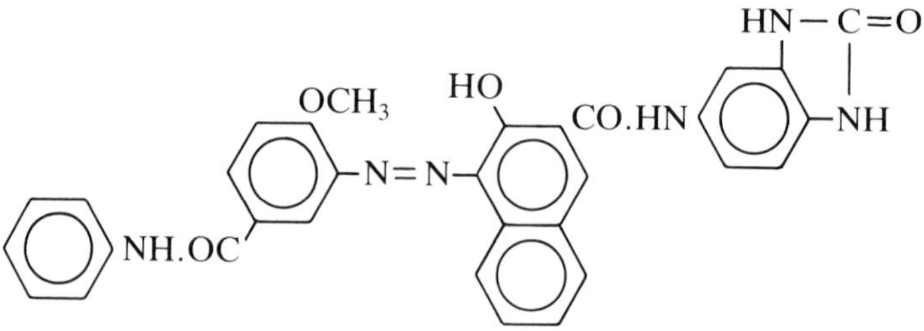

Naphthol Rubine F6B

(CI Pigment Red, 184)

Properties
It is a bluish red — very similar pigment to CI Red 146 but slightly stronger and more resistant. It has good resistance properties to acid, alkali and soap, but poor lightfastness.

Uses
Packaging liquid inks. It has good baking stability and so is suitable for metal decorating inks.

Structure

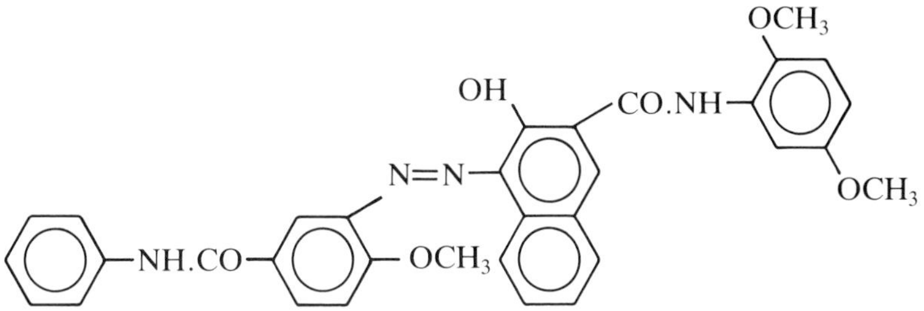

Benzimidazolone Carmine HF4C

(*CI Pigment Red 185 No. 12516*)

Properties
This is a process magenta of very pure clean colour, very resistant and superior to Red 52 and 57. However rather more expensive. Excellent quality pigment.

Uses
As a process magenta where stringent properties are necessary. Resistant to acid, alkali, most solvents, plasticizers, grease, wax and soap. Poor–medium lightfastness. Particularly poor in tint. Used widely in paste inks and metal decorating inks.

Structure

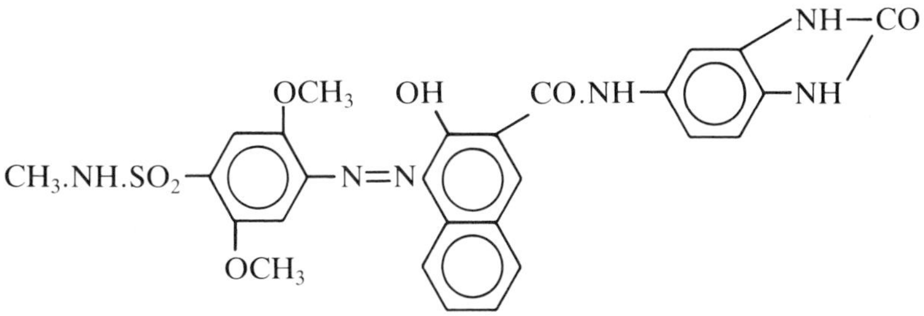

Rubine Red 6B

(*CI Pigment Red 200 No. 15867*)

Properties
Slightly bluer version of Red 52 and 57. It has Poor acid, alkali, solvent

and soap resistance. Excellent brightness, colour, purity and tinting strength.

Uses
It is used where, at similar cost and properties, a bluer version of 4B (Red 57/52) is required. Works well in liquid and paste printing inks.

Structure

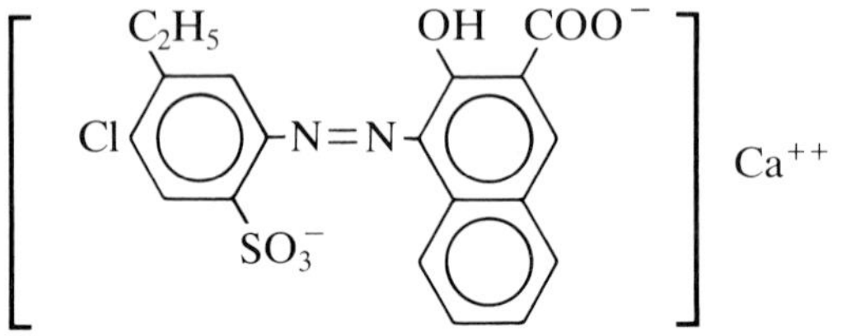

Quinacridone Magenta B

(*CI Pigment Red 202*)

Properties
This is a high performance pigment, the most lightfast of its shade. It is very similar to R122 at full strength but much redder in tint and has excellent resistance properties. Very expensive.

Uses
Very high quality applications which require permanency of all colour properties. Usable in all systems. Care in formulation is necessary to avoid poor flow characteristics.

Structure

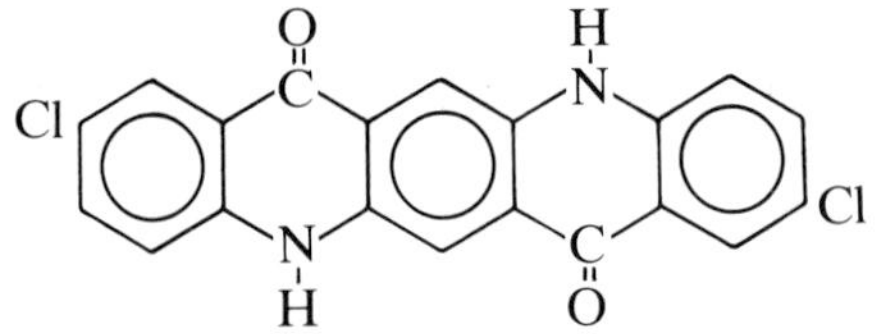

Benzimidazolone Red HF2B

(*CI Pigment Red 208 No. 12514*)

Properties
A very bright medium shade red with good heat stability and light fastness. Acid, alkali, grease, and soap resistant, and relatively stable to most solvents. Transparent with high tinting strength.

Uses
Liquid packaging inks and metal decorating inks particularly to achieve strength in tint at reasonable cost.

Structure

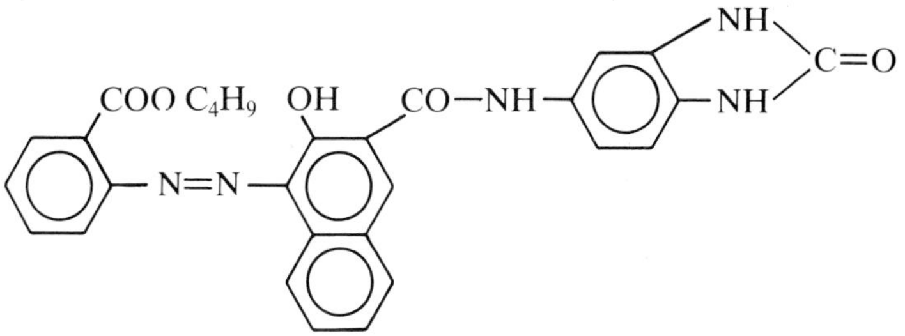

Naphthol Red F6RK

(*CI Pigment Red 210*)

Properties
Stronger, bluer and cleaner version of PR 170. Good resistance properties resistant to acid, alkali, soap and wax. Good heat resistance but poor lightfastness particularly in tint.

Uses
Packaging liquid inks and lacquers. Care is needed for lightfast specification inks.

Structure

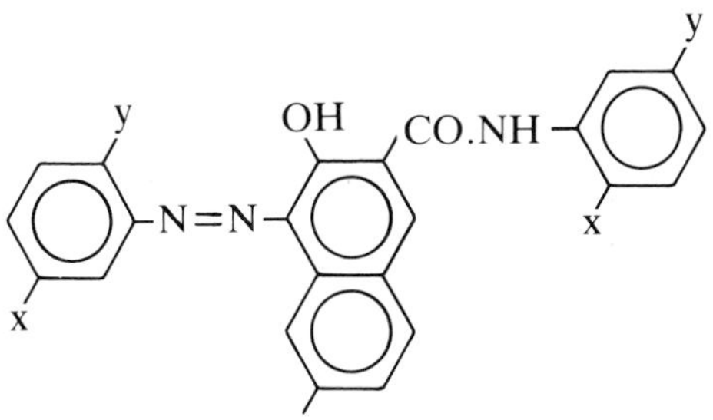

x and y not identified

Azo Magenta G

(*CI Pigment Red 222*)

Properties
Very highly resistant bluish red. Excellent lightfastness and good resistance properties. Strong in tint.

Uses

It is generally used in liquid packaging inks in applications where high specification for this shade is required.

Structure

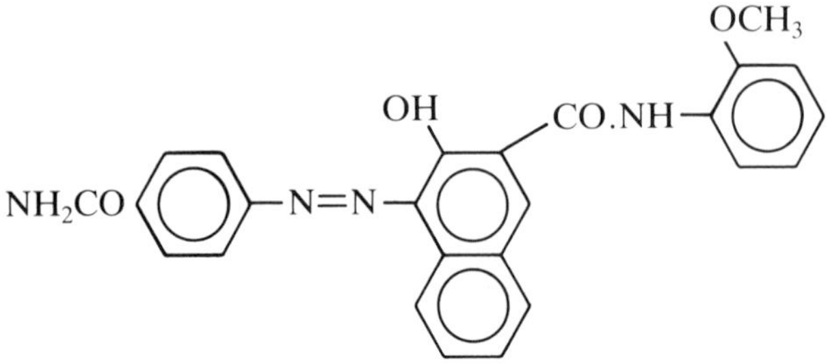

Anthraquinone Scarlet

Perylene Red Y
(*CI Pigment Red 224 No. 71127*)

Properties

A transparent very yellow shade pigment with excellent tinting strength and fastness properties. Gives dull tint colours with exceptional lightfastness both at full strength and in tint.

Uses

Security printing inks and other high specification inks not requiring alkali resistance.

Structure

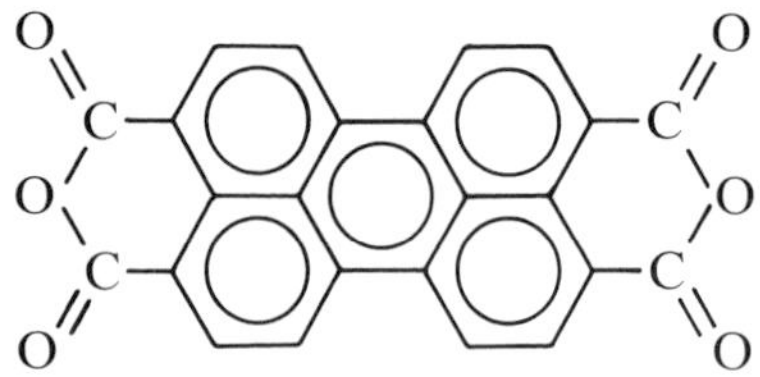

Quinacridone Violet

(*CI Pigment Violet 19 No. 73900*)

Properties

Various shades from yellowish red to bright violet. High performance pigment. Resistant to everything, very lightfast and heat resistant. The yellower shades are more opaque. The most lightfast organic blue shade red.

Uses
Any system where its superb properties are required. Typical of high performance blue shade reds, it is very expensive.

Structure

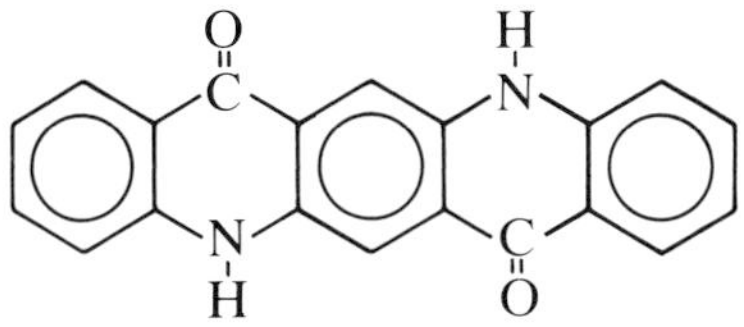

Benzimidazolone Bordeaux HF 3R

(*CI Pigment Violet 32 No 12517*)

Properties
Very bright violet red pigment. Highly transparent chemically resistant to everything except some organic solvents. Very heat resistant but has only limited lightfastness.

Uses
It is widely used as soapfast standard in this shade for both liquid packaging and metal decorating inks. Particularly good in foil lacquers where lightfast dyes can not be used.

Structure

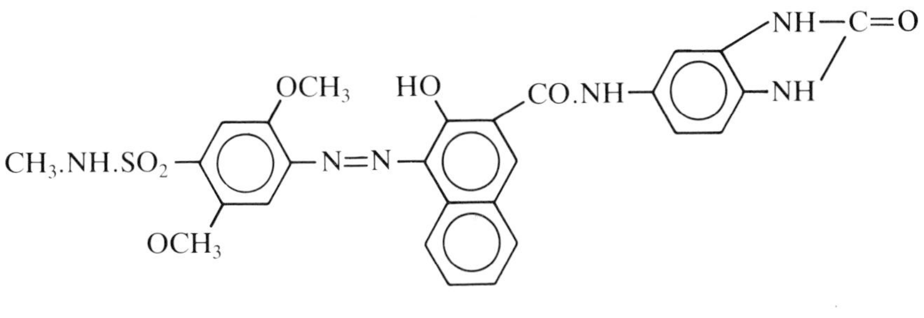

4.4   GREEN PIGMENTS

PMTA Deep Green

(*CI Pigment Green 1 No. 42040:1*)

Properties
A bright bluish green with high strength and clean undertone. It has fair lightfastness but tends to darken on exposure to sunlight. It has poor resistance to alkali, soap and strong solvents, but is unaffected by paraffin wax. Typical of PMTA type pigment.

Uses
It is suitable for all types of printing ink, especially liquid and water based inks where a bright intense hue is required.

Structure

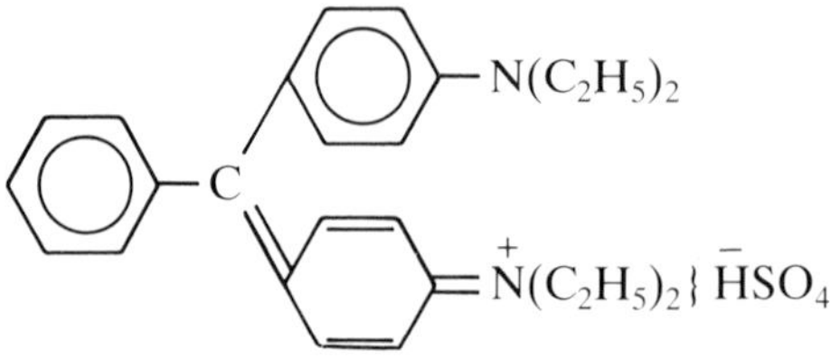

No. 42040:1 (CI Pigment Green 1) is the phosphotungstomolybdic acid salt of the above.

PMTA Vivid Green

(CI Pigment Green 2) No. 42040:1 (Pigment Green 1) and No. 49005:1 (Pigment Yellow 18)

Properties
It is a blend of two PMTA colours. A very bright emerald green of outstanding brilliance; a very strong colour. It has poor fastness to light, heat and alkali.

Uses
It is used in many printing inks to achieve this shade but is limited in use due to lack of resistance properties.

Structure

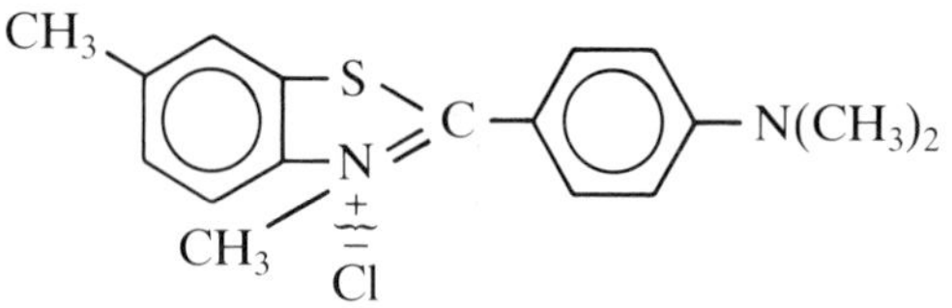

Pigment Yellow 18 is the phosphotungstomolybdic acid salt of the above.

Phthalocyanine Green

(CI Pigment Green 7 No. 74260)

Properties
It is a bright green resistant pigment *not* prone to crystallisation or flocculation like its blue counterpart; a Blue shade multi-purpose green

with superb properties. Resistant to everything. Very lightfast and heat resistant. Excellent transparency, tinting strength and colour strength.

Uses

Major green standard used in all types of printing inks. Very useful in tints as there is no lack of resistance properties. Careful formulation is necessary at high pigmentation levels in order to avoid poor flow.

Structure

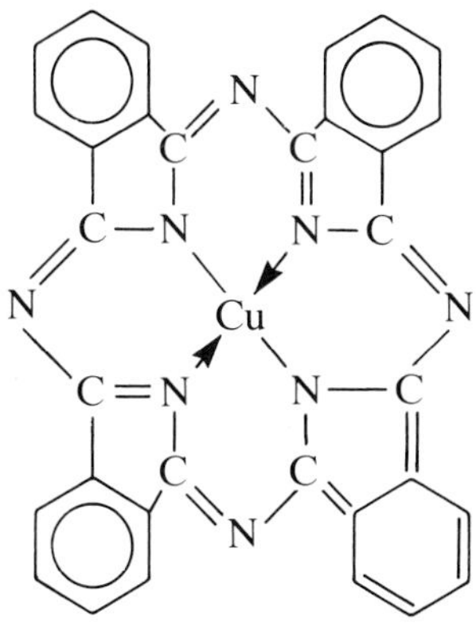

Basic
phthalocyanine
molecule

Chlorinated

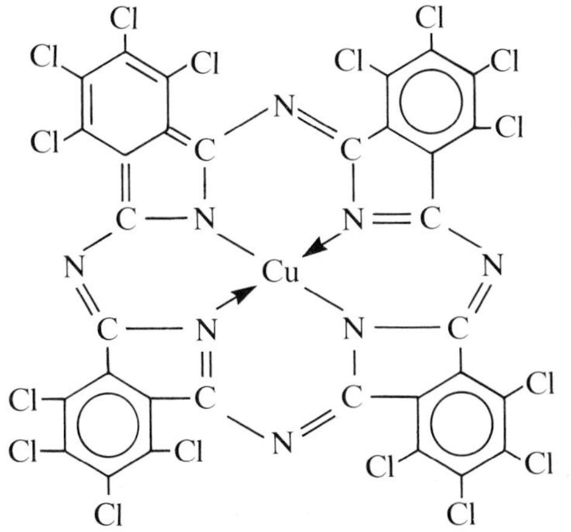

Chlorinated with
16 chlorine atoms to
give CI Pigment
Green 7

Phthalocyanine Green

(*CI Pigment Green 36 No. 74265*)

Properties
Augments Green 7 to cover the colour range of the phthalocyanines from blue shade to yellow shade (the more bromine the yellower the tone). High transparency, clean and bright, *very* resistant in all properties.

Uses
All types of printing inks. Any use for resistance properties as Green 7. It is more expensive than Green 7.

Structure
The bromine atoms vary from four to nine.

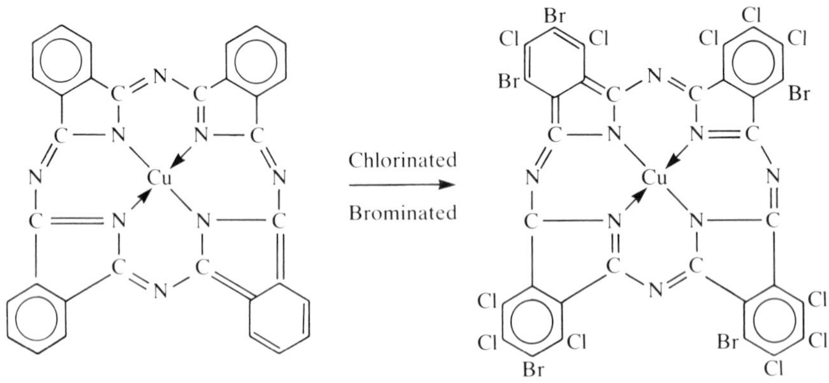

4.5   **BLUE PIGMENTS**

PMTA Victoria Blue, PMTA Brilliant Blue

(*CI Pigment Blue 1 No. 42595:2*)

Properties
A bright reddish blue of high tinctorial strength and purity of hue. Good lightfastness for strong colours subject to darkening on exposure and falling off sharply on reduction. Readily affected by polar solvents. It has good heat stability.

Uses
Used in most types of ink including water-based inks. Makes superb dispersions but rather short flowing inks. High cost and poor resistance do not curtail its use, as the shade cannot be obtained by combining other pigments, Products under this heading include the phosphotungstic molybdate salts of CI Basic Blue No. 7. Like the double salts (PTA and PMA) and copper ferrocyanide complexes, they have price advantages but poor lightfastness.

CI Basic Blue 7 formula

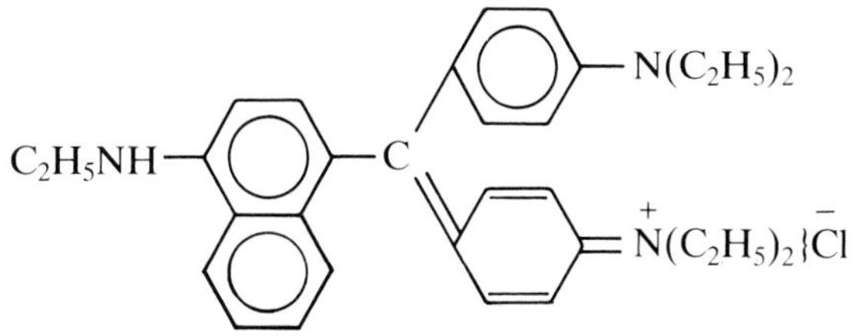

C_2H_5NH—[structure]—C[structure]—$N(C_2H_5)_2$

$\overset{+}{N}(C_2H_5)_2\}\overset{-}{Cl}$

Phthalocyanine Blue alpha form

(*Pigment Blue 15 No. 74160*)

Properties
The solvent-unstable form of alpha Phthalo Blue. *Very* resistant properties. Lightfast, heat stable resistant to acids, alkalis, solvents, plasticizers, waxes, greases, and soap. Blue 15 is the reddest and strongest Blue 15 type. The other pigments in this series are all modifications to correct any instability problems with this pigment.

Uses
Some usage in news inks, but probably the least used form of the Blue 15s in inks.

Structure

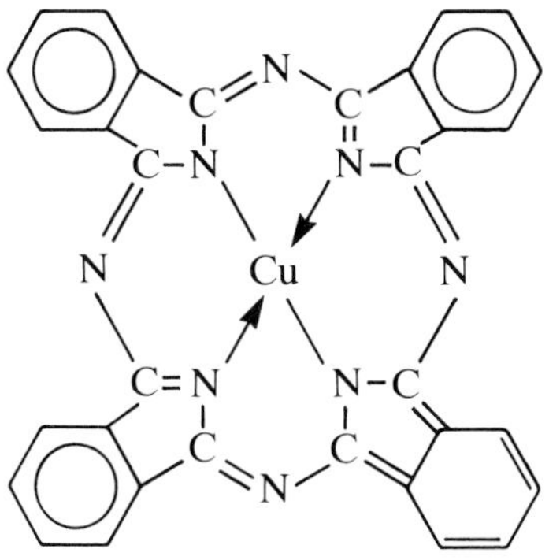

[Structure of copper phthalocyanine with central Cu atom]

Phthalocyanine Blue Alpha solvent-stable.
Phthalo Blue NC Alpha

(*Pigment Blue 15:1 No. 74160 or No. 74250 (mono chloride version)*)

Properties
Bright reddish blue excellent resistance properties. More heat resistant and solvent stable than Blue 15.

Uses
The most commonly used alpha reddish shade blue. The most cost effective and versatile red shade blue for use in all systems. Accounts for about 25% of the B15 types usage. Chosen for stability and shade.

Phthalocyanine Blue NCNF alpha form: solvent-stable, non-flocculating

(CI Pigment Blue 15:2 No. 74160 or 74250)

Properties
Bright reddish blue more extensively treated than 15 or 15:1, giving a non-flocculating version.

Uses
Improved stability version of 15:1. Higher cost restricts its use. Accounts for more than 5% of Blue 15 usages.

Phthalocyanine Blue NC Beta form: solvent-stable

(CI Pigment Blue 15:3 No. 74160)

Properties
Bright greenish blue. Completely resistant to acids, alkalis, solvents, plasticizers, greases, paraffin wax and soap. Also solvent stable, heat resistant, and very lightfast. Transparent, strong tinting power, very clean colour, very permanent. Slightly weaker than alpha forms.

Uses
The perfect cyan for process printing. Usable in every type of system and ink. Accounts for *over* 50% of the Blue 15 production.

Phthalocyanine Blue B NCNF

(CI Pigment Blue 15:4 No. 74160)

Properties
This is a bright greenish blue treated to prevent flocculation and is a useful cyan in systems where 15:3 does not work well and appears weak. All the resistant properties of the Blue 15s.

Uses
Suitable for many liquid ink resin systems. Used more in liquid inks applications than 15:3. Slightly more expensive but with increased stability. Very clean bright green shade colour. More stable in tint and strong solvent liquid ink formulae. Accounts for about 18% of Blue 15 usage.

Phthalocyanine Blue Epsilon form

(Pigment Blue 15:6 No. 74160)

Properties
This is the only very stable red shade Phthalo Blue. It has the alpha shade and does not change to the beta shade on heating. It has all the admirable Phthalo Blue resistance properties.

Uses
In cases where very clean, bright red shade blue is needed, mainly liquid ink application where shade stability can be more of a problem. Expensive. Accounts for 2–3% of the total Blue 15 production.

Phthalocyanine Blue (copper free)

(*CI Pigment Blue 16 No. 74100*)

Properties
It is a bright greenish blue — the greenest Phthalo Blue obtainable. It has excellent resistant properties but is less heat stable than Blues 15:1, 15:2, 15:3 and 15:4.

Uses
Can be used for most purposes but it is more expensive than 15:3 or 15:4. Its use is restricted to materials where a metal-free product is specified.

Structure

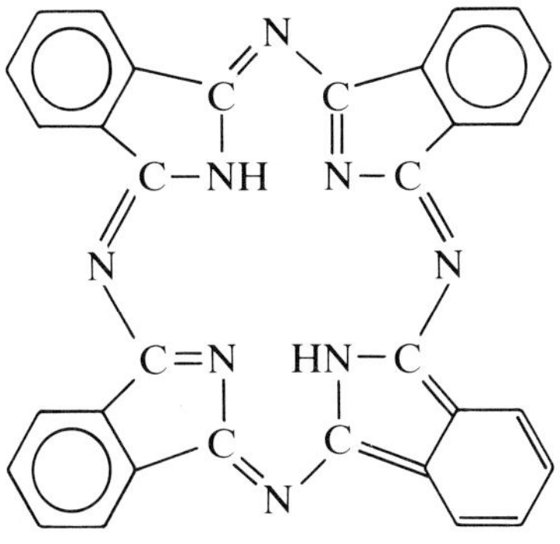

Milori Blue, Iron Blue, Bronze blue, Prussian Blue, Chinese Blue

(*CI Pigment Blue 27 No. 77510*)

Properties
A range of strong blues from dirty-reddish tone (Prussian) to a cleaner greener tone (Chinese). A lot of surface bronzing which can vary considerably both in shade and intensity, depending on drying conditions, degree of dispersion and the amount of binder in the recipe.

Iron blues possess excellent lightfastness, heat resistance, oil, fat, wax and acid resistance. They are attacked by alkalis and decompose readily. This renders them unsuitable for water-based systems, soap wrappers and fertilizer bags.

Uses
Iron blues are available in various grades to suit all types of printing ink. Their low cost and high strength make them popular pigments. They are difficult to disperse and should not be left mixed and unmilled as spontaeneous combustion can occur.

Tints are unstable on storage as the inorganic pigment reduces to ferro-ferrocyanide, thus turning green and dirty. They need little or no extra driers in oxidation drying ink vehicles. Iron being a transition metal catalyses the polymerisation of the vegetable oils. These blues have some limitation compared with phthalo blues but in the USA they still are used as the publication gravure standard in four-colour sets.

Formulae
$KFe''' [Fe''(CN)_6] x.H_2O$ or $K[Fe''Fe'''(CN)_6] x.H_2O$.
Note: Potassium may be replaced by sodium or ammonium or mixtures of both. Often referred to as potash blue or soda blue.

Ultramarine Blue

(CI Pigment Blue 29 No. 77007)

Properties
A red shade clean blue with low tinting strength semi-opaque to transparent dependent on surface treatment and particle size distribution.

Highly resistant to the effects of light, heat, alkali, wax and solvents, but dilute acids cause fading. Uncoated grades are sensitive to water.

Uses
Mainly used in screen process, diestamping and water-based inks. As they are hydrophilic special coated grades are available for lithography. In oil inks they are difficult to disperse, and do not transfer well on three roll mill rollers unless care has been taken in formulation to prevent piling. Suitable for food packaging but darken in the presence of copper plates or cylinders and under acid conditions on external exposure. FDA approval in certain foods and cosmetics.

Formula
$Na_{6-8}Al_6Si_6O_{24}S_{2-4}$.

Alkali Blue G, Reflex Blue 2G

(CI Pigment Blue 56 No. 42800; CI Pigment Blue 18 No. 42770:1; CI Pigment Blue 61 No. 42765:1)

Properties
Tinctorially the strongest blue pigments manufactured. CI Blue 56 is the

strongest and most alkali resistant. CI Blue 61 and 18 have poor alkali resistance. Red shade to green shade blues with a strong red-blue sheen, poor lightfastness, reasonable heat resistance, and good acid resistance. Other properties only moderate. These pigments retard oxidation drying in oil-based systems.

Uses
These pigments have traditionally been supplied as the flushed form. Either aqueous or oil based. Recently, however, powder supplies are becoming available giving the formulator more flexibility with his recipes. The powder form is very cost effective for flexo water-based inks giving excellent strength for this type of colour. The flushed form is still used in large quantities for paste inks and publication gravure inks. Mainly used as a toner in black ink (including news ink). A cheap and strong useful colour.

Structure

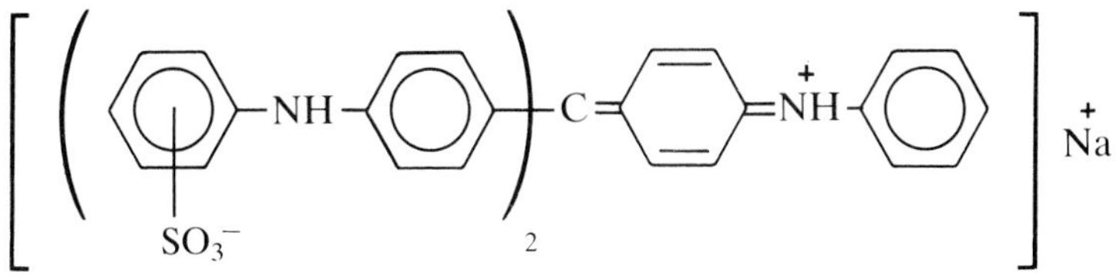

CI Blue 18

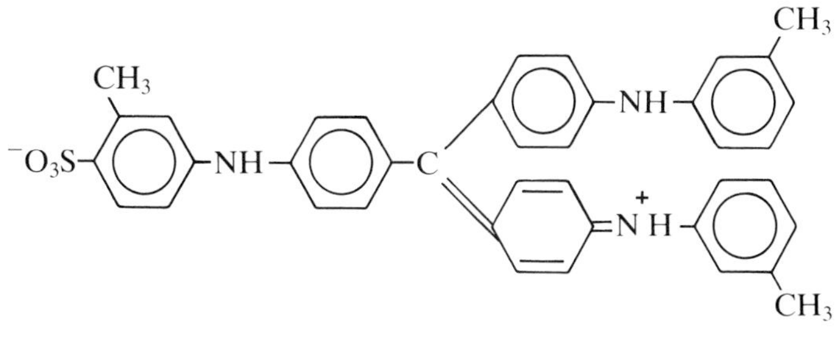

CI Blue 56

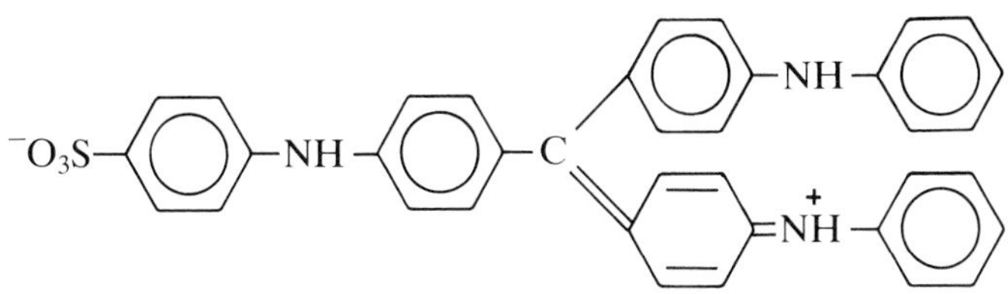

CI Blue 61

Indanthrene Blue, Anthraquinonoid Blue

(CI Pigment Blue 60 No. 69800)

Properties
Red shade transparent blue of good strength. Extremely resistant to light, heat, steam sterilization, acid, alkali, solvents, soap and waxes. Redder shade than Phthalo and very clean.

Uses
Can be used in all types of ink but is expensive; selected where outstanding fastness is the dominant requirement for example in gravure inks for soap or detergent packages. It disperses well but has high oil absorption which results in fairly short inks.

Structure

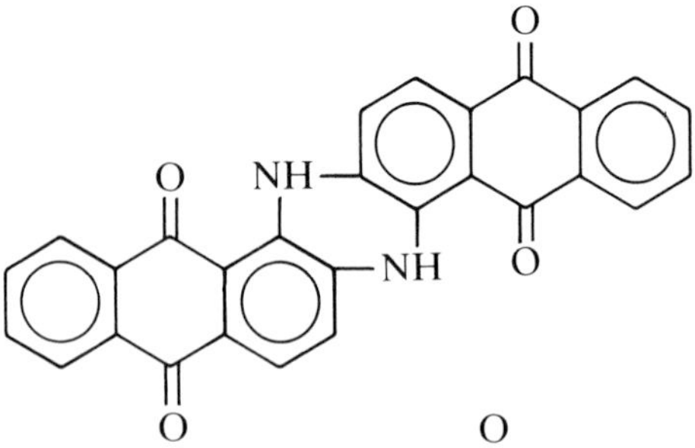

CI Pigment Blue 60

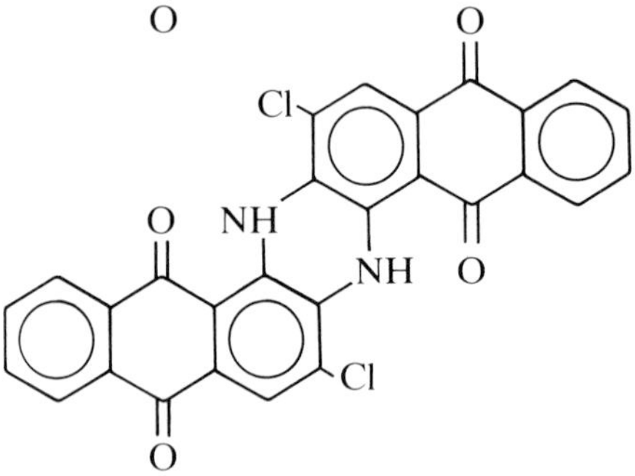

CI Pigment Blue 64

Note: The chlorinated grade is much redder and is classified as CI Pigment Blue 64 (CI Vat Blue 6, No. 69825).

Victoria Blue CFA, Copper Ferrocyanide Blue

(CI Pigment Blue 62 No. 42595x)

Properties
Royal blue coloured, very bright pigment not as bright as PB1 (the PMTA pigment) but a slightly more resistant and better value pigment. Useful for many applications.

Uses

Because of its shade and price often used instead of the PMA or PMTA toners. Good for gravure and water-based flexo packaging inks.

Structure

4.6   VIOLET PIGMENTS

PMTA Rhodamine

(*CI Pigment Violet 1 No. 45170:2*)

Properties

Strong clean magenta shade, which is often used in inks as a blue-shade red. Exhibits high purity of hue and tinctorial strength. Darkens on exposure to light and bleeds in water, solvents, oils and soap.

Uses

Declining in use. It is suitable for liquid inks including water based, but unsatisfactory for lithography or high temperature steam processing. Letterpress prints can be varnished or lacquered only under special conditions. CI Pigment Violet 2 is very slightly bluer and stronger, and has similar fastness.

Structure

Pigment Violet 1 is the phosphotungstomolybdic acid salt of the above.

Note: Double salts of this dye are made, the most popular being phospho-molybdate and silico-molybdate complexes.

PMTA Violet, Methyl Violet

(CI Pigment Violet 3 No. 42535:2)

Properties
A bright bluish violet with very high strength. It has good resistance to light and heat. No bleed in water or aliphatic hydrocarbons but bleeds in most other solvents and in soap. Suitable for steam processing. The double salts and the copper ferrocyanide complex give duller, stronger shades than the triple salt. They darken rapidly on exposure to light.

Uses
The triple salt is used in letterpress and lithographic inks. The cheaper double salt forms are used extensively in liquid inks and for toning blacks.

Composition
The phosphotungstomolybdic acid salt of CI Basic Violet 1 (Methyl Violet), No. 42535. The copper ferrocyanide complex is CI Pigment Violet 27. No. 42535:3.

Ultramarine Violet

(CI Violet 15 No. 77007)

Properties
Brightish violet to red violet — very high colour purity similar to Blue 29. Good lightfastness and heat stability.

Uses
Metal decorating inks to tone white baking enamels to avoid yellowing.

Formula
$H_2Na_{4-6}Al_6Si_6O_{24}S_2$

Dioxazine Violet (RL) Carbazole Violet

(CI Pigment Violet 23 No. 51319)

Properties
A tinctorially strong, red shade purple of high transparency. Superior resistance to heat, light, steam sterilization, acids, alkalis, solvents, fats and soaps.

Uses
Although an expensive pigment, it offers special advantages in formulation. It is used in all the printing processes for applications that require a stable purple, with excellent fastness properties. Also used for reddening phthalocyanines without loss of properties. Care in formulation is need to avoid poor flow.

Structure

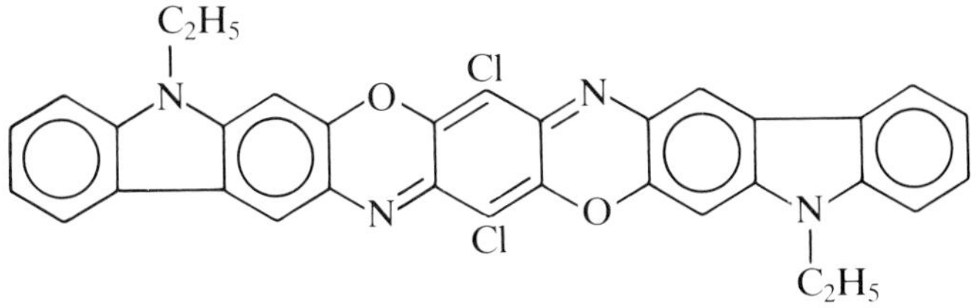

Crystal Violet (CFA)

(CI Pigment Violet 27 No. 42535:3)

Properties
Very strong and bright colour — more economical than PMTA toners (Violet 3). Not quite as clean but better properties. Poor alkali and soap resistance.

Uses
Mainly used in water-based flexo packaging rules. Sometimes used in gravure inks where the poor fastness properties can be accepted.

Structure

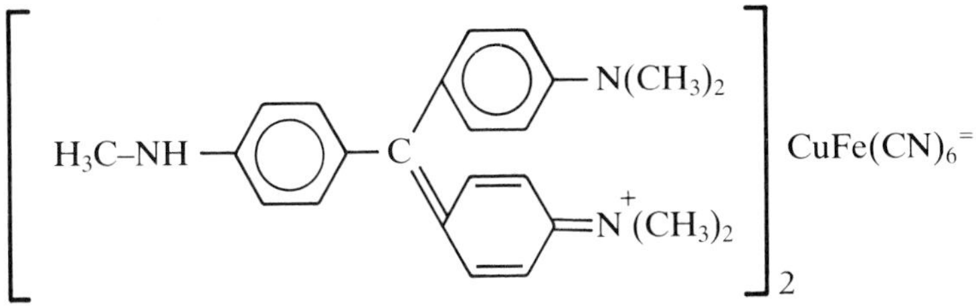

N-N-dimethylaniline oxidised with cupric chloride and converted to the copper ferrocyanide complex.

Dioxazine Violet B

(CI Pigment Violet 37 No. 51319)

Properties
Very high performance colour resistant to everything. Lightfast and heat resistant. Exceptional colour intensity redder than its counterpart Violet 23.

Uses
Used in a wide range of inks, where its exceptionally good fastness properties are required. Its excellent lightfastness is valuable in wall-covering inks.

Structure

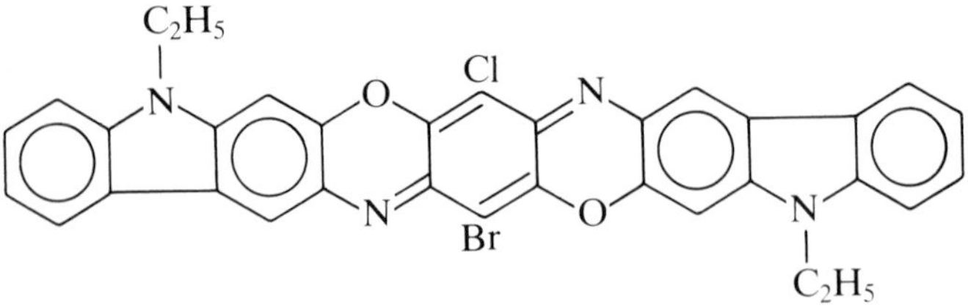

Benzimidazolone Bordeaux HF 3R

(CI Pigment Violet 32 (listed under reds)) (See p. 139).

Thioindigoid Red (listed here as it is a violet shade)

(CI Pigment Red 88 No. 73312)

Properties
High quality pigment. Very clean colour, excellent fastness properties.
Slightly soluble in ketones.

Uses
Used for security inks. An excellent red shade violet sometimes used with
CI Pigment Red 202.

Structure

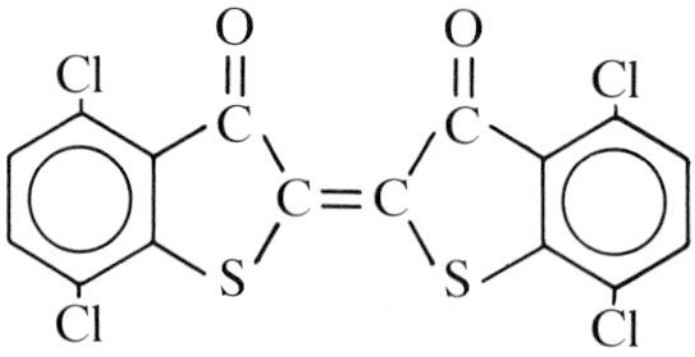

4.7   BROWN PIGMENTS

Brown iron oxides

(CI Pigment Browns 6 and 7 No. 77491 (77492) (77499))
Synthetic brown iron oxide, natural brown iron oxide, burnt sienna.
(CI Pigment Reds 101 and 102 No. 77491)
Synthetic red iron oxide, natural red iron oxide, red haematite.

Properties
Resistant to everything. Exceptional lightfastness and stability. Strong UV
absorber. Water resistant.

Uses
The natural oxides are difficult to grind, and their use is restricted to formulations where the very low cost and good fastness properties are required. The synthetic materials are finer and find a much wider usage. Used in inks for imitation wood grain laminates.

Empirical Formula
In all cases Fe_2O_3 (over 97%).

Diazo Brown

(*CI Pigment Brown 23 No. unassigned*)

Properties
Moderately high performance pigment. Acid and alkali resistance good. Some bleed in soap and organic solvents. Excellent grease, fat, and paraffin wax resistance. Very good lightfastness and heat resistance.

Uses
Tarnish-proof bronze printing inks, lacquers, and vinyl wall covering inks where the lightfastness is needed. More expensive than iron oxides but easier to disperse. Excellent tinting strength.

Structure

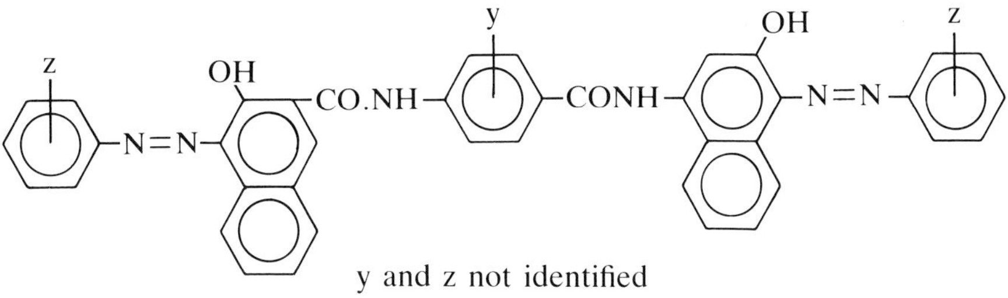

y and z not identified

Chromium Antimony Titanium Buff Rutile

(*CI Pigment Brown 24 No. 77310*)

Properties
Buff-coloured slightly abrasive (Rutile) pigment. Outstanding resistant properties. Exceptional lightfastness and heat resistance (over 300 °C) Low tinting strength slightly dirty shades.

Uses
Used in any system where coarse texture and poor tinctorial strength can be tolerated.

Basic Composition

TiO_2	80–90%
Cr_2O_3	1–6%
Sb_2O_3	8–12%

Benzimidazolone Brown HFR, Fast Brown HFR

(*CI Pigment Brown 25 No. 12510*)

Properties
Very transparent, very resistant chemically (some organic solvent bleed) good lightfastness and heat resistance to 220 °C. Readily dispersible in ink vehicles.

Uses
Where transparency is needed to achieve good effects on foil and a metal free specification on various films (particularly metallised). The very high heat, and other resistant properties make this particularly suitable for laminating inks and stoving metal decorating inks.

Structure

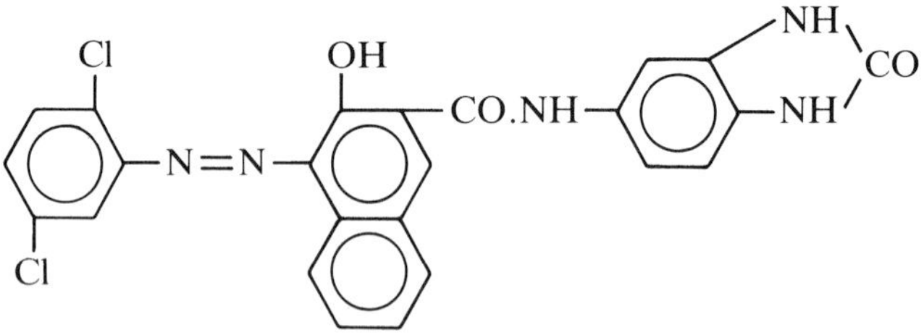

4.8   BLACK PIGMENTS

Vegetable Black (Lamp Black)

(*CI Pigment Black 6 No. 77266*)
Nitro Pigment

Properties
Soft black with bluish undertone, greyer and coarser than the channel blacks. It is inert but because it has a large surface area, it tends to absorb organic colouring matter, therefore it should not be added to tints.

Uses
Oldest pigment known, vegetable blacks are used as matt blacks. They improved the strength and colour of diestamping and copperplate inks. They have now been almost entirely replaced by furnace blacks, unless cost is a factor.

Carbon Blacks (Channel and Furnace)

(*CI Pigment Black 7 No. 77266*)
Most important black pigments. All types of carbon black are included under the above classification. They are pigments of extremely fine particle size, with colour varying from a grey masstone with blue undertone, to a dense jet black masstone with brown undertone,

dependent on the method of manufacture. Non-rubber varieties, used in printing inks, are produced in pellet, bead, densed or fluffy form. They are chemically inert, extremely fast to heat, light, acids, alkali, solvents and soap. Most grades however do contain a small proportion of volatile material, the effect of which may become apparent on dispersion.

Furnace Black

Almost all carbon black is today produced by the furnace process, and most of this production is allocated to the rubber compounding industry for vehicle tyre manufacture.

Some is produced by the slow vertical process in which oil is sprayed and burned under controlled conditions. The carbon is collected by cyclonic or electrostatic methods. Finer blacks are manufactured by a faster vertical process, where hot gas and oil are combined under pressure to disintegrate or crack. Carbon Black produced by this process is not exposed to air and therefore absorbs little oxygen. Newer types of treated furnace black have replaced channel black in most applications. They undergo a further process which bonds them with oxygen and other surfactants to obtain required surface characteristics.

Channel Black

Now produced in only a few locations throughout the world as it requires natural gas, which is more valuable as fuel than as a source of Carbon Black. Natural gas consists of a mixture of gaseous hydrocarbons, predominantly methane. It is burnt in a series of small luminous flames which impinge on a water-cooled inverted-U-shaped channel. A layer of fine Carbon Black is deposited. The particles are continuously removed by scrapers and conveyed to beading, densing and packaging plant. They have a dense black appearance with brown undertone. Channel blacks possess good wetting characteristics. They have a high oil absorption caused by their porosity and very large surface area. This wetting is aided by the low pH that channel blacks normally have, caused by the formation of a volatile absorbed layer of between 10 and 15% by weight, including moisture. Oxygen absorbed during manufacture can be chemically bonded to produce 'long flow' channel blacks having optimum colour, density, flow, gloss and dispersion. A disadvantageous effect of Channel Black in inks is the tendency to absorb both driers and alkali blue toners over a period of time.

Lamp Black

Creosote and tar residues are burned in a hooded flue, with a limited amount of air, to a smoky, sooty gas. It was formerly allowed to settle in cooling chambers, to be swept and bagged. Today it is normally collected by more modern methods. It has a large particle-size compared with channel black and it flocculates to form chain-like structures.

Formula of carbon blacks

Carbon	90–99%
Volatiles	1–10%

Almost pure carbon.

Black iron oxide, magnetic iron oxide, synthetic black oxide

(CI Pigment Black 11 No. 77499)

Properties

A dense matt brown-shade black of high specific gravity and low oil absorption. It is resistant to heat, light, alkali, solvents, soap and wax but it dissolves in acid. Cheap, strong UV absorber. Converts to red iron oxide at 150 °C.

Uses

The ordinary black iron oxide has been used in both copperplate and diestamping inks. It imparts the flow and clean wiping properties necessary for such inks, but is tinctorially weak.

Magnetic iron oxide is required in inks which print well-defined characters on cheques, for automatic cheque processing. When dry, prints using such inks will have a permanent magnetic response to reading apparatus. Better dispersability in water than Black 7. Synthetic grades are stronger and bluer than natural grades.

Composition

Ferrous-ferric oxide (FeO-Fe_2O_3). Occurs widely in nature as magnetite, the magnetic oxide of iron. The grades used in inks are synthetic products.

4.9   WHITE PIGMENTS AND EXTENDERS

Zinc White, Chinese White

(CI White 4 No. 77947)

Properties

Clean white of medium opacity, fine ground particle size. Pure grades are FDA approved. Strong UV absorber. self-fungicidal material, neutralises acidic oxidation products. Prone to livering in acidic vehicles. Discolours with copper compounds. Yellow when hot, white when cold.

Uses

Widely used as a white extender in tin printing and many other ink systems. It improves flow and levelling.

Formula

ZnO	99–99.7%
Volatiles	0.3–1.0%

Lithopone

(CI Pigment White 5 No. 77115)

Properties
Chemically inert pigment available in easily dispersable micronised forms. Opaque white extender. Decomposed by acids.

Uses
Gaining popularity again as an extender. More popular on the continent than in the UK. Useful in all systems where opacity is required.

Formula

$$BaSO_4 \quad 60-72\%$$
$$ZnS \quad \ \ 28-40\%$$

Titanium dioxide Anatase and Rutile

(*Pigment White 6 No. 77891*)

Properties
Most important opaque white pigment currently in use; accounts for over 80% of all CI whites used (including all extenders). It has replaced nearly all other whites. Resistant to everything, very lightfast and heat resistant. Inert material. Various grades are available with coatings of zinc, aluminium or silica oxides to achieve various properties.
Two major grades: Rutile (harder, bluer, more opaque than anatase)
Anatase (less opaque, more stable except in tint, softer texture).

Uses
Every ink system. Anatase is preferred in gravure ink systems giving less cylinder wear. Almost 100 grades/types are available. Generally the purer, less extended, non-coated grades are used for cheaper water-based flexo inks, screen process inks, and some gravure whites. Coated grades are for higher quality more exacting work giving higher gloss results. Used in decorating, screen and liquid ink base coats.

Formula and composition

$$TiO_2 \qquad\qquad 80-99.9\%$$
$$(Al, \ Si \ or \ Zn)O \quad 0-20\%$$

Zinc sulphide

(*CI Pigment White 7 No. 77975*)

Properties
Opaque white pigment. Resistant to everything but acid. Heat resistant and colour stable. Not as opaque as TiO_2 but less after yellowing. Great for abrasion resistance so a good extender for TiO_2. Greater UV reflectance than TiO_2.

Uses
Printing inks of all systems except acid vehicles. Metal decorating systems

gives pure clean baking enamels. Gaining favour on cost as an extender for TiO_2.

Formula and composition

ZnS	97%
ZnO	< 0.3%
Solubles	< 0.4%
Moisture	< 0.2%
Residue	3%

Calcium carbonate, whiting

(CI Pigment White 18 No. 77220)

Properties

White (dull) cheap extender now used to reduce the cost of inks without affecting their printing qualities. Precipitated calcium carbonate may contain up to 3% stearate or resin coating. Both the coated and uncoated grades have superseded other extenders as tinting media. They are bulky but impart good working properties and a fairly glossy finish. Varies from opaque to semi transparent.

Uses

Coated grades are used for letterpress, lithographic and gravure inks, whilst both coated and uncoated grades are used in aqueous inks. Screen process inks normally contain uncoated grades or whiting (the natural form of calcium carbonate) to obtain a matt finish. Copperplate inks also contain whiting for its wiping qualities. Selection of the correct grade aids dispersion, flow and press performance. Calcium carbonate is not recommended for use with high acid value vehicles such as moisture-set as it reacts to liberate carbon dioxide. It also reacts with alkali sensitive pigments like Milori Blue (B27) so care is needed in formulation. Calcium carbonate is also used extensively as a 'dry spray' power to prevent set-off. Ground mineral versions, once milled to a good dispersion, give results with outstanding gloss and transparency but are appreciably more costly than the very basic forms available.

Production

Calcium carbonate ($CaCO_3$) occurs naturally in many countries as chalk or limestone which is pulverised to produce whiting. It is produced synthetically by slaking quicklime and exposing the slurry to carbon dioxide. It is also obtained as a by-product in water purification.

Formula composition

	Natural	Synthetic
$CaCO_3$	95–98%	98–99%
$MgCO_3$	< 1–3%	< 1–1.5%
SiO_2	< 0.5%	< 0.1%

$(Al, Fe,)O$	$<0.5\%$	$<0.3\%$
H_2O	$<0.2\%$	$<0.2\%$

China clay, bentonite, kaolin clay

(CI Pigment White 19 No. 77004)

Properties
FDA approved for coating use, white to pale grey powder. Chemically inert and inexpensive. It has a higher oil absorption than most extenders except alumina.

Uses
Moisture set and low cost letterpress inks, some gravure inks. High oil absorption grades can be used as a thickener. Useful because of its inertness and cost in a wide range of applications.

Formula and composition
$Al_2O_3.2SiO_2.2H_2O$.

Natural hydrated aluminium silicate, it occurs in many parts of the world. It contains impurities of calcium, magnesium, iron, quartz and mica. It is cleaned and segregated into grades by levitation in settling tanks.

	Hydrous	*Calcined*
Al_2O_3	37–45	42–45
SiO_2	44–53	51–53
$(Fe, Ca, Mg)O$	12–14	1

Blanc Fixe, Process White

(CI Pigment White 21 No. 77120)

Properties
Clean and white extender. Excellent resistance properties to everything also good lightfastness and heat stability. It exhibits a high diffuse continuous reflectance of light giving a pure white, which helps to retain the original brilliance and tone of colours mixed with it.

Uses
Mainly in paste ink where its tendency to matt down can be prevented by the resin solids of the ink system. Recent innovations in the control of particle size have resulted in the introduction of grades suitable for liquid ink (Blanc Fixe micro). It has the highest density of any extender so considerable amounts by weight can be added to an ink without loss of printing strength. It has FDA status due to its low solubility.

Composition

$BaSO_4$	97–99%
SiO_2	0–2%
Residues	0–1%

Alumina hydrate, light alumina hydrate, Lake White, Transparent White

(CI Pigment White 24 No. 77002)

Properties
White, soft textured extender. Quite transparent compared with all the others. Resistant to everything except dilute HCl. Heat stable and light-fast. Can absorb the drier in oxidising ink systems.

Uses
Improves flow of litho inks, and the gloss of a wide range of paste and liquid inks. It is a reasonably priced extender and is easy to disperse. FDA approved.

$$Al_2O_3.O\text{-}3SO_3.3H_2O \text{ or } 5Al_2O_3.2SO_3.xH_2O.$$

Talc, French Chalk

(CI Pigment White 26 No. 77718)

Properties
Pure talc is the standard 1 on the MOH scale (where Diamond=10). Lustrous white very fine powder. Chemically inert lightfast and heat stable up to 850 °C, cheap and useful, but many versions can reduce gloss unless the binder system is adequate to compensate it.

Uses
In quality litho inks in particular gives good excess tack removal and improved rub resistance to inks.

Composition
A hydrated magnesium silicate locally produced from talc, soapstone or steatite.
   Can vary in make-up:

	Pure	*Average*	*Bad*
MgO	31−32	25−30	20−70
SiO_2	60−64	52−56	15−40
CaO	1	2−5	2
Fe_2O_3	1	1−3	1
Al_2O_3	1−2	1−2	5
Residues	4−5	4−10	

Silica

(CI Pigment White 27 No. 77711)

Properties
Transparent white light fluffy matt pigment with a high oil absorption. Completely inert, unaffecting the resistance properties of inks. Only the synthetic version is used for ink.

Uses
Flatting, anti-settle, and bodying agent for any ink system. Perfect for reducing tack, but at the cost of a lower gloss level. It can gel when added to non-polar solvents. Anti-mist additive.

Composition

SiO_2	98%
Other oxides	0.1%
Ignition loss	2.0%

4.10   PEARLESCENT MATERIALS

Titanium treated micas — pearlescents

(CI White 6 and White 20) (No. unassigned)

Properties
Synthetic effect, non-metallic shades from silver to gold. FDA approved. Chemically inert. Transparent with a high refractive index. Often needed between 10 and 25% pigment loading. In transparent media they produce effects due to the multiple reflection of light which match the lustre of mother of pearl. Stable to 800 °C.

Uses
In all cases where bronze or aluminium cannot be used and the substrate is not foil or metallised film, i.e. soap wrappers on paper. Usually screen, offset, flexo or gravure packaging inks. Very large choice of shades with the golds achieved with oxide blends. Care must be taken to ensure the binder content of the ink is high enough to provide complete fixing of the pearlescent to avoid poor results given in underbound systems.

Composition
Blends and co-precipitates of TiO_2 (rutile) and mica and variations.

Possible analysis (%)	
TiO_2	40
Al_2O_3	25
K_2O	6
SiO_2	27
H_2O	2

4.11   METALLIC PIGMENTS

Aluminium

(CI Pigment Metal 1 No. 77000)

Properties

Brilliant silvery powders (called linings) or pastes which have high hiding power. Some tend to tarnish on exposure and they will not resist alkali or acid. They are inert in hydrocarbon solvents.

Uses

Non-leafing grades are used as linings or pastes in all types of ink, either alone or with another colour to produce a 'metallic' colour effect. Many hundreds of ready-made pastes are available on the market these days suitable for all types of manufacture. One of the latest innovations is water-stable silvers for aqueous flexo use.

Copper powder, bronze powder

(CI Pigment Metal 2 No. 77400)

Properties

Brilliant reddish gold to greenish gold. Copper or copper–zinc alloys prepared for printing use. Available in many grades differing in particle size, leafing and covering power. Coarser powders generally are brighter. Finer powders are matter and more opaque. Special non-tarnish grades for all types of printing are now available already pasted into different oils, solvents, or media. They are sensitive to heat, humidity, and either strong acid or alkaline media. Stabilised paste forms should be used when possible.

Uses

Decorative printing inks, particularly for packaging, mostly on paper.

Manufacture

Powder flaked by ball or stamp milling with stearic acid for leafing grades or other fatty acid lubricants for the non-leafing grades. The shade depends on the composition of the alloy.

Composition

Colour	Copper	Zinc
Copper	100%	0%
Pale gold	90%	10%
Rich pale gold	85%	15%
Rich gold	80%	20%
Green gold	70%	30%

4.12   FLUORESCENT PIGMENTS

Various yellows, oranges, reds and greens and a single blue. CI numbers not assigned.

Properties

Powders created by pulverising solutions of fluorescent basic or reacted dyes in resins. Therefore they can be unstable, not very heat resistant and able to be ground away in dispersion equipment. Dazzlingly bright colours giving a fluorescent effect in daylight. Mainly in the yellow orange and red portions of the spectrum. They appear most brilliant in the dark with a UV source of radiation to illuminate them. Fluorescent dyes/daylight pigment powders have the property of converting short-wave visible radiation into longer wavelengths giving very brilliant colours. Blue, indigo and violets are extremely rare for daylight fluorescence.

Uses

Any type of printing advertising and packaging where this effect is required. Their use was confined until the 1960s to screen inks for poster work, but more recently, all the properties have been improved, so by the late 1970s fluorescent pigments were used in litho, letterpress, flexo and gravure systems.

Their stability has improved and they can be heat resistant up to 200 °C in some cases, and lightfastness has also been considerably improved. Due to the range of inks for printing purposes most reds, yellows, greens and orange fluorescent pigments are each available in about eight to ten different forms. The applications of each are different and care must be taken to choose the grade need in a specific case. Useful for safety labels, greeting cards, price changes and brilliant tri-chromatic inks.

4.13   GENERAL PROPERTIES OF PIGMENTS

Tables 4.5–4.7 are based upon test results and information supplied by the main producers of pigments. Often one supplier produces several grades of pigment with a single colour index reference. These may differ in relative density, oil absorption and opacity. They may have special properties such as improved lightfastness, water resistance, and insolubility in various reagents.

The information tabulated here is intended only as a guide for selection. It must also be appreciated that the tables give estimated average test results and do not state extremes of variation. Performance in a test can depend upon a variety of factors, including vehicle composition, treatments such as rosination, metallic salt and laking complexes, surfactants, etc. as well as tinctorial strength, particle size and batch to batch variations.

It should be remembered that small differences can be significant when assessing test results; these include the percentage of pigment, the type of vehicle, hardness of drying, film thickness and possibly the substrate. Any deviation from the specified test procedure could also alter the performance and comparative test results.

With the exception of lightfastness, test results have been expressed on a 1 to 5 scale which is a cypher form of the following ordered set.

1 = very poor;
2 = poor;
3 = fair;
4 = good;
5 = excellent.

Table 4.5 Average properties for tests to BS 4321 : 1969

Colour Index Pigment No.	Name	Bulk value	Conc. paste (%)	Lightfastness Blue Wool Strong	Tint	Water	Meths 74 op	Mixed solvent	Alkali	Soap	Detergent	Cheese	Fats and oil	Waxes	Special comments
Yellow 1	Arylide G	65	45	7	5–6	5	3–4	2–3	5	4–5	5	5	3–4	1–2	
Yellow 3	Arylide 10G	65	45	7	6	5	3–4	3	5	4	5	5	3–4	2	
Yellow 4	Arylide 13G	65	45	7	3–4	5	2–3	2	5	3–4	4	5	2	2–3	
Yellow 5	Arylide 5G	65	45	7	6–7	5	2	2–3	5	3	4–5	5	3–4	2	
Yellow 12	Diarylide AAA	72	40	3–4	2	5	5	2–3	5	5	5	5	3	2–3	
Yellow 13	Diarylide AAMX	80	40	5–6	4–5	5	5	4–5	5	5	5	5	5	4–5	
Yellow 14	Diarylide AAOT	69	40	4–5	3	5	5	4–5	5	5	5	5	5	5	
Yellow 17	Diarylide AAOA	77	37	6	4–5	5	5	4	5	5	5	5	3–4	5	
Yellow 34	Lead chromes	16–20	80	6–7	4D	5	5	5	1	2–3	5	3	5	5	Food and toy restrictions. Darkens
Yellow 55	Diarylide AAPT	70	45	6	4	5	5	5	5	5	5	5	5	5	
Yellow 65	Arylide Yellow RN	62	48	7	3–4	5	2	2	5	4	4	3–4	3–4	2	
Yellow 73	Arylide Yellow GX	60	50	7	4	4–5	3–4	2	5	4	4	3–4	3–4	2	
Yellow 74	Arylide 5GX	70	42	6–7	5	5	4–5	2	5	4–5	5	5	3	2	
Yellow 83	Diarylide AADMC	72	40	6	5	5	5	4–5	4–5	5	5	5	5	3–4	
Yellow 97	Arylide Yellow FGL	52	56	7	3	5	4	4	5	5	5	5	5	3–4	
Yellow 98	Arylide Yellow 10GX	42	65	7	4	5	5	4	4	5	5	4	4	3–4	
Yellow 101	Brilliant BGF	42	65	1–2	1	5	4–5	4	4	5	5	5	2	3	
Yellow 111	Arylide Yellow F46	40	65	7	4–5	5	5	4	5	5	5	4	4	4	

Note: 1 = poor, 5 = excellent. Blue Wool Scale 1 = Poor, 8 = excellent. D = darkens.

Colour Index Pigment No.	Name	Bulk value	Conc. paste (%)	Lightfastness Blue Wool Strong	Tint	Water	Meths 74 op	Mixed solvent	Alkali	Soap	Detergent	Cheese	Fats and oil	Waxes	Special comments
Orange 5	Dinitroaniline	67	45	6–7	4–6	5	4–5	3	4–5	2–3	2–3	3–4	3	2–3	
Orange 13	Pyrazalone	68	40	5	3–4	5	4–5	3–4	5	3–4	5	5	3–4	3–4	
Orange 16	Dianisidene	84	30	5–6	2–3	5	5	5	5	4	5	5	5	2	
Orange 34	Diarylide	70	40	6–7	4–5	5	4–5	4–5	5	3–4	5	5	3–4	3–4	
Orange 36	Benzimidazolone HL	80	32	6–7	6–7	5	5	5	5	5	5	5	3–4	3–4	
Orange 46	Ethylake Red C	55	45	3–5	2–3	3–4	2–3	3	2	5	5	5	5	3–4	
Orange 48	Quinacridone Gold	75	30	7–8	7–8	5	5	5	5	5	5	5	5	5	Food and toy Restrictions
Orange 51	Pyranthrone	62	40	7–8	7–8	5	4	4	4	5	5	5	5	5	
Orange 60	Benzimidazolone HGL	75	30	7–8	7–8	5	5	5	5	5	5	5	5	5	
Orange 61	Tetrachloro-iso-indolinone	40	48	8	4–5	5	5	5	5	5	4	5	5	5	

Note: 1 = Poor; 5 = excellent (except lightfastness), Blue Wool Scale 1–8; D = darkens).

C.I.	Pigment													Notes
Red 1	Para	70	45	7	2-3	1-2	2	1-2	2	1-2	2	2	2	1-2	
Red 2	Naphthol F2R	65	45	6-7	4-5	5	4-5	4	5	4-5	4	5	2	3	
Red 3	Toluidine	68	45	7	5	5	2-3	1-2	5	2-3	5	5	3	2	
Red 4	Red R	64	45	5-6	3-4	5	4	2	4	2-3	4	3	3	3-4	
Red 5	Carmine B	77	35	7	5-6	5	4	1-2	4-5	2-3	4-5	5	4-5	4-5	
Red 8	Naphthol F4R	71	40	5	3	4-5	4-5	3-4	5	4-5	5	5	4-5	4-5	
Red 9	Naphthol LF	67	45	6-7	4-5	5	4-5	1	3	4-5	3	3	3	2-3	
Red 10	Perm Red FRL	67	45	6-7	4-5	5	4-5	2-3	3-4	3	3-4	5	3	3-4	
Red 12	Bordeaux F4R	77	36	5-6	5-6	5	4-5	2-3	3-4	4	3-4	4	4-5	3-4	
Red 17	Naphthol-mid	74	40	5-6	4-5	5	5	1-2	4	3-4	3-4	5	1-2	1-2	
Red 22	Naphthol YS	68	34	6	1	5	5	1-2	5	3-4	3-4	4	3-4	1-2	Toy restrictions
Red 23	Naphthol Dark	83	30	5-6	1	5	4-5	1-2	4	5	5	5	5	5	Food and toy restrictions
Red 48:1	Barium 2B	52	48	3-4	1-2	4-5	5	4-5	3-4	5	4	3-4	2-3	5	Food and toy restrictions
Red 48:2	Calcium 2B	62	45	5-6	4	5	3-4	3-4	3-4	1-2	4	5	5	5	
Red 48:3	Strontium 2B	50	45	3-4	3	5	5	3-4	2-3	2-3	3	4-5	4-5	5	Food and toy restrictions
Red 48:4	Manganese 2B	63	45	6-7	5-6	5	4-5	3-4	2	2	3-4	4-5	4-5	5	
Red 49	Sodium Lithol	66	45	2-3	2-3	4-5	2-3	3-4	3-4	2	2	3-4	4	3-4	Food and toy restrictions
Red 49:1	Barium Lithol	67	45	2	1-2	5	3-4	4	2	2	3-4	4	3-4	4-5	
Red 49:2	Calcium Lithol	64	45	1-2	2	5	4	4-5	4	2-3	3	3	3-4	5	Good flow
Red 52:1	BON Ca	67	40	3-4	3-4	5	4-5	2-3	1	2-1	4-5	4-5	4-5	5	Food and toy restrictions
Red 53:1	Lake Red C	57	45	3-4	2	5	4	4	2-3	3	3-4	3-4	4-5	5	
Red 57:1	Rubine 4B	68	42	4-5	2-3	4	3-4	3	1-2	4	4	4	4-5	4-5	
Red 60	Scarlet 3B	45	55	4-5	4D	4	4-5	1	1-2	4	1	5	5	4-5	Darkens
Red 63	BON Maroon	67	45	5-6	3	5	3	3	3	4	3	3	5	3-4	Blues, darkens
Red 81:1	PMTA Pink	47	45	3-4BD	2-3BD4	3	2-3	4	1-2	4	4	4	5	5	
Red 88	Thioindigoid Red-Violet	58	40	7	4	5	5	5	5	4	5	5	4	5	Food and toy restrictions
Red 90	Phloxine	40	48	2	1-2	3	3	3	1	2-3	2-3	3	4	4	Food and toy restrictions
Red 104	Molybdate Chrome	18	80	6-7	5-6	5	5	5	1	4	4	5	3		Food and toy restrictions
Red 108	Cadmium Red	22	75	7-8	5-6	5	5	4-5	4-5	4-5	4-5	3	5		Food and toy restrictions
Red 112	BON Arylamide	66	45	6-7	6	4-5	2-3	3-4	5	4	4	5	4	4-5	
Red 122	Quinacridone	60	45	7-8	7	5	5	5	5	5	5	5	5	3	
Red 146	Carmine FBB	70	45	6	7	5	4	3-4	5	5	5	5	5	5	
Red 169	Cu Fe Magenta	67	40	5	2-3	4	3	1-2	4	1-2	2-3	2-3	4-5	4-5	
Red 170	Naphthol FSRK	81	32	7	4	5	3-4	3	5	5	5	5	5	5	
Red 176	B. Carmine HF3C	80	29	6-7	5-6	5	4	3-4	5	5	5	5	5	3-4	

Table 4.5 continued

Colour Index Pigment No.	Name	Bulk value	Conc. paste (%)	Lightfastness Blue Wool Strong	Tint	Water	Meths 74 op	Mixed solvent	Alkali	Soap	Detergent	Cheese	Fats and oil	Waxes	Special comments
Red 184	Napthol F6B	85	26	6	4	5	3–4	3–4	5	5	3–4	3–4	4	4	Darkens on exposure
Red 185	B. Carmine HF4C	82	31	6	5	5	3–4	3–4	5	5	5	5	5	5	Darkens on exposure
Red 200	Rubine 6B	40	50	4–6	2–3	3	1–2	1–2	2–3	1–2	1–2	4–5	5	5	
Red 202	Quinacridone	40	45	7–8	6–7	5	5	3–4	5	5	5	5	5	5	
Red 208	Benzimidazolene HF2B	85	28	6–7	5–6	5	3–4	4	5	5	4–5	5	5	5	
Red 210	Napthol F6RK	76	36	6	5	4–5	3–4	3–4	5	5	3–4	3–4	3–4	5	
Red 222	Azo Magenta G	78	33	6	5	5	3–4	4	5	5	3–4	5	5	5	
Red 224	Perylene Red Y	35	55	8	8	5	5	5	3–4	3–4	4–5	4–5	4–5	4–5	
Violet 19	Quinacridone Red	55	50	7–8	6–7	5	5	5	5	5	5	5	5	5	
Violet 32	B Bordeaux HF3R	97	18	6	5–6	3–4	5	5	5	5	5	5	5	5	

Note: 1 = poor; 5 = excellent (except lightfastness). Blue Wool Scale 1–8, B = bluer, D = darkens.

Colour Index Pigment No.	Name	Bulk value	Conc. paste (%)	Lightfastness Blue Wool Strong	Tint	Water	Meths 74 op	Mixed solvent	Alkali	Soap	Detergent	Cheese	Fats and oil	Waxes	Special comments
Blue 1	Brilliant PMTA	59 / 49	45	3D	2D	4	2–3	2–3	3–4	3–4	4	4	4	4	Darkens on exposure
Blue 3	Sky Blue PMTA	60 / 66	45	4D	2–3D	4	3	3–4	3–4	4	4	4	4	4	Darkens on exposure
Blue 15	Phthalocyanine	65	45	7–8	6–7	5	5	5	5	5	4–5	5	5	5	
Blue 16	Phthalo Copper Free	40	40	7–8	7	5	4–5	4–5	5	5	5	5	5	5	
Blue 18/56/61	Alkali	56	40	2–3	1–2	5	4–5	3	3	2–3	5	5	5	5	
Blue 24	Blue Lake	42	45	1–2	1	4	1–2	2–3	1	1	4	2–3	4	3–4	
Blue 27	Iron Blue	53	60	7–8	5–6*	5	5	5	1	2	4	5	5	5	*Decomposes
Blue 29	Ultramarine	70	75	7–8	7	4–5	5	5	4–5	4–5	5	2–3	4–5*	5	*Affected by rancid fats
Blue 60	Anthraquinone	53	40	7–8	6–7	5	5	4–5	5	5	5	5	5	5	Good heatfastness
Blue 62	CuFe Blue	70	35	4–5	3–4	4–5	3	3	2–3	2	2	2–3	2–3	5	
Green 1	Deep PMTA	50	50	3D	1–2D	5	2–3	2–3	2–3	1–2	4	5	5	5	Darkens on exposure
Green 2	Vivid PMTA	46	50	3D	1–2D	5	2–3	2–3	3	2–3	3–4	5	5	5	Darkens on exposure
Green 7	Phthalocyanine	48	45	7–8	6–7	5	5	5	5	5	5	5	5	5	
Green 36	Phthalocyanine YS	35	30	8	7–8	5	5	5	5	5	5	5	5	5	

Violet 1	Rhodamine PMTA	49	50	2–3D	1–2D	4–5	1–2	1	2	1	2–3	3	5	4–5	Darkens on exposure
Violet 3	Violet PMTA	49	50	3D	1–2D	5	1–2	1	4	2	4–5	4	5	4–5	Darkens on exposure
Violet 15	Ultramarine Violet	40	70	5–6	5	4–5	5	5	2–3	5	3–4	2–3	5	3–4	Heat resistant
Violet 23	Dioxazine	70	40	7–8	6–7	5	4–5	5	5	5	5	5	5	5	
Violet 27	CuFe Violet	80	30	4–5	3–4	4–5	2–3	5	2–3	2	2	3–4	5	3–4	
Violet 37	Dioxazine	70	30	7–8	7–8	5	5	5	5	5	5	5	5	5	

Note: 1 = poor; 5 = excellent (except lightfastness), Blue Wool Scale 1–8; D = Darkens.

Table 4.6 Black Pigments

Types of Carbon Black	Relative prices	Toptone shade and gloss	Flow properties in ink formulation	Ultimate particle size (millimicrons)	Concentrated past (%)	Volatile material (%)	pH of aqueous slurry	Fields of application	Special properties, etc.
Treated oil furnace	B	Blue and glossy. Gloss level up to ordinary channel blacks	Excellent. Vastly superior to the original furnace blacks	20–30	Approx. 53	Approx. 1.0	7.5–8.0	Letterpress and lithographic inks	
Treated oil furnace	C	Blue and matt	Very poor	25–60	Approx. 43	Approx. 1.0	8.0–10.0	Gravure and flexographic inks. News inks, magazine and cheap letterpress	High oil absorption. Difficult to disperse
Treated gas furnace	D	Grey/blue and matt	Low oil absorption, long flow	60–200	Approx. 53	Approx. 1.0	8.0–10.0	Cheap extender pigment	
Long flow channel	A	Brown and glossy	Very good. This permits higher loading of pigment	10–30	Approx. 53	12–15	3.5–4.0	High quality letterpress and lithographic inks	Glossiest type almost unavailable
Lamp Black	B	Grey/blue and matt	Low oil absorption, long flow	60–200	Approx. 70	Approx. 1.0	8.0–8.5	Cheap extender for matt inks	Extremely low oil absorption and very weak tinctorially

A–D represents price scale, A being most expensive.

Table 4.7 White pigments and extenders

CI Pigment White No.	Description	Bulk value	Concentrated paste (%)	Oil absorption	Transparency or opacity	Soap resistance	Alkali resistance	Paraffin wax bleed	Special properties
4	Zinc oxide	18	75	12	Semi-transparent	Good	Good	None	Not acid resistant
5	Lithopone	35	70	8	Opaque	Good	Good	None	Not acid resistant
6	Titanium dioxide	24	75	20	Opaque	Good	Good	None	
7	Zinc sulphide	33	70	15	Opaque	Good	Good	None	Not acid resistant
18	Calcium carbonate	38	55	30	Transparent	Good	Good	None	Not acid resistant
19	China clay	38	60	60	Semi-transparent	Good	Good	None	
21	Blanc fixe	26	80	18	Transparent	Good	Good	None	
24	Alumina hydrate	55	50	60	Transparent	Poor in alkali and soap	Poor	None	Not acid resistant
26	Talc	23	55	40	Transparent	Good	Good	None	
27	Silica	39	70	300	Transparent	Good	Good	None	

Bulk value
This is equivalent to 100 divided by the relative density of the pigment. The numbers indicate relative volume per unit mass.

Concentrated paste
The maximum percentage of pigment in 28 poise stand oil or its equivalent to give a handleable printing ink base or concentrate.

Methods of test for printing inks

BS 4321 : 1969 (ISO 2834–2844–1974) has been taken for the basis of these tests. A précis of such tests is given for guidance.

Test method 1 — Preparation of standard print — solid and 70% tone print. Defined volume of ink per unit area.

Test method 2 — Fastness to daylight — Blue Wool Scale ratings, 1 to 8; (1) Very low; (2) low; (3) moderate; (4) fairly high; (5) high; (6) very high; (7) excellent; (8) exceptional.

Test method 3 — Fastness to water — print between filter papers, 1 kg weight/24 hours.

Test method 4 — Solvents — Fastness;
(a) 74 op meths, or
(b) Acetone   10% by volume
     Ethyl acetate   40% by volume
     Ethylene glycol monoethylether
                      10% by volume
     Meths 74 op   30% by volume
     Tolune   20% by volume
5 minutes immersion in test tube.

Test method 5 — Alkali resistance — between filter papers soaked in 5% sodium hydroxide solution; 10 minutes/1 kg weight.

Test method 6 — Soap resistance — between filter papers saturated with 1% soap solution for 3 hours/1 kg weight.

Test method 7 — Resistance to detergent — between filter papers saturated with 3% detergent solution for 1 hour/1 kg weight.

Test method 8 — Resistance to cheese — in contact with crust and fresh slice for 3 days at 20 °C.

Test method 9 — Resistance to edible oils and fats — in contact with solid fat for 24 hours at 20 °C plus 1 hour at 4 °C in refrigerator to facilitate removal of specimen. Edible oils between soaked filter papers for 24 hours/ under kilogram weight in water-saturated atmosphere at 20 °C.

Test method 10 — Resistance to waxes — 5 minutes in melted wax, drip back through margin.

Test method 11 — Results from this test, the resistance of printing inks and prints to spices, have not been included because of the variety of spices available.

Note: For more information the complete standard test methods should be studied. All tests should be carried out exactly as specified in BS 4321 : 1969.

Table 4.8 gives details of the benzimidazolone pigments. While these all

Table 4.8 Benzimidazolone pigments

	Fast Yellow H2G	Orange HL	Red HF2B	Fast Red HFT	Fast Brown HFR	Carmine HF 4C	Bordeaux HF 3R	Carmine HF3C	Fast Maroon HFM
Colour Index No.	Y120	036	R208	R175	B25	R185	V32	R176	R171
Appearance (shade)	Yellow	Orange	Red	Reddish brown	Brown	Carmine	Dark red	Carmine	Dark brown
Density (g/cm^2)	1.47	1.60	1.38	1.42	1.49	1.39	1.39	1.38	1.39
(lb/gal)	12.4	13.4	11.6	11.9	12.4	11.6	11.6	11.5	11.6
Oil absorption (lb/100lb)	34	32	34	35	39	40	40	36	36
Optical properties									
Brightness	G	G–F	G	F	G	G	G	G	F
Tinctorial strength	G	G	G	G	G	G	G	G	G
Hiding power	G–F	G	F	F	G–F	F	F	P	F
Bleed resistance									
Water	E	E	E	E	E	E	E	E	E
Ethanol	G	E–G	E	E	E–G	E	E–G	E	E–G
Methyl ethyl ketone	G–F	E–G	E–G	E–G	G	G	G	G	E–G
Mineral spirits	G–F	E–G	E–G	E–G	E–G	E	E–G	G	E–G
Xylene	G	E	E	E	E–G	G	G	E	E
Dioctyl phthalate	E	E	E	E	E–G	E	E	E	E
Linseed oil	G	E	E	E	E–G	E	E	E	E
Chemical resistance									
Alkali	E	E	E	E	E	E	E	E	E
Acid	E	E	E	E	E	E	E	E	E

Processing								
Dispersibility	G	G	G	G	G	G	G	G
Heat stability (baking)	G	G	G	G	E	E	E	E
Lightfastness Exterior								
Full strength	E–G	G–F	G–F	G	E	G–F	G–F	G
Tint	G–F	G–F	G–F	G	E–G	G–F	G–F	G
Interior								
Full strength	E	E	G	E	E	G	G	E
Tint	G	G	G	E	E	G	G	E–G
Durability weathering	E–G	E–G	G	E	E	G–F	G–F	G
Toxicity rating	E	E	E	E	E	E	E	E

E = excellent; G = good; F = fair.

belong to the same chemical group the range of hues extends from yellow through orange and red to violet.

It is becoming increasingly difficult to develop and introduce new pigments. Not only are the research costs very high, but the time-scale of the evaluation of a new product now extends over several years adding substantially to the costs before any return can be expected. It may well be many years before an another entirely new chemical type of pigments, like the benzimidazolones above, becomes available. Meanwhile modifications of existing pigments to meet specific needs will continue to be developed by the pigment manufacturers.

Section II: Dyestuffs

Dyestuffs are largely used in liquid inks, coatings and lacquers. Special disperse dyes are used in both liquid and paste inks for heat transfer printing. In this application the choice of printing process is of minor importance compared with the subsequent operation of transferring the dyestuff from paper to the textile by sublimation. Other dyestuffs used in paste inks include acid dyestuffs for double tone and invisible inks, but by far the largest use has been in cheque security inks. Basic dyestuffs in the free-base form dissolve in fatty acids to form oil-soluble toners and can be used as toners for black inks.

In order to distinguish between the various types of dyestuff used increasingly in printing inks it is advisable to consult the Colour Index published by the Society of Dyers and Colourists jointly with the American Association of Textile Chemists and Colorists. Dyes are classified for specific usage and may be included in more than one entry; for example, food dyes include acid, solvent and natural dyes as well as pigments. This is followed by a classification of dyes by chemical composition and a comprehensive alphabetical index.

The classification used in the Colour Index is:

Acid dyes	Leather dyes
Azoic dyes	Mordant dyes
Basic dyes	Natural dyes
Developers	Oxidation bases
Direct dyes	Pigments
Disperse dyes	Reactive dyes
Fluorescent brighteners	Solvent dyes
Food and drug dyes	Sulphur dyes
Ingrain dyes	Vat dyes

Table 4.9 shows the Colour Index classification by chemical composition.

4.14   ACID DYES

This term applies as much to an application classification as to a chemical class. The name 'acid dye' was originally used to indicate the presence

**Table 4.9   Colorants arranged strictly on the basis of
their chemical structure**

	CI Nos	Page
Nitroso	10 000–10 299	4001
Nitro	10 300–10 999	4003
Azo		4009
Monoazo	11 000–19 999	4013
Disazo	20 000–29 999	4139
Trisazo	30 000–34 999	4277
Polyazo	35 000–36 999	4325
Azoic	37 000–39 999	4345
Stilbene	40 000–40 799	4365
Carotenoid	40 800–40 999	4375
Diphenylmethane	41 000–41 999	4377
Triarylmethane	42 000–44 999	4379
Xanthene	45 000–45 999	4417
Acridine	46 000–46 999	4431
Quinoline	47 000–47 999	4435
Methine and polymethine	48 000–48 999	4437
Thiazole	49 000–49 399	4441
Indamine and indophenol	49 400–49 999	4443
Azine	50 000–50 999	4445
Oxazine	51 000–51 999	4459
Thiazine	52 000–52 999	4469
Sulfur	53 000–54 999	4475
Lactone	55 000–55 999	4503
Aminoketone	56 000–56 999	4505
Hydroxyketone	57 000–57 999	4508
Anthraquinone	58 000–72 999	4511
Indigoid	73 000–73 999	4593
Phthalocyanine	74 000–74 999	4617
Natural organic colouring matters	75 000–75 999	4623
Oxidation bases	76 000–76 999	4641
Inorganic colouring matters	77 000–77 999	4651
Intermediates index		4691
Elementary formula index		4865

of sulphonic acid or other acidic groups in the molecular structure. However, since acidic groups are present in other types of dye, their presence is not a sufficient distinguishing factor. They are regarded as anionic dyes, soluble in water, mainly insoluble in organic solvents. Selected dyes are soluble in alcohol, ketones and esters. They give very bright hues, with lightfastness ranging from poor to very good.

Chemically they are azo, anthraquinone, triphenylmethane, azine, xanthene, ketonimine, nitro and nitroso compounds. An example, eosine, follows.

Eosine

(*CI Acid Red 87 No. 45380*)

Properties
A bright yellow-red hue, with a marked yellow green fluorescence which provides a typical example of dichromatism. It is readily soluble in water and other reagents. Lightfastness is between 1 and 2 on the Blue Wool Scale.

Uses
It is used in water-based fugitive cheque inks, invisible inks in painting books and in dyeing paper.

Structure

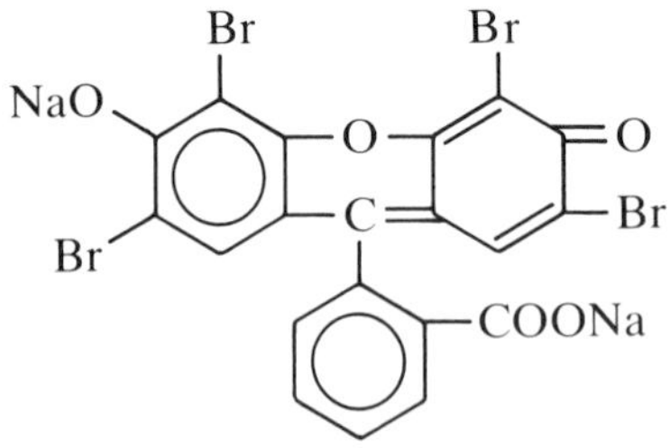

The most popular acid dyes in current use in ink manufacture are:
Acid Yellow 3, 5, 17, 23, 36, 54, 73, 121, 157, 194, 204 and 236.
Acid Orange 3, 7, 10, 142 and 144.
Acid Red 18, 52, 87, 88, 143, 221, 289, 357 and 359.
Acid Green 1, 16, 26 and 104.
Acid Blue 1, 7, 9, 15, 20, 22, 93, 129, 193, 254 and 285.
Acid Violet 9, 17, 90, 102 and 121.
Acid Brown 101, 103, 165, 266, 268, 355, 357, 365 and 384.
Acid Black 47, 52 and 194.

4.15 BASIC DYES

The first dyes produced from coal tar were basic dyestuffs. These cationic dyes have maintained popularity with the printing ink industry due to their brilliant shades and high tinctorial strength. However, their lightfastness is poor and curtails much of their usefulness. They are soluble in water and alcohol, but have low solubility in most other organic solvents.

These dyes are often used in conjunction with a mordant or laking agent such as tannic acid, usually in amounts equivalent to the weight of the dye. This lakes the dyes *in situ* and not only fixes the dry ink, but also makes it more water resistant and helps the dye to develop its full colour and strength.

Some of the more widely used basic dyes are described below.

Auramine

(CI Basic Yellow 2 No. 41000)

Properties
A bright lemon yellow with poor lightfastness used in flexographic inks.
The dye is dissolved in a suitable solvent in the presence of laking agents.
The wax and water resistance of prints containing auramine is very much
a function of the chemical formula of the laking agent. Tannic acid and
dimethyl salicylic acid are extensively used in this role.
    Note: The free base is CI Solvent Yellow 41 No. 41000:1.

Structure

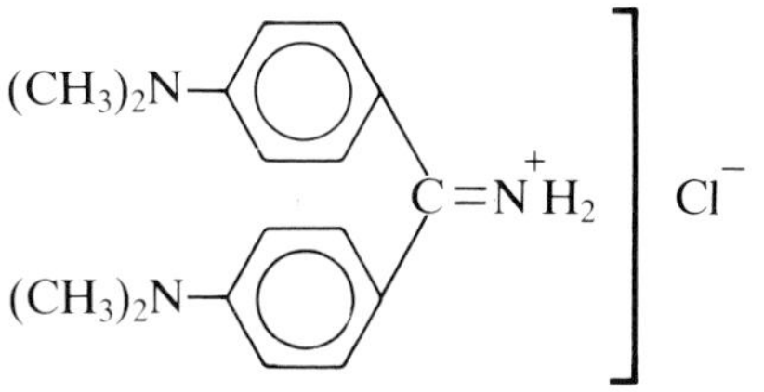

Severely declining in use in favour of Basic Yellow 37, Auramine Basic
Yellow 2 has been named as a suspect carcinogen.

Diarylmethyaniline Yellow

(CI Basic Yellow 37 No. 41001)

Properties
Strong brilliant yellow dyestuff. Decolourises immediately on contact
with strong alkali. Very soluble in ethyl alcohol and even much more so
in the presence of a laking agent. Has replaced Bright Yellow 2 in most
applications.

Uses
Comic inks, cheap paper flexo printing wrappers, foil lacquers not
requiring any lightfast specification. Some water reducible dye toners to
strengthen/brighten pigmented ink systems; again dye content will fade
out on prolonged exposure to daylight.

Rhodamine 6G

(CI Basic Red 1 No. 45160)

Properties
Very clean bright scarlet red, with a greenish fluorescence when dissolved
in water, but with a yellow fluorescence in alcohol. It has poor light-

fastness and properties similar to those described for auramine. It is used as the basis for PMTA and copper ferrocyanide salts.

Structure

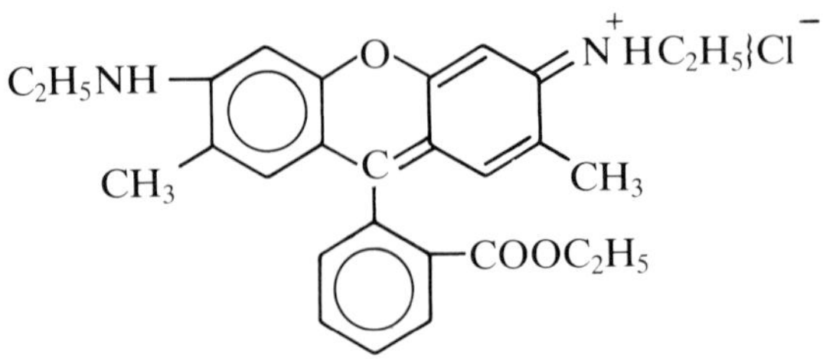

Rhodamine B

(CI Basic Violet 10 No. 45170)

Properties
Dirty and strong compared with Basic Red 1. A bright magenta red with yellow fluorescence, soluble in water, alcohol, and glycol ethers. It has poor stability to heat and light.

   Note: The free base of Rhodamine B is CI Solvent Red, No. 45170:1. The PMTA acid salt is CI Pigment Violet 1 No. 45170:2.

Structure

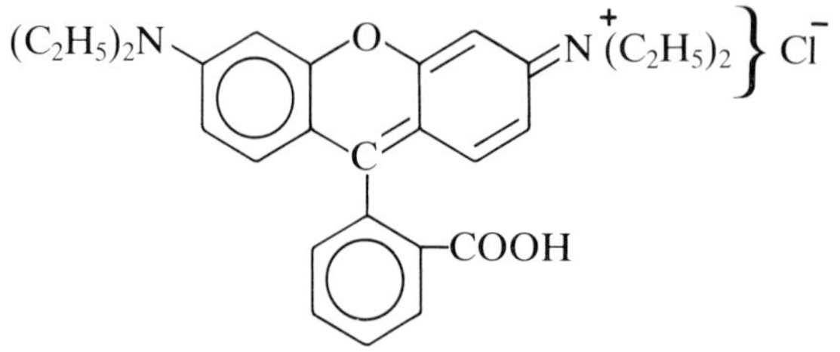

Victoria Blue

(CI Basic Blue 26 No. 44045)

Properties
Bright royal blue, very clean, soluble in water, and slightly soluble in cold water. It has poor fastness to light, tending to darken and become dull.

Uses
Used as dyestuff in flexographic inks. Further use of this dye is as the free base (CI Solvent Blue 5 No. 42595:1) and the PMTA salt (CI Pigment Blue 1 No. 42595:2) which is used extensively in all types of printing ink.

Structure

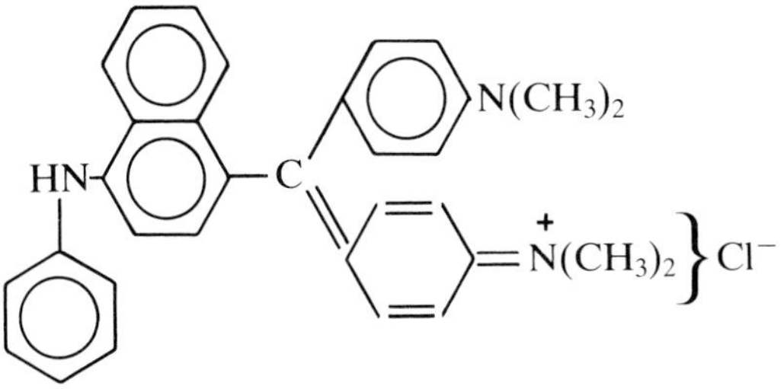

CI Basic Blue 26 (*Bright blue*)
(CI Solvent Blue 4) is the free base.
(CI Pigment Blue 2) is the phosphotungstomolybdic acid salt.

Methyl Violet

(*CI Basic Violet 1 No. 42535*)

Properties
Strong bright bluish violet, soluble in alcohol and water, but has poor
lightfastness, and turns black on exposure.

Uses
Used in flexographic inks for paper bags. Wider used in the following
forms, to obtain improved light stability:
CI Solvent Violet 8 No. 42535:1 is the free base.
CI Pigment Violet 3 No. 42535:2 the PMTA acid salt.
CI Pigment Violet 27 No. 42535:3 the copper ferrocyanide salt.

Structure

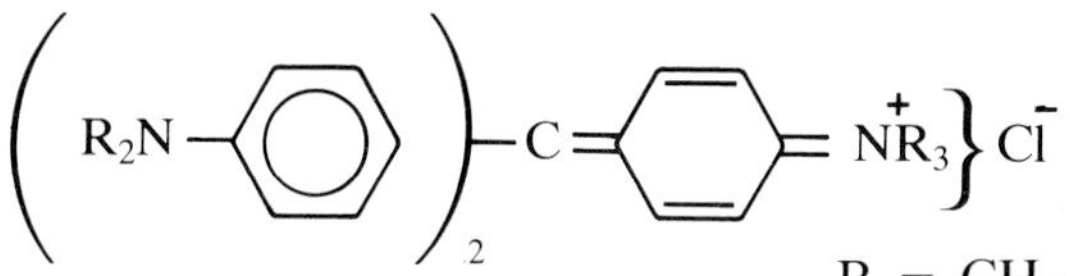

$R = CH_3$ or H

Other basic dyes frequently in use are:
Basic Yellow 13, 28 and 65.
Basic Orange 1, 2, and 59.
Basic Red 1:1, 2, 14, and 28.
Basic Violet 2, 3, and 11:1.
Basic Blue 1, 3, 5, 7, 8, 9, 11, 55 and 81.
Basic Green 1 and 4.
Basic Brown 1.
    For detailed information reference should be made to the Colour Index.

4.16 SOLVENT DYES

Often known as metal complex dyes. These are dyes in which solubility in an organic solvent or solvents is a characteristic physical property. They are drawn from a number of chemical groups and include what are sometimes described as oil dyes or fat dyes. The chemical constitutions associated with solubility in organic solvents includes acid dyes of the azo chromium complex, xanthene classes and the base salt form of certain basic dyes. sulphonic acid groups are usually absent, but dyes soluble in alcohols include salts formed between sulphonated dyes (usually sulphonated azo or phthalocyanine and organic high molecular weight bases such as diaryl-guanidine, mono and di-cyclohexylamine and di-isoamylamine).

The properties of these dyes are strongly influenced by their solubility in a particular solvent, so it is usual to use mixed solvents to obtain the optimum solvency and properties. The range of extremely bright colours with fair lightfastness makes them popular in a number of applications like golds on foil and uses where better lightfastness than that of a basic dye is required.

CI Solvent Yellow 19 No. 13900:1

Properties
A green shade yellow which has good lightfastness. It is heat stable up to 170°C and is soluble in alcohol and ester solvents, nitrocellulose, spirit and synthetic resin lacquers.

Uses
It is used for flexographic inks as well as wood staining. However, prints with this dye have poor water and alkali resistance.

Derivation
Chromium complex derived from:

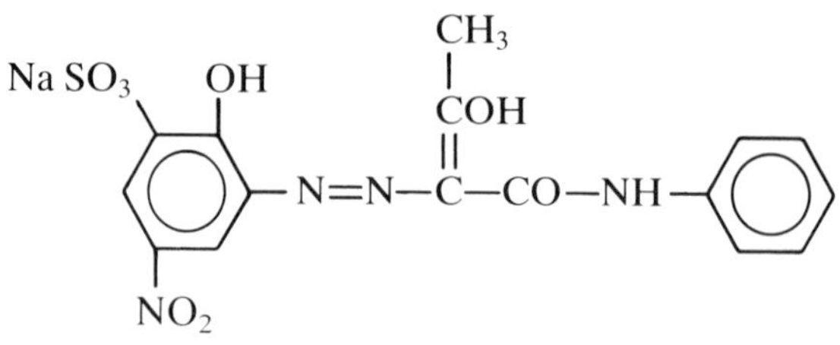

CI Solvent Orange 45 No. 11700:1

Properties
Transparent yellow shade orange, soluble in methylated spirits, diethylene glycol, esters and ketones. It has fairly good lightfastness, plus heat resistance up to 180°C.

Uses
Outlets include flexographic inks, gravure inks and lacquers based on nitrocellulose or spirit soluble resins used to simulate gold on film and foil.

Derivation
Chemically it is a cobalt complex derived from:

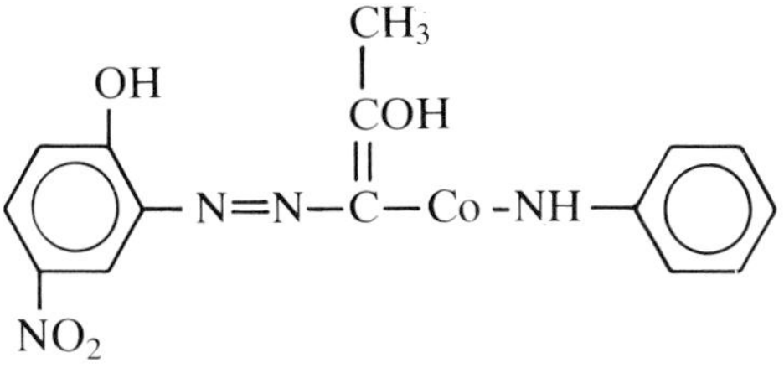

$$\text{OH} \quad \overset{\displaystyle CH_3}{\underset{\displaystyle \overset{\|}{C}OH}{|}}$$

(benzene ring with OH, NO$_2$ substituents) $-N=N-\overset{CH_3}{\underset{\|\,O}{C}}-Co-NH-$ (benzene ring)

CI Solvent Red 8 No. 12715

Properties
A blue-shade red with good solubility in alcohols, ethers, esters and ketones. Heat stability up to 220 °C and excellent lightfastness.

Uses
Suitable for gravure and flexographic inks and for use in stoving lacquers and two-pack polyester systems.

Derivation
Chemically a chromium complex derived from:

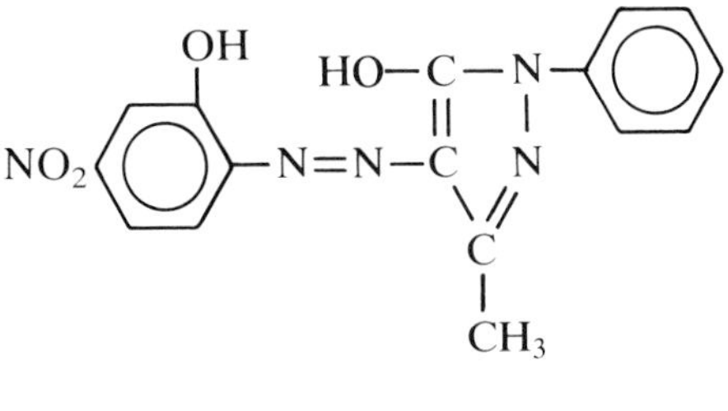

$$NO_2-\text{(ring, OH)}-N=N-C \quad HO-\overset{\|}{C}-N-\text{(ring)} \quad N \quad \overset{C}{\underset{CH_3}{|}}$$

Induline (spirit soluble)

(*CI Solvent Blue 7 No. 50400*)

Properties
Reddish blue to reddish navy, fair lightfastness, heat stable to 100 °C, insoluble in water, soluble in ethanol, also paraffin wax, oleic and stearic acid.

Uses
Recent policy in the UK is to exclude this material from printing inks

because of its toxic impurities, such as 4-amino-diphenyl. Used in stamping inks. Solutions of induline base in fatty acid were used as toners in cheap letterpress, news and gravure black inks.

Nigrosine (spirit soluble)
(*CI Solvent Black 5 No. 50415*)

Properties
Bluish black, heat stable to 150 °C. Good lightfastness, soluble in alcohol.

Uses
Used for flexographic inks and black marking inks. Sometimes referred to as 'aniline black'.

Nigrosine (free base)
(*CI Solvent Black 7 No. 50415:1*)

Properties
Black dye usually dissolved in a fatty acid, especially oleic or stearic acid to form the oleate or stearate. Fairly good lightfastness. Heat stable to above 180 °C. Slightly soluble in aromatic solvents.

Uses
Used in gravure inks as well as typewriter ribbons, carbon paper and stamp-pad inks.

General

Hundred of solvent dyes are available and they are gaining in usage. Some have very good lightfastness, and yellow and reds with Blue Wool Scale figure of 7 to 8 are available. The fact that these dyes exhibit good compatability with a large range of resins has made them particularly suitable for flexo and gravure coloured lacquers. Their solubility in some solvents varies considerably and solvent blends (including water) can prove extremely useful in ensuring completely solution at concentrations which give very strong results. These can be extremely expensive and so blending of colours to achieve the required shade should always take this into consideration.

4.17   DISPERSE DYES

The name disperse dyes was given in 1951 to a group of water insoluble azo, diphenylamine and anthraquinone dyes by the Society of Dyers and Colourists because they were sold as dispersions. Modern techniques allow them to be produced as dry redispersible powders.

Nearly all disperse dyes are primary, secondary or tertiary amines of three main types: amino azo benzene, aminoanthraquinone and nitrodiaryl amines. None of these contains solubilising sulphonic acid groups. A limited number are similar in structure to vat dyes.

Their main use in the printing industry is in heat transfer inks for printing on textiles, especially on to synthetic fibres. It is referred to either as heat-transfer printing, or sublistatic printing. The dyes are dispersed in a printing vehicle and are printed on paper by letterpress, lithography, flexography, gravure or screen printing. They are subsequently transferred to the fabric under heat and pressure. The dye sublimes and its vapour phase penetrates the fabric where it condenses. Bright, saturated colours result relative to the substantivity between dye and fibre, implying chemisorption or chemical bonding to the fibres.

There are hundreds of dyes of this type, and many of them are used in commercial heat transfer printing. Four disperse dyes which could form the basis of a trichromatic series are described below, and are typical of other dyes of this class. It should be noted that disperse dyes do not necessarily transfer with the same strength or hue on every fibre even when the optimum temperature and pressure for the particular substrate are applied, also that mixtures of disperse dyes can, under certain conditions have a catalytic effect which may cause premature fading. Sublistatic transfers can also be printed to transfer images to thermoplastics, metal, carpets and plastic bottles. Sometimes the dye vapour is made to cross relative large gaps (1 cm) with the assistance of vacuum.

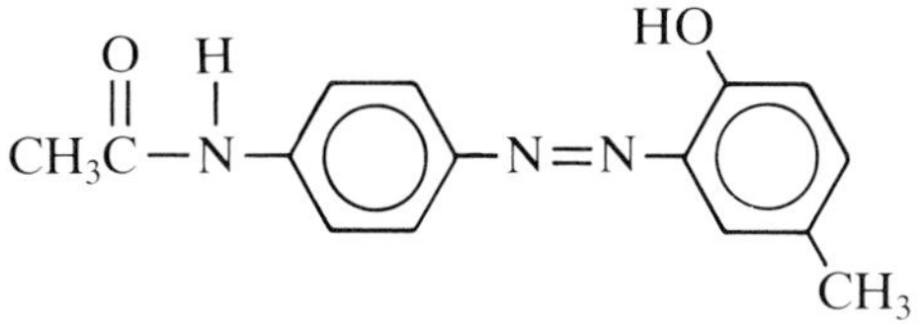

CI Disperse Yellow 3 No. 11855

Properties
A bright medium yellow hue that becomes slightly greener and more saturated when transferred to fabric. It has a very good fastness to light and is resistant to water, washing, wet and dry rubbing, acid and alkaline perspiration and dry cleaning when assessed according to the relevant BS or ISO tests.

Uses
It is used in inks for the five major processes to produce heat transfers on to polyamide, nylon, polyester and acrylic fibres, etc.

Structure

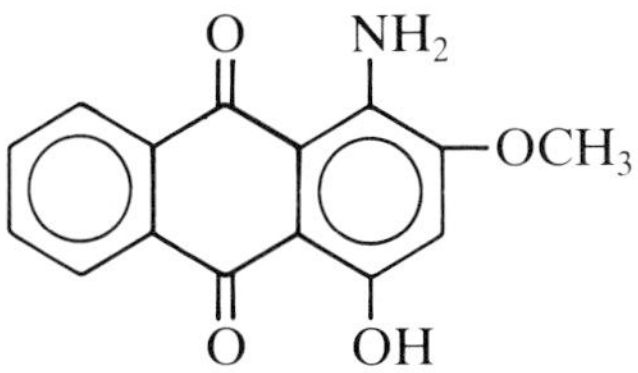

CI Disperse Red 4 No. 60755

Properties
A bright scarlet anthraquinone dye which has excellent lightfastness, washing, dry cleaning and perspiration fastness.

Uses
Although expensive, it is used for heat transfer process colours and can be used in inks for any conventional printing process.

Structure

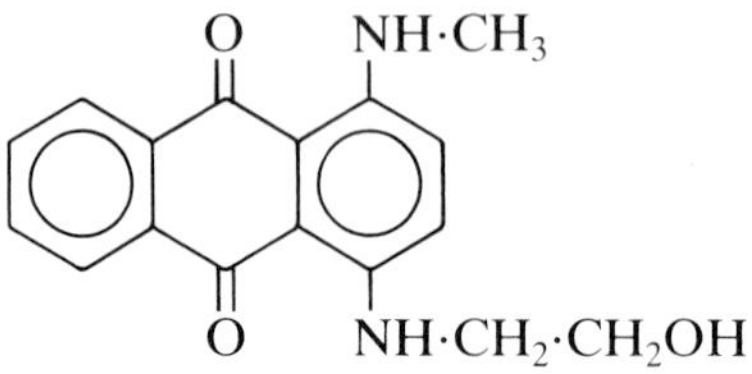

CI Disperse Blue 3 No. 61505

Properties
This is a clean blue-green shade dye. When printed and transferred to fabric, it gives a shade similar to phthalocyanine blue. However its lightfastness is only fair and there are stronger blue dyes which may be used instead or blended with it. It does not transfer well on to nylon, but it does transfer to other synthetic fabrics. Its fastness to washing, rubbing and perspiration is excellent, but it must be dry cleaned with care.

Uses
Transfers using this dyestuff can be produced by all conventional methods of printing. The degree of transference will depend upon the film thickness of the print, the nature of the substrate and the transfer conditions. Excess of dye transferred will give worse fastness results than when the optimum amount is used: an upper limit to colour strength is thus set by the requirement for washing fastness.

Structure

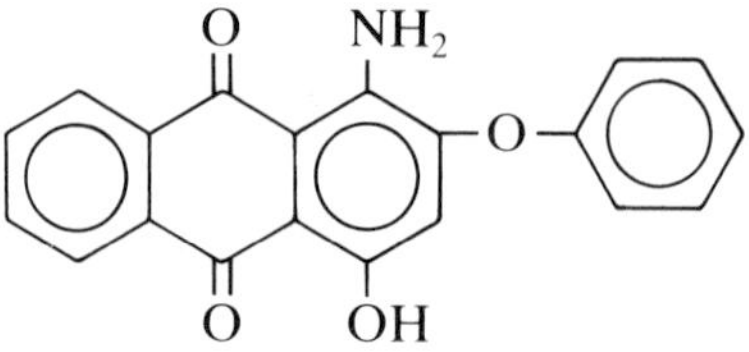

CI Disperse Red 60 No. 6075

Properties
Brilliant rose pink dye, with excellent lightfastness, excellent resistance to water, washing, acid and alkaline perspiration, wet and dry rubbing and dry cleaning.

Uses
Ideal colour for trichromatic transfer printing by the recognised major printing processes.

Section III: Oils

Oils in one form or another are probably the oldest raw material in the printing ink industry, and yet they still have an important role in the formulation of printing inks and varnishes. Many oils are treated and purified or modified with other polymers to meet modern requirements so that they bear little resemblance to earlier products except that they are still derived from the same vegetable, animal or mineral sources. (Table 4.9).

In the past it has been convenient to classify oils under three headings, drying, semi-drying and non-drying, dependent on the degree of unsaturation of the raw oil, defined by its iodine value. As individual measured results differ from season to season, and from source to source, quoted iodine values are usually averaged figures. There has been some confusion in various parts of the world as to how these values should be interpreted for the purpose of classifying oils. Borderline cases are often found under different headings in different catalogues.

4.18   DRYING VEGETABLE OILS

Chemically they are glycerides or triglycerides of fatty acids varying from those that are completely saturated (no double bonds) to those that contain three or more double bonds linking carbon atom pairs in the fatty acid compound. This characteristic of an unsaturated compound is represented by the symbols—CH = CH—.

When there is more than one double bond associated with neighbouring carbon atoms —CH = CH—CH = CH— it is referred to as a conjugated system. Vegetable drying oils are characterised by their power to absorb oxygen from the surrounding air and to form elastic films or skins, dependent upon whether or not complete polymerisation takes place. Polymerisation is defined as 'the chemical union of two or more molecules of the same compound uniting to form a single molecule'.

The amount of oxygen absorbed is in the main proportional to the iodine value of an oil, which is regarded as a measure of the extent of unsaturation of the fatty acids present, which in turn indicates the drying power of an oil. Iodine value is defined as the weight in grams of iodine that will combine with 100 g of oil. The recommended procedure is laid down in BS 684 (see Chapter 14).

The nature of the fatty acids present in a drying oil, and the proportions of each, largely determine its drying performance. For convenience they are normally divided in two classes: saturated fatty acids (e.g. palmitic, stearic and oleic acid) which play little part in drying, and unsaturated fatty acids such as linoleic acid, linolenic acid and its conjugated isomer elaeostearic acid etc. These play a vital part in the drying, colour, resistance and durability of ink vehicles subsequently manufactured from particular oils.

A high iodine value is an indication of rapid drying ability of an oil, but it is not the sole criterion for quick drying. A detailed knowledge of the actual positions of the double bonds in a molecule is essential for interpretation of iodine value because it indicates how subsequent polymerisation will take place.

With the exception of linseed oil, raw oils are seldom used unmodified in ink formulations. Refined and treated linseed oil is used alone or as an adjunct to petroleum distillate in many letterpress and lithographic ink formulations. The largest outlet, however, is in the manufacture of resins, varnishes and ink vehicles (see Chapter 12).

Linseed oil

Linseed oil, in the form of refined and treated oils, bodied and blown varnishes, also in conjunction with modifying resin or as an integral part of a resin system, is a prime constituent in several important classes of paste ink.

Linseed oil is obtained from the seeds of flax (*Linum*), cultivated in Argentina, USA, Canada, India, Russia, and to a limited extend in Australia and New Zealand.

Each seed contains between approximately 32–40% weight of oil. It is not economical to extract it all and some is left in the residual meal which is used to feed livestock.

Two methods of expelling oil from the seed are (a) the mechanical process and (b) the solvent extraction process. In the former, seed is first reduced to a suitable condition by crushing on a bank of five steel water-cooled rollers mounted one above the other. The crushed seed is then cooked to yield the optimum amount of oil at 80–90 °C if it is to be extracted by hydraulic pressure, at 110 °C if expellers are to be used. Meal from the cooker is moulded into shape, wrapped in a cloth and subjected to hydraulic pressure up to about 315 kgf/cm². The oil is then drained into tanks. Alternatively, mechanical screw presses called expellers force the seed through a conical grating, the oil exudes from the side and the feed cake is extruded from the end.

Solvent extraction is a semi-continuous process. It gives the highest oil yield but the residue is less acceptable as cattle food. In this process the seed is crushed and is mixed with solvent (usually petroleum ether or trichlorethylene). Solvent and oil percolate from the vessel. The solvent is removed by distillation and recycled.

Raw linseed oil contains a small percentage of impurities in varying amounts. They include free fatty acids, phosphatides, carbohydrates, waxes and suspended mucilage. Unless they are removed the oil 'breaks' on heating rapidly to 300 °C, i.e. it darkens and precipitates the impurities to form a gelatinous mass. On standing for long periods these impure materials precipitate by the absorption of atmospheric moisture to form 'foots'. Over a period of months or years these sink to the bottom of the tank allowing the removal of clear oil from above. Originally this was the method of preparing linseed oil for making artists' oil and varnishes hence the name 'stand oil' which today is given to bodied linseed varnishes in the surface coating industry.

A common method of refining linseed oil is the acid process. The vegetable oil is treated with sulphuric acid, allowed to settle overnight and the clear oil is syphoned off. This is washed thoroughly with hot water and allowed to settle once more. It is finally separated from the water and is vacuum dried. Acid refining does not remove all the fatty acids and produces refined oils of relatively high acid value. Such oils are used for their excellent dispersing qualities where the free acidity contributes to pigment wetting.

By far the most important method is alkali refining. Besides removing the break and colouring matter this process saponifies the fatty acids to form soaps, which can also be separated. Aqueous caustic soda is sprayed in and stirred until an emulsion forms. It is heated rapidly to break the emulsion and to encourage separation between oil and soapstock. The oil is thoroughly washed several times, allowed to settle and vacuum dried. There are also continuous processes in which separations are conducted by high speed centrifugation, instead of allowing water to settle out batch by batch.

The main constituent of linseed oil is linolenic acid which has three isolated double bonds ie $CH_3—CH_2—CH = CH—CH_2—CH = CH—CH_2 —CH = CH—(CH_2)_7—COOH$. the other unsaturated fatty acid is linoleic acid which confers colour retention on heating. Linoleic acid has 18 carbon atoms with isolated double bonds on the 9th and 12th carbon atoms. $CH_3—(CH_2)_4—CH = CH—CH_2—CH = CH—(CH_2)_7COOH$.

Linseed oil has always been the oil most used in ink and paint manufacture. It is also the most versatile. It is a straw coloured oil which without driers takes almost a week to dry as a thin film on glass.

Until very recently vast quantities were bodied by heating to 305 °C in open or closed pots. These were marketed in a wide range of viscosities to the ink industry as lithographic varnish.

Heat bodied linseed oils are sometimes called lithographic varnishes, a name which is little used in Europe, but is still used in the USA. They are today made in enclosed kettles or pots, which are made of various metals or alloys, particularly stainless steel which is ideal for bodying very pale oils. Kettles are made with interchangeable configurations that include facilities to blow air through the oil, blanket the surface with oxygen or an inert gas, or to heat the oil in a vacuum. Catalysts of metal powders, gases, inorganic and organic reagents are sometimes used to accelerate the bodying process. Strict control of temperature, heating rate and time ensures that the intended oil viscosity is obtained. Control is most important in the later stages of heat bodying as viscosity increases at an accelerating rate towards gelation.

Besides using raw linseed and lithographic varnishes, the inkmaker has several other linseed oil derivatives which he can include in formulations. They include boiled oils whose purpose is to increase the drying power without unduly altering the viscosity. They are manufactured by heating the raw oil in a steam kettle to 120–140 °C. The oil is stirred and occasionally air is blown in once it becomes hot, and driers in the form of oil-soluble organic salts of cobalt or manganese and lead are added.

Blown oils are produced by blowing large volumes of air through the oils at temperatures in the range of $100-140\,^{\circ}C$. They are a clear chestnut brown, faster drying with a higher viscosity and have improved wetting, dispersion and flow qualities. Inks containing blown oils must be formulated with care as they have a higher acid value and stronger odour. They are sometimes thickened by heating to $290\,^{\circ}C$ for use as ink vehicles; as these are liable to body up or gel on storage, viscosity is usually kept well below 200 poise.

Linseed oil is very versatile. Its fatty acids are used in the manufacture of alkyd resins, and refined linseed oil is used for making varnishes. It is sometimes treated while still retaining the character of an oil, e.g. maleinised oils, where about 10% of maleic anhydride is incorporated to form an adduct. Subsequently this adduct can be neutralised to make it soluble in water.

It is also possible to produce urethane oils by reaction of the mono- and diglycerides of a drying oil with a di-isocyanate so that it becomes soluble in alcohol. This also gives faster drying times and better water resistance. However, in addition to thermal decomposition hazards, some of these can yield small quantities of isocyanates on a warm running press or mill. This product could be toluene di-isocyanate (TDI) which can cause allergies — in particular skin sensitisation or asthmatic symptoms — in confined spaces without adequate air extraction apparatus. In the UK, inkmakers are barring TDI from any formulation; therefore treated oils cannot be used without a thorough investigation.

Tung oil (China wood oil)

The source of tung oil is the seed from a nut grown on shrubs in North and South America, Africa and the Far East. The shrubs are named *Aleurites fordii* and *A. montan*. Oil from the latter, being cultivated at a higher altitude, contains less elaeostearic acid and bodies more slowly, the gel time being about $20-30$ minutes at $276\,^{\circ}C$ against $11-15$ minutes according to the heat test of Appendix J in BS 391:1962.

The seeds are encased in hard nuts which are usually roasted to open them. They are then crushed and ground followed by either mechanical or solvent extraction similar to that for linseed oil. Tung oil has a pale straw colour with a slightly higher viscosity and refractive index than other vegetable drying oils. It dries slowly to form a soft, wrinkled, lustreless film. The raw oil has a characteristic odour, sometimes described as 'hammy' and it continues to release odorous by-products during drying.

Like oiticica oil, it is known as a conjugated oil becuase about 80% of the total fatty acid component is elaeostearic acid. Three conjugated double bonds, which are very reactive sites for polymerisation, occur in the molecular structure of elaeostearic acid:
$$CH_3(CH_2)_3{-}CH=CH{-}CH=CH{-}CH=CH{-}(CH_2)_7{-}COOH.$$
The other 20% of the fatty acid component is made up of oleic and

various saturated fatty acids in the ratio of approximately 2:1. For every three fatty acid molecules there is one molecule of glycerol in the structure of true triglyceride oils. Tung oil can be heated rapidly to a temperature of 290–310 °C to ensure that it has the ability to dry to a clear glossy film. This can be a delicate operation as it starts gelling at about 275 °C after seven minutes. This gelation can be retarded or prevented by adding a small quantity of linseed oil, glycerin or rosin. When properly heat-bodied it becomes the most water and alkali resistant of the drying oils.

Because tung oil gives a tough and glossy finish with rapid drying and excellent resistance properties, it is usually used as a component in oleoresinous varnishes, particularly with pure phenolic or alkyd resins where it gives a rub-proof film that is resistant to water, alkali and soap.

This makes it a good vehicle for lithographic inks for soap and detergent cartons. Other applications are in gloss overprint varnishes and metallic inks. It is normally omitted from inks to be used on wrappers for packaging food on account of odour when drying.

Oiticica oil

The only other conjugated oil normally used in ink manufacture is oiticica which is extracted from the nut of a tall tree *Licania rigida.* The tree grows wild mainly in the tropical climate of NE Brazil.

Shortly after extraction the oil sets to form a buttery mass, but this is stabilised to a permanent liquid by heating at 200 °C for 2 hours.

Liquid oiticica oil resembles tung oil in several respects. It dries rapidly to a wrinkled matt finish and forms a gel when heated to 275 °C. It has a characteristic odour and is greener than the pale straw colour of tung oil.

The principal fatty acid is *d*-licanic acid, which has three conjugated double bonds, but is unique in having a keto group in its formula:

$$CH(CH_2)_7 — CH = CH — CH = CH — CH = CH — (CH_2) — CO — (CH_2) — COOH.$$

Oiticica oil is used in the manufacture of oleoresinous varnishes that are quick drying and have good gloss and adhesion properties, rubproofness, water resistance and alkali resistance.

It gives a yellower more brittle film than tung oil and is sometimes combined with dehydrated castor oil, fish oil or soya bean oil to overcome this. Its usage to a large extent depends on price fluctuations relative to tung oil.

It is similar to tung oil in that it can be used to produce letterpress and lithographic inks that dry hard on non-absorbent substrates, and give scratch-proof prints on film and foil.

Dehydrated castor oil

Unmodified castor oil is a non-drying oil derived from the seed of the shrub *Ricinus communis* and as such was used in duplicator inks, typewriter ribbon inks and compounds for producing imitation water marks. It has been used as a plasticizer in liquid inks.

Castor oil mainly consists of the triglycerides of ricinoleic acid. This fatty acid contains one double bond and a hydroxyl group. The latter, together with an adjoining hydrogen atom, can be removed by dehydration. This is often carried out by heating to 280 °C in the presence of a catalyst, usually sulphuric acid, using a vacuum or inert gas to prevent oxidation and to remove volatile decomposition products. The reaction produces two isomers of linoleic acid, 9:12 linoleic acid (unconjugated) and a 9:11 (conjugated) isomer of linoleic acid, usually in the ratio of 3:1.

Dehydrated castor oil is pale and slightly more viscous than other drying oils. It has a slight smooth lingering characteristic odour. The earlier oils had a residual tackiness but this has been largely overcome.

Like other drying oils it can be bodied to produce varnishes, but these are seldom used on their own. Blown DCO is used for wrinkle finishes.

Dehydrated castor oil is mainly used in gloss varnishes to give pale coloured films that have good colour retention on air drying or stoving. It is ideal for metal decorating inks because of its pale colour and excellent flexibility. Because of low residual odour, DCO is sometimes used in food wrapper inks. By far the largest usage is in the manufacture of air drying and stoving alkyds for white roller coatings and inks. Such fluid polyesters are referred to as DCO alkyds.

4.19   OTHER OILS

Segregated oils

These are prepared by treatment of drying oils with a solvent such as liquid propane, furfural or sometimes a solvent mixture of liquid propane and furfuraldehyde. The solvent partitions the oil in two layers.

Both layers contain oil fractions of differing iodine value, one higher and one lower than the original oil. They are then separated and the solvent is removed from each by distillation. The segregated oil with the higher iodine value gives faster air drying times than the original.

Marine oils

Fatty portions of fish and the blubber of whales are processed to yield the so-called 'fish oils'. They have a more variable composition and a wider range of fatty acid components than vegetable oils. Exploitation of these oils is limited by their high odour and colour, but they yield flexible, glossy films. To improve their drying speed and reduce odour some have been segregated, but industry has been slow in accepting them. High cost and conservation effectively prohibit further adoption of marine oils by industry.

Semi-drying oils

Almost exclusively used as components in synthetic resins, mainly alkyds. Alkyds modified with semi-drying oils are usually more flexible, are often lighter in colour and have better colour retention than drying oils.

The main semi-drying oils are tobacco seed oil, soya bean oil, safflower oil and sunflower seed oil. Of these the most important is soya bean oil which finds increasing scope in pale alkyds for tin printing inks.

Soya bean oil

Today soya bean cultivation is of considerable importance. It produces the largest amounts of natural oil throughout the world, and is grown on all continents.

It comes from beans which are the seeds of a legume *Soya japonica* or *Glycine hispida*. These beans are a valuable source of protein. The oil is extracted by extrusion. The amount extracted is fairly small, about 20%, to allow for a proportion to be retained in the resulting foodstock.

It contains a high proportion of linoleic acid, and oleic acid with a typical odour and pale colour. It can be refined and bodied by conventional methods.

Alkyds based upon soya bean oil, either alone or with other modifications, give excellent colour retention and adhesion in tin-printing inks. They are also used in overprint varnishes and web-offset inks, despite their slower drying qualities. Their pigment wetting ability, especially with carbon blacks, is good.

Non-drying oils

Mineral oils

Mineral oil in various forms is used as a vehicle in printing inks or as a solvent or diluent in the manufacture of ink vehicles. Its use and development will increase with the exploitation of the North Sea oil fields.

A wide range of products is available. Mineral oil viscosities range from spindle oil, and oil just slightly thicker than a solvent, to heavy bodied oils. Colours range from pale straw to dark brown.

They are the products of their higher boiling fractions evaporated from petroleum at temperatures above 350 °C at atmospheric pressure. They consist of varying percentages of aromatic, naphthenic and paraffinic hydrocarbons, with a small sulphur content ranging up to 4%. The pigment-wetting qualities of mineral oils depend upon the relative ratio of each hydrocarbon present. High percentages of aromatic and paraffinic hydrocarbons give poorer pigment wetting properties as well as poorer miscibility with resins. They are used alone or in conjunction with rosin, some rosin derivatives and hydrocarbon resins in newspaper inks and cheap jobbing blacks. Oils such as spindle oil are a diluent or solvent for ester-gum, resinates, isomerised rubber, phenolic and maleic resins in quick-setting inks.

News ink oils

The essential requirement for a news ink oil is that it should be inexpensive. Also it must show good wetting and dispersion characteristics so that it can produce an ink which has sufficient flow properties

Table 4.9 Source, uses, composition, physical and chemical constants of oils

Oil	Origin	Chemical composition of fatty acids in the triglycerides				Relative density 15.5°C	Acid value (mg KOH/g oil)	Saponification value (mg KOH/g oil)	Iodine value (cgI/g oil Wijs)	Hydroxyl value[†]	Application
		Sat. fatty acids (%)	Oleic acid (%)	Linoleic acid (%)	Other acids (%)						
Linseed	Seeds of flax plant. Canada, India, USA. Argentine	10	20–25	14–16	Linolenic acid: 50–56	0.932–0.935	0.5–9.0	185–195	177–183	0	Oleoresinous varnishes, long oil alkydresins
Tung	Seeds of *Aleurites fordii*. USA, Japan, Malawi	4–5	4–10	5–9	Linolenic: 0–3 eleostearic: 77–83	0.940–0.943	0–4.0	190–196	160–180	0	Oleoresinous varnishes, alkyd resins
Oiticica	Seeds of *Licania rigida*. Brazil	10–13	4–7	0	*d*-Licanic acid: 82 other acid: 4	0.968	2.0–4.0	185–194	140–150	0	Oleoresinous varnishes, alkyd resins
Dehydrated castor	Synthetically from castor oil	1–4	7	62	9–11 Octadec-adienoic acid: 29	0.929–0.940	0–6.0	188–194	130–150	Variable about 20	Gloss varnishes with pale colour. Alkyd resin
Fish (whale and cod liver)	Whale blubber, fatty body tissues, liver, etc.	Complex mixture of acids containing 14, 16, 18, 20 or 22 carbon atoms				0.917–0.925	0–10.0	185–195	110–190	0	Oleoresinous varnishes, segregated oils, alkyd resins
Tobacco seed	Seeds of tobacco plant. USA	12	27	61	0	0.928–0.931	0–5.0	190–195	135–148	0	Alkyd resins
Soya bean	Seeds of *Glycine hispada*. Japan, USA	10–16	22–30	50–55	Linolenic: 5–9	0.925–0.927	0.5–4.0	190–195	126–130	0	Alkyd resins
Sunflower seed	Seeds of sunflower. USA	7	38	55	0	0.924–0.928	0–5.0	188–195	123–135	0	Alkyd resins
Tall	By-product	2–10	35–50	23–35	Rosin acids; 0–40	0.910–1.00	2.0–3.0	160–200	120–150	0	Alkyd resins
Safflower seed	Seeds of *Carthamus*. Europe. USA	11	25	64	0	0.925–0.928	0–2.0	186–194	140–150	0	Alkyd resins
Coconut	Fruit of *Cocus unifera*. India, Ceylon, S. Sea Islands	90	5–8	2–5	0	0.920	0–10.0	250	9	0	Alkyd resins
Castor	Seeds of *Ricinus communis*. Europe, Asia, USA	4–5	5–6	4–10	Ricinoleic acid: 77–90	0.950–0.970	0.5–4.0	180–187	79–86	146–166	Plasticiser. Preparation of Turkey Red oil

*The only oil listed in this table which is not a triglyceride is Tall oil. This oil is a mixture of fatty acids and rosin acids of variable constitution.
†The hydroxyl value is the number of milligrams of potassium hydroxide required to neutralise the acetic acid capable of combining by acetylation with 1g of the oil or fat.

without flying or misting, together with good binding qualities, adequate penetration and speedy setting.

The oils that fulfil these requirements are dark brown mineral oils, some of which are fairly viscous, but for normal newspaper black inks a viscosity of between 8 and 10 poise at 21 °C (70 °E) is recommended. Such a viscosity gives a reading of about 4000 seconds on a Redwood No. 1 and 400 seconds on a Redwood No. 2 viscometer, which are the instruments normally used in the UK for measuring the viscosity of lubricating oils.

The more viscous oils are measured on similar instruments at temperatures of 60 °C (140 °F) and 93.3 °C (200 °F). The flash point for such oils ranges from about 159 °C (350 °F) to above 200 °C (400 °F).

Test methods vary with suppliers and are covered both by the Institute of Petroleum (known as IP Standards) and the American Society for Testing Materials (ASTM Standards).

Mineral oils used in inks for newspaper production have fairly high aromatic contents, which may be as low as 22% but can be double that figure. The balance is made up of naphthenes and paraffins as well as small percentages of other unsaturated polycyclic and olefinic hydrocarbons. News ink oils have to be produced at very low cost and are made in specialised refinery processes using selected feedstocks of lower petroleum distillates or residues, which give products with a relatively low aniline point.

As these are products with a high flash point they do not present handling or storage problems. Recent tests have indicated that there may be some health hazards associated with these oils, especially in the fine mists around the rollers of a fast running press. Certain oils therefore are no longer used. Precautions should be taken to avoid prolonged skin contact with any aromatic petroleum oils.

Non-drying vegetable oils

Non-drying vegetable oils are valuable as plasticisers, lubricants and as components of synthetic resins. The most important is castor oil.

Castor oil

Castor oil is derived from beans which are seeds of the shrub *Ricinus communis* grown in tropical areas. Oil is obtained by cold pressing and solvent extraction, the finest grades being selected for pharmaceutical use. Ricinoleic acid forms from 80 to 87% of the fatty acid portion (averaged) in the castor oil triglyceride molecule. The chemical formula of ricinoleic acid is:

$$CH_3(CH_2)_5CH(OH)CH_2 - CH = CH - (CH_2)_7COOH.$$

Castor oil is used in the manufacture of DCO by dehydration (refer to p. 188) and Turkey Red oil which is produced by sulphonation. Castor oil can also be bodied by blowing or oxidising. Its main use is as a plasticiser for use in plasticising alkyds, also in duplicating inks, typewriter ribbon inks, for making imitation watermarks and for calibrating viscosity, instruments. It has good pigment wetting qualities and is sometimes used as a bridging agent, having both alcohol and oil miscible parts in the principal fatty acid molecule.

Section IV: Resins

The term 'resin' is used to describe non-crystalline solid materials or liquids of relatively high molecular weight. Such materials often lack a precise melting point. Resins used in printing inks contribute to the properties of hardness, gloss, adhesion and flexibility in an ink. They are binders for pigments.

Resins are of two types, natural and synthetic. A few naturally occurring resins are still used in the preparation of ink vehicles and varnishes, usually after some chemical modification. Most natural resins are obtained from living vegetable matter although some are in fossilised form. Shellac is an exception, its origin being the secretion of an insect. The chemical structure of natural resins is complex and in many cases has not been entirely proven.

Synthetic resins are prepared by polymerisation involving condensation or addition reactions between relatively small molecules. In many cases the composition of synthetic resins is known with considerable accuracy because they contain fewer molecular types than most natural resins. The chemical and physical properties of synthetic resins are closely related to their chemical structure and composition. Uniform synthetic products can be repeatedly manufactured in contrast to naturally occuring resins where the feedstock composition governs the final product. A great variety of modification is available for each class of synthetic resins and for this reason such resins can only be discussed here in general terms.

In the past few years some natural and a few synthetic resins have been superseded. The decline was caused by the limited nature of natural resources and the changing economic climate, which made the processing and distribution of small quantites very expensive. The future outlook is that conservation, plus health and safety considerations, will encourage the use of aqueous systems as well as those derived from vegetable products. This will accelerate the decline in demand for certain materials which have served the printing ink industry well in the past.

Chemical and physical properties of resins with their usual applications in printing ink formulations are mentioned, together with a description of the origin or preparation of each resin. An indication of chemical structure is outlined where possible.

For further reference to current materials, the Surface Coating Resin Index published every third year by the British Resin Manufactuers Association should be consulted. This association has recently published useful guidelines for the safe handling of resins.

4.20 NATURAL RESINS

Rosin

Rosin is obtained from pine trees throughout the world. Wood rosin is obtained by crushing the stumps of pine trees after they have weathered for a number of years, then extracting rosin and rosin oil with a petroleum solvent. Gum rosin is obtained by tapping the living trees, sometimes with the aid of artificial stimulation to accelerate the flow of exuded sap.

Wood rosin and gum rosin are similar, except that gum rosin generally has a higher melting point and wood rosin tends to crystallise from alcohol solution more readily. Rosin is amber in colour with a low melting point at about 60 °C by the capillary method and up to 80 °C by the ring and ball method. Its saponification value is 172 and its acid value is 168–172.

Rosin is regarded as an inexpensive resin. Its relative price has risen slightly over the last decade, but there has been little decline in its use. It is ecologically sound to produce rosin from living sources, as opposed to fossil sources and tree stumps, making gum rosin a regenerative starting material for printing ink resins.

Rosin has been used in alcohol based lacquers, also in combination with mineral oils in some comic publication and newspaper inks. Rosin has a strong peptising action in breaking down gels. Otherwise it has limited use in inks in its natural state. Its main outlet is as raw material in the production of chemically modified rosin derivatives, such as dimerised and polymerised rosin, limed rosin (rosin soaps or metallic resinates), esters of rosin including ester gum, rosin/maleic adduct, modified fumaric adduct, also as a modifying resin in cresylic, phenolic and alkyd resins. Cobalt, lead and manganese salts react with rosin to give the appropriate resinate driers.

Rosin consists of approximately 90% acids and 10% neutral material. The principal acid, abietic acid, converts to laevo-pimaric acid when heated.

Abietic or laevo-pimaric acids, as can be seen from their chemical formulae both possess two types of reactive centre, namely conjugated diene and carboxylic acid groups.

Abietic Acid   Laevo-pimaric Acid

Rosin has the tendency to crystallise from solution. Varnishes made from rosin do not dry well, and possess a residual tack. These disadvantages are least apparent in rosin/tung oil varnishes.

Rosin soaps
Limed rosin
Metallic resinates

The high acid value of rosin may be lowered by reacting calcium acetate or lime or zinc oxide or a combination of them with rosin or polymerised rosin. (see Table 4.10). A metallic resinate with a considerably higher softening point is obtained; the increase in softening temperature is dependent on the metal content; they are inclined to give brittle films, but they have relatively rapid solvent release, and a colour similar to rosin. They are soluble in hydrocarbons but not in alcohol.

They are used in low cost gravure inks as well as in heatset inks. Wetting of pigments is good and they prevent some basic pigments from livering. Resinates can also be used as a peptiser in heat bodied drying oil varnishes to prevent gelation.

Rosin or polymerised rosin is heated to 235–260 °C. Up to 7% lime (or calcium acetate) or up to 9% zinc oxide plus acetic acid is added slowly with constant agitation, and steam is blown through. Sometimes both calcium and zinc are added, usually in a ratio of 3:1 Zn:Ca, but still keeping the metal content of the resin below 10%.

Ester gum and other esters of rosin

Ester gum is the triglyceride derived from rosin by direct interaction between rosin and glycerol. It is an amber coloured resin with a softening point of approximately 85–90 °C (B & R). It is soluble in aromatic and aliphatic hydrocarbons, drying oils and most of the usual lacquer solvents with the exception of alcohols. Varnishes made with ester gum and drying oils have good pigment-wetting properties and find applications in most ink systems. Its low acidity means that it can be used in conjunction with metallic pigments. In liquid inks careful formulation is required to reduce tack.

It is prepared by heating rosin in a stainless steel kettle to approximately 260 °C, then slowly adding glycerol. The paleness of colour is preserved by the stainless steel reaction vessel, also by stirring and by conducting the esterification in an inert atmosphere of carbon dioxide or nitrogen. The temperature is held at 260 °C until the required drop in acid value (usually to below 10) is obtained. The theoretical requirement of glycerol is 10%, but an excess of 2–5% is recommended to balance any losses that occur during the reaction.

The general formula for the glycerol ester of rosin is:

$$
\begin{array}{l}
R.COO\!\!-\!\!CH_2 \\
\quad\quad\quad\ | \\
R.COO\!\!-\!\!CH \\
\quad\quad\quad\ | \\
R.COO\!\!-\!\!CH_2
\end{array}
$$

where R represents the rosin acid.

Table 4.10

	Percentage metal	Softening point (°C) Ring and Ball	Acid value (mg KOH/g)
Calcium resinate	2.4	117–122	95
Calcium resinate	3.4	140–145	85
Calcium resinate	6.2	155–165	40
Zinc resinate	4.8	130–135	100
Zinc resinate	9.0	140–150	35

Rosin and modified rosins can also be esterified with other alcohols and esterifying agent, for example pentaerythritol.

The pentaerythritol ester of rosin is similar to traditional ester gum in appearance, but it has a higher softening point of 112 °C (B & R) and quicker solvent release. It also has better water and alkali resistance. For these reasons it is much more popular than the glycerol ester although it is more expensive.

Its chemical formula is:

$$
\begin{array}{c}
CH_2{-}OOCR \\
| \\
RCOO{-}CH_2{-}\!\!-\!\!-C{-}CH_2{-}OOCR \\
| \\
CH_2{-}OOCR
\end{array}
$$

where R = rosin acid.

Pentaerythritol esters of dimerised and polymerised rosin acids have even higher melting points, give harder films and have faster rates of solvent release from an ink film. They find widespread application in letterpress and lithographic inks and are gaining use in liquid ink formulae.

Maleic resins and esters

A wide range of rosin maleic and maleic modified resins are commercially available. Maleic anhydride reacts with rosin according to the Diels–Alder synthesis. Properties vary with the molar ratio of rosin to maleic anhydride and with the manufacturing procedure. They can roughly be divided into three main classes (Table 4.11).

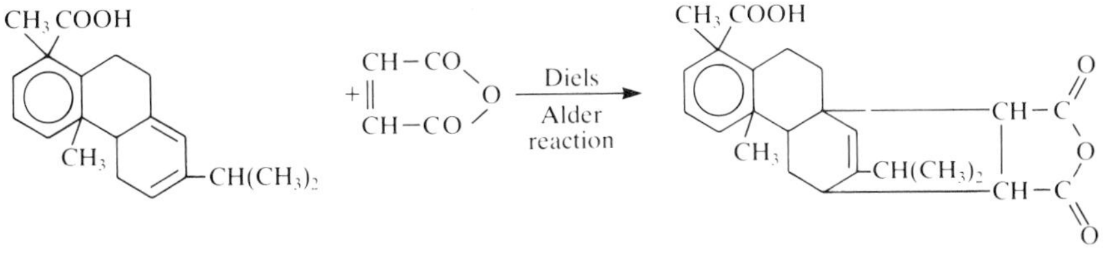

Laevo-pimaric acid   Maleic anhydride   Maleic resin (high acid value)

Table 4.11

	Softening point (°C)	Acid value (mg KOH/g)
Glycol-soluble maleics	140–150	280–320
Spirit-soluble maleics	140–160	90–130
Hydrocarbon-soluble maleics	120–160	Less than 25

To alter the solubility and reduce the acid value of the rosin maleic adduct, maleic modified resins are prepared by esterifying carboxylic groups with a polylol such as glycerol or pentaerythritol during an *in situ* preparation.

Besides modifying the qualities of high acid value rosin maleic resins, esterified maleic modified resins find wide scope for use in inks. High acid value maleics are used in steam set, moisture-set and water reducible letterpress inks, also for water-based flexo inks, in conjunction with an amine or ammonia to secure their solubility.

Medium acid value maleics have applications in alcohol-based gravure and flexographic inks, in combination with polyamides, nitrocellulose or ethyl cellulose, for printing on polymer films and foil. They are also used in water miscible letterpress inks.

Hydrocarbon-soluble maleic resins are very pale. They are used in letterpress and lithographic gloss inks, tin printing inks, heatset inks and overprint varnishes.

Dimerised and polymerised rosins
In order to raise the softening point of rosin it can be dimerised or polymerised by heating with a suitable catalyst. Its unsaturation makes it possible to polymerise to the dimer stage, which results in a high softening point, 150–170 °C and a lower acid number of 148. Dimerised rosin has increased compatibility with other resins and varnishes; it is less susceptible to oxidation due to greatly reduced unsaturation. It can also be used at high concentration in hydrocarbons without risk of crystallisation.

Further polymerisation from the dimer stage gives a polymerised rosin which is very pale in colour. It has a lower melting range near 77 °C by capillary method and 95 °C with ring and ball. It has an acid value of 148 mg KOH/g similar to dimerised rosin.

Its main applications are in cheap photogravure inks and gloss over-lacquers. It has also been used to improve the gloss, dispersion and working properties in letterpress black inks. It has also been used in heat-set inks.

The chemical constitution of polymerised rosin is not completely elucidated but reaction takes place at the olefinic centres. A dimerised rosin may have up to 80% rosin acid dimers, the other 20% being monomeric rosin acids and neutral material.

Rosin-modified fumaric resins

Fumaric acid is the isomer of maleic acid, therefore rosin modified fumaric resins are classified together with maleic resins. In practice fumaric acid is used in place of maleic anhydride to produce rosin fumaric resins with high acid values between 250 and 330 mg KOH/g and higher softening points, 140–150 °C.

They are used in flexographic and gravure inks, in combination with other resins for non absorbent substrates and with glycols for moistureset inks where hardness and gloss are required.

Their chemical structure is thought to be:

in which the carboxyl groups A and B are *trans* to one another.

Shellac

The importance of shellac in industry has rapidly declined due to shortage and high prices. At the same time the developments in acrylic emulsions and dispersions have been quickly accepted by flexographic inks formulators to fill the gap for water- and alcohol-based inks.

Shellac has an acid value of 65–80, a hydroxyl value of 260–280 and an iodine value of 15–25. Its melting range is 75–85 °C. Relative density is 1.15–1.20, and its water content is between 2 and 16%. It is soluble in methylated spirits and isopropyl alcohol. It is compatible with a wide range of resins and polymers including rosin, rosin esters, ethyl cellulose, cellulose nitrate etc. It is not compatible with polymers or co-polymers or vinyl acetate.

Shellac has been widely used in the preparation of flexographic inks, and in spirit varnishes. The addition of an alkali or amine produces a sodium or amine salt which renders it water soluble so that it can be used for water-based inks.

Shellac is the flaked form of purified lac. Lac is the natural secretion of the insect *Laccifer lacca kerr*. The lac is dissolved in alcohol, filtered and allowed to dry. Bleached shellac is prepared by dissolving the resin in sodium carbonate solution, allowing the impurities to settle and bleaching the liquid with sodium hypochlorite until the solution reaches a standard colour. The resin is then precipitated with sulphuric acid, washed, dried and powdered.

Other shades of shellac include garnet lac, lemon lac and orange lac.

Shellac contains from 3 to 5% wax. It can be dewaxed by solvent extraction, or the wax may be precipitated from an alcoholic solution by the addition of water.

The formula of shelloic acid is:

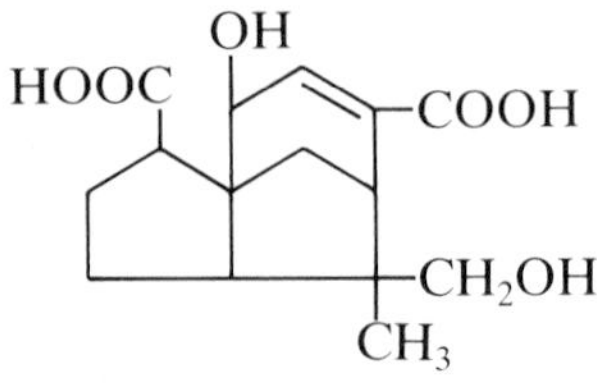

According to the Shellac Export Promotion Council for Indian Shellac, the original lac substance has the folllowing constitution:

Aleuritic acid, $HO(CH_2)_6CH(OH)$—$CH(OH)(CH_2)_7COOH$, m.p. 101 °C	46.0
Shelloic acid, m.p. 206 °C	27.0
Kerrolic acid, a tetrahydroxy palmitic acid, m.p. 132 °C	5.0
Butolic acid, a monohydroxy-pentadecanoic acid, m.p. 54 °C	1.0
Wax esters, derived from wax acids and wax alcohols	2.0
Neutral compounds of unknown composition	7.0
Polybasic acid esters of unknown composition	12.0
	100.0

Manila copal

Natural resin with softening point 122–132 °C (B & R), $d = 1.07$; acid value 120–130; iodine value 115–130. Soluble in alcohols, esters, ketones and glycols. Made oil soluble by cooking at 310 °C for 90 minutes. Main use was in alcohol varnishes for flexo inks, also in lacquers. Used in conjunction with ethyl cellulose for hardness and water resistance. Almost entirely replaced by synthetic resins.

Obtained from the tree *Agathis alba*. Many grades were available. The softer grades were obtained by tapping and the hard ones were fossilised. Copal resins consist of complex mixtures of monobasic acids, particularly those containing 8–10 carbon atoms.

Asphalts

This is a term which covers a large number of resin mixtures of complex compounds consisting of carbon, hydrogen, sulphur and nitrogen which are formed either as natural mineral compounds or as the residue from distillation of petroleum, coal tar or rosin.

They include asphaltum, pitches and bitumen which vary considerably depending on their origin or processing conditions. They range from hard brittle resins to those which are plastic or even fluid at room temperature.

Those found as minerals have such names as Gilsonite, Rafaelite, Malthas and Grahamite. These are referred to as bitumens, according to the British Standards Institute who has many specifications covering them.

They are dark solids, with fairly high softening points, and are soluble in hydrocarbons, but not always soluble in drying oils. In solution they make varnishes which vary from yellow to dark brown. Materials derived from the distillation of coal tar, rosin or petroleum are defined as 'pitches' or 'residual pitch'.

They are widely used in photogravure, letterpress and heatset black inks. The yellow undertone of such inks is usually counteracted by addition of dye toners.

Starch and dextrin

Starch is a commonly occurring natural material, which is synthesised in plants. The main sources of starch are tapioca, maize, wheat and potato.

It consists of amylose and amylpectin, in ratios of about 1 to 4, dependent upon the source. Starch is a stable white powder at room temperature. It is insoluble in cold water, but when an aqueous suspension is heated to 60–80°C, and allowed to cool, it forms a gel. Colloidal solutions of starch react with iodine to give a characteristic blue coloration.

Starches with a large particle size are used as anti-set-off powders in sheet-fed printing; treated water-swellable starches have also been used for anti-set-off in lithographic inks. Finely ground potato starch is used to add body and to eliminate excessive tack in platen inks.

Dextrin is partially hydrolysed starch produced by heating starch alone or in the presence of hydrochloric acid. This converts it to a yellow-brown colour, which forms a viscous gummy liquid when dissolved in cold water. It replaced gum arabic in anti-set-off solutions. It is also used as an adhesive for stamps, etc. and as part of a vehicle system for water-colour inks and cheque book inks.

Gum arabic

The main use for gum arabic was in lithographic fountain and plate 'gumming up' solutions. Gum arabic, being hydrophilic, helps to keep the non-image areas ink repellant. (Supply shortages have shown that polyacrylamides can equally well fulfil this function.) To a lesser extent gum arabic is used in water-based inks to reproduce copies of water colour paintings.

It is obtained from trees of the acacia species in Africa, especially in the Sudan. It consists of organic polybasic acids, related to sugars, with an estimated molecular weight of 250 000. It is readily soluble in water, but dries to give fairly brittle films.

4.21 SYNTHETIC RESINS

Pure phenolic resins

Pure phenolic resins are used in conjunction with tung oil to produce varnishes for letterpress and lithographic inks. Their pigment-wetting qualities are good and a small amount is sometimes added to modified phenolic varnishes to improve wetting and stability. If the non-reactive type is added to flushing vehicles it improves their performance. They have a fairly low softening point 80–130 °C and are soluble in aromatic hydrocarbons, but only partially soluble in aliphatic distillates. Tung oil phenolic varnishes have excellent drying speeds and give very tough glossy films, with good alkali resistance and adhesion properties. The most serious drawback of some pure phenolic resins is their tendency to yellow on exposure to light.

Oil soluble, oil reactive pure phenolic resins are prepared by heating a para substituted phenol with an aqueous formaldehyde solution under reflux in the presence of an alkaline catalyst. The para substituent must be at least a C_4 hydrocarbon chain or a phenyl group. If the substituent group is less than C_4 the resin will not be oil soluble. After the reaction has gone to completion, the water is removed by vacuum distillation. Properties of the final product depend on the phenol used, the catalyst, molar ratio of reactants and the plant operating conditions.

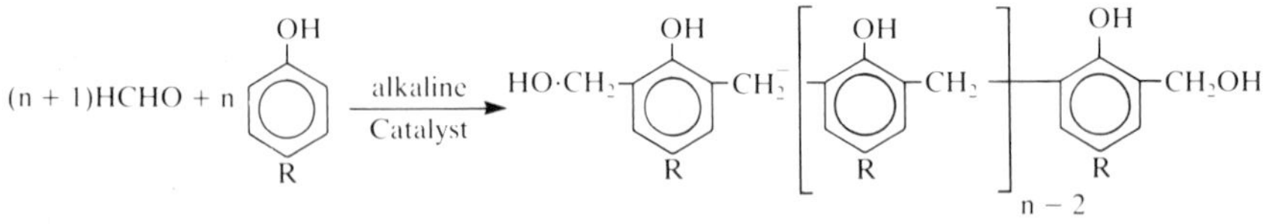

R = aryl, or at least a C_4 alkyl group   Oil-soluble pure phenolic resin
n = degree of polymerisation

If the catalyst used is an acid the product will be similar to the structure shown above but without the reactive methylol groups, as shown next:

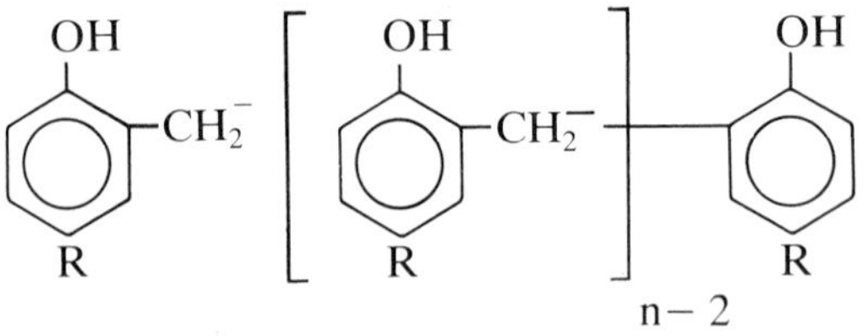

Pure phenolic resins prepared with acidic catalyst can be dissolved in drying oils, especially tung oil, without reaction occurring. Solution of the resin takes place between 200 and 250 °C.

The pure phenolic resins prepared with an alkaline catalyst and which contain the reactive methylol groups can be combined with tung oil. A reaction takes place accompanied by rapid bodying of the varnish. The reaction involves the methylol groups of the pure phenolic resin and the olefinic bonds in the tung oil.

Rosin-modified phenolic resins

Rosin-modified phenolic resins have widespread use in all types of ink vehicle system and there are many grades commercially available. They have limited solubility in aliphatic distillate which makes them ideal for use in quick-setting varnishes and gels. Generally they impart tough, glossy finishes with good chemical resistance. They are compatible with alkyds and other resins to produce high gloss inks with good rub resistance. They are also used in heatset and web-offset inks for their solvent release properties, and in screen inks for their handling and weathering resistance. Rosin-modified phenolic resins are sometimes used in toluene based gravure inks. It is also possible to obtain alcohol-soluble grades that have an acid value between 80 and 120 mg KOH/g, making them useful film formers in alcohol- or water-based inks.

Rosin (80–90 parts) is reacted with an oil reactive pure phenolic resin (10–20 parts) at 150 °C. The temperature is then increased to 250 °C in the presence of glycerol or other polyol in order to esterify the rosin acids. The functional properties of the modified phenolic resin depend on many factors including the chemical identity of the substituted phenol, the polyol used, plant operation conditions, etc. It is interesting to note that the methylol groups of the pure phenolic resin react with the unsaturated centre of the rosin acid in preference to the carboxylic acid groups in rosin.

Several chemical species are present in the final product. One possible structure for a pentaerythritol ester of a rosin-modified phenolic resin is given below.

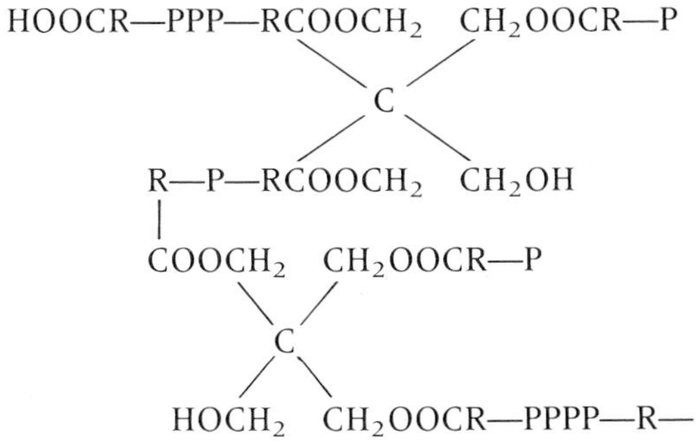

$$HOOCR—PPP—RCOOCH_2 \quad CH_2OOCR—P$$

$$\diagdown C \diagup$$

$$R—P—RCOOCH_2 \quad CH_2OH$$

$$| \quad COOCH_2 \quad CH_2OOCR—P$$

$$\diagdown C \diagup$$

$$HOCH_2 \quad CH_2OOCR—PPPP—R—$$

R = Rosin acid component
P = Substituted phenolic grouping, e.g. where R' = aryl or at least a C_4 alkyl group.

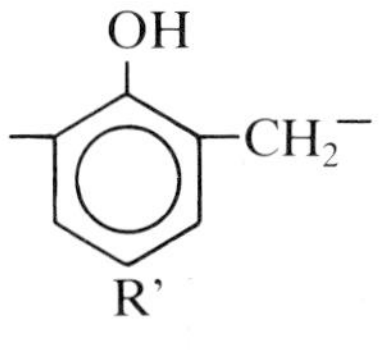

Rosin modified phenolic resins are soluble in oils and aromatic hydrocarbons. They usually need heating to quite high temperatures to achieve solution. They also reduce the gelation rate of tung oil when the oil is heated to high temperatures.

Many resins of this class are commercially available. Their properties depend on the structure of the initial pure phenolic (or cresylic resin), the esterifying polyol, the molar ratios of reactants and the physical conditions of manufacture. The acid value and softening points of two typical rosin-modified phenolic resins are given in Table 4.12.

Alkyd resins

The use of alkyd resins in printing ink has been influenced by increasing usage in other surface coating industries and the ability of manufacturers to produce 'tailor made' resins for specific purposes. They have become a primary resin in the formulation of paste inks, especially as they are obtained at workable viscosities without the addition of solvents (Table 4.13).

One important function in both paste and liquid inks is as a controllable resin plasticiser, where it is possible to combine control of drying rate and setting time with flexibility and adhesion. The high gloss of many alkyds ensures that they are a main film-forming constitutent in quick-setting inks for letterpress and lithography. Their solubility in a variety of solvents enables them to have important applications in screen inks. When non-drying alkyds are used they make excellent plasticisers in liquid ink systems based upon nitrocellulose. Tin printing inks rely upon alkyds as their main film former for colour retention, gloss, adhesion, steam retort processing, drawing flexibility and varnishability.

Their compatibility with other resins is excellent provided that the solvents or diluents used in combination with the other resins, or in subsequent formulation are carefully selected. A certain amount of solvent incompatibility can sometimes be used to advantage, for example in two phase quick-setting ink vehicles.

Alkyd resins are manufactured by esterification of a polyhydric alcohol with a polybasic acid, which makes them members of the polyester family. The majority of polyesters are produced for the plastics industry. Generally alkyds include a combination of drying oils and fatty acids that modify the polyesters to allow them to cross-link and become a pliable film. These film forming properties are dependent on the quantity, type

Table 4.12

Esterifying polyhydric alcohol	Softening points (°C) Ring and Ball	Acid value (mg KOH/g)
Glycerol modified	125–150	20
Pentaerythritol modified	145–175	22

Table 4.13 Typical alkyd resins

Type	Oil length	Polyol	Dibasic acid	Percentage dibasic acid	Solvent	Percentage solids	Acid value (mg KOH/g)	Viscosity (poise) 25°C
Linseed	72	Pentaerythritol	Isophthalic acid	20	—	100	12	150–250
Linseed	67	Glycerol	Phthalic anhydride	22	—	100	10	800
Linseed	76	Pentaerythritol	Terephthalic acid	16	—	100	12	250–350
Dehydrated castor	65	Glycerol	Phthalic anhydride	25	—	100	12	Viscous (unspecified)
Linseed/tung (styrenate)	36	Glycerol	Phthalic anhydride	20	Xylene	60	15	30–40
Soya	64	Pentaerythritol	Phthalic anhydride	23	—	100	12	450–520
Soya	52	Glycerol	Phthalic anhydride	32	White spirit	40	20	20–30
Castor	42	Glycerol	Phthalic anhydride	41	Xylene	50	50	10–18

and nature of the modifying oils, fatty acids or acid anhydrides which form them. The processing methods influence the nature of the alkyd end product.

For convenience alkyds are classed as drying or non-drying depending on the amount and the type of drying oils incorporated. These include linseed, tung, dehydrated castor, soya and tall oil. They are also classified as short, medium or long oil alkyds depending on the amount of oil present. Below 50% oil is a short oil alkyd. Normally the higher the oil length, the more soluble is an alkyd in aliphatic hydrocarbons. Most alkyds used in printing inks are long oil alkyds, containing more than 65% of the oil or fatty acid component.

Styrenated and vinyl toluenated alkyds are prepared by reacting styrene and vinyl toluene respectively with an alkyd in the presence of hydrogen peroxide.

Uses

This brief guide to the use of alkyds in oil-based inks, is in three short sections:

(1) Pigment interactions.
(2) Hard resin interactions.
(3) Film-forming properties.

It is possible to predict levels of performance in these areas, but there is no substitute for practical examination of materials within the framework of local formulation principles.

Pigment interactions

No alkyd exists with all its molecules having the same molecular weight; a molecular weight distribution is always observable. Frequently more than one molecular species is involved. A mean figure is derived from the molecular weight data which may have double or triple peaked distribution. The fact that the polarity or the chemical activity of individual species may differ, increases the possibility of such variations. This, together with batch blending, complicates attempts to predict performance. Surface properties of pigments, together with polarity and viscosity data for a given alkyd type, are compared to indicate potential wetting performance. The measurement of flow properties of standard pigments in a selection of alkyds augments such predictions. Pigmentation levels are chosen together with strict controls on dispersion, ageing, etc. Variation in flow properties at differing pigment volume concentrations provides much useful information. As the viscosity of an alkyd is raised, improvements are found in its versatility such that a wider range of formulation options is available.

In general, high hydroxyl value, low viscosity phthalic alkyds show very high wetting potential irrespective of the pigment type. Isophthalic alkyds are generally found to be more specific in their wetting performance. Good wetting for carbon black is frequently quoted. Terephthalic alkyd systems are also notable in this.

Styrenated and vinyl toluenated alkyds have been found rather unsatisfactory as sole ink vehicle and are often used in conjunction with other film formers.

Hard resin interactions

The distribution of physical and chemical properties described above applies equally generally to the behaviour of alkyds in the presence of hard resins. Peptising/solubility characteristics of a varnish prepared using an alkyd in conjunction with an oil to solubilize a hard resin vary with the amount of polar material present. Viscosity/concentration data combined with measurement of solution clarity provide a basis for ink vehicle studies which should also include measurement of gloss, tack, solvent-release and setting speed. The fact that gloss vehicles and grinding media are prepared using differing proportions of hard resin to alkyd shows that pigment carrying properties are also significant. If a pigment selectively absorbs components from the ink vehicle, it can affect the balance of hard resin to alkyd and upset tolerances in a quick-setting ink vehicle system.

Further effects arising from the loss of low molecular weight materials on to adsorbent particles should be considered, especially where four colour process inks are concerned.

Film-forming properties

Short to medium oil length alkyds that incorporate drying or semi-drying oils are used in both air drying and stoved systems. As the functionality of the polyhydric alcohol is increased (at the same oil length) drying speeds increase. As the oil length for the same functional alcohol is increased, so the flexibility of the dry film improves.

High acid value alkyds (those above 12 mg KOH/g) with a high viscosity will have a tendency to react with basic pigments, but give better wetting, good solvent release and faster setting.

In general the progression from ortho- to iso- to terephtalic alkyds, gives shorter drying times and a reduction in film flexibility, although much depends on the presence of hard resins and on relative concentrations.

Styrenated and vinyltoluenated alkyds yield tough films with short drying times and rapid setting in appropriate varnishes.

Hydrocarbon resins

Continual interest is being shown in hydrocarbon resins. Since exploitations of the North Sea oil fields, demand and usage in them has heightened. The varying cost of oil products has sometimes slowed the expected encroachment on use of other resin systems. Hydrocarbon resin is conventionally regarded as an inexpensive resin for cheaper letterpress and lithographic, and gravure inks. The advantages of chemical inertness and pale colour outweigh its poor wetting and solvent release properties in certain applications. The solubility of higher melting point hydrocarbon resin combined with its alkali resistance makes it suitable for use in soap and detergent carton inks. The softening points range up to 140 °C and acid value is less than 1.0 mg KOH/g.

Polystyrene resins and copolymers

Polystyrene is readily soluble in a wide range of solvents including aromatic hydrocarbons. It gives a water-white hard brittle film, with excellent insulation properties. However, its bad marring and poor adhesion to non-absorbent substrates limit its use in liquid inks and lacquers.

Polystyrene resins can be co-polymerised with maleic anhydride to give a range of powdered resins with a variety of acid values and viscosities. These can be made water soluble and have some applications in water-based inks.

Styrene is also polymerised with acrylonitrile and butadiene to yield products that are invaluable in the production of synthetic rubber rollers. Polystyrene emulsions can be used in liquid inks either alone or in conjunction with shellac, etc., however, they possess the typical styrene odour.

The monomer is produced from benzene, by reaction with ethylene in the presence of a catalyst to produce ethyl benzene and by subsequent dehydration to produce styrene.

Styrene is polymerised by one of four techniques: bulk, solution, suspension or emulsion. It is possible to control the molecular weight of the polymer by means of transfer agents such as mercaptans. The molecular structure is as follows:

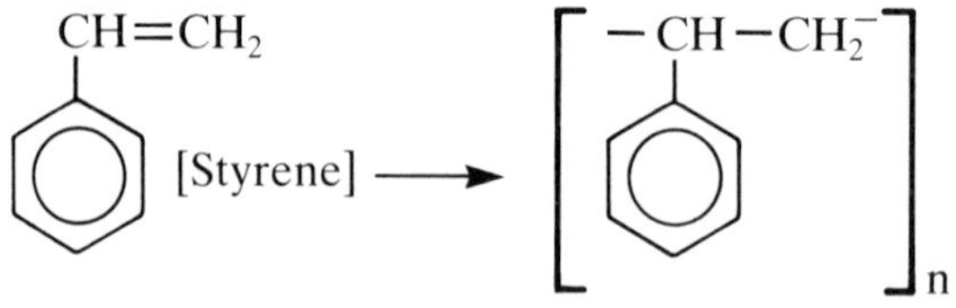

where $n =$ the degree of polymerisation.

The grades of polystyrene which are of interest for printing ink formulation have molecular weight averages from 800 to 15 000 above which the solutions become too viscous for practical purposes and 'string' badly. The melting point varies from 60 to 104 °C, increasing in accordance with the molecular weight. Acid number and saponification values are below 1.0 mg KOH/g. The emulsion technique is not used for making solid resin, due to soap residue which impairs electrical resistance and optical clarity, but it does produce a valuable polystyrene latex which is gaining usage in water-based surface coatings.

Terpene resins

Clear, pale inert resins with low softening points. On economic grounds they are usually limited in their application to gloss lacquers and letterpress or lithographic overprint varnishes. They are soluble in higher ketones and alcohols, aliphatic, aromatic and chlorinated hydrocarbons to form films which are inert to dilute acid and alkali. They are also compatible with most other resins used in printing inks.

Terpene resins are manufactured by the cationic polymerisation of alpha-pinene and beta-pinene monomers using Lewis acids.

Silicone resins

These have limited use in the printing ink and roller industry at present, due to high cost, but may assume a more prominent role in future. They consist of monomethyl, dimethyl, monophenyl, diphenyl, methyl phenyl, monovinyl and methylvinylchlorosilanes, and silicone tetrachloride. Resin hydrolysis is carried out batchwise or continuously, followed by the removal of acid, washing and condensation.

They are normally supplied in a solvent and can be baked to a hard film. They have been used in paints and coatings to obtain weather durability and gloss retention. Dichlorosilanes have the ability to adhere to almost any solid including polymers such as Teflon. Copolymers have also been introduced by blending and copolymerising with alkyds and acrylic resins.

Silicone polymer-coated litho plates have the advantage that they do not need a fount solution or any water at all to maintain ink free non-image areas in lithographic printing (see Chapter 2).

Silicone elastomers based on linear high molecular weight polymers such as polydimethylsiloxanes are used as silicone rubber compounds in roller manufacture.

Alkylated urea formaldehyde resins

Unmodified urea formaldehyde resins are used as moulding powders in plastics. By modifying them with alcohols it is possible to produce resins that are soluble in certain solvents.

Alkylated urea formaldehyde resins are compatible with nitrocellulose, epoxy resins and short oil alkyds. They find widespread application in two-pot high gloss lacquers for paper, film and foil, also in flexographic and photogravure inks where chemical inertness of the finished film is essential.

They can be cured at elevated temperatures without prior addition of a curing agent and therefore find use in stoving finishes.

Preparation
The first stage of preparation involves condensation of urea with aqueous formaldehyde in carefully controlled proportions keeping the pH between 7 and 8. The reaction products are:

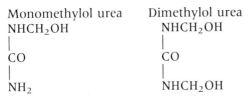

Monomethylol urea
$NHCH_2OH$
|
CO
|
NH_2

Dimethylol urea
$NHCH_2OH$
|
CO
|
$NHCH_2OH$

The reaction is carried out by heating the mixture under reflux. Some polymerisation through the methylol groups takes place at this stage and during the following stage of alkylation of the methylol groups. This is carried out by adding an alcohol, usually butanol, to the mixture which is further heated. Finally the excess alcohol/water/formaldehyde azeotrope is distilled over, leaving the urea formaldehyde resin in the reactor.

$$—NHCH_2OH + ROH \rightarrow NHCH_2OR + H_2O$$
alkylated product

R = propyl, butyl, octyl, etc.

The alkylated or etherified resin is easily soluble in aromatic hydrocarbons and is usually reduced with xylene.

Chemical and physical properties

The resins may be cold-cured by mixing with acid curing agent before application or may be cross-linked thermally when inert durable films are produced. The structure of the cross-linked resin is complex and description is beyond the scope of this book. The main reaction responsible for the cross-linking involves ether linkage with hydroxy groups which may come from another urea-resin molecule, an epoxide, or an alkyd resin.

$$—CONHCH_2OBu + HO \rightarrow —CONHCH_2O— + BuOH$$

The lacquers so produced are light in colour but they have poor pigment wetting properties.

Alkylated melamine formaldehyde resins

The applications are similar to those of alkylated urea formaldehyde, and when cured, melamine resins have fairly similar properties, with slightly better chemical, heat and water resistance. They form slightly more brittle films than urea coatings and are therefore formulated in conjunction with alkyd resins. Melamine formaldehyde resins are normally supplied as solutions in xylene or butanol. The cross-linking reaction is similar to that described for alkylated urea formaldehyde resins. The rate of cross-linking with an acid curing agent is faster than with urea resins.

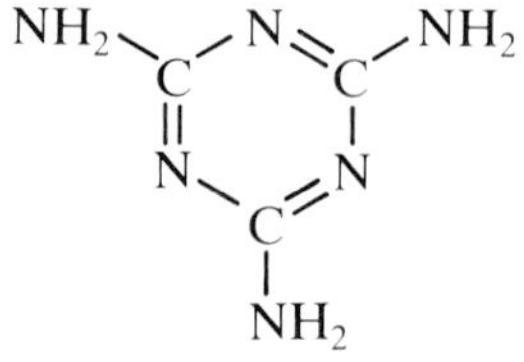

Melamine

Theoretically six methylol groups can be introduced on each molecule of melamine, as in hexamethylol melamine. In practice a commercial product consists of mixtures of tri, tetra, penta and hexamethylol derivatives. These methylol compounds are subsequently etherified with an alcohol, usually butanol.

Commercial products contain many chemical species. Butylated trimethylol melamine has the formula:

$$BuOH_2CHN-C \overset{\displaystyle C}{\underset{\displaystyle N}{\Vert}} \overset{\displaystyle\Vert}{\underset{\displaystyle C}{\Vert}} C-NHCH_2OBu$$

NHCH$_2$OBu

Polyamide resins

A large family of polymers that includes the amide grouping (—CO—HN—) in the main polymer chain. The largest class of polyamides is the 'Nylon' group of polymers having high molecular weight and poor solubility. They are produced for extrusion and moulding. Lower molecular weight species known as reactive polyamides are used in combination with other polymers. Thermoplastic polyamides, referred to as non-reactive, are soluble in solvents used in liquid inks. These last two general polyamide types referred to as fatty polyamides, interest the inkmaker. The reactive types are used in coatings and solventless inks. The soluble types are used in conjunction with other resins to improve film-forming properties.

The functional groups of polyamides are amino groups and the hydroxyl groups distributed in the ratio of one to one.

Chain terminators and blocking agents are commonly used to direct the condensation towards more soluble and compatible products.

The polyamides used for nylons are based upon the formation of a linear polycondensation of a diamine and a dibasic acid.

$$n[HOOC-R-COOH] + n[H_2N-R'-NH_2]$$
$$\rightarrow HO-[-CO-R-CO-HN-R'-NH-]_n-O + (2n-1)H_2O$$

They are manufactured either by self-condensation of amino acids or by thermal polymerisation of lactams. The first process involves reaction of diamines with dicarboxylic acids, whereby interaction takes place between a carboxyl group and an amino group with the elimination of water. For example, Nylon 66 is the polyamide formed with hexamethylene diamine $H_2N-(CH_2)_6-NH_2$ and adipic acid $HOOC-(CH_2)_4-COOH$. A preliminary step of salt formation, usually in an aqueous solution is often carried out to control the polymer chain and avoid termination.

In the case of polymerisation of carprolactam or higher lactams by thermal polymerisation the polymer chain formation is controlled by use of additives or adjustment of reaction time, temperature and pressures. High molecular weight polyamides are either extruded or moulded. Extruded filaments are woven or knitted into fabrics. The ink-maker is becoming more aware of these materials as they often form the substrate for transfer printing of textiles.

Reactive polyamides are obtained by the reaction of di- or polyfunctional amines with polybasic acids originating from unsaturated vegetable oils, usually in the form of dimers or trimers. Reactive polyamide resins contain reactive amino or carboxyl groups and can be prepared using an excess of the amino reagent or the carboxyl reagent. However, it is more usual for one of the reagents to be difunctional, while the other is trifunctional. When reacted in the appropriate proportions with dimer acid, they yield high molecular weight polymers containing free amino or carboxyl groups.

Example
If a dimeric fatty acid such as dilinoleic acid reacts with diethylene triamine on a mole to mole basis, one free amino group will remain for every unit of the polymer chain. This is shown in simple schematic form as follows:

$$n[HOOC - R - COOH] + n[H_2N - R' - R' - NH_2]$$

$$\underset{NH_2}{|}$$

$$\rightarrow HO - \left[- CO - R - CO - HN - R' - \underset{\underset{NH_2}{|}}{NH} - \right]_n$$

$$- O + (2n - 1)H_2O$$

Alternatively reacting a tricarboxylic acid with a diamine produces a resin containing free carboxyl groups.

$$n[HOOC - R - COOH] + n[H_2N - R' - NH_2]$$

$$\underset{COOH}{|}$$

$$\rightarrow HO - \left[- CO - R - CO - NH - R' - NH - \right]_n$$

$$\underset{\underset{COOH}{|}}{}$$

$$- O + (2n - 1)H_2O.$$

The amino-containing reactive polyamides have received most attention. They are capable of reacting with a number of materials, the most important being epoxy compounds. In some of these reactions using primary and secondary amino groups, the reactive portion is the hydrogen atom not the nitrogen atom.

A combination of epoxy and reactive polyamide resins is ideal. The polyamide portion of the molecule confers toughness, adhesion and resilience, whereas the epoxy resin imparts durability, good wetting, low shrinkage and low curing temperature. Copolymers so formed are thermosetting.

Amino containing polyamides react with acrylic resins, increasing the total unsaturation of the compound. They also react with diacrylates to bring about cross-linking between molecules, with acids to form amine salts, anhydrides to form amides and aldehydes to form methylene linkages. In conjunction with phenolic resins they produce thermosetting products having the inherent toughness and resilience of polyamide combined with the chemical and solvent resistance of phenolic resins. These polyamide resins have a low molecular weight and are usually liquid at room temperature. They form a basis for two-pot systems and other solventless inks.

The thermoplastic polyamides of higher molecular weight (2000–9000) have important rôles in liquid ink technology. They are produced by condensing dimerised vegetable oil acids with a suitable polyamino compound by blending at medium temperatures and reacting until the water can be distilled off.

Thermoplastic polyamides are solid amber coloured resins with softening points ranging from 90 to 150 °C. They can roughly be divided into classes below and above molecular weight 4000. Those with a molecular weight average of 4000–7000 normally require an alcohol/hydrocarbon co-solvent blend although they are soluble in aromatic hydrocarbons and higher alcohols. They are generally compatible with nitrocellulose. Those of mean molecular weight 2000–4000, have improved alcohol solubility and readily blend with nitrocellulose. Compatibility with shellac, rosin, maleic, phenolic and ketone resins is excellent. When incorporated into flexographic and gravure inks they confer excellent adhesion to treated polyethylene, polypropylene, polystyrene and films coated with polyvinylidene chloride, as well as other difficult substrates.

High gloss, resistance to fats and grease, heat sealability, fast solvent release and clean printing are also obtained. In overprint varnishes they provide scratch resistance in addition to the above properties. Other uses for polyamides include hot melt adhesives and heat seal coatings for flexible substrates such as film, foil or paper. Such coatings also provide a good grease and water barrier for packaging.

Aqueous polyamide dispersions have been used in adhesives, heat seal and barrier coatings when blended with rubber latices, acrylic or vinyl emulsions. Reactive polyamide colloidal dispersions can be used for curing epoxy emulsions in water-thinnable coatings.

Polyimide resins

Polyimides were introduced commercially in 1961. They are polymers incorporating the imide group —CONCO— in the main chain and are produced by a two-stage reaction between an acid dianhydride and a diamine. A typical synthesis is the exothermic reaction between pyromellitic dianhydride and 4,4'-diaminodiphenylether to form a poly (amic acid).

The second stage is the cyclisation of the poly (amic acid) with subsequent dehydration to form an aromatic polyimide. These resins are used mainly for high temperature and high insulation coatings.

Polymers of this class which are of interest to the ink-maker are those which are more closely related to the polyamides and are known as poly (amide imides).

Poly (amide imide) resins

These resins are prepared by the reaction of excess diamine with diacid dichloride to form a low molecular weight amine-capped polyamide

which is then reacted with pyromellitic dianhydride in dimethylacetamide to form a poly (amide-(amide-acid)). It is possible to achieve this
in three ways. The synthesis can be conducted homogeneously by adding
pyromellitic anhydride to the reaction mixture of diamine and diacid
dichloride in dimethylacetamide. Alternatively, the reaction of excess
diamine with pyromellitic dianhydride to form a low molecular weight
amine-capped poly (amide-acid), which is reacted with diacid dichloride
to form poly (amide-(amide-acid)) as described above. These various
reaction mixtures can each be cast into poly (amide(amide-acid)) film
which is then converted to poly (amide-imide). A third method for preparation of amide-imide polymers consists of reaction of trimellitic
anhydride derivatives, such as the acid halides or acid ester, with a
diamine (as illustrated). The poly (amide-(amide-acid)) is formed, which
upon conversion yields a poly (amide-imide).

Such polymers have recently been used to give good adhesion, hardness, and resistance properties to flexographic inks, also as resins for
obtaining easily dispersible pigmented chips for gravure inks.

Chlorinated rubber

Chlorinated rubber is available as a fine white powder, with up to 67% of
chlorine by weight. Several grades are available, usually the higher the
chlorine content the lower the viscosity of the solution obtained. It is
soluble in aromatic and chlorinated hydrocarbons, esters, ethers and
ketones, but insoluble in aliphatic hydrocarbons and alcohols. It is compatible with the majority of plasticisers, also hydrocarbon resins, rosin
maleic, modified phenolic, acrylic, medium oil alkyds and long oil alkyds.
It can also be dissolved in drying oil vehicles at $100-110\,°C$ and in high
boiling solvents at temperatures of up to $130\,°C$. It disintegrates and chars
at $135-140\,°C$. Clear solutions yellow on exposure to UV light. Chlorinated rubber has a relative density of 1.6.

It is used in toluene-based gravure inks, and in publication ink formulations based on resinates to increase drying rates, gloss and rub resistance. Adhesion to plastic film and foils is good. Chlorinated rubber has
good barrier properties, it lowers water permeability and aids resistance to

plasticiser migration in PVC films. Other applications include use in paper coatings, heat seal coatings, textile stencil inks and as an additive to alkyds in screen process systems to improve solvent release and film forming properties.

Natural rubber in the form of a solid, a solution or latex is chlorinated to give an approximate formulation of $(C_{10}H_{11}Cl_7)_n$.

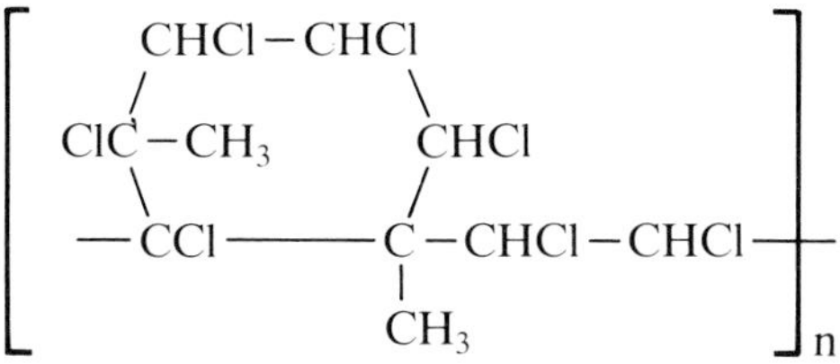

$$
\left[
\begin{array}{l}
\text{CHCl}-\text{CHCl} \\
\quad/ \qquad\qquad \backslash \\
\text{ClC}-\text{CH}_3 \qquad \text{CHCl} \\
\quad \backslash \qquad\qquad / \\
-\text{CCl}\!-\!\!-\!\!-\!\text{C}-\text{CHCl}-\text{CHCl}- \\
\qquad\qquad\quad | \\
\qquad\qquad\quad \text{CH}_3
\end{array}
\right]_n
$$

Cyclised rubber (isomerised rubber)

Several viscosity grades of the solid resin are available ranging from light to dark brown. It is also offered as a solution in white spirit or high boiling petroleum distillate. Cyclised rubber is readily soluble in aliphatic or aromatic hydrocarbons and esters. It is compatible with drying oils and minerals oils, most oil varnishes, ester gum, resinates, cellulose derivatives and hydrocarbon resins, also some maleics, modified phenolics and certain alkyds. It is not compatible with alcohols, ketones, many short oil alkyd resins, etc. Adhesion and chemical resistance are very good and it works well in lithographic inks in conjunction with isophthalic alkyds. The softening points vary from 110 to 140 °C.

It is used in letterpress and lithographic inks, for rapid setting on paper and good rub resistance when dry while remaining 'open' on the machine. Small amounts added to lithographic inks make them more hydrophobic. Their application has been limited by spraying and flying, but with recent grades this can be overcome by correct formulation. They are used in gravure and screen process inks, in combination with other resins, where the printed surface has to have good rub resistance.

Natural rubber is treated with an acid catalyst to reduce the degree of saturation. A rubber isomer forms which is regarded as a cyclic product.

The most probable structure is a polymer in which the repeating units have a structure such as:

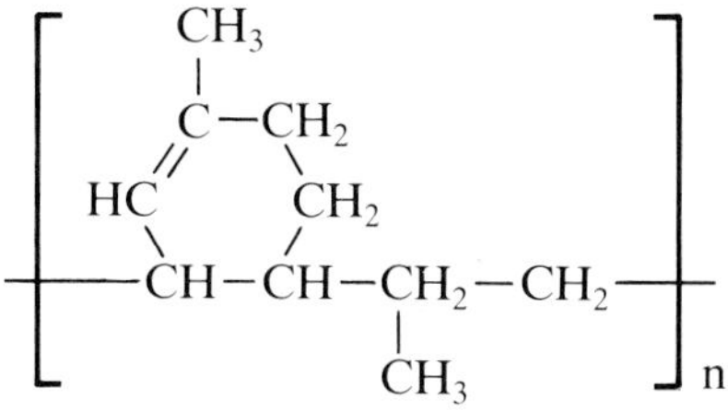

$$
\left[
\begin{array}{l}
\qquad\quad \text{CH}_3 \\
\qquad\quad | \\
\qquad\quad \text{C}-\text{CH}_2 \\
\qquad \nearrow\!\!/ \qquad \backslash \\
\text{HC} \qquad\quad \text{CH}_2 \\
\quad \backslash \qquad / \\
-\text{CH}-\text{CH}-\text{CH}_2-\text{CH}_2- \\
\qquad\qquad | \\
\qquad\qquad \text{CH}_3
\end{array}
\right]_n
$$

where n = degree of polymerisation, but more complex structures have been demonstrated.

Vinyl resins

Polyvinyl acetate
Vinyl acetate is produced by reacting acetylene with acetic acid using a catalyst.

$$CH \equiv CH + CH_3COOH \rightarrow CH_2 = CH - O - \overset{\overset{\displaystyle O}{\|}}{C} - CH_3$$

Vinyl acetate may be polymerised by bulk, solution, suspension or emulsion techniques using peroxide catalysts, giving a head to tail structure.

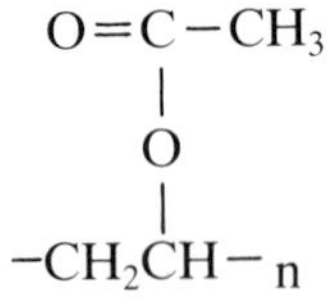

$$
\begin{array}{c}
O = C - CH_3 \\
| \\
O \\
| \\
-CH_2CH-_n
\end{array}
$$

where n = degree of polymerisation.

It is frequently copolymerised with 15–20% of other monomers including alkyl acrylates, fumarates and maleates. Except where the solid resin is required, it is polymerised in water to yield latices used in water-based paints and adhesives.

Many viscosity grades of the solid resin are commercially available. The high molecular weight and high viscosity grades give the toughest films which have more water resistance and higher heat resistance than films of lower molecular weight grades. The bulk polymer is a transparent solid with a characteristic odour of vinyl acetate monomer. The softening points vary from 32 to 86 °C depending on molecular weight. The resins are soluble in most solvents, except aliphatic hydrocarbons. Compatibility with other resins is not good but they are partially compatible with nitrocellulose, phenolic resins, and chlorinated rubber. The relative density is 1.18 and acid value zero.

Used in flexographic transparent lacquers for foil and in heat sealing lacquers. Polyvinyl acetate has poor pigment wetting and is not usually pigmented.

Polyvinyl alcohol
Polyvinyl alcohol cannot be prepared from its monomer vinyl alcohol which has not been isolated in the free state. It is commercially obtained by controlled hydrolysis of polyvinyl acetate.

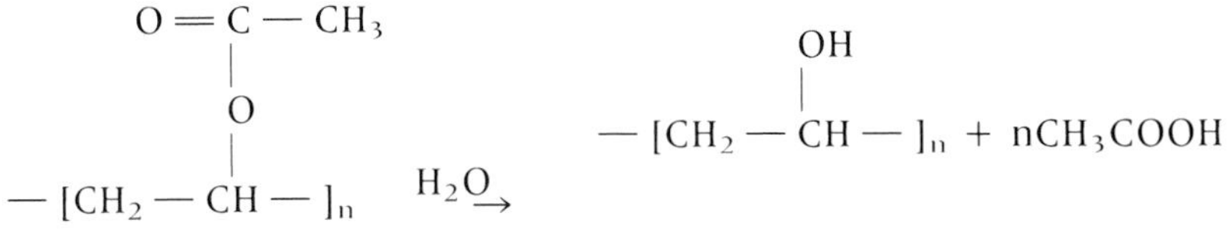

$$O = C - CH_3$$
$$|$$
$$O$$
$$|$$
$$- [CH_2 - CH -]_n \quad \xrightarrow{H_2O}$$

Polyvinyl acetate

$$OH$$
$$|$$
$$- [CH_2 - CH -]_n + nCH_3COOH$$

Polyvinyl alcohol

where n = degree of polymerisation.

Several viscosity grades are available. All grades are white powders with relative densities between 1.25 and 1.35. The resins, which are essentially netural, are soluble in water but insoluble in solvents except those solvents with hydroxy groups such as glycols, amides and amines. The hydroxy groups in polyvinyl alcohol are reactive and many derivatives are possible. Some grades have a small percentage of residual acetyl groups and these are more soluble in organic solvents. The use of polyvinyl alcohol resins in inks is limited to making prints that are sensitive to water as in tamper-proof security inks. They are also used in litho plate photosensitive coatings.

Polyvinyl butyral

Prepared by the reaction of butyraldehyde with partially hydrolysed or completely hydrolysed polyvinyl acetate (polyvinyl alcohol) under carefully controlled conditions.

A simplified chemical structure is:

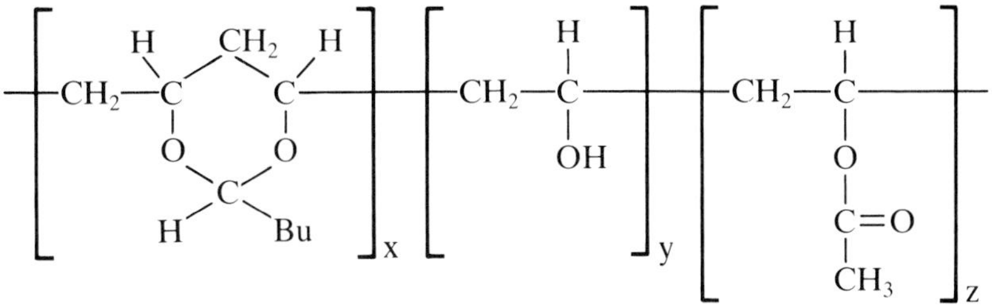

The three components are in controlled proportions and are 'randomly' distributed along the molecule.

The percentages of polyvinyl butyral, polyvinyl alcohol, and the polyvinyl acetate in most commercial products are approximately 80, 17 and 2.5.

Polyvinyl butyrals are supplied as a white powders which used to have a characteristic rancid odour precluding their use in inks for food wrappers. However, more purified products are now available which are more suitable. The solubility of the resins depends to a large extent on the proportion of functional groups present. They are generally soluble in alcohols, esters, ketones, glycols and glycol esters. Non-solvents such as aromatic hydrocarbons are frequently used as diluents. Films of polyvinyl butyral are very flexible. The resins are compatible with shellac, epoxides, maleic and acrylic resins, ethyl cellulose and nitrocellulose. They can be cross-linked with dialdehydes, reactive pure phenolic resins and isocyanates. Polyvinyl butyrals have a relative density of 1.1 and acid value of zero.

They have excellent adhesion to metal, glass, plastics, paper and textiles and are used in combination with other resins to modify flexographic and gravure inks. They improve the key of the ink film to the substrate, gaining use again in liquid ink applications.

Ketone resins

Very pale transparent resins ranging from almost water white to pale straw; they have excellent resistance to ultra violet light, and are chemically inert with very low iodine and saponification values. Ketone resins are neutral (acid and hydroxyl values zero). They have good solubility in acetates, alcohols, ketones and castor oil but limited solubility in aromatic hydrocarbons. They are insoluble in aliphatic hydrocarbon solvents and glycols and water. They have wide compatibility with vinyls, cellulose, acrylics, short-oil alkyds, and in small amounts remain compatible with chlorinated rubber and polyamides. They also possess good pigment-wetting properties.

They are used in conjunction with other resins to improve adhesion, mainly in lacquers as well as high gloss flexographic and photogravure inks. They are included in nitrocellulose lacquers and white coatings to increase hardness and durability.

Condensation of cyclic ketones with formaldehyde in the presence of an alkali catalyst gives rise to the formation of methylol cyclohexanones by substitution of the α-methylenes. Further alkylation of the methylol groups occurs as in urea formaldehyde resins (see p. 207).

Acrylic resins

The ranges of acrylic and methacrylic polymers are based upon monomers of acrylic acid $CH_2 = CH—COOH$ and methacrylic acid:

$$CH_2 = \underset{\underset{COOH}{|}}{\overset{\overset{CH_3}{|}}{C}}$$

These are easily polymerised or copolymerised with other comonomers because of their highly reactive double bonds and miscibility with oil-soluble and water-soluble monomers. Polymerisation can be carried out by bulk, solution, suspension or emulsion techniques using a variety of catalysts.

The general formula for an acrylic homopolymer is:

$$\left(CH_2 - \underset{\underset{R}{|}}{\overset{\overset{CO \cdot OR'}{|}}{C}} \right)_n$$

where R represents H or CH_3, R' represents H, alkyl, alkoxy, alkenyl or an aryl radical, and n = degree of polymerisation.

Various acrylic esters are available through reaction of acrylic acid with the appropriate alcohols. Acrylic esters readily polymerise under the influence of heat, light or peroxide catalysts. These and their co-polymers from the bases of acrylic elastomers which can be vulcanised or cured to yield oil-resistant rubbery polymers. Synthetic rubbers based upon polyethylacrylate or polybutylacrylate copolymerised with about 10% of acrylonitrile are of interest to roller manufacturers.

Poly-acrylic acids and poly-methacrylic acid are often considered together because of their similarity. Both are soluble in water. They are prepared by direct polymerisation of the appropriate monomer.

$$nCH_2{=}CH{-}COOH \rightarrow \left[\overset{\displaystyle COOH}{\underset{}{{+}H_2C{-}CH{-}}} \right]_n$$

poly-acrylic acid

$$n{-}\ CH_2{=}\overset{\displaystyle CH_3}{\underset{}{C}}{-}CH_3{-}COOH \rightarrow \left[\overset{\displaystyle COOH}{\underset{\displaystyle CH_3}{{+}H_2C{-}CH{-}}} \right]_n$$

poly-methacrylic acid

Acrylic resins are characterised by their clarity, chemical inertness, very good lightfastness and resistance to yellowing on stoving.

Their softening points vary widely depending on chemical formula and molecular weight.

Because the range of acrylic resins is so large examples can be found which are soluble in almost any solvent system used in modern printing inks.

Most types are soluble in gravure ink solvents but grades are available which are soluble in alcohols and both low and high boiling aliphatic distillates. Many homopolymers and copolymers of fixed composition are available in more than one viscosity grade. Acrylic resins can be formulated to be compatible with most resins used in printing inks.

Acrylic resins are used in flexographic, photogravure and tinprinting inks and high gloss lacquers. They impart good chemical resistance and improved key to the substrate. The odour of residual monomer from most acrylic resins is sufficient to exclude them from inks for food wrappers, though odourless grades can be made by purging.

The ability of acrylic monomers to polymerise with each other leads to widely differing applications using numerous combinations, for example those with thermosetting acrylic copolymers for stoving finishes. These are often terpolymers, the first for hardness and rigidity, the second for flexibility, the third providing pendant reactive groups for cross-linking with epoxy resins, amines, butylated urea and melamine resins, etc.

These polymers also undergo reactions characteristic of carboxylic acids. Recent interest has been shown in the neutralisation of aqueous solutions with basic materials such as triethanolamine or ammonia which cause viscosity changes demonstrating the presence of differing molecular conformations.

Ordinarily aqueous solutions of these polyacids have low viscosity because the polymer is tightly coiled, being only slightly ionised. As the pH is raised, more carboxyl groups become ionised. Mutual repulsion of the charges forces the polymer chain to uncoil. In diluted solution extended coils result in higher viscosities (subsequently the viscosity would become lower if hydrochloric acid were added to lower the ionisation).

The above means that solutions with high solids content can be supplied as stable fluids. When the pH is lowered they become useful vehicles for liquid inks which can then be adjusted for correct viscosity with water or water/isopropanol mixtures.

Such polymers when converted to their alkaline salts are soluble not only in water, but alcohols and glycols. These are used in flexographic inks and in water-reducible letterpress applications.

Water-soluble homopolymers and copolymers of acrylamide are non-ionic. They are resistant to bacterial attack and have been used for gumming up litho plates.

Photopolymerisation of acrylic esters can occur both with and without sensitizer. It can also be initiated by high energy radiation. They are primarily used in UV curing inks because they are stable, yet capable of rapid polymerisation.

The large range of acrylic esters available, together with the numerous photo initiators used in conjunction with them, involve several mechanisms the theories of which are still not well understood.

The use of acrylic esters as the main vehicle in UV printing ink systems is well established. When used in suitable formulations they impart the ink making, printing, drying and end-use qualities needed. Pigmentation problems have been overcome satisfactorily, and special particle sizes of carbon black are selected. It is necessary to compromise between wanted tinctorial strength and unwanted absorption of the UV radiation required to photopolymerise the acrylic ester.

Epoxide resins

Epoxide resins are characterised by the presence of more than one ethylene oxide or epoxide group per molecule.

$$R - \underset{\diagdown \quad \diagup}{CH} - CH_2$$
$$\underset{O}{}$$

The name epoxy resins currently being used for this group of resins outside Europe, is gradually becoming universal. They are linear polymers of low to medium molecular weight requiring a co-reactant to obtain three-dimensional cross-linking. (Exceptions to this are the high molecular weight resins with n greater than 18).

Prepared by the reaction of epichlorhydrin with diphenylolpropane (bisphenol A) in the presence of sodium hydroxide. The degree of polymerisation of the product will depend on the molar ratio of the reactants and the physical conditions prevailing. The general formula is:

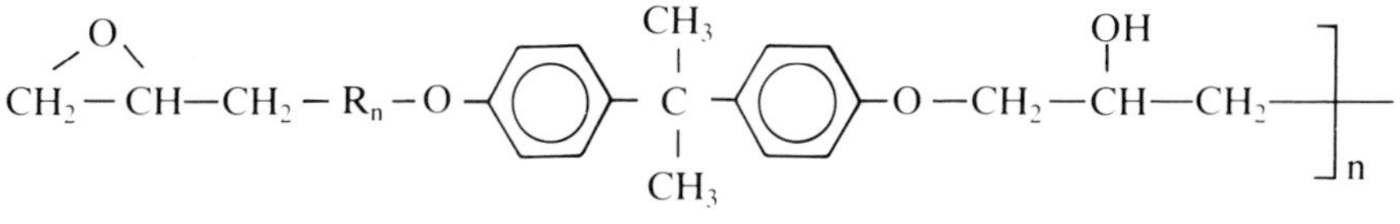

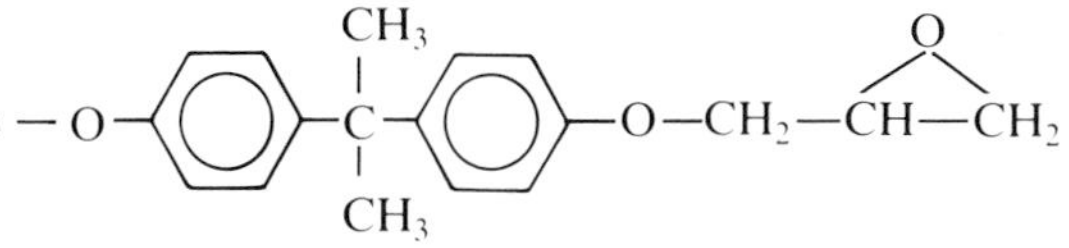

Chemical and physical properties will vary with the mean value of n, the molecular chain length, which may be obtained by using: $(n + 1)$ mol. bisphenol A + $(n + 2)$ mol. epichlorhydrin.

The lowest molecular weights are obtained when $n = 0$, as illustrated in Table 4.14.

Epoxide resins are used extensively in the electronics, paint and lacquer industries. In printing ink manufacture they are used in two-pot lacquers and inks where extreme product resistance is essential.

Liquid epoxy resins are used in solventless coating systems and are crosslinked through the epoxide groups with aromatic and cycloaliphatic amines.

The solid resins are pale in colour, without appreciable odour. Their brittleness increases with increasing molecular weight. Epoxide films are fairly resistant to alkalis and promote good adhesion to many substrates. They are soluble in methyl ethyl ketone, methyl isobutyl ketone, methyl cycohexanone, diacetone alcohol, glycol ethers and many co-solvent systems. They are compatible with certain pure phenolic, urea-formaldehyde, melamine-formaldehyde, polyamides and vinyl resins, but incompatible with vegetable oils, alkyd resins, and cellulose and rosin derivatives. Epoxide resins can be cross-linked to form inert flexible films with excellent adhesion by reaction with:

Table 4.14 Raw materials

$n =$	Melting range (°C)	Epoxide equivalent	Molecular weight	Relative density (20 °C)
0	Liquid at room temperature	175–210	380	1.167
2	64–76	450–525	900	1.206
3.7	95–105	870–1025	1400	1.156
8.8	125–132	1650–2050	2900	1.147
12	140–155	2400–4000	3750	1.190

(1) Amines or reactive polyamides, when reaction takes place with the epoxide group.
(2) Organic acids, when esters are formed at the hydroxyl and epoxide groups.
(3) Isocyanates, when reaction takes place at the hydroxyl group.
(4) Phenol, urea and melamine-formaldehyde resins.

Epoxy esters based on drying and non-drying fatty acids can be modified with styrene, vinyl toluene or rosin. Like alkyds, long oil esters can be thinned with aliphatic hydrocarbons. Water-soluble esters are also available.

Metal decorating inks are often protected with overprint varnish based on epoxy esters. Besides conferring good abrasion and processing resistance in such things as food and chemical containers, they protect the print from any contents spilled on the outside.

Polyisocyanates and polyurethanes

Polyurethanes are polymers obtained by the reaction of polyisocyanates with a polyhydroxy compound such as polyethers, polyesters, castor oil or glycols, etc. or with water and a stabiliser.

A great number of polyurethanes are used in the manufacture of flexible foams, rigid foams, adhesives and surface coatings as well as in roller manufacture. The most widely used systems are based upon two components. The main component is the polyisocyanate, a typical example of which is toluene 2,4-diisocyanate, obtained by the reaction of phosgene on toluene 2,4-diamine, which has been reacted with trimethylolpropane to reduce the health hazard. It will then remain stable until the curing agent (a polyol) is added. One-component systems are available. These use water as a curing agent, or are modified with drying oils or by producing a 'blocked' isocyanate which requires heat to reactivate the isocyanate groups.

Their use in inks is restricted by their effect on printing rollers, and health and safety considerations which put strict limits and controls on the amount of free isocyanates permissible. The screen process is the main outlet. Although polyurethanes obviously contain urethane groups, they also contain a multiplicity of other groups as well as aromatic rings. This results in a very cömplex range of possible structures. The fundamental structure of a polyurethane derived from a dihydroxy compound HO—R—OH and a diisocyanate OCN—R'—NCO can be given thus:

$$\left[-\text{R} - \text{O} - \overset{\overset{\displaystyle O}{\|}}{\text{C}} - \text{NH} - \text{R}' - \text{NH} - \overset{\overset{\displaystyle O}{\|}}{\text{C}} - \text{O} - \right]_n$$

A water-diluable polyurethane can be prepared by reacting a polyisocyanate with 2,2-di (hydroxymethyl) carboxylic acid, then reacting this with ammonia or an amine.

Note: Current legislation and practice should be carefully studied, before this class of resins is used.

Nitrocellulose, N/C (Cellulose nitrate CN)

Nitrocellulose has wide applications in flexographic inks for paper, polymer films, foil and board, usually in combination with other resins such as maleic, polyamide or acrylic. Provided that suitable plasticizer is used, inks based upon nitrocellulose have excellent scratch and rub resistance. It is heat resistant and will withstand high heat-seal temperatures. Also used in overprint lacquers and in foil 'washing' formulations.

Cellulose nitrate (strictly the nitric ester of cellulose and not a nitro compound) is made by treating purified cellulose $(C_6H_{10}O_5)_\chi$ with excess nitric and sulphuric acids. The degree of nitration is controlled by selection of acid strength, time and temperature of nitration. When nitration is complete the product is centrifuged and washed with water. The term nitrocellulose is the standard name applied to industrial grades of cellulose nitrate.

Various viscosity grades are prepared by partially hydrolysing the nitrated product with water, under pressure, when the molecular weight of the material is reduced. Most of the nitrocellulose products used in printing ink manufacture today are 'dense grades'. These are in the form of small square platelets manufactured by replacing the water with ethanol, isopropanol, normal propanol, by washing prior to the densing operation. The final product usually contains 70% solid cellulose nitrate, with 30% of solvent or plasticizer damping.

Cellulose is composed of β-glucose anhydride units joined in a head to tail fashion. Each unit has three hydroxyl groups which can be nitrated. In practice never more than 2.25 hydroxyl groups per unit are nitrated, which corresponds to 12.2% nitrogen in solid cellulose nitrate.

Handling and storage precautions: alcohol-damped or water-damped nitrocellulose should be stored in a cool place with the lids kept firmly on containers to prevent loss of water or alcohol. If this is not done the nitrocellulose becomes dry and extremely dangerous to handle. It is advisable to ensure that half-empty containers are not left lying about.

The legal minimum of damping is 25% below which nitrocellulose is classified as an explosive within the meaning of the Explosives Act. Drums of solvent damped nitrocellulose should be opened and processed only in buildings which comply with the Highly Flammable Liquids and Liquefied Petroleum Gases Regulations 1972.

The structure of a low nitrogen content nitrocellulose product is shown below. (10.7% nitrogen roughly corresponds with two nitrate groups per unit).

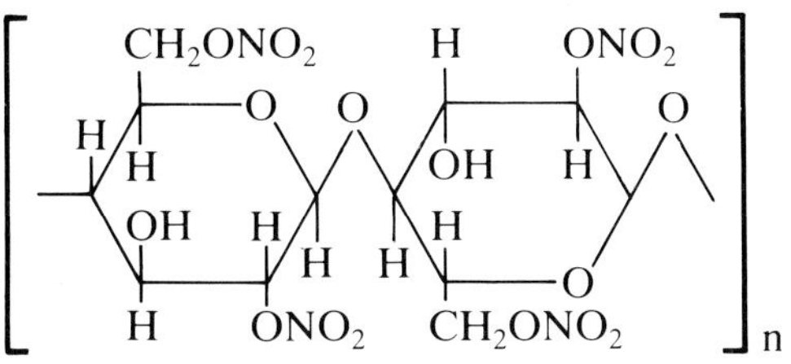

Grades of commercial nitrocellulose are characterised by their nitrogen content (Table 4.15).

For each nitrogen content there are several viscosity grades available characterised by the molecular weight distribution after hydrolysis.

Nitrocellulose is a white fibrous material with excellent thermal stability at room temperature. At elevated temperatures it decomposes violently with the evolution of brown nitrous oxide fumes. Ignition point is 180°C. It is compatible with shellac, rosin derivatives, epoxides, short oil alkyd resins, etc. and with most common plasticizers. The main plasticizers used are phthalates such as dioctylphthalate, and citrates. The film-forming and keying properties of nitrocellulose are outstandingly good.

Table 4.15

Classification	Percentage nitrogen	Suitable solvent
Low nitrogen content	10.7−11.2	74 op methylated spirit
Medium nitrogen content	11.2−11.8	Mixture of alcohols and esters
High nitrogen content	11.5−12.2	Esters

Ethyl cellulose

A cellulose ether manufactured by the reaction of ethyl chloride with alkali cellulose under carefully controlled conditions whereby some hydroxyl groups are replaced by ethoxyl groups (OC_2H_5). By varying the molar ratio of reactants before etherification, grades are produced with ethoxyl content varying from 44 to 50%, that is 2.2−2.6 ethoxyl groups per anhydroglucose unit. The reaction is symbolised as:

$$R—O—Na + C_2H_5Cl \rightarrow R—O—C_2H_5 + NaCl$$

(where R represents the cellulose radical).

Products are divided into four types according to the degree of substitution. Most requirements for ink-makers are met by grades within the 47.5−49% ethoxyl content range which are often referred to as the N-type. A number of viscosity grades ranging from 4 centipoise to 30 poise based upon 5% by weight in an 80/20 toluene ethanol solution are marketed in powder form. Ethyl cellulose is supplied as a colourless odourless white granular powder, which is non-toxic and chemically inert. It has a relative density of 1.14 and softening points are between 150 and 170°C depending on the degree of ethoxylation and the viscosity grade. Being chemically more inert than other chemical modifications of cellulose, it shows better alkali resistance and heat resistance and is less affected by sunlight. It also has moderate resistance to acids.

Solubility varies with the degree of ethoxylation. Grades with a lower ethoxyl content (45.5–47%) are used in plastics and films, whereas some of the high ethoxyl grades tolerate high dilution with aliphatic hydrocarbons, and have good solubility in many solvents including aromatic hydrocarbons, alcohols, chlorinated hydrocarbons, esters, glycol ethers, ethers and ketones. Co-solvent blends are used to match the solubility parameter more closely. Ethyl cellulose is compatible with rosin, glycerol esters, coumarone, gilsonite, shellac, some alkyds, including styrenated alkyds, also with maleic and phenolic resins, waxes and drying oils. It is incompatible with short oil alkyds, acrylates, methacrylates, polyvinyl chloride, polyvinyl acetate, polystyrene and nitrocellulose.

Ethyl cellulose can be plasticised with a large variety of plasticizers. The plasticized films are not as flexible as the corresponding nitrocellulose film, nor can they tolerate such a large percentage of plasticizer.

Lacquers based on ethyl cellulose have excellent toughness and flexibility. In flexographic and gravure formulations it has been used in combination with other resins for printing on paper, aluminium foil and polymer films. Emulsions and suspensions of ethyl cellulose are also used for coating paper, leather and textiles.

Ethyl hydroxyethyl cellulose (EHEC)

Mixed ethers of cellulose, similar to ethyl cellulose, but with wider solubility characteristics. Manufactured by treating alkali cellulose with ethyl chloride and ethylene oxide water under carefully controlled conditions.

No simple formulation of its chemical constitution can be given because the structure is dependent on the molar ratio of reactants. Ethyl hydroxy ethyl cellulose is a mixed ether containing:
(1) ethoxy groups as described for ethyl cellulose; and
(2) β-hydroxyethoxy groups by addition of ethylene oxide.

$$ROH + \underset{\underset{\displaystyle CH_2 — CH_2}{\diagdown}}{\overset{O}{\diagup}} \rightarrow ROCH_2CH_2OH$$

where R represents the cellulose unit.

The further reaction:

$$ROCH_2CH_2OH + \underset{\underset{\displaystyle CH_2 — CH_2}{\diagdown}}{\overset{O}{\diagup}} \rightarrow$$
$$ROCH_2CH_2OH_2CH_2OH$$

is possible.

Three different viscosity grades are produced and classified as high, low or extra low. They are marketed as a white odourless powder that is insoluble in water, methanol and diacetone alcohol and only partially soluble in ethanol and isopropanol. It dissolves in a wide range of aromatic and aliphatic hydrocarbons to form a hazy solution, which clears on addition of between 2 and 10% of ethanol, butanol or isopropanol. It is also soluble in chlorinated hydrocarbons, MEK, MIBK, and some esters. It is compatible with zinc calcium resinate and other rosin-based resins.

Ethyl hydroxy ethyl cellulose is used mainly as a film former in screen inks and to impart toughness and scuff resistance to gravure inks. It is also used as a flatting agent in some varnishes, and as a bodying agent, especially in heat-set inks.

Cellulose acetate propionate (CAP)

Mixed esters of cellulose, very similar to cellulose acetate butyrate. The choice of procedure employed in manufacture depends upon the composition of the ester desired. It is possible to introduce amounts of propionic acid rather than butyric acid into the acetic and acetic anhydride esterification mixture to produce a uniform product containing both acyl groups. All ratios of acyl groups can be made by varying proportions and procedures, although there is far less scope than with the butyrate. Commercial grades usually contain 38–47% propionyl.

CAP is similar in appearance to cellulose acetate, with a melting range of 188–210 °C and low odour. It has excellent solubility in a wide range of solvents; some grades are soluble in alcohol/water mixtures.

Good compatibility with acrylic resins, phenolics, polyvinyl acetate and maleic resin. Because of its high hydroxyl content it can be reacted with melamine and urea formaldehyde to give long pot life, low temperature curing inks. Easy pigment wetting of dry powders or wet presscake.

Used in flexographic, gravure and screen inks to give fast solvent release and good adhesion — useful in laminating inks. Also used as a clear overprint varnish or coating which can be applied over wet ink films. Inks formulated with cellulose acetate propionate have excellent resistance to heat, chemicals, solvents and penetration of oils or grease. Also recommended for use on paper to be recycled.

Cellulose acetate butyrate (CAB)

Cellulose ester containing more side chains than cellulose acetate, whereby some residual hydroxy groups are present.

Pure cellulose is esterified with acetic and butyric acids or more often anhydrides in the presence of a small proportion of sulphuric acid. The proportions of acetyl to butyryl may be varied from high to low. Therefore, the nature of this material will depend upon the degree of esterification of the cellulose and the molar ratio of acetyl to butyryl groups. Many grades of greatly differing composition and properties are available. They are produced as a sweet smelling white powder. The butyryl content varies between 17 and 37% with corresponding decrease in acetyl content which varies from 30 to 13%. Increase in butyryl groups increases the flexibility, moisture resistance, solubility and resin compatibility, but it lowers the softening point and hardness. The softening points of CAB vary from 120 to 240 °C. Hydroxyl contents range from less than 1% to almost 5%. Relative densities range from 1.17 to 1.26. Grades with the higher butyryl content are soluble in ketones, esters, methylene chloride, diacetone alcohol, and various co-solvents, whereas those with a lower butyryl percentage have a more limited solubility. The grades having a higher hydroxyl content are more soluble in diacetone alcohol and glycol ethers. Cellulose acetate butyrate resins are compatible with many resins particularly acrylic, phenolic, polyester, vinyl acetate, some alkyds and urethanes, and nitrocellulose. They are also compatible with a large range of plasticisers. Lacquers based on cellulose acetate butyrate have good clarity, gloss, resistance to UV light, gloss retention and solvent resistance. They are used in gravure and screen inks, as well as forming excellent media for the dispersion of pigmented chips by two roll milling or extrusion, but are generally limited to a role of additive resin in the film former, because excess amounts give a mild odour to prints.

Sodium carboxymethyl cellulose (CMC)

A white granular powder which is readily soluble in water or aqueous mixture of the lower alcohols and ketones. Addition of acidic materials causes precipitation. It is compatible with gum arabic, polyvinyl alcohol, starch, hydroxyethyl cellulose and certain urea formaldhyde resins. CMC can be plasticised with the usual water soluble plasticisers, such as glycerine, glycols, ethanolamines, etc.

Note: In the food industry 'cellulose gum' is used to designate purified CMC for use in foods.

Commerically it is manufactured by treating alkali cellulose with sodium monochloracetate. The amount of sodium monochloracetate is controlled such that 0.3–1.2 (usually 0.7) sodium carboxymethyl groups are introduced per unit of β-glucose anhydride. Thus, each unit of product will usually have 2.3 hydroxyl groups. Several viscosity grades are available.

Chemical constitution

The basic formation reaction is:

$$RONa + ClCH_2COONa \rightarrow ROCH_2COONa + NaCl$$
$$\text{Sodium carboxymethyl derivative.}$$

where R represents the cellulose radical.

Mainly used as a protective colloid and thickening agent for water based inks and to 'gum up' lithographic plates.

It is also used as a sealer to control the porosity of paper so that it gives more gloss and consumes less ink.

Section V: Solvents

The term solvent is widely applicable to a great number of solid, liquid and gaseous substances. The concepts of solvent and solute are interchangeable, so it is necessary to refer to a definition of solution when stating what a solvent is. A solution is a stable separation of molecules of one substance by admixture with molecules of one or more other substances. Ink-makers frequently make solutions by mixing substances (using heat if necessary) which may not spontaneously intermix on the molecular scale, but which subsequently remain in solution once mixed. Substances which spontaneously separate are non-solvents for each other: physical interactions between molecules determine the readiness with which substances intermix under given conditions of concentration and temperature.

A generalised rule that 'like dissolves like' is useful as a guide but it must be appreciated that the large number of possible differences between molecules weakens the concept of likeness as a useful criterion. Difference in size (molar volume), polarity, dispersion forces, hydrogen bonding, and enthalpy and entropy changes on mixing must all be accounted for by a complete theory of miscibility.

Solvent power is the most important factor in considering the usefulness of a solvent. Solvent power is not a general, abstract property, but a specific one relative to what is to be dissolved. The often met requirements that some solutions must have a high film-former content at a relatively low viscosity, while solvent must separate entirely from the printed film on evaporation drying, lead back to consideration of the degree of likeness of film-former and solvent. In polymer solutions of high concentration, liquids which possess the highest solvent power yield solutions with the lowest viscosity, and this property can be used as a yardstick in assessing the power of a particular solvent relative to a particular polymer. Conversely in solutions of low concentration, the solvent producing the greatest intrinsic viscosity has the highest solvent power for the polymer.

Solvents which have a high hydroxyl content are strongly polar and have a high dielectric constant, whereas hydrocarbon and other solvents have a low dielectric constant, being non-polar.

Because the equilibrium of physical interactions between molecules in a mixture of solvents is a complex mean of the component interactions, it often happens that two or more non-solvents for a polymer can be mixed in proportions such that the blend dissolves the polymer. This is the phenomenon of co-solvency. When a non-solvent for the principal resin is added to an ink it is conventionally referred to as a diluent or reducer. In fact diluents sometimes form a co-solvent blend with the ink vehicle solvents; a small change in proportions may result in greater solvent power of the blend for the film-former, but large additions of diluent cause precipitation in extreme cases. A converse of the co-solvency effect is possible. Examples exist where two solvents for a polymer form a non-solvent blend for the same polymer because a critical solvent–solvent interaction is far stronger than the same solvent–polymer interaction.

The solubility parameter is a measure of cohesive energy density (CED), proportional to the energy required to vaporise $1 \, cm^3$ of a liquid. The square root of CED was given the designation 'δ' by Hildebrand. A further analysis of this measure in terms of dispersion forces (δ_d), polar forces (δ_p) and hydrogen bonding forces (δ_h) exists. Charts and three-dimensional treatments of solubility parameters for solvents, plasticizers and resins can be of considerable assistance, and are therefore strongly recommended as an area of further study. It is outside the scope of this section of the *Ink Manual* to elaborate on theories of compatibility and solubility; but three accessible summaries are given in refs 1–3.

Rate of evaporation is next in importance after solvent power when selecting solvents for inks. Slow solvents (those of low volatility) are necessary for press-stability of letterpress and lithographic inks. Controlled evaporation rates are needed in inks that dry by evaporation. The evaporation rates of solvents in a blend vary with the components, the concentrations and temperature. Physical interaction between species of molecule alters both the effective and the relative volatility of components in a blend. Volatility at a given temperature is largely determined by vapour pressure and heat of vaporisation.

A solvent mixture of definite composition, having a constant boiling temperature (which may be higher or lower than that of all components in the mixture) is called an azeotrope. The term *balance* is applied to solvent blends having a composition that remains constant as the blend evaporates. A solvent blend that is balanced for evaporation drying at room temperature usually differs from one that is balanced for heat assisted drying (around 60 °C), and both blends normally differ from the azeotrope if one exists. It is possible for mixtures to have vapour pressures such that the blend has a flash point lower than any of its constituents. Other properties that have to be seriously considered when selecting a solvent for use in printing inks include residual odour, flammability, toxicity, purity and colour. The chemical constitution of pure solvents is of less importance than that of drying oils or resins (except in consideration of solvent power), since solvents do not usually take part in chemical reactions in ink vehicle manufacture. However, the presence or absence of water is significant in certain cases.

The properties of some solvents are tabulated on page 236. They are arranged in the following chemical groups for convenience: (A) hydro-carbon: (i) aliphatic; (ii) naphthenic; (iii) aromatic; (B) monohydric alcohol; (i) aliphatic; (ii) alicyclic; (C) glycol; (D) glycol ether; (E) ketone; (F) ester.

The chemical formula for each solvent is given (where possible), with commercial preparation, boiling range, relative density, flash point (closed cup, unless otherwise stated), colour, the main type of resin which it usefully dissolves and finally miscellaneous properties and applications.

Recent legislation now requires manufacturers to make available to all users details of hazards and precautions for the handling of their products, including solvents. Manufacturers have co-operated in a sensible manner and heed should be taken of their recommendations for any precautions needed. All solvents used in printing inks can be used with safety, if the precautions prescribed by factory or industrial health and safety committees are taken.

It is the user's responsibility to see that care is taken in marking and handling of materials, that sufficient ventilation is installed and where necessary protective clothing is provided and *worn*. Any risks that exist must be minimized, as nearly all solvents have an effect on the human body, serious effects being dependent on the amount of solvent and the time of exposure to it. Large doses in a short time cause acute poisoning usually by affecting the nervous system. Smaller dosage over a long period of time can cause sensitisation or chronic damage to organs.

Exposure to solvents is influenced by volatility of solvents which depends upon concentration, vapour pressure and temperature, or by physical contact with the solvent itself by immersion or ingestion.

The human body reacts in many ways. Swallowing solvent or inhaling concentrated vapour can result in poisoning and causes vomiting, halucinations, unconsciousness or death. Immersion of the hands in solvents is undesirable as toxic action may become possible by transfer to the mouth or absorption through the skin into the blood stream. Skin complaints can be caused by solvents, due to personal allergy to a particular solvent or by a degreasing of the skin which leads to irritation or infection. Barrier creams and the use of lanolin or hand cream after washing are a preventative measure in that they restore the grease level that protects the skin from bacterial attack.

It should also be remembered that when heated or running on a warm machine, 'high boiling' solvents behave very much like 'low boiling' ones being used at ordinary temperatures.

It should be noted that many technical grades of solvent are not pure chemicals, but contain isomers and sometimes homologues, which it may be uneconomic, and even undesirable to remove. This results in boiling ranges instead of boiling points and applies particularly to hydrocarbon solvents. In the aromatic series, terms such as 'toluol' were used to describe the less pure grades of toluene, and British Standards exist, such as BS 805:1977 for toluenes, defining three different grades. The chemical names benzene, toluene and xylene are now standard for referring to aromatic hydrocarbons. Three grades of xylene are specified in BS 458:1977.

4.22   HYDROCARBON SOLVENTS

Low boiling petroleum distillates — aliphatic

Various low boiling petroleum distillates are used in gravure, flexographic and screen inks. They are produced by fractional distillation of petroleum, allowing manufacturers to sell selected brands, with specified boiling range and flash point. Distillates may be close-cut, in which the initial and final boiling points are within 20 °C of each other, or they may be medium cut or wide cut solvents where this difference may be over 100 centigrade degrees. Products that are refined to a standard specification are known as Special Boiling Point solvents, usually referred to as SBP with number suffixes such as 1, 2, 3, 4, 5, 6 or 11, which refer to standard grades, but not to an arrangement in sequence of solvent properties. They are non-polar, with low dielectric constants and, with the exception of SBP 6 and 11, all have flash points below 23 °C (73 °F).

Evaporation rates vary considerably between grades. This has become an important consideration in maintaining a working atmosphere with adequate ventilation. The aromatic content, such as small amounts of toluene, in SBP solvents ranges from 1 to 10%. A description of the naphthenic grades is given below.

Low boiling aliphatic hydrocarbon distillates are good solvents for rosin, ester gum, resinate and hydrocarbon resins.

White spirit
White spirit is a higher boiling petroleum fraction than SBP6. It is cheap, evaporates fairly readily and is ideal for cleaning purposes. It is used as solvent or diluent in a few types of screen inks though its rate of evaporation is too slow for gravure and too fast for letterpress. It has an aromatic content of up to 19% and is a solvent for ester gum, isomerised rubber, hydrocarbon resins, phenolics, maleics and EHEC.

Paraffin oil (kerosene)
A water white distillation fraction of petroleum, with a higher, wider boiling range than white spirit. It has a distinctive odour, due to its volatile 'tail', although odour-free grades are obtainable. Its aromatic content is approximately 8%. It is very cheap. Besides being used for cleaning, paraffin oil is sometimes used as a diluent in letterpress, lithographic and screen inks. The following resins are soluble in paraffin to some degree: maleics, rosin, modified phenolics, hydrocarbon resins, ester gum, calcium zinc resinates, cyclised rubber.

High boiling petroleum distillates — aliphatic

Many close-cut high boiling petroleum fractions are marketed, the most important being those with a 20 or 30 degree gap between their initial and final boiling points. They range between 230 and 320°C, above which boiling point they are usually classified as mineral oil.

They are the most popular solvents for all types of letterpress and lithographic inks. With them it is possible to control evaporation and to obtain the press-stability needed in all types of printing machine including web-offset and heatset machines. Many of them are branded products made to suit all types of requirement. Some distillates are low odour grades with an aromatic content of 2% or less, others have aromatic contents up to 24% enabling resin/solvent varnishes to be produced. They are usually water-white when supplied but deteriorate by photo-chemical reaction to quickly become pale or dark straw coloured.

Petroleum distillates are good solvents, for rosin, ester gum, hydrocarbon resins, calcium zinc resinate, phenolics, maleics, cyclised rubber and EHEC.

Hydrocarbon solvents — naphthenic

The solvency power of petroleum distillates is largely governed by the proportion of aromatic hydrocarbons they contain. By hydrogenation of the aromatics found in hydrocarbon solvents it has been possible to produce a whole new range of solvents known as the naphthenic solvents. This range almost duplicates the various solvents in the aliphatic range above. Hydrogenation of the aromatics largely retains and sometimes increases their general solvency powers, lessens their odour and reduces aromatic, sulphur and olefin contents. This has meant that they have a more acceptable TLV (Threshold Limit Value, referring to the vapour concentration which may adversely affect health). With changes in legislation which may be introduced within the EEC, these solvents are likely to supersede many of the aliphatic and aromatic solvents at present in use.

The low boiling fractions have improved solvency characteristics with no increase in toxicity so they can be used in liquid inks as a replacement or partial replacement for some low boiling aliphatic and aromatic solvents. They are referred to as naphthenic fractions or some identification is added to the SBP number. Higher boiling fractions suitable for paste inks have already been accepted by the ink industry as 'low emission solvents'. The latest developments have low odour coupled with a high KB value and an aromatic content of 0.5% or less. This makes them suitable for food wrapper and packaging inks.

Aromatic hydrocarbons

Toluene (toluol)
The pure solvent is derived from the fractional distillation of light coal tar oil or the catalytic cracking of petrochemicals. 'Toluene' is also the name given to commercial mixtures of which various grades are available as described in BS 805:1977. Its odour is objectionable and the vapour is narcotic.

Toluene is extensively used in gravure inks for which special low odour grades are available. Care must be taken to ensure solvent release to prevent sticking and blocking of the printed reel.

It is a solvent for rubber, chlorinated rubber, melamine, urea formaldehyde, phenolics, ethyl cellulose, resinates, ester gum, polyvinylacetate and polystyrene.

Xylene (xylol)
Slower evaporating solvent than toluene consisting mainly of mixtures of three isomers of di-methyl benzene. Also derived from coal tar or petroleum, its flash point is 74°F but its slow evaporation rate (equal to half that of toluene) limits it to use in gravure proofing and sheet-fed gravure inks. However, its excellent solvency power for many resins makes it a useful solvent in the removal of dried ink films.

Its solvency is similar to toluene, except that it is a poor solvent for polyvinyl acetate, but it will penetrate PVC and treated polythene.

High boiling aromatic solvents

Mostly single ring aromatic structures with side chains or alkyl groups attached, containing small amounts of paraffins and naphthenes. A number of proprietary brands are available, with differing boiling ranges and flash points. They are normally water white or slightly yellow with a sweet odour characteristic of aromatic solvents.

They are used mainly in roller coating finishes and screen inks. They have KB values from 60 to 100 and are solvents for common resins such as ester gum, alkyds, maleics, modified phenolics, urea and melamine formaldehyde, chlorinated rubber and polystyrene.

4.23   ALCOHOLS

Methylated spirits

Ethyl alcohol (ethanol) is not used as pure anhydrous alcohol in commercial manufacture of printing inks. Industrial methylated spirits is essentially ethanol C_2H_5OH and water, to which a denaturant such as methyl alcohol has been added.

Two grades of industrial methylated spirits (IMS) are used extensively in gravure and flexographic inks. They are 64 op and 74 op (op denotes over-proof. Proof spirit contains 57.1% alcohol by volume (49.3% by weight) and has a relative density of 0.92 at 15.6 °C, 60 °F).

Industrial methylated spirits 64 op (density 0.821) contains 9.7% by weight of water, while 74 op IMS (density 0.797) contains 1.4% water.

Industrial methylated spirits as specified in BS 3591:1963 consists of a mixture of 95 volumes of the ethanol/water mixture with 5 volumes of acetone-free wood naphtha approved by the Commissioners of Customs and Excise as a denaturant. The mixture must conform to the Methylated Spirits Regulations 1962.

BS 3591:1963 further specifies maximum residue on evaporation, non-opalescent miscibility with distilled water, and limits to alkalinity or acidity when tested by specified methods.

Extensively used in flexographic and gravure inks and overprint lacquers, IMS is used in preference to toluene in inks for food wrappers because of its much lower residual odour.

The following resins are soluble in methylated spirits: shellac, some maleics, some phenolics, melamine formaldehyde, urea formaldehyde, PVAL, low nitrogen cellulose nitrate, PVB, cyclic ketone and many polyamides. Basic dyestuffs are also soluble in methylated spirits.

Normal propanol (*n*-propyl alcohol)

A very pure solvent, used because it has a higher boiling point than meths or isopropanol. It is miscible with water in all proportions and with many other solvents. It has a characteristic sweet odour.

n-Propanol is used in flexographic and gravure inks and overprint lacquers.

Resins that are soluble in methylated spirits are soluble in *n*-propanol, also it is a partial to good solvent for polyamides.

Isopropanol (isopropyl alcohol or secondary propyl alcohol

A colourless alcohol C_3H_7OH with chemical and solvent properties similar to ethyl alcohol. It is supplied industrially in two grades, sometimes called the isopropyl solvents. As a first grade it contains 99.7%

alcohol B.pt. 82.3 °C, (1PS 1) the second is a water azeotrope containing a minimum of 87% isopropanol B.pt. 80.3 °C (1PS2). It is miscible with water, also it is used as a latent solvent in conjunction with ketones and esters in nitrocellulose lacquers, aiding their solvency and tolerance of hydrocarbon diluents.

It is solvent for many natural and synthetic resins including ester gum, rosin, shellac, ethyl cellulose, spirit soluble phenolics, amino resins, PVAC and PVB. Like *n*-propanol it is only a partial solvent for polyamide resins, but exhibits cosolvency for polyamides in conjunction with hydrocarbon diluents.

Normal butanol (*n*-butyl alcohol)

A slowly evaporating alcohol of high purity consisting essentially of butan-1-ol, $CH_3—CH_2—CH_2—CH_2—OH$, miscible with most solvents, partially miscible with water and possessing excellent solvency power.

Used in metal coatings, some gravure and screen formulations but has considerably more residual odour than *n*-propanol or isopropanol, which precludes it from use in food wrapper inks. Used in conjunction with kauri gum to produce the 5% solution for measuring KB value. Widely used as a latent solvent in nitrocellulose and acrylic based lacquers. Among other resins that are soluble in butanol are alkyds, urea and melamine formaldehyde, shellac, ester gum, PVAC and metallic resinates. Basic dyestuffs, oils, fats and waxes are also soluble in butanol.

Specification requirements of *n*-butanol are described in BS 508:1966.

Alicyclic alcohols

Cyclohexanol
An oily liquid with a strong but not objectionable odour. It has a fairly high flash point and boiling range. A solvent for cellulose ethers, ester gum, shellac, low viscosity silicones and polyvinyl chloride. It has a fairly limited use in screen inks. It is miscible with oils and hydrocarbon solvents.

Methyl cyclohexanol
An oily liquid with an odour similar to cyclohexanol. Its solvency power is less than that of cyclohexanol for the resins named above and its uses in ink manufacture are similarly limited.

4.24   GLYCOLS

Ethylene glycol

Ethylene glycol is a colourless odourless somewhat viscous solvent which is very hygroscopic, miscible with water, alcohol and some ketone solvents. It is used in moistureset inks and water-reducible letterpress inks. The relatively rapid evaporation rate leads to poor press stability.

Ethylene glycol is a solvent for gelatin, dextrin zein and maleic or fumaric resins of high acid value.

Propylene glycol

A colourless odourless solvent similar to ethylene glycol in most respects, but is considered as absolutely safe for use in foodstuffs and medicines. It is recommended for use in food packaging inks and formulations for marking foodstuffs.

Its solubility is similar to ethylene glycol though it is slightly more organophilic.

Hexylene glycol

A colourless liquid with no odour or taste used in the soft drink, confectionery and cosmetic industries. It is slightly slower in evaporation rate than ethylene glycol, is miscible with vegetable oils and many organic solvents. Its use is somewhat limited by its high cost. The purified grade can be used in moisture-set inks for food wrappers. Its solubility is similar to ethylene glycol though it is more organophilic.

Diethylene glycol

Colourless liquid more viscous and hygroscopic than ethylene glycol. Miscible with water alcohols, acetone, glycol ethers and esters. Water-reducible letterpress inks formulated with diethylene glycol have improved press stability, the solvent being more organophilic and less volatile then ethylene glycol, but they have slower setting speeds. Odour level is low.

It is a solvent for high acid value maleic and fumaric resins, nitrocellulose, zein and shellac. It is currently under scrutiny for toxicity.

Dipropylene glycol

A solvent which is very similar to diethylene glycol. It is not as safe for use with foods as propylene glycol, but has lower toxicity than ethylene or diethylene glycol. It is used in moisture-set food wrapper inks and other similar applications.

Triethylene glycol

A stable high boiling water-white hygroscopic liquid, miscible with water and ethanol, but not soluble in hydrocarbon solvents. Its use is similar to that of diethylene glycol but inks are slower setting and possess excellent press stability.

It is a slightly better solvent than diethylene glycol especially for high acid value maleic and fumaric resins, nitrocellulose, zein and shellac.

Glycerine

A clear, colourless syrupy liquid with faint odour and sweetish taste, highly hygroscopic, being able to absorb up to 50% of its weight of water. It is miscible with water, alcohol, glycols and ketones and will dissolve oils, fats and acid dyestuffs. It is used in special water-soluble letterpress inks, e.g. cheque inks, and is a constituent of gelatine rollers. It is too hygroscopic a solvent for formulation of stable moisture-set inks. It can be used as a plasticiser, and it is used extensively in resin manufacture as an esterifying alcohol. Gelatine, sugars and gum arabic are all soluble in glycerine.

Ethoxy propanols

Propylene glycol mono-methyl ether (methoxy propanol) (MP) is a colourless liquid and the first of this type of solvent to be introduced into the UK in the early 1970s. It is now replacing methoxy ethanol (methyl cellosolve) in many applications as it presents less toxicological risk. It has similar solvency characteristics to methyl cellosolve, but is inferior to ethoxy propanol for solvency and wetting.

Propylene glycol mono ethyl ether (ethoxy propanol) (EP) is the nearest ethylene oxide derived substitute for ethoxy ethanol. It has very similar properties without as yet any doubt regarding toxicity. Ethoxy propanol is slower evaporating than methoxy propanol but is a strong solvent with good wetting and retarding properties. It is now gaining recognition as the replacement for ethanol (cellosolve).

Propylene glycol mono methyl ether acetate (MPA)
Propylene glycol mono ethyl ether acetate (EPA)

The acetate esters of methoxy and ethoxy propanol, introduced under many trade names over the past 5 years, have replaced the ethylene glycol derivatives. These are equally good solvents for many resins used in ink formulae and particularly useful as powerful solvents with low volatility.

4.25 KETONES

Acetone (dimethyl ketone)

One of the most widely used industrial solvents with relatively low cost and high solvency power. It is a colourless flammable liquid with a characteristic odour and low boiling range. Miscible with water, hydrocarbons and natural oils in all proportions. Occasionally used as a rapid evaporating solvent in flexographic and gravure inks and in lacquers, though it is too volatile for most applications.

Powerful solvent for cellulose acetate and nitrocellulose, ethyl cellulose, CAB, CAP, PVAC, PVC, ester gum and many other natural or synthetic resins. It also penetrates polythene and methyl methacrylate.

Methyl ethyl ketone (butanone-2)

Colourless solvent with a higher boiling point than acetone and less soluble in water, it has its own distinctive odour and is completely miscible with castor and linseed oils and with most organic solvents. Its main application is in gravure inks, the evaporation rate being about a third that of acetone at room temperature. It is often used as a substitute for ethyl acetate in nitrocellulose lacquers. Its solvency power is similar to acetone with the exception of cellulose acetate for which it is not a very good solvent.

Methyl iso-butyl ketone (hexone)

Water white solvent with a bland ketonic odour, almost immiscible with water but miscible with most solvents and oils. Used in gravure and screen inks, having an evaporation rate of only about one-eighth that of acetone. Solvent for nitrocellulose, ethyl cellulose, CAB, PVC, PVAC, vinyl copolymers, acrylics, chlorinated rubber, polyurethane, phenolic and epoxy resins.

Cyclohexanone (sextone)

Colourless liquid having a strong characteristic odour. Only slightly soluble in water, but a powerful solvent for oils and fats, miscible with hydrocarbons. It is used in special screen inks. A good solvent for most polymers, it will penetrate polythene and soften perspex.

Methyl cyclohexanone (sextone B)

This solvent resembles cyclohexanone but is slightly slower evaporating, colourless to pale yellow, with a strong odour described as similar to peppermint and acetone. It is used in special screen inks and lacquers. Its solvency action is similar to cyclohexanone.

Isophorone

Excellent ketonic solvent in the high boiling range, colourless but with a strong odour due to impurities present, only slightly soluble in water but miscible with most solvents. Solvent for oils and fats. Used in metal decorating coatings and screen inks. Regarded as a powerful solvent for nitrocellulose and vinyl resins, also for many other natural or synthetic resins.

Diacetone alcohol

Clear colourless liquid which yellows on standing. It is a ketonic alcohol, normally classified with the ketones. It has a faint odour and is miscible with water, many solvents and some oils including castor oil. It is not compatible with high-boiling aliphatic hydrocarbons. Used mainly in flexographic inks. Because it is slow evaporating it assist 'flow out', giving glossly films, especially with nitrocellulose. Diacetone alcohol is also used in some roller cleaner formulations. It is a good solvent for cellulose acetate and nitrocellulose, PVAC, shellac, zein and basic dyestuffs.

4.26   ESTERS

Ethyl acetate

One of the best low boiling solvents for nitrocellulose, water white with

Table 4.16 Solvents

Name	Chemical formula	Preparation	Boiling range (°C) Initial	Final	Relative density	Flash point (°F)	(°C)
ALIPHATIC HYDROCARBONS							
Low boiling petroleum distillates. Some examples are given below:	No simple formula. Always complex mixtures of compounds. Mainly of the paraffinic type. The percentage of given molecular species is dependent on the nature of the parent feedstock	Fractional distillation of petroleum			Varies according to manufacturer's specification		
SBP 1			40	170	0.680	−50	
SBP 2			70	94	0.708	−40	
SBP 3			100	120	0.740	20	− 6
SBP 4			40	160	0.710	−50	
SBP 5			90	105	0.728	0	−17
SBP 6			140	160	0.768	83	28
SBP 11			154	168	0.776	101	38
White spirit (the US equivalents are Stoddard Solvent and Mineral Spirit)	No simple formula. Always complex and variable mixtures of paraffinic, alicyclic and aromatic hydrocarbons	Fractional distillation of petroleum	150	190	0.78	102−120	38−48
Paraffin oil (kerosene)	No simple formula. Always complex and variable mixtures of paraffins with smaller quantities	Fractional distillation of petroleum	180	250	0.78−0.80	145−150	62−65

of alicyclic and aromatic (approx. 5%) hydrocarbons

High boiling petroleum distillates	Complex mixtures of higher molecular weight hydrocarbons, especially paraffinic hydrocarbons. The aromatic content is variable (8–18%)	Fractional distillation of petroleum	240	320	0.82–0.83	>150	>65
High boiling naphthenic distillates	Low odour, low emission mixtures. Very low aromatic content (0.2–2.0%)	Hydrogenation of selected petroleum fractions	210	310	0.76–0.82	>150	>65

AROMATIC HYDROCARBONS

Toluene (Toluol, Toluole)	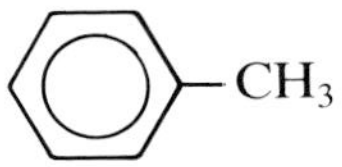 [structure: benzene ring with CH_3]	(1) Distillation of coal tar (2) Petrochemical prepared by catalytic processes	105	111	0.850–0.873	45	7
Xylene (Xylol, Xylole)	Essentially a mixture of the *ortho*, *meta* and *para* isomers of dimethyl benzene 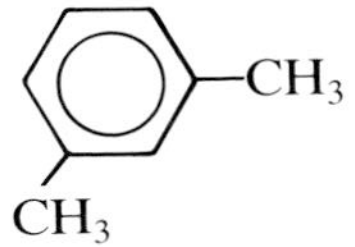 [structure: benzene ring with two CH_3]	(1) Distillation of coal tar (2) Petrochemical prepared by catalytic processes	136	145	0.887	78	25
High boiling aromatic solvents	Single ring aromatic solvent with side chain (Aromatic content 95–99.5%)	Catalytic process and blending	170	270	0.900–0.990	>200	>93

Table 4.16 continued

Name	Chemical formula	Preparation	Boiling range (°C) Initial	Final	Relative density	Flash point (°F)	(°C)
MONOHYDRIC ALCOHOLS **(i) Aliphatic alcohols**							
Methylated spirits (ethyl alcohol, ethanol)	Essentially CH_3CH_2OH with traces of methanol and water	Catalytic hydration of ethylene and fermentation of starch	77.5	79.0	0.806	55–60	12–15
Normal propanol (n-propanol, propanol-1, n-propyl alcohol)	$CH_3(CH_2)_2OH$	Prepared by Fischer-Tropsch process	97	98	0.804	59	15
Isopropanol (Isopropyl alcohol, propanol-2)	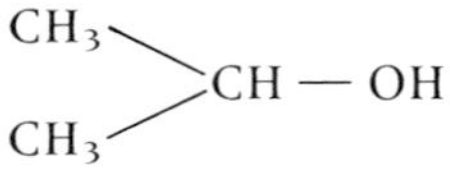 $CH_3 \backslash CH_3 / CH - OH$	Catalytic hydration of propylene	79.5	82.5	0.784	53	11
Normal butanol (n-butanol, butanol-1, n-butyl alcohol)	$CH_3(CH_2)_3OH$	Catalytic hydrogenation of crotonaldehyde	114	118	0.813	95	35
(ii) Alicyclic alcohols Cyclohexanol (Hexalin)	(ring structure: CH_2, CH_2, CH_2, CH_2, H_2C, H_2C with CH_3OH)	Catalytic vapour phase hydrogenation of phenol	158	166	0.963	154[a]	68

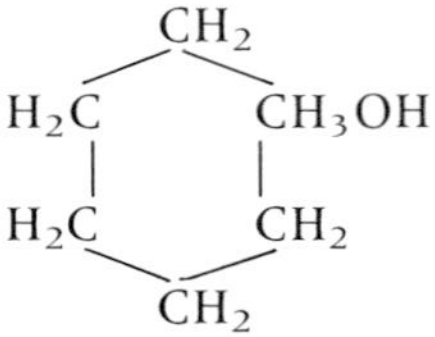

Name	Formula	Method of preparation					
Methycyclo-hexanol	Isomeric mixture of: (i) 2-Methylcyclohexan-1-ol. (ii) 3-Methylcyclohexan-1-ol. (iii) 4-Methylcyclohexan-1-ol. Each isomer has *cis*- and *trans*-stereoisomers.	Catalytic vapour phase hydrogenation of commercial cresol	165	170	0.920 –.0930	145	62
GLYCOLS							
Ethylene glycol (glycol, ethane-1,2-diol, 1,2-di-hydroxy-ethane)	CH_2 — OH | CH_2 — OH	Catalytic hydration of ethylene oxide	196	198	1.113	230	110
Propylene glycol	CH_3CHCH_2OH | OH	Catalytic hydration of propylene oxide	187	189	1.037	225[a]	107
Hexylene glycol (2-methyl-pentane-diol)	OH OH | | H_3C — C — CH_2 — C — CH_3 | | H CH_3	Catalytic hydrogenation of diacetone alcohol	197	199	0.924	245[b]	118
Diethylene glycol (Digol)	CH_2CH_2OH O CH_2CH_2OH	Controlled catalytic vapour phase reaction between ethylene oxide and water	244	248	1.117–1.120	290	143

Table 4.16 Continued

Name	Chemical formula	Preparation	Boiling range (°C) Initial	Final	Relative density	Flash point (°F)	(°C)
Dipropylene glycol	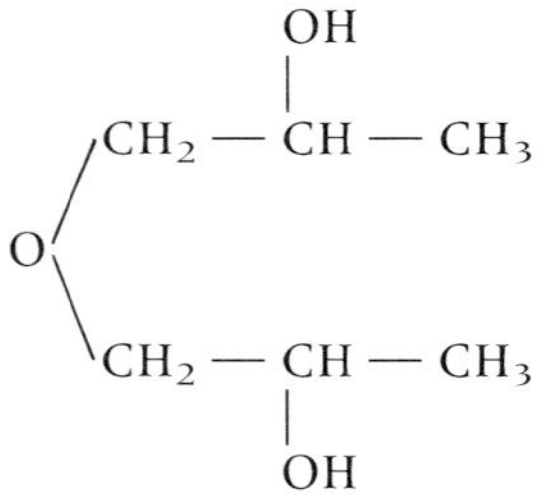 $CH_2-CH(OH)-CH_3$ and $CH_2-CH(OH)-CH_3$ linked via O	Catalytic condensation and hydration of propylene oxide	230	234	1.025	244[a]	117
Triethylene glycol	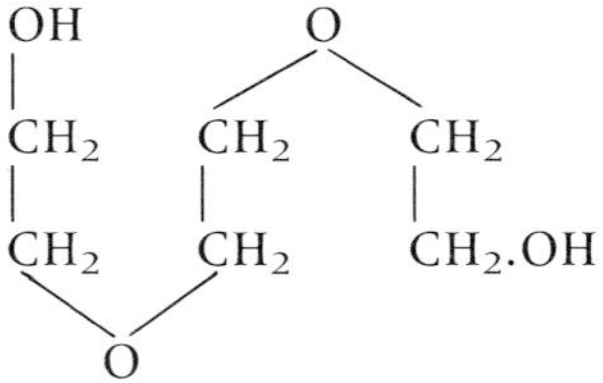 $OH-CH_2-CH_2-O-CH_2-CH_2-O-CH_2-CH_2.OH$	Controlled catalytic vapour phase reaction between ethylene oxide and water	278	280	1.125	330[a]	165
Glycerine	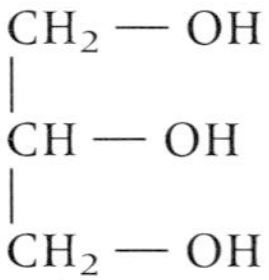 CH_2-OH / $CH-OH$ / CH_2-OH	Saponification of naturally occurring triglycerides, e.g. animal fats and vegetable oils	287	293	1.26	320	160
GLYCOL ETHERS Ethylene glycol monomethyl ether (methyl glycol, 'Methyl Cellosolve' 'Methyl Oxitol')	CH_2-OCH_3 / CH_2-OH	Catalysed addition of methanol to ethylene oxide	120	126	0.963–0.966	102	38

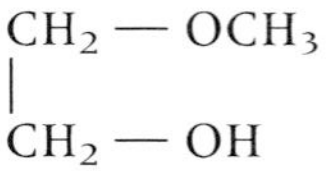

Name	Structure	Method of manufacture					
Ethylene glycol monoethyl ether (ethyl glycol, 'Oxitol' 'Cellosolve')	$CH_2 - O.CH_2CH_3$ $\mid$ $CH_2 - OH$	Catalysed addition of ethanol to ethylene oxide	130	136	0.928–0.934	120	48
Ethylene glycol monobutyl ether ('Butyl 'Cellosolve', 'Butyl Oxitol')	$CH_2 - O.CH_2CH_2CH_2CH_3$ $\mid$ $CH_2 - OH$	Catalysed addition of n-butanol to ethylene oxide	170	172	0.902	165[a]	73
Diethylene glycol monoethyl ether (Carbitol, Ethyl Digol, 'Dioxitol')	$CH_2CH_2O - CH_2CH_3$ O $CH_2CH_2 - OH$	Catalysed reaction of ethanol with ethylene oxide	200	204	0.995	205[a]	96
Diethylene glucol monobutyl ether ('Butyl Carbitol', 'Butyl Digol', 'Butyl Dioxitol')	$CH_2CH_2 - O(CH_2)_3CH_3$ O $CH_2 - CH_2 - OH$	Catalysed reaction of butanol with ethylene oxide	229	231	0.953	240[a]	115
Propylene glycol mono methyl ether ('Methoxy Propanol')	$CH_2 - OCH_3$ $\mid$ CH_2 $\mid$ $CH_2 - OH$	Catalysed addition of methanol to propylene oxide	120	125	0.962	95	35
Propylene glycol mono ethyl ether ('Ethoxy propanol')	$CH_2 - O - CH_2CH_3$ $\mid$ CH_2 $\mid$ $CH_2 - OH$	Catalysed addition of ethanol to propylene oxide	129	135	0.896	108	42

Table 4.16 Continued

Name	Chemical formula	Preparation	Boiling range (°C) Initial	Final	Relative density	Flash point (°F)	(°C)
KETONES Acetone	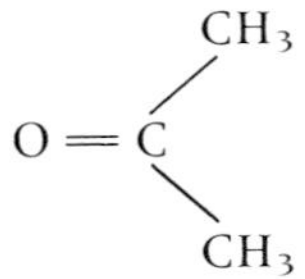 $CH_3-C(=O)-CH_3$	Mainly by the vapour phase catalytic dehydrogenation of isopropanol	55.5	56.5	0.79	1.0	−17
Methyl ethyl ketone (butan-2-one)	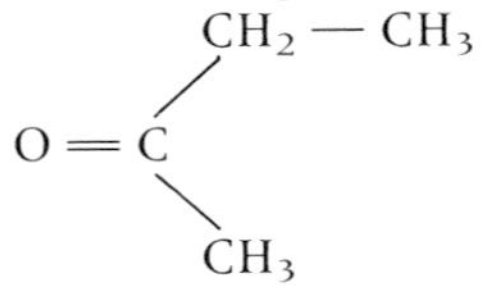 $CH_3-C(=O)-CH_2-CH_3$	Vapour phase catalytic dehydrogenation of sec-butanol	79	80	0.805	26	− 3
Methyl isobutyl ketone	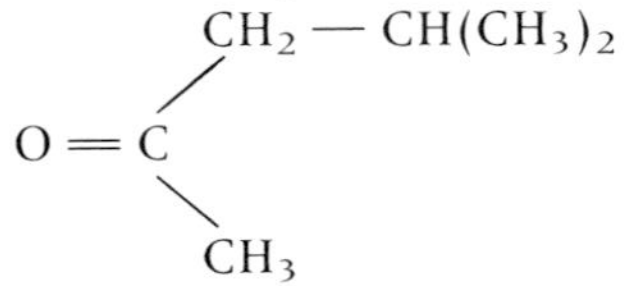 $CH_3-C(=O)-CH_2-CH(CH_3)_2$	Vapour phase catalytic hydrogenation of mesityl oxide	114	116	0.800–0.801	56	13
Cyclohexanone (Sextone)	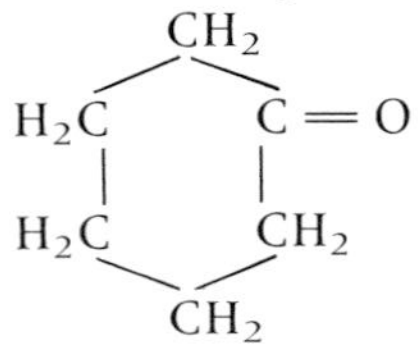 cyclohexanone ring ($CH_2-C=O$, CH_2, H_2C, H_2C, CH_2)	Vapour phase catalytic dehydrogenation of cyclohexanol	151	152	0.948	117	47
Methyl cyclo-hexanone (Sextone B)	Isomeric mixture of 1, 2 and 3 methyl cyclo-hexanones	Vapour phase catalytic dehydrogenation of methyl cyclohexanol	165	175	0.914	117	47
Diacetone alcohol	$CH_3-\underset{OH}{\underset{|}{C}}-CH_2-\overset{O}{\overset{\parallel}{C}}-CH_3$	Alkaline condensation of acetone (Aldol change)	158	166	0.931	55	12

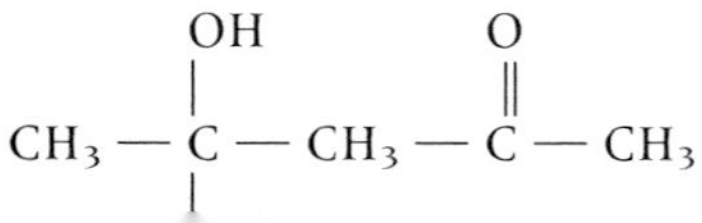

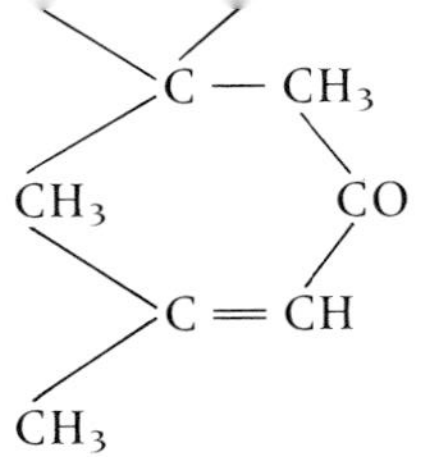

condensation
of acetone

ESTERS	Structure	Preparation	B.p.	Density	[a]	[b]
Ethyl acetate	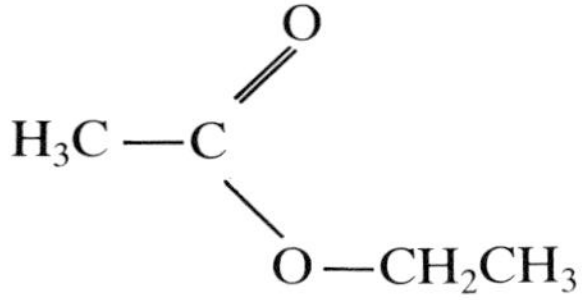 $H_3C-C(=O)-O-CH_2CH_3$	Esterification of acetic acid with ethanol	76– 79	0.902–0.904	26	– 3
Isopropyl acetate	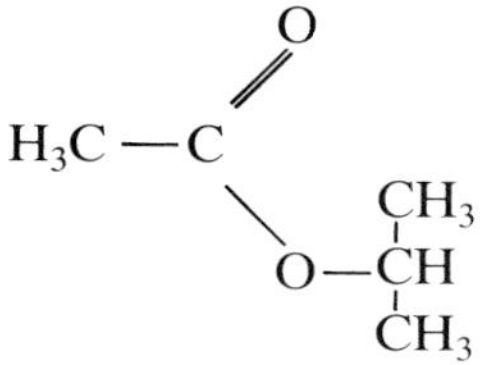 $H_3C-C(=O)-O-CH(CH_3)CH_3$	Esterification of acetic acid with isopropanol	85	0.872	62	16
n-Butyl acetate	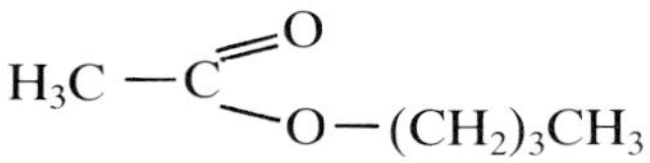 $H_3C-C(=O)-O-(CH_2)_3CH_3$	Esterification of acetic acid with n-butanol	124– 128	0.883	99	37
n-Propyl acetate	$H_3C-COO-(CH_2)-CH_3$	Esterification of acetic acid with normal propanol	100– 105	0.89	58	14
Ethoxy propanol acetate	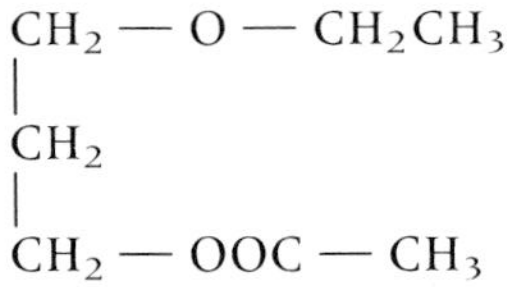 $CH_2-O-CH_2CH_3$ / CH_2 / $CH_2-OOC-CH_3$	Esterification of ethoxy Propanol with acetic acid	155– 160	0.941	130	54

[a] Open cup.
[b] Pensky Martens.

strong fruity odour. Partially miscible with water (8% w/w at 20 °C) but miscible with castor and linseed oil, and hydrocarbon solvents. Its solvent power is increased by adding a small quantity of alcohol. Extensively used in flexographic and photogravure inks and in lacquer formulations as a fast drying solvent. Its evaporation rate is slightly below that of acetone at room temperature. Solvent for NC, EC, CAB, PVAC, polystyrene, ester gum and maleics. Ethyl acetate penetrates polythene, perspex and PVC.

Isopropyl acetate

Properties very similar to ethyl acetate but its relative evaporation rate at room temperature is about half and its water tolerance is less.

N-butyl acetate

Colourless liquid with fruity acetate odour. Not miscible with water, but miscible with oils and many solvents including hydrocarbons. Used in small amounts in flexographic inks, gravure inks and overprint lacquers as well as some screen and metal decorating inks. Because of its excellent solvent properties and slow solvent release it gives good gloss and reduced solution viscosities. The evaporation rate is about one-sixth that of ethyl acetate. Good solvent for nitrocellulose, hydrocarbon resins, rosin, ester gum, chlorinated rubber, vinyls, polystyrene and acrylates. In the presence of 20% butanol, n-butyl acetate will dissolve shellac and some alkyds.

N-propyl acetate

Water white solvent with fruity odour. Gained recognition, particularly with the demise of cellosolve, as a retarder with good solvency and no residual odour. Now an important liquid ink solvent.

Section VI: Plasticisers

The main function of a plasticiser in an ink film is to make the dried print more flexible and pliable. Plasticisers need to be essentially non-volatile in their final ink format. Inks that dry by evaporation, especially on non-porous surfaces, tend to be brittle, and flexing can crack the film and rupture the adhesion. Plasticisers help to supply elasticity to an ink film and allow it to bend or crease without failing. They do this by acting as solvents for the film former polymer molecules, maintaining the exact degree of mechanical freedom required.

Many chemical classes of compounds are used as plasticisers in printing ink manufacture. They are mainly viscous liquids but more solids are being used these days as they plasticise without migration problems and effect better heat resistance than liquid ones. Certain plasticisers are incorporated to promote specific properties in the final ink film such as:
   (a) increased gloss;
   (b) deep freeze resistance;

(c) to prevent blocking;
(d) extra flexibility;
(e) increased adhesion to difficult substrates;
(f) less discoloration at high temperatures.

Very careful formulation is necessary in the initial choice of plasticiser type, and the percentage incorporated into the final ink recipe in order to achieve the optimum balance of ink properties for an efficiently safe and workable plasticised dry ink film. The major use of all these types of plasticisers is in liquid inks, paste inks having in-built plasticised resins unsuitable for liquid ink use.

The main families of plasticisers are:

Abietates	Phthalates
Adipates	Polyesters
Benzoates	Polyol esters
Butyrates	Ricinoleates
Citrates	Sebacates
Epoxidised compounds	Stearates
Phosphates	Sulphonamides

The major suppliers in the plasticiser market concentrate more on other industries than ink, and the most knowledgeable people tend to be in the films and plastics industries. A large amount of published information is available on the subject of plasticisers mainly because of the possible toxicity and hazard levels in contact (after migration) with foodstuffs.

Many data sources and references are available for study, nearly all published in the 1970s. These include the NPIRI *Raw Materials Data Handbook*, Volume 2, which gives a comprehensive study of 246 known plasticisers on the market.

Sucrose acetate iso-butyrate (SAIB) $C_{10}H_{16}O_9(CH_2OOCCH_3)(CH_2OOCC_4H_9)$

FDA status 2520; boiling point: decomposes at 288°C; freezing point = N/A; flash point 260°C; relative density 1.146.

White powder, solid plasticiser compatible with nitrocellulose, CAB ethyl cellulose, polystyrene and p.v. acetate used as a plasticiser in inks to be laminated and when the level is carefully controlled does not affect the strength of the seal. Insoluble in water.

Triethyl citrate (TEC) $CH_2C(OH)CH_2[COOCH_2CH_3]_3$ ## Tributyl citrate (TBC) $CH_2C(OH)CH_2[COO(CH_2)_3CH_3]_3$

(Well-known trade names: Citroflex 2 (TEC); Citroflex 4 (TBC))
FDA status: TBC 2520; TEC 101, 2005, 2514, 2526, 2548, 2550, 2559, 2569, 2571. Boiling point TBC 170°C; TEC 127°C Freezing point: TBC 62°C; TEC 46°C. Flash point: TBC 185°C; TEC 155°C. Relative density: TBC 1.042; TEC 1.136.

Clear liquid plasticisers compatible with a very wide range of resins. TEC is compatible with CAB, cellulose acetate, N/C, ethyl cellulose, PV acetate, PVB and vinyl chloride acetate. TBC is compatible with all the above and polystyrene and chlorinated rubber.

They impart excellent gloss to N/C systems but, as with all liquid plasticisers the amount used should be carefully considered and migration tests checked with food, even though TEC is in GRAS category (generally recognised as a safe foodstuff chemical). Incompatible with water.

Epoxidised soya bean oil, $C_{57}H_{76}O_{18}$

FDA status — not considered harmful: 2005, 2514, 2520, 2526, 2531, 2548, 2550, 2559, 2571, 2572; boiling point over 400°C; freezing point +5°C; flash point 310°C; relative density 0.993.

Light yellow viscous liquid compatible with N/C, E/C chlorinated rubber and vinyl chloride and vinyl chloride acetate. Popular to plasticise vinyl systems and stabilise them, but useful in many applications due to its cellulose compatibility. Immiscible with water.

Tables of FDA regulated plasticizers
Applicable Sections in the Federal Code of Regulations
(Title 21, Part 121 except for one in Part 8)

8.300	Diluents in Color Additive Mixtures
121.101	Substances Generally Recognized as Safe (GRAS)
.1059	Chewing Gum Base
.1164	Synthetic Flavoring Agents
.2005	Substances for Which Prior Sanctions have been Granted
.2506	Industrial Starch — Modified
.2507	Cellophane
.2511	Plasticizers in Polymeric Substances
.2513	Polyethylene Glycol (mean MW 200-9500)
.2514	Resinous and Polymeric Coatings
.2519	Defoaming Agents Used in the Manufacture of Paper and Paperboard
.2520	Adhesives
.2526	Components of Paper and Paperboard in Contact With Aqueous and Fatty Foods
.2530	Reinforced Wax
.2531	Surface Lubricants Used in the Manufacture of Metallic Articles
.2534	Animal Glue
.2535	Textiles and Textile Fibers
.2537	Anti-Offset Substances
.2548	Zinc-Silicon Dioxide Matrix Coatings
.2549	Melamine Formaldehyde Resins in Molded Articles
.2550	Closures With Sealing Gaskets for Food Containers
.2553	Lubricants with Incidental Food Contact
.2556	Preservatives for Wood
.2557	Defoaming Agents Used in Coatings
.2559	Xylene-Formaldehyde Resins Condensed With 4,4'-Isopropylidenediphenol-Epichlorochydrin Epoxy Resins
.2562	Rubber Articles for Repeated Use
.2567	Water-Insoluble Hydroxyethyl Cellulose Films

.2569	Resinous and Polymeric Coatings for Polyolefin Films
.2570	Ethylene-Vinyl Acetate Copolymers
.2571	Components of Paper and Paperboard in Contact with Dry Food
.2572	Polysulfide Polymer-Poly-epoxy Resins
.2576	Cross-Linked Polyester Resins
.2577	Pressure-Sensitive Adhesives
.2591	Semirigid and Rigid and Modified Acrylic Plastics
.2592	Rosins and Rosin Derivatives

Phosphates

Recently out of favour in general ink applications (despite good gloss level achievements) due to their migration properties and possible toxic hazards.

Phthalates

Liquid

Di methyl phthalate (DMP) $C_6H_4(COOCH_3)_2$
Di butyl phthalate (DBP) $C_6H_4(COOC_4H_9)_2$.
Di isobutyl phthalate (DIBP) $C_6H_4[COOCH_2CH(CH_3)_2]_2$
Di iso octyl phthalate (DIOP) $C_6H_4(COOC_8H_{17})_2$
Di octyl phthalate (DOP) $C_6H_4(COOC_8H_{17})_2$

Solid

Di cyclo hexyl phthalate (DCHP $C_6H_4(COOCH(CH_2)_4CH_2)_2$

FDA status

DMP	2520	2576	[2591]					
DBP	2520	2576	2507	2514	2526	2562	2567	2571
DIBP	2520	—	2507	—	—	—	2567	—
DIOP	2520	[2005]	[2548]	2514	2526	[2550]	[2559]	2571
DOP	2520	[2005]	[2548]	2514	2526	[2550]	[2559]	2571
DCHP	2520	[2511]	2507		2526		2567	2571

Boiling point

DMP	282 °C	DIOP	235 °C
DBP	340 °C	DOP	232 °C
DIBP	327 °C	DCHP	212 °C

Freezing or melting point

DMP	− 26 °C	DIOP	− 50 °C
DBP	− 45 °C	DOP	− 50 °C
DIBP	− 45 °C	DCHP	+ 62 °C

Flash point

DMP	149 °C	DIOP	215 °C
DBP	171 °C	DOP	210 °C
DIBP	174 °C	DCHP	215 °C

The phthalates are the largest group of plasticiser, there being 15 major phthalates used in the industry. Ease of handling and the low cost of the liquid grades account for their large usage. They are compatible with nearly every resin system; DCHP for instance, can be used with any resin:

Dimethyl phthalate (DMP)   All, but limited in chlorinated rubber and vinyl chloride

Dibutyl phthalate (DBP)   All, but limited in CAP and cellulose acetate

Dibobutyl phthalate (DIBP)   All, but limited in CAP and cellulose acetate

Diisooctyl phthalate (DIOP)   All, but cellulose acetate — limited with CAP, PVB, chlorinated rubber and acrylic

Dioctyl phthalate (DOP)   All, but cellulose acetate — limited with CAP, PVB, chlorinated rubber and acrylic

Dicyclohexyl phthalate (DCHP)   All

Polyesters

A range of plasticisers, varying widely in chemical composition which can be of use in specialised formulations.

Polyol ethers

Glyceryl triacetate (Triacetin $CH_2CHCH_2(COOCH_3)_3$) is the most well known in this class and is the most acceptable. Generally regarded as safe.

It is the only member of this group with some solubility in water, a *rare* property in plasticisers.

Compatible with all forms of cellulose resins CAP, CAB, NC, EC, CA, acrylics and poly vinyl acetate. Boiling point 258 °C; freezing point −50 °C.

Ricinoleates

Butyl ricinoleate

$C_4H_9OOCC_{10}H_{18}CHOHC_6H_{13}$ sold as Felxricin P3). Boiling point 220 °C; freezing point −23 °C; flash point 207 °C.

Insoluble in and immiscible with water, FDA approved 2519, 2520 and 2535. Compatible with and suitable with CAB, NC, EC, poly vinyl acetate, PVB and vinyl chloride acetate.

Glyceryl tri ricinoleates (castor oils)

$Ch_2CHCH_2[OOCC_{10}H_{18}CHOHC_6H_{13}]_3$ (sold as castor oil std. and crystal.). Melting point −15 °C. No boiling point; flash point 290 °C.

Insoluble in water used in liquid and paste inks. FDA status 2514, 2526, 2531, 2535, 2548, 2550, 2553, 2557, 2559, 2562, 2571.

Colour usage restricted to suitable inks.

Stearates

For example, butyl stearate, $CH_3(CH_2)_{16}COO(CH_2)_3CH_3$. Melting point 19 °C; boiling point 350 °C; flash point 188 °C.

Usage gradually dropping in favour of more easily handlable materials.

One of the oldest plasticisers known in the trade, compatible with CAB, NC, EC, chlorinated robber and polystyrene. Now less used because of properties and compatibilities.

Section VII: Waxes

Waxes of many chemical types are incorporated in most letterpress, lithographic, screen, gravure and flexographic inks principally to impart mar-resistance, improved slip and water repellancy properties. They are the main constituent of carbon copying inks and other 'hot melt' inks, where printing is done at an elevated temperature and drying is by solidification.

Small additions of wax or wax paste are known to reduce tack without appreciably increasing the flow of letterpress and lithographic inks. Waxes are normally introduced into inks in one of two ways. Wax of a controlled fine particle size can be mixed or ground into the batch along with pigments or may be introduced during the final blending operations. A second way of incorporating them into formulations is to compound them into 'wax media' by dispersing or in some cases melting them into varnishes and/or solvents and adding these to the ink.

It is generally accepted that the non-rub qualities imparted by an individual wax are a function of both the particle size and the hardness as well as the melting temperature of any particular wax. It is also generally accepted that wax particles that are fine enough to show no adverse manufacturing or printing problems, but which stand just proud of the dried printed film tend to give the maximum scuff resistance. However, with the heat and movement imparted by the friction of constant rubbing under pressure, particles of the ink film can ball up and mark unprinted areas. It is therefore essential that the use of waxes be kept in perspective. With a hard drying flexible film the quantity of wax needed to obtain excellent scuff and mar-resistance is minimal. With a poorly formulated vehicle system, more waxes are added to improve rub resistance, which in themselves will have an adverse effect on ink flow and transference, gloss, hardness and resistance characteristics.

Addition of waxes to inks almost invariably decreases their gloss and increases their drying time in both oil and liquid inks, hence the ink must be formulated in the knowledge that there is an optimum amount to give the best compromise between non-rub properties and any detrimental effect such as decreased gloss. Proofing and checking on a rub tester (see p.743) under controlled conditions is necessary with each change of ink formulation or substrate.

Various greases, many of which may be regarded as plasticising wax, are used in letterpress and lithographic inks for control of rheological properties. The two most widely used, petroleum jelly and lanolin, are included in this section.

Certain wax compounds and solutions added to liquid inks, which are printed on non-porous substrates tend to migrate to the surface of the film. This helps to prevent blocking as well as giving the printed film some slip and scratch resistance. It is possible to produce ink films from which Scotch tape can be peeled off easily, thus masking the anchorage measurement at the substrate/ink interface. This can be due to the wax floating to the top surface or to exudation of other mobile components.

Synthetic waxes, defined as those chemically produced, are now the most popular ones used in the ink industry; these are therefore given prime place, whereas natural waxes are those produced by plants or animals and include the fossil and petroleum waxes, all of which need some physical or chemical refining before they are suitable for use. The use of natural waxes has declined as they have been largely replaced by polyethylene waxes.

Since many proprietary waxes are blends of both natural and synthetic products, it is difficult to accurately describe their constituents. Some pure waxes can be identified by the type of fluorescence they give when exposed to black light from a UV lamp. This test should be carried out in both solid and liquid state, also upon a solution of the wax.

4.27   SYNTHETIC WAXES

Polyethylene waxes

These are waxy hard white solids, normally supplied in powder or pellet form. They are odourless, tasteless, and can be used in inks or coatings that have direct contact with foodstuffs. They are chemically inert, and contribute towards the chemical resistance properties of an ink film. They are soluble in a large number of common solvents and oil/resin blends at elevated temperatures, but are precipitated from solution on cooling to room temperature, when the polyethylene wax assumes the form of a very fine dispersion.

Polyethylene waxes are incorporated in all types of printing inks. They are usually added in the form of 'non-rub' or 'slip' media which are fine dispersions of the wax in the solvents, oils and resins etc., of that particular type of ink formulation in which it is to be incorporated. For example, for use in a letterpress or lithographic ink the wax is melted in solvent, or in a mixture of solvent and resin, or in linseed stand oil. The dispersion is obtained by cooling the mixture in specialised equipment. Today the printing ink formulator has a choice of many polyethylene waxes and it does not automatically follow that a wax which can be formulated into a medium imparting excellent rubproofness to a letterpress or lithographic ink can be formulated into a first class gravure or flexographic ink wax medium. Hence different waxes are used in differing types of ink, in order to achieve maximum rubproofness. Generally speaking the most effective type of polyethylene wax for slip properties in an ink film is the high density, high melting point isotactic polymer. This and some of the softer waxes are now available in fine powder form with a particle size of between 2 and 4 μm. It is possible to incorporate these waxes both at the pre-grinding stage, or by mechanical dispersion into finished inks. These waxes are very abrasion resistant if an optimum quantity is used, between 0.25 and 2.0%.

Emulsifiable grades of polyethylene wax are also commercially available. These usually are waxes in which some carboxyl groupings have been introduced by partial oxidation of the hydrocarbon. An emulsion is produced by forming salts with basic compounds when the aqueous mixture is stirred vigorously at temperatures above the melting point of the wax. More recently the basic materials have been already incorporated in the commercial product, when the material is truly self-emulsified.

Polyethylene is produced by the polymerisation of ethylene $H_2C = CH_2$ to form a chain structure of $(-CH_2 - CH_2 -)_n$. Many commercial grades are available with molecular weights between 1500 and 4000.

The lower the mean molecular weight, all factors being equal, the lower the melting point of the wax. They vary from the high density high melting point isotactic polymer, which is produced by the low pressure Ziegler process in which an aluminium trialkyl and a transition metal chloride is the catalyst combination, to the low density low melting point atactic polymer, which is produced by the high pressure (up to 1000 atmospheres) free-radical process using oxygen as the source of the propagating free radicals. High density polyethylenes have a relative density of approximately 0.94–0.96, at 24°C. Low density ones are approximately 0.90 at 24°C.

Chain branching, which is present in varying amounts in all commercial polyethylene waxes produced by the high pressure process, as demonstrated below, has the effect of lowering the melting point of the wax.

Polymer with chain branching:

$$-CH - CH_2 - CH_2 - CH_2CH_2 -$$
$$|$$
$$CH_3$$

Polymer without chain branching:

$$-CH_2 - CH_2 - CH_2 - CH_2 -$$

Polytetrafluoroethylene

Polytetrafluoroethylene — PTFE — is produced under a variety of trade names, i.e. Teflon (Du Pont), fluon (ICI), Hostaflon (Hoechst), etc. It exhibits useful properties over the widest temperature range of any known polymer. It is uniquely non-adhesive and antifrictional and will withstand a temperature of 320°C. As yet there is no known solvent for this material. The above properties are all ideal in theory for an anti-set-off powder for an ink which must possess a high degree of rub resistance. The coefficient of friction of solid polytetrafluoroethylene decreases with increasing load.

The relative density of such powders is 2.1 and the mean particle size 4 or 5 µm. Suitable for all types of printing inks, but especially ideal for heat set inks, where the temperature of the drying apparatus does not permit polyethylene waxes to be used effectively. It can also be stirred into finished inks to improve their rub and scuff resistance. The relative cost of PTFE is prohibitively high for many applications. It is formed by the pyrolysis of chlorodifluoromethane at high temperature to yield tetrafluoroethylene.

$$2CHClF_2 \rightarrow CF_2 = CF_2 + 2HCl$$

This is subsequently polymerised at high pressure.

Halogenated hydrocarbon waxes

Many different products are obtained by the halogenation of hydrocarbon polymers, mainly with chlorine.

Paraffin may be chlorinated by replacing hydrogen atoms in the teminal CH_3 groups with chlorine atoms and yields a viscous liquid at 40% chlorine, a soft solid at 60% and a hard solid wax-like material at 70%, when it has a melting point of 90°C (194°F) and a density of 1.64. It is an inert material with a low dielectric constant and is soluble in hydrocarbons, ketones, esters, and drying oils. It is compatible with ethyl cellulose, nitrocellulose, alkyds, urea formaldehyde, vinyl and polystyrene resins.

They find limited used as plasticisers in resins, as moisture proofing agents and fire retardant additives.

Other chlorinated olefins and polyolefins are manufactured, and they also range from soft wax-like products to brittle resinous solids. Some are blended, some are marketed as solutions in aromatic solvents. They give excellent adhesion in liquid inks applied to plastics films and foil and are used in specialised heat seal formulations.

They are produced by passing chlorine through the molten material or through solutions or suspensions in carbon tetrachloride. No simple chemical formulation can be given as the product depends on the type of feedstock and the degree of chlorination. Like chlorinated solvents and many other materials, they can form thermal decomposition products which could prove dangerous. The relevant safety data sheets should be consulted before use.

Fatty acid amides

A fatty acid amide is described as an organic acid derivative in which the OH group is replaced by an amine group NH_2 to form a long chain amide which is a waxy material. The three fatty acid amides which are used in printing ink formulations are:

$CH_3(CH_2)_7CH = CH(CH_2)_7CONH_2$ Oleamide
$CH_3(CH_2)_{16}CONH_2$ Stearamide
$CH_3(CH_2)_7CH = CH(CH_2)CONH_2$ Erucamide

Fatty acid amides are wax-like to the touch, vary in colour from pale yellow to dark brown, have melting points between 68 and 109°C, and are soluble to varying degrees in most solvents. They also blend well with waxes, but odour can be a problem if the refined versions are not used.

Fatty acid amides are used as rub resistant and slip improving additives in flexographic and gravure inks and over-lacquers. They minimise set-off and blocking, and improve the adhesion of subsequent colours. Stearamides and oleamides are used as solubilisers for oil-soluble dyestuffs. In paste inks a small amount of oleamide will improve pigment wetting. In letterpress inks stearamide will reduce stringiness without lowering viscosity and will help reduce misting problems. Often less than 1% is used.

Stearamide increases the polarity of non-polar oils and helps stabilise oil-in-water emulsions.

Fatty acid amides are chemically neutral compounds and are prepared by passing ammonia through molten fatty acid in a special reactor:

$$R.COOH + NH_3 = R.CONH_2 + H_2O$$

where R is an alkyl or alkenyl group in which the length of the carbon chain varies from C_{12} to C_{18}.

4.28   PETROLEUM WAXES

Petroleum waxes are extracted from crude petroleum. There is a variety of crude feedstocks from different world sources, which may have up to 10% wax content. Petroleum waxes are obtained by distillation of oils and subsequent refining. They are roughly classified by type dependent on the oil content and treatment.

In printing inks the petroleum waxes are used to reduce the tack of letterpress and lithographic inks, as slip agents, especially the micro-crystalline waxes, and as components of carbon-copying inks.

Petroleum jelly is used as a medium to reduce tack and to improve setting or increase penetration into paper or board. It is also used to retard the speed of drying of letterpress or litho inks on the press-rollers.

The petroleum waxes are composed almost entirely of complex mixtures of paraffinic hydrocarbons. Paraffin wax consists of paraffins between $C_{18}H_{38}$ and $C_{32}H_{66}$. Microcrystalline wax consists of paraffins between $C_{34}H_{70}$ and $C_{43}H_{88}$. Petrolatum is a mixture of paraffinic and olefinic hydrocarbons in which the carbon chain length of the molecules varies between C_{16} and C_{32}.

Produced by distillation of petroleum at a temperature of about 300°C a wax oil mixture distils over. On chilling, the wax crystallises from the wax-oil mixture and is removed by filter-pressing. The solid residue in the filter is known as slack wax and contains about 65% of paraffin wax, from which commercial paraffin wax is extracted by further refining. Petrolatum or petroleum jelly is extracted from a higher boiling distillate of crude petroleum. The refining of the various waxes is usually by treatment with solvents of various kinds, including liquid sulphur dioxide, nitrobenzene, furfural and methyl ethyl ketone. The waxes are further refined by decoloration with Fuller's earth, charcoals, activated carbons or clay, and are then recrystallised.

Slack wax

A soft paraffin wax with up to 35% oil content and melting points between 49 and 60°C (120–140°F).

Scale wax

An intermediate paraffin wax between the slack wax and the higher refined grades with large flake-like crystals. Colour ranging from white to

yellow depending on the amount of oil retained which maybe up to 6%.
Melting point 49−57°C (120−135°F).

Fully refined paraffin wax

Sometimes referred to as paraffines or block wax. These are hard waxes
with a coarse crystalline structure. Colour translucent white to slightly
yellow according to the degree of refining. Relative density (15°C) 0.916,
melting point from 45−74°C (118−155°F). Molecular weight 350−420.

At room temperature it is more soluble in aromatic hydrocarbons than
in aliphatic solvents. At elevated temperature paraffin wax is also soluble
in drying and non-drying oils, rosin, rosin esters, phenolics, maleic and
alkyd resins and will not crystallise out on cooling if less than 2% is
present. It is only partially soluble in alcohol.

Petrolatum or petroleum jelly

A mixture of liquid and solid hydrocarbons with a grease-like consistency.
It often exhibits a green or blue fluorescence. Relative density (15°C)
0.820−0.860. Melting point 45−74°C (113−165°F).

Microcrystalline waxes

High melting waxes with a very fine crystallite particle size in contrast to
the coarse crystallites of the paraffin waxes. They have an average molecular weight of 490−800 and they exhibit higher viscosities when molten.
The residual oil content varies from 0.3 to 13% and the colour from
yellow to white. Melting range lies between 66 and 93°C (150−200°F).
Relative density at 15°C is 0.930.

Ceresin wax

Originally ceresin was an alternative name for ozokerite, but today the
name is applied to blended hydrocarbon waxes of variable composition
and melting point. The mixture may consist of petroleum waxes only or
petroleum and mineral waxes with small amounts of carnauba wax to
raise the melting point. Ceresins are used for similar purposes to paraffin
waxes.

Montan wax

Extracted from bituminous lignite or shale, montan wax has a colour
ranging from very dark brown to yellow, which usually relates to the
amount of de-resination. It also has rather an oily odour. Crude montan
wax has melting point 80−83°C (176−181°F); acid value 23−28;
saponification value 78−92, whereas bleached montan wax has a melting
point of 72−77°C (158−170°F), acid value 93−107 and saponification
value 95−143. It is soluble in aromatic and chlorinated hydrocarbons
when cold, but when hot it is soluble in aliphatic hydrocarbons. It contains a large amount of organic acids and can easily be emulsified. It also
blends well with other waxes.

Besides finding some use in carbon paper inks and rubber manufacture it is used in the manufacture of synthetic wax esters of high molecular weight. Montan wax is obtained from brown coal by a solvent extraction process. The crude extract usually contains 70% montan wax, the remaining components being resin and bitumen. The montan wax is then purified by a further solvent extraction process. An ester is formed when the de-resinated montan wax is subjected to pyrolysis.

Montan esters

The almost pure montan wax, essentially the ester montanyl montanate $C_{27}H_{55}COOC_{28}H_{57}$, is first hydrolysed, and the alcohol is oxidised by a chromic sulphuric acid mixture to montanic acid, $C_{27}H_{55}COOH$. This acid is the raw material for most of the synthetic ester waxes, which have a higher melting point than that of montanic acid.

The most common glycols used in the esterifying reaction are ethylene and butylene glycols, when the molecular weight of the wax is greatly increased. These glycols are used in various combinations often in the presence of small amounts of saponified wax and fatty acids to give the desired properties of commercial wax blends (including the IG and Gersthofen waxes).

The ester waxes, partially saponified ester waxes, and special blends of wax, fatty acids, and other additives vary in colour from almost black, through a continuous colour range to pale yellow. Their melting points vary from 70 to 104°C (158–219°F) and their acid and saponification values from 6 to 85 and 80 t0 170 respectively.

Synthetic waxes based on montan wax are used to improve the rheological properties and to control the rub resistance and water repellency of lithographic and letterpress inks.

4.29   NATURAL WAXES

Generally natural waxes are complex mixtures of fatty acids, esters, alcohols and hydrocarbons. The acids consist mainly of mixtures of saturated fatty acids, general formula C_nH_{2n+1}, ethylenic acids, C_nH_{2n-1} COOH, and many dibasic acids, etc. where n is any even number between 12 and 32. The esters are formed by reaction of the wax acids with higher alcohols from C_{12} to C_{34}. The hydrocarbons are saturated paraffins containing between 19 and 31 carbon atoms. Unsaturated hydrocarbons are rarely found in natural waxes.

Beeswax

Beeswax is a yellowish white to a yellowish brown in colour, the yellow being due to pollens carried by the bee. It has a melting point of 62–64°C (143–147°F) with a solidifying point of 61°C (142°F), acid value of 19, iodine value of 10, saponification value of 96–100, and a relative density of 0.96 at 15°C. At room temperature it is a waxy solid with a tacky surface. It is soluble in ether, chloroform, carbon tetrachloride and vegetable oils. It is insoluble in water, only partially soluble in ethyl alcohol, and insoluble in mineral oil at room temperatures. When heated it blends well with paraffin waxes.

Beeswax is sometimes used in letterpress process inks, especially in the first colour down, where it helps the trapping of superimposed colour. It is also used in mixed oil/wax compounds for reducing set-off or reducing hard drying of a film.

Beeswax is obtained from the honeycomb of the honey bee *Apis*. The wax is secreted by the worker bee and consists of (approximately): wax acid esters, 71%; cholesteryl esters, 1%; alcohols, 1%; wax acids, 14%; hydrocarbons, 11%; moisture and impurities, 2%. The molecular weight of pure beeswax is approximately 570.

Carnauba wax

Hard and brittle waxes that range from straw yellow to brownish grey depending on the grade. Average constants for carnauba wax are acid values 2.5−5.0; iodine value 10−13; saponification value 79−85; softening point 82−84°C (180−183°F), relative density 0.999.

Carnauba is used extensively to elevate the melting point of other waxes, especially paraffin wax. Its solubility at room temperature is low, but increases rapidly at high temperatures. The wax tends to crystallise or precipitate out of solution on prolonged heating, retaining the solvent and retarding its evaporation. In inks it is regarded as a wax which does not float to the surface after printing; as such it has been used in overprinting varnishes, gloss inks and gold media. When dispersed in drying oil in conjunction with other waxes it imparts improved slip to the dried ink film, reduces the tack and improves the setting. Small quantities are also used in carbonising inks.

Brazil is the main country of origin of carnauba wax. It is obtained from the leaves of the South American palm (*Copernicia cerifera*). Each leaf produces about 5 g of wax which is scraped from the leaves as a powder, melted to remove coarse impurities and cast into blocks.

Carnauba wax consists mainly of esters of long-chain alcohols (C_{24} to C_{34}) with free fatty acids (C_{15} to C_{50}), and small amounts of resins, hydrocarbons, lactides, etc. An approximate composition is: wax acid esters 84%; free fatty acids, 3%; hydrocarbons, 2%; other materials, 10%; moisture 1%.

Miscellaneous natural waxes

Of the various other natural waxes which have occasionally been used in printing inks, a large number have been superseded by the synthetic waxes. They were used for solvent and oil retention as well as surface slip. Although their use has declined in recent years, more interest is now being shown for the future of some natural products. Some waxes from this group are described under the headings below.

Spermaceti wax
Soft white wax with good slip characteristics, soluble in hot toluene, melting point 43−47°C (110−117°F). Obtained from the sperm whale.

Shellac wax
Off-white with a tinge of pink, melting point 72–74°C (162–166°F). Obtained from the secretion of the lac insect (*Carteria lacca*) in India by dewaxing shellac resin.

Ouricury wax
Hard brittle yellow wax with melting point 83–85°C (181–187°F). Was used as a substitute for carnauba wax. It is obtained from the leaves of the palm *Attalea excelsa* in the Amazon region. Almost entirely used domestically and exported only on request.

Japan wax
Hard pale ivory coloured wax melting point 45–51°C (113–124°F). Used in letterpress and letterset (dry offset) inks to prevent 'crystallisation' and to aid superimposition. Classed as a vegetable tallow extracted from the seeds of a small tree (*Toxicodendron*) cultivated in the orient.

Candelilla wax
Brownish hard wax, easily powdered. Melting point 68–70°C (154–158°F) with good slip properties. Obtained from the stems of *Pedilantus pavonis* and other related Euphorbia, which are weeds found in North Mexico.

Lanolin
Anhydrous lanolin is wool grease that varies from almost white to dark brownish yellow. It is a semi-solid plastic sweet smelling mass with a melting point of 36–42°C (97–108°F), acid value 5–15, saponification value 82–120, iodine value 15–36 and relative density 0.940.

Widely used in the ink industry as a tack reducer in letterpress inks and varnishes. It also helps setting and stops crystallisation in wax compounds. A small amount is incorporated in cold carbonising inks. Ink with pigments of high oil absorption benefit from the use of lanolin for wetting properties and for clean working. It also helps lithographic inks to take up sufficient water to improve their printing properties.

Hydrous lanolin can be produced by kneading anhydrous lanolin with water, until a soft white ointment-like material is obtained containing 25–30% of water.

Lanolin is extracted from the water-insoluble portion of wool grease produced in sheep's wool. Its extraction is an important by-product industry of the wool-producing countries.

The grease is extracted by washing or scouring the wool first with alkali and then with soap solution. The grease is centrifuged to remove dirt or left for the dirt to settle. Finally it is treated with hot water and sulphuric acid to separate the grease as anhydrous lanolin.

Wool grease differs from other natural waxes in containing a high proportion of esters of sterols together with the usual long chain alcohol esters. An average composition is: cholesteryl and isocholesteryl estolidic esters, 33%; free sterols 1%; fatty alcohol acid esters, 48%; hydrocarbons, 2%; free alcohols, 5%; free fatty acids, 3%; lactones, 6%; other materials, 2%.

Section VIII: Driers

Catalysts used to promote oxidation of the drying oils used in letterpress and offset lithographic inks. In the presence of these, oxidation proceeds rapidly and the ink films dry hard in a few hours.

The most widely used catalysts are the inorganic salts and metallic soaps of organic acids.

Recently more complex organic derivatives of metals have been introduced to boost the effect of conventional driers.

Metallic driers are used in two forms:
(1) As oil-soluble soaps, which comprise the liquid driers.
(2) As dispersions of inorganic salts in oils, which comprise the paste driers.

4.30   LIQUID DRIERS

Liquid driers are prepared by the conversion of suitable organic acids to their heavy metal salts and soaps. These are soluble in oils and/or petroleum solvents to form liquids or soft pastes which readily mix with oil in inks.

The metals are almost invariably available as salts of complex fatty acids, either singly or as mixtures. The fatty acids include:
Octoic fatty acids to form octoates;
Rosin fatty acids to form resinates;
Naphthenic fatty acids to form naphthenates;
Tall oil fatty acids to form tallates;
Linseed fatty acids to form linoleates.

The chemical constituents of the organic acid component can influence dry print odour as well as the effectiveness of the drier in an ink formulation at a given percentage of metal. The metal contents range between 3 and 18% except for lead and zirconium which go up to 24%, or even 36%. An addition of between 0.5−4% of driers is the amount normally necessary to achieve adequate drying in printing ink films. Factors which affect the efficiency of drying of such films are the surface characteristics and the pH of the paper, stacking procedures and atmospheric conditions. It is a well-known fact that inks dry more rapidly at higher temperatures whereas a high relative humidity slows down the oxidative drying rate, especially when the pH is low.

Cobalt

The most powerful drier, also the most popular (see Fig. 4.1). Violet in colour until oxidised when it becomes brown. It tends to discolour tints and whites. It is readily soluble in organic acids, so may be affected by the fountain solution on lithographic presses, by leaching out from the ink into the fountain. Conversely cobalt acetate can be dissolved and introduced into the fount as a drying activator.

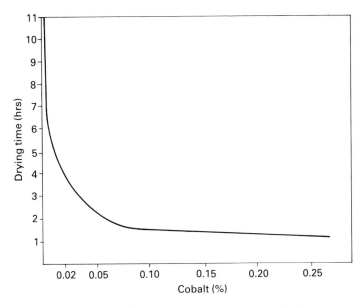

Fig. 4.1 Graph showing the relationship between cobalt content and drying time

Manganese

Less vigorous catalytic action than cobalt. Brown in colour, affects whites and tints less than cobalt and is unlikely to leach out in the fountain solution. By itself only suitable for tin printing inks, otherwise it needs heat to behave as an activator. Often used in conjunction with other metal soaps.

Lead

A slow acting drier of pale colour. Use limited by the common specification for freedom from lead. It is frequently used in conjunction with manganese soaps.

Cerium, zirconium and lithium

Driers of medium efficiency and pale colour. They have replaced lead in many vehicles and are usually used mixed or in conjunction with cobalt or manganese soaps.

Calcium and zinc

Ineffective as catalysts, having only one valency state, and are not used much in printing inks except in some whites. Calcium is used in low odour systems to modify the catalytic action of transition metal driers.

Iron
Useful in tung oil varnishes

4.31 PASTE DRIERS

These are prepared by grinding organic salts of lead and manganese in linseed oil varnishes. Lead acetate and manganese borate are commonly used. It is also possible to prepare pastes by flushing, but most formulae are confidential. The percentages vary but are of the order of 40% lead acetate, 8% manganese borate. Small amounts of soluble manganese driers may also be added, but cobalt is seldom used in paste driers.

The slow-drying characteristics of paste driers give a receptive ink surface for superimposition, therefore they are usually used in multi-colour process inks. Cobalt driers are inclined to yield dried prints which are not receptive to overprinting. Paste driers, although they dry inks slowly, do dry uniformly throughout the film.

Inorganic peroxides have been successfully used as components in paste driers, but great care has to be taken in respect of *fire* during their production. These are very useful in lithographic ink, where the acidity of the fount solution, combined with the pH of the vehicle will liberate oxygen which is absorbed along the chain of the polymer.

To obtain the optimum drying effect in oil based inks, it should be noted that some pigments such as the PMTA salts require more driers than normal. Others like bronze blue of phthalocyanine require less, because they accelerate oxidative polymerisation. The use of lead in driers is being phased out; lead is being replaced by other metals such as zirconium.

Section IX: Miscellaneous additives

4.32 CHELATING AGENTS

These are chemicals which react with multivalent metal ions, to form a chelate. This is a type of atomic cocoon produced by the formation of a closed ring of atoms which are attached to a central multi-valent ion, usually metallic. Normally this phenomenon is due to sharing with the central ion a lone pair of electrons from oxygen or nitrogen atoms in the chelating compound.

Chelating agents are used for locking up unwanted metal ions, to form extremely stable insoluble or soluble chelates. By chemically binding metal ions in additives, or trace metals in compounds, it is possible to improve shelf life, colour, clarity and stability. Chelating agents used in water-based systems are able to improve wetting through softening the water by locking up and rendering inactive the iron, calcium and magnesium salts.

Also by the use of chelates, it is possible to control the release of metal ions over a prolonged period or during a chemical reaction. Chelates have also proved useful in suppressing a reaction until an elevated temperature is reached. They are also used in the ink industry for cross-linking hydroxyl containing polymers.

Some important chelating agents are ethylene diamine tetra acetic acid (EDTA) and its sodium salts, nitrilotriacetic acid salts, sodium salts of diethylene triamine acetic acid, heptonates, di-methyl glyoxime and its sodium salts, as well as the alkanolamines.

A typical example is the chelation of isopropyl titanate by triethanolamine to form triethalolamine titanate, liberating isopropyl alcohol, a solution in which they are then marketed. The titanium is bonded and when liberated reacts to produce a thixotropic structure in emulsion systems.

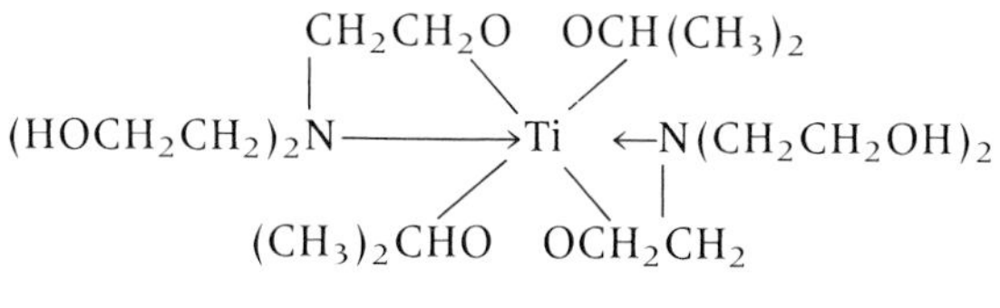

$$(HOCH_2CH_2)_2N \longrightarrow Ti \overset{\overset{\displaystyle CH_2CH_2O \quad OCH(CH_3)_2}{\diagdown \quad \diagup}}{\underset{\underset{\displaystyle (CH_3)_2CHO \quad OCH_2CH_2}{\diagup \quad \diagdown}}{}} \leftarrow N(CH_2CH_2OH)_2$$

4.33 ANTIOXIDANTS

Antioxidant molecules react with free radicals formed during the autoxidation process and continue to act until all the antioxidant molecules are exhausted. Consequently, any antioxidant incorporated into an ink will delay the initiation of oxidative polymerisation drying.

The most common antioxidants are the naphthols, substituted phenols, a variety of oximes and the aromatic amines. Quantities necessary are only a fraction of 1% of the final formulation.

Press stability of lithographic tin printing inks is usually improved by the incorporation of a small quantity of antioxidant which is lost during the initial part of the stoving schedule. The oxidation process then proceeds normally.

Numerous antioxidants are marketed for industry and many of them bear manufacturers' trade names. Some of the most effective are:

Eugenol (extract of oil of cloves)
Hydroquinone
Pyrocatechol
Guaiacol (methyl catechol)
Butylated hydroxytoluene
Butylated hydroxyanisole
Methyl ethyl ketoxime
Butylaldoxime
Cyclohexanone oxime
Typical symbolic formulae are:

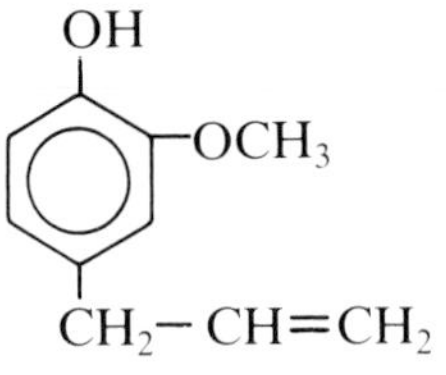

OH

$-OCH_3$

$CH_2-CH=CH_2$

Eugenol

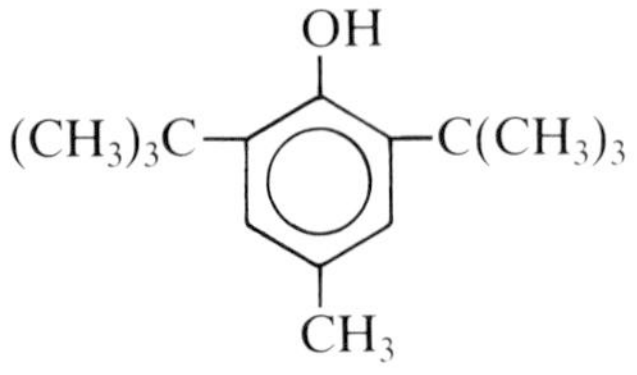

OH

$(CH_3)_3C-$ $-C(CH_3)_3$

CH_3

Butylated hydroxytoluene

4.34 SURFACTANTS

The name surfactant is given to substances which have the property of being adsorbed at surfaces or interfaces, thereby reducing the surface or interfacial tension. Soaps and detergents act this way in aqueous systems.

This action can be measured and demonstrated by taking the contact angle (xyz) of a drop of pure water on a non-absorbent surface (a). The addition of a very minute quantity of surfactant decreases the angle of contact considerably at the same time increasing the interfacial area (b) (Fig. 4.2).

The same concept of wetting and spreading applies both to oleo resinous and to aqueous systems. A contact angle less than 90° enables the vehicle to wet and disperse pigments. Wetting greatly influences the physical and printing characteristics of an ink. Numerous surfactants are available, many of them with trade names. The choice is large so it is left to the ink formulator to select the best surfactant, and to include in his formulation the optimum amount to obtain the highest tinctorial strength from pigments. Incorrect use of surfactants can result in decreased adhesion to some substrates, and can lower the water resistance of a print. Surfactants can promote emulsions; therefore care must be taken that they do not also cause emulsification of lithographic inks. Selection must be made with knowledge of the properties of the surfactant, the formulation objectives and the precise amount needed to attain those objectives efficiently. This approach discourages indiscriminate use which can be harmful and costly.

Surfactants normally contain polar and non-polar groupings in the same molecule. It is the ionic or active portion of the molecule which divides surfactants into anionic (which have an active negative charge) cationic (where the positive charge predominates) and non-ionic classes. Amphoteric surfactants have both positive and negative centres of charge (Table 4.17).

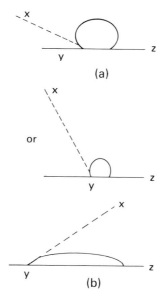

Fig. 4.2 The effects of adding surfactants

Anionic surfactants

Active constituent — anions (negative charge). These attract the positive charge of other large molecules, leaving a negative charge at the surface, causing excellent wetting in aqueous systems. Anionic surfactants include common soap and are usually alkali metal, ammonium or substituted ammonium salts of long chain fatty acids. The non-polar portion is the fatty acid hydrocarbon chain, while the carboxylate anion and its associated cation are the polar portion. Thus when a polar pigment is dispersed in a non-polar vehicle, the polar portion becomes adsorbed on the surface of the pigment as a mono-molecular layer, giving the pigment a negative charge, and leaving the non-polar part pointing outwards into the vehicle, to which it is attracted. In this way interfacial tension is lowered, so better wetting and hence dispersion of the pigment is achieved. An electrical double layer around the pigment particle also serves to reduce flocculation tendencies.

Cationic surfactants

The reverse of the above, i.e. positive charge attracting negative charge. Ideal for dispersing a pigment in a vehicle, the pigment assumes a positive charge and disperses well in non-aqueous systems. Cationic surfactants include quaternary fatty ammonium halides, acetates or sulphates.

Non-ionic surfactants

These have no residual electrical charge. Molecules in which part is lyophilic (solvent-liking) and part is lyophobic can behave as surfactants although they are non-ionic. To emulsify fats in water, a lyophilic (hydrophilic) polar polyethylene oxide chain is joined with a hydro-carbon chain which associates with the fat.

Amphoteric surfactants

As these contain central and balanced positive and negative charges, they exhibit behaviour of either anionic or cationic surfactants according to prevailing conditions.

The concept of hydrophile–lippophile balance (HLB) is of practical importance and HLB values may be assigned by a simple test.

Chemical formulae of some well established surfactants are given below in the table.

For a more detailed description of surfactants the reader is advised to consult a standard text book on the subject. *The Encyclopaedia of Surface Active Agents* by Sisley and Wood [4] is particularly recommended.

A catalogue of surface active agents in the United Kingdom listing chemical and brand names, data and applications has recently been published [5].

Table 4.17 Surfactants

Anionic	$R.\overline{COO}\ M^+$	*Fatty acid salt* R = Any fatty alkyl or alkenyl group M = NH_4, Na, K, etc.
Cationic	$R \overset{+}{\longrightarrow} N(CH_2)_3 X^-$	*Quaternary ammonium salt* R = Fatty alkyl radical X = Halogen, HSO_4, CH_3OO
Non-ionic	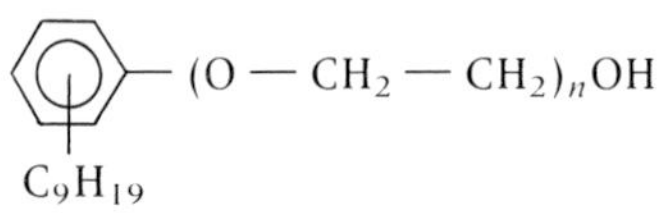 $\text{(benzene ring)} C_9H_{19} - (O - CH_2 - CH_2)_n OH$	*Polyethoxylated nonylphenol* n = 5 to 12
Amphoteric	$R \overset{CH_3}{\underset{CH_3}{\overset{+}{-N}}} - CH_2COO$	*Alkyldimethylbetaine*

4.35   DEODORANTS AND REODORANTS

Many of the materials used in the manufacture of printing inks have a strong odour. Some can be extremely unpleasant, causing nausea. Highly volatile solvents are probably the worst offenders for producing un-wanted odour, but even high-boiling printing ink distillate odours can be very disagreeable on a printing machine which has become warm. Other materials have odours which may not be strong, but which are persistent, for example some of the styrene and acrylic monomers. The volatile by-products of oxidation drying or of prolonged storage, also give rise to unwanted odour if trapped within printed matter. It is required to mask these unwanted odours, reduce their intensity and convey a clean, healthy and pleasant fragrance without altering the efficient functioning of the ink or having adverse effects on the printed products.

Reducers and proprietary sundries such as blanket washes, sometimes contain specially added re-odorant compounds. These are intended to make the use of the materials pleasant, to enhance sales appeal, and to disguise the ingredients, thereby adding security to the formulations.

A large range of natural and synthetic chemicals have, for many years, been used in other industries, notably food and cosmetics to mask, alter or intensify the odours of the product. Among these are included amyl and methyl salicylate, vanillin and blended derivatives of essential oils.

It should be noted, however, that few deodorants or reodorants are as persistent as the odour to be masked, therefore, on ageing, it is possible for a formulation to revert to its original aroma.

Some manufacturers of products increase their sales by inclusion of a reodorant in the ink on the package in such a way that it conveys to the purchaser a desire to buy. Compounds of lavender, leather, peppermint, carnation, antiseptic, cedarwood, citron, vanillin, etc, are available.

Microencapsulated perfumes are sometimes used as paper coatings. When scratched these pervade the surrounding area with a fairly concentrated fragrance. They have been incorporated into inks and overprint varnishes but further development work is necessary, especially on application, so that the odours are not released prematurely.

4.36   PURE CHEMICALS

The main classes of pure chemicals used in printing ink formulations today are:
(1) alkalis,
(2) organic acids and acid anhydrides, and
(3) polyols.
The most common members of each class are described below together with some relevant physical properties and their applications.

Alkalis

The major use of alkalis in printing inks is in the formulation of water-based or water-reducible letterpress, gravure, and flexographic inks. They act simply by converting the highly acidic and fairly insoluble resins such as shellac, fumaric, and maleic resins to salts which are very much more hydrophilic and soluble than the parent resin. Excellent varnishes can thus be prepared from these resin salts using water or one of the lower more hydrophilic glycols, especially ethylene glycol, as the principal solvents.

The quantity and type of alkali used has to be calculated so that the resulting dried ink film does not remain too water soluble. The alkali used normally evaporates or penetrates into the stock leaving the original insoluble resin system on the surface as film former.

It is also possible to use alkalis to alter the complete physical structure of a vehicle. Polymers and co-polymers containing acrylic and methacrylic acids are produced in high-solids emulsion form at a very low viscosity. By raising the pH and neutralising, these resins become water soluble; at the same time the particles become ionised, and form a

Table 4.18

Name of alkali	Chemical formula	Boiling point (°C) at 760 mm Hg	Miscellaneous comments
Caustic soda	NaOH	Not applicable	Rarely used since the inks tend to remain permanently water soluble as alkali cannot leave ink film. Inks have excellent press stability
Ammonium hydroxide	NH_4OH	Not applicable Evolves NH_3	Industrial grade is 35% solution in water. Sp. gr. = 0.88. Its extreme volatility gives ink films with excellent water resistance but poor press stability. Obnoxious odour has limited its use
Morpholine	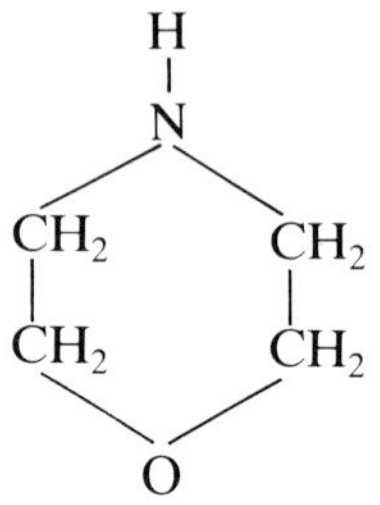 (structure shown)	128	Most volatile of the commonly used liquid amines
Monoethanolamine	$HO-CH_2-CH_2$ | NH_2	171	Similar to morpholine only inks more stable on press and prints remain water soluble for a longer period after printing
Diethanolamine	$HO-CH_2-CH_2$ | NH | $HO-CH_2-CH_2$	270	Functional chemical and physical properties intermediate between monoethanolamine and triethanolamine
Triethanolamine	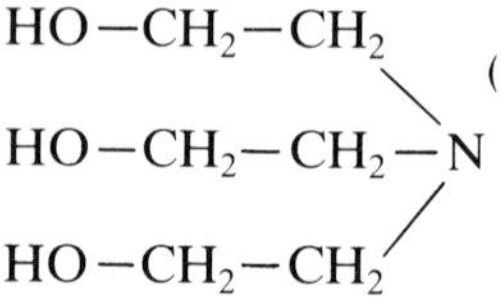 $HO-CH_2-CH_2$ $HO-CH_2-CH_2-N$ $HO-CH_2-CH_2$	279 (at 150 mm)	The high boiling point gives inks with excellent press stability. Maximum water resistance of prints is only achieved after a long period of time

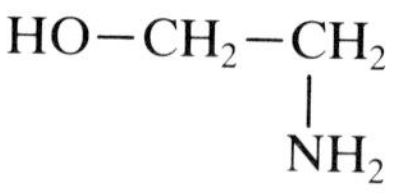

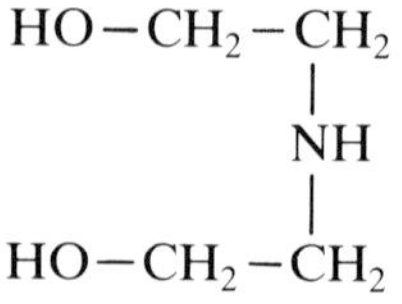

Morpholine structure:

$$\begin{array}{c} H \\ | \\ N \\ \diagup \quad \diagdown \\ CH_2 \qquad CH_2 \\ | \qquad\qquad | \\ CH_2 \qquad CH_2 \\ \diagdown \quad \diagup \\ O \end{array}$$

completely different configuration due to forces of repulsion, thereby increasing the viscosity.

These high solids emulsions can be of great assistance to the ink formulator.

The most commonly used alkalis, together with their chemical formula, boiling point, and miscellaneous points of interest are given in Table 4.18.

Organic acids and acid anhydrides

These are used, in the main, as raw materials for resin preparation.

A list of the common acids and acid anhydrides, together with their chemical formula, melting point, miscellaneous comments and application is given in Table 4.19.

Table 4.19

Name of organic acid or acid anhydride	Chemical formula	Melting point (°C)	Miscellaneous comments and application
Benzoic acid	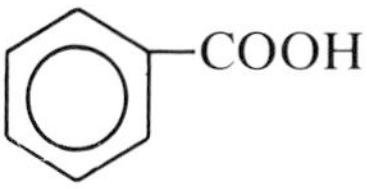	122	Finds use in medium and short oil alkyds as a viscosity reducer. It also gives fast drying alkyds. Alkyd resins with benzoic acid have limited application in letter-press and lithographic inks.
Phthalic anhydride	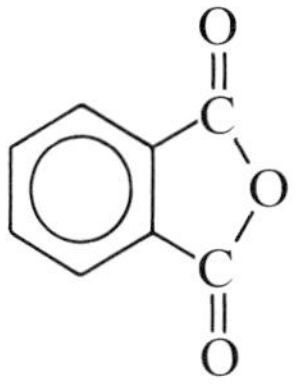	130	This is the commonest and most economic of the dibasic acidic materials used in hard resin and alkyd resin manufacture by reaction with hydroxy groups when polyesters are formed. See Chapter 12. Alkyd resins with phthalic anhydride have widespread application in letterpress, lithographic, tin-printing and screen inks.
Isophthalic acid	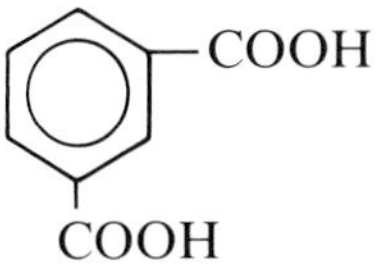	341	Application is similar to phthalic anhydride when resins with properties such as faster rate of drying, improved flexibility excellent pigment wetting properties, etc., are obtained. Its high melting point makes manufacture of the resins more difficult than with phthalic anhydrides.
Terephthalic acid	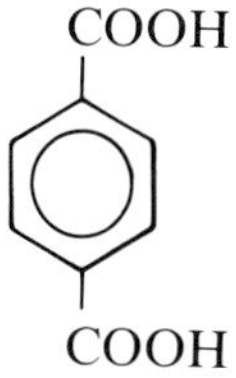	Sublimes at approx. 300	Application similar to isophthalic acid. Formulation of alkyd resins with this acid is more critical than with isophthalic alkyds. The high cost of this material limits its use.

Maleic anhydride

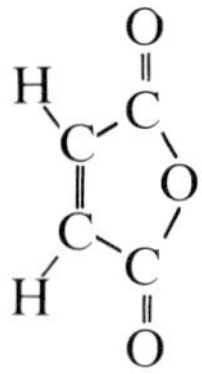

$$\begin{array}{c} H \\ \diagdown \\ C \\ \parallel \\ C \\ \diagup \\ H \end{array} \begin{array}{c} O \\ \parallel \\ C \\ \diagdown \\ O \\ \diagup \\ C \\ \parallel \\ O \end{array}$$

57   This material as well as being dibasic possesses an olefinic double bond which can take part in olefinic type polymerisation and Diels-Alder reactions with dienes. Its main uses are: (1) polyester formation with polyols which are often cross linked with an olefinic monomer (*e.g.* styrene) using a peroxide catalyst; (ii) Diels-Alder reactions with rosin derivatives when resins of high acid values are obtained. These can be made water or glycol soluble by reaction with bases; and (iii) Diels-Alder reactions with unsaturated natural oils when the water soluble "maleinised oils" and alkyds are obtained on treatment with bases.

Fumaric acid

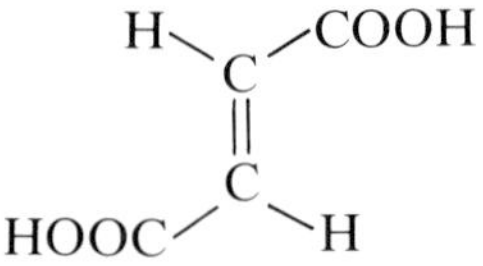

$$\begin{array}{c} H \diagdown \quad \diagup COOH \\ C \\ \parallel \\ C \\ HOOC \diagup \quad \diagdown H \end{array}$$

299–300 sublimes   Application and reactions similar to maleic anhydride. Fumaric acid cannot form an anhydride because the carboxylic groups are in the trans position. Under very vigorous dehydration conditions, however, it is converted to maleic anhydride.

Polyols

The most common polyols used in the manufacture of synthetic vehicles for the printing ink industry (see Chapter 9) are:

Glycerol

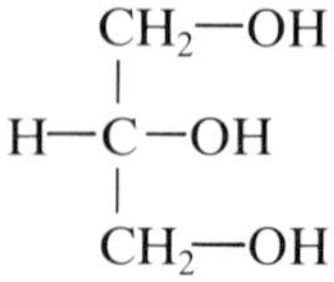

$$\begin{array}{c} CH_2{-}OH \\ | \\ H{-}C{-}OH \\ | \\ CH_2{-}OH \end{array}$$

Neopentyl glycol

$$\begin{array}{c} CH_3 \\ | \\ HO{-}CH_2{-}C{-}CH_2{-}OH \\ | \\ CH_3 \end{array}$$

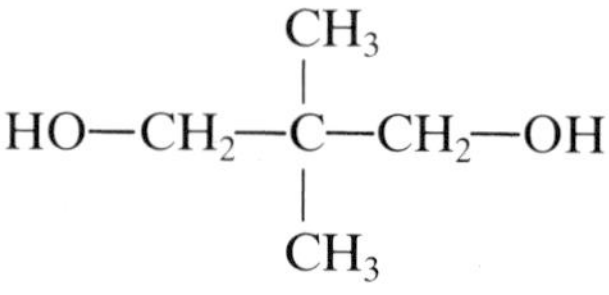

Trimethylol propane

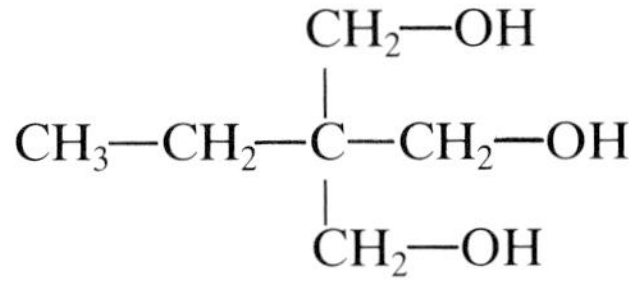

$$CH_3-CH_2-\underset{\underset{CH_2-OH}{|}}{\overset{\overset{CH_2-OH}{|}}{C}}-CH_2-OH$$

Trimethylol ethane

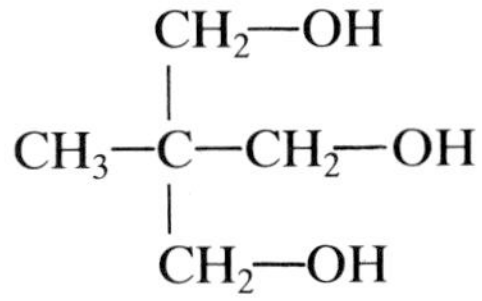

$$CH_3-\underset{\underset{CH_2-OH}{|}}{\overset{\overset{CH_2-OH}{|}}{C}}-CH_2-OH$$

Pentaerythritol

$$HO-CH_2-\underset{\underset{CH_2-OH}{|}}{\overset{\overset{CH_2-OH}{|}}{C}}-CH_2-OH$$

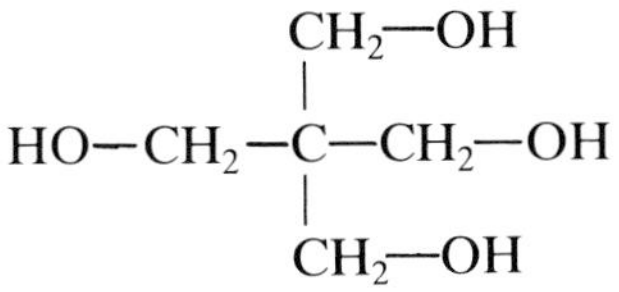

4.37 DEFOAMING AGENTS

Foaming is a condition which might happen when a monomolecular film of a surfactant which is fairly insoluble in the bulk of the liquid is present on the surface of the liquid. Elastisity of the liquid surface can cause foaming with a detrimental effect, if not controlled prior to printing. Ideally foaming should be inhibited by formulating to ensure that the surface tension is kept low. This is often difficult because of the surfactants used in the formulation of inks or in the ingredients used to make them. If foaming does occur it increases rapidly with agitation. Prevention is better than cure but where foaming does occur it should be dealt with as soon as possible.

Defoaming agents act in one of two ways:

(1) By being a good solvent for the surfactant. This causes the monomolecular film, so essential to foaming, to be dispersed into the bulk of the liquid. When this happens the tendency towards foaming is greatly reduced.

(2) By adding a small quantity of a material which is immiscible with the system, and which drastically lowers the surface tension of the system. This reduction in surface tension causes existing bubbles to burst and creates a physical condition such that stable bubbles and foams cannot form.

This latter type of defoamer is nearly always added in a form which makes it readily dispersible over the surface, for example as a liquid or an emulsion.

There are many defoaming agents and they are usually classified as silicone or non-silicone, the former being the most popular type. Silicone fluids and emulsions are often used as defoaming agents. There are fluids for non-aqueous media and for emulsions ranging from 10 to 30% silicone in water-based systems.

The silicone fluids most commonly used in the printing ink industry are of the polydimethyl siloxane type having the following structure:

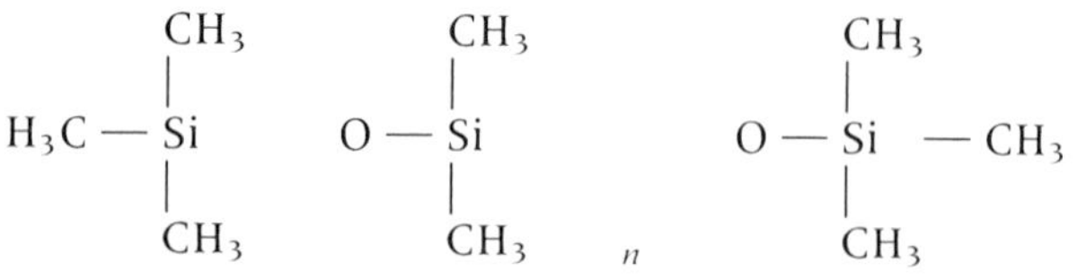

$$H_3C - \underset{\underset{CH_3}{|}}{\overset{\overset{CH_3}{|}}{Si}} \quad O - \underset{\underset{CH_3}{|}}{\overset{\overset{CH_3}{|}}{Si}} \quad O - \underset{\underset{CH_3}{|}}{\overset{\overset{CH_3}{|}}{Si}} - CH_3$$

The chain length (n) may be from 1 to 2000 and the viscosity of the oil will be dependent upon this.

When $n = 2$, viscosity will be 0.65 centistokes at 25°C, whereas when $n = 2000$ the viscosity $= 1 \times 10^6$ centistokes (i.e. 10 000 stokes) at 25°C.

Antifoam agents must be used with discretion as it is easy to over-estimate the amount necessary and to initiate other printing faults such as pinholing or reticulation.

4.38 LAKING AGENTS

In both water-based and ethanol-based flexographic dye inks, laking agents are needed to stabilise the basic dye in solution.

In ethanol inks the laking agent is usually a tannic acid derivative and combination with an acidic resin like shellac or maleic can ensure a stable ink system. Properly formulated dye inks can give increased colour strength and more surface 'sit up' with good resistance to wax and water/steam bleed.

Tannic acid

The main laking agent used for water-reducible flexographic dye inks. Now becoming expensive as its production from the natural sources in remote Third World countries is gradually dropping. This scarcity has pushed ink companies to look even harder for a substitute.

Tannic acid substitutes

A number of synthetic chemicals are now available as alternatives to tannic acid. These are cheaper but not quite as effective as pure tannic acid. They are, however more easily soluble in alcohols.

They have gained wide recognition for commercial reasons of cost and availability. They tend to be used in the alcohol/alcohol: water basic dye systems for flexographic work. Mixed with acidic resins, the inks are strong, very bright and will penetrate less into the substrate. The amount of laking agent used *must* be carefully checked in each formula to ensure the resistance properties required for that formula are attained. Basic dye solutions can be made in small concentrations with acidic resins alone but this should be checked for properties carefully as small quantities of laking agents will always improve the fastness properties.

Section X: Raw materials for radiation curing systems

One of the major growth areas of recent years, especially in the paste ink field, has been in radiation curing inks. These are inks, and clear lacquers whose components react when exposed to UV light, or when passed through an electron beam, to cure instantly to a solid polymer. This is achieved via a free radical reaction. In UV curing inks this is started by photoinitiators, in electron beam curing by the electrons themselves. The composition of the inks is described in chapter 10.

Radiation curing inks are basically formulated in the same way as any other ink — that is to say, they are composed of pigment, binders, diluents and additives necessary for specific applications. In these inks the binders are generally acrylates of some sort. The diluents are also acrylates and non-volatile. Pigments are as used elsewhere in printing inks, although careful selection and testing is necessary. The additives in addition to the normal waxes and surfactants, include photoinitiators and photo-activators in the case of UV inks, inhibitors and in some cases additives to improve printability. The additives, of course, especially the amines used as photoactivators, must be carefully selected to be insoluble in water for lithographic inks and varnishes.

A brief discussion of each class of component follows. For a more detailed examination of the subject there are now several excellent text-books available [6, 7].

4.39 PIGMENT SELECTION

Many of the pigments used in normal lithographic inks are suitable for use in UV systems. It is, however, essential to test each pigment very thoroughly indeed. Many pigments will react with the vehicle to give poor storage stability. Others may slow the cure unacceptably, while still others may change shade or burn out completely.

A reasonable guide to dark storage stability is an accelerated ageing test in an oven at 50 or 60°C. When compared with an ink of known stability the time taken to polymerise will allow a fair estimate of the shelf life of the ink to be made. Unfortunately it cannot be assumed that all pigments of the same chemical class are alike in stability since different manufac-turers may apply different surface treatments, or even different degrees of washing. Even batches from the same supplier may vary.

Metallic inks containing bronze and aluminium powders should be avoided since gellation may occur rapidly, if not explosively, and careful formulation is necessary to avoid this.

Many manufacturers now supply lists of their pigments which have been found to be suitable in UV systems. These should be treated as a guide but full tests should still be carried out.

Cure rate can be very dependent on the pigment used. This is particularly true of black inks, and of other dark colours. Carbon Black not only absorbs UV light, but also acts as a free radical scavenger. It can therefore have an inhibiting effect out of all proportion to its concentration.

In general, to avoid generating too much heat in the dispersion stage, pigments should be selected for ease of dispersion, although those which have been surface treated to enhance this should be examined closely, in case this treatment interferes with the stability.

4.40 PREPOLYMERS

The resins currently used as binders in radiation curing inks and lacquers are predominantly acrylate esters. Those commonly available include epoxy polyester and urethane acrylates, together with oil modified versions of these. They are used singly or in combination as the binder in inks or as the high viscosity resin component of a lacquer or varnish. Since the undiluted form may be at too high a viscosity for convenient use many are supplied already let down in an appropriate monomer.

Epoxy acrylates

These are used widely in inks and lacquers for most applications. They are the acrylated form of bisphenol A diglycidyl ether and its derivatives. They are relatively cheap, have a good cure rate and display good gloss, chemical resistance and reasonable adhesion. They tend to brittleness, especially in lacquers, and are often combined with a more flexible component. Oil modified versions have been supplied for pigment-wetting purposes, soya bean oil being a popular modification. This can, however, slow the cure rate, and leave ink films with an aftertack with remains permanently. Again, therefore, they tend to be used in combination with other prepolymers rather than singly.

Polyester acrylates and unsaturated polyesters

The term polyester covers a wide range of chemicals, and a correspondingly wide range of polyester acrylates are available. This variety makes it possible to select fairly accurately the desired properties. Many monomers fall into this category.

Unsaturated polyesters are typically cheaper products, and have tended to be used in applications such as wood finishing. Some more specialized products are finding limited use in both inks and varnishes, however.

Urethane acrylates

These are suitable for most applications, and are characterized by better adhesion and greater flexibility than epoxy acrylates or cheap polyesters but have a correspondingly high price. Again a wide range, both aliphatic and aromatic are available.

Recent requirements for lower odour and taint levels have led to 'stripped' versions of these monomers becoming available. These do have a much lower odour level, but as a consequence of more processing, a higher cost (see Table 4.20).

Table 4.20   Prepolymers

Bisphenol A epoxy acrylate

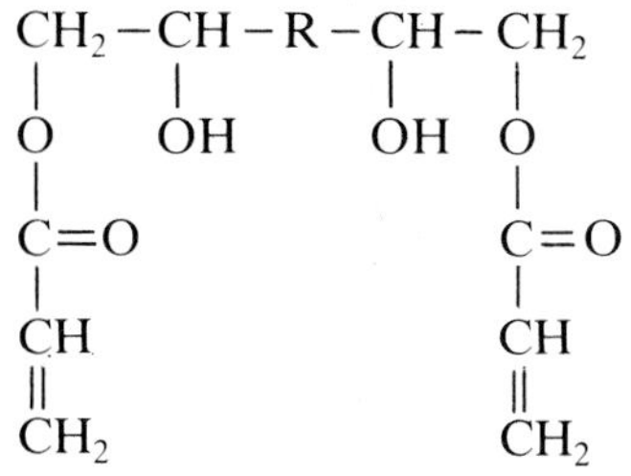

$$CH_2-CH-R-CH-CH_2$$

$$
\begin{array}{cccc}
| & | & | & | \\
O & OH & OH & O \\
| & & & | \\
C=O & & & C=O \\
| & & & | \\
CH & & & CH \\
\| & & & \| \\
CH_2 & & & CH_2 \\
\end{array}
$$

where R is

$$\left[O-\bigcirc-\underset{CH_3}{\overset{CH_3}{C}}-\bigcirc-O-CH_2-\underset{OH}{CH}-CH_2 \right]_n$$

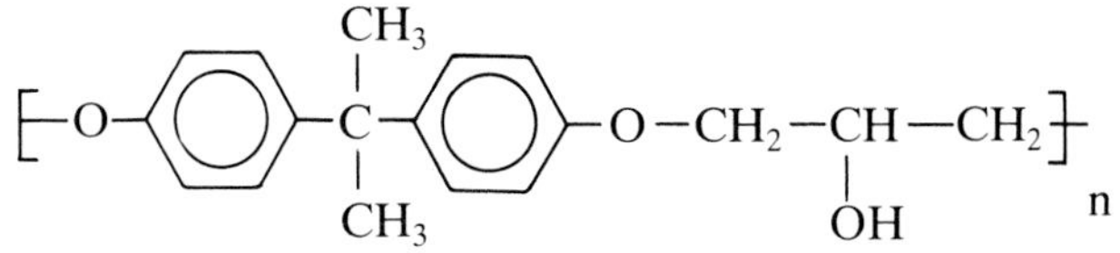

Polyester and Polyether acrylates

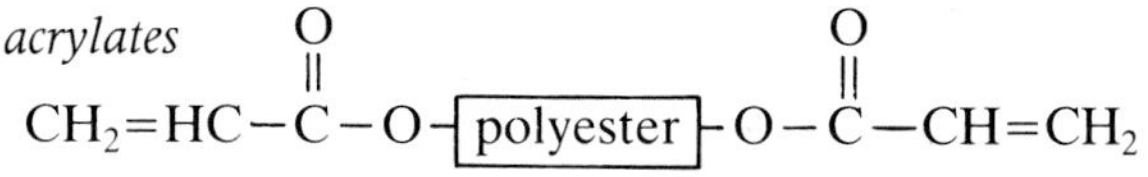

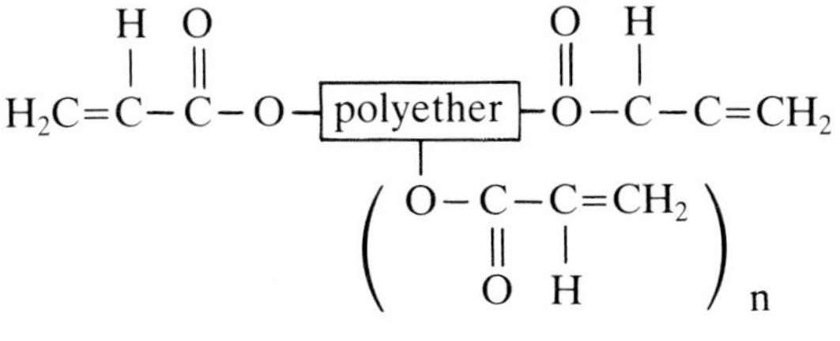

$$CH_2=HC-\overset{O}{\overset{\|}{C}}-O-\boxed{\text{polyester}}-O-\overset{O}{\overset{\|}{C}}-CH=CH_2$$

$$H_2C=\overset{H}{\overset{|}{C}}-\overset{O}{\overset{\|}{C}}-O-\boxed{\text{polyether}}-O-\overset{O}{\overset{\|}{C}}-\overset{H}{\overset{|}{C}}=CH_2$$

$$\left(\begin{array}{c} O-C-C=CH_2 \\ \| \ \ | \\ O \ \ H \end{array} \right)_n$$

Urethane acrylates

$$CH_2=\underset{Y}{\overset{}{C}}-R-O-\overset{O}{\overset{\|}{C}}-\overset{H}{\overset{|}{N}}-R^1-\overset{H}{\overset{|}{N}}-\overset{O}{\overset{\|}{C}}-O-R-\underset{Y}{\overset{}{C_2}}=CH_2$$

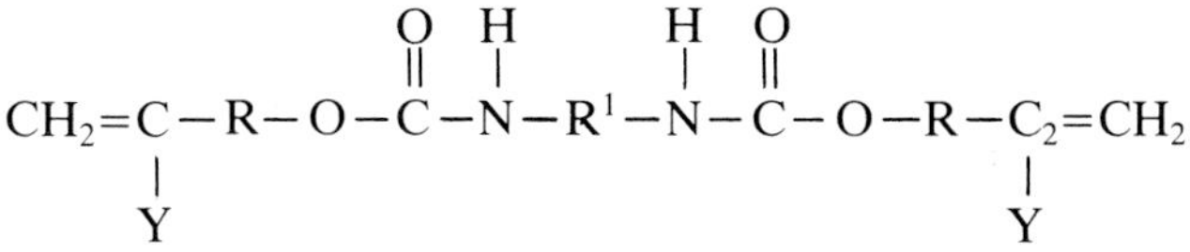

where R is $-\overset{O}{\overset{\|}{C}}-O-CH_2-(CH_2-OH)_n$ $n = 1$ or 2
Y is H or CH_3
R^1 is such that $OCN-R^1-NCO$ can be TDI, HMDI, IPDI, MDI

4.41   REACTIVE DILUENTS

The selection of appropriate monomers is governed by the functionality of the monomer, its cutting power, and its health and safety aspects. Many of the irritation problems associated with early UV formulations were caused by the use of irritant monomers, which at the same time were all that was available. It is an unfortunate fact that many of the most efficient diluents in terms of reactivity and adhesion have turned out to be among the more irritant. The monomers dealt with below meet the SBPIM specification for a Draize figure lower than 3.5, although this can vary with the individual supplier.

Of the monofunctional monomers currently in use n-vinyl pyrollidone is perhaps the most widely used. It is an excellent diluent with a relatively high reactivity. Many monofunctional monomers have a high odour level in the uncured state, and are therefore unacceptable in any large quantity.

Perhaps the most common monomer, certainly in those approved by the SBPIM is tripropylene glycol diacrylate (TPGDA). It is used as the major component in roller coating lacquers, at a reduced percentage in more viscous clears, and as a tack reducer in inks. Attention should be paid in inks to its lithographic properties.

Dianol diacrylate, or its analogues (dianol is a registered trade name of Akzo Chemie), have a much higher viscosity than TPGDA. They do have a very low Draize rating, and excellent film-forming and plasticizing properties.

In those areas where health and safety are not such important considerations 1,6-hexanediol diacrylate (HDDA) is perhaps the best difunctional monomer available for adhesion and reactivity. It has a high Draize figure, however, which precludes its use in the UK in printing inks, although it, and some other monomers which are suspect for health and safety reasons, are still used in other applications.

Trimethylol propane triacrylate (TMPTA) has been the most popular trifunctional monomer until recently when it too was placed on the SBPIM exclusion list. It has good lithographic behaviour, a high reactivity and a reasonable viscosity. Recent developments in monomers have been aimed at finding a less irritant substitute with similar properties. Of these the most promising are the ethoxylated or propoxylated versions of TMPTA or glyceryl propoxy triacrylate (GPTA). Of similar viscosity to TMPTA these have lower Draize values, but reduced reactivity, or poorer lithographic behaviour, or a higher cost, but are generally acceptable substitutes. All these triacrylates convey a degree of brittleness to a lacquer film because of their high cross-linking ability.

Tetra acrylates are rarely used, but again suffer from brittleness (see Table 4.21).

4.42   PHOTOINITIATORS

In UV curing inks these are chemicals which become excited by UV radiations, forming free radicals, which in turn react with the vehicle of the ink or varnish to begin polymerisation. The term is also used to include proton donors such as amines. The basic mechanism of photoinitiation and photochemistry in general is well documented, and can be found in many standard textbooks.

Benzophenone is one of the cheapest, most widely used photoinitiators. It requires a proton donor, such as amine, to be present in order to yield radicals easily. This synergism, two products being much more efficient than either one alone, is a common feature of photoinitiation in UV inks. Benzophenone is readily soluble and is itself a solvent, and so is useful as a viscosity modifier in both inks and lacquers. Its main drawback is its odour, which, while not unpleasant, is very strong. Not all the benzo- phenone content of an ink is used up in a reaction, and of what is used some readily combines to form other products with high odour levels.

Much work was done using benzophenone in combination with Michler's ketone (4,4-bis dimethyl amino benzophenone). This proved a most effective system, but the health risks associated with Michler's ketone have caused it to be withdrawn from use, and as yet no effective substitute has been found.

Several other photoinitiators work by a process of hydrogen abstrac- tion, requiring a proton donor for effective initiation. These include benzil, derivatives of benzophenone and a number of thioxanthones, of which isopropyl thioxanthone finds the widest use. These do tend to be more expensive than benzophenone to achieve the same cure rate.

Other photoinitiators operate without the need of a proton donor. These simply undergo fragmentation into reactive species under UV light. Those commercially available at the time of writing go under a variety of trade names are mostly benzil ketals or acetophenone derivatives. Again breakdown products of some of these can cause odour problems, although these are less severe than those with benzophenone.

A wide range of amines, and acrylated amines are available as proton donors for UV systems.

The combination of photoinitiators should be carefuly chosen to provide the best cure rate at the minimum cost. It is possible by choosing the wrong combination to cause bloom to appear on lacquers, or to achieve a brittle surface cure. Often after a certain cure rate is achieved no further addition of photoinitiator will improve it. It may be convenient to make compounds of initiators as liquids for easier ink manufacture. (See Tables 4.22–4.24)

4.43 ADDITIVES AND INHIBITORS

It is necessary to incorporate a small amount of inhibitor into radiation- cured inks and lacquers to guard against premature gellation. Hydro- quinone and its derivatives are commonly used for this purpose at very low percentages.

The shelf life of inks can be considerably extended, even when adequ- ately inhibited, by careful processing and storage. Inks exposed to high temperatures even for a short time may start to gel, and the reaction will

Table 4.21 Reactive diluents

Monofunctional monomer

n-Vinyl pyrrolidone

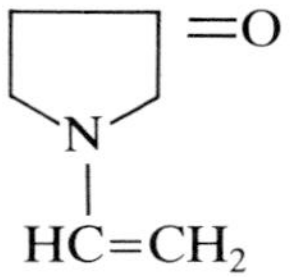

$$HC{=}CH_2$$

Difunctional monomers

Tripropylene glycol diacrylate (TPGDA)

Dianol* diacrylate (DDA)

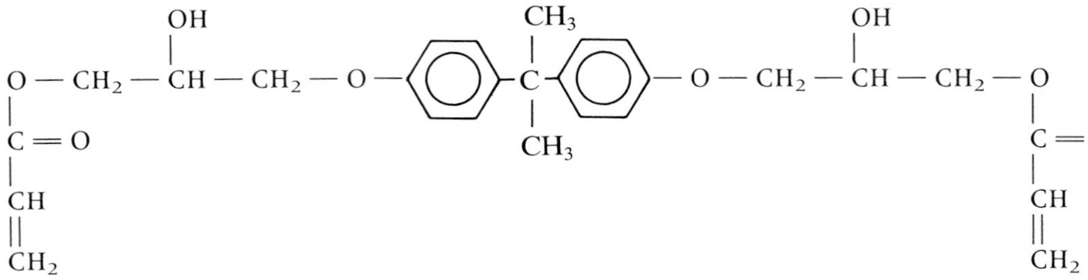

*Trade name

Trifunctional monomer
Trimethylol propane
   triacrylate (TMPTA)

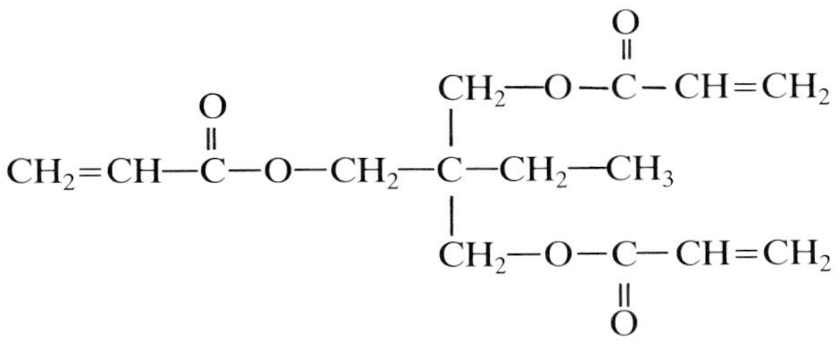

$$CH_2=CH-\overset{\overset{\displaystyle O}{\|}}{C}-O-CH_2-\overset{\overset{\displaystyle CH_2-O-\overset{\overset{\displaystyle O}{\|}}{C}-CH=CH_2}{|}}{\underset{\underset{\displaystyle CH_2-O-\overset{\overset{\displaystyle O}{\|}}{C}-CH=CH_2}{|}}{C}}-CH_2-CH_3$$

The molecular weight of some reactive diluents is increased to lower skin irritation. This is achieved by alkoxylation of the polyol with ethylene or propylene oxides prior to acrylation.

Example
Alkoxylated glycerol triacrylate
   e.g. OTA 480

Table 4.22 Aromatic ketone initiators used with a proton donor

Chemical name	Structure	Commercial name
Benzophenone		—
4-Phenyl benzophenone		Trigonal 12
Isopropyl rhioxanthone		Quantacure ITX
1-Hydroxycyclohexyl acetophenone (50%) and benzophenone (50%)		Irgacure 500

Table 4.23 Photoactivators. Proton donors used with aromatic ketone initiators

Chemical name	Structure	Commercial name
Triethanolamine	$(HO — CH_2 — CH_2 —)_3 N$	TEA
Methyl diethanol-anime	$(HO — CH_2 — CH_2 —)_2 N CH_3$	MDEA
Ethyl-4-dimethyl aminobenzoate	C_2H_5OOC ⟨ ⟩ $N·(CH_3)_2$	Quantacure EPD Nuvopol EMBO

2-(n-butoxy)
ethyl-4-dimethyl-
amino benzoate

$C_4H_9OC_2H_4.OOC$ — ⬡ — $N(CH_3)_2$

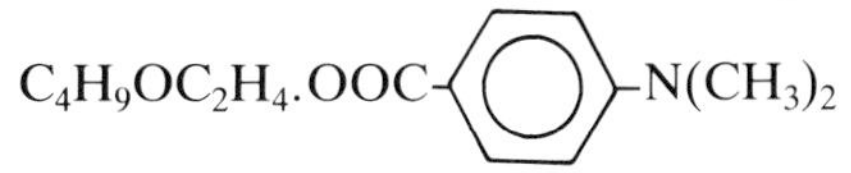

Quantacure BEA

2-ethyl hexyl-
p-dimethyl-
aminobenzoate

$$\begin{array}{c} H \\ | \\ C_4H_9-C-CH_2.OOC-\text{⬡}-N(CH_3)_2 \\ | \\ C_2H_5 \end{array}$$

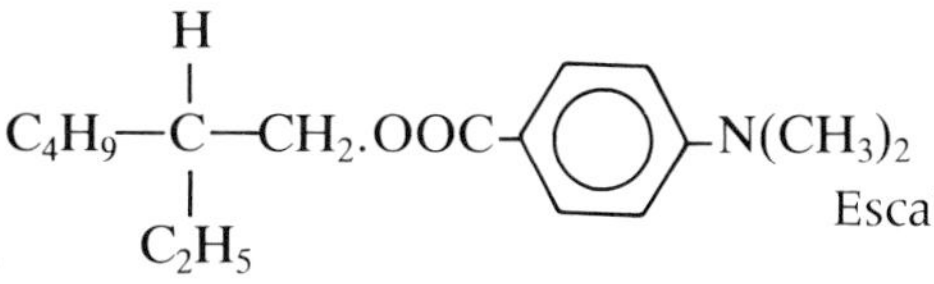

Escalol 507

**Table 4.24 Acetophenone, benzoin and benzil ketal initiators
used without proton donors**

Chemical name	Structure	Commercial name
2,2-Diethoxy acetophenone	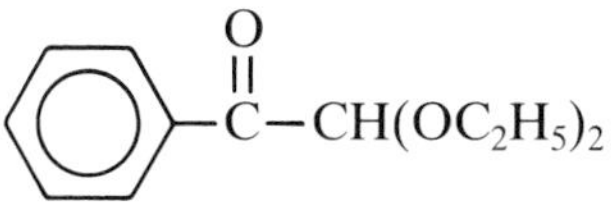 ⬡—C(=O)—CH(OC$_2$H$_5$)$_2$	DEAP
1-Hydroxy cyclohexyl acetophenone	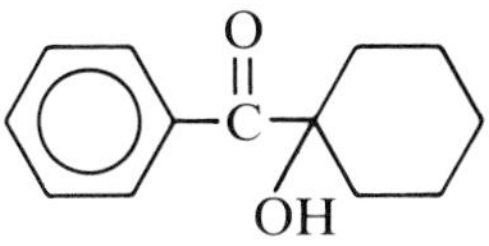 ⬡—C(=O)—C(OH)⬡	Irgacure 184
α, α-Dimethyl-α-hydroxy acetophenone	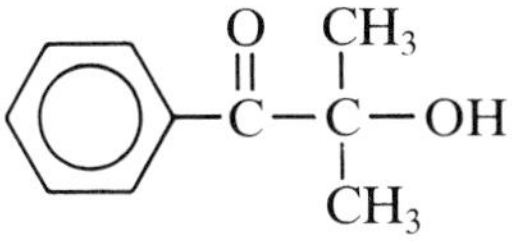 ⬡—C(=O)—C(CH$_3$)(CH$_3$)—OH	Darocur 1173
Benzoin	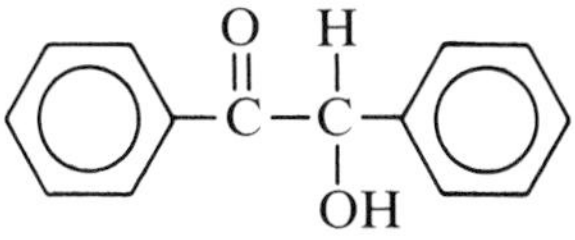 ⬡—C(=O)—C(H)(OH)—⬡	—
Benzoin ethers	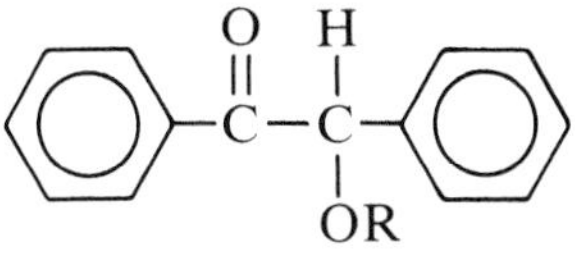 ⬡—C(=O)—C(H)(OR)—⬡	Esacure EB3 Trigonal 14 (Butyl ethers)

R = ethyl, butyl or isopropyl

Table 4.24   Continued

Chemical name	Structure	Commercial name
Benzil ketals		Irgacure 651 (Methyl ether)

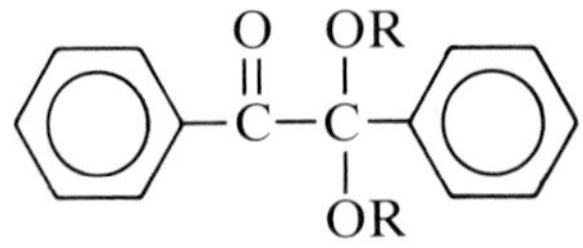

R = methyl, ethyl, butyl or isopropyl

| 2-Methyl-1-[4-(methylthio)phenyl]-2-morpholino propanone-1 | 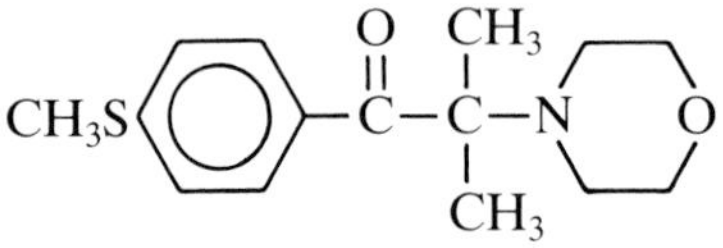 | Irgacure 907 |

continue from there. Containers should have an air space left at the top since oxygen helps inhibit these reactions.

Surfactants are much the same as those used in ordinary inks and lacquers, although some specialized acrylated products are available. As with all the other components each should be checked in the system in which it is to be used to establish compatibility, and any effect on shelf life. Some will also degrade rapidly under UV light.

Tables 4.20–4.24 summarize the specialist raw materials used in radiation curing systems.

Section XI: Health and safety at work

For many years, the British printing ink industry has taken the lead in establishing standards for freedom from toxic materials in inks for food packaging use. A totally different facet of safety has, however, developed with the introduction of the Health and Safety at Work, etc. Act in 1974.

The Act places two duties on manufacturers:

(1) To ensure as far as is practicable that the products are safe in use;
(2) To provide information on how the products may be used with safety.

At the same time, an employer is required to ensure that working conditions are free from hazard and that employees are suitably trained. Finally, the employee has a responsibility to act in a responsible manner.

It may be argued that these practices have been adopted by all conscientious employers and manufacturers for many years. In common with all legislation, this Act is designed to bring the unsatisfactory few up to the standards of the majority. It is doubtful whether such an Act is necessary in the printing ink industry. However, it is now law and a good deal of time and money is being spent to prove that the manufacturer is observing the Act.

The practical effects are largely as follows: ·

(1) When any new raw material is being used in inks or coatings for the first time, its effect on the human body should be investigated. Good examples are the many new materials being introduced in the UV curing inks.

(2) If it is thought that any hazard may exist, special handling advice must be given.

(3) General advice on the recommended methods of handling printing inks should be given to all ink users so that they and any new-comers to the industry can be suitably instructed.

One important change has taken place, although not specifically prompted by this Act, the removal of lead compounds from printing inks. A pro-gramme of action has been agreed by the Society of British Printing Ink Manufacturers which resulted in the discontinuation of the use of lead driers and lead-based dry colours from July 1978.

REFERENCES

1. Boielle, *Printing Technology* 100–6 (Dec. 1972).
2. Tawn, *JOCCA* **1**, 71–4 (1969).
3. Sears, Unit 22, Federation Series on Coatings Technology, pp. 29–34 (Apr. 1974).
4. Sisley, and Wood, *The Encyclopaedia of Surface Active Agents*. Chemical Publishing Co.
5. Hollis, Gordon L. *Surfactants UK*. Tergo Data Publications Ltd (July 1976).
6. Holman, (ed.) *UV and EB Formulation for Printing Inks, Coatings and Paints*. Sita (1984).
7. Phillips, *Sources and Applications of UV Radiation*. Academic Press (1983).

CHAPTER 5
Letterpress inks

During the last decade, the use of letterpress as a major printing process has shown a continual decline.

Nevertheless, it continues to be used to produce significant amounts of general printing throughout the world.

Jobbing printing, continuous stationery and corrugated cases are examples where the process is still used in this country, although on a diminished scale, since printing by alternative processes offers advantages of quality and output. A considerable number of newspaper titles are also still printed by rotary letterpress and are probably the largest daily print run to take place on a regular basis, in the UK. While general letterpress printing is likely to continue to have a place in the world market, it is unlikely that the process will have a long term future for newspaper printing, in view of current trends towards the purchase of web offset and flexographic presses.

5.1 NATURE OF THE PROCESS

The fundamentals of the letterpress process, together with a description of the configuration of the most widely used presses, have been dealt with in Chapter 2. The fact that the ink is applied to the paper from a raised or 'typographic' surface, gives the process all its characteristic properties. The printing surface can be 'hard' such as type metal alloy, and copper or zinc blocks, or 'flexible' as rubber stereos and polymer plates. The materials used to manufacture letterpress inks do not generally affect the various metallic surfaces of the formes, although there may be differences in wetting characteristics. Polymer plates, however, may be affected by the materials used in some types of ink, and care has to be taken when formulating inks to be used with these. It is the method by which the ink is applied to the printing surface, and the printing speed which has the most influence on the formulation.

Platen presses

Platen presses have relatively poor ink rolling power compared to other types of presses, and printing takes place from a rigid forme which is parallel to the paper at the moment of impression. The ink will need to have a sufficiently low viscosity to distribute over the short roller train, but must not be so long flowing that it will spread down the sides of the relief characters. At the same time, the ink should exhibit sufficient tack to transfer clearly under a reasonable level of impression if the embossing effect of type on paper is to be kept to a minimum.

The type of work printed on such presses is normally short runs of jobbing work, hand bills, invitation cards, dinner menus, etc. where a fast turn-around of the finished work is often required.

Flat-bed presses

These presses have increased ink rolling and distribution power, compared to platen presses, and the printing forme area can be larger. The presses are suitable for printing on a wide range of substrates, from newsprint to coated papers. The inks used, therefore, can have a wide range of properties and drying rates, and compared to the inks used on a platen press, they usually have a higher viscosity, increased tack, and may be slightly weaker.

Rotary presses

The web of paper is printed in a continuous operation, usually at high speed and this necessitates the use of thinner inks than are required for either platen or flat-bed presses. These presses are suitable for printing long runs, with ink drying by penetration, e.g. newspapers.

There are three methods of ink feed to the printing unit and each makes demands on the rheology of the ink used.

(1) *Duct feed.* This consists of a rotating roller in an ink fountain, the flow of the ink being controlled by a flexible blade. This system can handle inks of relatively high viscosity provided they have sufficient flow not to hold back in the duct.

(2) *Ink rail feed.* Ink is pumped through tubes to an ink rail which is set parallel and close to the first inking drum. Separate tubes supply each 'column width' and enable the ink feed to be easily regulated. This system, which is fully enclosed, requires an ink which is free-flowing and has a low yield value.

(3) *Page-Pak system.* Similar to the ink rail feed system, but each column width is supplied by an individual pump unit. Inks with low viscosity and extremely good flow are necessary.

Appearance of letterpress print

A well-printed copy produced by the letterpress process should have even solids, clean type and the minimum of distortion to the printed substrate caused by the impression pressure. In practice, however, this is not always the case. Where presses are incorrectly adjusted or where unsuitable papers are used, undue printing pressure is necessary in order to obtain a smooth print. The pressure may be such that the substrate is embossed on the reverse side of the print, when a halo appears around the edges of the solids. These characteristics of squash and embossing are present to some degree in all typographic printing and are the usual means of identifying the process.

Mottle or the uneven appearance of the solids occurs mainly when using platen presses, and this is more pronounced when transparent colours are used. Newspapers printed by the rotary letterpress process can also be identified by the general background greyness of the copy which is normally caused by second impression set off.

5.2   GENERAL CHARACTERISTICS OF LETTERPRESS INKS

To obtain the best from the process, letterpress inks should conform to the following basic principles:
  (1) It is always desirable to print with the thinnest film of the strongest and stiffest ink that the paper and the machine will tolerate. If the ink has insufficient colour strength, high film weights of ink are needed and excess squash, mottle and set off, are likely to result. At the other extreme, if the ink is too strong, it may be difficult to obtain the required strength, while at the same time producing an even print which 'bottoms' the paper.
  (2) The faster an ink is required to run, the less tacky it has to be. At high speeds, stiff inks do not transfer readily on the inking rollers, while an ink with little or no tack may cause the inking rollers to skid, with the resulting loss of ink transfer.
  (3) Printing by the letterpress process requires very high pressures, and inks must be formulated in a way which minimises the unwanted effects of such pressures.
  (4) The ink should not dry on the press during normal daily stoppages and in some instances, for much longer periods. In most press rooms, it is not economic to allow time to clean the inking rollers, either during or at the end of the run. The ink should therefore return to normal working properties, in the shortest possible time, after a break in production.
Observance and recognition of these principles will produce inks which suffer the minimum distortion and squash, at the moment of impression, and give the highest quality of print.

Drying mechanisms

The very smallest apart, most letterpress presses have a relatively long roller train to distribute the ink and to apply an even film over the area of the forme. Hence, letterpress inks should not be formulated using volatile materials since these would evaporate during the running of the press and lead to ink drying on the press with the consequent loss of transfer. Furthermore, this could also lead to an increase in the tack level of the ink, leading to picking problems on substrates with a low surface structure and strength. Apart from inks which are formulated for label machines, which have some form of assisted drying, when a proportion of 240–260°C petroleum distillate can be included, 260–290°C petroleum distillate is normally the lowest fraction that can be employed.

Letterpress inks dry either by:

(1) Penetration, e.g. newspaper inks;
(2) Oxidation, e.g. foil inks;
(3) Combination of both, e.g. quick-set inks, or
(4) A shift in chemical balance, e.g. water-reducible inks.

Inks which dry by UV radiation have also been formulated, notably for small label presses, and these are considered in detail in Chapter 10.

Penetration

Penetration 'drying' occurs where part of the vehicle is a low viscosity material such as mineral oil or petroleum distillate. This component penetrates into the substrate leaving a pigment/oil phase on the surface, which enables the print to be handled a short period after printing. In the case of conventional newspaper inks, the ink film does not dry further and can be removed by light pressure.

Cheap general-purpose inks used to print hand-outs and similar promotional work on uncoated papers can also be formulated to dry by penetration. They usually include bodied drying oil or resinous varnish to impart some oxidation drying characteristics and to give the print a degree of rub resistance.

Oxidation

Inks drying by oxidation alone are used to print impervious substrates, such as foil boards or certain plastics, and print tests should be carried out on the substrate to be used to ensure that the level of adhesion is sufficient for the 'end usage' requirements.

The drying process is catalysed by the metallic driers described in Chapter 4, in detail. Inks drying by this process alone sit on the surface of the substrate and the prints may require an anti-set-off spray, to avoid problems in the stack.

Combination drying

The combination of penetration and oxidation/polymerisation form the most widely used drying process for inks used on platen and cylinder presses for sheet-fed printing on the harder uncoated and coated papers and boards. They are formulated to have two phases. The low viscosity component penetrates the pores of the substrate immediately following printing, to set the ink, leaving the high viscosity vehicle/pigment phase on the surface which dries by oxidation. Depending on the rate of setting of this type of ink and the tack levels of the drying film, the prints may require a minimum of application of spray powder to prevent set off.

Precipitation drying

Until the early 1960s, 'Moisture set' inks were widely used for printing corrugated cases. These inks dried by the precipitation of the resin content of the varnish when in the presence of water. The water could come from three sources: the atmosphere, the moisture present in most papers and boards, or from an applied jet of steam.

In spite of many advantages, the early inks suffered from press stability problems, tending to dry rapidly under conditions of high humidity and to lack flow properties. To overcome these problems, water-reducible inks were offered to the trade and found a ready acceptance. These inks were, and still are, based on a high acid value resin dissolved in glycol, but the formulations also include sufficient alkali to neutralise this acidity and give an ink which is itself alkaline. Such inks can be reduced with water and still give good flow, hence their press stability and transfer properties show an immediate improvement over the moisture set types. The interaction between a wet print of the alkaline ink and the acidic paper, shifts the pH (or $\log_{10}$ hydrogen ion concentration) towards the acid side of neutral and the resin precipitates as before, to form a solid mass of resin and pigment, giving a dried print.

Care has to be taken to ensure that the level of alkali added to the ink is sufficient to make it water reducible but excess alkali leads to drying, and consequent rub resistance problems on some substrates. This is especially so if the print is subjected to wet conditions, for example, a container transporting flowers. The measurement of pH therefore forms an important quality control test for this type of ink. The advantages of using these inks are as follows:

(1) The drying can be extremely rapid, although it depends on the substrate and its pH.
(2) As no oxidation drying is involved, prints with a low odour and taint level are obtained.
(3) They can be printed on normal letterpress equipment without elaborate extras, although the rollers must be compatible with the ink system and those made from rubber are preferred. Care should also be taken to choose a photopolymer letterpress plate which will not be affected by glycols.
(4) Water can be used to wash up the press, and except in rare cases, highly flammable solvents are not required.

5.3   PHYSICAL PROPERTIES

Because of the large variation in types of press, printing speeds, and ink drying requirements, involved in the letterpress process, the physical properties of the inks used vary considerably from the long flowing, low viscosity inks used for high speed rotary newspaper presses, to short, relatively high viscosity paste inks, for use on platen machines.

Quality control

While the details of quality assurance methods are dealt with in Chapter 14, some consideration should be given to those tests, which are particularly relevant to letterpress inks. As for all inks, a thorough testing of the raw materials, before they are used, helps avoid major variations in the final product. Letterpress inks can be manufactured either by dispersing the pigments directly into the vehicle or by letting down previously made 'pigment bases' with additional vehicle and additives. Both ink or base can be manufactured by batch on a 'triple roll mill' or by 'bead-milling' where continuous production methods can be used. Testing the 'final' product will be carried out on each batch or by regular sampling, where continuous production methods are used. Routine quality assurance tests will be made for hue, colour strength, setting, drying etc., which are similar to those used for offset inks.

Test methods more specific to the letterpress process are outlined under the headings below.

Viscosity

The viscosity of sheet-fed letterpress inks is usually measured by the standard test on a 'falling bar' viscometer. Since the majority of the inks are non-Newtonian in character, an estimated yield value will be found by extrapolation to zero shear. High yield values may cause hanging back in the duct while inks with too low a yield value may give problems at the printing plate.

Rotary news inks which have comparatively low viscosities are measured on a variable speed rotational viscometer. The difference between the indicated viscosity at the various rates of shear will also give an indication of the yield value and thixotropic nature of the ink.

Dispersion

All inks are tested using a standard fineness of grind gauge, in the normal way. Low readings can be equated with excellent dispersion and maximum colour development, while high readings are an indication that possible press problems of filling in of the halftone areas of the forme and filter blockages may occur. A poorly dispersed ink may also give undesired print properties including poor gloss and high rub off.

The appearance of the ink under a medium power microscope is also useful as a confirmation of the degree of dispersion. In practise, while the letterpress process allows more latitude than the other processes, most inks for general printing will be produced to give a grinding reading of $6-8$ µm and inks for newspaper printing may be slightly higher.

Rub resistance

The rub resistance of most articles printed on letterpress sheet-fed presses will be assessed in accordance with British Standard *Methods for Measuring the Rub Resistance of Print* BS 3110 : 1959.

When considering newspaper inks, a distinction must be made between rub resistance and 'print smear', or the removal of the pigment/oil residue from the substrate surface by light pressure, which was mentioned above as a characteristic of penetration drying. The use of the usual rub resistance tests for this type of ink is too severe and the results obtained do not relate to the end usage of the newspaper. The resistance of the ink to smearing is more relevant and this is usually measured subjectively by rubbing a finger over the print and observing the condition of both finger and print.

Cloth and paper may also be used to determine the smear resistance of the printed paper. This test must be regarded as subjective, since there are considerable variations between the individuals conducting the test. Nevertheless, it is widely used in the industry as an indication of the drying/setting of newspaper inks.

Strike-through

'Strike-through' is the degree that an ink penetrates into the paper, and is an important element in the control of newspaper inks. It is often assessed by inspection of the reverse side of the standard drawdown test for colour, but the degree of wetting out experienced by the inks and their relative viscosity, has to be taken into account, when making a judgement. A better test is to print a known volume of ink under standard conditions on newsprint and measure the optical density of the reverse side of the print after a set time using a spectrophotometer.

The term 'show-through' is sometimes confused with 'strike-through' but is more properly used to describe the appearance of the print from the reverse side, due to the transparency of the paper used, rather than ink penetration.

Tack

The standard test methods will be used but, in general, the results will be in the lower range, those for rotary news inks being just sufficient to prevent roller-slip on the tackmeter. Inks for platen presses, which require sufficient tack to transfer from the stereo to the paper, will give readings in the medium range dependent on the tackmeter used and its calibration. If colours are being superimposed wet-on-wet, tackgrading becomes very important, in view of the heavier ink films laid down by the letterpress process.

Flow

The flow of letterpress inks can be assessed by using either a flow plate or a strip of paper. Whichever method is used, a preset volume of ink is applied by means of an ink pipette or syringe and allowed to flow in a vertical position for a fixed period of time. Dependent on the type of press and ink duct, newspaper inks will usually flow between 4 and 14 inches (10–35 cm). Flat-bed and 'knifeable' inks will probably require a flow length of less than 1 inch.

Misting tests

In view of the health and safety aspects of printing newspaper inks in fast running rotary presses, these tests have become highly important on the development and control of formulations. A standard test can be carried out using a tackmeter running at high speed and measuring the amount of ink transferred by visual or optical density methods.

Most of the ink transferred is caused by ink sling rather than ink mist and if more detailed tests are required it is necessary to set up a small roller system to simulate the roller chain on the press. This is positioned in an airtight enclosure and the air extracted through a filter or particle size counter. The amount of ink mist and fly can then be expressed volumetrically.

Residue

Also of particular importance to the control of newspaper inks is a test for extraneous material, which may cause blockages in the pump and press filters. Relatively small amounts of material may cause problems in these due to the large volume of ink passing through them.

A weighed amount of ink is reduced in viscosity with solvent and passed through a fine sieve. The residue remaining on the sieve is weighed accurately and examined under a microscope. With good housekeeping during the whole manufacturing process, the amounts detected by this very stringent test will be very small indeed.

5.4   RAW MATERIALS

Pigments

The range of pigments that can be used in letterpress inks is comprehensive and the process does not impose the stringent demands of 'ink-fount' balance of lithography. The trend over the recent years has been to restrict the number of pigments used in most types of ink, and letterpress is no exception. Wherever possible, pigments and other raw materials, will have common use over the whole range of ink manufacture.

Due regard must be paid to the avoidance of certain pigments in inks designed for food wrappers and also those which are to be applied to toys or graphic instruments.

The film weights associated with letterpress printing are comparatively high and therefore the level of coloured pigmentation is lower than the offset process. It is often necessary to include transparent, or semi-opaque, extending pigments, such as Calcium Carbonate (CI Pigment White 18) or Blanc Fixe (CI Pigment White 21) in the formulation if the balance between colour strength, working properties and cost, is to be achieved. This is because the amount of coloured pigment used in the formulation, to obtain the strength required, is insufficient to give the ink enough body and viscosity. Reducing the colour strength by the addition of varnish alone, would tend to increase the set off properties of the ink or reduce the viscosity to a level where it would not print properly. The addition of extender helps adjust the pigment/vehicle ratio and gives the flow, transfer and setting properties required to print evenly. Since extender pigments are usually relatively inexpensive, their addition helps reduce the ink raw material cost.

The inclusion of some opaque pigment, where the specification allows, is also recommended as it reduces the effect of the uneven solids obtained by the process.

Water-reducible inks are often used on natural kraft stocks and need to have some opacity to hide the colour of the substrate. Until their use in this country was discontinued by member companies of the SBPIM in most inks in 1978/9, chrome pigments found a wide use in yellow, orange, scarlet and green shades and gave inks with good opacity and brightness. Combinations of Rutile Titanium White (CI Pigment White 6) with a range of other pigments, are now used to give the best result possible, although in some cases, loss in brilliance of colour has had to be accepted.

Some care has to be taken over pigment choice since the glycols used in the inks are relatively powerful solvents. The chemical structures of diethylene and dipropylene glycols give them some of the properties of ethers and some of alcohols, and these will cause certain pigments to bleed. PMTA pigments are among these, the basic dye being extracted with some change in hue and loss of lightfastness. Calcium 4B Toner (CI Pigment Red 57) also causes stability problems and inks based on this tend to 'liver' or thicken up.

Coloured inks for newspaper printing need to have good contrast of hue when printed on newsprint to give maximum impact and therefore brightness of colour is essential. Headlines and seals are often printed using light to mid scarlet shades, so Lake Red C (CI Pigment Red 53 : 1) finds a wide usage, although other pigments with some opacity can be beneficial, e.g. Chlorinated Para Red (CI Pigment Red 4).

The carbon blacks used in newspaper ink formulation are usually rubber grade, intermediate particle size, furnace blacks. They may be supplied in fluffy, in pelletised, or in an oil-beaded form. The carbon must have good tinctorial strength, have good ink-making properties, be free from abrasive materials, but above all, must be economic. The particle size and oil absorption vary according to the manufacturer and the form in which they are supplied. The properties imparted to the ink by the different particle size of carbon blacks are shown in Table 5.1.

The ink properties required often cannot be achieved with the use of a single grade of Carbon Black and it is likely that different grades will be blended to give the optimum results.

This type of pigment will also find a use in the formulation of the cheaper range of letterpress blacks for platen and cylinder presses, although for magazine printing higher grades of black with even finer particle size (20–30 millimicrons) are required.

Table 5.1   Carbon Black

	Particle size	
	SMALL	*LARGER*
Colour	Brown shade	Blue shade
Strength	Higher	Lower
Rub Resistance	Better	Slightly worse
Dispersion	More difficult	Easier
Flow	Shorter	Longer

Induline dyes were used to tone newspaper inks in the UK until 1977, when they were withdrawn owing to the possibility of health risks associated with their manufacture. The use of pigmented toners as a replacement to give a blue undertone to the ink is nearly always precluded on economic grounds for rotary letterpress inks for newspapers, but they are used in other types of letterpress inks when costs allow; Reflex Blue (CI Pigment Blue 18) is widely used for this purpose.

Resins

Drying systems
The earliest drying systems used in letterpress inks were based on linseed oil varnishes. The linseed oil was usually heat polymerised to give a range of viscosities between 15 and 240 poise. With the advent of presses with faster printing speeds, quicker drying systems were developed by reacting modified or refined linseed oil with a hard resin, such as a rosin ester or rosin-modified phenolic resin. In current varnishes, the drying oil component has been largely replaced by the use of long oil alkyd resins, and these, with the addition of phenolic, maleic or hydrocarbon resins, give the wetting, drying and gloss required in each particular formulation. The viscosity of this type of resin–oil varnish will usually be adjusted by the inclusion of a high boiling range petroleum distillate (see Solvents below), which will also have an influence on the setting of the ink. The inclusion of a suitable gellant may also be required if the formulation needs a structure. An ink based on this type of varnish should have a reasonable level of setting on most coated and uncoated substrates but, to increase the speed of setting, part of the gloss varnish is often replaced by a quick-setting varnish based on a higher solvent content.

Careful formulation produces a varnish which does not skin unduly in the container, has good press stability, but rapidly sets on the paper by solvent penetration. The highly resinous phase of the ink remains on the surface and oxidises and polymerises to form a hard film.

For non-absorbent substrates, where the drying depends almost entirely on oxidation–polymerisation, the addition of tung oil into the formulation may be required to give the speed of drying and the hardness required.

Non-drying systems
The majority of all non-drying systems especially for use on rotary presses for newspaper printing are based on mineral oil, although they may contain a minimal amount of hard resin to give a degree of rub resistance. The resins used will be drawn from the same classes as used for other types of inks, i.e. phenolic and hydrocarbon resins, etc. but the grades chosen must be soluble in mineral oil and have low cost. Since there is no requirement for fount tolerance, certain grades of zinc and calcium resinates can also be used. The mineral oils used can vary in viscosity and aromatic content depending on the source and nature of the treatment used to refine the oil.

Recent legislation has directed the suppliers of mineral oils to classify their products in accordance with the degree of treatment and refining given to the oils.

Solvent-refined oils
The oils are treated during their manufacture to remove the aromatic content. They do not normally carry any HASAW warning labels.

Hydro-refined oils
This grade of mineral oil is partially treated and may require a warning label dependent on the degree of treatment.

Aromatic extract oil
Aromatic extract oils contain a high aromatic content and usually require warning labels.

Water-reducible systems
The main resin used in early formulations was shellac, but in view of the increased cost of this material, alternatives have become more widespread. The most common being maleic and fumaric resins. The grade of resins used differ from those in 'oil' based varnishes in having a high percentage of maleic or fumaric acid, and in being unesterified giving the high acid value (200–300 mg KOH/g) required for solubility. Other resins, such as unmodified alkyds and some grades of acrylics, can also be used. The resins are usually dissolved by heat into the glycol fraction and some of the alkaline material such as monoethanolamine, triethanolamine or urea can be incorporated at this stage, the balance being added in the ink formulation.

Solvents/reducers

As noted above, the high boiling point petroleum distillates used will mainly have an initial boiling point of above 260°C, with a limited use of 240/260°C fraction. These are not strong solvents and some quick set varnishes require the inclusion of a 'bridging' agent to increase the solvency power of the system.

High boiling point alcohols have been used for this purpose and, since they are more effective than distillates, only small additions are necessary to reduce viscosity and improve flow and press stability. Some care has to be taken to avoid the loss of these 'bridging' agents through the heat generated during the milling process, otherwise the solvent balance will be altered with consequent problems of drying on the press. In non-drying systems and those drying systems where some loss in setting can be tolerated, 0.5 poise 'spindle' oil makes a useful reducer and also stabilises the system.

Mineral oils have been dealt with in the previous section, when used as the major ink component, but small additions can also be made to other types of formulation, to lower the viscosity and as an aid to increase penetration drying.

In water-reducible inks, the main solvents used will be diethylene or dipropylene glycol, but small quantities of water may also be included as a diluent where the formulation can tolerate it without the loss of stability.

Higher molecular weight glycols can also be used, in small quantities, to improve press stability, although the more normal method of doing this is to increase the level of pH, but without causing drying problems on the substrate.

Additives

Additives are those materials which are present in a formulation to promote:
  (1) the drying properties;
  (2) the rub resistance;
  (3) the press stability;
  (4) the low mist character; and
  (5) the dispersion and rheology of the ink.

Driers
The chemical crosslinking of a 'drying' ink system is promoted by the addition of 1−2% of certain organic soaps of metallic elements. Cobalt, manganese, cerium and zirconium compounds are the most common types of drier present in letterpress inks and may be included in liquid or paste form. They may be used either alone or as a mixed drier to achieve the end result required. The use of excess cobalt and manganese driers can lead to press drying problems without resulting in a harder print surface. Since there is no water present from the fountain solution as is the case of lithographic printing, the use of perborate driers is hardly effective in letterpress inks, which would have to rely on the moisture present in the substrate or atmosphere.

Rub resistance
The addition of waxes can improve the rub resistance of a drying and semi-drying ink system, either by making the ink film tougher or by giving it extra slip. Although petroleum-based waxes, for example, paraffin and montan, and natural waxes, for example carnauba, have been used in the past, most modern formulations will rely on polyethylene (PE) wax to improve rub and scratch resistance. Polytetrafluoroethylene (PTFE) wax will be used to reduce the friction level of the ink film surface and impart 'slip' properties.

Waxes will often be added to the formulation as 'pastes' or 'dispersions' with the ink vehicle used, but the introduction of 'micronised' grades has reduced the need for this extra processing. Blends of PE and PTFE waxes may be used and some mixtures are being marketed by suppliers.

Care has to be taken, if the end use involves delayed superimposition of a second colour, UV varnish or laminate film, to avoid the use of excessive amounts of wax paste in the first colour, as this interferes with the 'lay' and keying of the subsequent film.

Press stability
Although most inks are required to dry on the substrate, unless a suitable balance is achieved, those drying by oxidation can also dry on the rollers

and forme of the press. The types of antioxidants used are detailed in Chapter 4 and the formulator will choose from these the material which gives him a dry print with a stable ink. Volatile anti-oxidants can cause rewetting problems in printed stacks, while non-volatile types can retard oxidation drying too far. Since these materials can be very effective in their anti-oxidant properties and only small amounts are required in the formula, they may be added in reduced forms, using as extender a high boiling solvent, oil or some of the vehicle, to minimise variations in the amount added during production. Other types of material which have found a use in letterpress formulations, to retard drying, have been greases such as yellow or white petroleum jelly, but these will appear mainly in non-drying or slow-drying types of ink.

The use of monoethanolamine, triethanolamine and urea, in water-reducible systems, not only solubilises the system by achieving chemical balance but any excess acts as a retarder.

Low mist additives

Many of these are available, the most common being the range of bentonite clays, which can be added as pastes, or with the newer grades, as a stir-in material. Since these also tend to reduce the flow characteristics of an ink, a balance between reducing the mist level and flow length has to be achieved.

Dispersion and flow aids

The wetting characteristics of an ink system will usually be obtained by careful choice of the resins and oils involved. For example, in rotary news inks and similar non-drying systems, the most common wetting agent is an asphaltum complex solution in mineral oil. The correct addition, which may only be 2 or 3%, but can be higher, will improve the dispersion of the pigment and give better flow without causing excessive strike-through or ink mist. Other materials which help pigment dispersion are being marketed and several suppliers are now offering ranges of hyperdispersants applicable to individual pigments for optimum effect.

5.5 LETTERPRESS INK FORMULATION

Letterpress inks will be formulated to be as versatile as possible, to run on a number of types of presses and substrates. Examples are given of a range of ink formulations suitable for the conditions stated, which can be modified to suit the other requirements of local conditions.

Platen ink for absorbent papers

Much of the work carried out on platen presses can involve absorbent papers and requires a quick turn around. An ink which penetrates the paper but remains open on the press is therefore advantageous and can be formulated on similar lines to a flat-bed newspaper ink, with higher viscosity and shorter flow:

Carbon Black (CI Pigment Black 7)	19.0
Reflex Blue	2.0
70 poise mineral oil	55.0
0.5 poise mineral oil	9.0
15 poise bodied linseed oil	10.0
Asphaltum solution	5.0
	100.0

Cylinder press ink for uncoated papers

An ink formulated for a more general range of uncoated papers will require to have some drying properties. The incorporation of some resin giving the ink a degree of tack and flow, and the addition of slightly more blue toner can also be expected:

Carbon Black (CI Pigment Black 7)	18.0
Reflex Blue	4.0
Long oil alkyd	16.0
0.5 poise mineral oil	5.0
Varnish[a]	55.5
10% Cobalt drier	0.5
Antioxidant	1.0
	100.0

[a] Consists of a hydrocarbon resin or modified phenolic resin cooked into a blend of linseed and mineral oils.

Quick-set inks for coated paper

This type of ink will usually be based on a varnish formulated from resin, drying oil and distillates as described in the section dealing with resins. The choice of resin and ratio of resin/oil blend to distillates are determined by the quality of the substrate and the properties required from the final print.

The higher the absorbency of the substrate, the higher the viscosity of the resin/oil phase needs to be, in order that chalking of the print does not occur:

Calcium Carbonate	8.0
β Phthalocyanine Blue (CI Pigment Blue 15.3)	12.0
Gloss varnish	49.0
Setting varnish	16.0
Polyethylene wax paste	5.0
Anti-set-off paste	3.0
Cobalt/manganese drier	1.0
280–320°C distillate	6.0
	100.0

Letterpress ink drying by oxidation

An ink drying by oxidation for use on foil board or other impervious substrates would be formulated on the lines of:

Carbon Black (CI Pigment Black 7)	20.0
Reflex Blue	6.0
Long oil alkyd	30.0
Modified phenolic/tung oil varnish	30.0
Polyethylene wax paste	8.0
260–290 distillate	4.0
Cobalt/managanese drier	2.0
	100.0

The small percentage of distillate shown can be tolerated, but higher amounts would tend to be trapped in the ink film and make it easier to remove from the surface.

Water-reducible inks

For letterpress printing on corrugated boxes, a typical formula would be:

Water-reducible red

Blanc Fixe (CI Pigment White 21)	10.0
Rutile Titanium White (CI Pigment White 6)	5.0
Lake Red C	14.0
Varnish	54.0
Diethylene glycol	8.0
Wax paste	5.0
Amine	4.0
	100.0

Varnish
High acid value

Maleic resin	50.0
Glycol	40.0
Amine	10.0
	100.0

The printing involved will mainly consist of solids, type and linework and the print usually has a semi-matt appearance. On this type of substrate, two colours can usually be superimposed, wet-on-wet without tack grading, but on more solid boards, tack adjustments with mixtures of water and glycols will be required. A low viscosity ink aids distribution on the rollers and the transfer properties from the rubber stereo to the kraft paper.

Process inks

The theory of four-colour reproduction was dealt with in Chapter 3. The use of the letterpress process is now hardly known in the UK for this purpose and BS 4160 : 1967 (European Reference CEI 12/66) has recently been made 'obsolescent' by the BSI. This standard specified the colour, colour strength, transparency, lightfastness and solvent resistance of the three primary colours — yellow, magenta and cyan. It also referred to a previous British Standard for letterpress process inks (BS 3020 : 1959) for a suitable black ink.

Process magenta letterpress

Calcium 4B toner (CI Pigment Red 57.2)	15.0
Polyethylene wax paste	3.0
Cooked quick-set vehicle[a]	32.5
Gloss quick-set vehicle[b]	28.0
Cobalt/manganese Drier	0.5
280–320°C distillate	20.0
Antioxidant	1.0
	100.0

[a] Based on modified phenolic resin, cooked into a long oil alkyd and linseed oil and let down in 280–320°C distillate.
[b] Based on modified phenolic resin, cooked into a long oil alkyd, let down with 280–320°C distillate and aluminium gelled.

Similar formulations for the yellow and cyan would contain 15% Diarylide Yellow (CI Pigment Yellow 13) and 14% β Phthalocyanine Blue (CI Pigment Blue 15:3) respectively, to give the correct chromacity figures when printed in accordance with the standard. Because of the thicker ink films laid down, tack grading is essential, when printing two colours wet-on-wet.

Newspaper coloured inks

The titles and late news items are sometimes printed on very small units which have very limited rolling capacity. Due to the high operational speed of these units, considerable ink spray can be developed. Coloured inks are also widely used in provincial papers to give impact to advertising. These are usually printed on page wide units which present the formulator with similar problems to that of the seal or headliner colours. Occasionally, if the amount of coverage of colour warrants, a complete unit will be used, but more usually, a half-deck, and the careful design of the page layout will provide the necessary colour.

A typical formulation would be:

Chlorinated Para Red (CI Pigment Red 4)	10.0
Extender pigment	10.0
Resin varnish[a]	40.0
Mineral oils	25.0
Long oil length alkyd resin	10.0
Wetting agent } Antioxidants, etc. }	5.0
	100.0

[a] A hydrocarbon resin or low melting point phenolic resin dissolved in mineral oil.

The amount of pigment/extender is varied depending on the properties of colour strength, etc. that are required and the ratio of resins, solvents and oils adjusted to suit the individual units being used.

Rotary black inks for newspapers

The formulations used for printing newspapers have to meet the following targets which require good tolerance for trouble-free running:
    (1) Low misting character;
    (2) Non-drying on press, but reasonable rub resistance on paper;
    (3) Easy delivery to the press;
    (4) Good transfer to the substrate;
    (5) Low linting character;
    (6) No adverse effect on stereos used;
    (7) Good penetration on newsprint but a minimum of strike through during the normal life of the paper;
    (8) Good de-inking properties;
    (9) Economical to use.
The colour strength is critical in formulating newspaper inks. If the ink is over-strength, reduced film weights are used, and this may give rise to poor coverage and uneven inking. Problems of linting and transfer may also occur if an insufficient volume of ink is being replaced on the forme and ink distribution roller chain. Conversely, with low-strength ink, problems of strike through, misting, flooding the stereo and marking off on path rollers occur, as high volumes of ink are transferred through the printing system to obtain the correct density.

The possible health risks that may be associated with the inhalation of ink mist have been monitored for a number of years by the printing industry.

According to the International Agency for Research on Cancer, the carcinogenicity of some grades of mineral oil may be linked to the presence of certain polycylic aromatic hydrocarbons (PAHs) of which Benz-α-Pyrene (BαP) has been one of the most widely studied.

In 1984, a subcommittee of the Printing Industry Advisory Committee of the Health and Safety Commission, produced an advisory leaflet, and recommended that a time-weighted average exposure to ink fly of 1.5 mg/m^3 was a realistic target for individual exposure. There is some evidence to suggest that the measurement of BαP can be used to indicate the level of total PAHs present in the ink and a target of 10 parts per million, present in the ink, has been suggested in the document.

The actual amounts of ink used per copy vary with the content of the edition and the absorption of the newsprint used. A guide to how much ink is used is: 1.7 g of rotary news ink per 1000 copies of a 32-page tabloid. (This is based on experience rather than a calculated formula.) While the amount of ink used per copy is extremely small, the total amount of ink used by a newspaper publisher can be over 3000 tonnes per annum. The cost of the ink is therefore a significant factor in newspaper production costs, and can be a limiting factor which has to be taken into account by the formulator, restricting the use of certain raw materials.

A further economic factor to be taken into account is the need to include a proportion of reclaimed material in the newsprint paper manufacturing process. Before this happens, the printed paper must be de-inked to maintain the final brightness of colour of the newsprint. The de-inking process consists of treating the paper with alkali to reduce the paper to fibre form allowing the ink to separate and float to the surface from where it is removed. The paper slurry is then neutralised and included in the fresh paper-making cycle.

While this process is reasonably easy with conventional rotary letterpress inks, harder drying formulations can prove more difficult to separate.

The formulations in Table 5.2 show the type of variations in formulae used to obtain the correct properties for various presses, plates and papers. They must only be regarded as starting formulations to give the technologist guidance. Individual printers require inks with properties specific to their needs, and each type of manufacturing plant will produce inks of slightly differing properties using the same materials. The formulations have been chosen to give similar print performance under the various conditions used.

Formula A: General-purpose low mist black
This ink is capable of running at very high speeds (50 000 cph) on a variety of presses and newsprints. It has a blend of mineral oils and flow aids which will give good density with other good print qualities of strike-through, and non-marking on the turner bars of the press. The anti-spray additives are added to give the ink low mist and ink sling properties.

Formula B: Ink rail
A differing blend of mineral oils and flow aids to give excellent print qualities to an ink which is thinner and longer flowing than Formula A. This is necessary to allow the ink to freely circulate on the ink rail distribution system.

Formula C: Page-Pak
This formulation has longer flow and is thinner than both the previous formulations because of the restrictions in the Page-Pak distributing systems.

Formula D: Keyless inking (indirect flexo)
The formulations of news inks for this type of ink metering system are similar to those for conventional units. The viscosity of the ink needs to be low enough to fill the cells of the etched roller and the flow sufficient for the inks to be readily pumped. The level of pigmentation is determined by the depth and shape of the cells of the etched roller.

Formula E: Flat-bed rotary
The flat-bed rotary formulation included in Table 5.2 would normally be supplied in tins or buckets for use on the slower flat-bed and rotary

Table 5.2

| | Formulae | | | | | |
	A General low mist	B Ink rail low mist	C Page-Pak low mist	D Keyless inking	E Flat-bed knifeable	F Better rub
Carbon blacks news ink grade	13.0	13.0	13.0	12.0	14.0	14.0
70 poise mineral oil					77.5	
9 poise mineral oil	68.0	67.0	71.5	61.5		
2.0 poise mineral oil		13.5		18.0		35.0
0.5 poise mineral oil	10.0		7.0		4.5	
Asphaltum solutions OR flow aid/surfactants	5.0	4.0	3.0	5.0	2.0	5.0
Kerosene/distillate	2.0	1.0	4.0	3.5	0.5	260/290°C 18 : 0
Anti-spray additives	2.0	1.5	1.5		1.5	
Resin varnish						28.0
Viscosity (poise)	20–25	14–18	11–14	15–25	100–150	15
Flow	5–7 inches (127–178 mm)	7–10 inches (178–254 mm)	8–10 inches (203–254 mm)	6–9 inches (152–228 mm)	½–1 inch (13–25 mm)	7–8 inches (178–203 mm)

presses. The relatively low ink usage and the position of the ducts is such that the ink is normally transferred to the duct manually using a palette knife. This type of ink is commonly referred to as 'knifeable'. The slow press speeds involved also required inks of a higher viscosity and shorter flow than previous formulations and the inclusion of anti-spray additives is not normally required.

The amount of Carbon Black in the formulation needs to be increased compared with high speed rotary presses in order to achieve similar print densities due to the different transfer properties of these inks.

Formula F

The formulation has been included to show that rub resistance of conventional news inks can be improved by the incorporation of an amount of varnish. This varnish usually consists of hard resin, typically hydrocarbon resin, mineral oil and distillate.

Use of refined oils

The formulations described so far have been based on mineral oil which contains an amount of aromatic content. This aromaticity determines the wetting of the carbon blacks and also affects the flow of the ink. However, in some instances, the HASAW regulations are directing the formulator to the use of more refined mineral oils. Mineral oils which have been hydro-refined or solvent-treated are available but as well as being considerably more expensive than the aromatic mineral oils, they do present the formulator with a considerable challenge to achieve comparable properties of viscosity, flow, and yield value. *For example:*

Carbon Black	11.5
9 poise solvent-refined mineral oil	78.0
Surfactants and flow aids	3.0
Anti-spray compounds	3.5
Distillate	4.0
	100.0

The formulation based on solvent-refined oils is superficially similar to that of the general low mist ink in Table 5.2. However, the grade and amount of Carbon Black may require to be changed in order to overcome the poor pigment wetting properties of the solvent-refined oils.

The flow aids and surfactants have to be chosen with great care to give the best results with the particular grade of refined oil chosen. An increased amount of anti-spray compound is required to maintain the same low mist properties as an ink based on non-refined oil.

Generally speaking, the press working properties of an ink formulated on highly refined mineral oils falls short of that obtained with the more conventional oils.

5.6 INK-RELATED PROBLEMS AND THEIR POSSIBLE SOLUTIONS

Many of the problems associated with letterpress printing are due to the relatively high film weights carried, which affect the setting and drying properties of the inks used and also contribute to the misting characteristics on rotary presses. Strike-through and 'filling in' problems can arise with the thin inks used on absorbent newsprints, while inks with too high a tack level can lead to 'linting'.

Mottle

Mottle shows itself as uneven colour, caused by uneven inking of the paper. It is most likely to occur on work printed on platen presses and when the ink is too weak to obtain the correct density with normal film weights. The excess ink which has to be carried, shows up the local variations of film thickness and this is particularly noticeable when transparent blues, greens and browns are being printed. Over-reduction by the printer to overcome a tack problem is a common cause of mottle and this can be helped by recommending a 'gel' reducer. Inks with some opacity are less prone to give this problem.

Set-off

The more an ink 'stands up' on the surface of the paper, the more it is likely to 'set-off' on to the reverse side of the following sheet. Therefore, set-off is increased by:
— Thick films of ink.
— Non-absorbent substrates.
— Smooth papers (which allow close contact between sheets).
— Highly resinous inks which dry to give an ink film with a high tack surface.
— Slow setting inks.
— Faster running presses.
— Large stacks of print.
Formulating the ink to have sufficient colour strength to achieve the density of colour required, at a low film weight, will therefore help reduce 'set-off'.

The formulation of inks containing resinous varnishes must ensure that the tack level of the ink film as it sets, does not rise sufficiently to allow it to adhere to the underneath side of the following sheet. Choice of the resin/solvent blend to give good solvent release is therefore important. Some rise in tack level will almost certainly occur and can only be prevented from becoming a problem by the use of anti-set-off sprays on the press.

Marking-off

Newspaper inks which 'set' on top of the substrate, will tend to 'mark off' on to the path rollers of the press. Usually a deposit of ink and paper fibres builds up on the rollers causing a background tint of the ink, while the prints will also exhibit poor rub properties due to the heavy ink film deposited. Keeping the ink as strong as possible, with a low viscosity and reduced resin content, helps overcome this problem, but needs to be balanced against the requirements of strike-through and rub resistance.

Second impression set-off

This is a special form of set-off associated with work which is 'perfected', that is to say, printed on both sides in close succession, as on newspaper rotary machines. When the second side is being printed, the first is face downwards on the second impression cylinder, and the print is transferred under pressure. It has become a particular problem with the use of water-based flexographic inks for newspaper printing, which have drying properties. The use of carefully chosen substrates and longer web leads between printing units, have helped reduce the problem which is still the subject of much technical investigation.

Drying and rub-resistance

Bearing in mind the 'drying' methods mentioned above, the recommendation of the correct ink for a particular substrate is highly important, especially since so many letterpress inks are formulated to be 'non-drying'. Inks of the 'quick-setting' type, which have 'oxidation drying' capability should be suitable for most requirements, but the use of extra wax paste and liquid driers may be necessary on harder stocks, provided this does not lead to the ink drying on the press during short stoppages.

During the production run of a newspaper, the presses will stop for various reasons. If the plates are being changed or for web breaks, the stoppage may be prolonged. The presses are so designed that the impression between printing forme, ink rollers and paper is not released when the press stops and they remain in contact with one another. The ink must not dry on the rollers as this could cause uneven inking. Traditional mineral oil based inks are quite satisfactory, but the search for harder drying, more rub resistant inks, can lead to problems.

Misting

Ink mist and sling are the two components of the phenomenon which is sometimes described as ink fly. The ink on a pair of fast rotating rollers is drawn out into fine filaments as the ink is transferred from one roller to the other. These filaments of ink split at each roller simultaneously and the central part is released into the atmosphere as a microscopic particle of ink. These particles are electrically charged and move away from the rollers into the atmosphere.

Long flowing inks are more prone to misting than short flow inks and the addition of 'low mist' additives is the usual cure, although a compromise with 'ease of delivery' to the press must be part of the specification.

Strike-through

The problem was stated in the section on quality control above, and is always present, to a certain extent, when printing on newsprint. It can be minimised by keeping the viscosity as high as the press and 'set-off' properties will allow. The use of refined grades of oil also helps reduce the problem because the contrast of the colour of the oil, with that of the substrate, is not so marked.

Filling-in

This occurs when the non-printing area becomes contaminated and begins to show on the printed copy. The main reason for this problem is the unsuitability of the ink for the paper. An ink with too high a tack value for the surface structure of the substrate being used, will cause the short paper fibres from the surface to be transferred to the inking system. These fibres will eventually adhere to the forme and 'fill in' the non-printing area until they themselves start to print. This is commonly called 'linting'.

An alternative cause is where a low viscosity ink does not contain sufficient cohesive structure and flows down the side of the 'relief' character, and is transferred to the substrate under the impression pressure.

Poorly dispersed ink can also achieve the same result, as the unground material 'fills in' the non-printing area.

Hanging back in the duct

Hanging back in the duct is usually associated with short flowing inks which are liable to 'set up' quickly and this may be due to over-pigmentation, the inclusion of too much wax paste or very short vehicle or oil. On rotary presses, with open ducts, the 'lint' from the fast moving webs easily contaminates the ink and unless good housekeeping is employed, can cause it to lose its flow properties.

Under normal circumstances, the ink should have sufficient flow to follow the rotating duct rollers during the press run, and in mild cases, where it 'hangs back', the addition of a bodied varnish or oil may be sufficient to overcome the problem. More severe 'hanging back' often requires complete reformulation.

Where bulk rotary newspaper black ink is concerned, the problem will almost certainly have shown previously, because of the need to pump these inks, often for long distances. An increase in the asphaltum complex solution, used in the formulation, will cure most problems where the ink has high yield value.

5.7 NEW DEVELOPMENTS

During the past year, several keyless inking systems have been installed and currently many more are being discussed. These, combined with the use of modern generation photopolymer relief printing plates, will offer challenges to the ink chemist in the development of suitable inks, giving improved results from what is basically, still the letterpress printing process. In the UK at least, developments of inks for general printing purposes, will be aligned with those for offset lithographic printing. The use of rotary letterpress for narrow web label printing is gaining ground from flexography. Some converters are adding rotary letterpress units to their flexographic operation. It is claimed that letterpress gives better consistency and higher print quality particularly on very small text and with fine screens. This will involve extended use of IR drying or UV curing inks, depending on the drying method chosen.

Photopolymer plates

Photopolymer plates are widely used by provincial newspapers, and this technology has recently been implemented by a number of national newspapers.

A new generation of plates has recently been introduced using advanced polymer chemistry, to give improved tone reproduction and smoother ink lay down. When printing has been carried out under carefully controlled conditions and using optimum ink and paper, results close to those obtained by the offset process have been achieved.

Keyless inking

During 1980, a project was initiated in the USA, by the American Newspaper Publishers Association (ANPA), to develop a new ink metering system for a lightweight press they were investigating. They required an inking system that would apply a consistent and reliable ink film over a wide range of press speeds. The system that was developed comprised an etched roller rotating in a trough of ink, the excess being removed by a doctor blade. The ink film is transferred to the printing forme by one or two rubber rollers. Because the ink feed system does not have control keys to adjust the amount of ink being used, the system has become known as the 'Keyless Inking System'. It is sometimes referred to as 'Indirect Flexo' because of the use of the etched roller that is also used on flexographic presses, and because this roller is not in direct contact with the printing plate.

A number of European press manufacturers showed interest and during 1981, Crabtree-Vickers joined the Newspaper Society and the *Leicester Mercury* to evaluate an ANPA inking system which was under test.

The inking system replaced an existing unit and following the success with this unit, Crabtree-Vickers have developed a series of 'Retrofit' units to their own design which have been specifically engineered to replace existing printing units on Crabtree, Wifag, MAN and Goss presses. Keyless inking has also been developed in Europe by Koenig & Bauer, who are incorporating the mechanism into new press designs.

Description of the unit

Ink is fed to the engraved drum by an ink chamber. This chamber is formed into the top surface of a full width heavy cast-iron stretcher which is the main structural member of the inking system (see Fig. 5.1).

Fig. 5.1 Keyless inking unit

The doctor blade is an adjustable carrier, fixed to the precision machined top surface of the stretcher to ensure it is perfectly flat and rigid, thus giving an even blade contact across the full width of the drum, with the lightest contact pressure possible. Ink is constantly circulated within the system and is taken from the reservoir at the base of the unit and fed to the chamber by a motorised pump, through a filter. When the unit is in operation, the excess ink on the etched drum is removed by the reverse angle doctor blade and returned to the ink reservoir.

The ink is delivered to the printing cylinder by two forme rollers. These rollers play an important part and need to have a rubber covering of between 30−35 shore hardness when using hard polymer plates.

The following advantages for this type of system that are claimed by the manufacturers are:

(1) Automatic even ink coverage which leads to reduce printed waste, and gives better solids and halftones.

(2) Lower energy consumption due to the fewer inking rollers required.

(3) The lack of any adjustment to the operation, when printing, eliminates the need for operator attention.

(4) Ink mist reduced to negligible levels due to the reduced number of inking rollers.

(5) The ability to benefit from modern technology inks (i.e. water emulsion and resin based) which gives sharper print, reduced rub-off and less strike-through.

Flexography

The introduction of keyless inking and the use of advanced technology polymer plates is an attempt to increase quality without the economic drawbacks of the web offset process. A further development has been the introduction of the flexographic process into newspaper production. Purpose-built flexographic units are being developed and have been built with the particular needs of newspaper production in mind. A number of installations are in production in Europe and the USA. Water-based inks are used for both technical and environmental reasons and they dry to a hard rub-resistant print.

As in packaging flexographic presses, paper impression needs to be released when the press slows to crawl speeds, to avoid paper stick. The etched roller must also continue to revolve when the web stops for edition changes, paper breaks, etc. An automatic ink wash-up system is also necessary should the press be completely stopped for all but short periods of time.

The use of water-based inks also presents unique problems to the ink formulator, and the plate, paper and press manufacturers. Early test runs have shown that they give rise to second impression set-off if the balance between ink drying, printing impression, and the distance between printing units, is not correct.

The use of water as the solvent in the ink, causes all but the most controlled newsprints to lint, which combined with any softening effect the water may have on the polymer plate, leads to 'filling in' and deterioration of the print quality.

The problems are being solved by co-operation between the various parties involved and the writer looks, with great interest, towards this process in the production of newspapers.

CHAPTER 6
Lithographic Inks

Chapter 2 of this manual has dealt in detail with the mechanics of the lithographic printing process. In this chapter, we will be looking at lithographic ink formulation and raw material selection but first of all, it is worth while to look at the type of printed material produced using the lithographic printing process.

Nature of the products

The lithographic printing process is very versatile and used to produce a wide range of printed matter. The bulk of this falls into two main categories: (i) publication and (ii) packaging (Fig. 6.1). Publication work ranges from newspapers and leaflets through magazines, catalogues and brochures to books.

Newspaper production around the world is dominated by the use of coldset inks that penetrate into the substrate, although the printing technique may vary. Rotary letterpress (see Section 2.1) is still widely used but web-offset lithography and more recently flexography are displacing rotary letterpress when new plant is installed.

Magazines, catalogues and brochures are other major products of web offset lithography, with heatset drying being a requirement for all but the lower quality publications on newsprint or similar stocks. Shorter run-length publication work is produced by sheet-fed litho although some letterpress production still exists. Additionally, web offset heatset is penetrating this market area as a viable alternative as economic run-lengths are continuously being reduced, particularly for the smaller eight-page presses, because of increasing automation and control of machine make-ready. At the other end of the spectrum, gravure is a major competitor to heatset for magazines, catalogues and brochures for mass circulation. (Fig. 6.1). The position between the two processes in this market area is constantly shifting as improved heatset quality and quicker, cheaper gravure cylinder origination challenge each other's traditional strengths. Other major publication products for lithographic printing are directories and books. For both, coldset web-offset is a common option, although many books are printed on sheet-fed perfecting presses (see Section 2.2). Letterpress still competes with lithography in this area of the print market.

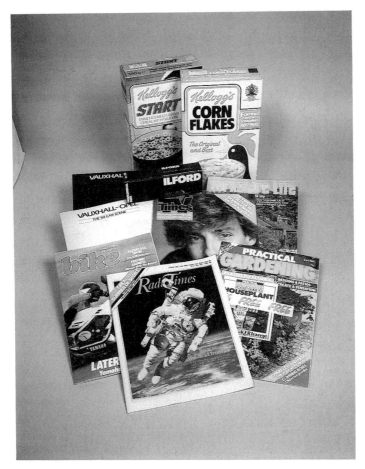

Fig. 6.1 The wide scope of lithographic printing is represented by packaging, periodicals and promotional literature

Litho-printed packaging is represented by the two major product groups of folding cartons and labels. Folding cartons are particularly diverse in their design and packaged product requirements. Their design range covers the spectrum from the aesthetic appeal of luxury cosmetic, spirits and tobacco packaging to functional strength for heavy mechanical components. Product-dictated performance needs a range from soap and detergent resistance on relevant cartons to low odour and no tendency to taint the contents with food and confectionery packaging. Sheet-fed litho printing is so successful in these areas that it is the dominant process for producing folding cartons. Gravure competes for some long run-length, fixed cut-off cartons, and a little letterpress printing remains, but otherwise sheet-fed litho is the automatic choice. The situation is similar for paper labels, both in terms of diversity of requirement and dominance of sheet-fed lithographic printing. However, in self-adhesive label production, flexo and UV rotary letterpress have become the major processes used on specialised narrow web presses. Nevertheless, it is worth noting here that the main use of litho printing is on paper and board substrates. Attempts to formulate a wide range of wholly satisfactory inks for film, plastics and foil substrates are impeded by litho restrictions. This is not to say that such materials cannot be printed by lithography, but achieving full ink adhesion on the stock is often more problematic than with gravure, screen or flexo systems.

Nature of the process

Lithography is a printing process that relies on a chemical distinction between image and non-image area rather than any physical relief differentiation. With the exception of driographic, or waterless plates (see Section 6.8) all types of lithographic plate require application of an aqueous fount solution to activate and maintain the distinction between ink-accepting image areas and ink-repelling non-image areas. To explain why films of ink and fount form in the respective areas requires a study of surface tension.

Surface tension is described as the internal pressure that restrains a liquid from flowing. Polar liquids such as water have a high surface tension, non-polar liquids have a low surface tension as does a litho printing ink. When a drop of liquid is placed on a solid surface the relationship between the surface energy of the solid and the surface tension of the liquid determines whether the droplet will spread across the surface or form globules. If the liquid is placed on a surface of lower surface energy than itself, it will form into globules and fail to wet the surface. Conversely, if the liquid has a lower surface energy than the solid surface, the liquid will spread across, or wet, the surface. If a liquid wets a surface, it is said to have a low contact angle; if it forms globules, it is said to have a high contact angle (see Fig. 6.2).

When applied to lithography, the distinction between image and non-image areas is sufficient to cause the following effects:

(1) aqueous fount solution spreads on the non-image area because of its low contact angle.
(2) aqueous fount solution in the image area is unable to form a continuous film because of its high contact angle, leaving the image free to accept ink.

In the absence of water, the image/non-image distinction is not, however, sufficient to stop ink, with its lower surface tension, from wetting both areas. Hence the need for water in lithography to generate a barrier between the ink and the non-image area. This highly simplified, theoretical explanation of the principle of lithography is based on a static situation. In practice, the situation is complicated because ink and fount are being continuously applied from roller systems to the printing plate. This dynamic situation poses the following questions:

(1) With its lower surface tension, why does the ink not displace the water in the non-image areas?
(2) What happens to the droplets of water in the image area?

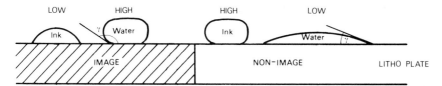

Fig. 6.2 Litho plate surface characteristics. The low contact angle (γ) of water in the non-image area allows wetting with water. Water's high contact angle in the image area causes it to form into minute globules

(3) Why does the ink not form a film across the top of the non-image fount layer?

(1) The attraction of fount solution in the non-image area is promoted by the use of hydrophilic colloids, such as gum arabic, activated with acid, which are present on the surface of the plate and incorporated into the fount solution to maintain replenishment during the print run. Such colloids adhere to the non-image surface and increase its hydrophilic nature such that displacement of the non-image fount film by ink does not occur.

(2) The high contact angle of fount solution on the image surface prevents the formation of a continuous film of damp. However, reticulated droplets of dampening solution do remain and could be expected to prevent ink transfer in the image areas that they cover. This is avoided with properly formulated litho inks since they are able to emulsify the tiny fount droplets into the ink film to give a continuous image. This limited attraction between ink and water and its ability to take up water are fundamental to the success of the litho process.

(3) There are number of factors at work which prevent the ink from forming a film across the top of the non-image fount layer. To understand this dynamic situation, it is necessary to look more closely at what is happening in the nip where ink forme roller, ink, fountain solution and plate non-image area come into contact. As we have seen, despite their differences in surface tension and chemical nature, there is a fundamental attraction between ink and water. In the nip, a composite film of ink and fountain solution is formed together with partial emulsification of water in the ink. At the nip exit, it is the relative viscosities of the various components that determines how the film splits. (The force required to split a liquid is directly proportional to its viscosity.) Thus the much lower viscosity fount solution will split in preference to the ink film, leaving a film of fount on the non-image area and transfering some droplets of fount onto the ink film which subsequently emulsify into the ink. Thus, according to this theory, the ink will not form a layer over the fount solution.

Having achieved clear and precise inking of the image on the plate, through the mechanisms described, the ink is then transferred to the stock to be printed via a resilient offset blanket (Fig. 6.3). This happens in all litho printing with the exception of the direct, or di-litho, printing of newspapers that is carried out on converted rotary letterpress news-

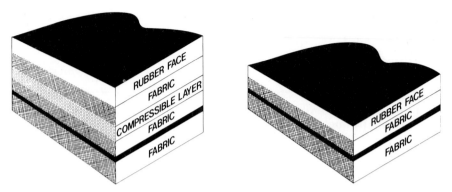

Fig. 6.3 Lithographic blanket construction: — compressible (left) and conventional

paper presses. For this reason the terms litho and offset are virtually synonymous in the printing industry. The use of a rubber blanket ensures good plate life and high quality transference even to substrates of poor smoothness.

Lithographic plates can be prepared at relatively low unit cost as base metal preparation and the chemistry employed in coatings are sophisticated but not unduly expensive. Processing of plates through exposure, development and press preparation involves simple techniques, often akin to photographic processing, and can be partially automated to ensure high productivity. Fine detail line and half-tone work lithographic printing is generally cheaper and quicker than comparable stages in the other major printing processes.

The method of image/non-image distinction on a litho plate permits high quality graphic reproduction. Fine detail in line and half-tone work can be maintained, particularly where fine-grained aluminium plates are used (Fig. 6.4). At the same time, the use of a resilient offset blanket ensures smooth transfer of large solid areas even on to lower quality paper and board stocks.

In summary, offset lithography is characterised by high quality and low origination cost and is particularly suited to text, solid and tone work on all the types of paper and board.

6.1 GENERAL CHARACTERISTICS OF LITHO INKS

Lithographic restrictions

The mechanisms involved in image/non-image distinction by ink and fount impose both chemical and physical restrictions on lithographic inks.

In order to achieve controlled emulsification, lithographic inks are generally produced at relatively high viscosity, typically in the range 100–300 poise. Viscosities below these characteristically 'paste' levels can be satisfactory under certain lithographic conditions but emulsification

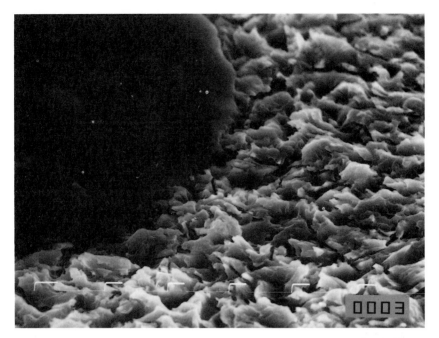

Fig. 6.4 Electron photomicrograph of a litho plate surface showing grained non-image area and half-tone dot image area (top left)

problems, in particular, may occur. This is because, even where there is no chemical affinity, two liquids become more readily emulsifiable together as their viscosities approach one another. Aqueous fount solution is a very low viscosity material and there exists a significant physical barrier to over-emulsification between typical litho inks and fount. It is very difficult to make low viscosity inks that are not too readily emulsifiable, and hence too readily miscible, with dampening solution. Such miscibility is, of course, totally inappropriate as it would negate the whole lithographic principle by destroying the distinction between fount and ink.

With these relatively high ink viscosities, a long ink distribution roller train is required. This is because a thin, even ink film must be delivered to the plate and substantial rolling and distribution is essential to break down the viscous ink to this state. Inevitably the use of such a roller system imposes its own restrictions on the ink. Evaporation of volatile materials from the ink must be severely limited as the ink is exposed for long periods in a high surface area to volume ratio on the distribution train. The necessary rolling and distribution work is achieved by using a system consisting of alternate metal and elastomeric-surfaced rollers (Fig. 6.5). The elastomers used here and for the offset blanket surface can be chosen from a wide variety of synthetic rubbers and polymers. Nevertheless, most readily available and suitably resilient materials are susceptible to attack from strong chemical solvents, particularly under the influence

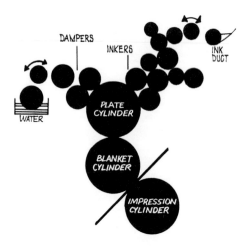

Fig. 6.5 Typical litho press roller distribution train

of the pressures and temperatures generated on an offset litho press. Such solvent attack leads to softening and swelling of the elastomers. The latter produces further physical stress in the form of increased pressures and softening renders the surface increasingly susceptible to physical dis-integration.

Ink viscosity and distribution requirements therefore prevent the use of significant proportions of both volatile materials and strong solvents. The inability to use powerful solvents in litho formulations naturally produces a severe restriction on the choice of binder chemistry, since only resins soluble in weak solvents can be considered.

Film thickness considerations

Offset lithography prints finer ink films than any of the other major printing processes. This is partly because the planographic litho plate has a relatively low capacity for carrying ink, and partly due to the extra film splitting involved in transferring ink from plate to substrate via the offset blanket. This latter effect is not always as great as it might at first appear since the performance of a modern blanket, assisted by poor ink adhesion to metal and a capillary-draw effect from the paper substrate, can lead to preferential ink transfer from plate to blanket to paper. This results in greater than 50% transfer at both nips. Nevertheless, ink film thickness on print is normally in the $1-3$ μm range.

It is therefore essential for litho inks that are required to print full-strength colours to be formulated with high pigmentation. Further, very careful manufacture is required to ensure satisfactory pigment dispersion for both strength development in the ink and freedom from agglomerates that will be significantly larger than the thickness of film transferred. Such large particles, if present in excessive numbers, can produce various adverse effects from piling on the blanket through (see Section 6.6) to poor gloss and rub resistance.

Thus, the low ink film thickness of lithographic printing influences both raw material selection, as discussed in Section 6.4, and ink manufacture, as covered in Chapter 12.

6.2   DRYING MECHANISMS

Restrictions of the process

As reviewed in Section 6.1, ink volatility at the temperatures encountered on the inking rollers (which may be as much as 15 degrees Centigrade above ambient due to frictional heating) must be minimal. Therefore, evaporation can only be employed as a significant drying mechanism for lithographic inks if heating to relatively high temperature is utilised. This is the technique employed in heatset drying of web-offset print where the web may be raised to temperatures in the range of 100–160°C through ovens 10–25 feet (3–7.5 m) long. Even here, evaporation is not the sole mechanism involved in ink drying, as will be discussed under the sub-heading of 'Quick-set mechanism' later.

Although lithography poses restrictions on the use of evaporation drying, many other drying techniques are suitable and utilised. These techniques employ both chemical and physical mechanisms, either singly or in combination. The major drying processes involved are:
(1) Penetration — physical;
(2) Oxidation — chemical;
(3) Quick-set — physical and chemical;
(4) Radiation — chemical.

Penetration

Penetration is the drying mechanism involved in the web-offset coldset printing of books and newspapers on absorbent uncoated paper stocks. If drying is defined as the conversion of the fluid ink to a solid, immobile film on print then strictly speaking the ink in this case does not dry. The ink remains fluid but becomes sufficiently entrapped in the interstices between the substrate fibres for its mobility to be severely restricted. Problems can be encountered if the ink film thickness exceeds the absorptive capacity of the paper or if there is insufficient time between printing and handling to allow full absorption. Both problems can be encountered in newspaper printing but the coldset penetration mechanism does allow the distinct advantage of high-speed production at minimum cost. Thus economics counter drawbacks in the area of marking.

Oxidation

Oxidation is the classical drying mechanism for lithographic inks, involving the oxygen-induced free radical polymerisation of unsaturated (drying) vegetable oils such as linseed and tung. It is a chemical process which can be catalysed further or accelerated by small amounts of appropriate metal, usually transition metal, driers (see Section 6.4).

The primary reaction is considered to be that of autoxidative polymerisation which proceeds by several stages:

(1) Peroxide/hydroperoxide formation;
(2) Decomposition to form free radicals, which initiate;
(3) Polymerisation;
(4) Termination.

Peroxide/hydroperoxide formation
This stage involves the attack by atmospheric oxygen at activated sites on the drying oil fatty acid chains. One such site is the methylene group adjacent to one of the carbon–carbon double bonds present in all drying oil molecules, e.g.:

$$-\overset{*}{C}H_2 - CH = CH -$$

$*$ = active methylene group.

If we consider linseed oil as an example of a typical drying oil, which has linoleic and linolenic acids as principal fatty acid components, the following reaction scheme has been proposed.

Taking linoleic acid as a specific example, which occurs naturally in the *cis* configuration:

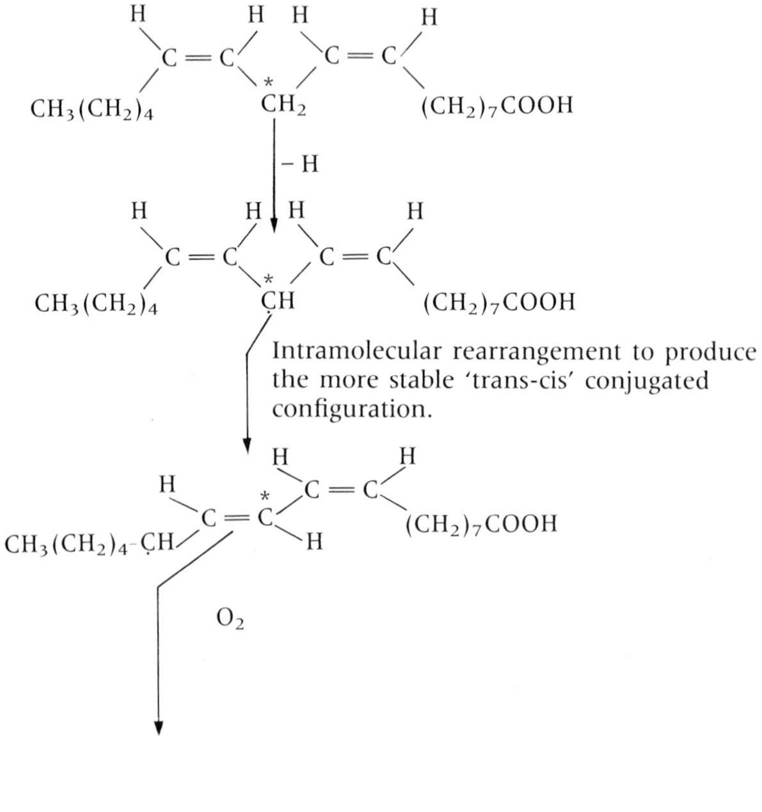
H
H H
H
C = C
C = C
*
$CH_3(CH_2)_4$
CH_2
$(CH_2)_7COOH$
– H
CH
Intramolecular rearrangement to produce the more stable 'trans-cis' conjugated configuration.
$CH_3(CH_2)_4$-CH
O_2

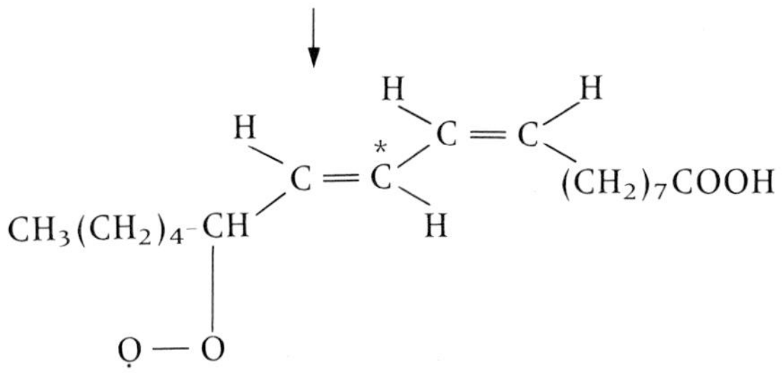

$$CH_3(CH_2)_4-CH \overset{H}{\diagup} \!\! C = C \overset{*}{\diagdown} \!\! \overset{H}{\diagup} C = C \overset{H}{\diagup} \!\! \diagdown (CH_2)_7COOH$$

with $O-O$ attached to the first CH.

This peroxy free radical can abstract a proton from an active methylene group on another fatty acid chain producing the following hydro-peroxide:

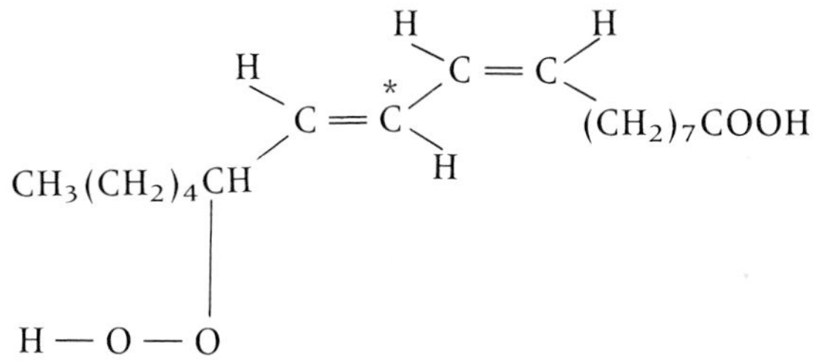

$$CH_3(CH_2)_4CH \overset{H}{\diagup} \!\! C = C \overset{*}{\diagdown} \!\! \overset{H}{\diagup} C = C \overset{H}{\diagup} \!\! \diagdown (CH_2)_7COOH$$

with $H-O-O$ attached to the first CH.

Note: The $-CH_2-$ group marked with an asterisk, one of three methylene groups adjacent to carbon–carbon double bonds in the molecule, is more reactive than the others because it is adjacent to two double bonds and is therefore the major reaction site.

Where conjugated structures are present, as in tung oil, direct oxygen addition to the conjugated system to form a 1, 4-cyclic peroxide is possible.

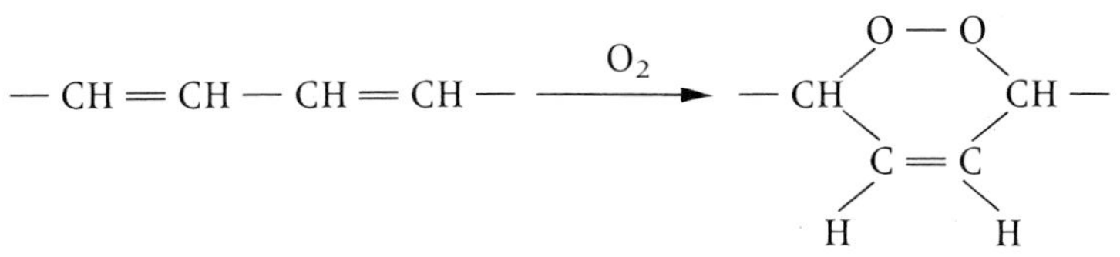

$$-CH=CH-CH=CH- \quad \xrightarrow{\;O_2\;} \quad -CH \overset{O-O}{\diagup \diagdown} CH- \;\; \text{with} \;\; C=C \;\text{(H, H)}$$

It follows, therefore, that any further oxygen addition to the conjugated hydroperoxide formed by the oxidation of linoleic acid will also give a 1, 4-cyclic peroxide.

Decomposition to form free radicals

Once hydroperoxides or cyclic peroxides have been formed, they can decompose into free radicals as follows:

$$ROOH \rightarrow RO^{\cdot} + {^{\cdot}OH} \qquad\qquad (1)$$
$$2ROOH \rightarrow RO^{\cdot} + ROO^{\cdot} + H_2O \qquad\qquad (2)$$

The following reactions are then possible:

$$RO^{\cdot} + RH \rightarrow ROH + R^{\cdot} \tag{3}$$
$$^{\cdot}OH + RH \rightarrow R^{\cdot} + H_2O \tag{4}$$

The $R^{\cdot}$ radicals formed in reactions (3) and (4) can react with atmospheric oxygen to form peroxides and hydroperoxides thus propagating the chain reaction.

Polymerisation

The radicals formed in the above reactions can add on to another molecule of drying oil, and these addition reactions can continue increasing the molecular weight until termination occurs. There is also more than one site for the growing radicals to attack and this produces cross-linked molecules. The increasing molecular weight and the cross-linking reactions cause the ink vehicle to become a solid material, encapsulating the pigment.

Termination

The $R^{\cdot}$ radicals formed in reactions (3) and (4) can also react with each other leading to the termination reaction:

$$R^{\cdot} + R^{\cdot} \rightarrow R - R \tag{5}$$

Other, minor reactions involving the other free radicals formed include:

$$R^{\cdot} + RO^{\cdot} \rightarrow R - O - R \tag{6}$$

and

$$RO^{\cdot} + RO^{\cdot} \rightarrow R - O - O - R \tag{7}$$

Nowadays, oxidation as the sole drying mechanism is normally only encountered in inks for printing impervious substrates such as foils and plastics. Even here newer drying techniques involving radiation curing have made inroads in the use of 100% solids drying oil-based formulations.

Quickset mechanism

Despite the upsurge over the last decade of radiation curing in lithography, the quickset mechanism remains the dominant drying process within the industry. It is employed extensively on all forms of paper and board. While quickset is usually associated with sheet-fed printing, the mechanism, in conjunction with evaporation, forms a part of the heatset drying process of web offset print. Quick-setting involves a physical process followed, over a longer period of time, by chemical drying.

The physical state is illustrated in Fig. 6.6. The quickset vehicle is composed of two phases which are of limited compatibility. One phase is a highly viscous solution of hard resin in drying oil, and the other phase is very low viscosity petroleum distillate. The resin/drying oil and distillate must be sufficiently compatible for the vehicle to remain as a stable, homogeneous fluid throughout ink distribution and transfer to the

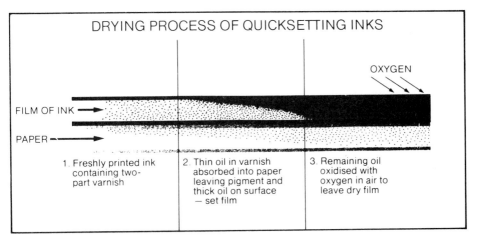

Fig. 6.6 Drying process of quicksetting inks — the principal method used by lithographic inks. It relies on rapid separation of thin mineral oil from the ink film followed by oxidation/polymerisation of drying oil

substrate on the press. However, once deposited as a thin film on an absorbent paper or board, a capillary-draw mechanism pulls the highly mobile distillate away from the rest of the ink. If the two phases are too compatible, the affinity of the resin/oil for the distillate will inhibit this penetration. However, if only limited compatibility exists the distillate will be separated from the ink and will be drawn into the interstices of the substrate coating or fibre network leaving the hard resin in drying oil (or alkyd) phase binding the pigment on the print surface. While not fully dry, the ink film reaches such a high viscosity that it loses mobility and ceases to transfer readily. At this stage the ink is said to be 'set' and will not mark the reverse of the subsequent sheet in a stack of sheet-fed work, unless the stack is subjected to hard lateral scuffing.

Setting may take from 2 minutes to over half an hour depending on the ink formulation, the printed film weight, the nature of the paper or board substrate and ambient conditions, principally temperature. Generally a short setting time is desirable, although this may not necessarily guarantee minimum set-off in the stack. This point is considered in more detail in Section 6.6.

After the print is set, oxidation drying within the drying oil or alkyd, and possibly the resin, leads to polymerisation and the formation of a three-dimensional cross-linked network of chemical bonds. Depending on formulation and ambient conditions in the stack, such as temperature, humidity and acidity, this chemical reaction is usually well advanced in 8–16 hours, although it may take a matter of days to reach completion. The result is the conversion of the set film into a hard, dry solid exhibiting the maximum potential rub, scratch and scuff resistance for the particular ink formulation and substrate combination.

Heatset inks are usually similar to quickset formulations with vehicles formulated to have very limited compatibility between the resin/alkyd phase and distillate phase. Again, a degree of compatibility is required to ensure adequate roller stability and transference but this must be limited to ensure the minimum tendency for distillate retention. By minimising distillate retention, removal of the low viscosity component by evaporation is promoted, but also, due to the capillary-draw effect of the paper substrate, some degree of setting by distillate penetration will take place. Thus the formulation of a very high viscosity film of pigment bound by resin/alkyd on the print surface is achieved by distillate removal to atmosphere (evaporation) and to substrate (penetration or setting). In a similar way to the quickset process, the resin and alkyd binder cross-link by oxidation drying. Generally, however, this extra degree of drying is rather incidental since the major physical demands on the print such as folding, cutting or sheeting occur immediately the ink leaves the oven and chill-roller station. At modern press speeds of up to 2000 feet/minute (600m/min) there is virtually no time at all between setting of the ink and running it over turner and folder arrangements.

Radiation curing

Radiation curing mechanisms are fully described in Chapter 10. It is sufficient to note here that free radical chain reactions are involved which produce a highly cross-linked dry film almost instantaneously. Substantial energy input is required, either in the form of highly energetic electron beams which possess energy in the appropriate form to initiate the chain reaction directly or, more commonly in the litho printing industry, in the form of UV radiation. UV radiation, which is an electromagnetic energy form, has to be converted into usable chemical energy by processes involving special so-called 'photoinitiator' chemicals incorporated into UV curing inks.

Drying and setting tests and measurement

Before leaving the fundamentally important areas of drying and setting it is worth considering the assessment of these properties in the laboratory. More detail is presented in Chapter 14, so at this stage only the guiding principles will be considered.

All setting checks involve taking a print in a controlled manner (with respect to substrate and film thickness) and subjecting it to contact under pressure against a virgin piece of stock at appropriate intervals throughout the likely setting period. Unless a temperature and humidity controlled environment is available, it is usual to conduct all tests directly in parallel against a standard ink of known performance.

Set-off in the stack may be assessed by the controlled building of a weighted stack on top of a print in contact with virgin stock over an appropriate period of time. As with setting, comparison against a known standard ink tested in direct parallel is usually required. This type of testing assesses the set-off produced both during initial setting and during the subsequent drying stages, which may sometimes involve periodic ink softening (see Section 6.6).

Two types of ink drying test are used for oxidation drying and quick-setting formulations. The first involves printing the ink on a relevant substrate and assessing rub resistance at appropriate intervals. The second involves producing ink films on impervious surfaces and periodically assessing them until they are touch dry. This latter technique isolates the the oxidation drying mechanism from any penetration or setting effects. However, if the film is freely open to the atmosphere there may be substantially greater availability of oxygen than exists in a stack of printed work. For this reason tests are sometimes employed where the ink film is stacked or sandwiched in some way to limit severely atmospheric contact. Oxidation drying is greatly influenced by ambient conditions of temperature and humidity, therefore all tests should either be done in a suitably controlled environment or against a standard ink of known performance. The latter approach is not wholly reliable since ink drying characteristics may change with ink ageing and it is therefore difficult to ensure that standards are truly of known performance in this respect.

6.3 PHYSICAL PROPERTIES

We have seen in Section 6.1 that the lithographic process dictates certain physical characteristics of litho inks. These can be described under the headings of rheology and tack. Chapter 13 covers in detail the theory of ink rheology and tack and the practical measurement of these properties. Consequently, these topics are only outlined in this section.

Rheology

Four main rheological quantities are relevant to litho inks:
  (1) Viscosity;
  (2) Yield value;
  (3) Thixotropy;
  (4) Flow.
Viscosity is defined as the ratio of shearing stress (shearing or deformation force per unit area) to shearing rate (velocity gradient of flow or deformation). Lithographic inks are substantially non-Newtonian as this ratio is not constant but is shear rate dependent. This is primarily due to the high content of dispersed pigment present and, indeed, unpigmented litho varnishes frequently approach Newtonian character. The non-Newtonian behaviour necessitates viscosity measurement over the full range of shear rates that litho inks can encounter in use. This range is very wide since very low shear rates are encountered in removing ink from the can and in the ink duct, whereas exceedingly high rates of shear exist in the roller nips of fast running presses. Consequently, it is not normally possible to measure viscosity under all relevant conditions on a single viscometer. A variety of techniques are employed. The two most common instrument types are the falling-rod viscometer for high shear rates, and the cone and plate viscometer for lower shear conditions (see Fig. 6.7).

One important feature of the non-Newtonian behaviour of litho inks is that they usually demonstrate a finite yield value. That is, a distinct shear

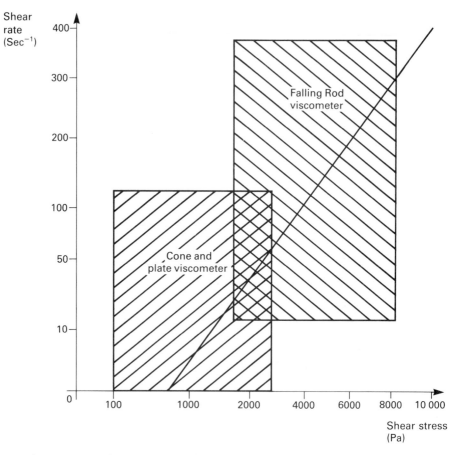

Fig. 6.7 Operating ranges of falling rod and cone and plate viscometers (courtesy of BASF Coatings + Inks Ltd)

stress or force is required before any deformation or flow takes place. It is difficult to obtain a direct measurement of yield value without sophisticated and expensive controlled stress rotational viscometers. Normally an estimation of yield value is made by extrapolating stress results obtained at much higher shear rates to zero shear rate, for example as determined using a falling-rod viscometer. As with any extrapolation, validity has a degree of uncertainty.

Lithographic inks are generally thixotropic as well as non-Newtonian. Their viscosity depends not only on the rate of shear at a given point in time but also on the previous shearing history of the sample under investigation. In everyday terms, the inks thicken on standing, and this 'structure' can be broken down by stirring. Thus, when a thixotropic ink is subjected to constant shearing, the apparent viscosity decreases with time until all the thixotropy that can be eliminated under the particular shear conditions has been destroyed, and a limiting value for viscosity is obtained. Thixotropic behavior is a major complication in the measure-

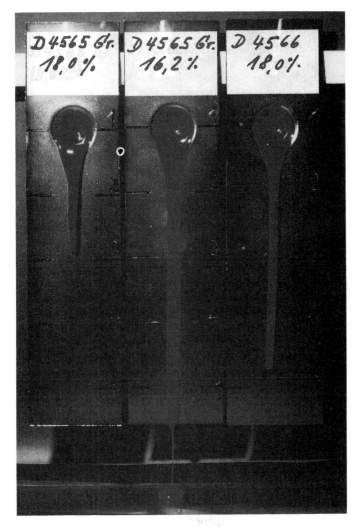

Fig. 6.8 The difference between a long-flowing ink (middle), a short-flowing ink (left) and medium flowing ink (right) (courtesy of BASF Coatings + Inks Ltd)

ment of litho ink rheology. Methods employed on the commonly used viscometers are usually specifically designed to eliminate the influence of thixotropy as far as possible. However, this results in a lack of quantitative information about this particular property. The major problem in attempting to carry out useful studies of ink thixotropy is that virtually any method for introducing a sample on to rheological instrumentation involves shearing actions that may destroy some of the thixotropic nature. Leaving samples on instruments for substantial periods while structure redevelops is a possible solution but hardly conducive to large-scale or frequent studies.

A simple study of ink rheology may be conducted by allowing an ink sample to flow under the influence of gravity down some form of inclined plane (Fig. 6.8). This technique enables a quantitative determination of flow to be made by recording distance travelled by the ink sample against time. However, flow in this sense is not a fundamental rheological property but rather a complex interaction of viscosity, yield value and possibly thixtropy, the latter depending on the technique used for sample presentation to the inclined plane.

In addition to viscosity, yield value, thixotropy and flow, all of which can be quantified, albeit with differing degrees of ease and reliability, purely subjective assessments of rheological behaviour are common in the litho inks industry. A number of terms are used to cover these subjective concepts, the more accepted being consistency or body, and knifeability.

Consistency or body relates to the 'feel' of a small sample of the ink when worked with a palette or taper knife on an ink slab. An experienced technician's assessment of consistency carried out in this manner can encompass relevant aspects of thixtropy, yield value, viscosity and flow, all integrated into a few seconds of various knife motions. This can be very useful for comparing an ink directly beside one of known performance. However, the inability to yield quantitative results as well as the requirement for considerable operator experience and skill, are major drawbacks for such an approach to rheological assessment.

Knifeability is an even simpler test, generally used only in the area of low viscosity web-offset inks supplied in pails or similar containers. The assessment merely consists of ensuring that the ink can be transferred out of the container on a broad knife without it being prone to excess dripping. As with consistency, the property is a complex interaction of thixotropy, yield value and viscosity, and the test suffers similar drawbacks because of its subjectivity.

All rheological properties are greatly influenced by temperature. For this reason all measurements and assessment must either be carried out on temperature controlled equipment, preferably operating in a controlled environment, or in a strictly comparative manner against reference samples of known performance.

Relevance of rheology

The relevance of rheology to press and print performance can be seen by considering the transport of the ink from the can or container to its final position as the printed image.

High viscosity, substantial thixtropy and high yield value can all produce problems in transferring the ink from the container to the duct of the press. This is particularly relevant where pumps are used to transport the ink from bulk delivery units or storage tanks to the press, unless the pumps are of the 'follower-plate' type which are able to move viscous or poorly-flowing ink by a basic high pressure pushing action. Conversely, low viscosity, minimal thixotropy and low yield value can pose problems of dripping with inks that are hand-fed, usually by some form of palette knife, from container to duct. A frequent problem area is coldset web-offset printing inks supplied in pails rather than in some form of bulk pumping system. Due to the demands of the presses and substrates used, inks are usually relatively low viscosity products and problems with retaining all the ink on the palette knife during feeding may be encountered if the ink has too low a yield value or if its thixotropy is very readily broken down. The problem may be much more than just nuisance value if ink falling off the knife lands on a newsprint web in such a way as to cause a web break.

Once in the duct, it is important that the ink flows readily into the gap between the duct blade and roller in order to begin its journey up the ink distribution train. This will not occur if the ink has a yield value above the relatively low shear stresses that exist in the duct, nor if there is too much thixotropy. Thixotropic structure may inhibit ink feed if it builds up significantly during the dwell time of the ink in the duct or if it was inadequately broken down when the ink was fed onto the press. Whether due to thixotropy or yield value, poor flow in the duct leads to inadequate or sporadic feeding of ink up the inking rollers, a condition known as 'hanging-back' (Fig. 6.9). This problem may be alleviated by manual agitation of the ink in the duct or by the use of automatic duct-agitators, which rotate as they traverse through the ink and are shaped in such a way as to push the ink towards the duct-gap. Clearly, both techniques break down thixotropy and serve to raise shear stresses above the ink's yield value. Again, a converse problem can exist, most commonly on newspaper presses, where low viscosity free-flowing ink drips from the sides of the duct and possibly even from under the duct blade. The situation is aggravated where printing units are supplied with ink even when they are not being used, as can happen in coldset web-offset production of newspapers on multi-unit presses. As with knifeability difficulties, higher ink yield value and more rapid thixotropic set-up reduce the problem although duct engineering is obviously an important factor.

As discussed in Section 6.1, a long ink distribution train is employed to roll out relatively viscous litho inks to an even film for application to the litho plate. While such substantial rolling power is a design feature of

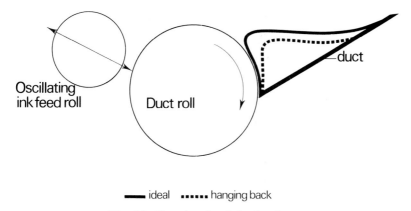

■ ideal ▪▪▪▪▪ hanging back

Fig. 6.9 Hanging-back in the duct

litho presses, it is still necessary for the ink to have the correct rheological characteristics to ensure even distribution. Too high a viscosity and poor flow can impede distribution, particularly where the image demands rapid ink replenishment in certain areas. From the inking rollers through to transference to the substrate, the rheological characteristics of the ink may be modified by the presence of emulsified fount solution. This must always be borne in mind when attempting to relate laboratory rheological studies with commerical press performance. The two properties are interrelated as emulsification is influenced to a large extent by viscosity (see Section 6.1). A large measure of the litho ink formulator's skill centres around producing inks which show rapid, but controlled, emulsification that does not inhibit distribution or transference. Chemistry, as well as rheology, has a major influence here, but generally lower viscosity inks emulsify more rapidly but may be closer to the danger point at which distribution and transference can be impaired.

Problems at the ink distribution stage may also be encountered if droplets of ink are expelled or 'fly' from the roller train. The droplets may be formed by cavitation and multiple string rupture during the film splitting that occurs at the exit side of an ink roller train nip. Cohesive, and even electrostatic, factors are believed to be involved in flying, but rheological properties also influence string formation. Low viscosity, and particularly long flow, appear to promote flying and the problem is often eliminated by incorporation of particulate structuring agents which increase viscosity and significantly restrict flow.

In the transference stages from plate to blanket to substrate the rheology of the ink, or more precisely that of the fount -in-ink emulsion, affects both the amount of ink printed and the fidelity of the reproduction. Generally, low viscosity and yield value will promote transference but these factors will definitely result in a greater physical dot gain in halftone printing (see Section 6.6). The latter phenomenon arises from the physical squashing of the fluid ink that inevitably results from the pressures existing in the plate/blanket and blanket/impression transference nips. Naturally, high viscosities and yield values give reduced flow, and hence less squash, at a given pressure.

Tack

Tack is a measure of the forces required to split a single film of ink into two. Such film splitting is influenced by rheological and adhesive properties, in addition to the internal cohesion of the ink. Various instruments have been developed to measure tack but all rely on the principle of measuring the force exerted on a roller as it splits a single ink film in a nip into two films at the nip exit. A subjective, comparative assessment of tack can be made using fingertips to separate a thin film of ink on a slab into two films, one portion remaining on the slab and the other adhering to the fingertip. A skilled technician, using repetitive dabbing of two inks with adjacent fingers, can make a surprisingly good comparison in this way. However, the method suffers the disadvantages of all such approaches which depend on individual skill and which do not produce a quantitative result.

Tack does not vary as rapidly with temperature as does viscosity. Nevertheless, there is an influence and all competent tack measuring devices incorporate appropriate temperature control of the roller system. As an empirical quantity, tack can be significantly influenced by film thickness and it is important that the volume of ink applied to tack measuring devices is carefully controlled. Even with temperature and film weight control, different tack machines can produce different results on given inks. Where the machines are not of a single type two main factors are involved:

(1) Since tack is an empirical quantity, there is no scientifically defined unit of tack. Various manufacturers of tack instruments have established their own arbitrary scales.

(2) Different designs involve different roller surfaces, geometries and speeds. These variants produce different interaction of the cohesive, rheological and adhesive components that contribute to tack. This accounts for the phenomenon where ink A is higher tack than ink B on one instrument, but B is higher than A on another. Such differences also prevent simple scale conversion factors which might otherwise be a solution to the problem of differing arbitrary scales.

Instruments of the same design from a single manufacturer would be expected to overcome these problems of inter-instrument comparability. In reality, agreement is not always perfect. It appears that this is due to the basic tack machine approach being very susceptible to influences from bearing wear, roller surface condition and other mechanical and frictional factors.

Despite all this, tack is a major influence on ink performance and currently available instrumentation, whilst not perfect, is fully satisfactory in enabling control of this parameter.

Relevance of tack

As a measue of the forces involved in ink film splitting, tack is relevant to all stages where distribution or transference takes place. Too high a tack can cause some form of rupture in the substrate since this may require less work than producing the very high forces necessary to split the ink film. In the most severe cases, part or all of the paper will be pulled on to the surface of the offset blanket and the press will have to be stopped and cleaned immediately. Where the tack is only marginally too high for the strength of the substrate a small degree of coating pick or fibre linting will occur at each impression. This will obviously impair the smoothness, and possibly the gloss, of the print and will eventually result in a deposit of fibre or coating on the blanket which begins seriously to affect print quality, necessitating a blanket wash.

From the point of view of minimising the risk of substrate disruption, particularly with weak, low-quality papers, it is clearly advantageous to use inks with as low a tack as possible. However, very low tacks can produce a variety of problems from the ink roller train right through to final print quality:

(1) Litho press distribution trains are usually engineered such that only alternate rollers are driven by the press motor. Every other roller relies on frictional contact with a press-driven roller to ensure rotation. Inks with very low tack may have insufficient cohesion to transmit the frictional forces across the nips, resulting in roller slippage and inadequate ink feed and distribution.

(2) In wet-on-wet printing on multi-unit presses it is necessary, in relevant image areas, for ink to transfer or 'trap' on top of a previously printed, wet ink film. If the inital colours printed are very low in tack they may be susceptible to an unacceptable degree of 'back-trapping' from subsequently printed, higher tack, inks. This will occur if less force is required to split the previously printed ink film from the paper than is necessary to split the subsequent ink from its blanket. Naturally, such poor trapping completely disrupts colour reproduction and may easily result in an unsatisfactory final print.

(3) The influence of rheology on physical dot gain in halftone printing was referred to in the previous sub-section. Studies have shown that dot gain is also dependent on tack. It is less easy to identify a simple explanation for this effect but it seems possible that the low cohesion of dots of low tack ink provides less resistance to the squashing action at the plate/blanket and blanket/substrate impression nips. Thus, excessively low tack inks may produce unacceptably high dot gain.

6.4   FORMULATING PRINCIPLES

We have seen in Section 6.1, how the nature of the lithographic process imposes certain fundamental requirements on the chemical, physical and strength characteristics of litho inks. Achievement of these required characteristics, and those dictated by typical end-uses of litho printed matter, will now be considered in terms of raw material selection and formulating principles. As with any ink, the formulations comprise:

(1) Pigments;
(2) Resin/vehicles;
(3) Solvents/diluents;
(4) Additives.

Pigments

As previously stated (see Section 6.1) offset lithography prints ink films of lower thickness than any of the other major printing processes. Since the final print is generally required to match the range of colour strengths available by other printing techniques, it follows that litho inks must be produced at the highest colour strengths. Hence pigments selected must be inherently strong and able to develop their strength when dispersed in litho ink vehicles. Also, they must not produce unacceptable rheology, in terms of high viscosity, high yield or excess thixotropy, even when incorporated at high loadings. Dispersibility to fine particle size is necessary, not only for strength development but also to ensure that the presence of particles substantially bigger than litho ink film thickness is limited. Overriding all these requirements, it is essential that the pigments chosen must be fundamentally insoluble in, and unreactive with, the aqueous or aqueous/alcohol fount solution employed.

The nature of the printed work produced by lithography can impose further restrictions on pigment choice. Transparency is required where ink films or halftone dot patterns are to be superimposed, either by wet-on-wet or wet-on-dry multi-colour printing. However, there are instances, particularly when printing on stocks of poor quality and whiteness, where a degree of opacity can be beneficial since it helps 'cover' or hide the less than perfect substrate. Publication printing rarely places any particular resistance demands on pigment selection, except for the excellent lightfastness demanded in poster printing, but packaging, both cartons and labels, can have stringent requirements in terms of chemical resistance, lightfastness and toxicity.

Cartons and labels generally require at least moderate lightfastness, in order to cover the occasions when merchandise is displayed in shop-windows, and may have more stringent lightfastness specifications when greater exposure is anticipated. Where such performance is required, poor lightfastness pigments are either avoided altogether, or are used in a careful balance with pigments of superior performance such that light exposure produces only a slight lightening or darkening rather than an unacceptable fade or shade-shift. Chemical resistance requirements are as diverse as the products packed or labelled with litho print. Major areas are soap and detergent resistance on relevant cartons, and solvent resistance on cosmetic labels. Solvent resistance may also be dictated, both in packaging and publication work, if the print is to be further processed by spirit varnish or lamination.

The major pigment types which have the right performance to comply with the fundamental demands of the litho printing process and which have established bulk use in litho ink formulations are:

(1) The process colour pigments (for four-colour halftone printing):
   CI Pigment Yellows 12 and 13.
   CI Pigment Red 57:1.
   CI Pigment Blue 15:3.
   CI Pigment Black 7 (usually toned with CI Pigment Blue 18).
(2) Other yellows:
   CI Pigment Yellow 17.
   The Hansa yellows (CI Pigment Yellows 1, 3 and 5) are restricted in use by their poor strength, rheology and transparency. However, their soap resistance and lightfastness characteristics lead to some use in relevant applications.
(3) Oranges:
   CI Pigment Orange 13.

CI Pigment Orange 34.
(4) Other reds:
   CI Pigment Red 2.
   CI Pigment Red 4.
   CI Pigment Red 48.2.
   CI Pigment Red 53.1.
   CI Pigment Red 81.
(5) Other blues:
   CI Pigment Blue 1.
   CI Pigment Blue 15 and 15.1
   Some CI Pigment Blue 27 (Iron Blue) is used but it is generally difficult to disperse adequately for modern litho requirements.
(6) Violets:
   CI Pigment Violet 1 (PMTA Rhodamine).
   CI Pigment Violet 3.
   CI Pigment Violet 23.
(7) Greens:
   CI Pigment Green 1.
   CI Pigment Green 2.
   CI Pigment Green 7.
(8) White:
   CI Pigment White 6 (Rutile grades).
(9) Extenders:
   CI Pigment White 18.

Resin/vehicles

Litho ink vehicles may be divided into two major classes. These are the oleoresinous (hard resin and drying oil alkyd) systems used to produce quickset, heatset and oxidation drying inks and the acrylate systems used in radiation curing inks. The latter are described in detail in Chapter 10. However, it is worth noting here the relevance to radiation curing vehicles of all comments in Section 6.1 and below, which relate to the emulsification, cohesion, surface energy and pigment wetting dictates of the litho process, and the requirements that they pose in both chemical and physical characteristics. In other respects this subsection is restricted to a consideration of oleoresinous vehicles, whether drying exclusively by oxidation or by the quickset mechanism:

Oxidation drying vehicle   *Quickset mechanism vehicle*

Hard resin(s)   Hard resin(s) ⎫ A
Drying oil/alkyd   Drying oil/alkyd ⎭

                                  Petroleum distillate} B

                                  Vehicle is formulated such
                                  that there is limited
                                  compatibility between A and B

We have already seen (Section 6.1) that the lithographic process places stringent demands on the chemical nature of the binder resins and the physical form of the vehicles based on these materials.

Chemistry

The resin chemistry used has to meet a number of fundametal require-ments:

(1) Solubility in weak solvents.
(2) Controlled water tolerance — neither fully miscible nor totally water repellent.
(3) Cohesive   ⎱ when used in appropriate weak solvent-
(4) Low surface energy⎰ based vehicle.

Hardly surprisingly, the range of basic resin types which comply with this demanding set of requirements is very limited. In fact, in the major area of oleoresinous systems the vast majority of the hard resins used fall into only two classes. These are the modified rosin ester and modified hydro-carbon groups of resins. The drying oil and alkyd components used, which can essentially be regarded as liquid resins, also represent a narrow range of chemistry. Linseed and soya are the dominant vegetable oils, either for direct incorporation into litho ink vehicles or for alkyd modifi-cation. The majority of alkyds employed are long oil modified isophthalic esters of either glycerol or pentaerythritol. This is not to say that other classes of resin and oils or alkyds cannot be found in litho ink formula-tions. Some, such as cyclised rubber resins for certain small offset applications and tung oil for very hard drying requirements, may be essential choices in certain formulating situations. However, it is a distinc-tive feature of litho ink vehicles that the range of basic chemistry avail-able to the formulator is severely limited. Attempts have been made to broaden this base and, for example, oil-compatible acrylic resins suitable for litho applications have been developed within the last few years. These resins have not made a major impact and it seems unlikely that fur-ture developments will achieve major departures in terms of novel litho vehicle chemistry at acceptable costs.

On this basis, the formulator of oleoresinous litho ink vehicles would seem to be provided with only a handful of raw materials with which to perform his task of producing a variety of vehicles for both sheet-fed and web-offset applications. While the pure chemist might be tempted to agree, the ink technologist will recognise that resin manufacturers have produced a bewildering range of resin grades, albeit on the basis of the restricted fundamental chemistry that we have already identified. Both chemical modifications, as in the phenolic or maleic modification of rosin esters, and molecular weight variants exist, and even minor changes have a marked influence on printability and print characteristics. One of the prime properties that can be altered is the solubility or compatibility of the resin with petroleum distillate. As discussed in Section 6.2, fine control over this parameter is required in both sheet-fed quickset and heatset systems in order to ensure an optimum balance between the conflicting demands of press stability and setting. Producing resins that have good solubility, for maximum stability, coupled with rapid viscosity rise on solvent release, in order to give the fastest possible setting, has been a major goal of the resin industry. The compatibility of a vehicle is not controlled by resin choice alone but is also influenced by oil and distillate types and proportions. The role of distillate will be considered in more detail later.

Vegetable oils and alkyds have a major solubilising influence on the hard resins used in oleoresinous litho vehicles. This effect is usually proportional to the alkyd or oil content and it is, therefore, customary to consider quickset and heatset vehicles in the context of their resin-to-oil (and/or alkyd) ratio. It must always be borne in mind that resin and distillate choice influence vehicle compatibility and hence the stability/ setting balance of the formulation. Nevertheless, when these factors are controlled, resin-to-oil ratio relates directly to the compromise achieved. Thus, high resin-to-oil ratio minimises compatibility and promotes setting at the expense of stability. The converse applies for low resin-to-oil. Other properties are influenced by compatibility and hence by resin-to-oil ratio. Gloss is a prime example, being generally improved by high vegetable oil or alkyd contents. Within the quickset mechanism, resin-to-oil ratios of vehicles may vary from 7 : 1 to 0.5 : 1, dependent on the choice of resin, oil and distillate and the end application of the vehicle. The latter may range from high-speed heatsetting to high-gloss, low speed sheet-fed.

Vehicle compatibility can be influenced by manufacture as well as by raw material choice and proportions. This effect is exploited in the production of so-called 'cooked' varnishes. In these the hard resin and drying oil components are heated together at elevated temperatures, usually in the 220−240°C range. Here appropriately modified rosin ester grades will undergo chemical reaction, largely involving ester-interchange, with vegetable oils. The reaction modifies the distillate tolerance of the resin/oil phase. In this way a resin which is incompatible in a given distillate grade may be cooked to a tightly controlled tolerance before crash-cooling and completion of vehicle manufacture by incorporation of the distillate.

Manipulation of vehicle components is also carried out in order to achieve the optimum chemical balance for the particular lithographic performance required. We have already seen in this subsection that hard resin and alkyd chemistry is severely constrained by the requirement for controlled emulsification. Even so, within the materials that are potentially appropriate, there are subtle variations in water tolerance that can make a critical difference to machine performance under adverse conditions such as high press speeds, poorly controlled dampening, and plates of poor image/non-image distinction. Within the limited range of chemistry that is available, combinations of materials sometimes enable improved lithographic compromises to be achieved. One of the classic examples of this is the combination of modified rosin ester and modified hydrocarbon resins in precise proportions to achieve particular water tolerance characteristics. The simple guiding principle in this area of vehicle formulating is that the hydrocarbon is more water repellent than rosin ester. In practice, much empirical work is normally necessary before the exact proportions of specific grades are defined such as to yield rapid, controlled emulsification with minimum impact on transference and distribution.

Physical form

The required viscosity for good litho performance can generally be achieved in oleoresinous vehicles by manipulation of the resin-solids content. Oxidation drying vehicle viscosity can be adjusted by altering the drying oil content and similarly quickset vehicles can be adjusted by varying the amount of distillate. Nevertheless, certain limitations apply since too low a resin-solids content may produce insufficient hardness when the vehicle is acting as the binder on the dried print. Even more important, too much distillate in a quickset vehicle can result in poor pigment wetting. This impedes dispersion and is likely to produce ink of unacceptable rheology — high viscosity and yield value and excessive thixotropy — even though the vehicle appears to be of appropriate viscosity. This is a particular problem in high resin-to-oil ratio systems for heatset and very rapid setting sheet-fed inks. Therefore, resin grades must be selected with due consideration of the vehicle viscosity that they will produce. In a given formulation where the alkyd and distillate types and contents are defined, the viscosity produced by a particular resin grade will be dependent on its inherent viscosity (as measured in strong solvent solution) and its compatibility, or distillate tolerance. This latter parameter influences vehicle viscosity in quickset systems since the vehicle is not a true solution. The resin is best regarded as being partially dissolved and partially in colloidal dispersion. Forces of association between the dissolved and dispersed resin components ensure that the vehicle is a stable fluid exhibiting bulk homogeneity. However, these forces also impair vehicle flow and thereby raise viscosity. Consequently, compatibility, or distillate tolerance, and vehicle viscosity are generally inversely proportional.

Careful raw material selection and formulation can yield vehicles which produce inks of the required viscosity for litho printing while maintaining resin solids that ensure good pigment wetting and adequate hardness on print. However, it is often difficult to achieve the optimum balance between viscosity and tack. In simple terms, litho ink vehicles based solely on appropriate hard resins, oils, alkyds and, possibly, distillates often yield inks which are rather too high in tack at optimum viscosity. Lowering the tack by distillate or vegetable oil addition produces low viscosity and may lead to emulsification and inhibition of distribution or transference. To maintain viscosity and reduce tack, the resin solids need to be lowered.

A frequently used technique for achieving this is to produce a structured gel varnish at reduced resin solids. The gel structure is created by introducing a material, most commonly some form of aluminium complex, which will increase intermolecular bonding, hence inhibiting flow and raising viscosity. Gel varnish production, despite having been established in the industry for decades, can be problematic since the reactivity of both the resin and gellant can vary. The result is not a simple increase in viscosity but rather the introduction of a yield value and thixotropy into the vehicle. This further complicates the rheological behaviour of inks based on these gels. Despite these drawbacks, gel varnishes are frequently used in litho ink formulation for fine balancing of tack and rheology.

Solvents/diluents

The volatility and solvency dictates of the process (see Section 6.1) restrict the litho ink formulator to the use of very weak solvent power, high boiling petroleum fractions. Indeed, few people outside the lithographic industry would regard these materials as solvents and in many ways they might be considered as diluents. The very low solvent power is sometimes boosted in a controlled manner by the use of small proportions of stronger solvents which may have alcohol or ester functionality. Tridecanol is an example of one such 'bridging' or 'co'-solvent. High molecular weight materials are chosen since volatility restrictions apply. While these are strong solvents to the litho ink chemist, they are still weak solvents in absolute terms as a result of their molecular weight and low functionality. Because of their higher solvent power compared to distillates, a small addition of bridging solvent will give a significant increase in compatibility. Thus, they can be useful for reducing viscosity with little effect on tack. They can also produce an improved balance between compatibility and phase separation, particularly in heatset web-offset inks and may, therefore, be used to promote stability and gloss with minimum sacrifice on setting.

Typical bridging solvent contents in litho inks may be of the order of 2% whereas distillate contents can be in the range 20–50%. Petroleum distillates are therefore major components of inks which dry by the quickset mechanism and produce subtle effects beyond their simple role as viscosity and tack adjusters. The two prime properties of distillates are their boiling range and aromatic content:

Boiling range

240–260°C   Heatset

260–290°C   Slow heatset/ultra-fast quickset

280–320°C   Quickset

Aromatic content

Conventional distillate	Above 15% aromatic	Highest solvent power
Low aromatic distillate	5–6% aromatic	Medium solvent power
Zero aromatic distillate	0–1% aromatic	Low solvent power

Volatility restrictions necessitate a high boiling range. However, slightly more volatile grades are acceptable in high-speed sheet-fed or web-offset printing where the ink dwell-time on press rollers is minimised. Boiling ranges are narrow or 'close-out' in order to achieve the optimum balance between press stability and phase separation. A broad boiling range can result in both poor stability, due to evaporation of the more volatile, lower-boiling fractions from the ink distribution rollers, and slow setting, caused by retention of poorly mobile high-boiling fractions.

Modern resin development has made it increasingly acceptable to use low or zero aromatic distillates to give adequate resin solids levels which would only have been possible with 'conventional' relatively high solvent power distillates a few years ago. There are additional advantages in using these weaker solvent power distillates. They often give a better balance of rapid setting with adequate stability because of the inherent solubility of the resin. In low odour and taint systems for food packaging, and where heatset oven emissions are subject to controls, the inks supplied must be formulated on the lowest aromatic content distillates available.

Before leaving this consideration of the solvent power of the distillates used in litho inks for drying by the quickset mechanism, it should be mentioned that aromatic content is not the sole controlling factor. Two different distillate grades of zero aromatic content may differ substantially in the proportions of naphthenic and olefinic unsaturates present. A grade which contains a high proportion of these materials will be a stronger solvent than one which is mainly or exclusively paraffinic in nature.

Additives

As with all inks, a wide range of additive components are employed in small proportions to modify various printability and print performance characteristics of the basic pigment, resin and solvent combinations. The main classes of additives used in litho formulations are:
 (1) Driers;
 (2) Waxes;
 (3) Antioxidants;
 (4) Anti-set-off compounds;
 (5) Litho additives;
 (6) Rheology modifiers.

Driers
The oxygen-induced polymerisation of drying oils and drying oil modified alkyds is accelerated by incorporation of small quantities of certain metal compounds. These drier catalysts are mainly based on transition metals that are able to exist in stable forms in more than one oxidation state. This provides reaction pathways which speed the oxidation reaction and enable a highly cross-linked film to form in a matter of hours rather than days. Metal soaps of long-chain carboxylic acids are usually employed because of their solubility in the oleoresinous vehicles found in oxidation drying and sheet-fed quickset inks.

With the demise in the use of lead for toxicological and environmental reasons, cobalt and manganese are the principal metals used today. Their soaps are normally used in the form of relatively dilute solutions in hydrocarbon solvent in order to assist handling and make weighing of small metal concentrations less critical.

Cobalt is probably the most widely used drier catalyst in oxidation drying systems at present due to its outstanding ability to accelerate the drying reaction. The catalytic activity of cobalt relies on repeated transitions from the Co^{2+} oxidation state to Co^{3+}.

We have seen in Section 6.2 that a key step in the oxidation drying mechanism is the formation, and subsequent decomposition, of hydroperoxides. These hydroperoxides decompose at elevated temeratures to give free radicals:

$$ROOH \rightarrow RO^. + {}^.OH$$

However, in the presence of small quantities of metals such as cobalt, this decomposition reaction proceeds readily at ambient temperatures. The reaction can be represented as follows:

$$Co^{2+} + ROOH \rightarrow Co^{3+} + RO^. + OH^-$$
$$Co^{3+} + OH^- \rightarrow Co^{2+} + {}^.OH$$

The free radicals generated during both the above oxidation and reduction reactions accounts for the increase in the rate of polymerisation seen when cobalt is used as a drier.

Generally, drying efficiency, in terms of speeding cross-linking and producing a harder final film, increases with drier content up to a limiting value. Additions above this level may give more rapid drying but can also result in soft films. This phenomenon is believed to be associated with the formation of large numbers of relatively low molecular weight polymers as opposed to a small number of very high molecular weight polymers by extensive cross-linking. Over-addition of driers can also result in press stability problems due to skinning, particularly on the roller ends where there is little take-off and so ink dwell time as a thin film is maximised. Both quantity and choice of driers can affect the formation of odorous oxidation by-products during the drying reaction and hence can influence the odour and tainting potential of the final print. Thus, drier combinations for inks to be used on food and confectionery packaging must be carefully formulated to minimise the formation and retention of oxidation by-products, such as volatile aldehydes, ketones and carboxylic acids, which are odorous and mobile.

Volatile aldehydes, ketones and carboxylic acids occur as by-products of oxidation drying of drying oils through decomposition of the peroxides and hydroperoxides formed as intermediates in these reactions. The following general scheme for production of these materials is:

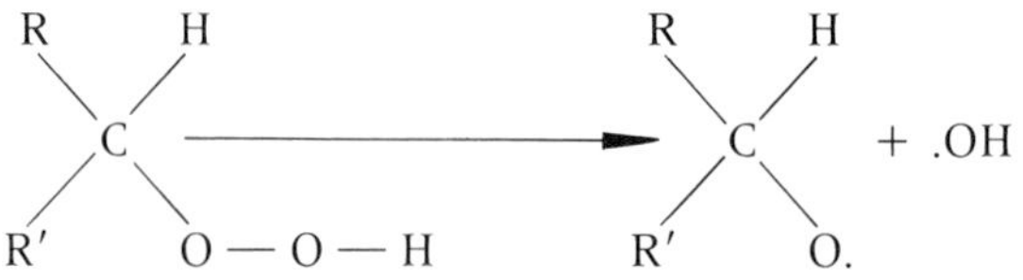

(1)

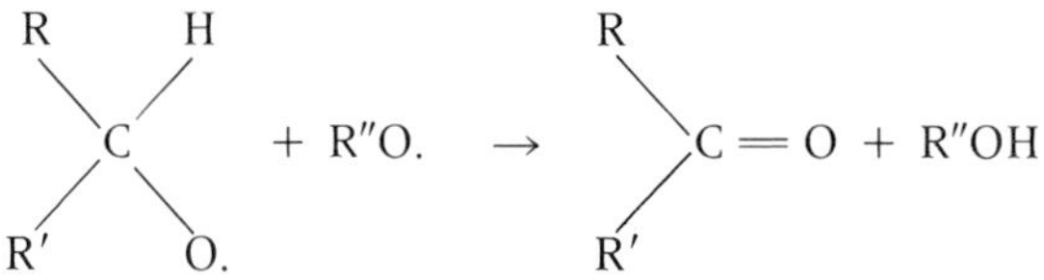

Ketone

(2)

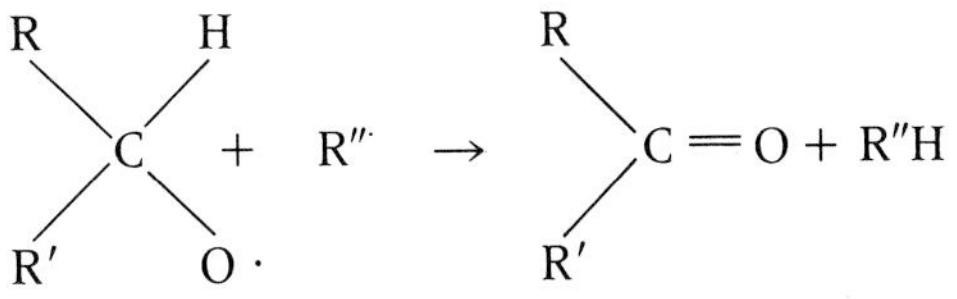

$$\underset{R'}{\overset{R}{>}}\underset{O\,\cdot}{\overset{H}{C}} + R''' \longrightarrow \underset{R'}{\overset{R}{>}}C=O + R''H \qquad (3)$$

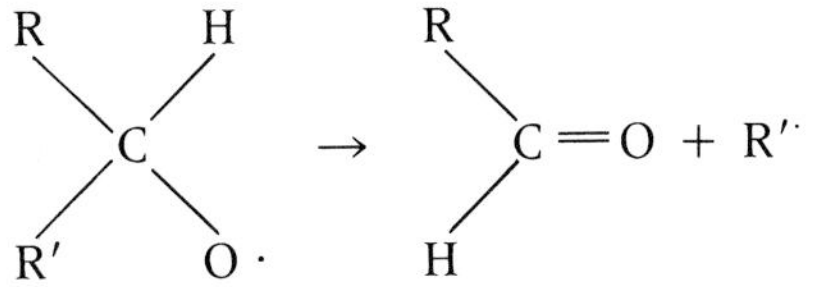

Ketone

$$\underset{R'}{\overset{R}{>}}\underset{O\,\cdot}{\overset{H}{C}} \longrightarrow \underset{H}{\overset{R}{>}}C=O + R'' \qquad (4)$$

Aldehyde

The aldehyde produced in reaction (4) can then oxidise further to produce carboxylic acids:

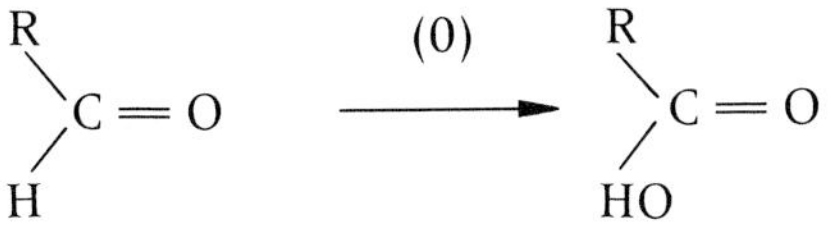

$$\underset{H}{\overset{R}{>}}C=O \quad \overset{(0)}{\longrightarrow} \quad \underset{HO}{\overset{R}{>}}C=O$$

To illustrate how such low molecular weight materials can be generated, the following reaction scheme, starting with the hydroperoxide of linoleic acid (see Section 6.2 for details of formation of this material) is proposed. The scheme follows reactions (1) and (2) above:

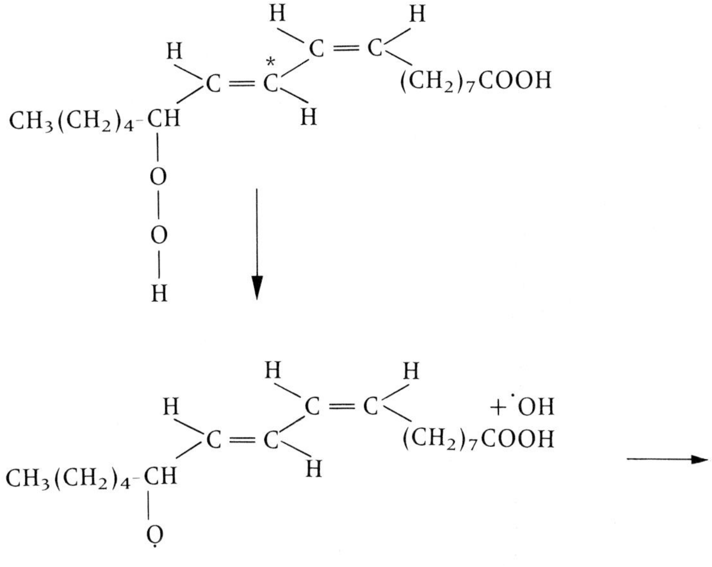

$$CH_3(CH_2)_4 - \underset{\underset{H}{\overset{|}{O}}}{\overset{|}{CH}} \; \underset{\underset{H}{\overset{|}{O}}}{} \quad \overset{H}{>}C = \overset{*}{C}\overset{<}{\underset{H}{}} \quad \overset{H}{>}C = C\overset{<}{\underset{(CH_2)_7 COOH}{\overset{H}{}}}$$

$$CH_3(CH_2)_4 - \underset{\underset{O}{\overset{|}{O}}}{\overset{|}{CH}} \quad \overset{H}{>}C = C\overset{<}{\underset{H}{}} \quad \overset{H}{>}C = C\overset{<}{\underset{(CH_2)_7 COOH}{\overset{H}{}}} + \cdot OH \longrightarrow$$

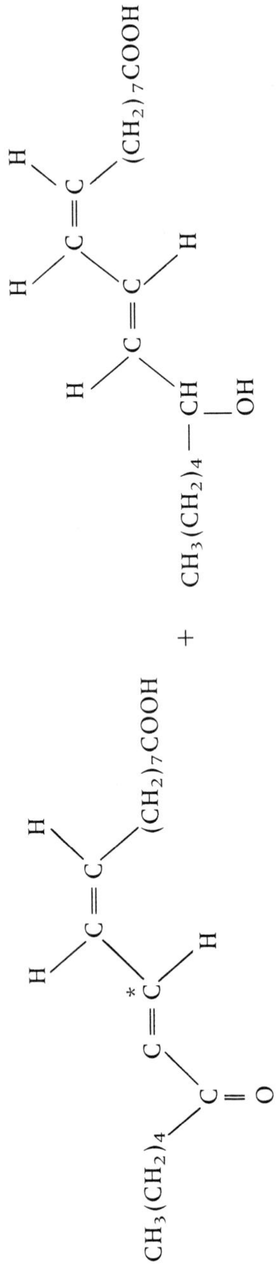

$$CH_3(CH_2)_4 \quad C=C \atop \underset{O}{\overset{H}{|}} \quad H \quad C=C \atop H \quad (CH_2)_7COOH$$

Ketone

$+$

$$CH_3(CH_2)_4 - CH \atop \underset{OH}{|} \quad C \atop H \quad H \quad C=C \atop H \quad H \quad C=C \atop H \quad (CH_2)_7COOH$$

Alcohol

The unsaturated ketone (molecular weight 246) produced by this mechanism is capable of further oxidation as follows:

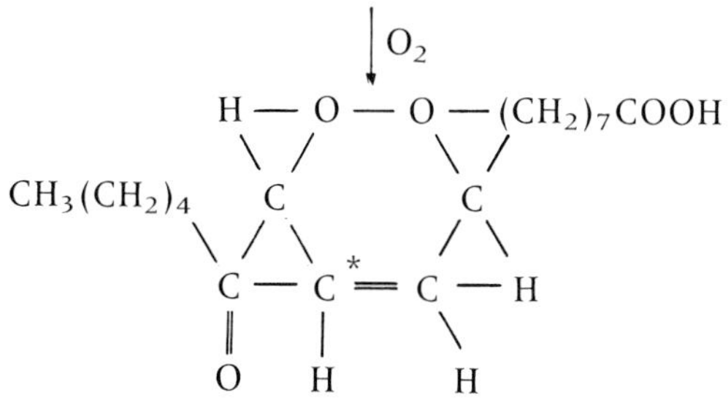

$$
\begin{array}{c}
\Big\downarrow O_2 \\
H - O - O - (CH_2)_7COOH \\
CH_3(CH_2)_4 \quad C \qquad C \\
C - C \overset{*}{=} C - H \\
\| \quad | \qquad \backslash \\
O \quad H \qquad H
\end{array}
$$

The cyclic peroxide then decomposes in the following way to give a diketone of molecular weight 84, and aldehyde of molecular weight 100, an acid of molecular weight 144 and a keto-acid of molecular weight 242.

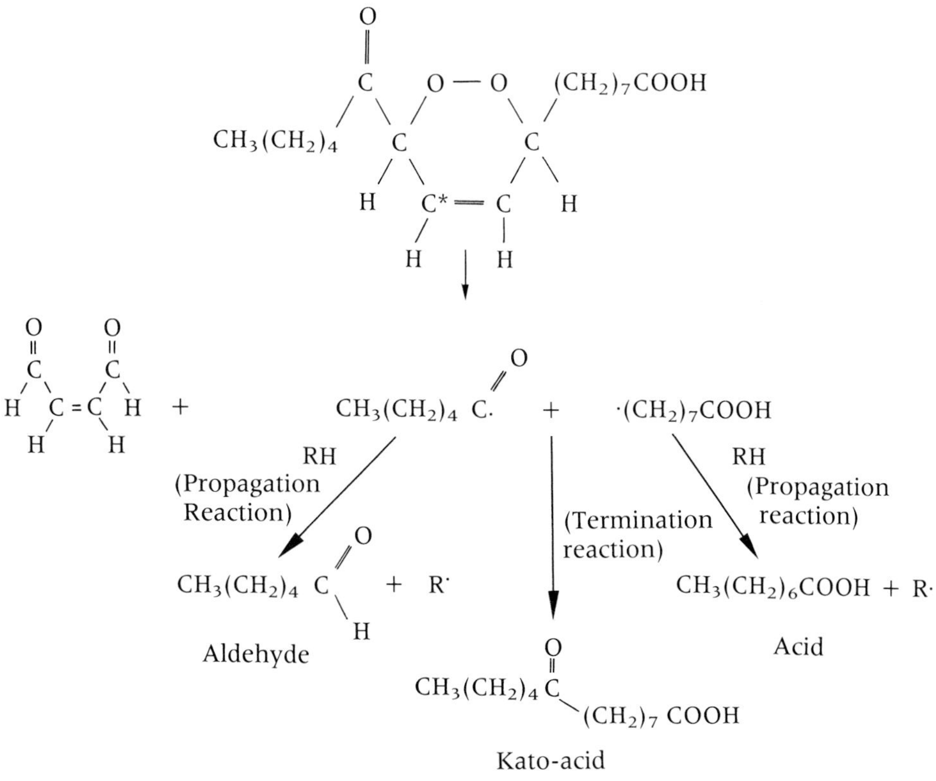

$$
\begin{array}{c}
O \\
\| \\
C \qquad O - O \qquad (CH_2)_7COOH \\
CH_3(CH_2)_4 \quad C \qquad C \\
H \quad C^* = C \quad H \\
| \quad | \\
H \quad H
\end{array}
$$

$$
\begin{array}{ccc}
O \quad O \\
\| \quad \| \\
C \quad C \\
H \quad C = C \quad H \quad + \qquad CH_3(CH_2)_4 \; C\cdot \qquad + \qquad \cdot(CH_2)_7COOH \\
H \qquad H
\end{array}
$$

RH   RH
(Propagation   (Propagation
Reaction) O reaction)
                                    ‖
(Termination
reaction)

$$CH_3(CH_2)_4 \; C \diagdown H \qquad + \quad R\cdot$$

Aldehyde

$$
\begin{array}{c}
O \\
\| \\
CH_3(CH_2)_4 \; C \diagdown \\
(CH_2)_7 \; COOH
\end{array}
$$

Kato-acid

$$CH_3(CH_2)_6COOH + R\cdot$$

Acid

The low molecular weight diketone, aldehyde and acid formed are potential sources of off-odours and taints; the diketone produced in this scheme is noted in chemical references as being a lachrymatory liquid which will obviously have significant potential to cause odour and possible taint problems.

Driers are not incorporated into inks which rely on penetration or evaporation to achieve drying, i.e. coldset or heatset web-offset formulations.

They are not applicable either to the entirely different chemistry of radiation curing systems. In UV inks the photoinitiator species play a similar role, but through totally separate mechanisms.

Waxes

As with inks for other printing processes, waxes are incorporated into litho inks to produce some or all of the following properties:

Slip   For in-line handling

Scratch resistance⎫
Rub resistance   ⎬   For end-use requirements
                              ⎭

The major chemical types of wax used in modern litho ink formulations are polyethylene (PE) for rub and scratch resistance, and polytetrafluoroethylene (PTFE), for good surface slip. Numerous different grades are available but generally these are molecular weight (and hence melting point) or particle size variants of these two chemicals. Some materials are now available which combine PE and PTFE, either by chemical reaction or by simple physical blending. These are designed to enable an optimum balance of surface slip and rub/scratch resistance to be achieved by incorporation of a single additive.

Low melting point grades promote surface slip but are usually inferior for rub and scratch resistance. Problems may also be encountered in ink manufacture if the wax is processed at any stage at a temperature near or above its melting point. Recrystallisation of the wax on subsequent cooling can result in the formation of wax particles of uncontrolled size which can adversely influence ink rheology, printability and final print characteristics.

Large particle size wax grades are particularly effective for both slip and rub/scratch resistance performance. This is because the particles protrude significantly from the ink film and thus the print surface is essentially wax rather than ink vehicle binder. However, the presence of such large particulate matter in the ink can produce printability problems, such as blanket piling, and print deficiencies, such as poor gloss.

Certain substrate requirements can make the choice of wax combination absolutely critical to the performance of the final print. A case in point has been the introduction of matt-coated papers and boards. The optically matt surface has generally been achieved by the incorporation of significant proportions of calcium carbonate extenders into the coating formulations. Unfortunately, this has tended to produce harsh and abrasive surfaces which give severe ink scuffing problems when print is subjected to face-to-face rubbing. Ink formulation changes cannot always completely overcome this substrate problem, although the utilisation of large additions of highly effective PE and PTFE grades to tough binder vehicles can alleviate the situation.

Waxes may be incorporated into litho inks either directly, as finely micronised powders, or via an intermediate wax paste dispersion. The latter route involves easier blending but destroys some formulating flexibility by dictating incorportion of the vehicle present in the wax paste. Micronised grades have improved substantially over recent years and now rarely produce blanket piling problems involving separation of the wax from the ink vehicle, unless too large a particle size grade is chosen or unless the wax is improperly incorporated into the ink.

Antioxidants

Antioxidants are used to control the oxidation drying potential of litho inks. The main types are as follows:

(1) Oximes, e.g. methyl ethyl ketoxime — volatile and only effective where atmospheric contact is restricted.

(2) Phenolic types, e.g. butylated hydroxy toluene (BHT) — moderate efficiency.

(3) Quinones, e.g. hydroquinone — very active, severely inhibits oxidation.

The use of antioxidants always involves a compromise between obtaining desirable long skinning times and retaining adequate drying potential for the ink on print. A number of non-skinning requirements may be encountered ranging from freedom from skin in the can through to full overnight stability on the inking rollers. The former may be achieved by the use of volatile oxime antioxidants which are largely lost from the ink by the time that it has transferred up the distribution rollers and on to the substrate. Thus, the desired non-skinning property may be formulated into the ink without significant deteriment to drying. At the other end of the scale, full overnight roller stability can only be achieved using a fierce antioxidant such as hydroquinone. Even with careful balance of the drier content, it is exceedingly difficult to maintain robust drying characteristics. The balance between non-skinning and good drying can be assisted by the incorporation of calcium perborate driers. These assist drying by the generation of oxygen in the presence of water. Thus, skinning is little affected whilst drying of the emulsified ink on print is boosted. However, even with the exploitation of such subtle techniques, it is difficult to arrive at inks which do not skin on the roller ends on a hot night but dry adequately in stacks stored in cold, damp warehouses in winter.

Anti-set-off compounds

The quickset mechanism (see Section 6.2) is a major boost to achieving set-off-free stacks of sheet-fed print. However, even the fastest quick-setting ink is likely to take a couple of minutes to set. In this time on a fast running press a significant weight of stack can build up on top of the sheet. Set-off during this period is avoided by the use of anti-set-off spray powder on the press and by careful adjustment of the delivery to ensure minimum frictional contact between the sheets. Anti-set-off spray is a nuisance in the press room and the printer is always looking for ways to minimise its use. Infra-red drying units have been a major help in this area, but the ink-maker is also able to play a part by the incorporation of anti-set-off compounds. These compounds are based on particulate materials such as silica or starch which have a particle size slightly greater than the printed ink film thickness. Thus, the particles protrude through the film and separate adjacent sheets, functioning as it were as 'internal spray powder'. Naturally, the exact size of the particulate material employed and its content in the ink must be carefully controlled to minimise adverse effects such as blanket piling, poor print gloss and poor rub resistance. Due to these limitations, anti-set-off compounds can never be total solutions but merely another tool in the reduction of set-off.

Litho additives

We have already seen that inks will only perform satisfactorily in the lithographic process if their physical properties and fundamental chemical make-up are correct. Additives cannot be used to produce adequate litho performance from an ink that is based on chemistry inappropriate to meet the emulsification, cohesion and surface energy requirements of the process. However, additives may be beneficial in controlling undesirable reactions between fount and ink that might, in certain circumstances, jeopardise the otherwise good machine performance of a basically sound formulation. One example of this is the use of soluble salts of tartaric acid and ethylenediamine tetra-acetic acid (EDTA), which are able to form complexes with soluble calcium ions which may be present with certain pigment grades, or which may originate from the coatings of paper and board substrates. If they are not removed, these calcium ions can react with other materials and create problems of roller stripping and ink-in-water emulsification. The former is caused by the formation of ink repellent deposits of insoluble calcium salts produced by reaction between calcium ions and certain acids used in fount solutions. Ink-in-water emulsification can result if calcium ions are free to form soaps with fatty acids present in some vegetable oil based vehicles.

Rheology modifiers

The use of aluminium gellants to produce particular rheological charac-teristics in litho ink vehicles has been referred to earlier. Ink rheology can also be manipulated by incorporation of additives directly at the ink manufacturing stage. Ideally, it would be necessary to have a series of materials, each being capable of producing a significant change in a single rheological parameter only, when added in small proportions. This would give the formulator maximum flexibility to achieve the exact balance of viscosity, yield value, thixotropy and tack that was desired. In reality, the additives available all produce more complex rheological effects and considerable care has to be exercised to ensure that the benefit in one area is not achieved at the expense of an unsatisfactory influence on another property. Such potential pitfalls can extend beyond rheological considerations into other properties such as gloss and setting. With these thoughts in mind, the major ink rheology modifiers may be listed as follows:

Montmorillonite clays

These are now available as stir-in grades for direct incorporation into inks, thus avoiding the polar activation and shear requirements that previously existed. Viscosity, yield value and thixtropy are all increased, generally with little influence on tack. Higher levels of incorporation can reduce gloss.

Fumed silicas

These are very fine particle size powders, which produce substantial increases in thixotropy and yield value but do not raise true viscosity to any great extent. They have little influence on tack. Gloss may be reduced, and indeed some silica grades are sold as matting agents.

Polyamides and aluminium chelates

These are supplied in liquid form for direct incorporation into inks, in addition to their more widespread use in gel varnish production. As with gelled vehicles, the effect involves increases in yield value and thixotropy rather than a simple rise in viscosity. Polyamides, in particular, produce a structure which is readily broken down when the ink is subjected to shear. A drawback with these additives when incorporated at ambient temperature is that structure may continue to develop over an extended period after ink manufacture. Generally, gloss and tack are little influenced by these materials.

Micronised hydrocarbon resins

Low molecular weight and melting point grades are available to increase tack by stir-in addition to inks. Some increase in viscosity usually also occurs. Large additions can produce slow setting, sticky inks which are prone to set-off in the stack.

As well as being useful for adding direct to inks, rheology modifiers such as montmorillonite clays and fumed silicas may be used to produce gelled reducers by incorporation into distillates. Gelled reducers are useful for cutting tack without too much impact on viscosity and are easier to mix into ink in the duct than are straight distillates.

6.5 TYPICAL INKS AND VARNISHES

The details of litho ink and varnish formulation are dependent on the conditions under which the products are to be printed and the end-use requirements of the print. The major distinctions are between sheet-fed and web-fed printing and between the main classes of substrate, namely low-quality papers, high-quality papers, carton boards and impervious substrates. Specific formulations are also usually employed for the small-offset presses found in instant print shops and in-plant printing departments.

Inks and varnishes for sheet-fed paper printing

The quickset mechanism is the standard drying process employed in the sheet-fed offset printing of paper substrates. The formulating approach shown below is used to produce process colours, blacks, colour blending system inks and specially matched individual spot colours:

Organic pigment[a]	18.0
Cooked quickset vehicle[b] ⎫	
Gloss quickset vehicle[c] ⎬	70.0[c]
Fast quickset vehicle[d] ⎭	
PE wax paste	5.0
Anti-set-off paste	3.0
Cobalt/manganese driers	1.0
280–320°C distillate	3.0
	———
	100.0

[a] For example, CI Pigment Blue 15 : 3, for a process cyan.
[b] Based on low solubility modified rosin ester cooked into linseed oil and let down in 280–320°C distillate.
[c] Based on high solubility, low melting point modified rosin ester and long-oil alkyd, aluminium gelled.
[d] Based on soluble modified rosin ester, used at high resin-to-oil ratio.
[e] The exact proportions of the various vehicles will depend on the desired balance of gloss, setting and machine performance.

Black inks of the highest density require the incorporation of specific wetting vehicles, generally based on low viscosity resins and alkyds to give high solids in the vehicle at moderate viscosity. This enables the vehicle to wet fully the substantial surface area associated with high loadings of fine particle size carbon blacks.

Good pigment wetting is essential for satisfactory ink rheology, particularly flow and transference, to the promotion of gloss and a true black jetness.

Quickset overprint varnishes for sheet-fed lithographic application to paper substrates are based on rather different formulations to anything usually encountered in vehicles for inks. A typical varnish would consist of the following:

Maleic modified rosin ester (pale colour)	30.0
Tung oil	15.0
Long-oil alkyd	15.0
PE wax paste	7.5
Cobalt driers	2.5
260–290°C distillate	30.0
	———
	100.0

The pale colour of the resin is necessary to minimise yellowing on top of non-ink-image areas and to avoid shade distortion of underlying colours. Tung oil is introduced to give a tough dry film for maximum rub and scuff resistance. As setting of overprint varnishes is often partially restricted by the ink films already present, oxidation drying is speeded by the use of high drier contents, and solvent separation is encouraged by the use of more mobile and slightly more volatile, lower boiling range distillate fractions. These latter generally give acceptable roller stability since the throughput of varnish on the press is usually high due to the relatively high film weights and coverage which are commonly applied.

Specific ink formulating approaches are utilised where there are particular substrate requirements, such as with label papers, or particular pigment characteristics, as in metallic inks, or where there is a distinct variation in the characteristics of the printing process, e.g. small-offset.

Sheet-fed label inks

Plain paper labels for the can, jar and bottle industries are generally printed on large format sheet-fed machines with a single sheet carrying many individual labels. The substrates used are normally low in base weight and coated on the front side only. Large sheet size, high ink coverage (in order to meet the visual impact demands placed on labels) and lightweight paper produce severe sheet curl problems unless specially formulated low tack inks are used. Adequate viscosity is preserved in such inks by the use of gelled vehicles and gelled reducer. A typical formulation would be:

Permanent Orange (CI Pigment Orange 13)	20.0
Cooked quickset vehicle[a] ⎫	60.0[c]
Gelled quickest vehicle[b] ⎭	
Long-oil alkyd	5.0
Anti-set-off paste	3.0
Gelled reducer	3.0
Cobalt/managnese driers	2.0
PE wax paste	7.0
	100.0

[a] Based on low solubility modified rosin ester cooked into linseed oil and let down in 280–320°C distillate.
[b] Based on soluble modified rosin ester. Aluminium chelate gelled.
[c] The ratio of the two vehicles depends on the exact viscosity/tack relationship required.

Metallic inks

Gold and silver inks are based on bronze and aluminium linings respectively. These linings are powders of a specific physical form which consist of extremely small, flat particles. In order to achieve a satisfactory metallic lustre at the low film weights printed by offset lithography, very high loadings of the finest particle size grades are utilised. Even then, optimum brilliance and lustre can only result if the particles are able to 'leaf' in the ink vehicle, i.e. orientate themselves such that the flat surfaces are overlapping and parallel to the substrate surface. Care is required in formulating metallic litho inks in order to ensure that the vehicle both permits leafing and retains adequate tack and rheology characteristics when very high proportions of the metal linings are present. Rheological stability and freedom from tarnishing are also major considerations because reactions can take place between the metal powder and ink vehicle components. A typical formulation of an offset litho gold suitable for sheet-fed printing on to paper and board would be:

Bronze lining paste[a]	50.0
Metallic quickset vehicle[b]	41.5
Cobalt driers	1.0
PE wax paste	7.5
	100.0

[a] Approximately 85–90% bronze lining in low sulphur content distillate. (The latter is important to avoid tarnishing.)

[b] Based on soluble modified rosin ester resin and long-oil alkyd in low sulphur content distillate. Stabiliser is incorporated into the vehicle for single-pack gold production.

Silver inks utilise a lower lining content than golds, due to the lower specific gravity and superior lustre and opacity of aluminium linings. A typical level of incorporation would be 20%. Also, stabiliser is not usually required in single-pack silver inks since aluminium linings are inherently less reactive with oleoresinous vehicles than bronze powders.

In order to preserve the leafing potential of the linings, all metallic inks must be manufactured by processes involving low shear only. Thus, slow mixing and blending are appropriate but high-speed stirring and three-roll milling are not.

Small-offset

The presses used in small-offset are designed for ease of operation, since they are often run by staff who are not fully trained litho printers, and for economic construction, in order to provide a low-cost facility for the duplication of type matter. Inking and dampening systems are frequently rather different from those employed on larger, sheet-fed presses and are often extensively integrated. The ink distribution system is restricted in rolling power. Additionally, a wide range of plates is used, extending

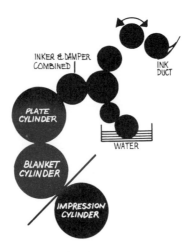

Fig. 6.10 Integrated inking system — small offset

down to very low cost grades which rely on a tough paper as the base support material. Such plates may have only a limited distinction between image and non-image areas.

These factors impose a unique set of requirements on the physical and chemical characteristics of small-offset inks. Primarily, the inks must show good distribution and transference, controlled emulsification even on integrally damped presses (Fig. 6.10) with their associated founts (which may contain glycols or glycerol), and high cohesion in order to prevent ink transfer to poorly protected non-image areas. The inks are usually required to remain non-skinning over extended periods because the presses are not routinely washed up, even at the end of the day or week, particularly where they are only utilised for printing black. This does not pose too many drying problems when the inks can be formulated to dry by penetration into uncoated paper stocks. Here rubber resin based inks can be used as well as inks based on oxidation drying vehicles with appropriately high additions of anti-oxidant. Where the role of small-offset presses is extended to include the printing of coated papers, oxidation drying systems with a very careful balance of driers and anti-oxidants must be utilised, usually to give extended non-skinning with relatively slow drying.

Typical formulations for oil-based and rubber-based small-offset blacks are as follows:

Carbon Black (CI Pigment Black 7)	20.0
Reflex Blue (CI Pigment Blue 18)	2.0
Oxidation drying vehicle[a]	70.0
Anti-oxidant paste[b]	2.0
Alkali-refined linseed oil	6.0
	100.0

[a] Based on modified rosin ester in linseed oil.
[b] Butylated hydroxy toluene dissolved in oleoresinous vehicle.

Carbon Black (CI Pigment Black 7)	20.0
Reflex Blue (CI Pigment Blue 18)	2.0
Micronised talc[a]	10.0
Cyclised rubber vehicle[b]	65.0
280–320°C distillate	3.0
	100.0

[a] Incorporated to control the severe flying tendency of cyclised rubber based system.
[b] Cyclised rubber resin dissolved in relatively high aromatic content distillate, at low resin solids.

Inks and varnishes for sheet-fed carton board printing

A significant proportion of sheet-fed offset carton printing is now carried out using UV curing systems. Ink and varnish formulations for this process are fully detailed in Chapter 10. This subsection will consider those formulations for board substrates which dry by the quickset mechanism. Although inks and varnishes produced for paper printing can be applied to board stocks, optimum performance to meet the stringent and unique demands of the folding carton packaging industry is usually achieved with specialised formulations. The major requirements are:

(1) Hard drying with maximum scuff resistance in order to avoid marking on filling, distribution or retail of the packaged goods.
(2) Very rapid setting to minimise set-off problems and the need for anti-set-off spray, aggravated by the substrate weight and its rapid development of pressure in the stack.
(3) Minimum odour and minimum tendency to impart off-flavours (taint) when used for food and confectionery packaging.
(4) Suitability of subsequent varnishing or lamination in relevant instances. (*Note*: If this requirement also applied to sheet-fed paper jobs, formulations similar to that outlined below can be used.)

Maximum scuff resistance is achieved in inks typified by the following:

Naphthol Red (CI Pigment Red 2)[a]	20.0
Cooked quickset vehicle[b]	50.0
Hard-drying vehicle[c]	20.0
Micronised PE wax	2.0
Micronised PTFE wax	1.0
280–320°C distillate	5.0
Cobalt/manganese driers.	2.0
	100.0

[a] Resistant red pigment for soap and detergent carton requirements.
[b] Based on low solubility modified rosin ester cooked into linseed oil and let down in 280–320°C distallate.
[c] Distillate-free oxidation drying vehicle based on modified rosin ester in tung and linseed oils.

This type of ink is relatively slow setting, particularly if the content of hard oxidation drying vehicle is further increased to meet particular rub resistance demands. Consequently, substantial use of a relatively coarse spray powder is normally necessary to avoid set-off. Where a less hard drying ink can be tolerated, substantial advantages in setting can be achieved by using a formulation as below, which is also appropriate as the basis for low odour and taint or varnishable requirements:

Rubine 4B (Ba salt) (CI Pigment Red 57:1)	20.0
Low odour quickset vehicle[a]	55.0
Hard drying quickset vehicle[b]	15.0
Micronised PE wax[c]	2.0
Micronised PTFE wax[d]	1.0
Low aromatics 280–320°C distillate	6.0
Cobalt/manganese driers	1.0
	100.0

^a Based on soluble modified rosin ester and soluble low odour modified hydrocarbon resin in long-oil soya modified alkyd and low aromatics distillate.

^b Oxidation drying vehicle. This is eliminated where minimum odour characteristics are demanded.

^c PE wax content is reduced in inks for subsequent varnishing or lamination.

^d PTFE wax is eliminated in inks for subsequent varnishing or lamination.

The key to success in this approach is very careful formulation of the low odour quickset vehicle. Resins must be chosen that have inherently low odour and good solubility in low aromatic distillates, coupled with rapid and effective solvent release in order to promote quick, hard setting. A careful chemical balance between resins is also essential to give optimum litho performance on fast-running carton presses.

Alkyd selection is critical in order to balance odour and taint properties with hardness of drying. This is because odour and taint problems can arise from the formation of volatile aldehydes and ketones as by-products of the oxidation drying process. Thus, the elimination of aromatic hydrocarbons by appropriate choice of distillate grade is not alone a guarantee of acceptable odour and freedom from tainting potential.

Overprint varnishes for sheet-fed lithographic application to food and confectionery packaging have to be formulated with due consideration for both these odour and taint factors and the colour and drying aspects described earlier. Thus, a typical formulation would be:

Maleic modified rosin ester (pale colour)	30.0
Long-oil soya alkyd	25.0
PE wax paste	7.5
Cobalt driers	2.5
Low aromatics 260−290°C distillate	35.0
	100.0

Ink for sheet-fed impervious substrate printing

Offset lithography is heavily orientated towards the printing of paper and board substrates which possess a significant degree of absorbency. This enables penetration and setting to play major roles in drying and adhesion in most sheet-fed work. However, a range of foil and plastic stocks are printed lithographically, either with oxidation drying or UV curing formulations. The latter are detailed in Chapter 10.

Oxidation drying inks for impervious substrates represent a difficult compromise between press stability and hard drying. They must be non-skinning in the duct and on rollers, plate and blanket, but free from set-off and must give good key and adhesion to the substrate. In practice, stability times are trimmed to the bare minimum in order to promote rapid, hard oxidation drying. Even so, anti-set-off spray powder and small stack heights are essential to avoid severe set-off and blocking. Bearing in mind the restrictions that apply to the chemistry of litho vehicles as discussed in Section 6.1, it is not possible to achieve key to plastic substrates by incorporating resins of similar chemistry to the plastic in the binder vehicle of the ink. Key and adhesion have to be achieved through oxidation drying producing a hard, yet flexible, polymerised film. Because of this lack of formulating flexibility, litho inks cannot readily cope with the wide range of impervious substrates that are printed by gravure, screen and flexo.

A further complication arises in the area of ink/water balance. Since the substrates are impervious they are unable to transport fount solution away from the impression area of the press in the way that paper and board substrates can. This increases the likelihood of over emulsification of the ink. As well as being a potential source of poor reproduction quality, such emulsification can also retard oxidation drying and provide another obstacle to satisfactory ink adhesion on print.

Despite all these potential pitfalls, correctly formulated oxidation drying inks in the hands of skilled litho printers regularly produce fully satisfactory work on foil boards and on plastic sheets. A typical formulation would be:

Phthalocyancine green (CI Pigment Green 7)	20.0
Oxidation drying vehicle[a]	70.0
Micronised PE wax	3.0
Micronised PTFE wax	1.0
Cobalt driers	3.0
Manganese driers	1.0
Alkali-refined linseed oil	2.0
	100.0

[a] Based on modified rosin ester and containing high proportions of tung oil. Must be completely distillate free. The chemistry of the vehicle is carefully optimised to minimise emulsificatioin that would affect transference or drying.

Inks for web-offset paper printing

The engineering advantage of handling a continuously running web rather than discrete sheets enables web-offset printing to cope with lower weight papers at significantly faster speeds than even the most productive sheet-fed press. This advantage can only be exploited with inks that are capable of printing and drying at high speeds. Thus, web-offset inks must be formulated to the correct chemistry and rheology for high speed, high shear lithographic printing and fast drying. On lower quality, highly absorbent papers such as newsprint adequate drying can be achieved very rapidly by ink penetration into the interstices between the paper fibres. This so-called 'coldset' mechanism achieves a relatively mar resistant print due to the severely restricted mobility of the ink amongst the fibres. On higher quality, more closed surface and coated papers insufficient penetration can take place to ensure that the print will not mark as it is handled in-line to the printing units through folders and cutters. In such cases heatsetting is used to achieve ink film solidification through distillate evaporation and penetration. The different drying mechanisms require different formulating approaches.

Coldset

A typical coldset formulation is illustrated by the following process cyan:

Phthalocyanine Blue (solvent stable) (CI Pigment Blue 15.3)	15.0
Hydrocarbon vehicle[a]	10.0
Process oil[b]	50.0
Montmorillonite clay-gelled process oil	20.0
280–320°C distillate	5.0
	100.0

[a] Low cost hydrocarbon resin in process mineral oil and distillate; used to assist pigment wetting.

[b] Broad-cut liquid petroleum fraction, significantly less refined than the close-cut distillates used in quickset sheet-fed and heatset inks. Unlike distillates, process oils possess some pigment-wetting capabilities.

Since coldset printing is used for relatively low quality, low cost publications, cost is a prime consideration in ink formulation, hence the use of hydrocarbon resin and process oil as the basic vehicle components. Where the dark colour can be tolerated in black inks, it is usual to utilise a cheap natural asphalt or bitumen resin, such as gilsonite, in place of hydrocarbon resin to boost the pigment-wetting capabilities of the process oil.

Some of the rheological requirements for coldset inks have already been discussed in Section 6.3. The inks are relatively low viscosity, due to the large proportion of process oil, and this assists pumping, distribution and penetration into the newsprint. However, a degree of structure is required in order to avoid duct dripping and handling problems, the latter being relevant only on presses where the ink is fed by hand. Tack is inherently low due to the low resin solids of the formulations, and this is essential for high speed printing on the weak surface strength papers used.

Heatset

Heatset inks are much more like quickset sheet-fed formulations. This can be seen from the following example:

Diarylamide Yellow (CI Pigment Yellow 12)	15.0
Fast heatset vehicle[a]	
Fast heatset gel vehicle[b]	75.0[d]
Slow heatset gel vehicle[c]	
Micronised PE wax	2.0
Micronised PTFE wax	0.5
Low aromatic 240–260°C distillate	7.5
	100.0

^a Based on a combination of modified rosin ester and modified hydrocarbon soluble resins at high resin:alkyd ratio in 240−260°C low aromatic distillate.

^b As (a) but gelled with aluminium chelate.

^c As (b) but on 260−290°C low aromatic distillate.

^d The exact proportion of the various vehicles will depend on the desired balances between tack and viscosity and between stability and heat-setting.

Resins are chosen for good solubility in low aromatic distillate but must also allow rapid solvent release for optimum heatsetting. A highly efficient wax combination is required, to achieve good surface slip immediately upon solidification of the ink film. This ensures that the print is protected adequately as it passes over chill rollers *en route* to the in-line folders and cutters.

6.6   INK-RELATED PROBLEMS AND THEIR POSSIBLE SOLUTIONS

There are five main areas where difficulties can arise with lithographic printing inks:

(1) Litho problems;
(2) Rheology problems;
(3) Tack problems;
(4) Setting and drying problems;
(5) Appearance problems.

Litho problems

Lithography runs into problems when ink transfers to the fount solution or to non-image areas of the plate; when it fails to retain a continuous film on the inking rollers, or when it builds up on rollers, plate or blanket. The names given to these various difficulties are, respectively, tinting, scumming, stripping and piling. All can occur in inks whose chemical and physical properties meet the demands of the lithographic process identified in Section 6.1. However, the exact detail of these characteristics may lead to litho problems under particular circumstances, as will be seen from the following descriptions.

Tinting

Tinting occurs when pigments, with or without other ink ingredients, become solubilised or emulsified into the fount solution. The fount takes on a weak coloration from the ink which transfers as a wash of colour on to the non-image areas of the print. This problem is usually caused by pigments which have been incompletely washed free of soluble components during their manufacture, or which contain materials that are reactive with certain vehicle components to form water emulsifiable soaps. It can therefore be avoided by using an alternative pigment grade. Tinting caused by ink-in-water emulsification resulting from calcium soap formation may be overcome by adding a complexing litho additive, as described in Section 6.4.

Scumming

Scumming occurs when non-image areas of the plate accept and transfer ink such that the print does not remain completely clear in these areas. Minute flecks of scummed ink are present on virtually all lithographic print. In fact, their presence represents the only simple way to distinguish lithographic print from high quality letterpress work. Obviously, scumming becomes a problem when ink transfer in the non-image areas becomes apparent to the naked eye or when it affects tonal rendition by increasing apparent density in halftone areas. It may be caused by a variety of factors associated with plates and their processing, with fount solutions, or ink characteristics, both chemical and physical.

When ink chemistry imposes a narrow range of water tolerance the chances of scumming are increased. Thus, if ink is too water repellent, dampening levels will have to be kept very low to avoid the fount from inhibiting receptivity and transfer of ink to sections of the image area (so-called 'water marking'). At these low settings there may be insufficient fount present to protect fully all non-image areas from accepting ink. Such problems are aggravated by irregular demands for fount and ink due to image layout across the width of the press. While there is control of ink feed to different lateral zones on all presses, any such control of dampening feed is very rare. Thus, it is usual for a standard fount application to apply across the plate regardless of image/non-image relationships. Variations in the latter have to be accomodated by ink−water balance tolerance if problems such as water marking and scumming are to be avoided.

The physical characteristics, of the emulsified ink can have a major impact on scumming. Problems may arise because the ink has too low a viscosity possibly caused by over-reduction or elevated temperature. Ink may then transfer over the top of the fount film on the non-image areas (see Section 6.1). Excessive emulsification may also lead to a greater attraction between emulsified ink and fount, leading to the same effect.

Most ink-related scumming problems are due to a complex mixture of the particular physical and chemical properties of the ink, often linked through emulsification. Sometimes it is possible to produce a solution by a simple increase in viscosity and tack. This increases inherent cohesion and reduces the degree of emulsification. Equally, there are occasions when a change in formulation is required to reduce water repellency and increase water tolerance of the ink, or to give a more cohesive emulsion.

Stripping

Stripping occurs when the ink ceases to transfer along all areas of the roller train surface. This can adversely affect ink feed to the image areas of the plate and can result in a print that is weak or of uneven density. The problem can be caused by poor setting or condition of the rollers, but when related to ink rather than machine, it is most often caused by a reaction between ink ingredients and fount components which can produce hydrophilic deposits on the rollers. This reaction and its control were discussed in the paragraph on litho additives in Section 6.4.

Piling

Piling is the name given to the build-up of excessive ink film thickness on rollers, plate or blanket. The immediate cause is a lack of distribution or transference in the relevant areas. The problem is visually apparent on the press and manifests itself on the print as a lack of continuous and even coverage of all image areas. This is normally first observed along the trailing edge of solids when the sheet is examined in the direction that it was printed. If the press is not stopped and relevant areas cleaned, the problem usually increases to the point where print quality is seriously impaired and physical damage is caused to the plate and blanket by the increased pressures in the areas of piling.

There are a number of potential causes of piling and these may sometimes be interconnected in a complex manner. Firstly, piling on the rollers, plate or blanket may occur due to emulsification having an adverse influence on rheology and tack properties and hence hindering distribution and transference. Alternatively, plate and blanket piling in particular may be caused by loss of distillate, with the resulting poor press stability causing the ink to dry up and lose transference. Both these causes essentially involve the ink as a whole piling. There are also a number of instances where the build-up involves a type of filtration process whereby most of the fluid components of the ink transfer, but hard, particulate material remains behind on the plate or blanket, bound by a small proportion of retained vehicle. Common examples of this latter phenomenon are piling of poorly wetted micronised waxes and piling of coarse or poorly dispersed pigment particles. Emulsification can make things worse by promoting de-wetting of particles that are not completely coated in vehicle. Finally, piling may involve significant proportions of components which originate from the substrate, and which have been removed by the influence of fount and ink. Thus blanket pile on web-offset presses is often found to contain large quantities of fibre lint from uncoated papers or fillers from poorly bound paper coatings.

With this diverse set of causes there is clearly no single solution which will overcome all instances of piling. Changes in the basic chemical and physial characteristics of the ink may be necessary if emulsification inhibits distribution and transference. However, piling associated with poor press stability may be overcome by addition of bridging solvent or by the use of a distillate grade of higher boiling range or stronger solvent power. Filtration type piling clearly requires improved ink manufacture, or a change in the grades of the relevant particulate materials used to ensure that satisfactory wetting of the particles with vehicle is retained, even when thin films are sheared and emulsified with fount. If materials originating from the substrate are a major component of piling, then ink is not the sole cause of the problem. However, it may still be possible to avoid the problem by ink modification, for example by reducing tack in order to reduce the force pulling fibres or coating from the stock at the point of impression.

Rheology problems

Rheology problems have been covered in Section 6.3 under the heading 'Relevance of rheology'. One of the major complications in litho ink technology is the interaction of rheology and emulsification, with each property being a major, but not sole, influence on the other. Thus, as has been discussed, rheology may be involved in complex ways in litho problems such as scumming and piling. With this background it is easy to understand why a considerable array of techniques has been developed to enable ink rheology to be manipulated. These range from straightforward viscosity reduction with distillate, through techniques for producing various degrees of structure in vehicles and inks, to flow promotion by incorporation of low viscosity alkyds or other wetting resins. With these tools the formulator is able to tailor litho ink rheology to overcome problems, whether these be directly caused by the physical properties of the ink or associated with emulsification.

Tack problems

Section 6.3 provides a summary of tack problems under the heading 'Relevance of tack'. Tack is controlled by the use of gelled vehicles and by incorporation of distillate or gelled distillate into the ink. Additives are also available which enable minor upward adjustment of the tack of a finished batch of ink that would otherwise have to be rejected and replaced with a new make of a modified formulation. This can be helpful where trapping problems are encountered. Consequently, inks can be manipulated through a reasonable tack range by additives or, if this is insufficient to overcome particular tack problems, they can be reformulated using different balances of gelled and ungelled vehicles and reducers.

Setting and drying problems

Lithographic inks generally represent a major compromise between the conflicting requirements of press stability and drying. The former demands that thin ink films must retain fluidity throughout the distribution and transference processes on the press, whereas the latter necessitates rapid formation of a hard, solid film at only slightly lower film thicknesses. Press stability considerations normally result in penetration, quicksetting and oxidation drying inks being retarded in some way to maintain printability. Thus, setting and drying characteristics may be jeopardised, particularly under adverse press or print conditions. In this context, high press speeds, high ink film thickness and substrates which are heavy or which restrict penetration or oxidation can all produce problems. Among the specific difficulties encountered are:

(1) set-off;
(2) chill-roller marking;
(3) turner-bar marking;
(4) carry-over piling;
(5) slow drying;
(6) poor rub/slip/scratch.

Set-off

Set-off occurs in sheet-fed printing when ink transfers from the print to the reverse of the adjacent sheet in the stack. It can be influenced by a whole range of substrate and machine factors, such as:

Absorbency of the stock	— lack of absorbency inhibits setting.
Smoothness of the stock	— smoothness promotes contact and hence set-off.
Weight of the stock	— heavier stock produces more pressure and hence more set-off in the stack.
Speed of the machine	— faster running gives a more rapid rise in stack pressure.
Adjustment of the delivery	— any lateral sheet movement increases marking.
Spray-powder setting	— insufficient spray-powder allows sheet contact.
Maximum stack height	— greater pressures exist in higher stacks.

However, ink performance may also be a prime cause of set-off. Problems can arise if the ink is too slow setting, or if it passes through a very sticky phase as it sets. Either characteristic can make the printed ink film prone to set-off as the stack is being built in the press delivery. Even an ink which sets rapidly and without passing through an excessively tacky stage can cause set-off in the stack if it is susceptible to 'sweat-back'. As oxidation drying gets under way in a stack of work heat is generated by the chemical reactions involved. Thus, temperatures are raised and the affinity of the set ink film for distillate fractions which have penetrated into the substrate may be increased. If this occurs, such solvent components may migrate back into the warm ink film, softening it and rendering it prone to set-off or 'sweat-back'.

All ink-related set-off problems are associated with the compatibility balance of the quickset vehicle. Slow setting speeds are a result of too great an affinity between the resin and the distillate in the same way that 'sweat-back' is associated with too much residual compatibility between the binder resin and the separated distillate. Where the ink has set satisfactorily initially, 'sweat-back' is most often due to the presence of a small proportion of stronger solvent components in the distillate. This problem may be avoided by the use of cleaner, tighter-cut distillates in the ink and ink vehicle. Slow setting problems may be reduced by the use of weaker solvents or an increase in resin-to-oil ratio to decrease compatibility. If the ink passes through a sticky stage this may necessitate a change in resin to a grade which rapidly builds viscosity, but without increase in tack, as distillate is lost in the setting mechanism. (Fig. 6.11).

Anti-set-off additives (see Section 6.4) help to separate sheets in the stack and can therefore reduce the tendency of inks to set-off. They will also assist in fully oxidation drying systems where no actual phase separation setting can occur. However, these latter inks generally require severe limitations on stack heights and appropriate use of spray powder to avoid excessive set-off.

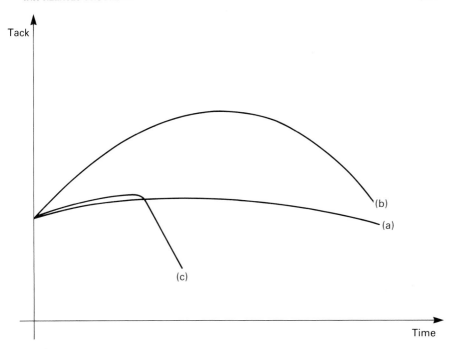

Fig. 6.11 Set-off and the influence of tack characteristics of an ink after printing: (a) slow setting, low tack rise — some set-off; (b) slow setting, high tack rise — severe set-off; (c) fast setting, low tack rise — ideal situation, minimal set-off

Chill-roller marking

On exit from the drying oven, heatset web-offset print is guided around a set of large-diameter cooled rollers. These chill rollers assist in the solidification of the ink film which has had distillate removed in the oven, but which is likely still to be molten due to the temperatures involved. Chill-roller marking will occur if the ink film is too tacky in its molten state or if it has insufficient surface slip to cope with any minor speed differential between the web and the rollers. The problem may be reduced by using higher resin-to-oil ratios or a higher melting point resin in the ink vehicle to yield harder, less plastic films. The use of minimum aromatic content distillates and a reduction in bridging solvent content also assist in avoiding chill-roller marking by reducing any tendency for retained solvent fractions to plasticise the ink film. The use of a wax combination that promotes surface slip in the molten film is also clearly beneficial.

Turner-bar marking

Web-offset print, whether coldset or heatset, is passed over small diameter turner-bars as it enters the folding station in line with the press.

If the ink film is inadequately set it may transfer on to the turner-bars and then be redeposited on non-image areas of the print. Such marking may be avoided by increasing the penetration of coldset inks or by improving the setting and surface slip characteristics of heatset inks as described in the preceding subsection. Penetration of coldset inks may be improved by simply reducing the viscosity of the formulation. If this is not possible due to lithographic or other machine restrictions, then a reduction in the resin content of the ink may be appropriate. Alternatively, as with all problems associated with slow setting or drying, an increase in colour strength such that a reduced film weight of ink may be carried can be beneficial.

Carry-over piling

So far all problems associated with setting that have been described have been to some extent involved with a lack of setting speed. Carry-over piling is a problem that can occur on multi-unit presses if the ink sets too rapidly on the particular substrate being printed. Even an ink that has a sufficiently suppressed tack rise on setting to avoid set-off will, nevertheless, go through some increase in tack as distillate separation commences. It is therefore possible for an ink printed on the first printing unit of a multi-colour press to be at a sufficiently high tack level as it passes through subsequent units for it to transfer back from the print to the blanket. Thus, print quality is impaired and a build-up of ink occurs on a later colour blanket. The problem may be overcome by slowing the setting speed of inks to be printed on early stations of a multi-colour press. This may be achieved by decreasing the resin-to-oil ratio of the vehicle or by some other technique to increase compatibility, such as addition of bridging solvent.

Slow drying

As with set-off, slow drying may be due to a range of factors:

Low temperatures	Oxidation drying, like other chemical reactions, is retarded by reduced temperatures.
Acid stocks	At high humidity (which is almost always present in litho printed stacks) acidity can severely retard oxidation drying.
Lack of oxygen	This can be particularly problematic in large stacks, especially if these are wrapped for dust protection.

The main ink-related causes of slow drying are inadequate oxidation potential in the vehicle, or lack of available drier catalysts. The former may be rectified by incorporating more oil or alkyd into the vehicle or by the use of fully drying materials such as linseed or tung oil in place of semi-drying oils such as soya. Ineffective driers may be due to inappropriate quantities being included in the formulation, or to an effect known as drier absorption. This involves pigments with a high surface area, notably carbon black, and prevents the drier being available for reaction with the vehicle since it is absorbed into the pigment surface. Drier absorption is linked to the age of an ink as a given ink's drying potential can deteriorate with time. Such drying problems can clearly be rectified by addition of further catalyst to the ink. When antioxidant has been added to extend skinning times, such a simple approach may be more difficult to implement. Careful balancing of drier and antioxidant types and proportions is essential to give adequate drying without jeopardising stay-open requirements.

Ink-formulating approaches can also assist in instances where poor drying is not solely an ink problem. Poor drying due to lack of oxygen availability can be boosted by the use of calcium perborate based driers which liberate oxygen when mixed with water, as occurs when fount becomes emulsified into the ink. Drying on highly absorbent substrates, such as cast-coated papers, may be impaired by vehicle penetration into the substrate or its coating. This then leaves insufficient binder on the surface and the pigment is prone to 'chalk' from the print. This gives the appearance of poor drying. The problem may be alleviated by increasing the resin and oil binder of the ink vehicle.

Poor rub/slip/scratch
Lack of rub and scratch resistance or inadequate surface slip characteristics may be associated with poor drying. However, it is also possible to have problems in these areas even when all possible oxidation drying has taken place. This is because dry litho ink films have a very high pigment-to-binder ratio and it may not be possible for the polymerised binder on its own to produce all the desired surface protection. Thus, various waxes and slip aids are used in litho ink formulations.

Consequently, it is possible to encounter rub, scratch and slip problems if the incorrect type and quantity of these additives is present. Rub and scratch resistance problems may be alleviated by using greater quantities, coarser grades or harder grades of PE and other waxes. Surface slip characteristics may be improved by appropriate use of PTFE wax or silicone slip aids.

Appearance problems

A litho print may fail to produce the desired appearance through lack of fidelity in the printing process or because of unevenness or poor gloss in the printed ink films.

Dot reproduction
Reproduction is most critical in halftone dot areas since variations affect both tonal balance and colour rendition. Dot gain, due to both optical and physical effects, is an inherent feature of offset litho printing on paper and board substrates. It is not, as such, a fault, provided that it occurs in a sufficiently controlled and predictable manner to enable appropriate allowances to be made at the colour separation stages (see Chapter 3). Optical dot gain is caused by light scattering under the printed dots in the surface layers of the substrate. Consequently, it is almost entirely controlled by the characteristics of the stock. Physical dot gain is a result of the squash or flow-out that occurs when the fluid ink dots are transferred from plate to blanket to substrate under impression pressures. It is there-

fore influenced by press, blanket and substrate characteristics as well as ink factors.

Inks must be formulated to minimise the degree of sideways spread during the transference stages. Rheology, in terms of viscosity and flow, and cohesion (tack), require appropriate control to maintain dot gain within identified tolerances. Further, the tinctorial strength of the ink is also important since variations in this property will be compensated for by adjustment of ink film thickness to maintain colour density. Thicker dots will naturally be more prone to dot gain through edge squash under printing pressures. Techniques for controlling the rheology, tack and colour strength of litho inks have already been covered in detail in preceding subsections of this chapter.

Poor lay

Lay is the term given in the printing trade to the smoothness and evenness of a printed ink or varnish film. Poor lay means that the solid areas of ink or varnish are not of a completely uniform film thickness. This may be caused by variations in the receptivity of the substrate surface or to inadequacies in transference and flow characteristics of the ink or varnish.

Receptivity problems may be an inherent deficiency of the given stock or, particularly in the case of over-varnishing, may be related to the surface characteristics of underlying ink films. An ink film which has too low a surface energy, either due to the nature of the polymerised binder or from the presence of low surface energy rub and slip additives, will cause the varnish to reticulate since its surface energy is greater than that of the underlying ink. This is a severe case of poor lay where the substrate surface is non-receptive to the varnish film. Sometimes the problem may be overcome by additions of surfactant materials to lower the surface tension of the varnish. Alternatively, it may be necessary to reprint the job using a different ink and substrate combination to ensure that the surface is more receptive to the varnish.

Borderline transference characteristics can lead to poor lay since some image areas may not transfer as well as others. Transference is dependent on both rheology and emulsification characteristics at the point of impression and other properties, such as press stability, have an impact on these. Therefore, there is clearly no single solution to overcoming poor lay due to borderline transference. However, factors which improve the compatibility of the vehicle, such as minor additions of oil, alkyd or bridging solvent, frequently produce sufficient change to eliminate this marginal deficiency.

Poor flow characteristics of the ink film initially transferred to the substrate can result in print showing poor lay. This is because completely even transfer is not always achieved and a certain amount of flow is required to enable a film of completely even thickness to form. Again, a variety of factors influence the flow characteristics of the printed, emulsified ink. A small addition of a flow promoting oil or alkyd may be sufficient to improve lay. However, if emulsification is severely restricting flow then it may not be possible to effect improvements to the lay without a significant adjustment of ink chemistry or rheology.

Poor gloss

Gloss, or the mirror-like reflection of the ink film, is a much demanded property for high quality print, particularly where it is needed to make a marketing impact. Since gloss is dependent on the formation of a smooth, continuous film it is strongly influenced by the absorptivity and surface smoothness of the substrate and the chemical and physical properties of the ink. Poor ink gloss may be caused by poor pigment wetting, poor transference and flow, or the presence of large particle size material. Incomplete wetting of the pigment particles by the ink vehicle will result in light scattering within the ink film which reduces gloss. It may also promote over-emulsification of ink with fount which will inhibit the transference and flow properties that are so important to ensure the formation of a smooth ink film on print. Particulate material which protrudes through the thin ink film will act as light scattering sites and will impair gloss.

A number of options are open to improve gloss by ink modification. Increasing the oil or alkyd content of the ink is beneficial in a number of ways. It improves pigment wetting and boosts compatibility. The more compatible vehicle has improved press stability and hence superior transference and flow on print. Use of a lower viscosity resin in the ink vehicle enables solids to be increased and thus promotes hold-out of a smooth ink film on the substrate surface. This gives superior gloss compared with an ink film that has penetrated extensively into the stock and which therefore allows light to be scattered from protruding paper fibres and coating particles. Choice of pigment and wax grades which are readily wetted and which break down to fine particle sizes in the ink manufacturing stages are also clearly beneficial. Within appropriate limits, gloss can also be promoted by a reduction in pigment content, and hence tinctorial strength, which results in a thicker ink film that boosts hold-out and flow. The disadvantage of this approach, and indeed of the use of more compatible vehicles, is that set-off becomes more problematic. If this is not controlled by appropriate stack handling techniques advantages in gloss may be lost, either due to marking in the stack or through the light scattering produced by excessive application of coarse anti-set-off spray powders. In this respect the only techniques that can improve gloss without detriment to setting are in the areas of pigment choice and manufacturing method to improve wetting and reduce the presence of over-size particles. Pigment flushes can be particularly advantageous for gloss inks because of their final level of dispersion and excellent degree of wetting.

6.7 RECENT AND FUTURE TRENDS

Lithographic printing is now a fully established, mature printing process. Consequently, change, both in the process and the ink technology that serves it, tends to be evolutionary rather than revolutionary. Relevant evolutionary trends can be identified in three main areas:

(1) the litho printing process;

(2) drying and print finishing techniques;
(3) litho ink raw materials and manufacture.

Litho printing

The major trend in the litho industry is towards higher printing speeds and larger presses. As yet only tentative steps are being taken in new approaches, such as anilox inking, and even more radical departures, such as driographic plates.

Productivity

The demand for higher productivity, heard throughout all manufacturing industries, produces a trend in litho printing towards higher speeds and larger machine sizes. Speeds are continually increasing in all areas of web and sheet-fed printing. Both the press manufacturer's maximum mechanical ratings and the average actual production printing speeds have increased significantly over the last decade. There are now numerous sheet-fed presses rated at over 10 000 impressions per hour (iph) (Fig. 6.12) and web-offset machines engineered for over 1400 feet/minute (425 m/min). Running speeds in excess of 9000 iph, sheet-fed, and 1400 feet/minute, heatset web-offset, are commonplace. Indeed, no speed under 2000 feet/minute (600 m/min) is now regarded as worthy of special note in the heatset industry and the talk is in terms of moving up towards the achievement of 3000 feet/minute (900 m/min). Sheet-fed presses appear to have been at their limit for sheet size for some time, but both wider and larger cylinder circumference web presses have been

**Fig. 6.12 A modern six-colour sheet-fed litho press
(courtesy of Perschke Price Ltd)**

developed over recent years. This has enabled standard 16-page heatset presses to be complemented with various 32- and 48-page machines. In both sheet-fed and web-offset the number of printing units in a given machine is also growing with the search for in-line productivity. Thus, five- and six-colour presses are now the automatic choice for new investment in the carton printing industry, and eight-unit heatset web offset presses are being installed to enable full colour printing on twin webs (see Fig. 6.13).

These trends in speed and size are encouraging news for the ink-maker because of the associated increased in litho ink consumption. However, while only an evolution of the lithographic process, they require a critical reappraisal of ink characteristics in such diverse areas as chemistry, rheology, dispersion and drying.

Requirements for improved drying and setting characteristics to allow modern presses to run at full speeds without producing problems at the end of the machine are readily appreciated. It may be less apparent that chemistry, rheology and dispersion need modification to cope with new conditions. The two prime aspects of litho press trends which impinge on ink technology are: (i) the higher shear conditions encountered at higher printing speeds and, (ii) the various modified dampening systems that have been devised by press designers to provide the rapid response required to minimise wastage on large, fast-running machines. The litho ink formulator is constantly striving to produce a chemical and rheological balance which will not break down under high shear conditions nor produce unsatisfactory emulsified states when the inks are run in con-

Fig. 6.13 Web-offset press (courtesy of Harris Graphics Ltd)

junction with an ever fuller range of continuous operation dampening systems.

New, finer levels of dispersion have been found essential to avoid piling problems since at high shear rates the presence of pigment agglomerates seems to render an ink particularly susceptible to coarse over-emulsification with subsequent loss of transference. Such problems are undoubtedly made worse by the increased throughput of ink through the press and by the trend to minimise wash-ups in order to maximise productive printing time.

Anilox inking

Despite remote and programmable controls for setting the ink ducts on litho presses this aspect of press make-ready remains time consuming and a potential source of print wastage. There is, therefore, clearly something attractive about an inking system which automatically delivers ink to the plate in line with the demands of the image layout. This is what occurs in the anilox inking of flexo plates. Attempts are under way to modify this approach for offset litho presses. This is not straightforward since anilox cylinders, like any engraved or etched system, work most readily with very fluid low viscosity inks. Conversely, the lithographic process demands inks of relatively high viscosity. It remains to be seen whether a satisfactory compromise can be achieved to enable anilox inking to operate on a range of litho presses.

Driography

The basic concept of driography is to eliminate the need for fount dampening by producing a plate where the non-image surface is inherently non-receptive to ink. The idea is not new but it has been relaunched from a practical point of view in the last few years with the availability of presensitised driographic or 'waterless' plates. The plates rely on a silicone polymeric surface in the non-image areas to maintain ink repellency. The attractions, in terms of the removal of dampening and hence the elimination of one of the major variables of the litho process, are obvious. Unfortunately, the system has so far not shown itself to be sufficiently flexible for widespread application in the litho industry. The plates are physically delicate, being particularly prone to scratching. More fundamentally, it has proved very difficult to eliminate ink receptivity completely in the non-image areas over prolonged runs. Thus, there is a tendency for a wash of colour to print in the background areas after a relatively small number of impressions. This 'toning' can be alleviated by very careful setting of the inking system and by the use of inks which have been appropriately modified to have minimum tendency to wet the silicone surface. However, where inks have to be adjusted to cope with multi-colour printing on lightweight substrates or where press condition or size makes fully accurate roller settings impractical, the lack of flexibility of the 'waterless' approach is exposed.

Drying and print finishing

Major trends in litho ink drying, such as the past introductions of heatset and UV, seem unlikely to be repeated. The only discernible new approach at present is the tentative introduction of electron beam as an alternative radiation curing approach to UV. The first EB installation on a commercial litho press was in the USA at the beginning of the 1980s. For the next 4 years or so this remained unique. However, new equipment is beginning to be installed, particularly for the web-offset printing of liquid packaging cartons which are cut and creased in-line. At the moment, EB equipment is very much more expensive than UV because of the massive electrical voltages employed and the requirement for shielding to prevent leakage of X-rays. Electron beam curing is also particularly susceptible to oxygen inhibition of surface cure and it is therefore necessary to blanket the curing chamber with nitrogen. This adds to design cost and complexity, particularly for sheet-fed applications, and introduces significant running costs in terms of nitrogen requirements. These factors, coupled with the ability of UV to produce satisfactory cure under most offset-litho conditions, seem destined to restrict EB to a limited number of installations in specialised areas.

Print finishing on offset litho machinery has become widespread in recent years and it appears that growth will continue in this area. Two main technologies are employed for application of the finishing varnish, namely aqueous (or water-based) and UV curing. Both types are generally applied as low viscosity systems and lithography is not employed. Application is via the dampers or by using a specialist coater or coater-dampener unit. Ultraviolet curing varnishes may also be applied dry-offset from the ink roller train. This latter route is not suitable for water-based systems since the emulsions employed do not have sufficient tack to drive undriven rollers in the ink train and are often insufficiently stable for such prolonged thin-film exposure. Where varnish application is via the print cylinder, a relief photopolymer plate may be used for pattern application. A similar, if rather cruder result, may be achieved by cutting the blanket. Otherwise application results in complete coverage of the sheet. This may give gluing problems unless specially formulated varnishes are used.

Generally, low viscosity varnish print finishing requires experience and skill in order to achieve a satisfactory result with minimum mess and press down-time. This is especially true when existing damping or inking systems are employed, as opposed to investing in new specialised units. The benefits, however, can be considerable. Ultraviolet varnishing can achieve a quality of gloss and protection that is only just short of lamination. Aqueous systems are not as physically impressive but they can meet certain gloss and resistance requirements economically, and their ready in-line application over wet oleoresinous inks on multi-unit presses produces a set-off free print without any spray powder requirements.

Ink raw materials and manufacture

Trends are currently apparent in both the raw materials available to the litho inkmaker and his options for manufacture. Major raw material advances are in the areas of pigments, resins and distillates.

Pigments

Pigment manufacturers have been working for many years to make their products easier to disperse. Significant advances have been made but it appears that further developments are possible. The benefit to the ink-maker is a reduction in energy consumption if the pigment powders can be fully dispersed under an appropriate mixer only, such that no milling or grinding of the ink is required.

Inkmakers as well as printers are always looking for ways to automate their production operations. One of the prerequisites for this is that the raw materials involved must be capable of being pumped and metered by some technique. To assist in this, pigment manufacturers are working to develop granular pigment forms which can be automatically transported and metered, and which can be readily dispersed in the ink mixing and milling stages. These granular forms also have an advantage in traditional, manual handling as they are substantially free from the dusting problems that arise with pigment powders. Thus, their desirability is not limited to ink production plants of the future.

Pigment flushes have been widely used in litho inks in the USA and Japan for decades, but their use in Europe has been very much more restricted. A variety of reasons can be put forward including availability, economics and technical flexibility. The development of a new generation of very high pigmentation flushes has given considerably improved formulating flexibility, and it now appears that economic and availability factors may be changing. There is, therefore, a possibility that European ink-makers may start to utilise these pigment pastes more widely. Certainly, the flushing process, which involves mixing aqueous press-cake with flushing vehicle, gives very good levels of dispersion, since the agglomeration that usually occurs on pigment drying is avoided. Thus, flushes can be particularly advantageous in gloss inks and inks for very high speed printing.

Resins

Resins manufacturers have succeeded in producing a range of grades suitable for litho vehicle applications which have excellent solubility in very weak solvent power distillates. The best of these grades also demonstrate rapid viscosity rise on distillate loss to give rapid setting. The next stage is the development of grades with a wider range of viscosity characteristics. Traditionally, fully distillate soluble grades, that do not require cooking into vegetable oils to produce compatible quickset varnishes, have been of low viscosity. This has meant that gelation has been necessary to give high viscosities while restricting resin solids and hence tack. Obviously, the development of high viscosity grades of resin which reduce the requirement for gelation will simplify varnish manufacture and control. However, at present it seems unlikely that the resin chemists will be able to eliminate all requirements for gelled varnish production.

Modifying hydrocarbon resins has been another major area of involvement for resin suppliers. Basic hydrocarbon resin chemistry is inert and rather too water repellent for optimum litho performance. Continuing efforts to develop a wider range of modified hydrocarbon resins seems likely, particularly as the basic cost of this approach can be low.

Distillates

Petroleum distillates are now available in zero aromatics rather than just low aromatics grades. Early products were exceedingly weak solvents and were often prone to 'waxing' (solidification into a wax state) at low temperatures. These properties have now been improved by increasing the proportions of naphthenic and olefinic unsaturates present with a corresponding decrease in the purely paraffinic character of the distillates. As well as being desirable for low odour and taint carton inks and for low emission odour heatset, the controlled solvent power of the zero aromatics grades can be advantageous for maximising solvent release and hence setting characteristics, with appropriate soluble resin grades.

Manufacture

The three-roll mill is the traditional machine for the grinding and pigment dispersion stages of litho ink manufacture. Increasingly, bead mills are being harnessed for their productivity advantages, particularly for production of standard process colours.

The goal of automating litho ink production has been referred to briefly earlier. Continuous, automatic ink manufacturing concepts are now being promoted by some equipment suppliers. The basic features are automatic feeding and metering of both liquid and powder materials, pre-mixing as a continous process with direct feed to a bead mill. The equipment technology appears to be available, but there is understandable caution regarding such installations in the litho ink industry since such machinery tends to be highly inflexible and so unable to cope with the vehicle and colour variants that are required by the diverse applications of offset lithography.

Summary

Offset lithography is now firmly established as the dominant printing process for paper and board substrates. It is both economically and technically fully competitive. In this situation it is unlikely that the process or the inks that serve it will undergo any revolutionary changes. Nevertheless, continued developments will be necessary to exploit the maximum benefits of the process. Increased productivity, both in ink production and in the printing industry, will be a major force behind the demands for litho ink modification and refinement.

CHAPTER 7
Gravure inks

The gravure process has become established as a major part of the printing industry. At the time of its invention, at the beginning of the twentieth century, it offered the means of high quality reproduction on paper coupled with high press speeds, which in due course gave it an advantage over the established litho and letterpress processes, especially for publication printing. In later years the process was to prove ideally suited for printing the rapidly developing non-porous substrates, principally coated and uncoated cellulosic films and aluminium foils. As the era of the plastic films emerged, flexographic printing also developed and combined with gravure to become the major sources of printed flexible packaging. Gravure printing has many other areas of application, including paper and board packaging materials for confectionery, foodstuffs and the tobacco industry. It gained favour in the printing of cigarette cartons because of its lower production costs. It also offered superior metallic golds when required. The lower costs can be attributed to the use of the in-line cutting and creasing operation which produces finished blanks from the reel. This is made possible by the drying characteristics of gravure inks. There are many non-packaging applications where the gravure process has been adopted. Postage stamps, football coupons and catalogue printing are typical examples. In the industrial and home areas, plastic laminates, wallcoverings, decorative wall panels, PVC cladding for buildings and PVC floor coverings, all represent products that are dependent on the ability of the gravure process to print consistent, fine tone reproduction.

From the description of the gravure process in Chapter 2, it is apparent that the system requires inks to be mobile, low in viscosity and rapid drying. In order to print, the ink must fill the cells of the gravure cylinder and after removal of the excess by the doctor blade it must be sufficiently fluid to transfer to the substrate at the point of impression, flowing out to give a smooth print.

For inks to have the fundamental fast drying properties necessary for the gravure process, the use of highly flammable and sometimes harmful solvents is required. Efforts have been made over recent years to reduce the harmful effects of solvents by improved press extraction and better ventilation. As relevant information has emerged, certain solvents have been eliminated from gravure ink formulations, being replaced by more acceptable alternatives.

When the characteristics and physical nature of gravure inks are studied, it becomes apparent that highly volatile solvents are fundamental to their successful performance.

7.1   GENERAL CHARACTERISTICS

Viscosity

Unlike the offset and letterpress processes discussed in previous chapters, the gravure process requires low viscosity inks. However, they are similar to offset and letterpress inks in that they contain resinous binders and pigment colourants. Gravure inks dry by the evaporation of highly volatile solvents and normally there is no oxidation or chemical change once the ink film is dry. It is this need for low viscosity, combined with rapid solvent evaporation, that determines the essential characteristics and performance of gravure inks.

The range of ink viscosities met with in gravure printing is generally between 15 and 25 secs. Zahn Cup No. 2 at 25 °C, dependent on the conditions which influence ink viscosity on the press. The main factors are:
  (1) the rheological characteristics of the ink;
  (2) speed of printing;
  (3) rate of evaporation of the solvent system;
  (4) shape and range of cell depths;
  (5) the design parameters;
  (6) nature of the substrate.
Printing at high speeds demands faster drying inks and hence lower viscosities, to enable the ink to flow rapidly out of the cells at the point of impression. Absorbent substrates require lower viscosities than non-porous surfaces, in fact some substrates require the addition of retarding solvents to enable an acceptable print to be achieved. It is not until all these conditions have been taken into account that the correct viscosity can be established.

Ink viscosity must be adjusted to meet several critical factors. Too high a viscosity results in an inadequate flow of ink from the cells, giving a pattern to the print in some areas that follows the cell walls, commonly known as 'screening'. Too low a viscosity results in a 'slur out' or 'halo' occurring on the trailing edge of the print, appearing as a thin film of ink printing beyond the limits of the design. It is important to note that any final strength and shade adjustments to an ink should only be made when the correct viscosity and printing speed have been determined.

Flow and rheology

As indicated previously, a gravure ink has to change from a low viscosity liquid to a solid in a matter of a few seconds. The ability to do this lies primarily in the rapid evaporation of the solvent content of the ink, followed by the resin attaining a tack-free state as soon as the solvent has left the ink film.

The solvent and resin components are the carrier or ink vehicle which carries the colourant whilst the resin binds the colourant on to the substrate. Pigments and dyes provide the colour, additives being used for a variety of purposes, particularly to improve press performance and enhance the characteristics of the printed ink.

Prior to considering the rheology of gravure inks it is necessary to understand the conditions prevailing on the gravure press.

The gravure process consists of a design engraved or etched in the form of minute cells into the surface of a cylinder or occasionally a plate. The cylinder A rotates in an ink duct or reservoir which is supplied with ink pumped from the ink box. Excess ink is wiped from the cylinder with a reciprocating doctor blade C, leaving the unetched area free of ink and a deposit of ink in each of the cells. The substrate F is passed over the cylinder with a rubber impression roller D forcing the substrate on to the surface of the cylinder at the nip B, thereby transferring the ink from the cells on to the substrate (Fig. 7.1).

It is important that the correct machine settings are maintained, the doctor blade must be set at the correct angle and pressure. Too much pressure and incorrect angling of the blade can lead to similar print defects to those described for incorrect viscosity adjustment. The impression roller must have the correct shore hardness for the type of substrate being printed, with only sufficient pressure used to ensure proper transfer of ink to substrate.

It is when the ink is transferred to the substrate that the rheological

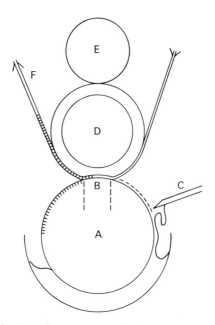

Fig. 7.1 The gravure printing system

properties of the ink influence the print quality and smoothness, i.e. lay characteristics; the ink must flow out of the cells to give as smooth and unblemished a print surface as possible.

On occasions this is not the case and close examination of the print shows that bubbling has occurred which subsequently leaves minute surface craters. Excessive pigment flocculation can occur, manifesting itself as irregular lay, characterised by 'crows' feet' like patterns in the print. Frequently the overall appearance of poor lay is a 'hills and valleys' configuration which is attributed to the ink splitting unevenly and failing to flow out before it dries.

Clearly, factors other than ink performance affect lay; substrate, machine speed and particularly the type of engraving, all contribute to the print quality achieved. Two traditional types of cylinder engraving are conventional (Fig. 7.2), and direct or hard dot (Fig. 7.3). In recent years mechanical methods, e.g. Helioklischograph, have been used to engrave cylinders, resulting in solids with smoother lay and fewer imperfections, when compared with similar designs using conventional engraving. This has proved helpful when printing difficult colours such as greens and browns (Fig. 7.4).

The printability and lay characteristics of a gravure ink can often mean the success or failure of a specific formulation. Unfortunately there is no exact way of predetermining if a combination of pigment, resin, solvent or additives, dispersed by a chosen method will produce an ink with acceptable lay.

The influence of components on characteristics

Gravure inks are composed of resins, pigments, dyes, plasticisers, waxes, additives and solvents, each playing an important role in the characteristics and properties of a particular ink.

Pigments and dyes

The prime purpose of pigments and dyes is to confer colour. Basic dyes are rarely used in gravure packaging inks due to their reactivity and poor lightfastness. Occasionally they are used in publication inks to brighten and strengthen the blue and red. Lightfast dyes are primarily used on aluminium foil where their transparency is complemented by the metallic reflectance of the foil. However, they are expensive in comparison with transparent pigmented chips which are more widely used.

The choice of pigments most suited to the gravure process can be complex and their effect on the behaviour of an ink can be critical. Ideally a pigment should be resistant to the solvents in the ink, and should not give rise to adverse flow properties. Pigments which produce very short inks generally give poor lay, due to the lack of flow into and from the gravure cells. A pigment must exhibit good printability characteristics in the chosen resin/solvent system.

As the pigment in most coloured gravure inks is the most expensive part of the formulation, the economics of pigment selection is of vital importance. The normal practice is for ink makers to have a preferred standard range of pigments against which alternatives can be assessed.

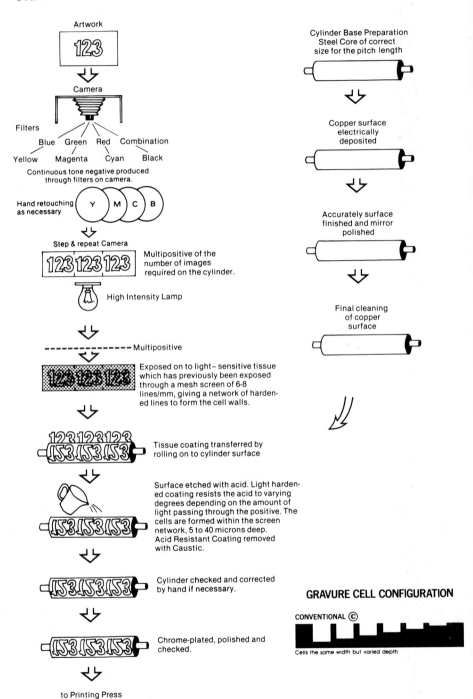

Fig. 7.2 A conventional gravure engraving (courtesy of Gravure Cylinders Ltd, Derby)

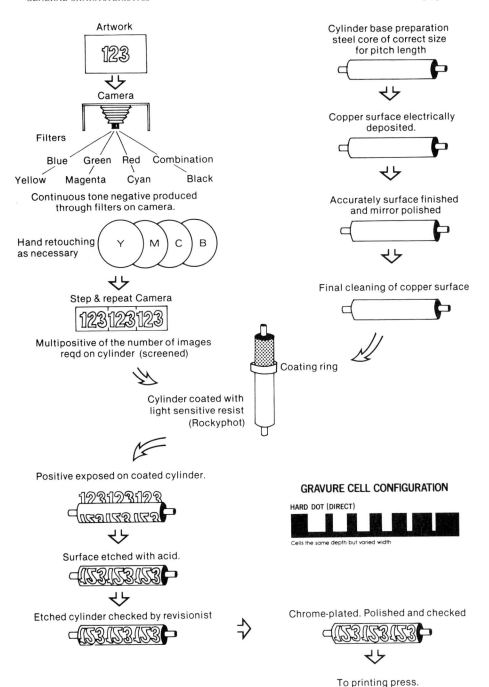

Fig. 7.3 Direct engraving (courtesy of Gravure Cylinders Ltd, Derby)

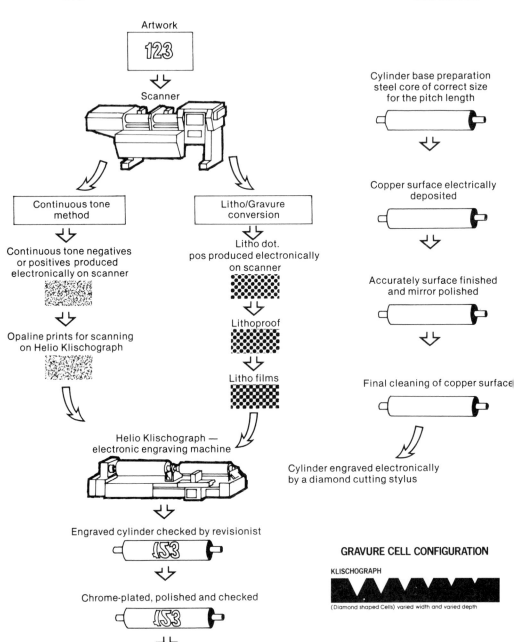

Artwork

Scanner

Cylinder base preparation
steel core of correct size
for the pitch length

Continuous tone
method

Litho/Gravure
conversion

Copper surface electrically
deposited

Continuous tone negatives
or positives produced
electronically on scanner

Litho dot.
pos produced electronically
on scanner

Accurately surface finished
and mirror polished

Opaline prints for scanning
on Helio Klischograph

Lithoproof

Litho films

Final cleaning of copper surface

Helio Klischograph —
electronic engraving machine

Cylinder engraved electronically
by a diamond cutting stylus

Engraved cylinder checked by revisionist

GRAVURE CELL CONFIGURATION

KLISCHOGRAPH

Chrome-plated, polished and checked

(Diamond shaped Cells) varied width and varied depth

To printing press

Fig. 7.4 Electronic engraving (courtesy of Gravure Cylinders Ltd, Derby)

Additionally to the basic requirements outlined above, consideration should be given to their tinctorial strength, ease of dispersion and gloss characteristics. Where it is envisaged that a pigment is to be used in a concentrated base scheme it must have good flow properties at high pigmentation.

In comparison with flexographic, offset and letterpress printing the gravure process can, when required, deposit fairly heavy film weights of ink, which are governed by depth of engraving and hence the volume of ink held by the cells. At press viscosity the maximum pigmentation for organic pigments is unlikely to exceed 15%, most shades being matched with 5–10% of pigment (by weight).

With inorganic pigments, of which titanium dioxide is the most common, higher pigment levels of 25–35% are common, particularly for backing whites on transparent films and aluminium foil.

Extenders such as china clay and precipitated calcium carbonate are used in many gravure inks to control the gloss level and improve the lay characteristics in formulations with an otherwise high binder/pigment ratio. In addition because of their low cost they can effectively be used to cheapen gravure inks. They combine this attribute with their ability to reduce tack, notably in publication inks. Occasionally silica-based pigments are employed as matting agents, although only limited amounts can be used before the flow properties of an ink are substantially impaired.

In many areas of gravure printing, pigments need specific properties to meet demands made on them during the processing of the print, e.g. printing of PVC wallcoverings and in the end use of the printed material, e.g. deep-freeze polyethylene bags.

Resins

It is unusual to find an individual resin that will impart all the desired properties to a gravure ink and therefore the formulator will usually select two or possibly three resins to achieve a combination giving the desired characteristics. Nevertheless, it is important to ensure that the main resin in the formulation has most of the essential properties. These properties should be:
(1) Adequate adhesion to the substrate.
(2) Good solubility in the preferred solvent system.
(3) Combination of good solvent release properties with the ability to dry to give tack-free ink films.
(4) Capability of providing the desired level of gloss.
(5) Good pigment-wetting properties and acceptable printability characteristics.
(6) Toughness to ensure adequate rub resistance of the print.
Secondary properties of the resin system will be largely determined by the end use of the ink. These might include heat resistance, product resistance, e.g. to soap and alkali, and low odour.

Solvents

Choice of solvent type is largely determined by the material to be printed and its end use. With the demand for low odour packaging by many end users, severe restrictions are placed on the solvents that can be used. For example typical solvents for food packaging would be chosen from ethanol, isopropanol, ethyl acetate, isopropyl acetate, acetone, methyl ethyl ketone, selected aliphatic hydrocarbons and as retarder certain glycol ethers.

As the toxicity of solvents is increasingly taken into account the use of those that are considered harmful is in decline. However, there are areas where solvents such as toluene and xylene are still used; the former often as a co-solvent with ketones in PVC inks and with alcohols for polyethylene inks. Football coupons, postage stamps and bottle labels are currently printed with toluol-based inks, advantage being taken of their superior printability in the fine tones. Inks for gravure publication printing can contain both toluol and aliphatic hydrocarbons, being formulated to give excellent tonal printing at high press speeds with the lowest possible cost. It is particularly important that the solvents used in inks for this process have excellent lubrication properties to reduce wear on gravure cylinders during long print runs.

Although water is not, as yet, used extensively in gravure inks, it has proved a very valuable solvent for overprint varnishes that avoid re-wetting problems when applied over alcohol based inks.

To summarise, solvents, albeit only a temporary part of gravure inks, play a vital role in their printability characteristics; their contribution must be to have good solvency for the chosen resin system and have an evaporation rate commensurate with the speed of printing. With increasing attention being paid to health and safety in industry, solvents should be utilised that are the least hazardous in their nature.

Additives

The overall performance of most gravure inks can be improved by the use of additives. They perform a wide variety of functions, some of which are essential to the sucess of the ink, while others enhance its characteristics.

Some additives are primarily functional during the manufacturing stage, e.g. pigment-wetting aids, others perform as stabilisers and bactericides ensuring storage stability. Many resin systems require the addition of chemical plasticisers or plasticising resins, in order to prevent ink drying in the cells on the press and to ensure satisfactory adhesion and flexibility on the substrate.

The surface characteristics can be considerably modified by additives which promote rub resistance and slip. The most common and probably best additives for these purposes are polyolefin, paraffin and amide waxes.

7.2 PHYSICAL PROPERTIES OF INKS AND THEIR MEASUREMENT

Dispersion

The process of dispersion of the pigment into the vehicle to produce a gravure ink can be achieved by several quite different manufacturing techniques. The subject of ink manufacture is considered in detail in Chapter 12, but the relationship of the dispersion process to the end properties of the ink is of great importance and will therefore be discussed further.

The options available to the ink manufacturer for dispersion of the colourant, range from the use of chips and pre-dispersed pigments to conventional dispersion equipment such as bead mills, ball mills, rod mills or even high-speed stirring.

The use of pigment chips and to a lesser extent pre-dispersed pigments, tends to give short thixotropic inks with very high gloss and transparency. Dissolution is usually carried out on a high shear mixer and this is normally sufficient to produce an ink with a dispersion of well below 10 mu as measured on the Hegmann gauge (see Chapter 14). In the event of poor dispersion a mill pass may be necessary but this is usually indicative of a poor quality chip either through an inadequate pre-dispersion technique or degradation/oxidation of the resin carrier.

There are other advantages in using chips, particularly with pigments that are difficult to grind, where advantage can be taken of the increased brightness and tinctorial strength the process achieves e.g. PMTA colours and lithol rubine 4B toners. The high cost of the pre-dispersion process, together with the shortness of flow which places a limit on ink strength attainable, means that such techniques are, however, only normally used when no other option is feasible.

Bead mill dispersion is now the most common process for the manufacture of gravure inks. Various modifications are available which enable different diameter bead sizes to be utilised. In general terms the smaller the bead size the better the dispersion but the shorter the flow and hence a compromise has to be reached. If glass beads are used then special care must be taken to ensure total separation of the beads from the ink to prevent the risk of cylinder wear.

Ball mills are becoming less popular due to their lack of batch size versatility and the relatively long times required to achieve dispersion. Ink flow, however, is excellent and high pigment loadings can be used.

Microflow mills give extremely high quality inks for gloss and strength but the resultant ink flow can be a limitation and the slow throughput rates mean that it is not a viable process for bulk gravure ink production.

High-speed stirring as a means of dispersion is little used although in theory titanium dioxide can be ground by this technique. The final ink, however, lacks the gloss of a bead mill dispersion and is dependent on consistent quality pigment to ensure that no oversized particles are present. A stirring technique is of course used for dyestuffs but this is a solubilising process as opposed to a dispersion technique.

Strength, hue, transparency and gloss

The transparency requirements for an ink vary with the substrate to be printed. For printing on foil or other metallised substrates and the 'pearlised' films, inks with a high transparency and gloss enhance the appearance of the substrate, whereas on paper a glossy transparent ink will appear weak in comparison to a matt opaque ink. In film printing, gloss is usually paramount for surface printing but of no value in reverse printing where the design is viewed through the substrate. Transparency is not of value since coloured inks are used in conjuction with whites.

Pigment strength is normally assessed by use of the drawdown technique and by 'bleaching' (see Chapter 14). The drawdown will also give an indication of the transparency of the ink if a black bar is used (see Chapter 14) and the hue can be compared on the undertone and the mass tone. A wide variation in hue should be avoided in any gravure ink colour match since this will lead to a change in hue at different ink film thicknesses.

In carrying out strength comparisons using the bleach technique only inks manufactured by the same dispersion process should be considered. This is becasue the particle size distribution produced by each dispersion process differs widely and any such comparison will lead to anomalous results on bleach strengths.

Relative strengths of inks produced by differing manufacturing processes must still be assessed since the pigment is normally the most expensive raw material in a gravure ink formulation. Reliance is therefore placed on the standard drawdown method with variations in gloss and bronzing produced by the different dispersion processes eliminated by overdrawing the inks with a matt varnish.

The gloss level achieved on a print is governed by the choice of raw materials, the dispersion process and the drying mechanism on the printing press. For example, an ink which dries rapidly on a hot drum to give a high gloss finish may not achieve the same gloss level when dried by hot air. This usually occurs when the last solvent removed is a non-solvent for the resin system causing a precipitation of the binder, and flocculation of the ink. In the case of drum drying, all solvents are evaporated together and hence this situation does not arise, but this drying technique has limitations due to the effect on the substrate.

High gloss levels, while normally demanded for surface printing for visual appeal, are not always desirable from a print quality viewpoint and in certain cases, e.g. wallpaper printing, from an appearance aspect. In process printing a high gloss first colour (usually yellow) will give inferior trapping of the superimposed colours to a matt or even semi-gloss colour. Poor trapping will detract from the overall visual appearance of the print and hence the first colour down must be carefully formulated to give a balance in between the gloss of the solid areas of this shade and the print quality of superimposed areas of colour.

Wallpaper printing will be discussed in greater detail later in this chapter but matt inks enhance the appearance of the print, especially after embossing, since they reduce surface reflection effects. The use of silica matting agents to produce such finishes can give rise to packing in the etch if the raw materials and the dispersion process are not carefully controlled.

The measurement of gloss is discussed in Chapter 14, but while the use of a gloss meter will enable a numerical figure to be assigned, the industry still relies heavily on the human eye. The difficulty with instrumental techniques is that surface defects in print quality are not detected whereas the observer will visually combine these with his judgements on gloss levels.

Viscosity

The importance of obtaining the correct printing viscostiy to achieve uniform lay has already been discussed in the previous section (7.1). In addition the solvent plays a role in acting as a lubricant between the cylinder and doctor blade thus reducing wear on these components. On high-speed presses the cylinder is moving at speeds of up to 500 m/minute past the doctor blade and the force which has to be applied to the blade to wipe the cylinder clean is highly dependent on the ink viscosity. In general terms the lower the ink viscosity the lower the force which has to be applied to the blade and consequently the wear rate is diminished (Fig. 7.5).

However, in practice there is a limit to lowering the ink viscosity since when a critical point is reached the wear rate actual begins to increase. The reason for this phenomenon lies in the belief that a microscopic layer of ink, which is not visible to the naked eye, actually passes beneath the doctor blade even in the non-etched areas of the cylinder acting as a lubricant between them. When the viscosity of the ink approaches that of pure solvent there is insufficient binder present to act as lubricant and in such cases the formulation requires to be adjusted to include additives to take on this lubricating role.

Wear on doctor blades and cylinders will in turn result in print defects and while blade wear can be easily corrected by fitting a replacement, wear on a cylinder can be costly and may result not only in press down-time, but, in the extreme, rechroming.

Since viscosity of the ink is such an important factor in the operation of the gravure press, an accurate, quick and reproducible method of measurement is necessary. Traditionally, printers have used flow cups for viscosity measurement at the press side (see Chapter 14) but while these are fast and reliable for inks with Newtonian flow properties they give false results with thixotropic inks. Hence, print defects due to over reduction of thixotropic inks with solvent are commonplace.

For these higher-bodied inks flow cups can again be used for viscosity measurement provided the inks have near Newtonian flow, but flow times of greater than one minute should not be exceeded. Alternatively, the Brookfield Viscometer can be used for all ink types since this records a viscosity at a particular set rate of shear.

Once the viscosity of the ink in the duct has been set it can be kept constant during the run by the use of an automatic viscosity controller (Fig. 7.6). Such equipment allows for solvent from a header tank to be metered into the duct ink to replace solvent loss by evaporation.

The low viscosity required for gravure printing can give rise to pigment settling on storage and therefore it is customary to supply inks to the converter at a viscosity higher than that used in the printing process. This

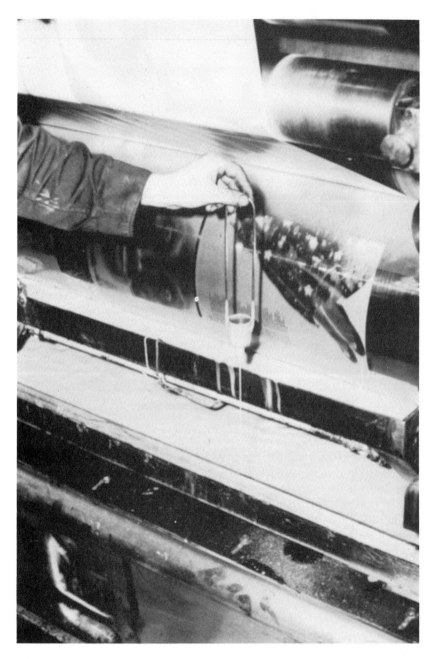

Fig. 7.5 A gravure unit with the duct open to allow a check on viscosity

Fig. 7.6 An automatic viscosity controller

has the additional practical and commercial advantages of lowering the transportation costs due to the lower solvent content.

It must be emphasised that the viscosity of all inks and varnishes will vary with temperature and in any set of measurements the temperature should not only be kept constant but recorded along with the viscosity reading.

Drying speed

The drying speed of an ink is of extreme importance. An ink which is too slow drying may give:

(1) trapping and pick off problems where two inks are superimposed on one another;

(2) marking on the turner bars;

(3) retained solvent in the final print;

(4) set off in the reel or stack.

In the alternative situation where the ink is too fast drying, printability problems can occur due to screening (especially in the fine tones) and perhaps surprisingly it may lead to retained solvent. This is because certain resin systems tend to skin rapidly and if this occurs prior to the solvent being released then the excess solvent becomes trapped within the ink film.

The drying speed of an ink is dependent on the choice of solvents, the resin system and the substrate. The evaporation rate of the solvent is an obvious factor, but equally important is the solubility of the resin in the solvent and the ease with which the binder will release this solvent. A resin which is soluble in a mixture of solvents but not in the individual solvents themselves, will release the solvents on the point of drying, faster than resin which is soluble in one solvent alone. The porosity of the substrate and the affinity of any substrate coating for the ink solvents will also affect the drying speed of the ink system.

Measurement of the drying speed is difficult and can only be truly carried out on the actual production press. The ink-maker will, therefore, normally resort to the Hegmann Gauge to compare the drying rate of a test ink against that of a known standard. This test, which is more fully described in Chapter 14, has the disadvantage of excluding the role of the substrate and to overcome this problem, a comparison with a known ink can be made by using drawdowns at a number of film thicknesses on the actual substrate to be printed. These are prepared using metal rods wound with steel wire of various diameters.

Tack

Ink tack is not important for gravure in the same way as for a lithographic ink where correct adjustment is essential not only for the substrate but also to allow one ink to trap over another wet ink. In fact a contrasting situation occurs in gravure printing since each ink must be dry before a second one is superimposed on it or pick-off and back-trapping will occur.

In gravure printing the ink formulator is solely concerned with the tack of the dried ink, especially in areas of two or more colour superimposition. Factors influencing tack include:

(1) choice of resin;
(2) retained solvent;
(3) heat retention.

Selection of the correct resin is important and this will depend not only on the stickiness or dryness of the resin at room temperature, but also on its solvent release characteristics and on its actual softening point, since retained heat in a web or in a stack may contribute to an apparent abnormal situation. This situation may be particularly prevalent in high-speed publication gravure printing where a high solids relatively high tack resin system is used for maximum gloss. Build-up of tack can then occur, especially in areas of superimposition. Such a problem can be overcome by inclusion of a low tack binder, e.g. Ethyl cellulose (EC) or Ethyl hydroxy ethyl cellulose (EHEC). Alternatively a pigmented extender medium can be incorporated but this will normally detract seriously from the gloss level of the print.

7.3   FORMULATING PRINCIPLES

The composition of gravure inks varies widely dependent on the choice of substrate, press parameters and end use of the finished print. A typical ink will, however, contain the following ingredients:

Pigment (or dyestuff)   4–12%
Extender pigment   0–8%
Resin   10–30%

Solvents   40–60%
Plasticiser   ⎫
Wax   ⎬   2–10%
Additives   ⎭

Pigments

The initial factor in pigment selection is that it should be chemically suitable for the end-use specification, e.g. an acidic pigment should not be adopted for a design requiring alkali resistance. Secondly, the pigment should disperse readily in the selected vehicle system giving an ink with near Newtonian flow properties, which when reduced to press viscosity will give good gravure printability. Thirdly, the pigment must exhibit good dispersion stability, both as the ink is supplied and also at press viscosity, since press returns may be stored for lengthy periods prior to reuse.

All pigments are treated by the pigment manufacturer to render them more easily dispersible in particular resin systems. Hence, even within the same chemical class (as defined by its Colour Index reference) the ink formulator will require several pigments to obtain inks with adequate dispersion and flow properties in the wide variety of resin systems at his disposal. For example it is rare for the same phthalocyanine blue pigment to be used in publication gravure ink as well as a nitrocellulose packaging ink.

The gravure process will frequently require the same ink to be printed as a solid and tonal colour and it is therefore imperative that where pigment blends are adopted they should be of a similar particle size. This prevents the problem of the finer particle size pigment being preferentially printed out in the shallow tones with a consequent variation in hue from the same ink in solid areas.

Below is an abridged list of pigments which are particularly suitable for gravure inks. Their properties are discussed fully in Chapter 4 and only points pertinent to the gravure process will be highlighted.

Yellows

Diarylide and monoarylide yellows are almost exclusively used, although caution must be exercised with the latter due to their solubility in toluene, poor fat and wax resistance and poor heatfastness, especially at tint strength.

Lead and cadmium yellows have virtually no application in the UK because of the regulations governing disposal of waste ink. However, in certain European countries and the USA they are still used on non-food applications for their superior printability, high lightfastness and lay characteristics.

Oranges

These fall into similar chemical classes to yellows and hence the same comments apply. In general oranges of Colour Index reference Pigment Orange 5, 13, 34 are used.

Reds

Lake red C, Permanent Red 2B and the Lithol Red 4B and 6B toners find wide application due to their high tinctorial strength at economic cost. Care should be exercised with the 4B, 6B toners if a waterbased overprint varnish is to be applied, due to their partial solubility in water. With the exception of the Manganese 2B toner, all the above pigments have poor–moderate lightfastness and if an improvement in this property is required then the para and toluidine reds can be utilised. These pigments, however, are highly variable in their ink making and chemical resistance properties and hence the permanent carmine grades are sometimes adopted, albeit at higher cost.

Finally, for brightness and cleanliness of shade PMTA grades also find specific applications.

Blues

Phthalocyanine, bronze and PMTA blues are the principal grades for gravure printing. Care should be exercised with bronze blue as it is notoriously difficult to disperse and can give rise to streaking and/or scumming if oversized particles cause a temporary slight lift of the doctor blade.

Violets

PMTA and indanthrene violet are almost exclusively used although the latter generally gives inks with poor flow characteristics and its use is avoided where possible.

Greens

Phthalocyanine and PMTA greens are the most common, though great care should be taken with the former on long print runs as it can cause excessive wear of the chrome cylinder.

Whites

Titanium dioxide (rutile grade) is the generally adopted pigment. Grades produced by the sulphate process are preferred as the chloride grades tend to give cylinder wear problems. For certain applications, even sulphate grade rutile pigments can cause excessive wear and then either the anatase grade or preferably zinc sulphide is utilised, albeit with a decrease in opacity.

Extenders

Alumina, calcium carbonate and china clay are the most commonly adopted extenders. Addition of these tends to aid printability and improve smoothness of lay, but they do have a detrimental effect on gloss. Since extender pigment particles are not readily coated by the resin in the dispersion process, they also tend to settle easily and any settlement is generally very difficult to restir into the ink system. The choice of type of extender depends on the binder system but calcium carbonate must not be used in the presence of an acidic binder or decomposition will occur.

Blacks
Furnace blacks are the most common though on occasions the more expensive channel blacks are used for their better flow and gloss characteristics.

Metallics
Bronze and aluminium powders are used but the vehicle system must be chosen carefully to provide a stable system.

Dyestuffs

In packaging inks solvent dyes are sometimes used in preference to pigments for optimum transparency on foil and metallised substrates. Their solubility depends to an extent on the chemical structure but they all tend to prefer polar solvents and in particular the lower alcohols. In contrast to the basic dyestuffs used in cheaper flexographic printing, most of these dyes exhibit excellent lightfastness.

Dyestuffs are also available which are soluble in toluene and these find some application in toning publication gravure inks though their use is limited by their poor lightfastness.

Resins

The general requirements of the resin will be apparent from the tasks it has to perform during the printing operation. The resin should be film forming, hard, amorphous, low odour, usually colourless or pale, transparent, and soluble in the solvents to be used in the ink. It should also assist the wetting of the pigment to ensure good pigment dispersion and stability. The viscosity of the varnish prepared from resin in solvent should increase as the resin solids increase and a thin film of varnish when exposed to air and heat should release its solvent quickly to give a tack-free homogeneous resin film. Clearly, the resin must impart adhesion of the ink to the substrate to be printed. On paper, resins can be chosen to give penetration into the substrate and thus a low gloss level, or less penetration giving higher gloss levels.

The melting ranges of resins vary considerably and in many instances, softening point is less critical than adhesion, rub resistance or lightfastness. Resins should not interact with the other constituents of the ink otherwise gelation usually occurs. Thus, basic pigments in the presence of an acidic resin will more often than not give problems. Acidic resins do help to prevent pigment flocculation in low viscosity inks at low resin solids. Under certain conditions, the resin can react with another specific constituent of the ink in a controlled manner. These are cross linkable resin systems, which may also be catalysed, and will be discussed later.

Resins are complex organic compounds of high molecular weight. Originally, the choice of resin was restricted to a limited number of processed natural products, some of which are still used, e.g. shellac, gilsonite and modified rosin derivatives. Synthetic resins have much more to offer the ink technician because they can be tailor-made to suit individual requirements. In general, they are higher melting and lower in odour than the natural resins.

The major classes of resin are examined below in relation to the gravure print process but for more general data the reader should consult Chapter 4.

Cellulose resins

Undoubtedly the most important resin in this class is nitrocellulose because of its excellent ink-making characteristics, wide compatibility with modifying resins and additives, excellent solvent release, low odour, good colour retention and heat resistance. Nitrocellulose is available not only in a range of viscosity grades but also with varying nitrogen content. This enables the solubility characteristics to be varied from alcohol–ester to alcohol—toluene blends or even to straight ester solvents. Compatibility with other resins is also affected by variation in the nitrogen level.

Nitrocellulose therefore finds wide application in the gravure packaging field and has only a few limitations which the experienced ink-maker cannot overcome. For example, in multicolour gravure printing, trapping of one nitrocellulose ink over another can be problematical and in extreme cases the use of an alternative resin system for one or two of the colours may be necessary. Another drawback is in biscuit roll-wrap applications where a water-based PVdC overlacquer is applied for maximum seal integrity at low pressures. Inks based on nitrocellulose cannot be utilised because of poor interlayer adhesion between the ink and PVdC overcoat. A further limitation is in metallic inks where the use of bronze and to a lesser extent aluminium pigments, can cause decomposition of the resin, resulting in the liberation of nitrogen dioxide gas.

Ethyl cellulose in contrast is more expensive and has relatively poor pigment-wetting characteristics and a tighter compatibility spectrum than nitrocellulose. It gives very tough, flexible ink films and will dissolve in both toluene and alcohols depending on the grade used.

Toluene gravure inks based on ethyl cellulose are noted for their fine half-tone printability and often there is an advantage in using at least one ink based on this resin to improve the tonal reproduction on films.

Both ethyl and ethyl-hydroxyethyl cellulose (EHEC) are useful resins for mixing with other resinous systems in order to remove the tackiness which is sometimes present during the drying of the ink film. This is particularly helpful on multicolour superimposition when picking can be a problem. In publication gravure inks EHEC is often used because of its solubility in aliphatic solvents.

Cellulose acetate propionate has found increasing use in the packaging field especially in conjunction with acrylic resins for printing on PVdC and acrylic-coated films where excellent low temperature adhesion is given. This resin blend can also be successfully used for PVdC overcoating where nitrocellulose cannot be utilised. Two grades of resin are currently available. The first, half-second propionate (HSP), has excellent heat resistance but limited application in colours due to its poor wetting properties with organic pigments; the second, alcohol soluble propionate (ASP), contains a higher hydroxyl content giving improved solubility in ethanol-ester (or even ethanol–water) blends. This latter grade shows superior wetting characteristics to the half second propionate but requires greater selectivity of pigments in comparison to nitrocellulose.

Polyamides

The use of polyamide resins for printing on polyethylene is well established. Historically the co-solvent grades, which are soluble in blends of alcohols and hydrocarbons, were used due to their excellent pigment-wetting properties, high gloss, rapid solvent release and good deep-freeze resistance. The trend towards hydrocarbon free inks for food wrappers has resulted in an increased use of the alcohol soluble grades. The slower solvent release of these resins means that for non-blocking properties to be achieved it is necessary to use the alcohol grades in conjunction with nitrocellulose.

The emergence of cold seal packaging has given a new lease of life to polyamide resins. Since the cold seal adhesive has to be applied in line with the printing stations, the ink/lacquer system is in direct contact with the adhesive during storage of the printed reel. Polyamide resins have been found to be practically unique in providing release to the cold seal adhesive while not impairing the strength of the bond formed when the adhesive is sealed to itself during the packaging operation.

Polyamides also find value in lacquers because of their high gloss and resistance to products such as soap and margarine. Typical end uses are on detergent cartons, confectionery and margarine wrappers.

Further applications are limited because of the low softening point of the resins and unsuitability for use in inks for adhesive lamination.

Acrylics

Apart from areas already discussed in blends with cellulose resins, and with vinyl resins, acrylics have little scope because of their poor pigment-wetting properties and low softening points. This latter property can, however, be used to advantage in formulating gravure heat seal lacquers, e.g. foil to polystyrene.

Solutions of carboxylated acrylic resins in aqueous alkaline media are the basis of water-based inks and lacquers and these will be discussed fully later in this chapter.

Vinyls

Vinyl resins generally give inks with poor flow, gloss, heat resistance and solvent release. Hence, applications are limited and their use is confined to printing vinyl wallpaper, polyester films for lamination and for vinyl washed foil or vinyl films. Pigment selection is highly important to give inks with reasonable flow and good gravure print characteristics.

As with acrylic resins the low softening point of vinyls allows some use in heat seal lacquers.

Hydrocarbon-soluble resins

The major application for such resins is in the publication gravure field. Calcium and zinc calcium resinates are soluble in both aromatic and aliphatic hydrocarbons giving high solid but low viscosity varnishes with excellent solvent release. In addition their low cost makes them an attractive proposition for publication gravure inks, but they do suffer from the drawback of forming powdery films which tend to give poor rub resistance.

Rosin modified phenol-formaldehyde resins have high gloss, good solvent release and better rub resistance in comparison to the resinates. However, they are more expensive than the resinates and their use is mainly confined to toluene-based inks because of their limited tolerance of aliphatic hydrocarbon solvents.

Hydrocarbon resins, while giving high solid low viscosity varnishes, are inferior to resinates for solvent release and are therefore only of limited value.

Gilsonite is a natural resin which is deep brown in colour and hence its use is confined to black/brown inks. Pigment-wetting properties are excellent and black/brown inks with excellent gloss and good solvent release can be obtained. Varnishes on storage do have a tendency to gel and deposit insoluble matter unless a small quantity of alcohol is included in the formulation. Gilsonite is incompatible with cellulose resins and its use should be avoided if these resins are present in the ink or reducing medium.

In packaging inks chlorinated rubber is used for high quality gloss inks on paper. As well as being soluble in aromatic hydrocarbons chlorinated rubber is soluble in esters, but totally insoluble in alcohols. The system has excellent chemical and abrasion resistance and its non-curl properties on paper make it an ideal basis for label work.

Modifying resins

A wide variety of chemical types of resin are used to modify the main vehicle system usually as a means to increase gloss or reduce cost. Typical of resins in this category are rosin modified maleics which can be produced with or without acidic functionality to vary the solubility into polar or non-polar solvents respectively. Typical use of such resins are in combination with nitrocellulose or chlorinated rubber (low acid value grades only), and since flexibility on film and foil is limited the main application is on paper and board. The resins have a typical rosin odour and care should be exercised in food packaging applications.

Ketone, aldehyde and styrene-allyl alcohol resins are all used to promote gloss improvement of nitrocellulose inks. Again all suffer limited flexibility on film and especially on foil and will detract from the heat resistance and solvent release of a nitrocellulose ink.

Shellac, despite its wide application in flexographic inks, is only rarely used in gravure inks because of its poor resolubility characteristics. This leads to resin packing in the gravure cell and a progressive loss in colour strength of the print.

Solvents

The solvents in gravure inks are temporary ingredients which are present purely as a means of applying the vehicle solids to the substrate by way of the printing unit. In theory the solvent is then eliminated, mainly by evaporation and/or absorption, and takes no further part in the properties of the printed film. However, in practice this can be an over-simplification and the consequences of trapped solvent will be discussed in more detail later.

The choice of solvents in a gravure ink is governed by:
(1) the resin system to be employed;
(2) the press speed;
(3) direct or offset gravure process;
(4) nature of the design;
(5) the substrate;
(6) end-use properties of the print;
(7) health and safety considerations.

The chemical nature of the resin system will often severely restrict the choice of solvents available to the ink-maker, and also with most resins a blend of solvents will give lower viscosity than a single solvent. Selection of solvent can be predicted by consideration of the solubility parameter, and the possible hydrogen bonding associated with a solution of the resin in the solvent. While this may appear complex, details of solubility parameters of individual resins can normally be obtained from the resin supplier and the ideal solvent balance can easily be predicted without the need for theoretical knowledge. Information obtained by this technique is particularly useful since it can predict where a non-solvent for a resin can be used in conjunction with a true solvent to give a lower viscosity solution.

Press speeds may vary from about 10m/minute on an in-line extruder and printer to several hundred metres per minute on a publication gravure press. The solvent must allow the ink to remain open during the printing stage until the ink has flowed out on to the substrate, but then be rapidly removed within the drying cycle. Hence, the evaporation rate of the solvent(s) must be geared to the length of time in between excess ink being removed by the doctor blade and the point of removal of the solvent. This time span is clearly governed by the press speed and also by whether a direct or offset gravure process is utilised.

The nature of the design itself also plays a major role in the speed of evaporation required in the solvent system. Solvent loss from ink in a shallow cell will be liable to cause drying in problems more quickly than the same solvent loss from ink in a deep cell simply due to the quantity of ink and therefore solvent present in the particular cell. Hence, the amount of tone work on a design will govern the evaporation rate of the solvents required in the individual inks.

Many print formats will involve inks with a high coverage and others with very low usage. The low usage inks may require little or no replenishing even on long print runs and any loss of solvent by evaporation in the duct must be replaced in order to retain a constant viscosity. While this is a simple task for single solvent inks it is a rare situation and in the more normal circumstances of mixed solvents, further additions must be made in accordance with the evaporation rate of the individual solvents. For example, as an ink based on a 9 : 1 mixture of toluene and ethanol evaporates, the solvent vapour consists of a 54 : 46 ratio of toluene and ethanol respectively. Hence, any solvent replenishment should be done with the later ratio rather than with the original ink solvent balance.

Substrate can have two main effects on the selection of the solvent system. Firstly, with permeable substrates the rate of absorption of individual solvents can affect the flow-out properties of the ink in between the point of impression and the ink losing its solvent and this can give rise to a variation in the smoothness of lay. Secondly, certain films and substrate coatings can be attacked by some solvents which can have an influence on the anchorage of the coating or even on the properties of the substrate itself.

Inks for food packaging in particular must not be tainted by the effects of residual solvent. Therefore, special attention must be given to the release characteristics of the solvents employed from the chosen resin system so as to give the minimum retained solvent levels in the finished print.

The new EEC labelling regulations introduced at the beginning of 1986 specify the hazard code label which must be attached to all ink systems. While this is not restricted to just the ink solvents these in the main are the constituents which dictate the label required on a particular gravure ink.

In recent years changes have been made to the Threshold Limit Values (TLV) of certain solvents which have reduced the maximum concentration of the solvent vapour allowable in the working atmosphere. Such changes have resulted in the virtual elimination within the UK of some traditionally used solvents, e.g. 2-ethoxyethanol and its corresponding acetate. However, alternative glycol ethers and esters whose TLVs have not been reduced can be used with similar results.

The most common solvents found in gravure packaging ink formulations for paper, film and foil are ethyl, iso-propyl and *n*-propyl alcohols, acetone, methyl ethyl ketone (MEK), ethyl, isopropyl, *n*-propyl and butyl acetates, methoxy and epthoxy propanols, toluene, aliphatic hydrocarbons and water. Publication inks are basically restricted to toluene and the aliphatic hydrocarbon solvents. Some properties associated with these solvents, together with a few other solvents which can be utilised, are listed in Table 7.1.

Key to risk and safety phases:
R 10 Flammable.
R 11 Highly inflammable.
R 20 Harmful by inhalation.
R 21 Harmful in contact with the skin.
R 22 Harmful if swallowed.
R 36 Irritating to eyes.
R 40 Possible risk of irreversible effects.

Table 7.1 Properties of gravure solvents

Solvent	Density at 4°C	Boiling point (°C)	Evap. rate (Bu Ac = 10)	Flash point (°C)	Hazard classi-fiction	General nature of risk	Risk phase Nos R	Safety phase Nos S
Acetone	0.79	56	115	−15(1°F)	—	Highly flammable	11	9 16 23 33
1,1,1,-trichloroethane	1.32	74	75	—	Class IIc	Harmful	20/22	2 25
Ethyl acetate	0.90	77	62	−5(24°F)	—	Highly flammable	11	16 23 29 33
Methyl ethyl ketone	0.80	80	57	−7(20°F)	—	Highly flammable	11	9 16 23 33
Hydrocarbon SBP 2	0.71	70–95	54	(−40°F)	Class IIa	Highly flammable, harmful	11 20/21 40	9 16 29 33
Isopropyl acetate	0.87	88	50	4(40°F)	—	Highly flammable	11	16 23 29 33
Hydrocarbon SBP5	0.73	90–105	35	(−7°F)	—	Highly flammable	11	9 16 29 33
Ethanol (74 op spirits)	0.79	75–81	33	13.9(55°F)	—	Highly flammable	11	7 16
n-propyl acetate	0.89	102	27.6	14(58°F)	—	Highly flammable	11	16 23 29 33
Isopropanol	0.80	81	23	12(53°F)	—	Highly flammable	11	7 16
Hydrocarbon SBP3	0.74	110–120	22	(15°F)	—	Highly flammable	11	9 16 29 33

Table 7.1 Continued

Solvent	Density at 4 °C	Boiling point (°C)	Evap. rate (Bu Ac = 10)	Flash point (°C)	Hazard classi-fiction	General nature of risk	Risk phase Nos R	Safety phase Nos S
Toluene	0.87	111	21	4 (40°F)	Class IIc	Highly flammable harmful	11 20	16 29 33
Water	1.00	100	18	—	—	—	—	—
Methyl iso-butyl ketone	0.80	118	16	14 (57°F)	—	Highly flammable	11	9 16 23 33
n-propanol	0.80	98	11	22 (72°F)	—	Highly flammable	11	7 16
Butyl acetate	0.88	126	10	21 (70°F)	—	Highly flammable	11	16 23 29 33
Propylene glycol methyl ether	0.92	121	8.3	36 (97°F)	—	Flammable	10	24
xylene	0.86	137	6	25 (77°F)	Class IIc	Harmful, flammable	10 20	24/25
n-butanol	0.81	118	5	29 (84°F)	Class IId	Harmful, flammable	10 20	16
Propylene glycol ethyl ether	0.90	132	4.9	49 (120°F)	—	Flammable	10	—
Propylene glycol methyl ether acetate	0.96	140–150	3.4	47 (116°F)	—	Flammable	10	—
Cyclohexanone	0.95	157	2.3	44 (111°F)	Class IIc	Harmful, Flammable	10 20	25
Methyl cyclohexanone	0.92	175	1.8	48 (118°F)	Class IIc	Harmful, flammable	10 20	25
Diacetone alcohol	0.94	166	1.4	52 (126°F)	—	Irritant, flammable	10 36	24/25
Di propylene glycol methyl ether	0.95	188	0.2	79 (174°F)	—	—	—	—

S   2 Keep out of reach of children.
S   7 Keep container tightly closed.
S 16 Keep away from sources of ignition.
S 23 Do not breathe gas/vapour/fumes/spray.
S 24 Avoid contact with skin.
S 25 Avoid contact with eyes.
S 29 Do not empty into drains.
S 33 Take precautionary measures against static discharges.

Ink additives

A wide variety of different chemical compounds are incorporated into gravure inks for a host of reasons. Generally levels of addition are small, usually below 5%, but even at 0.1% the effect on printability must be carefully examined.

Slip agents must be incorporated with great care as silicone materials usually give lay defects and micronised waxes if not finely and uniformly dispersed can give packing in the etch or even streaking problems. Overprint properties can also be affected by the choice of wax or slip aid.

Dispersion and surface active agents can be included for easier pigment wetting and also to prevent long-term pigment (or extender) settlement by imparting a weak gel structure to the ink.

Other additives such as epoxides, amines, phenolics, organic acids are incorporated as stabilisers, antioxidants, etc. to prevent ongoing side reactions which may occur on storage of the ink or finished print.

Anti-pinhole agents are used in inks for printing on the nitrocellulose-wax coated cellulose films (e.g. MS film) in order to lower the surface tension of the ink and allow a smooth pinhole-free coating to be obtained. Other films such as the corona discharge treated polyolefines will often require additives like titanium acetyl acetonate to ensure good adhesion of a nitrocellulose ink system.

A comprehensive list of additives cannot be attempted within the scope of this section but further details of specific materials will be given in the section on ink formulations.

7.4   INKS AND VARNISHES FOR SPECIFIC END-USE APPLICATIONS

Before discussing individual ink systems it is important for the reader to understand the basic ink requirements for the different substrate types, the end use properties, and details of certain after processing techniques such as lamination, PVdC coating and cold seal application.

Gravure printing of paper and board is relatively straightforward using in the main opaque coloured inks directly on the substrate. In contrast inks for foil and metallised substrates utilise transparent coloured inks to emphasize the metallic nature, but if a white is used then it should be formulated with maximum opacity to prevent the greyness of the foil showing through. In cases where a gloss opaque white is required on foil a primer or wash coat under the white ink may be necessary.

As a means of enhancing the brightness and strength of the colours, multicolour designs on film are usually printed with a backing white. In surface printing the white is printed first and provides the vital property of base adhesion to the film. The colours may solely be printed over white, but occasionally an overdraw can mean that adhesion of the colours to base film is also required. In reverse printing the colours are printed first, followed by a backing white and hence all inks must have adhesion to the base film. The gloss level of inks for reverse printing is of no importance since the design will be viewed through the film.

The end-use property requirements of gravure inks will vary widely but all inks must show good basic adhesion to the substrate, good flexibility or resistance to cracking and good resistance to scuffing.

Inks for packaging applications will have to withstand tests designed to reproduce packaging and storage of the finished product. Tests for packing line simulation will involve testing for slip and in certain cases heat seal resistance. Increases in packing line speeds mean that slip control is vital and on heat sealing, not only must the ink not be plucked from the surface, but it must also rapidly release from the sealing jaws. To this end inks now contain additives specifically aimed at providing hot jaw release.

Product storage tests will be varied ranging from odour and taint tests to resistance tests for soap, alkali, oils, fats, waxes, light and deep-freeze to name but a few. For details on all the above test methods the reader should consult Chapter 14.

In addition the Society of British Printing Ink Manufacturers has published lists of raw materials which they recommend should be excluded from use on immediate food wrappers. This list primarily covers pigment, plasticiser and solvent selection. Futhermore print for a specific end use, e.g. bundling tissue for tobacco wraps, may be subject to even more stringent safety regulations.

End property testing on inks for non-packaging applications can also be very stringent. For example inks for wallpaper and melamine laminates will have to withstand a whole variety of tests to simulate marking, staining or rub by a wide range of different chemicals.

Lamination

Lamination is a means of bringing two or more substrates together by either the use of adhesive, wax, heat or an extrusion process. In adhesive lamination (see Fig. 7.7) the design is normally reversed printed on film and the adhesive is coated over the ink system. Any solvent is then removed prior to bringing the second web into contact, often by the use of a hot nip.

The adhesive can be solvent- or water-based or even a 100% solids system which can be applied via a special applicator. The solvent-based adhesives still predominate and fall into two main classes:

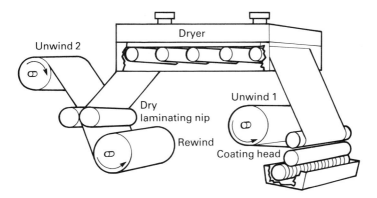

Fig. 7.7 The process of dry bonding adhesive lamination

(1) Polyurethane adhesives which are available as single-pack moisture curing types or two-pack reactive systems, the latter providing
the best results for high specification demand, e.g. boil-in-the-bag
and retortable pouches. The chemical nature of these products
means that only water-free ester and ketone solvents can be used
as true solvents, though limited dilution can be made with aliphatic
hydrocarbons.

(2) Polyether type adhesives cured with an epoxy resin which can be
used in alcohol solutions have found increasing application since
they are isocyanate free and give less rewetting when applied over
most common laminating inks. This is solely due to the use of
alcohol solvents as opposed to the more aggressive ester and ketone solvents in polyurethane adhesives, and is especially applicable to in-line print and lamination when the underlying inks
have not had time to harden off before application of the adhesive.

Few guidelines can be given in formulating inks for adhesive lamination.
Clearly the ink must show a good level of adhesion to the base film but
even then this does not guarantee that good bond will be given on
adhesive lamination. Polyamide resins should be avoided as poor bonds
will result and PMTA pigments can, under certain circumstances, give a
bleed in contact with the adhesive.

In extrusion lamination (see Fig. 7.8) two films are bonded together by
extruding a thin layer of polyethylene (or similar material) on to one
surface and nipping in a second layer of substrate while the polyethylene
(or extrudant) is still molten. The process can be applied to two identical
or quite dissimilar substrates.

The best results are obtained when the resin system of the ink has an
affinity for the extrudant. Hence, nitrocellulose inks modified with
polyamide or shellac usually suffice for polyethylene extrusion lamina-
tions.

In general terms bond strengths are inferior to those in adhesive

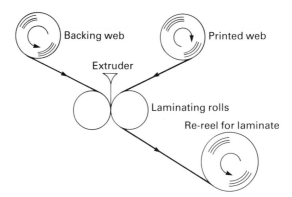

Fig. 7.8 The process of extrusion lamination

lamination, but can be further improved by application of a primer over the inks prior to extrusion. Volatile components such as solvents or water in the printed web will reduce the strength of the finished bond due to a bubbling effect on application of the hot extrudant.

Extrusion coating is a similar process to that of extrusion lamination but instead of a second web being brought into contact, the extrudant is simply passed over a chill roller.

Wax lamination is basically similar to extrusion lamination but in this case the extrudant is molten wax, usually paraffin wax. The vast majority of this work is in foil-paper laminates for the confectionery and snack food markets. While in adhesive and extrusion lamination, printing is carried out before lamination (to reduce wastage) the printing of foil-paper wax laminates is often done after the webs have been combined to prevent the problems of printing unsupported foil. However, to guard against poisoning of the foil surface by chemicals in the paper the foil surface must be printed or at least primed within hours of the lamination process taking place.

Heat lamination is now only rarely used and it utilises a web printed with a thermoplastic heat sealable ink or varnish which is brought into contact with a second web before passing through a hot nip under pressure. The process is shown in diagrammatic form in Fig. 7.9.

PVdC coating

Aqueous PVdC coatings are applied primarily to obtain better seal integrity of the package and also to promote gloss levels. The extra barrier properties given by the layer of PVdC are normally superfluous especially when a PVdC-coated film is used.

The coating can be applied either in-line with the gravure printing operation. Drying of the PVdC coating is an important factor since if excess moisture is present in the final film then poor adhesion to the underlying inks results. Hot air dryers are the most effective with the

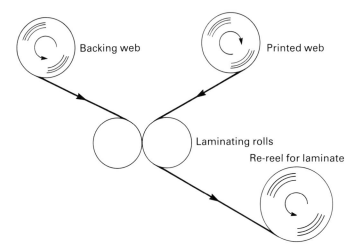

Fig. 7.9 The process of heat lamination

facility to increase the web temperature gradually to prevent surface skinning of the coating which can give rise to trapped water or blistering of the surface layer.

Nitrocellulose-based inks have been shown to be unsuitable for PVdC coating giving poor interfilm adhesion on long-term storage of the final print. Pigment selection is also an important factor since interlayer adhesion can be affected (e.g. with bronze blue pigments) and those giving a water bleed (e.g. some 4B and 6B toners) will discolour the PVdC during the coating operation.

Cold seal applications

Cold seal packaging has increased rapidly during the 1980s, offering the advantages of higher line speeds, less wastage and no side effects on the end product due to the heat sealing process. In this process the design is surface printed on the substrate in the normal manner and then the web is reversed and a cold seal adhesive applied to the back of the substrate in pattern form. The adhesive must be printed in line for registration purposes, and is formulated to seal to itself at the packaging stage.

Cold seal adhesives are all based on natural rubber latex, modified with varying amounts of synthetic latices to adjust their properties to suit the end-use requirements. After application the adhesive is in direct contact with the inks within the reel and hence no interaction between the two must take place during the storage period. Typical problems which can occur are:

(1) Poisoning of the adhesive by set-off of the inks on to the adhesive or vice-versa.
(2) Migration of additives, e.g. waxes or slip agents from the ink or varnish into the adhesive again causing a poisoning effect.

(3) Chemical interaction by components in the inks with those in the adhesive. For example metallic pigments will react with the ammonia in the latices.

(4) Bleed from pigments or dyes into the adhesive and hence risk of contamination of the product being packaged.

Publication inks

Once a major source of publication print, the last 10 years has seen the decline of the gravure process in favour of web offset, certainly where runs below 500 000 copies are concerned.

Publication gravure inks are formulated to achieve the lowest possible cost; the main demands are on their press performance. Once printed their prime function is to provide text and illustrations, merely requiring adhesion to the substrate and a moderate degree of scuff resistance and lightfastness.

Paper in publication printing

In publication printing the nature of the paper surface plays a critical role in the print quality achieved. The gravure printing process is dependent on direct contact between ink and paper at the point of impression. Because the surface of a gravure cylinder is smooth, the ink cannot be easily forced into any indentations in the paper. In line printing where the engraving is relatively deep, lack of contact with very small areas of paper are not necessarily too critical, as there is sufficient ink in the adjacent cells to flow out into these areas. However, in the tonal areas of publication printing there is insufficient ink in the adjacent cells for this to happen. Therefore, if the paper has an uneven surface with a predominance of indentations then undersirable speckle occurs (Fig. 7.10).

In efforts to improve the quality of paper produced for gravure publication printing, paper manufacturers have resorted to a number of tests designed to indicate the degree of speckle that might be expected from a particular quality or batch of paper. For the test to be relevant all other possible variables have to be eliminated so that at the point of testing any change in the degree of speckle is due to the surface of the paper only.

This can be achieved by using a mechanically driven reel fed laboratory proofing press which must have a pneumatically operated impression roller, a reciprocating doctor blade, with a rewind sufficiently distanced from the point of impression to provide an adequate web carriage. Additionally the machine should have an effective speed control and an inking system which will enable consistent ink viscosity control to be maintained during the tests.

The test inks should be of proven quality consistent with a typical publication ink formulation. Black ink is normally chosen as it provides the greatest contrast between ink and paper. The cylinder used for the test is engraved to a depth which represents the tonal values where speckle is the most critical.

A standard paper is selected which has a predetermined and acceptable

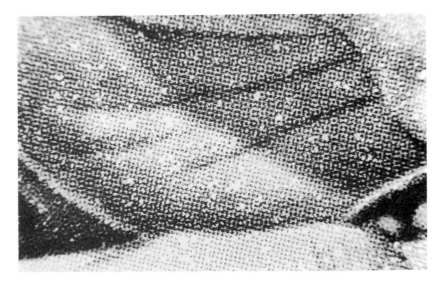

Fig. 7.10 The effect known as dot skip or speckle

level of speckle. The standard paper and test paper are attached adjacently to each other on the web and run through the press at the same time. The machine is stopped before the papers reach the rewind reel. By visually counting the number of speckles on the test paper, over a given area of print, its suitability for gravure printing can be assessed against the number of speckles showing in the same area of the standard print. The test should be repeated a number of times, to assess different areas of the test paper, i.e. both sides, centre and edges. By logging the results, numerical data can be accumulated to give both development and batch records.

Gloss and absorbency

Given inks of acceptable gloss characteristics the gloss level of a print will depend on the quality of paper selected. Papers ranging from absorbent newsprint, highly calendered mechanical paper through to good quality blade coated papers are used in different areas of publication printing, with each type of paper having a different absorbency charactertistic.

Only the first colour down, usually the yellow, prints directly on to the paper surface, subsequent colours print not only on the paper but also on a dry ink film of varying thickness and coverage.

Lower ink viscosities are required for absorbent papers to enable the interstices between paper fibre and filler to be penetrated, thereby covering the whole of the surface of the substrate before ink flow is prevented by solvent evaporation.

With highly absorbent papers, the penetration of the ink into the paper results in poor gloss and dullness of colour, particularly with the red and the blue. On calendered mechanical papers the gloss and colour brightness is greatly improved, as the paper fibres and fillers have been compressed to form a less absorbent surface on which the ink penetrates to a lesser degree.

Publication gravure printing with the highest gloss and colour brightness, though not necessarily speckle free, is achieved on blade-coated papers. This is due to the coating of titanium dioxide and method of application which produces a glossy white surface with a low degree of absorbency thereby enhancing the gloss and brightness of the inks.

Papers that are excessively absorbent allow penetration of the ink vehicle which can cause undesirable 'strike through'. A similar problem occurs when the ink vehicle has a refractive index close to that of cellulose, which renders the print area more transparent.

Formulation requirements

As indicated earlier the greatest demands on publication inks are made during the actual printing operation. Prior to use, the ink should have viscosity stability during storage. Due to the large quantities involved, inks are often kept in tanks ready to be pumped to the machine room. On the press they should have good lubricity and wiping properties to enhance cylinder life. They should exhibit good pigment suspension at low viscosities and be tolerant to solvent shock. When printed there should be immediate and rapid solvent release, followed by the formation of a tack free ink film, thus avoiding set off on the turner bars. The ink should print with a sharp definition and provide a gloss level consistent with the type of substrate being printed.

The choice of pigments for the four process colours normally employed in gravure publication printing is dependent on them fulfilling the pigment-orientated properties outlined above. The yellow pigment should have a degree of opacity especially when used on less than white absorbent papers. The pigment chosen should be a neutral yellow to ensure bright greens, but not too light or greenish to produce weak scarlets and reds. The rubine or magenta should be as bright as possible particularly in the tones, thereby enabling reasonable mauves and violets to be printed. The blue should be midshade and again as clean as possible. The choice of black depends principally on the preferred tonal hue; most publication houses choose a neutral or slightly blue black.

A number of alternative resin systems can be used for publication printing. The majority of inks are based on zinc/calcium resinates with ethyl hydroxy ethyl cellulose (EHEC) used to improve hold out and reduce surface tack. Occasionally cyclised rubber is also used, detracting less from the gloss than EHEC. However, both these resins are expensive and can only be used to a very limited degree. To assist in reducing tack it is normal to use either china clay or precipitated calcium carbonate as extenders; care in selection is required to ensure that colour brightness and gloss levels are not adversely affected. Resinous plasticisers with extremely good wetting properties, can be employed to ensure pigment suspension at machine viscosities.

Other approaches to publication ink formulation have been made with the use of phenolic resin systems, which because of their more complex molecular structure are capable of exhibiting better hold out and gloss. However, their inferior solvent release and drying characteristics, have reduced their potential for modern publication printing, where machine speeds of up to 500 m/per minute are common. Well-dispersed polyethylene wax is usually used to achieve the necessary rub resistance.

Choice of solvent systems is extremely limited in publication printing, normally a combination of toluene and aliphatic hydrocarbons with a very small percentage of xylene. In some cases printing houses prefer 100% toluene, especially where solvent recovery is installed and the printing units can be effectively sealed to prevent solvent loss to the surrounding atmosphere. Where a solvent system with a low aromatic content is preferred, the overall solvent power can be enhanced by the introduction of a small percentage of *n*-propanol. This is particularly helpful when the formulation contains EHEC.

Typical formulations would be:

Conventional:

Diarylide Yellow (CI Pigment Yellow 12)	6.0
Calcium carbonate	6.0
EHEC resin	2.0
Zinc/calcium resinate	37.0
Polyethylene wax	1.0
Resinous plasticiser	3.0
Aliphatic hydrocarbon solvent	20.0
Toluene	22.5
Xylene	2.5
	100.0

Low aromatic:

Rubine Red (CI Pigment Red 57)	7.5
Calcium carbonate	5.0
EHEC resin	2.0
Zinc/calcium resinate	37.0
Polyethylene wax	1.0
Resinous plasticiser	3.0
Aliphatic hydrocarbon solvent	39.0
n-propanol	2.0
Toluene	2.5
Xylene	1.0
	100.0

Easily dispersible pigments for publication gravure inks

The earliest pigments of this type appeared on the market over 20 years ago. Strictly speaking they were not pure pigments but resinous dispersions achieved by mechanical means and had a pigment content of more than 50%. Their main characteristic was their ease of dispersion by conventional milling and in some cases, high shear stirring was sufficient to achieve adequate dispersion. Dispersion was achieved as the resin component dissolved, but their use was limited by their high initial cost and they were superseded in the main by the use of selected pigments designed specifically for use in publication gravure inks using conventional dispersing techniques.

Over the years developments in pigment technology have taken place to a high degree. Pigments are coated with surface active agents, hyperdispersants, resins and other chemical compounds during their manufacture resulting in products which are ideally suited for the high speed production of publication inks. Although pigment costs may still be higher than a conventional pigment, the overall production cost of the ink is beneficial.

Techniques of production using these newer pigments vary since some are fine powders which disperse easily by triturating with solvent, e.g. toluol, while granulated dust-free types require mixing at high pigment loadings to produce shear. They have strong deflocculating properties and the mixes produced then require only the addition of varnishes and additives to produce the finished ink.

Generally speaking, inks produced using the yellow, cyan and magenta easily dispersible (ED) pigments yield inks that are higher in gloss, brighter, cleaner in hue and more transparent than those produced with conventional pigments.

Inks for catalogue printing

The ink requirements for catalogue printing are somewhat different from those outlined above. The coated paper normally used is highly calendered with a rather more brittle surface than magazine papers. This means there is less ink penetration into the surface and a reduction in mechanical bonding.

In addition to high gloss, inks are required to have tough flexible films with good rub resistance and printability. As there is little assistance from the substrate, rapid solvent release and hard film forming properties are essential to ensure adequate drying. Because of the nature of the paper, press speeds are usually lower than those used for magazine printing.

Pigment selection for catalogue inks is to some extent dependent on the requirements of individual printers, but in general the pigments are similar to those used in other areas of publication printing. A suitable set of inks can be formulated using diarylide yellow, phthalocyanine blue, calcium 4B and a blue-toned carbon black pigment. As high gloss level is essential only low concentrations of extender are possible.

Unlike weekly magazines, catalogues have to withstand considerable handling over varying lengths of time. This sometimes calls for resins that have tougher film-forming properties, than can normally be achieved with resinate/EHEC or resinate/phenolic combinations. A particularly suitable resin system is chlorinated rubber modified with a phenolic resin, which gives the toughness and flexibility required. Rub resistance can be achieved by using carefully selected well dispersed polyethylene waxes. Careful plasticisation is required to ensure a correct balance between flexibility and tack.

The solvent system is usually toluene or possibly toluene/ester.

Typical formulation of a catalogue ink would be:

Phthalocyanine Blue (CI Pigment Blue 15.3)	8.0
Chlorinated rubber	15.0
Phenolic modified resin	10.0
Resinous plasticiser	5.0
Polyethylene wax	1.0
Toluene	51.0
Ethyl acetate	10.0
	100.0

Packaging inks for paper and board

Labels

The use of labels in packaging is widespread and varied, normally the preprinted label is attached to the product container to identify its contents. Unlike printed wrappers, labels do not have close contact with the product, which is normally packed in glass, plastic, metal or some other impervious material. This allows, within certain limitations, a fairly wide choice of solvent and resin combinations to be utilised for label printing inks. When printed labels are produced for returnable bottles, the ink film should be penetrable during the washing process, which utilises a dilute solution of caustic soda, thus allowing easy removal of the label.

Pigment selection will be dependent on colour requirements. A reasonable degree of lightfastness is obviously desirable and some labels may require product resistance. Where high levels of gloss are important those pigments specificially developed for high gloss in liquid inks should be utilised. Normally, ball milling or bead milling is the method of manufacture. Advantage can be taken of the low viscosity grades of chlorinated rubber to help promote gloss.

Typical fast-drying formula for labels:

Rubine 4B (CI Pigment Red 57)	5.0
Lake Red C (CI Pigment Red 53)	7.0
Chlorinated rubber	12.5
Phenolic resin	20.0
Resinous plasticiser	4.0
Dioctyl phthalate	3.5
Polyethylene wax	1.0
Toluene	27.0
Ethyl acetate	10.0
Aliphatic hydrocarbon	10.0
	100.0

Some label printers have a preference for alcohol/ester-based inks choosing to avoid toluene for environmental reasons. This reduces the choice of resin systems considerably and normally means the use of nitrocellulose, maleic and ketone resins.

Non-curl properties and easy wash-off can be achieved by correct plasticisation and the choice of a suitable maleic resin. The main problem is to formulate suitably glossy inks that are sufficiently hard drying, especially when the label is not varnished. The use of specially developed, easily dispersed pigments give reasonable results. But as these pigments are a limited range, it is often found to be expedient to use nitrocellulose chips as a means of equalling the gloss of the chlorinated rubber-based inks.

Typical formula of a chip-based non-curl label ink:

50% Rubine 4B nitrocellulose chip	10.0
50% Lake Red C nitrocellulose chip	14.0
Maleic resin	10.0
Dioctyl phthalate	3.5
Wax dispersion	5.0
Ethanol	40.0
Ethyl acetate	14.0
Selected glycol ether	3.5
	100.0

A third option for labels are co-solvent polyamide based inks, where advantage is taken of their natural non-curl properties and potentially high gloss levels. There are some limitations to their use due to the high degree of resistance to weak alkalis, which prohibits them being used on wash off labels. Also ink films based on polyamide tend to weld when ram punched or guillotined, therefore they cannot be used on the edges of the labels.

Over recent years there has been a small but significant increase in the use of metallised paper for labels. There have been problems associated with ink adhesion on this substrate, mainly due to the oxidation of the aluminium surface. Lacquering immediately after the vacuum depositing process has been completed has alleviated the problem to great a extent. However, as the surface is basically non-absorbent, problems can occur with ink trapping as is the case with aluminium foil and certain films. In these circumstances it is useful to use a co-solvent polyamide backing ink, overprinted with nitrocellulose inks.

Typical co-solvent polyamide ink:

Phthalocyanine Blue (CI Pigment Blue 15.3)	10.0
Co-solvent polyamide	22.0
Nitrocellulose	3.0
Polyethylene wax	1.0
Isopropanol	12.0
Toluene	37.0
Isopropyl acetate	8.0
Ethanol	7.0
	100.0

Metallic label inks

Many labels require the use of metallic inks as part of the overall design. The various shades of bronze inks are particularly popular, usually they are the last colour to be printed in a multicolour design and often overprint one or more of the preceding inks.

Aluminium-based inks are used less frequently than bronze inks on labels. Primarily they are used as background to give a metallic effect to overlying colours. Care is required in the selection of a suitable grade of aluminium, if this is the case, to ensure interlayer adhesion. A typical formula for a bronze medium into which bronze powder would be added at 30–40% by weight is:

Chlorinated rubber (medium viscosity)	30.0
Dioctyl phthalate	10.0
Wax dispersion	7.0
Stabilizer	1.0
Toluene	52.0
	100.0

and a bronze ink:

Bronze powder (gravure ink lining)	40.0
Acrylic resin	20.0
Wax dispersion	5.0
Stabilizer	1.0
Toluene	27.0
Isopropyl acetate	7.0
	100.0

Paper wrapper inks

With paper wrappers being used predominantly for food, confectionery and toiletry packaging, the choice of raw materials in gravure inks is largely determined by their suitablity for these end uses. It is of paramount importance that the dried ink film is free from odour or potential taint that might be derived from retained solvent or odorous resins. Although inks are not intended to come into direct contact with foodstuffs, they should not contain pigments or dyes which are likely to bleed into food constituents, for example animal fats, vegetable oils and in the case of chocolate products, cocoa fat. Migratory materials should be avoided where possible and harmful pigments, chemicals and additives should not be used.

For guidance on this subject the Society of British Printing Ink Manufacturers has a standard booklet *Printing Inks for Use on Food Wrappers and Packages* and also a list of materials that should be excluded from inks for food wrappers.

After the exclusion of harmful pigments and certain dyes, the possible choice of colourants for wrappers is wide and largely dependent on their end use. As indicated earlier specific product resistance is often a factor, with standard tests available to predetermine a pigment's suitability. Soap wrappers for example require inks that are resistant to 2% sodium hydroxide solution and also to a 25% toilet soap gel.

Because of its low odour characteristics, coupled with its fast solvent release properties, nitrocellulose has been generally accepted as the ideal film former for paper wrappers. It can be used with plasticiser only or modified with carefully selected maleic and ketone resins.

Ethanol with small quantities of esters and glycol ethers are the preferred solvent combinations. When the label is not varnished, nitrocellulose chips may be required to achieve an acceptable gloss level. Otherwise ball milled or bead milled inks can be used, with the overprint varnish providing gloss. As with labels, inks for paper wrappers should be adequately plasticised to prevent curling during the packaging process. One of the inherent problems with varnishing paper wrappers for foodstuffs is the choice of solvents for the overprint varnish. Having formulated the inks on low odour nitrocellulose and ethanol, there is little point in introducing odorous solvents such as toluene and ester solvents into the overprint varnish. These would not rewet the inks and would provide an acceptable gloss level, but there would always be a danger of retained solvent leading to possible taint. Alcohol based varnishes can cause rewetting and glossy versions are often prone to tackiness. If they are used to overprint inks containing PMTA pigments or dyes they can quickly become discoloured and tint the unprinted areas of the paper.

It was the introduction of water-based overprint varnishes in the late 1960s that largely overcame this problem. Initially vinyl chloride vinylidene copolymer emulsions were used, they satisfied all the main demands of gloss and low odour, and because of their high water and low alcohol content did not rewet the inks. The main drawback was their high temperature of coalescence, which meant that only machines fitted with extended drying capable of reaching a temperature of 120 °C could use this type of varnish successfully.

Following these came the acrylic-based varnishes which require a much lower coalescing temperature. With a wide range of acrylic solutions, dispersions and emulsions becoming available, the formulator is now able to provide water-based varnishes for paper wrappers with a wide range of properties.

A typical ink for a food wrapper would be:

Calcium Rubine 4B (CI Pigment Red 57.1)	10.0
Nitrocellulose	15.0
Maleic resin (low odour)	5.0
Wax dispersion	2.5
Dioctyl phthalate	5.0
Ethanol	54.0
Ethyl acetate	5.0
Glycol ether	3.5
	———
	100.0

Water-based overprint varnish, using specially selected low odour materials:

Acrylic resin solution	28.0
Wax emulsion	12.0
Acrylic resin dispersion	40.0
Wax dispersion	5.0
Anti-foam	1.0
Isopropanol	8.0
Water	6.0
	100.0

Inks for cork tipping

The first requirement for cork-tipping inks is that their constituents should be fully FDA approved. They also should be manufactured under very carefully controlled conditions to avoid possible contamination.

Pigmentation is achieved with specially selected iron oxides and titanium dioxide, all chosen for their inert properties.

Cork-tipping inks are normally based on nitrocellulose combined with a specially selected plasticiser, which when pigmented is tack free and insoluble in human saliva.

Carton Inks

This is one of the areas in which gravure printing competes with offset litho and comparisons in print quality are often made. For this reason the lay characteristics of a carton gravure ink is of critical importance. Also many of the cartons for well known brands of chocolate confectionery and cigarettes, have considerable coverage of a colour with which their name is associated. Therefore particular attention is given to the print quality to ensure that it is smooth and unblemished.

Before discussing inks it is worth considering the role played by the engraving on the gravure cylinder and its influence on lay. With some shades, dark greens, browns and deep blues, it is an advantage when printing on board to have an engraving of greater depth than normal, this ensures good ink coverage and flow out reducing the tendency for mottle to occur. It also enables the ink maker to include a reasonable amount of tianium dioxide in the formulation to give increased opacity and improved lay.

Recent developments have shown that Helioklischograph engraving can also reduce the lay deformities typical of certain colours.

In common with other packaging inks for food and tobacco, low odour is of prime importance for cartons printed for these purposes. The inks are similar in characteristics to those discussed for paper wrappers where comparable requirements apply.

If inks with acceptable lay are to be formulated then careful attention must be given to pigment selection, not only individually, but in the combinations chosen to match a particular shade. For example browns are often best matched from an orange or red which when darkened with black has the nearest hue to the shade required. Bronze blue and yellow (PY 83) are a useful combination with black for matching dark greens. Sometimes, despite their cost, PMTA pigments can be used to advantage to achieve good lay.

Almost invariably, cartons are varnished and as with paper wrappers water-based overprint varnishes have become increasingly important. All cigarette cartons and many chocolate boxes are overwrapped with transparent film which is heatsealed directly over the printed surface. Obviously this requires the inks to be heat resistant and the overprint varnish to have heatseal release properties against the particular film employed.

In addition to heat resistance properties, the water-based overprint varnish should be formulated to give an acceptable gloss level and also be based on hard drying acrylic resins, which have the ability to release water from the drying varnish film as rapidly as possible. In this respect the amine content should be critically evaluated and, if tolerable, ammonium hydroxide should be considered. Sometimes the best overall results can be realised by a carefully balanced combination of two amines.

Heat-resistant carton ink:

Rubine 4B (CI Pigment Red 57.1)	10.0
Orange 5Y (CI Pigment Orange 5)	2.5
Titanium dioxide (PW 6)	7.5
Nitrocellulose	13.5
Maleic resin	5.0
Wax dispersion	5.0
Dioctyl phthalate	5.0
Ethanol	42.0
Isopropyl acetate	7.0
Glycol ether	2.5
	100.0

Heat seal resistant varnish:

Hard acrylic resin	15.0
Isopropanol	20.0
Water	15.0
Amine or ammonium hydroxide	2.0
Wax emulsion	5.0
Acrylic emulsion	35.5
Wax dispersion	5.0
Release agent	1.5
Anti-foam	1.0
	100.0

When metallic inks are required for carton printing the overall design has to be carefully considered. If the area of metallic print is small and if the ink has to overprint other colours, it is likely that the best printability will be achieved with a toluene-based acrylic resin type of ink described previously. If the ink is required to print directly on to the board then an ester-based acrylic metallic ink should perform satisfactorily.

Problems can occur when a large area of gold is required to overprint a nitrocellulose based colour and, because of the low odour requirements, a toluene system cannot be used. The best approach is to use a vinyl modified acrylic with a carefully balanced ester solvent system, suited to the machine speed and its drying capacity.

The choice of materials for bronze inks is discussed in detail in the section dealing with metallic inks.

Metallic bronze gravure ink for cartons:

Bronze powder (Superfine lining)	40.0
Ester soluble vinyl resin	2.5
Acrylic resin	17.5
Wax dispersion	3.0
Stabiliser	1.0
Ethyl acetate	10.0
Isopropyl acetate	16.0
Glycol ether	10.0
	100.0

Foil inks

Aluminium foil is widely used in packaging both unsupported and as a laminate to paper and board. Having been used for a wide variety of packaging outlets over many years, aluminium foil is now being challenged by metallised papers and films, the former because of lower cost, the latter on product appeal.

Virgin foil

When unsupported foil is printed it is normally untreated, the inks being required to key directly to the virgin aluminium surface. During the annealing process any rolling oil left on the surface is removed. When exposed to the atmosphere aluminium oxidises, rendering ink adhesion more difficult. Normally the foil is sufficiently tightly wound to prevent this happening too quickly.

One of the problems when printing virgin foil, especially the very thinnest gauges, is to achieve adhesion and flexibility of the ink film without leading to subsequent blocking. This can be particularly hazardous where colours overdraw or where the nature of the design creates a build-up of ink in certain areas. This can lead to excessive pressures in those areas in the rewind reel thus increasing the danger of set-off or blocking.

Many colours printed on foil are transparent, advantage being taken of the metallic lustre of the aluminium. The required transparency is usually obtained by the use of pigmented nitrocellulose chips and occasionally lightfast dyestuffs, or a combination of both. As with paper wrappers the choice of pigments and dyes is fairly extensive. The most common use of lightweight foil is for chocolate and confectionery wrappers where the colourant will need resistance to vegetable oil, cocoa butter and possibly lactic acid.

Providing the foil is free of oil and oxidation, adhesion to the surface can be achieved with plasticised nitrocellulose modified with polyvinyl butyral. The plasticiser content should be balanced carefully to ensure sufficient flexibility without incurring a soft or sticky ink film that could subsequently set-off or block. Adhesion on foil is usually assessed by the self-adhesive tape test and crinkling the printed foil between finger and thumb.

Solvent choice is limited to those that are suitable for food packaging as discussed previously. Particular attention should be paid to the use of any retarding solvents, as these can remain in the ink film giving a false degree of flexibility, which is not enduring. Also their presence increases the danger of blocking and residual odour.

Formulation for lightweight aluminium foil:

50% pigmented nitrocellulose chip	25.0
Polyvinyl butyral	4.0
Dioctyl phthalate	4.0
Wax dispersion	3.0
Ethanol	54.0
Isopropyl acetate	10.0
	————
	100.0

Not all chip-based inks key to foil satisfactorily; certain pigments can be difficult if the overall chip content exceeds a certain level. Typical examples are Bronze Blue and Lake Red C; it may well be the dispersing aids used during the manufacture of the chips that are responsible for the problem.

Harder grades of foil laminated to polyethylene are used to package medical tablets individually. Hard foil is produced by varying the annealing process and is of a heavier gauge than that used for confectionery wrappers. A thin film of polyethylene is coated on what is to be the reverse side of the foil by hot melt application and is used to heatseal two strips of foil with the tablet sandwiched inside. The encapsulation is done at very high speeds using crimp heatsealing and to achieve the required throughput the sealing temperature is between 180 and 200 °C.

The heavier grades of aluminium foil are usually referred to as capsule foil and are primarily used for milk bottle tops and polystyrene cream and yogurt tub closures. Bottle tops are printed by both the gravure and flexographic process. The main requirement is for wet rub and water resistance and when printed gravure either vinyl-based or nitrocellulose-based inks can be used. Yogurt lids are normally printed with plasticised nitrocellulose and are heatsealed to the tub by means of a heatseal varnish which is applied on the reverse side of the foil. A modified vinyl chloride/vinyl acetate resin combined with a suitable soft acrylic is used, which is designed to give a sufficiently firm bond when sealed, but allow the closure to be easily peeled from the tub.

Foil laminates
Foil to paper laminates are to be found in the food, confectionery, tobacco and toiletry industries. To prevent poisoning and oxidation of the foil

after lamination it is coated with a wash of nitrocellulose, shellac and, in some cases, vinyl. Therefore the ink no longer has to key to aluminium but to a resinous base which enables a wider range of inks to be employed.

When heat resistance is not essential, advantage can be taken of employing polyamide or acrylic backing whites overprinted with nitrocellulose colours, thus ensuring good trapping. Some difficulty can be met with the rewetting of nitrocellulose on foil, both as the basecoat and when inks of this type overprint each other. With inks the solvent sensitivity of the nitrocellulose can be reduced by modification with a hard acrylic; the solvent system however will have to be predominantly ester based. Inks of this type plus plasticised nitrocellulose, can be used where heat resistance is required and also for soap wrappers.

Shellac-washed foil rarely presents a problem with rewetting due to its tendency to insolubilise on ageing. Because of this alcohol and ester based inks can be printed without difficulty.

Wax laminated wrappers are commonly used for chocolate biscuits. Preprinting on virgin foil, then lamination with hot melt wax, requires inks with a reasonable degree of heat resistance to prevent picking off immediately after lamination. Prelaminated material must be printed within a few hours of lamination to prevent poisoning of the foil surface, thus making ink adhesion difficult. Heat resistant polyamides ($> 120\,°C$) can be used in conjunction with nitrocellulose-based inks. Care is required in pigment and dye selection to ensure there is no bleed into the wax.

Foil board laminates

Printing foil board laminates was at one time almost exclusively done by sheet-fed litho. The foil was usually coated with a gold tinted nitrocellulose lacquer. After printing by offset the sheets were then varnished, often as a separate operation. A similar approach was followed when the gravure process was originally adopted to print cigarette cartons on foil board laminate. Using a sheet-fed machine the gold lacquered foil/board was printed in one operation. Problems were encountered with printability and adhesion on the gold nitrocellulose base lacquer.

It is now more customary to print on foil/board laminates which have been coated with shellac. Often the darker colours are printed directly on to the substrate with the light colour, usually a gold, printed last also playing the part of an overprint varnish.

Modern reel-fed machines with in-line cutting and creasing demand inks that are hard drying and flexible. The adhesion and flexibility should be sufficient to withstand a $180\,°C$ fold followed by ten firm finger rubs on the ink surface. All cigarette cartons are overwrapped with film requiring the inks to be heatseal resistant. The surface slip of the final overprinting colour should be carefully adjusted to meet the requirements of the high speed cigarette packing machines.

To achieve the adhesion and heat seal resistance requirements, nitrocellulose modified with melamine resin are suitable combinations. On occasions where there is superimposition of the darker colours the base colour can be hardened with an acrylic resin to reduce rewetting. Mainly

nitrocellulose chips and occasionally lightfast dyes are used as colourants.
For example:

50% Yellow 83 nitrocellulose chip	3.0
50% Orange 5 nitrocellulose chip	2.0
Nitrocellulose	20.0
Modifying resin	10.0
Wax dispersion	5.0
Ethanol	46.5
Isopropyl acetate	10.0
Glycol ether	3.5
	100.0

Inks for polyethylene film

The vast majority of polyethylene film is converted by the flexographic process because of either the simplicity of the design or the length of run which makes the gravure process commercially unviable. However, designs of a more complex nature still utilise the gravure method or alternatively the offset gravure process which can allow a blend of flexo and gravure printing stations.

Conventionally co-solvent polyamide resins have been used in gravure inks for polyethylene because of their excellent pigment-wetting properties, good solvent release characteristics, good water resistance and not least they enable a high gloss print to be obtained. The trend to remove hydrocarbon solvents from packaging inks has meant a change to the use of alcohol-soluble polyamide resins. Unlike their co-solvent counterparts the alcohol-soluble grades have inferior solvent release and water resistance properties and modification with nitrocellulose is essential to provide inks of the desired quality.

For heat-resistant applications polyamide resins cannot be employed and inks based on nitrocellulose modified with polyurethanes are used. These systems will be discussed more fully under the section 'Inks for treated polypropylene films'.

Regardless of the resin system employed, pigment selection plays a vital part in the determination of the water resistance properties of the final ink film and the particular pigment selected will not necessarily be the same for the above resin systems.

In polyamide inks PMTA pigments (especially PBl. 1) should be used with caution so as to ensure there is no bleed into the base film on storage. Certain pigments based on transition metals may react with polyamide resin to give severe odour problems. For example Manganese 2B toners and copper bronze are both liable to give odours, whereas copper phthalocyanines are satisfactory since the copper has already formed a highly stable chemical complex.

Polyamide inks frequently gel at low temperatures and the ability to recover to a fluid system on warming varies widely from one resin to another. In co-solvent resin systems a small quantity of water greatly facilities the ease of gel recovery despite having no effect on the original temperature at which gelation first occurs. Water has no effect on alcohol soluble inks, gel recovery being adjusted by resin selection.

Ink based on co-solvent polyamide resin:

Phthalocyanine Blue (CI Pigment Blue 15.3)	12.0
Co-solvent polyamide resin	25.0
Antioxidant	0.5
Isopropanol	32.0
Toluene	14.0
Aliphatic hydrocarbon	12.0
Water	2.0
Amide wax	1.0
Polyethylene wax	1.5
	100.0

Ink based on alcohol-soluble polyamide resin:

Phthalocyanine Blue (PBl. 15.3)	12.0
Alcohol-soluble polyamide resin	13.0
Nitrocellulose	18.0
Dioctyl phthalate	2.0
Antioxidant	0.5
Ethanol	40.0
Ethyl acetate	7.5
n-propyl acetate	5.0
Amide wax	1.0
Polyethylene wax	1.0
	100.0

Inks for treated polypropylene films

Treated polypropylene films have continued to find increasing use in gravure packaging outlets. The homopolymer film (i.e. OPP) is still little used despite being the lowest cost since it is unsuitable for heat seal operations and the corona discharge treatment loses its effectiveness with time. Treated co-extruded polypropylene films are mainly used for heat resistant applications while the 'pearlised' grades have attained tremendous growth in cold seal outlets. The pearlescent effect in these films is produced by air entrapped in the film rather than by use of mica pigment. The insulating properties of these air bubbles makes heat sealing difficult within the dwell time on a packaging line.

For co-extruded films, inks based on nitrocellulose modified with polyurethanes and titanium acetyl-acetonate (TAA) are used for both surface and reverse printing as well as for lamination. Titanium acetyl-acetonate acts as the adhesion promoter to the treated side of the film but has a drawback in that it may also react with the pigment to give an unstable ink. The same system can also be employed for printing on the treated side of homopolymer polypropylenes though higher levels of the adhesion promoter may prove necessary with a consequent risk of ink instability and high residual odour.

White ink for treated side of co-extruded polypropylene:

Titanium dioxide (CI Pigment White 6)	30.0
Nitrocellulose resin	11.0
Dioctyl phthalate	3.0
Polyurethane	8.0
Polyethylene wax	1.0
Erucamide	1.0
Ethanol	27.0
Ethyl acetate	16.0
Titanium acetyl-acetonate (TAA)	3.0
	100.0

While the above ink system is widely adopted both in the UK and in other countries, it has the drawback of relatively poor gloss levels for surface printing. Special two-pack reactive white ink systems for use with conventional overprinting colours have therefore been developed to overcome such a deficiency. The white can be based on epoxy-amine, polyol-isocyanate or alkyd-melamine chemistry, though the latter is little used because of the evolution of formaldehyde in the heat seal process during the packaging operation. Iso-cyanate systems are no longer used in the UK for health and safety reasons but they are still frequently adopted within other countries. The epoxy-amine system is therefore preferred in the UK and has the additional advantage that if used in conjunction with cellulose acetate proprionate-acrylic colours it can also be suitable for PVdC overcoating.

Cold seal packages will normally use an overprint varnish to give release to the adhesive within the printed reel though in certain cases, dependent upon the design, the release properties can be built into the inks themselves. Polyamide varnishes frequently contain antioxidants either as an additive in the varnish or in the resin itself to reduce the risk of unwanted odours and the antioxidant, if phenolic in its chemical nature, will react with transition metals to form a pink-coloured complex. Hence, if a TAA-based ink system is used in conjunction with a polyamide varnish then extreme care must be taken to ensure no discoloration occurs on storage of the printed reel.

Since in cold seal work there is no heat resistance requirement, polyamide inks can be employed with polyamide varnishes to give base adhesion to the treated polypropylene film. While this might appear to be an attractive solution to the problem, however, other difficulties still have to be overcome. The use of alcohol soluble polyamide resins throughout can give rise to excessive tack levels due to the poor solvent release and an all co-solvent resin situation can lead to undesirable retention of hydrocarbon-based solvents. A combination between the two resin grades for the inks and varnish is therefore preferred. Finally, it is important to ensure that the chosen ink–lacquer combination gives the required slip characteristics and since the amide waxes normally used with polyamide resins can migrate into and poison the cold seal adhesive, such additives must be avoided. For this reason micronised polyethylene or polypropylene waxes are used to give the correct coefficient of friction levels to suit the packaging requirements.

Red ink for pearlised OPP			*Cold seal release lacquer*	
Calcium 2B toner	10.0		*for pearlised OPP*	
Alcohol soluble			Co-solvent polyamide resin	30.0
Polyamide resin	12.0		Ethanol	34.0
Nitrocellulose	10.0		Aliphatic hydrocarbon	34.0
Ethanol	52.0		Micronised PE wax	2.0
Ethyl acetate	14.5			———
Micronised PE wax	1.5			100.0
	———			
	100.0			

Coated polypropylene films

The PVdC coated polypropylene films are heat sealable and have excellent gas and moisture barrier properties. Modification of the PVdC latex with a methacrylate can broaden the sealing range of the films with only minimal effect on the barrier properties. In recent years pure acrylic coated films have become available which have the advantage of lower cost but inferior barrier properties and this has been followed by combinations of one side acrylic and one side PVdC and pearlised grades.

The coatings on all the above films are affected by hydrocarbon solvents and high ester content inks which can destroy adhesion of the coating to the base film. In addition the acrylic coatings are sensitive to glycol ether solvents and the slower ester solvents (eg. *n*-propyl acetate) leading to very high solvent retention and ink set-off.

Coated polypropylenes are primarily used where the cheaper uncoated grades have insufficient barrier properties or where the seal format of the latter is unsatisfactory. In addition the ability to overcoat the inks with PVdC emulsion enables very high bond strengths to be obtained on front/front (i.e. A/A) and back/back (i.e. B/B) seals. Here A and B refer to the front surface and reverse surface respectively. Such seal compatibility can only be realised on the treated films by coating both sides of the film with PVdC and this is usually impracticable. In or out of line PVdC coating of the inks is used particularly on biscuit roll wraps where the bunch seals and the undulations of the pack on the side seam enable only low sealing pressures to be used.

For majority of end applications cellulose acetate propionate (CAP) in combination with acrylic resins are used as these offer excellent print adhesion at low web temperatures and hence minimise press registration problems. Such systems give low retained solvents and are suitable for both surface and reverse printing, PVdC overcoating, lamination and with a polyamide overlacquer, cold seal applications. The major problems with these inks are on fine typework and tone areas where the rapid solvent release can cause premature drying. Retarding solvents are not a straight-forward solution due to the coating sensitivity described above and in case of extreme difficulty, and where the print format allows, a nitro-cellulose system may be preferred.

Yellow ink for printing on coated polypropylene film:

Diarylide Yellow (CI Pigment Yellow 12)	8.0
Alcohol soluble propionate	10.0
Acrylic resin	6.0
Ethanol	45.0
Ethyl acetate	30.0
Polyethylene wax	1.0
	100.0

Cellulose films

Due to the high energy costs involved in the manufacture of cellulose films they are continuing to decline in use in contrast to the increasing popularity of the polypropylene based products. For convenience the films can be divided into three categories:

(1) Single side coated films, using either a PVdC or nitrocellulose coating. The uncoated side (designated PT) must be printed and then laminated to another film to protect it from moisture. Inks are relatively simple, being based on plasticised nitrocellulose but high in ester solvents to prevent excessive retention of alcohol-based solvents.

Green ink for reverse printing on PT film lamination:

Diarylide Yellow (CI Pigment Yellow 13)	7.0
Phthalocyanine Blue (CI Pigment Blue 15.3)	2.5
Nitrocellulose (high nitrogen grade)	14.0
Dioctyl phthalate	5.5
Polyethylene wax	1.0
Ethyl acetate	70.0
	100.0

(2) Two side coated nitrocellulose films are available varying in the content of wax and plasticiser in the coating to alter the properties of the film. While plasticised nitrocellulose inks offer excellent adhesion to these films great care must be exercised in the resin to plasticiser ratio to prevent blocking. The high wax content in the film coating also causes the ink to pinhole and the surface tension of the ink must be adjusted, usually by the addition of a phenolic resin.

White ink for surface printing on N/C coated films:

Titanium dioxide (CI Pigment White 6)	30.0
Nitrocellulose	12.0
Dioctyl phthalate	7.0
Phenolic resin	1.0

Polyethylene wax	1.0
Ethanol	30.0
Ethyl acetate	19.0
	100.0

(3) Two side PVdC coated cellulose films form the third grouping with the PVdC being applied via an aqueous emulsion or solvent-based varnish. Ink systems are similar to those described for PVdC coated polypropylene films though since the coatings have better anchorage to the cellulose base retarding solvents can be used in moderation.

Polyester films

The high tear and impact strength of polyester film makes it particularly suitable for packaging both heavy and abrasive goods and its high heat resistance lends the film to boil-in-the-bag and retortable pouches. Only grades which are PVdC coated or have a special chemical coat can be heat sealed and therefore in the majority of applications lamination to polyethylene or foil and polyethylene is necessary.

For reverse printing and lamination on uncoated polyester a vinyl ink must be used since other resin systems give poor bond strengths either initially or after a period of time. It is especially important to select the pigments carefully for the vinyl carrier in order to obtain maximum flow and aid printability characteristics. Solvent retention can also be a problem especially with ketone/ester based laminating adhesives and alcohol soluble grades are therefore preferred as virtually no rewetting of the ink is then given.

On the rare occasions when inks are required for surface printing polyamide inks can be used for non-heat resistant applications. For heat resistant packs inks based on nitrocellulose modified with polyurethane and TAA, as for co-extruded polypropylene film, can be adopted.

For both surface and reverse printing of the PVdC-coated grades inks on CAP–acrylic systems can be used as previously described for PVdC-coated polypropylene films.

Red ink for reverse print and lamination on uncoated polyester film:

Barium 2B red toner (CI Pigment Red 48.1)	9.0
Vinyl chloride/vinyl acetate/vinyl alcohol co-polymer	10.0
Ethyl acetate	40.5
n-propyl acetate	40.0
Polyethylene wax	0.5
	100.0

Nylon films

Nylon is primarily used for its thermoforming properties for example in wrapping cheese and in bacon pouches. The nature of the surface of nylon film appears to vary with the source of manufacture and therefore unlike most films a simple ink recommendation cannot be made and ink systems differing widely in choice of resin systems have been adopted.

Vinyl films

Ink systems for printing on vinyl films differ not only in choice of resin system but also in pigmentation types dependent on the plasticiser content. For unplasticised grades (UPVC) either nitrocellulose–acrylic or CAP–acrylic systems can be adopted without any special pigment limitations. However, for plasticised vinyl films not only does the ink-maker need to take special care on the adhesion and non-blocking properties of the resin system, but also pigments which do not cause bleed into plasticiser have to be chosen. For further details the reader should consult the latter section on inks for vinyl wallpaper.

Metallised films

Metallised films have been included as a separate section because of the special problems associated with the printing of such films. To date metallised PVdC-coated cellulose, co-extruded polypropylene and polyester films have been made commercially available. Difficulties have arisen since the adhesion properties of inks vary not only with the age of the films after metallising but also vary from batch to batch of film. Hence, it is virtually impossible for the ink-maker to guarantee that an ink system will adhere satisfactorily to any batch of film. The phenomenon which causes such variable ink receptivity is still not readily understood and attempts to classify the behaviour as due to surface oxidation or migration have not been substantiated by chemical evidence.

Polyamide-based inks do show a greater tolerance to film variation than most other resin systems and hence for cold seal applications such inks can be recommended provided reasonable precautions are taken by the printer. For heat-resistant applications individual ink manufacturers have formulated their own specialist systems which cope in variable degrees with the film inconsistency, and for extremely difficult cases primers have even also been utilised.

Wallcoverings

A variety of processes are used to print wallcovering, with gravure, flexo and ink emboss predominating where surface printing once reigned supreme. Many manufacturers use all three processes and it is not uncommon to see combinations of them used to print the same reel of wallcovering.

A modern trend is for printers to carry the same range of inks for printing vinyl-coated papers and plain paper by both gravure and flexo. This is done to reduce the range of inks they have to stock. Normally the ink maker supplies a pallette of colours along with reducing mediums, pearlescent bases and specials such as non-tarnish golds. Colour matching is carried out on site by the printer.

Inks for paper

Gravure inks for paper are normally expected to withstand the sponge-ability test without varnishing after embossing and the washability test after varnishing. As the base paper plays an important role in resistance to these tests, it is sometimes difficult to judge whether a failure is due to paper or ink.

The pigments used for wallpapers must have a minimum lightfastness of 5 on the blue wool scale at tint strength. The resin system should be flexible, but tough enough to withstand cold embossing and when modified with suitable wax additives, provide an ink film with excellent rub and mar resistance. As tonal printing is common on wallpapers an essential feature is that inks should have excellent printability. Toluene is usually the preferred solvent with chlorinated rubber and phenolic resin being used to give properties described above.

It is important that the ink should print to give a matt finish, even where there are overlying colours. This is to prevent distorted light reflections, especially when the wallpaper is viewed from an oblique angle. The mattness of the inks is normally controlled by the use of transparent white extenders.

Typical formula for a gravure wallpaper ink:

Phthalocyanine blue (CI Pigment Blue 15.3)	10.0
Calcium carbonate (CI Pigment White 18)	12.5
Chlorinated rubber	12.0
Modified phenolic resin	17.5
Resinous plasticiser	2.5
Wax dispersion	7.5
Toluene	38.0
	100.0

Vinyl-based varnishes can be applied by air knife coatings or in-line gravure application when a washable standard of paper is required.

Vinyl coated wallcoverings

Two types of inks are used on vinyl-coated papers for gravure printing, those which are for gravure printing only, with a solvent base of methyl ethyl ketone (MEK) and toluene, also the dual purpose, but more expensive type, which can be printed by both gravure and flexo and has an ester solvent base.

As the processing and end use are identical, the choice of pigments for both types of inks is comparable. Their selection is governed by the specific requirements of the production process and the overall specifications demanded of vinyl wallcoverings. All pigments must have a lightfastness of 5 on the blue wool scale at tint strength. They must be fully resistant to bleeding in plasticised PVC and be heat resistant at 200°C during the embossing process. As pigments with these properties tend to be very expensive, particularly oranges, reds and violets, care should be exercised in pigment selection to achieve the greatest economy of use.

The resin systems for the gravure and dual purpose inks are not identical but have much in common, being based on vinyl chloride/vinyl acetate copolymer with acrylic for the former and modified vinyl resin plus acrylic for the latter.

As indicated above, most vinyl wallcoverings are heat embossed at about 200°C. In addition to enhancing the appearance of the wall-covering, the heat and pressure integrate the ink into the surface of the vinyl. This process is usually sufficient to ensure that correctly formulated inks will pass the scrubbability test.

When printing on vinyl, it is an advantage if the solvent system softens the substrate slightly, thereby giving the ink 'bite' and ensuring good adhesion. With straight gravure inks this is achieved by a correct balance of MEK (the strong solvent) and toluene (the diluent). If the formulation has excess ketone there is a danger of softening the vinyl too much leading to screening. With the dual purpose inks the solvent power of esters for vinyl is less than ketones and therefore they can be used without diluents.

Gold printing is normally done using simulated golds, where aluminium-based inks are tinted with transparent yellows, oranges and scarlets. These are used to avoid the tarnishing which eventually occurs with bronze powder based inks.

Background mica finishes are based on gravure printed inks based on pearlescent pigments. These can be plain silver pearl or tinted pastel shades, giving a pleasing lustre to superimposed colours.

Typical formula of a vinyl wallcovering gravure ink:

Organic pigment (heat and bleed resistant)	10.0
Extender	7.0
Vinyl chloride/vinyl acetate copolymer	6.0
Hard acrylic resin	12.5
Wax medium	7.5
Methyl ethyl ketone (MEK)	23.5
Toluene	28.5
Butyl acetate	5.0
	100.0

Typical gravure/flexo wallcovering ink:

Organic pigment (heat and bleed resistant)	12.0
Extender	5.0
Modified vinyl resin	8.0
Hard acrylic resin	10.5
Wax medium	7.5
n-Propyl acetate	57.0
	100.0

Expanded polyethylene

Expanded polyethylene is a third substrate type used for wallcoverings. Problems with long-term adhesion are associated with polyamide-based inks. Specially formulated inks have been developed to overcome this using an ester-based resin system. The expanded polyethylene base gives a wallcovering that is distinctly warmer and softer to the touch than either vinyl- or paper-based products.

Decorative wall panels

This is one of the few processes where offset gravure printing is exclusively used. It is principally designed to print sheets of hardboard or plywood, either to produce a woodgrain effect or simulate bathroom or kitchen tiles.

The sheets of hardboard are fed along a roller train where initially a base coat is applied and stoved dry. The offset gravure unit is situated above the roller or blanket which then transfers the ink on to the hardboard as it passes underneath. Unprinted ink is wiped from the offset roller after printing and when the system is idling. The inks dry by solvent evaporation, assisted by heat absorbed by the board during the earlier stoving operation. Finally an acid cured precatalysed varnish is applied by curtain coat to protect the print.

Pigments should exhibit a high degree of lightfastness, iron oxides were originally used for woodgrain designs. Nowadays organic pigments have gained favour due to their less abrasive nature.

Nitrocellulose, ketone resins, ethyl cellulose and polypale resin can be employed as ink binders.

The solvent system must be sufficiently slow drying to ensure that the ink remains fluid when it is transferred to the rubber roller and subsequently to the substrate. Solvents should be selected with care as they must not swell or distort the rubber roller unduly.

Melamine laminate inks

Gravure printed laminated plastics are used to produce decorative panels and sheeting for household furnishings, particularly in kitchen and bathrooms where exposed surfaces have to be highly resistant to water and heat. They are produced by printing paper which has been impregnated with either melamine formaldehyde or urea thermosetting resin.

The laminate is produced by laying the thoroughly dried print over layers of unprinted impregnated paper to a thickness slightly greater than the laminate required. A sheet of pure thermosetting resin is overlayed on the print. The stack is sandwiched between highly polished metal plates and subjected to a pressure of 600–1000 lb/square inch for 45–90 minutes at a temperature of 150–200°C. The temperature and pressure compress the stack inducing the thermosetting resin to flow and cure to produce a solid plastic. Finished laminates should resist boiling water for two hours without delamination.

Pigments should have a high degree of lightfastness and have an acceptable resistance to colour change and development during the lamination process. There is a major difficulty in producing consistent colour matching to standard patterns, especially with multicolour tone work. A final assessment against the standard can only be made after a test laminate of the print has been produced. As each test takes nearly an hour the overall colour-matching time can be lengthy. Normally pigment selection is from phthalocyanine blues and green, naphthol reds and certain diarylide yellows e.g. PY 83.

The resin system used for laminate inks should be compatible with the type of thermosetting resin used in the laminating process. Polyvinyl butyral resins are used for solvent-based inks with styrenated acrylic systems being preferred for water-based products.

The use of additives such as waxes and silicones should be avoided if possible as they are a source of blistering and delamination.

Low pressure laminates

The majority of wood finishes used on modern furnishing are some form of gravure printed laminate. Low pressure laminates are printed on paper with a woodgrain design then coated with an acid catalysed lacquer to give protection. The print is adhesive laminated to block or chipboard. Apart from reducing costs, one of the main advantages of this method is that the grain matching required with natural wood finishes is eliminated. This is particularly important if part of the furnishing is damaged during manufacture, when a printed matching replacement panel can be used. With natural wood grains this is not always possible.

For home use heavy gauge plasticised PVC is surface printed with either woodgrain or decorative designs, given a light emboss then reverse coated with self-adhesive and protected with a release film.

Stamps

The paper used for stamp printing is a smooth highly calendered coated paper with low absorbency. Designs may call for single colour or multi-colour with overlaying tone and line work involving very fine detail.

Pigments should be carefully selected for their printability properties in the chosen resin system. They should have a lightfastness of 5 on the blue wool scale and be resistant to household cleansing agents and bleaches. They should be well dispersed during manufacture to ensure problem-free printing. Filtration of the finished ink requires care to ensure there are no large particles present which might give rise to streaking.

Care is required in both the formulations and manufacture of gravure inks for stamps to ensure they have excellent printability, rub and mar resistance. Choice of resin system is of great importance, as it should be tough and flexible and have excellent resolubility in the solvent system to maintain consistent tonal reproduction. Both phenolic resins modified with low viscosity ethyl hydroxy ethyl cellulose and ethyl cellulose/maleic resin systems are used successfully with toluene as the main solvent. Any waxes should be finely dispersed and used sparingly.

Speciality systems

Metallic inks

For many years ready mixed metallic gravure inks were of minor interest to ink manufacturers, as gold and silver inks were mainly blended on site by the printer.

It was the development of frame board for cigarette cartons using metallic bronze inks to match the existing gold lacquered foil board laminates, that introduced the need for premixed inks. Printers lacked the facilities to mix the large quantities of ink involved.

Inks for this purpose require careful formulation and must have the following basic properties:

(1) Be viscosity stable; although a slight increase in viscosity as the bronze powder is wetted out is normal, long-term stability should be achieved.

(2) Have good solvent release coupled with a low level of odour after printing. The use of toluene should be avoided.

(3) Exhibit a high degree of brightness even after prolonged storage as the frame board is often part of a foil board laminate pack.

(4) The ink should print with maximum coverage and minimum striation, for which all the components are important including the correct solvent balance.

(5) The ink film should be tough, flexible and have good rub resistance. However, as complete rub resistance is difficult to achieve, frame board is normally given a light coat of varnish.

The choice of resin systems for frame board golds can only be established after stability tests have been carefully evaluated. Resins should be neutral in nature and non-reactive with the copper content in bronze powders. An increase in viscosity, dulling of the bronze when printed, and in extreme cases a green or blue tint in the medium are all indications of instability. Suitable resins are certain grades of acrylics with or without selected vinyl resins as modifiers.

With toluene being excluded and the adverse effect on storage stability of some alcohols, the normal choice of solvents for frame board golds, are esters with selected propylene glycols as retarders.

An example of a frame board gold would be:

Bronze powder	40.0
Acrylic resin	16.0
Wax dispersion	3.0
Stabiliser	1.0
Ethyl acetate	10.0
Isopropyl acetate	25.0
Glycol ether	5.0
	100.0

Bronze inks for film

When printed on transparent film, bronze inks are normally backed with a colour to enhance the overall effect. Reds are particularly pleasing with the paler shades of gold, with greens and blues complementing the rich shades. The choice of resin system will largely depend on the type of film to be printed. If the resin selected to meet end use requirements has limited stability with bronze powders, then it is possible to premix them just prior to printing, disposing of the overs at the end of the run.

When polyamide resins are used as binders, care should be taken to ensure that sufficient antioxidant is incorporated into the formulation to prevent reaction between the resin and the bronze powder. Incorrectly balanced formulae can lead to objectionable odour rendering the printed material worthless.

Aluminium-based inks

These are used in two different ways. On the one hand as a self-colour providing a metallic silver backing for superimposed inks. On the other hand as a means of providing coloured metallic effects when blended with transparent organic pigments.

To achieve the best results for both uses, careful selection of the aluminium base is needed. It is necessary that an aluminium gravure ink film should provide a surface on which superimposed inks can achieve acceptable adhesion.

A very fine particle size aluminium is required to enable the formulator to produce metallic colours with pleasing decorative appeal. These inks are particularly suitable for printing gift wrappers, labels and cosmetic cartons. Poor results are obtainable on absorbent stock, due to the preferential absorbance of the smaller pigment particles leaving the large alumium particles on the surface. Ideally only suitably coated papers and boards should be used.

Transparent yellows and orange/reds are of particular interest, as from these and aluminium a wide range of simulated golds can be produced. Correctly formulated inks of this type can compete for visual effect with bronze-based inks. Because they have a lower specific gravity they also have a greater mileage and when formulated correctly can be used to coat board on which it is possible to print and obtain good adhesion with selected offset inks.

Advantage is taken of simulated golds to produce copper-free inks which may be an essential requirement for printing wrappers for certain types of foods, e.g. prawns.

When bronze-based metallic inks are used on wallcoverings, there can be a reaction between sulphur-based compounds and the copper content in the powder resulting in the formation of black copper sulphide and subsequent print spoilage. Additionally, because of the presence of sulphur oxides in the atmosphere, particularly in heavy industrial areas, bronze based golds used on wallcoverings tend to tarnish quickly. Simulated golds are free from this problem maintaining their brightness over the natural lifespan of the wallcovering.

Typical formula — vinyl wallcovering simulated gold:

Vinyl Yellow chip base (CI Pigment Yellow 83)	10.0
Vinyl Orange (CI Pigment Orange 34)	5.0
Aluminium base	15.0
Vinyl resin	7.0
Acrylic resin	10.0
Wax dispersion	5.0
MEK	20.0
Toluene	23.0
Methyl isobutyl ketone	5.0
	———
	100.0

Pearlescent inks

The use of pearlescent pigments to replace pure mica has been a growing practice in the gravure printing of wallcoverings. Their advantage is that they are more flexible and provide a considerably more cohesive surface than the old dextrin bound micas, thus ensuring adequate adhesion of superimposed colours.

Pearlescent pigments are titanium dioxide coated with mica and are dispersed into the gravure media by stirring. There are different shades available in different particle sizes, the most popular being the silver or pearl white grades. Pleasing effects can be obtained by small additions of tinting colours to produce, pinks, light blues and greens.

Also of interest, although somewhat expensive, are interference pearl pigments which reflect different colours from different light sources. They also exhibit a change of colour when printed over a black or dark background.

Cosmetic and soap packaging designs take advantage of pearlescent inks to enhance customer appeal. As with wallcoverings the pearlescent print often forms a background to the whole design. Overprinted colours are usually transparent thus allowing the pearl effect to be transmitted.

Typical formula — pearlescent ink for wallcovering would be:

Pearlescent pigment	20.0
Vinyl resin	10.0
Acrylic resin	15.0
Wax dispersion	5.0
MEK	20.0
Toluene	25.0
MIBK	5.0
	100.0

Fluorescent inks

Fluorescent gravure inks are not widely used; often their main role is to emphasise some point of sale feature. However, there has been a consistent use of this type of ink on detergent packs. Occasionally they are used to good effect on wrapping paper and they can be effectively utilised on wrappers for childrens sweets.

Being resin based, fluorescent pigments are adversely affected by strong solvents which swell the resin giving inks with poor flow. Aliphatic hydrocarbons are suitable solvents but they limit the choice of resins considerably; cyclised rubber, vinyl toluene alkyds and certain specially formulated acrylics can be used.

Achieving a satisfactory dispersion with fluorescent pigments is not easy and unless the pigment is adequately wetted out, hard settlement can occur. As an alternative to pigments, soluble dye toners offer a means of formulating inks on a more conventional and wider range of resins such as nitrocellulose, polyamide and cellulose esters using alcohols, esters, toluene and MEK solvents. This enables films, foils and papers to be printed with resin systems that have the requisite properties.

Where possible, fluorescent pigments should be used as single colours to ensure the maximum brightness. Additions of small amounts of conventional pigments can be used to adjust shade, but this is inevitably accompanied with a corresponding loss of flourescence.

It should be noted that the lightfastness and fat resistance of these types of pigment is generally poor. However, it is possible to improve the light-fastness by backing the fluorescent ink with an ink based on conventional pigment of similar shade, thereby minimising any colour change. Also a degree of protection can be achieved by overprinting with a suitable lacquer.

7.5   PRINTING INK FAULTS

As with other printing processes, the identification of some faults can be difficult and on occasions time consuming and expensive to resolve. Most faults manifest themselves as some form of irregularity in the print appearance or as a blemish on the unprinted area of the substrate. Some may be due to incorrect adjustment of the machine or ink and disappear when the correct conditions are attained. Others are more persistent and need further careful investigation.

Streaking

Gravure printing is unique in that on occasions it suffers from this particular fault. As the term implies it occurs when the doctor blade fails to wipe the cylinder cleanly allowing a narrow streak of ink to pass under the blade. If the machine is printing at the time, the streak is deposited on the unprinted area of the substrate. The reason for the streak can often be determined by its position and its nature. A heavy streak near the ends of the cylinder is often caused by ink skin which can form in the corners of the unit, eventually becoming dislodged and finding its way under the doctor blade.

If the resulting degree of print spoilage is unacceptable, the machine should be stopped, the skin removed from the duct and the ink carefully filtered. Any draughts impinging onto the ink should be excluded. Should the skinning and subsequent streaking persist, reformulation of the ink may be required to eliminate the problem.

During the printing of paper and board, fibres and particles of coating find their way into the ink. Where hydrocarbon-based inks are concerned the fibres and coating remain finely dispersed in the ink system and rarely give rise to streaking problems.

With alcohol-based inks there is a tendency for the fibres to form nibs which can lodge under the doctor blade resulting in streaking. Fine filtering of the ink can reduce the problem but if streaking persists then either a change of substrate or ink may be the only solution. Streaking may be aggravated by an ink that has a tendency to scum and therefore to remove loosely bound fibres or coating from the substrate.

The most difficult type of streaking to eliminate appears as a very fine streak, often only a few inches in length, originating from the print area. The streak is too short to be caused by a solid particle lodged behind the doctor blade. Normally, filtering of the ink has little lasting affect on the phenomena. The most likely causes are from pigment dispersion, especially hard pigments, e.g. bronze blue, agglomerates of wax particles, small particles of gelled ink or even static. On occasions the addition of a reducing medium with a high pigment extender content can eliminate the problem.

Non-polar hydrocarbon solvent based inks are occasionally affected by static originating from the friction of the substrate as it passes through the machine. Its effect is to produce spiky protusions around the edge of the printed area. Antistatic additives, highly polar in nature, normally overcome the problem.

Scumming

In gravure printing the occurence of scumming is fairly rare. It results when the doctor blade fails to wipe cleanly by leaving a thin film of ink on the non-printing surface of the cylinder, which in turn is deposited on the substrate.

Although appearing perfectly smooth to the naked eye, it is possible for the surface of a polished chrome cylinder to be sufficiently porous as to retain a thin film of ink which cannot be removed by the doctor blade. Often inks containing highly dispersed pigment are prone to scum. Under these conditions, and as with some cases of streaking, additions of extender medium enable the doctor blade to wipe the cylinder more cleanly.

Scumming can also occur when the surface of the cylinder appears to be faultless. This happens when there is an attraction between the surface of the cylinder and certain pigments and a cure can only be achieved by changing the ink. This type of scumming may be partially due to the humidity conditions prevailing at the time.

Lay characteristics

Not all ink problems are related to the doctoring of the cylinder. In packaging printing in particular a solid print area should have an unblemished smooth appearance.

Before any assessment of the lay characteristics of a gravure ink can be made, the ink must be adjusted to the correct viscosity at the machine speed of the print run. Screening, which often shows at the leading edge of the print should be corrected by further dilution of the ink, while haloing which occurs at the trailing edge should be dealt with by increasing the viscosity slightly or on occasions adjusting the doctor blade setting or reducing the pressure on the impression roller.

If during the print run there is a persistent recurrence of what appears to be screening, especially if it is repeated in the same area of the print, then either drying in or packing in has occurred.

Drying in is basically due to the resin system failing to rewet rapidly enough in the solvents, so that the ink retained in the engraving after printing gradually forms an inert layer at the bottom of the cells that is too viscous to combine with the ink in the duct as the cylinder rotates. The drying in will tend to occur in the shallower areas of the engraving and can usually be remedied by additions of a strong solvent or the judicious use of a suitable retarder.

If the adjustments discussed above do not solve the problem then it is likely that the rather more serious fault of packing in has occurred. Two reasons for packing in are possible.

(1) Similarly to drying in, the resin system lacks resolubility in the solvents. It may not be the total resin combination at fault, as it needs only one of the resins to have poor resolubility so that it lodges in the cells preventing ink transfer. Normally the cells require vigorous scrubbing to clean them. Retarding the ink or adding stronger solvents has no affect. Shellac is an example of a resin which behaves in this way.

(2) Equally troublesome, but fortunately not very common, is packing in due to excessive pigment flocculation. When the design has a solid area of print, pigment flocculation would show as unacceptable reticulation, but where there is only fine type, the lay characteristics cannot be assessed. Under these circumstances the flocculated pigment tends to pack into the cells, eventually completely blocking them. Pigments which have large irregular shaped particles are prone to this fault, e.g. phthalocyanine blues. Additions of retarder or any other additive has little effect. The cylinder has to be cleaned with a soft wire brush and the ink reformulated on alternative pigments.

Having avoided the above faults the lay of a gravure ink can still be the subject of criticism. Ideally the ink film should be free of defects, show no reticulation and have an even coverage over the substrate. On occasions this is not the case and if modifications to the ink on the press fail to remedy the situation and if the substrate is not the source of the problem, then reformulation of the ink may be necessay. Sometimes additions of extender medium can improve the lay while the ink is on the press, however, the amount that can be added is limited by its effect on colour strength and gloss.

Improvements to lay can sometimes be achieved by changing or modifying the resin system. This may not always be practicable due to other requirements of the ink. Usually the most likely means of improving the lay is by using alternative pigments, or improving the method of dispersion.

Moiré pattern

In process printing a phenomenon can occur where a regular pattern appears across an area of tonal printing distorting the visual appearance of the reproduction. This is known as the moiré pattern and is due to the screen size and angle being identical on each of the engravings used to produce the print. Normally at least one of the cylinders has to be remade with a different screen configuration to overcome the problem.

Rewetting

When alcohol/ester-based inks with the same resin system, particularly nitrocellulose, are overprinted on each other on non-porous substrates there is a tendency for rewetting and hence screening to occur. Occasionally this can be overcome by thinning the base colour to a lower viscosity thereby reducing its solids. The screening results from the ink in the cells of the second colour rewetting and combining with the dried ink of the first colour. Thus at the critical moment of contact the ink in the cells increases in viscosity sufficiently to prevent flow out.

The problem is associated with the ease of rewettability and the rate of viscosity increase of the resin system. Often the problem can only be solved by using inks on different solvent/resin systems to avoid rewetting. Acrylic or polyamide resin based inks can be used as alternative systems for the overprinting of nitrocellulose inks.

The problems discussed so far have dealt with faults which are evident at the time of printing. Unfortunately there are others which only manifest themselves at some stage after printing.

Blocking and set-off

When printing on non-absorbent, flexible substrates the formulator is faced with the problem of producing an ink which has good adhesion and flexibility, yet is sufficiently hard drying to prevent blocking in the reel.

Lightweight unsupported foil is a difficult substrate, especially if there is overprinting which leads to the build up of excessive pressure in specific areas in the rewind reel. Achieving adhesion on foil normally requires the correct balance of plasticiser and resin. Too little plasticiser and the ink will flake, too much and there is a danger of set-off or blocking. Particular care has to be taken to avoid set off when the printed material is destined to directly wrap confectionery or foodstuffs. Some converters use cotton wool soaked with acetone to check if set-off has occurred on the reverse side of the substrate.

Likewise there is a possibility of set-off or blocking with flexible films which have solvent sensitive coatings, as these can retain solvent absorbed from the ink.

Immediately after printing, coated films are usually heated to a sufficiently high temperature to ensure adequate keying of the ink, and to avoid blocking it is important that the chill roller effectively cools the print prior to rereeling. Both blocking and set-off are serious faults, as they normally render the printed material unusable, leaving both the ink-maker and printer facing a difficult situation.

7.6   FUTURE DEVELOPMENTS

Future trends in the ink industry are often initiated by restrictions imposed by legislation or prevailing economic and ecological factors. For example, the USA governmental legislation to reduce solvent emission created the need for a water-based ink development programme which has stimulated interest not only in the USA but in virtually every country worldwide. In the seventies the effect of local ecological pressure was seen in the UK causing the withdrawal of lead and cadmium from printing inks, while in many other countries such pigments are still in common use.

New developments are also created as a result of novel approaches, either from the ink formulator himself or by the converter or end user of the final print. These ideas can lead to totally new technology or more often new applications of existing concepts. In the following sections various probable approaches for the future will be discussed, some of which may come to fruition while others may not prove to be technically feasible.

Economic needs

In discussing economic factors the aim is to reduce the cost of the end product and many different courses of action are available which inevitably affect the ink manufacturers recommendations.

(1) ink raw material costs;
(2) ink manufacturing costs;
(3) higher printing speeds;
(4) higher packaging speeds;
(5) redesign to use lower cost substrates.

Raw material cost reduction is an on-going situation affecting all ink makers and does not warrant further discussion in this section. However, reduction of manufacturing costs can be achieved not only by increasing throughput rates but in the case of inks by ensuring that the maximum strength is obtained in the dispersion process, since the pigment is usually the most expensive ingredient within the formulation.

Higher printing speeds often mean formulation changes to reduce tack levels, maximisation of solvent release characteristics and lower effective web temperatures for adhesion (due to shorter time dwell in the drying oven). The significant increases in press speeds which can be achieved with modern-day presses can even require a complete change of the basic ink vehicle.

The constant demand for higher packaging speeds has seen the evaluation of first cool seal wrappers and then cold seal packages. The change from a heat seal wrap to cold seal requires complete redesign of the ink system and future trends indicate more wrappers will move towards this technique. In addition, for smooth transit on packaging lines, slip levels of inks and varnishes are critical and the inkmaker will need to exercise even more control in future years in measuring the coefficient of friction values.

A reduction in cost of the end product can often by achieved by advancements made in ink formulation techniques. For example the use of an absorbent board instead of a more expensive coated board can result in savings even though a sealing coat provided by the ink manufacturer is necessary in the first unit of the press. Similarly savings could be made if the properties associated with a coated propylene film could be achieved with an uncoated film. Such challenges await the technology advance from the ink-maker and until these developments reach fruition the end user must wait patiently for his resultant cost savings.

Ecological considerations

In earlier sections of this chapter the trend away from hydrocarbon solvents and the elimination of certain glycol ethers has been discussed. The USA has already seen the introduction of legislation on solvent emission and similar controls could be introduced in European countries in the future. Ink manufacturers must, therefore, continue to strive for solvent-less gravure systems and in particular the areas of water-based and UV curing systems will come under the microscope.

Water-based inks and varnishes

Water-based varnishes as discussed earlier are already in commercial use for relatively simple specification end uses. However, water-based gravure inks are only rarely used and in general fail to meet the technical standard of their solvent counterparts. For absorbent stock the drying speed of water-based inks is just about acceptable with the current state of the art, but on non-absorbent substrates there is a severe curtailment of press speeds. In order to solve this problem advancements are required in stronger inks which will print from shallower etch depth cylinders, together with a greater knowledge of the optimum drying mechanisms.

The technology of water-based inks has remained unaltered in its general principle of using an acidic resin which is solubilised in water by formation of an alkaline salt with a relatively volatile amine. On drying, the amine is liberated to re-form the water insoluble acidic resin. A comparison of water-based and solvent inks in relation to the gravure process can now be made.

In a normal solvent ink the solvents are balanced so as to be sufficiently non-volatile for minimal evaporation to occur from the time at which the doctor blade removes the excess ink from the cylinder to the point of impression. After impression, transfer of ink to the substrate and time to achieve uniform lay, the solvent must evaporate prior to rewind or stacking. By comparison, in a water-based ink the solvent is replaced not just by the water, (which acts as the solvent) but also with the volatile amine required to neutralise the acidic resin. Water, while being relatively non-volatile and not causing problems due to evaporation from the cell can create difficulties at the final drying stage dependent on the porosity of the substrate. Furthermore, the surface tension characteristics of water are totally different to those of conventional solvents and uniform flow out of the ink from the cells to the substrate is more difficult to achieve. In a water-based ink the amine rather than the solvent influences press behaviour. Premature release of the amine before the point of impression will give precipitation of the resin in the cell and since the ink is only weakly alkaline it is unlikely that this will be resolubilised within the next revolution of the cylinder in the ink duct. Hence, as with a too volatile solvent system, an incorrect choice of a too volatile amine can lead to packing in the etch and a screening phenomena on the print. Alternatively a less volatile amine can lead to the final print not achieving water resistance because all the amine has not been released.

In order to provide water inks which will dry more quickly, resins have been developed which contain a dispersed resin phase within a solubilised polymer. This reduces the amount of amine to be released at the point of drying and increases the solids content of the ink and therefore reduces the quantity of water for evaporation. The disadvantage of such polymers, however, is that the risk of drying in the cell is increased and resolubility of the dispersed phase cannot be achieved.

The introduction of acidic groupings on to a resin limit the choice of resin systems available for waterbased inks, and the versatility to the ink-maker in coping with the adhesion requirements of the many different substrate types is equally limited. In turn, the pigment industry has yet to provide pigment coatings which enable the ink-maker to produce concentrated high gloss, good flowing, dispersions necessary for quality gravure work.

These points have been introduced to the reader to show some of the difficulties facing the inkmaker and the challenges which will have to be overcome before waterbased gravure ink systems rival their solvent counterparts on an equal technical basis.

Ultra violet curing of inks and varnishes

Ultra violet curing inks are now commonplace in letterpress and lithographic printing but have yet to make a significant impact in gravure technology. The difficulties facing the ink-maker will be examined in order to show the problems to be overcome before UV curing gravure systems are a practicable proposition.

Lithographic systems for UV curing inks contain the following ingredients:
— pigment
— oligomer
— diluent
— photoinitiator
— additives (e.g. waxes, inhibitors, etc.).

The oligomer replaces the resin in a conventional lithographic ink and the diluent replaces the solvent. However, in UV curing inks the diluent is also photoreactive and is crosslinked on irradiation with the oligomer and so the system is effectively a 100% solids ink. The photoinitiator is present to generate the free radicals necessary to start the polymerisation process.

This technology is equally applicable to lithographic varnishes by omitting the pigment component and it has also been further extended to fairly low viscosity coating varnishes by inclusion of a low molecular weight diluent at the expense of some or all of the oligomer. Cure speeds are, however, slower since a greater degree of crosslinking is required to build a final polymer of sufficient molecular weight to achieve the necessary hardness and rub characteristics. The higher toxicity and skin irritation figures of the low viscosity photoreactive diluents has limited progress to such an extent that less than a handful of the currently known diluents are classed as suitable for use by the Society of British Printing Ink Manufacturers (SBPIM).

In considering gravure application of UV curing systems an immediate advantage is that no evaporation will take place in the gravure cell since no volatile component is present. Hence, previous problems due to packing in the etch and resultant screening will be totally eliminated. However, even at low viscosities flow out remains a severe problem and while silicones can be successfully used in coating varnishes to reduce surface tension levels other problems of crawling and crazing are then often apparent if these additives are used in gravure.

The low viscosities required for gravure application can be achieved with some difficulty for lacquers, but for pigmented systems even greater problems arise in meeting the required goal.

New technology must therefore be developed before UV gravure ink systems prove a viable proposition for the conversion industry and even then further challenges still have to be overcome. In lithographic printing for instance trapping of one colour over another in the wet state is achieved by tack grading of the inks (see Chapter 6), whereas in the gravure process this is not feasible. The reasons for this are quite simply that one wet ink would pluck off the previous ink and that a web-fed gravure press requires the ink to pass face down over a turner bar prior to passing through the next unit. Inter-unit drying would therefore be necessary to at least tack dry each colour prior to a final curing process.

Ink film weight applied by the gravure process is higher than that achieved by lithography and in order to achieve full through cure at the commercial press speeds associated with gravure printing new polymers diluents and photoinitiators may have to be designed. In addition the problems of adhesion to the base substrate have also to be solved as well as satisfying the demands of the end user.

It is therefore unlikely that UV curing inks will be commonplace in the near future but the use of UV curing gravure varnishes for overprint applications may be an alternative proposition due to the high gloss levels which can be achieved; also the prevention of re-wetting of underlying inks by using UV water-based varnishes.

New concepts

Of all the areas of development new concepts are one of the most difficult to predict being a product of the ink formulators' own ingenuity. Solvent systems will still occupy a considerable amount of development activity with one goal being to create novel effects to stimulate market appeal. Typical examples are the production of new three-dimensional effects on wallpapers by use of expandable microspheres or real wood effects on melamine laminates by introduction of a three-dimensional wood grain.

On the food packaging side, further efforts will be made to reduce retained solvent levels in the wrapper to reduce the risk of taint. The introduction of a date stamp on all perishable goods will also prompt ink-makers to search for new ways in which the coding can be printed and then only activated at the time of packaging. This will replace the use of crude printing systems for date coding at the packaging stage.

In summary, changes in the gravure process are unlikely to be of a revolutionary nature but new ink concepts will be steadily introduced to satisfy the requirements of the changing world.

CHAPTER 8
Flexographic Inks

The process of printing now known as flexography originated during the latter part of the nineteenth century. This process which was first called aniline printing used simple inks that were water or alcohol solutions of coal tar dyes derived from aniline oil, the substrate being paper generally for bag making.

Commercial printing started in the early part of this century and the method was primarily used on the paper bag making machine as a tail end printer, the print being hand synchronized with the bag-producing operation. Around the time of World War I press manufacturers began to make equipment to satisfy the needs of printers and paper converters. Since the equipment was somewhat basic at that time a number of users manufactured their own aniline printers to link up with bag machines.

The name flexography and its adjective flexographic were adopted much later in 1952 after the process had made considerable progress with improved print quality, press design, inks and a much wider range of work and substrates. The original name aniline together with aniline inks had chemical and toxicological association with coal tar products which had been superseded. The process required a changed image not tied to its old bag-making origins. The original definition of the term flexographic became 'a method of rotary letterpress printing which employs rubber plates and rapid drying inks'. Since 1952 flexography has made further rapid strides and is used to print an increasing share of packaging and flexible packaging material to which it is ideally suited.

The range of materials for different end applications printed by the process is extensive. The principal materials are for packaging. In addition to the original paper bags the products include polyethylene bags and sacks, reels of paper, board, foil and films of all types. The reels can be converted either in-line as bags or boxes or subsequently used as slit or trimmed reels on various packaging machines to wrap or shape around food and other packaged items.

As well as packaging, the process is widely used to print stationery materials, tickets, coupons, forms, paper-backed books, comic papers and more recently newspapers. Other applications are small labels on specially designed machines for adhesive, gummed and heat sealable labels. Disposable products are printed and examples of these are paper cups, kitchen towels, cloths and toilet tissue. Wallpaper and plastic wall coverings, gift wrap substrates, decorative materials, strappings and adhesive tapes are also printed together with a number of other products.

Process fundamentals

The flexographic printing process is described in detail in Chapter 2 of this manual. Basically similar to rotary letterpress printing it uses a raised printing surface made of a flexible material to transfer an ink image to the substrate. Being flexible this surface is able to transfer a good image even to rough substrates. The ink is contained in a duct and metered into a thin film often by the use of two rollers (Fig. 8.1). The first roller (A) called the duct or fountain roller can be metal or rubber and is mounted in the duct partly immersed in ink; this roller rotates normally at a slower speed than the second roller (B) called the forme or transfer roller. The second roller can be made of rubber or metal but not of the same material as the duct roller. On modern presses the transfer roller is an engraved anilox metal or ceramic roller and is likely to be fitted with a doctor blade for better ink film thickness control. Originally the pressure between these two rollers was the only means of film thickness metering.

The ink film thus controlled is transferred by contact to the plate or stereo roller (C) which in turn transfers the ink image to the substrate web which is held against an impression cyclinder (D) normally made of rubber (Fig. 8.1).

The improvements made by a number of different methods to control the ink film carried, have over the years transformed the process. The use of engraved rollers, of trailing and reverse blades, the ability to vary the ratio of rotation of the duct and transfer roller, more refined engineering, improved register control for stack presses and the use of central impression machines have all contributed. Added to these are developments in alternative stereo materials which have stimulated counter developments

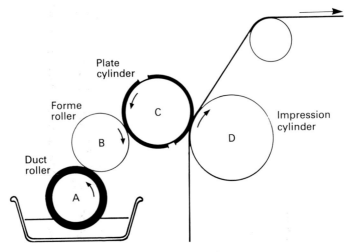

Fig. 8.1 A typical flexographic printing unit

with rubber stereos, inks, drying and web handling, etc. all of which have allowed the flexographic printing method to continue to improve its quality and make the dramatic growth of recent years.

In general the improvements in design, particularly those which have affected ink application and drying, have allowed the ink-maker to widen his choice of materials that will print by the process and thus increase the possibilities of making suitable inks for difficult substrates with increased end-use resistance properties. There are, however, still a large number of flexographic machines of the older type giving good service that do not have many sophisticated features and the ink formulator should remember that some of the latest inks while perfectly satisfactory for a modern press may cause problems on some old machines. In general press design is as important to formulation as considerations of substrate and end use.

8.1   GENERAL CHARACTERISTICS OF THE INKS

There are a number of fundamental properties required of flexographic inks that are determined by the nature of the printing process and the conditions and forces met during the act of printing together with the requirements of the printed article and the use to which it is put. The flexographic process did not originate and was not developed over the course of time in isolation. Considerations of ink performance, properties and limitations have played their part in the concept and improvement of the process alongside other essential associated products used by the printing industry. Flexographic ink technology has also made considerable progress over the years and continues to advance. Ink consists of three main constituents: colourant, binder and solvent, and the variation and choice of these components, their properties and interactions will be covered in detail later in this chapter. An examination of the general characteristics of the ink and how these relate to the requirements of the process and its products will assist in an understanding of the basis and act as a stimulus for the development of ink technology.

Fluidity

A necessary property of a flexographic ink on the press is that it be a free-flowing liquid and that it should remain in this state until deposited on the substrate. The ink as supplied may require reduction with solvent to adjust the viscosity for printing. Most presses are equipped with a reservoir that is sited below the level of the printing unit. The ink is pumped into the duct which is fitted with an overflow to keep the ink at a set level consistent with the position of the inking roller. The surplus ink is thus under constant recirculation by pumping and gravity fall, requiring a fluid state. The action of inking rollers and nips and the doctor blades also requires fluidity and on many presses the agitation can be considerable.

During recirculation of the ink through the unit, contact with the atmosphere in the duct and on the rollers causes some of the solvent to evaporate with a consequential increase of ink viscosity. This will cause an increase in film weight on the stereo as will be seen later, and it becomes necessary to replace the loss by the addition of a suitable solvent. This can be done manually by viscosity checks during the run adding solvent to correct or better still by the use of automatic viscosity control equipment.

Dye-based inks satisfy the fluidity requirement best of all since the solid content of the ink, colourant and binder, tend to be low when compared to other inks. Evaporation of the solvent moreover with dye inks does not cause significant increases in viscosity (which is already close to that of solvent) and the losses are noticed more by an increase in colour strength of the print, due to a higher ratio of dye to solvent. In contrast to organic solvent inks, water-based systems have the advantage of losing little solvent by evaporation and thus require less adjustment.

Viscosity

Ink viscosity control is important in flexography particularly as a means of maintaining the print density during the run. The actual printing viscosity chosen for a particular job will depend upon a number of factors among which are press speed, substrate, type of metering, temperature, solvent mix and print thickness required. In practice satisfactory print can be achieved over a wider range of viscosity than is the case with the gravure process. The image as deposited on the substrate is faithfully reproduced from the stereo after contact and within reasonable limits a satisfactory print will result.

Printers tend to use viscosity adjustment and the addition of solvent as a means of controlling colour strength of the print. As solvent is added to the ink there are two effects: firstly, a reduction of colour concentration due to the increased volume, and secondly, the lower viscosity of the diluted ink will give rise to lower sheer forces applied during metering and a decrease in film weight will result. This second effect is more apparent with simple roller nip metering.

Press speed will also influence the viscosity requirement of the ink. As speed increases sheer forces will also increase and the film weight of ink carried will correspondingly increase. A greater volume of solvent will need to be added to maintain the same print density. This is well known by printers and can cause problems since slow running during proofing at the start of a job will require the use of a thicker ink than is used on the run to give equivalent print density.

The control of colour strength and ink film weight by viscosity is a perfectly satisfactory method but care should be taken to avoid the extreme limits. Over-dilution may cause the print to suffer in appearance and lose abrasion resistance and the viscosity reading is of less significance as this moves closer to that of the solvent. At very high viscosities print definition and dirty working on fine type can occur and viscosity control is difficult due to wide variation as solvent is lost during the run. The use of a medium (non-coloured ink) to weaken excessively strong inks and the addition of a concentrated toner to a weak ink are ways of keeping the viscosity within the closer limits necessary for good quality printing particularly with halftone.

Transfer

The more sophisticated ink metering becomes with reverse angle bladed anilox rollers, etc, the less do press speed and sheer forces affect film weight but with more simple metering the transfer properties of ink will alter the weight carried. Dependent upon the choice of ingredients used in the ink, tack properties and wet adherance to rollers, stereo and substrate will vary extensively with each formulation and these will affect the amount of ink transferred to the substrate. Of the ink components the binder has the most influence on these properties and increases in concentration and molecular weight increase transfer. Solvent and pigment have a lesser influence although high pigment loading while conferring extra colour strength may affect it adversely.

The transfer properties of different formulations and resins can be compared in the laboratory by a side by side application of two inks using a hand anilox proofer, first ensuring that the strength and viscosity are equal. The transfer property of an ink has importance in formulation and the use of the hand anilox applicator for colour matching can be a better method of laboratory assessment than other hand applicators not related to flexographic ink metering.

Colour and strength

The colour and strength of a flexographic print will be determined by the ink film thickness applied to the substrate and the type and concentration of colourant used in the formulation. The wet ink film thickness will, in practice, vary over a wide range, apprxomately 2–15 μm, dependent upon press, metering, substrate and not least the working practice of the printer. On average the film thickness is lower than with gravure and a stronger ink is required to obtain a satisfactory coverage.

Since ink film thickness varies widely it is difficult to be precise when deciding colour concentration during formulation. The normal practice is to supply an ink at higher viscosity and of greater strength than is likely to be required on the press to allow adjustments to be made with solvent and, in some cases, a reducing medium. Experience and knowledge of the equipment and practice used by different printers can also be usefully considered when formulating inks of suitable strength and colour.

Stereo composition

Materials used to manufacture stereos have a considerable influence on the choice of solvent used in the ink. Those in use are natural and synthetic rubbers of different kinds and photopolymer compositions. Plastics have also been used. In general both water and ethanol will be suitable for all materials but modern flexographic requirements with many substrates and different end uses for the print lead to the use of many other solvents to widen the range of resins beyond those that are soluble in water and ethanol alone. A number of these solvents can swell or dissolve certain stereo composition materials. Table 8.1 gives an indication of the suitability or otherwise of different solvents for some of the more common materials in current use. The table should be only used as a general guide since both photopolymer and rubber compounds will vary in their solubility properties. Many inks are based on a blend of different solvents and a particular solvent may be used in a formulation at low concentration with no adverse effects, even though on its own it may be aggressive to the stereo. Stereo manufacturers are usually pleased to provide data on solvent resistance for their compounds.

The restrictions placed upon the selection of certain solvents particularly with the use of the more soluble photopolymer materials can create a limitation on the choice of binder for the ink. This may conflict with requirements for adhesion on difficult substrates or end use resistance. These conflicting demands cannot always be simply resolved by ink formulation (Table 8.1).

Table 8.1 Solvent resistance of stereo materials

Solvent	Natural rubber	Nitrile rubber	Butyl rubber	Photopolymer plate
Alcohols				
Ethyl alcohol	R	R	R	R
Iso-propyl alcohol	R	R	R	R
n-propyl alcohol	R	R	R	R
Esters				
Ethyl acetate	N	N	R	N
Iso-propyl acetate	N	N	R	N
n-propyl acetate	N	N	R	N
Ketones				
Acetone	C	N	R	C
Methyl ethyl ketone	N	N	R	N
Methyl cyclohexanone	N	N	R	N
Glycol ether				
Propylene glycol mono ethyl ether	R	R	R	R
Aromatic hydrocarbons				
Toluene	N	C	N	N
Xylene	N	C	N	N
Aliphatic hydrocarbons	N	R	N	N
Water	R	R	R	R

R = recommended; C = caution; N = not recommended.

Print characteristics

Ink will have a basic colour requirement and in addition may need to be glossy, matt, opaque or transparent dependent upon the design, substrate or job. The appearance of the print with good definition for halftone and type and good lay on solids are important considerations. The improvements in the process have helped ink printability although good print design that takes account of the strengths and weaknesses of the process are equally necessary. There can be conflicts on film weight requirements of solids and halftone and when these are on the same stereo, difficulties can result. Ink characteristics will also affect print quality and desirable features are good flow, wetting and good resolubility properties. Dye-based inks satisfy these needs and strong colours with high resolubility are easy to achieve.

Dye inks have limited fastness and most inks are now based on insoluble colourants and more care during manufacture and formulation is required to make pigmented inks with good printability. This is normally achieved but problems can be met when the formulation limits are stretched, examples being (a) excessive pigment loading reducing flow characteristics and (b) use of low solubility binders with high end-use resistance properties leading to dirty working.

After deposition on the substrate the ink is required to dry very rapidly this being most acute on a central impression multi-colour press where the gap between print units is only a few inches and at high press speeds the time allowed for drying is a fraction of a second. Since flexography is unsuitable for wet on wet printing, drying, by penetration with absorbent substrates or, otherwise, by solvent evaporation has to be achieved instantly between colours. Thus modern drying equipment with high velocity heated air has become normal on most printing machines. Solvents used in the ink need to be very volatile but a balance must be struck between satisfactory drying of the print and premature drying on the stereo which would otherwise affect print quality. Printing water-based inks on absorbent materials will most easily satisfy these requirements.

Overprinting colour on colour causes most drying problems as there is a tendency for solvent in the overprinting colour to resolubilise the first down colour and become trapped into the double layer which is less easily dried. With water-based inks overprint colour drying can also be a problem but in this case the water in the second layer penetrates more slowly through the first ink layer before it is absorbed by the substrate.

The choice of solvent in the ink and the diluent used on the press have a significant bearing on drying and blends of solvent may be used to achieve the desired result. The binder used will also influence the evaporation of solvent as resins have different solvent release properties. Some drying will occur on the stereo and it is important that this redissolves on the next revolution or poor image definition and dirty working could result.

Adhesion and end use

Consideration of the substrate to be printed is of fundamental importance to formulation and this is particularly the case with non-permeable substrates where ink adhesion properties are critical. With paper printing adhesion presents no problems since the ink is able to penetrate the paper surface and thus obtain purchase. In the case of non-absorbent substrates adhesion must be achieved by other means and chemical and physical bonds and wetting between the surface of the substrate and the ink are important.

Of all the ink constituents the binder is the most important for adhesion considerations and suitable resins will be needed to confer adhesion to the particular substrate. Each type of substrate will present different problems and sometimes a combination of resins may be needed to give both adhesion and a balance of other ink properties. In general the more inert the substrate surface the more difficult it will be to obtain adhesion.

The end-use conditions of the print have a particular bearing on formulation when making inks for packaging. There may be requirements during the packing or processing for resistance to heat sealing, non-adherence to reverse printed lacquers or adhesives, or for particular surface slip characteristics. The print may have to resist packaged products or storage conditions involving resistance of the ink against water, oils, fats, soap, detergents, in warm or deep-freeze conditions or other materials associated with the end use.

The choice of colourant will depend on the specification for bleed and fastness and generally dyestuffs will be excluded from work with a high requirement. The binder selected will need to have suitable resistance properties and many packaged products will soften certain resins resulting in a loss of adhesion or transfer of colour. Solvents will not normally affect resistance properties although resistant binders generally need stronger solvent mixtures for their solubility.

8.2 PHYSICAL PROPERTIES OF FLEXOGRAPHIC INKS AND THEIR MEASUREMENT

The measurement of the physical properties of flexographic inks and an understanding of how these properties are affected by the choice of ingredients and their interaction is a large part of ink technology. These measurements are ideally conducted alongside a study of ink performance on the press and in this way a relationship of properties and resultant performance can be built up by the ink formulator and subsequently used to improve inks and solve problems. The measurement of print resistance properties must also be related to the conditions of end use and care should be taken when setting specifications not to build in unnecessary margins of safety as this can lead to excessive restrictions on raw materials and print qualities may suffer as a result.

Viscosity and dilution measurement

Inks are normally supplied at a viscosity which is higher than needed for printing for two main reasons, firstly to allow for adjustment on the press by the printer and secondly for improved shelf stability of the ink. The importance of viscosity during printing has been discussed and the normal method for measurement is the efflux flow cup. This is a convenient method and is satisfactory for low viscosity liquids but can give false results with thixotropic inks. On the press it is helpful to allow the ink to circulate in the pump for a while before taking the final viscosity reading as this will tend to break any ink structure.

For the ink formulator a solvent dilution curve will yield more information than a simple viscosity measurement of the ink after manufacture. This can easily be carried out by taking readings at different dilutions and plotting the result as a graph. The graph shown gives typical results for some different ink types. Generally the steeper the curve the less solvent required to reach print viscosity and the stronger will be the resultant print (Fig. 8.2).

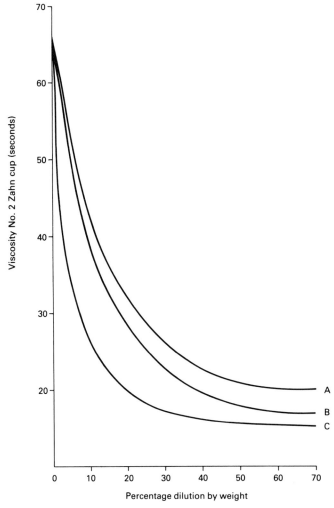

Fig. 8.2 Dilution–viscosity curve: (A) polyamide-based ink; (B) cellulose nitrate based ink; (C) water-based ink

The tack of a wet ink can have a marked effect on printing properties and this will be more pronounced at high press speeds. While for oil inks it is relatively easy to take measurements with a tackmeter this is difficult with thin volatile flexographic inks. The effect of tack can be observed, however, using the hand anilox proofers and will show as an increase in pull through and strength.

Colour and strength testing

The colour and strength of a formulation must first match the agreed standard and this level must be maintained with subsequent batches. The ink film weight carried by the stereo is the governing factor affecting colour strength and some experience or estimation of the likely print weight will be needed when the colour is matched. A laboratory anilox hand applicator is normally used to simulate the printed result but less dilution with solvent will be required by hand proofing than with printing due to the relatively slower speed of hand application.

Once the colour and strength has been established maintenance of later batches to the same colour will be a more simple task. Side by side applications against the standard sample can be made and corrections carried out but viscosity differentials must be minimised or this will affect the result. Strength comparisons can be made by the White Reduction test — by diluting one part of the colour with 20 parts of a compatible white and comparing the result with the same test on the standard sample. In addition to colour strength, opacity must be considered and it should be noted that opaque inks tend to look stronger than those which are transparent.

Batch samples of flexographic inks can and do tend to deteriorate on standing particularly when they have been opened several times. It becomes difficult over a long period to maintain a master standard sample and tests tend to be carried out against the last batch sample. This can lead to colour drift and it will be necessary to retain also proofs to avoid this problem. Colour matching and batch passing can be carried out with the aid of a colour matching computer but it will still be necessary to pay attention to proofing control although this could avoid the problem of standards.

Drying

Water-based inks dry mainly by penetration of water into the substrate and partly by evaporation. The speed of drying is dependent upon the substrate absorbency and the composition, i.e. the resins, bases and emulsions selected for the formulation. The inks dry too rapidly for this to be measured easily in the laboratory although a comparison can be made against an ink of known performance by testing a drawdown made on the paper with one's finger as they dry.

With solvent-based inks there are a number of factors which will influence the drying, the most important of these being the evaporation rate of the solvents used and the solvent release characteristics of the binder. Some resins will release certain types of solvent more rapidly than others dependent upon their chemical make-up. A particular resin may release a slower evaporating solvent in preference to a faster solvent thereby reversing the expected drying rate of each. The hardness of the binder will also affect the point at which the ink feels dry and tack free, even though solvent may still be present in the ink film. High concentration of pigment and extender can improve drying, keeping the ink surface open without skinning to assist evaporation. Various additives and plasticizers can affect solvent release.

Drying properties can be compared in the laboratory by making comparative drawdowns and checking the relative time for the surface to become tack free but as stated solvent may still be contained in the ink. When testing it must always be remembered that on a press heat is normally applied thus reducing the time taken for a print to dry. Under laboratory conditions it is difficult to record results or make comparisons when drying is very rapid as is the case with heat assisted drying.

Testing of the final state of solvent retention after printing is normally carried out by the use of gas-liquid chromatography (GLC) and the results are expressed as the amount of solvent present in each square meter of print. The use of this equipment can be very informative in ink drying research.

Prints which are poorly dried can give rise to serious problems in printing, these being solvent odour contamination in the reel and print blocking due to retained solvent leaving the ink in a semi-solid state and prone to adhere to the unprinted side of the web. On certain coated films a ghost image may form on the unprinted side caused by solvent attack.

Dispersion state

Pigmented inks exist as a dispersion and the state of dispersion is dependent upon the formulation and the process of manufacture. The normal method of measurement is the grinding gauge which consists of a metal wedge graduated from 0 to 25 μm the ink being drawn down the wedge with a blade until pigment particles begin to form scratches, this being taken as the particle size. Care should be taken that the test is conducted prior to the addition of waxes, etc. since these will interfere with the result. Other indications of the dispersion state are gloss, colour development and transparency and these are monitored. Alterations in settings controlling the conditions on milling equipment will give rise to variations in quality of dispersion and careful monitoring will be needed to avoid batch variations. Chip-based inks are more consistent from batch to batch than with other methods of high dispersion. Chipping can, however, induce loss of flow due to the high dispersion state achieved.

While processing plays a major part in producing a dispersion there is a considerable influence and interdependence between the binder, pigment and solvent. All are equally important to the final result particularly with respect to the dispersion stability. Finding the optimum conditions and formulation balance for maximising dispersion can be time consuming with many experiments required. There is little doubt that bulk manufacture of single pigment bases is to be preferred.

Stability of the dispersion is important, both with neat ink as supplied and diluted ink as used on the press. Storage stability can be assessed by an accelerated test at 50 °C for one week and dilution stability by comparing a diluted sample left overnight with a freshly diluted sample.

Adhesion and end use

The adhesion of an ink to the substrate being printed must be of a satisfactory level for the life of the printed article. Normally the level of adhesion will not change after printing but occasionally on some substrates certain chemical type inks systems will need a short time for full adhesion to develop. On plastic films adhesion is often improved if some heat is applied to the film while drying. Also heat may aid wetting or cause fusion of film coatings and ink resins. Once adhesion is achieved it is normally stable and should not deteriorate with time. On the rare occasions when this does happen it is usually associated with a change of state of the substrate or ink components.

The most common test for adhesion is the adhesive tape test and although it has the virtue of being simple to carry out, there are a number of variables. Attempts have been made to standardise the tests and test conditions but it should be remembered that the test is dependent upon the adhesion of the tape to the ink surface as well as adhesion of the ink to the substrate. The result should therefore be viewed alongside other tests, i.e. scuff resistance and scratch resistance, before a final conclusion is reached.

With most presses being reel-to-reel, prints tend to be subjected to very high pressure after rewind. With non absorbent substrates where the ink remains on the surface this pressure can cause blocking and it is necessary to ensure that film inks have good block resistance. There are a number of tests in the laboratory to simulate reel pressures by placing a test print under a hydraulic ram, screw clamp or weight for a period of time and observing if the print will block. Sometimes an elevated temperature is used and this will increase the severity of the test but the temperature chosen must not affect the substrate or its coating or the results could be invalid. In practice 50 °C with most coated films would be a limit.

Much packaging is heat sealed and sealing is often applied in the printed area of the pack. With reverse printing there will be no direct ink-to-heated-jaw contact but, with surface printing, ink is required not to melt or stick to the heat seal jaw. Test heat sealers are available and the formulation must be such as to give the required properties. Surface-printed ink is also required to have good rub resistance as packaged goods will be handled and tend to rub together in transit. There are a number of standard rub testers available.

Flexographic printing is widely used for packaging applications and the properties required of the print to resist packaging, packaged products and storage conditions can be diverse. It is common for the ink to be printed on the outside of the substrate without the protection of an over-lacquer. Many packaged products can have a powerful solvent or chemical action on the print and not all packaging materials are good barriers to migration or seepage of wrapped substances through to the ink layer.

Ink manufacturers will select their raw materials and formulations to give the necessary properties being sought, but it is important that the ink-maker be advised of the intended use of the ink and of any special conditions or specifications. Data will have been previously assembled by ink technicians of the resistance properties of raw materials and particular formulations but it is often necessary to carry out specific tests involving the substrate and the packed product or its active ingredients. On this basis recommendations or advice on ink selection will be made. A whole range of tests are devised for ink and print to simulate the various conditions of use and under ideal circumstances a specification is drawn up between the packer, converter and ink-maker to agree on test methods and standards which are acceptable to all. Within this specification the ink, print and package can be tested at all stages of manufacture and problems avoided. There are cases where the degree of resistance provided is left to the judgement of only one of the parties involved. It should be remembered that the final performance of the pack will be dependent upon all of the component parts and the conditions of printing and conversion.

The measurement of the physical properties of ink are clearly of prime consideration during batch testing of inks with quality control and assurance testing. All inks will be tested for colour, strength, adhesion and dispersion and dependent upon the ink product type specific fastness or end-use tests can be included. The decision regarding the extent of testing will be governed by the importance of the test, the associated risks and the time available during the manufacturing cycle. Every ink cannot be tested for all possible conditions and a sensible choice will need to be made of what constitutes a reasonable cover for good control. Systematic manufacture of a rationalised range of known specification products makes for a satisfactory resolution of this compromise.

Testing forms very much a part of the ink making responsibility and is carried out to ensure that mistakes during making and blending are detected. If the final stage of manufacture is a mixing, blending, operation on the premises of a printer similar testing will be necessary for the same reason. Normally a simple test procedure can be recommended.

8.3 FORMULATING PRINCIPLES

As previously mentioned, flexographic inks consist of three main ingredients, colourant, binder and solvent, each having a different function but dependent on the others. Additives of various types are also used to impart other properties and as modifiers for the basic components to widen the range of properties available.

In considering the selection of raw materials to manufacture different inks the properties of each raw material must be known. Manufacturers will supply data of fastness properties, solubility, compatibility and general chemical constitution and this will assist the selection. In practice the properties of raw materials will be affected by the other constituents in the ink. For example, the fastness of a pigment may be improved by the presence of certain resins used in the formulation. It will be necessary to compile data for different ink systems which will show modified properties from that of the raw materials themselves.

The formulation of the ink will be influenced by a large number of factors associated with the use of the ink and the printed job. These would include: colour standard, finish required, type of press, ink metering, drying equipment, stereo composition, substrate, print design, end use, after processing of print and any specification requirement. In addition there will be considerations of ink manufacturing equipment, batch size and cost.

It is normal to classify inks into different named product types according to formulation, substrate usage and properties rather than all inks being bespoke for each job. Most users will want the inks they use to have the widest possible application covering several substrates and end uses, this to avoid the proliferation of many ink types held in stock. Distinct product types with known and defined performance properties and limitations will make the selection of the correct ink easier without misunderstandings. Products can also be manufactured in a planned manner and with larger batches of fewer inks, better control can be exercised both with raw material stocks and QC testing. The printing and packaging industry is, however, subject to rapid changes and these may require modified ink properties. Therefore flexibility must also be preserved.

In order to consider the principles of flexographic ink formulation an examination of the basic raw material ingredients and their behaviour in the inks will be covered. Details of the properties, origin and chemistry of raw materials will be found in Chapter 4 of this book.

Colourant — dyes

The most widely used dyes in flexographic inks are the basic dyes which are salts of cationic dye bases (commonly hydrochlorides, sulphates, etc.). Without modification they are very soluble in water and have some solubility in alcohol, the alcohol solubility being improved considerably when the dyes are 'laked' with tannic acid or synthetic mordants. Colour strength is improved by laking and dependent upon the choice of mordant solubility in water is greatly decreased. Laking of the dye will improve lightfastness, although this remains relatively poor, and bleed resistance to wax, oil, water and glycerine will also improve. Properties will vary dependent upon the mordant used and the optimum dye/mordant ratio will also vary as will colour development and strength. Price considerations are important but with wide strengths variations the lowest cost mordant will not necessarily be the most economic. Phenolic and other high acid value resins are also used to lake dyes and these can reduce the amount of mordant required together with improving transfer properties of the ink. Other modifications are made for inks used on kitchen paper rolls to improve resistance to domestic chemicals.

Dye inks have near Newtonian flow and give low viscosity bright colours with high strength and, on the press, printing is clean with good drying at high speed. The inks are used for bags, waxed paper bread wrappers and decorative wrapping paper. Semi-pigmented inks are made by mixing dye with white or coloured pigment. The applications of dye inks are for prints where lightfastness is not required and their tendency to migrate and bleed will cause no problems. Generally these inks are not used on plastic films. The basic dyestuffs most commonly used are:

Basic Yellow 37	Victoria Blue
Magenta	Rhodamine G
Malachite Green	Methyl Violet
Rhodamine 6G	

Lightfast dyes which are mainly metal complexes are available that have improved fastness to light, fats, water and other chemicals. Solubility and strength are much lower than with the basic dyestuffs and the manufacturers' literature must be consulted with respect to solubility and compatibility. Inks for foil and labels are made with these dyes and they have been used for plastic films but there is a tendency to migrate under some conditions. It should be remembered that their lower strength can require higher concentration of dye and being soluble the dye will become part of the binder so that the film-forming characteristics must be considered.

Examples of these dyes are:

Solvent Yellow 19	Solvent Orange 45
Solvent Red 102	Solvent Blue 37

Colourant — pigments

Pigments used in flexographic inks have similar specification requirements to those used for other processes. Irrespective of the properties required by the end use of the print, suitable pigments will be chosen for their wettability and dispersion characteristics in the various solvent and resin systems which are used. Generally what is being looked for is ease of dispersion with stability on storage, good flow at high pigment concentration and maximum colour strength at reasonable cost. Properties of gloss, opacity and transparency will be important for certain applications. The flow and dispersion characteristics will be dependent upon pigment type but pigments of similar chemical constitution vary considerably depending upon the processing conditions used by the pigment makers and the various additives and treatments of the pigment during manufacture.

Flexographic inks will vary widely in the solvents used (water to aliphatic hydrocarbon) and the many resin types for the different ink ranges. While it is desirable to use the same range of pigments for all inks the dispersion and rheological behaviour will differ with each solvent/binder system. In order to optimise the properties the pigments selected will tend to vary somewhat for each range. Other inks may contain additives and materials which are unstable in combination with certain pigments and these factors will also have to be considered.

The following is a list of the main pigments which are particularly suitable for flexographic inks:

Pigments	*Colour Index*
Monoazo yellows	PY 3, 5, 98
Diaryl yellows	PY 12, 13, 14
Pyrazolone Orange	PO 13
Permanent Red 2G	PO 5
Lithol Rubine 4B	PR 57
Rubine 2B	PR 48
Lake Red C	PR 53
Lithol Red	PR 49
Permanent Red R	PR 4
Iron oxide	PR 101
Phthalocyanine Green	PG 7
Phthalocyanine Blue (β form)	PB 15.3
Ultramarine	PB 62
Permanent Violet	PV 23
Titanium dioxide	PW 6
Carbon Black (Furnace/Channel)	PB 7
PMTA pink, green blue, violet	PR 81, PG 1, PB 1, PV 3
Copper ferrocyanide dye complexes	PR 169, PG 45, PB 62, PV 27

Extenders	
Calcium carbonate	PW 18
China clay	PW 19
Blanc fixe	PW 21

Lead chromates have not been listed due to their decreasing use on grounds of toxicity. When permitted they are, however, good flexographic pigments giving free-flowing opaque inks. Other pigments to the ones listed would be required for increased specifications for example soap wrappers and in general the choice of pigment will be influenced by the end use of the print. When considering resistance properties many inks will be made using a combination of different pigments in order to achieve the desired shade. It is therefore important to remember that the resistance of the resultant mixture tends to be governed by the least resistant component.

Extenders are used in flexographic inks particularly when tints or weak shades are required but in general with the low film weights carried by the process they are not extensively used.

Bronze and aluminium powders are also used to make metallic shades and sometimes these are blended with colourants to produce particular effects. When aluminium powder is used with coloured pigments or dyes the resultant mixture is described as a polychromatic. It is convenient for the printer that metallic inks be supplied as a ready made mixed dispersion but care will need to be taken with the choice of resin, solvent and additives to avoid instability problems which, at their worst, can involve evolution of gas and heat generation. Where there is any doubt, inks are supplied as a two-part mixture, powder and varnish or paste and varnish, to be mixed together just prior to printing.

Resins

The function of the resin is to act as a carrier for the pigment or dye and bind the colourant to the surface being printed giving adhesion. There are a large number of resins available and the main restriction is that any resin used should be soluble in solvents which do not adversely affect stereo materials. This can be a severe restriction with some photopolymer stereos. The range of resins available has in fact increased with resin development and the variety of resins within each chemical type has also increased in recent years. The ink technician needs to keep up to date with these developments and to be familiar with the properties of the variations available. The general requirements for the resin are therefore solubility in flexographic solvents, adhesion to substrates, pigment wetting, good solvent release, good film-forming properties, low odour, pale colour and compatibility with other resins. All resins vary in their fulfilment of these properties and a resin may have good adhesion properties but be a poor film former or another resin have poor solvent release but good pigment-wetting characteristics. To overcome these difficulties and, because the perfect resin does not exist, blends of resins are commonly used in ink formulation and in this way a balanced range of properties can be incorporated into the ink. The more resins that are used in a formulation will, however, increase the possibilities of incompatibility and often increase the need for powerful solvents to bridge the solubility requirement of all the resins used. Compatibility and solubility tests will be essential during formulation.

Shellac

Early inks were restricted to a very limited range of natural resins and shellac became the principal one used. Shellac has good pigment-wetting properties, solubility and adhesion to a large number of substrates. It is also compatible with dyes and will make good water-based inks when the resin is saponified with a base. Shellac has, however, slow solvent release properties and low heat resistance and is now very expensive. The resin, although still used, has largely been replaced by alcohol soluble maleic and phenolics in pigmented paper and dye-based inks, although on other substrates the film-forming properties of these substitutes are not as good as shellac and other resins are used.

Nitrocellulose

The most common resin employed in solvent-based flexographic inks is nitrocellulose which is cheap and readily available and also has many other desirable features including good solvent release, low odour, good pigment wetting, heat resistance and very wide compatibility. This compatibility allows the possibility of making inks for nearly all substrates by modification with other suitable resins together with plasticizer, and the properties of nitrocellulose inks can thus differ widely dependent upon the particular modification. In these formulations nitrocellulose will normally confer pigment wetting, drying, heat resistance and film-forming properties, the other resins helping gloss and adhesion. Nitrocellulose inks can be further improved in gloss, strength and transparency by the use of chips made from pigment, nitrocellulose and plasticiser. This can be a useful method of giving the extra gloss required on film and foil printing. Various nitration and viscosity grades of nitrocellulose are available and the so-called spirit soluble grade allows inks to be made with a low ester content which is important for photopolymer plates. The main limitations of the resin are its incompatibility with basic dyes, the reaction with bronze powder and the necessity to avoid water contamination either in pigments, solvents or from the atmosphere, since blushing can occur during drying when water is present.

Ethyl cellulose

Ethyl cellulose is also used in solvent-based inks and this resin shares with nitrocellulose good heat resistance, solvent release, and film-forming properties with low odour. Flow and pigment wetting is however inferior and high viscosity is a limitation. Dye compatibility is good and the resin is widely used to reduce tack and improve drying with these inks. Chips and other pigment dispersions are available based on ethyl cellulose and although gloss is good and the dispersions carry a lower risk of fire during manufacture and storage their high viscosity and cost restrict their usefulness.

Cellulose acetate propionate

The third member of the cellulose family used in solvent inks is cellulose acetate propionate (CAP). It is similar to nitrocellulose in many of its properties but compatibility is more limited and more ester is required for solution. The primary use is in inks for PVdC coated films where it is used alone or with acrylic resin to give low odour, heat resistant inks with good adhesion. There are two main versions available, one being more alcohol soluble than the other, and unlike nitrocellulose, the presence of water causes no problems. CAP resin has rather a low tack level and this gives inks which print best at the upper speed range. Chips are available and with careful formulations free-flowing inks with good gloss can be made.

Polyamide resins. These are widely used in flexographic inks and they have good adhesion to many substrates. Pigment wetting is good and glossy inks can be made without chips by judicial selection of pigments and process. The resin has a slight odour and low heat resistance.

Combinations with nitrocellulose are common. Many grades are available and these vary in solubility and compatibility and in general the more alcohol tolerant the resin is, the softer it becomes. Early polyamide inks required the use of aromatic solvents and the resins were described as co-solvent types. Polyamide inks can now be made with ethyl alcohol as the main solvent and these are suitable for photopolymer plates. They dry to give a slippery waxy surface. The resin is not suitable for laminating inks.

Acrylic and methacrylic resins in recent years have found increasing use in flexographic inks over a wider area of formulations than before. The range of resins in the acrylic family is large and the diversity of properties in both solvent systems and water-based inks becomes ever wider as new developments become available. In solvent inks the resins are used primarily for their adhesion properties although the solubility of some of the resins can be difficult requiring a careful balance of solvent in the ink. Solvent release properties will also vary with resin type and this must also be allowed for when formulating. Mainly acrylic resins are used in combination with cellulose esters or nitrate in inks for PVdC lacquer applications or heat resistant surface printing. The resins with their good adhesion properties are used for lamination inks and as binders for metallic pigmented systems of different types. Other uses include PVC film and inks for adhesive tape printing. With solvent systems care must be exercised with the higher molecular weight acrylics as these tend to string, causing dirty working.

In water-based inks acrylics were first used as a substitute for shellac, which had become too expensive, and are used either as a sodium or amine salt in solution or as an emulsion in the majority of inks for sacks and corrugated cases. Much development in water-based film inks involves acrylic resins and they are also finding increasing use for water-based overprint varnishes and coatings which are begining to rival solvent varnishes in some fields.

The use of acrylic resin has become extensive in solvent based systems. The technology is somewhat complex in aqueous systems. The technician will need to catalogue properties, performance and modifications of the many types in order to obtain the best use of these important resins.

Ketone resins are used in flexographic inks to assist gloss or aid adhesion as they have good solubility and are inert. Low odour versions are available although the film-forming properties of ketone resin is poor.

Another resin used is *polyvinyl butyral* which is a good film former and its high adhesion properties and good compatibility make it suitable for many inks including laminates, although heat resistance is poor. Other resins are *polyvinyl acetate, polyvinyl chloride* and the co-polymers of these which find their main use in PVC film inks and other speciality applications. There are also a large number of additional resins used as minor components of inks to impart special properties.

Solvent

The two other main constituents of the ink, colourant and binder are both solids and therefore the prime function of the solvent is to convert the ink into the fluid form which makes it capable of being printed. The solvent is thus a carrier for the other ingredients but must be easily removed from the print by evaporation and by penetration into the substrate.

The choice of solvent is governed by a number of factors. The resin used in the ink will have specific solubility characteristics and as is often the case mixed resins with different solubility parameters will demand particular solvents or solvent mixtures. With flexography there will always be a restriction of choice of solvent determined by the nature of the stereo materials used. Some printers may specify the use of certain solvents which may of course restrict resin choice and also with inks for food packaging certain solvents may be proscribed as their presence as retained solvents in print after drying, may represent an unacceptable risk of odour or taint even in small quantities.

Solvent choice can also be influenced by colourant and with dye inks this is obvious since the solvent must dissolve the dye. With pigmented inks the solvent can affect flow characteristics and a particular pigment may give good free flow in one solvent and poor in another. A good example of this is the behaviour of diaryl yellows which are satisfactory in alcohol but give poor flow in ester solvents. There is also evidence that dispersion, stability and colour strength development are influenced by flow and the effect of pigment/solvent affinity. Solvents also affect adhesion to substrates and with coated films or soluble substrates a partial solubility of the surface by the solvent will assist ink adhesion. With non-soluble substrates and uncoated films, solvents can also affect adhesion by increasing or decreasing wetting of the surface and adhesion is normally at its best when wetting is of a high order.

Most inks contain mixtures of solvents and care will be taken to produce a satisfactory balance for solubility of resins and for the other considerations. During printing more solvent will be added by the printer and the volume used can equal the original solvent content. Therefore the total composition must be considered when formulating and re-commending diluents. During printing, solvent will be lost by evapora-tion from the duct and it is well know that solvent vapours will consist of classical mixtures determined by vapour pressure, effect of solvents on each other, resin composition of the ink and atmospheric conditions. Solvent added during the run should therefore ideally be of the same constitution as the evaporating mixture or imbalance can occur and the result could mean instability of the ink with possibilities of precipitation of resin, mattness of the print, increase in viscosity, loss of flow or adhe-sion failure. In a press unit with a good coverage of print, the rate of ink usage is such that constant replenishment with fresh ink is necessary and there is little chance of the problems described. With little ink usage, however, on a unit with a small area of type or tone where the same ink constantly recirculates through the unit for many hours without fresh ink being added, special attention will need to be paid to the make up solvent to ensure a balance is preserved. The ink formulator should try as far as possible to use a balance of solvent with suitable evaporation rates to avoid evaporation imbalance particularly with inks for halftone print-ing, a classical case of low usage application. The best solution in these extreme situations is to simplify formulation drastically and use only one solvent provided that a satisfactory ink for the substrate and end use can be made with these restrictions.

The principal solvents used in flexographic inks are: water, ethyl alcohol (methylated spirit) isopropyl and normal propyl alcohols, ethyl, isopropyl and normal propyl acetates, propylene glycol mono ethyl ether and propylene glycol mono methyl ether. In certain cases aliphatic and aromatic hydrocarbon (SBP, toluene) or ketone solvents (acetone, MEK, MIBK) may be used for some types of ink but the suitability of the stereo composition must be considered.

Additives

These materials are incorporated into inks to impart particular properties to improve the inks or overcome problems and meet specifications. The use of additives is increasing and a wide range of different types have become available and there is no doubt that the increase of available properties which additives give has to some extent compensated for the various restrictions of choice of raw materials in recent years.

During manufacture of ink the use of wetting agents will assist the dispersion of pigments and can improve gloss, colour development and stability but wetting agents tend to be selective with pigments and ink types. Flocculation problems can be reduced using surfactants. Also flotation, which is the tendency of different pigments used in the same formulation to separate causing streaks and marks on the print, can be overcome by careful selection and use of the optimum level of surfactant. Many pigments will contain dispersion aids added by the pigment manufacturers and these can be more effective in one type of ink than another. Normally manufacturers will not publish details of these additives but generally will give comprehensive details of the use of pigments in various systems.

Plasticisers are used to give flexibility to inks when printing on films and foils and many plasticisers will improve adhesion properties and drying. Excess plasticiser must be avoided however as the melting point of thermoplastic resins will be reduced and blocking could be a risk. Excess plasticiser, migrating to the print surface, can also lead to odour and taint problems on the packaged product. Various stabilisers which prevent reactions between ink components and inhibit breakdown of resins and reduce corrosion of metal containers are used. Antioxidants can be added to inks and varnishes based on polyamide, and also incorporated into chips where air contact during storage may be a problem with particular resins or resins in combination with certain pigments.

Waxes are very common additives and will be present in most inks to improve rub, scuff and scratch resistance and to improve slip properties. Polyethylene and polypropylene waxes are the most popular as these do not inhibit drying to any great extent since they are not soluble in solvents. Compounds of wax and varnish can be made by milling, but fine powder forms are also available for direct addition to the ink. One disadvantage of wax addition is reduction of gloss. There are a number of other materials which will also improve scratch and rub resistance and also reduce the coefficient of friction, among these are fatty amides, silicone fluids, talc and stearates. Most of these materials migrate to the surface of the print and this can reduce the tendency of inks to block. It should also be noted that there can be side effects caused by certain materials in particular circumstances, examples being loss of seal properties with film and cold seal adhesives, wetting and adhesion problems on varnishing and laminating.

Water-based inks and varnishes need more additives than solvent systems and antifoams, fungicides, waxes, surfactants, wetting agents, transfer aids, coalescing agents, anti-blocking compounds and slip additives are all used for different properties. Anti-foam is usually required as foaming can be troublesome both during manufacture of the ink and on printing presses particularly those with high agitation. Foaming may be difficult to overcome and there is a temptation to use excessive antifoam agents which can cause craters and uneven print. Some anti-foams contain hydrocarbons, which will attack stereo rubbers and therefore care should be taken in their choice.

Adhesion promoters can affect the shelf life of the ink and the range of different materials which will impart special properties is increasing. All of these materials will need attention by the ink technician to assess and monitor their use and any resulting side effects.

8.4 INKS AND VARNISHES FOR SPECIAL PURPOSES

Flexography originated as a process to print packaging materials in the form of paper bags. While the scope of the process has widened considerably with many substrates and products being printed, packaging has still remained the most significant area of printing for the process. Given the relative ease and low cost of flexographic plate origination, the possibilities of design changes at short notice and improving standards of printing and ease of handling of flexible materials, it is not surprising that flexography attracts an increasing share of packaging work. As previously mentioned there are, however, a large number of non-packaging applications, most of these being for paper products of diverse use. In recent years wall coverings have been extensively decorated by flexography and more recently still in a number of countries newspapers are also being printed by the process using water-based inks.

The formulation of inks and varnishes is normally governed by the needs of the substrates being printed and the end use of the print. Each class of substrate and its end use will be considered in respect of the formulation of inks and it will be seen that at present these are normally based on pigments and organic solvents. Initially and because of their special nature two types of inks will be considered: (a) dye-based inks and (b) water-based inks.

Dye-based inks

The dyes can be dissolved by stirring but solubility and laking are improved if some heat is used. Denatured ethyl alcohol (methylated spirit) is the solvent and dependent upon the mordant which is used this can contain up to 6% water which can improve solubility although an excess of water will precipitate the laked dye. After the inks are made and allowed to stand at room temperature there will be a tendency for a small deposit to settle which should be discarded as this may cause problems on the press when printed.

Additions of resin solutions may be made, e.g. shellac or alcohol soluble maleic, to improve transfer properties and wax or glycerine resistance (glycerine being used to improve the flexibility of paper for sweet wrappers). If further resistances are required phenolic resins and others can be used. Ethyl cellulose in small quantities is sometimes added to reduce tack and also to increase drying speed. Glycol ethers may be incorporated to retard the ink and because they are good dye solvents they improve clean working on the press by keeping the type open during printing.

A typical formulation for a paper ink is:

Basic Yellow 37 dye	8.0
Victoria Blue dye	4.0
Tannic acid (or other mordant)	20.0
Varnish (60% alcohol soluble maleic)	5.0
Glycol ether	4.0
Methylated spirit (64 op)	59.0
	100.0

Most dyes require a mordant level equal to their own weight but rhodamine B and Basic Yellow 37 need twice this quantity to give maximum water and wax resistance. Colour matching with dye inks is similar to other inks but the formulator must remember that high dilution during printing is common and proofing with the anilox hand roller is difficult due to the low viscosities of these inks, therefore a knife-down or squeegee applicator gives better results dependent on the substrate and its absorbency.

Dyes may be blended with white to make a semi-pigmented ink but the use of a pigment will require more binder to stabilise the dispersion and prevent settlement. However, simple bag machines are unable to cope with the resulting increased viscosities and tack. Dyes can also be incorporated into coloured pigmented inks to increase colour strength but the dye may interact with some pigments and binders and care must be taken to ensure compatibility. Pigmented water-based inks can also be strengthened using dye solutions but the alkaline nature of the inks precludes the use of a mordant and the print is prone to bleed under damp conditions. Some of the dyes also have poor stability and compatibility with alkaline resin suspensions.

Basic dyes are not normally used in inks for film printing because of their bleed and migration properties although some early polyethylene inks were made with dye and titanium dioxide pigment. These inks were, however, superseded by pigmented polyamide inks which have better fastness.

Flexographic inks based on lightfast dyes are employed for aluminium foil printing and these will give bright transparent effects. Inks for metallised substrates and coloured lacquers for overprinting silver inks are also made with lightfast dyes. These dyes do not require a laking agent and by consulting the manufacturer's data a selection of suitable dyes can be found that are compatible with several binders. Lightfastness can be good and resistance properties are better than with basic dyes although there is still a tendency for migration. Film inks can be made with these dyes although pigment would be more economic and have better fastness.

Water-based inks

The traditional water-based ink for paper and board consists of a pigmented resin system, the resin being first dispersed into water. While a number of natural resins exist which are soluble in water such as casein and gum arabic they will also remain soluble after printing, which is not acceptable. It is therefore necessary that the resin be capable of being dispersed but then become insoluble once printed. One method of achieving this is to add an alkali to a suitable resin and converting this to a soap or resin/water dispersion. Traditionally shellac was used but has been largely superseded by carboxylated acrylics although other resins are also employed.

Water-based inks have several advantages, the first being that due to their low evaporation rate the inks are stable on the press without drying on the stereo or in the duct giving constant viscosity. Water does not attack stereo materials including photopolymers and the inks present no fire hazards and give economies by the use of water as a reducer and cleaning solvent.

A problem for the printer is the disposal of waste ink and wash-up solution which tend to be generated at a larger volume because water is freely available. Such wastes cannot be discharged into the drains because of pollution and have no recovery value, the alternatives being expensive water treatment plant or the cost of dumping.

The most extensive area for use of water-based inks is for multiwall paper sacks and corrugated case printing where non-flammability and fast drying on the highly absorbent substrates are an advantage. Good water resistance shortly after printing is obtained and also good level of rub resistance essential for the case and box-making machines.

The choice of alkali for resin solution will be made depending on a number of factors, these being cost, equivalent weight, print water fastness, press stability and drying speed. Properties for three alkalis are given (Table 8.2) to illustrate how these may vary.

Resins may be used in solution form. Acrylic resins are often supplied

Table 8.2

	Cost	Equivalent weight	Water fastness	Press stability	Drying speed
Caustic soda	Low	40	Poor	Good	Fast
Triethanolamine	Medium	149	Fair	Good	Slow
Ammonia	Low	17	Good	Poor	Fast

as a resin in water emulsion allowing a solution to be formed instantly by the addition of the selected alkali to the correct pH. Alternatively the emulsion can be pigmented without alkali making an ink which will have faster drying and better resistance properties than the solution type. This will be at the expense of press stability and makes press wash-up difficult particularly when the ink is allowed to dry. Some inks are made as a mixture of the solution and emulsion form to try to arrive at a balance of properties. This may involve mixed acrylics and a limited amount of alkali.

Pigments have to be selected for stability in water based inks and those which react in alkaline systems or show settlement must be rejected. To give improved rub resistance the use of a wax compound will be needed. Too much slip on inks for paper sacks may cause instability in the stack. Water-based inks are prone to foaming problems and anti-foam additives of the silicone type or organic defoamers are used although too much may cause poor lay, wetting problems and 'fish-eye' effects.

The manufacture of water-based inks is relatively easy and three main methods are used:

(1) Grinding of dry pigment into varnish;
(2) Mixing of stabilised pigment water pastes with varnish;
(3) Use of pigment press cakes with power mixing or milling.

A strength advantage is given by the use of pastes but the presence of wetting aids and glycols which are present in the pastes may be troublesome. A small quantity of organic solvent incorporated into the inks sometimes improves tack and lay but solvent content must be strictly limited in case of fire risk. Typical water-based flexographic ink formulations are as follows:

Red water-based ink

Lithol red pigment (CI Pigment Red 49)	18.0
Acrylic/alkali water varnish	60.0
Polyethylene wax compound	4.0
Isopropyl alcohol	4.0
Water	13.9
Silicone anti-foam	0.1
	100.0

Blue water-based ink

50% Phthalocyanine Blue water paste	24.7
Acrylic emulsion	50.0
Water	20.0
Monoethylamine	2.0
Polyethylene wax compound	3.0
Organic defoamer	0.3
	100.0

Extender pigments may be incorporated into the ink to reduce costs but since many of these inks are used on natural kraft or coloured substrates white pigment will be added to improve opacity.

Water-based inks may be used for other paper printing and there are a number of applications where their non-flammability or low odour make them attractive, examples being confectionary wrappers, labels, fat wrappers, wallpaper and newspaper printing. The inks can be of the solution or emulsion type and a wide range of properties can be given. Many of the solvent-based flexographic inks now used for paper could be replaced by water-based inks albeit with some loss of drying speed on overprinting or on coated papers. Gloss is also lower.

Varnishes based on water are becoming more common and with gloss and heat resistant versions becoming available. The main area proving most difficult is cold seal release varnishes. Generally the varnishes are based on acrylic resins with modifications and highly volatile bases such as ammonia are used to give increased drying and lower retained odour in the print.

Much work is being carried out to make water-based inks for films and with certain film substrates good results have been achieved, e.g. polyethylene. Unlike paper inks, film inks based on water have poorer drying and show other disadvantages when compared with solvent systems and tend to be used only when solvent is restricted due to fire hazards or pollution constraints.

Pigmented inks for specific substrates

Paper and board

In formulating pigmented solvent-based inks for paper substrates the main concerns are in producing inks having satisfactory colour and strength with good printing properties together with good drying and rapid solvent dilution characteristics. Adhesion to board and paper surfaces is not normally a problem with the possible exception of glassine and heavily coated papers. Most inks must be dry rub resistant unless an overvarnish, which may be matt or glossy, is used.

Flexographic printed papers are used for labels, packaging and decorative wrappers, wallpaper and other applications already mentioned and in the ink the choice of pigment must satisfy the resistance requirements associated with the end use. General wrapping and bag inks do not demand lightfast pigments except the normal avoidance of those with poor lightfastness in pastel shades since this will reduce lightfastness even further. Pigment used for waxed papers and soap wrappers would need to be insoluble in the respective materials in each case.

The principal binder for these inks is nitrocellulose which has good pigment wetting and can be pigmented either by direct grind methods or, where increased gloss and strength are required, by using chip dispersions. Nitrocellulose resin gives good drying and since pigmented inks are not absorbed as readily into paper, as is the case with dye inks, faster drying is required. Maleic resin is often used with nitrocellulose in the inks and the main solvent is methylated spirits with esters or glycol ethers to aid solubility. Plasticizers and waxes may be incorporated to improve flexibility and rub resistance, and other binders or modifiers may be used for special applications or to give particular properties.

Typical formulations of inks for paper are:

Phthalocyanine blue (CI Pigment Blue 15.3)	14.0
Titanium dioxide (CI Pigment White 6)	6.0
Maleic resin varnish	16.0
Nitrocellulose varnish	38.0
Wax compound	4.0
Plasticiser	4.0
Methylated spirits	11.0
Isopropyl acetate	7.0
	100.0

or a formulation using chips:

N/C 2B red toner chip	24.0
Isopropyl acetate	10.0
Methylated spirits	40.0
Maleic resin varnish	12.0
Polyethylene wax compound	4.0
Nitrocellulose varnish	10.0
	100.0

Heat transfer papers

Flexographic printing for the heat transfer process has decreased in recent years although it is still carried out for particular textile applications mainly for synthetic fabrics. The inks are printed on a carrier web (usually paper) and are formulated with disperse dyestuffs which sublime at high temperatures on to fabric held in contact with the web. Print quality has to be of a high standard and both solvent and water based inks can be used. The print on the carrier web must not rub and the resin used should not interfere with the subliming properties of the dyestuff. The selection of the dyestuff and binder is influenced by the type of fabric to be transfer-printed and the specification of the final print.

Film inks

Inks for films have different requirements from those for paper. There are two main differences. Firstly the substrates being non-permeable will cause the ink to lie on the surface. As adhesion is not assisted by penetration as is the case for paper, it will therefore be dependent on the physical and chemical attractions between the ink and substrate surface. In some cases, fusion of ink and film coating or solvent attack on the base film or coating occurs. Secondly, because the substrate is transparent it is possible to print on the face or reverse side of the film and it is normal to use a white ink underneath the colour in the former case and to back the colour with white in reverse printing, the white being used to opacify the film and increase the contrast of the print. With hydrocarbon-based films such as polyethylene or polypropylene, then only the treated side is printed.

Reverse printing has the advantage that the colour is viewed through the film which increases gloss and brightness as well as giving protection to the print against abrasion and rub. Laminates will normally be printed this way if the lamination is carried out after printing. Films which are reverse printed need to be very transparent and cloudy substrates such as extruded polyethylene are unsuitable. In recent years there has been a move away from reverse printing for non-laminated food packaging towards face printing to avoid any possibility of contamination contact of packaged foodstuffs by the ink or retained solvent. Reverse prints are still being carried out but are mostly back lacquered or with an inner ply to limit the possibilities above, occurring.

In discussing inks for films it will be assumed that these are face printed unless otherwise stated or dealing with laminates. With face prints, adhesion of the white ink, if used, is paramount since this will be in direct contact with the substrate. It may be possible to use a common range of coloured inks on several substrates by taking advantage of the adhesion properties of different whites which are suitable for each substrate, the white being used as a key coat. The ink-maker will advise when such possibilities exist.

Cellulose films

Regenerated cellulose films were the first packaging films used and although still extensively printed they have given way to various plastic films in many areas of use for several reasons: price considerations, changing packaging demands and the more widespread use of deep freezing. Cellulose film is available in three main types: uncoated, nitrocellulose coated and polymer coated. Some of the more common films are listed below:

P	Plain cellulose films, uncoated
PS and QHS	Nitrocellulose coated semi-permeable, sealing
DMS	Nitrocellulose coated one side with moisture proof, sealing coating
MXDT	PVdC barrier coated one side, sealable
MXXT/S	Solvent PVdC barrier coated both sides, sealable
MXXT/A and MXXT/W	Aqueous PVdC barrier coated both sides, sealable
WS	PVC/PVA coated both sides, sealable

Uncoated and single side coated films

Uncoated P film has reasonable fat and grease resistance but poor moisture resistance. It is not used to the same extent as the coated films but can be used for twist wraps, adhesive tapes and some food packaging. The film can give problems to the ink-maker since adhesion can be difficult with higher ink film weights such as overprint areas and this tends to get worse with ageing of the print. Solvent retention can be a problem and certain pigments may bleed into the glycol type plasticisers contained in the film. Blocking is also a problem and tests should be conducted to avoid this when formulating. Polyamide inks can be used for P film, also inks based on nitrocellulose or ethyl cellulose for applications where grease resistance is required. Heat resistance is not normally needed.

Single side coated films DMS and MXDT are primarily used for polyethylene extrusion or adhesive laminates with printing being applied to the uncoated P side of the web. Similar ink problems to those mentioned with P film printing exist but adhesion to the substrate and compatibility with adhesives is of basic importance for a good laminate. Polyamide inks are unsuitable for adhesive lamination due to a tendency of polyamide to interfere with two component adhesive reactants. Polyamide/nitrocellulose ink can however be used for extrusion laminates.

Nitrocellulose coated films
There are a number of cellulose films coated with nitrocellulose most of which are heat sealable. PS and Q type films are permeable to moisture and are used for packaging bakery and meat products. Although the films are sealable, it is unusual for the seal to be in the printed area. Polyamide inks may be used which have the necessary deep-freeze resistance properties. Pigments which are prone to migration should be avoided as these may bleed.

MS film is designed to have heat sealing, moisture and gas barrier properties as well as being impervious to fats and grease. The film is used for packing products which require a moderate degree of protection from moisture. It has also been used for many years, as unprinted film for cigarette overwrapping. The main problem for the ink-maker is that wax contained in the coating causes wetting problems with the ink, which may crawl on the surface before drying. Pinholes which appear on the printed films, vary according to the source of the film. Formulations to overcome this will contain both resins which wet the surface and anti-pinhole compounds to reduce surface tension. Ethyl cellulose helps as it sets the ink before crawling while non alcohol solvents aid wetting. MS films vary in their tendency to give pinholing and over reduction of the ink with solvent will increase the problem.

Adhesion to MS film is less of a problem as it is possible to fuse the film coating and the ink together by heating the print while drying.

Inks which are heat resistant are, however, more difficult to fuse. Solvent retention can be a problem and alcohol solvents can be absorbed more readily than other solvents. Heat resistant inks are often required for MS film if the print comes in the seal area and nitrocellulose is the normal binder used together with a modifying resin and plasticizer. It will be necessary to conduct tests to ensure that the formulation will not block and has satisfactory adhesion and flexibility particularly when inks are superimposed.

A typical formulation for an MS film ink is:

Calcium 2B pigment (CI Pigment Red 48.2)	14.0
Titanium dioxide (CI Pigment White 8)	4.0
Nitrocellulose varnish	40.0

Shellac varnish	9.0
Polyethylene wax compound	3.0
Anti-pinhole varnish	3.0
Plasticiser	3.0
Isopropyl acetate	7.0
Methylated spirits	17.0
	100.0

Alternatively the ink could be formulated from nitrocellulose pigmented chips to give more gloss.

PVdC co-polymer coated film
PVdC coated cellulose film has improved gas and moisture barrier properties compared with MS film. The film is available in two main grades with the polymer coating being applied during manufacture from either a solvent solution (MXXT/S) or from an aqueous dispersion (MXXT/A and /W), the latter grade being less affected by strong solvents contained in the ink and having better barrier properties. MXXT films are widely used on food products which require its high barrier protection e.g. biscuits and some snack foods, although coated polypropylene and also co-extruded polypropylene films have replaced MXXT in many cases. The films are heat sealable and this will affect ink formulations since the seal is often in the printed area. Biscuits may be packed while still hot and the ink must not melt.

Polyamide resins have good adhesion to MXXT but poor heat resistance and although this may be improved by the addition of nitrocellulose it is difficult to achieve sufficient heat resistance for direct ink application in seal area. A number of ink additives can damage PVdC coatings either by affecting heat sealing properties or moisture vapour transmission rates. Tests will need to be conducted if additives of chemical active type are used.

Acrylic resins have traditionally been used to give adhesion to PVdC coated films and when these resins are combined with cellulose esters or nitrate, heat resistant inks can be formulated and since there are a large number of acrylic resins available the selection should be made on a balance of properties. These would include:
  — high alcohol tolerance;
  — low odour and a low solvent retention;
  — good compatibility with cellulose-based resins;
  — good pigment wetting;
  — adhesion to PVdC surfaces
The main problems with the use of these resins in flexographic inks is the relatively high ester solvent requirement for solubility and a tendency to string. Photopolymer plates are also difficult to use with the solvents needed. With surface printing it is possible to formulate a white ink on an acrylic/cellulose basis with good adhesion and use other heat resistance inks, for example nitrocellulose colours for overprinting. This will solve the problem of using photopolymer plates with the colours, together with heat resistance and clean printing of fine type or halftone.

A basic breakdown formulation for heat resistant white ink is:

Titanium dioxide (CI Pigment White 6)	32.0
Cellulose acetate propionate	10.0
Acrylic resin	5.0
Plasticiser	2.0
Ethyl acetate	15.0
Normal propyl acetate	10.0
Methylated spirit	25.0
Polyethylene wax powder	1.0
	100.0

Isopropyl acetate is avoided due to its tendency to be retained by PVdC coatings after drying.

Pinholing of the type associated with MS film is not a feature of MXXT films although occasionally wetting problems can be encountered which may be a function of the coating or its components. MXXT films (as is the case with other coated films) are subject to changes as manufacturers seek to improve their products for storage, sealing or packaging performance and these changes may influence ink behaviour in unpredictable ways. The ink manufacturer will need to keep abreast of such changes by constant liaison with film makers and users to make necessary adjustments or recommendations as required and hopefully avoid problems before they occur.

PVdC overlacquering
Prints on MXXT films in common with prints on coated polypropylene and other PVdC coated substrates may be overlacquered with aqueous PVdC emulsions to improve sealability, barrier properties or give added protection and gloss to the print. The overlacquer can be applied in line or as a separate operation and will be heated to coalesce the emulsified PVdC and to evaporate the water contained. With such applications both white and coloured inks need to have an affinity for PVdC. Adhesion to the film coating and receptivity of the print surface for the lacquer without poor wetting and crawling are necessary attributes. Additionally inter-layer bond strength must be high to resist tape tests and possible breakdown of heat seals.

Under these circumstances the overprint colours as well as the white need to be formulated with these considerations in mind and nitrocellulose overprint colours as previously described for non-lacquered surface printing are unsuitable. Colours similar to the white formulation given will be more suitable and tests should be made to check the performance of any formulation used. When considering the pigment choice for PVdC overlacquered colours, note should be taken of the highly acidic nature of these emulsions which may cause certain pigments to bleed.

Polyolefin films
There is a wide and expanding range of films in this group and together they now represent the largest use of packaging films. All of these materials are important with respect to the flexographic process and for

the purpose of covering the ink requirement they are placed into three main categories, these being: uncoated, co-extruded and coated. The following list gives the main films at the time of writing under these categories:

Uncoated polyolefin

LD polyethylene	Low density polyethylene may contain additions of ethylene vinyl acetate (EVA) or linear low density (LLD) polymers
HD polyethylene	High density polyethylene
OPP polypropylene	Oriented polypropylene
Cast polypropylene	

Co-extruded polypropylene

Co-extruded polypropylene	Three ply, polypropylene inner, modified polypropylene outer
Pearlised polypropylene	Three ply, voided inner, modified polypropylene outer

Coated polypropylene

PVdC coated polypropylene	Aqueous PVdC coated both sides
Acrylic coated polypropylene	Coated both sides
Differential coated polypropylene	Acrylic one side, PVdC other side
Vinyl coated polypropylene	PVC/PVA coated both sides
Coated pearlised	Various acrylic or PVdC coatings

Polyethylene and uncoated polypropylene films

The different versions of polyethylene together with cast and oriented polypropylene are normally treated for flexographic printing as ink adhesion is difficult if not impossible on untreated film by this process. Treatment is carried out by the corona discharge process and when only one side of the film is treated ink adhesion is possible without increasing the risk of blocking on to the reverse untreated side. If the film is treated both sides avoidance of blocking is more difficult particularly if each is of a different level as the dry print will tend to be attracted towards the side with the highest treatment. Treated film will normally show a surface wetting value of between 36 and 40 dynes which will give satisfactory adhesion. If the treatment is excessive however, a lower print–water resistance can result due to improved wetting of the polyolefin surface by water which can penetrate under the print. With polypropylene films treatment is more of a problem than is the case with polyethylene and some of its effectiveness can be lost with age. One disadvantage for the packer is that all treated polyolefin films do not heat seal as easily as untreated films.

The traditional resin used for polyethylene inks is polyamide although shellac was used at one time. Polyamide of the co-solvent type employs solvent mixtures of iso- or normal propyl alcohol with aliphatic hydro-carbons and good printing inks can be made from chips with low blocking tendencies, good water and deep freeze resistance, robust adhesion and high gloss. The main disadvantages are a tendency to gel in the cold and the presence of hydrocarbons which make the inks unsuitable for photopolymer plates and offer a potential risk of solvent absorption by sensitive foods.

There has been a steady move towards inks which are alcohol dilutable and hydrocarbon free and these have become the most common type in the UK. Most of these inks are still based on polyamide but of the alcohol soluble type which is softer than co-solvent polyamide, therefore nitrocellulose is added to increase drying. Inks of this type can be formulated for polyethylene which have all of the desirable properties of the co-solvent systems and also do not gell, are faster drying, suitable for polymer plates and require less solvent for dilution to reach press viscosity, thus giving stronger prints. Choice of resin, pigment and additives are all important in achieving a good result.

Polyamide inks can be modified with rosin-based resins to improve their adhesion and water resistance although compatibility must be considered. Waxes are used to improve scuff resistance and fatty amides may be incorporated to adjust slip characteristics. The film itself may contain several additives and the combination of these together with those in the ink will need careful control to achieve the desired result. Pigments should be chosen to give good flow in the binder and solvents used and for other properties such as water resistance since these are affected by some pigments. Polyethylene is a low barrier material and the use of pigments which may bleed or migrate through the film must be avoided or thoroughly tested. Some pigments may cause oxidation of polyamide and should be used only with a suitable antioxidant or avoided altogether.

Cast polypropylene film which is used for garment packaging because of its high clarity has somewhat improved barrier properties than polyethylene but because of the possibilities of lower treatment levels the alcohol dilutable inks may need some modification to improve their adhesion. In other respects similar inks are used.

A typical formula for a co-solvent ink:

Organic pigment	12.0
Co-solvent polyamide resin	20.0
Maleic or phenolic resin	3.0
Normal propyl alcohol	22.0
isopropyl alcohol	22.0
Aliphatic hydrocarbon	15.0
Polyethylene wax compound	5.0
Fatty acid amine	1.0
	100.0

A typical starting formulation for an alcohol dilutable hydrocarbon free ink:

Organic pigment	12.0
Alcohol soluble polyamide resin	22.0
Nitrocellulose (dry weight)	4.0

Methylated spirits	29.0
n-propyl alcohol	18.0
n-propyl acetate	10.0
Polyethylene wax compound	4.0
Fatty acid amide	1.0
	100.0

Alternatively chips could be used with both formulations although gloss is quite good by direct dispersion methods.

Polyethylene is also used in heavy gauges to make sacks to pack fertilisers, animal feeds and garden and agricultural products. Polyethylene sacks are stored in the open and the inks used must resist both weathering and the chemicals which are packed and have good lightfastness. A testing programme will have to be conducted over many months to ensure that both pigments and binders resist simulated weathering conditions before such inks are used. It should also be remembered that the substrate is equally important and controls must be carried out by all parties involved.

Co-extruded polypropylene (COEX) film
Co-extruded polypropylene films have in recent years become important for packaging of many products including crisps, snack foods, chocolate coated products and confectionery. Moisture barrier properties of these films are similar to coated films but gas barrier is inferior. COEX film consists of a central polypropylene core with the outside layers modified using polyethylene or other polymer groups to have a lower melting point, thus giving good sealing properties without film distortion.

Polyamide inks give good adhesion to the treated surface but, since the film is heat sealable, inks resistant to heat will be required for many applications and due to the nature of the products packed grease resistance is also desirable. Because of the difficulties in satisfying these requirements, early systems consisted of a chemically reactive two-pot white or primer based on polyester–isocyanate chemistry overprinted with conventional heat resistant inks. These gave way some years ago to nitrocellulose ink systems modified with a number of resins and chemical adhesion promoters of different types. These nitrocellulose inks give satisfactory heat and grease resistance for most applications and print well by flexography and are diluted with alcohol/ester solvents but gloss level although good is somewhat lower than their two-pot predecessors.

Co-extruded film may be used in cold seal applications and in this case non heat resistant inks would be suitable; overprint varnishes may also be used and these applications will be discussed later.

Pearlised co-extruded polypropylene is the latest co-extruded substrate and has made fast progress in similar market areas to the plain co-extruded films and also by the employment of special conversion methods used for biscuit packaging. These films have a voided polypropylene core which gives the pearl effect but also some loss of barrier properties and internal film strength. As much of the film is used on cold seal packs, inks with adhesion but low heat resistance, such as polyamide systems are satisfactory. The main concerns are suitable appearance for the particular job together with low odour, good drying and non-adherence of the ink and over-varnish to the cold seal adhesive. When heat resistance is required, similar inks to those described for plain co-extruded films are suitable.

Pearlised films have found an increasing use for biscuit packaging because of their attractive appearance, but due to the low barrier properties of the film a PVdC lacquer is often applied over the print and to help heat sealability; the lacquer may also be applied to the reverse unprinted side of the web. An ink will therefore be needed to both adhere to the uncoated treated surface of the film and be receptive to the aqueous PVdC over lacquer. Conventional flexographic inks are not suitable and will only perform one of the functions required. One possibility is to use a suitable print primer on the substrate overprinted with inks of the acrylic/cellulose ester type already described for use with coated films but in view of the many layers of coatings and inks an additional layer of primer is not desirable.

A method in use at present to solve these problems is to use a two-pot white of the epoxy type which will adhere to the base substrate and is also suitable for PVdC overlacquering this gives good inter-layer bond for heat sealing. More recently various one-pot white inks have been developed which will perform the same function, but these can have slightly lower bond to the base substrate when measuring heat seals made from the pack. Both the two-pot and one-pot whites can be overprinted with the PVdC lacquerable colours as just mentioned. It should be noted that converter-applied PVdC overlacquering on uncoated polyolefin films involves complex ink technology which is still under development and skills in conversion and printing to handle the various coatings, etc. often involving two printing or coating operations for a particular job.

Coated polypropylene film
PVdC coated polypropylene was the first established and is today the most extensively used coated plastic film. Suitable for biscuit wrappers and a wide range of other packaging the substrate has excellent gas and moisture barrier properties and in common with other polypropylene based films has good low temperature stability.

Heat resistant low odour inks are used to surface print the coated film and when the print is over-coated with aqueous PVdC lacquers the ink must be receptive to this coating application. Inks previously discussed for MXXT films are also suitable for printing this film although two main differences between the film types should be noted. Coated polypropylene has lower heat resistance than MXXT and the inks therefore need to show good adhesion at the lower drying temperatures used during printing. Secondly, there is a tendency for the coated polypropylene film to absorb strong solvent such as hydrocarbons and esters and as well as being retained in the prints these solvents can reduce the bond between the base film and its coating.

Acrylic-coated polypropylene is used for similar packaging applications to PVdC coated films. The film is a high barrier sealable type and ink adhesion is not normally difficult and a wider range of inks can be used that with PVdC coated films. Solvent retention by the acrylic coating and the possibilities of blocking do, however, demand care in the choice of solvents used in the ink as esters and hydrocarbons will solubilise acrylic resins.

Differentially coated polypropylene films with an acrylic coating one side and PVdC the other have become available during recent years mostly in Europe and both plain film and voided core versions are available. These films have good barrier properties and make very good heat seals as both sides are compatible for sealing. The manufacturer recommends that printing is carried out on the acrylic side and similar ink recommendations are made as for acrylic-coated films, also the same precautions need to be taken with respect to solvents.

Other plastic films

Polyvinyl chloride (PVC) films
PVC films vary in their polymer composition and can be unplasticized or contain different quantities of plasticiser. Flexography is not the main printing process although vinyl adhesive tapes, labels and wallcovering materials are all printed by the process together with certain other packaging. Co-polymer resins of polyvinyl chloride and polyvinyl acetate are difficult to use in flexographic inks because of the ketone and ester solubility requirement of this resin type, which in other ways, i.e. adhesion, excess plasticiser absorption, is ideal. Compromise formulations of cellulose with acrylic resins or other binders have to be used and careful formulation with comprehensive testing should be carried out to achieve a balance of adhesion, plasticiser tolerance and non blocking Solvent retention can also be a problem particularly on the unplasticised grades.

Pigments have to be chosen for non-migration into the film and non-bleed in the film plasticisers. Adhesive tape inks should avoid colourants which cause breakdown of latex adhesives. With inks for wallcoverings there are additional requirements of good lightfastness together with high temperature stability for heat embossing work.

Nylon films
Nylon films have high strength but poor moisture and modest gas barrier properties. They are used in laminates for thermoforming and in deep draw vacuum packs and various boil in the bag applications. Good adhesion is obtained with polyamide formulations such as those given for polyethylene but inks for laminates are more difficult. Boil in the bag work is also difficult as the substrate is permeable to water and two pack catalysed ink systems may need to be used. Much film particularly oriented nylon is unprinted and retailers apply their own adhesive labels.

Polyester film
These films are tough heat resistant and have very good barrier properties and are used in laminate complexes, pharmaceutical packaging, boil in the bag and liquid pouches and for carbon copy film.

In its uncoated form polyamide and vinyl inks have good adhesion and more inks are suitable if the film surface is corona discharge treated by the converter. Primed/treated versions of these films are available to which ink adhesion is less difficult and various modified nitrocellulose inks can be used for surface printing or with laminates. Polyester films are also coated with PVdC and this allows the use of similar inks to those described under MXXT film.

Metal and metallised substrates

Aluminium foil
This material gives the most light and barrier protection properties of all the flexible packaging materials and this together with its reflective appearance has led to foil being chosen where these properties are of high priority. The surface of aluminium tends to oxidise in the atmosphere rather quickly and adhesion becomes difficult. Virgin foil in the reel will not oxidise readily due to the exclusion of air and the presence of traces of lubricating oils left from the foil rolling process during manufacture. When foil is laminated to paper or board, however, it becomes necessary to wash coat the foil with a dilute lacquer containing nitrocellulose, shellac or vinyl resin. This wash coat acts as a primer for ink adhesion in addition to preventing oxidation.

In the virgin or washed form adhesion with polyamide, nitrocellulose or specialised formulations is not difficult. Inks are required to be flexible and resistant to the packed products and with confectionery and chocolate wrappers, low odour level is essential. Resistance to blocking must be carefully tested particularly with virgin foil as pressure in the reel can be severe.

Bright transparent effects can be given using dye inks as already mentioned but for non-bleed results and high resistance specifications pigmented inks which are transparent can be formulated using selected pigments and chips.

A typical formulation for a transparent ink for aluminium foil is:

Nitrocellulose pigmented chip (60% pigment)	20.0
Maleic or shellac varnish	10.0
Alcohol soluble polyamide	4.0
Nitrocellulose varnish	10.0
Micronised polypropylene wax	0.5
Plasticiser	3.0
Normal propyl alcohol	20.0
Methylated spirits	22.5
Normal propyl acetate	10.0
	100.0

Metallised substrates
An increasing number of substrates are now being metallised using a thin film of aluminium by the vacuum metallising process. The metal can be deposited directly on to certain base substrates and coatings or alterna-

tively using a primer which is applied before processing. Among the materials thus metallised are paper and board, cellulose films, polyester and coated or uncoated polypropylene. Metallisation will reduce light penetration of packs but does not offer the barrier properties of aluminium foil and the effectiveness of the barrier is more dependent on the base substrate although metallisation does improve barrier properties.

Ink adhesion in general on metallised substrates is more difficult than is the case with the base substrate itself and dependent upon the source this becomes progressively more difficult with time partly due to oxidation of the aluminium. Fresh material should therefore be used where possible. Another variable is batch consistency and preliminary tests for adhesion are advisable before production runs commence.

Metallised paper is relatively less difficult and polyamide inks and other formulations are used. Tape resistance if required may be difficult, however, and tests should be carried out on the substrate. Generally on film substrates printing is carried out on the metallised side of the web and although this is more difficult for adhesion the metallic effect is better. Inks which are non-heat resistant can be formulated most easily but when heat resistance is required complicated specialised formulations are needed. One solution is to use a print primer which will allow over-printed conventional heat resistant inks to adhere but printers will prefer inks that do not need priming.

Inks for lamination

Laminates are formed from two or more substrates which are held together and thus combine the properties from each substrate. Although printing may be carried out on the laminate complex this section will cover only those laminates where the ink is sandwiched between two layers. The possibilities of film combinations which can be laminated together is extensive although the majority of laminates employ polyethylene as the inner layer for heat sealability. There are several methods of lamination and three main types will be considered:

(1) adhesive lamination;
(2) extrusion lamination;
(3) extrusion coating.

Adhesive lamination is carried out using either solvent-based or solventless adhesive and these may be of the one-pot non-reactive or two-pot catalysed type, the adhesive being normally but not always applied to the printed web. Solvent from the adhesive is first evaporated and the two webs brought together between nip rollers under pressure and sometimes these rollers are also heated this being dependent upon the adhesive used. With reactive adhesives a few days may be required before the bond between the laminated layers is developed.

The ink-maker's main concern is the nature of the substrate, the properties of the adhesive and the lamination conditions, the unprinted web being less important to the inkmaker. Substrates which can be printed and adhesive laminated include DMS, MXDT, MS, MXXT, polyester, nylon and also uncoated and coated polypropylene of the different types. The requirements for ink are good adhesion to the base substrate and compatibility with the adhesive. It is difficult to be too specific regarding ink formulation because of the many substrates involved and adhesives used, but in general resins that give poor bonds are polyamide, ethyl cellulose and PVA while those which find use in formulations are acrylics, PVB, vinyl and various combinations and modifications of nitrocellulose and cellulose esters. The choice of plasticiser is important and often chemical adhesion promoters are used. Wax compounds and materials which give slip or abrasion resistance to surface-printed inks should be used after thorough testing. It is frequently also found that matt inks are more receptive to adhesives than gloss inks. Pigments are not normally a problem although PMTA types may bleed in adhesive solvents.

With the many substrates and adhesives used there is an obvious advantage if one ink type is suitable for several different laminates. It becomes necessary for the ink-maker to be very specific in the recommendations made regarding which substrates and which adhesives have been tested and found to be satisfactory. Lamination bond failure is only discovered after printing is completed and the laminate is formed. Checks which ensure satisfactory final results are not easily made during the production stages.

Extrusion laminates are formed by sticking two webs together by extruding a thin film of hot polyethylene on to one web and nipping the films together while the polyethylene is still molten. The web may be of the same composition or different dependent upon requirement. Extrusion coating is made in a similar manner but in this case the molten polyethylene is extruded on to the printed side of a single web which is then passed over a chill roller.

Ink formulations need to have a good affinity for polyethylene; polyamide and shellac inks in combination with nitrocellulose give good results. Bond strength is generally not as high as with adhesive lamination and the inks need good solvent release properties since retained solvent affects the strength of the bond.

Miscellaneous products

Metallic inks
These inks are made from aluminium or bronze powders mixed with a suitable resin solution. The particle size of the metallic powders used in flexographic inks are larger than pigment particles and normally are leafing grades which give maximum reflectance and coverage. When first mixed into a varnish brilliance will be at its best but rub resistance may take a day to develop to its maximum with some loss of brilliance, this being due to 'wetting out' of the metal with resin and solvent. Inks may be supplied ready mixed but the binder choice will be restricted to those that are neutral and give storage stability. With other less stable binders the inks will be supplied as a powder or solvent paste together with a varnish and the mixing is carried out prior to printing.

Aluminium powders are more stable with resins than bronze powders but gold shades can be obtained with aluminium by mixing with a varnish that contains some dye or finely dispersed pigment. These mixtures are, however, less bright than with a bronze powder based ink but have the advantage of being copper free which may be necessary for certain food wrapper specifications. The type of resin binders for metallic inks are dependent on the substrate and end use but good leafing of the metal flake has to be considered. Over concentration of the powder should be avoided as ink film strength can be affected together with the adhesion of any overprinting colours. High powder concentration should also be avoided in lamination inks. Suitable resins are polyamide, ethyl cellulose, PVA, acrylics and cellulose esters. Nitrocellulose can be stable in some aluminium-based inks but bronze powders are reactive and care must be exercised. Stabilisers can be added to help stability in certain cases.

Typical formula for gold and silver shade inks are:

Gold ink

Bronze powder	30.0
PVA varnish (25% solids)	55.0
Wax compound	5.0
Methylated spirits (64 OP)	10.0
	100.0

Silver ink

Aluminium powder	16.0
Polyamide varnish (40% solids)	60.0
Wax compound	4.0
Normal propyl alcohol	15.0
Aliphatic hydrocarbon	5.0
	100.0

Halftone inks

The use of halftone and colour process printing has increased dramatically in recent years and is made possible by better printing equipment and printing plates. Good quality inks will give satisfactory results with half tone but most manufacturers offer specialised inks for two main reasons: (a) to ensure that each colour gives the correct hue and is of the clean shade that is required for process work and can be used as a standard for making plates; (b) to ensure that the ink has good solubility with few solvents (preferably one) to avoid solvent balance difficulties which can arise with low ink usage over long periods, as can be the case with half-tone printing. High resistance specification inks with low solubility resins and complex solvent mixtures are unsuitable and will not give the clean printing that is essential for good reproduction.

Polyethylene coated board

Polyethylene coated board which is printed before extrusion has been discussed under laminates but when printing is carried out on the outside extruded layer the polyethylene is first treated by corona discharge. Both printed forms are used extensively for milk and other liquid packaging to make cartons of different types.

Flexibility of the ink is only required to facilitate folding and the main concern for the ink-maker is abrasion, rub resistance and adhesion together with meeting the specifications regarding resistance to the liquid packed. Suitable inks are made from nitrocellulose which has the necessary toughness and these are modified to improve adhesion and rub properties. Pigment choice is governed by the requirements for non-bleed in milk and other drinks.

Two-component reactive inks

The properties of prints produced from conventional evaporation drying inks are related to the properties of the resins employed. With flexography the choice of resins that may be used is restricted by the limited range of solvent allowed and in recent years these constraints have become more severe. Systems which incorporate a chemical reaction after the solvent has evaporated can give a 'cured' film which shows improvements in heat, solvent and product resistances with more gloss and adhesion than a conventional ink, thus offering the possibility of overcoming difficult print specification problems. The principal types are now usually based on epoxy-amine chemistry and isocyanate cured systems are no longer in favour.

These inks also have disadvantages:

(1) The need to mix two components together before printing;
(2) Limitations of pot life after mixing which can create waste ink;
(3) Formation of an insoluble deposit on the stereo or part of the press if allowed to dry out;
(4) In certain cases odour may be a problem if the print is not properly cured.

Given these difficulties two-component inks are used for certain applications where conventional inks will not satisfy the resistance or adhesion requirements. One way of reducing the problem is to use a two-component white overprinted with conventional colours and under these conditions the resistance properties of the colours are also improved.

Daylight fluorescent inks

Flexographic printing with daylight fluorescent inks has not increased in recent years although some of the problems with poor strength have been resolved with the availability of better pigments and soluble fluorescent toners.

The inks can be used on paper substrates but when film prints are made it is better to print over a white as the inks are transparent and contrast is improved. Single-colourant formulations are best as mixing two colours has the effect of quenching the fluorescence. Both organic solvents and water-based inks can be made using the binders contained in the toners or with simple resin systems. By using the coloured inks to overprint suitable white inks different substrates can be accommodated.

Overprint varnishes

The majority of flexographic printing on paper, foil and film is carried out without over-varnish, since this is an additional cost and requires the availability of a spare unit on the press. Certain work will specify or need a varnish to enhance the print or perform a certain function. Varnishes may also be used to increase resistance performance of the print which cannot be achieved any other way and examples of these are paper cups and plates and also some labels.

The most recent application need for varnishing is the increased use of cold seal adhesives for packaging. These adhesives are based on latex and are applied to the reverse side of the web and function by contact. Since adhesion to the face side of the web is to be avoided a release varnish is applied over the print. Although it is possible to formulate inks which perform a similar function, total ink coverage would be required as any area left uncovered would allow the adhesive to stick to the substrate. Total coverage is the exception and over-varnishing has thus become normal.

Polyamide resins are used in these varnishes and the aim is to achieve good release, lowest odour and good drying. Tests will need to be made over a period to ensure that release properties are maintained. The use of additives for slip and abrasion resistance will need to be thoroughly tested to ensure that they do not interfere with the properties of the adhesive.

Specialised varnishes may be used to give particular properties, e.g. heat-sealing varnishes to be applied over a reverse print to aid sealing and barrier varnishes which will require high resin solids to obtain maximum film weight. Two-pot varnishes give good barrier properties.

Water-based varnishes are increasingly being used to replace organic solvent-based systems and in many cases similar results can be obtained. Some of the problems that are met with in water-based pigmented systems are easier to overcome with varnishes and it is expected that the trend will continue.

8.5 INK-RELATED PRINTING PROBLEMS AND POSSIBLE SOLUTIONS

A number of problems can occur during printing and these may be a consequence of many factors including substrate, press design, ink and operating conditions. Most problems will be overcome by experienced printers making adjustments as necessary although sometimes it may not be possible to do so with the equipment or materials available.

The following problems are discussed which may have an ink origin but it should not be assumed that these are always ink related as usually there are a number of possible causes not all of which are discussed. A change of ink to one with different properties may overcome a particular problem even though the fundamental cause may lie elsewhere.

Colour and print appearance faults

Colour and strength drift may occur during a printing run. Viscosity or machine variations may be the cause but a badly dispersed or formulated ink may be unstable after dilution giving rise to pigment settlement or flocculation. Loss of gloss may be noticed and the condition may be aggravated if the solvent balance becomes upset due to evaporation from the duct leading to resin precipitation. Good processing and care in selection of materials and binders with good mutual compatibility and solvent blends with controlled evaporation to retain balance should avoid such problems.

Mottle and poor lay can be the result of over-reduction of the ink but formulations which have poor flow tend to produce prints with lay faults. Over pigmentation or poor pigment/vehicle wetting may be a cause and this should be corrected during formulation. If the problem is caused by the ink not wetting the substrate the addition of an alternative solvent or a better wetting binder may improve the lay properties.

Inks with poor flow or bad wetting may cause a 'ghost image' which is the appearance of a fainter unwanted image in a printed area and is the result of local ink starvation on the transfer roll from the previous impression.

Printability faults

Dirty working is probably the most common printing fault with flexography. When the problem is due to ink, poor resolubility on the edges of the stereo is the main reason allowing dried ink to build up which enlarges the stereo giving it a ragged thick edge which shows on the print. The use of a slower solvent or retarder will minimise the problem although binder solubility and low speed of solution are fundamental causes. Inks that need high resistance properties may necessitate the use of a binder which has poor resolution in flexographic solvents and this may give rise to dirty working.

Pinholing can occur on coated films, particularly MS film, due to poor wettability of the surface giving rise to ink crawl after impression and before drying. The addition of anti-pinhole compound may overcome the problem by improving wetting and sometimes an alternative mix of diluent that includes a good wetting solvent will also help. Surface properties of substrates can vary and the printer may increase the ink film weight or pre-heat the web which also helps.

Pinholing or bad wetting may also be evident on overprinting colour on colour due to the non-wetting of the dried ink surface by the next colour. Some ink additives and waxes can cause this problem.

Stereo swelling is normally caused by solvent attack and absorption on stereo materials. This problem is fundamental to the choice of both solvent and stereo composition. Changes consistent with the properties of these are required as soon as the problem is detected.

Foaming of water-based inks may be attributable to the press recirculation system causing high turbulence of the ink and thus aerating the liquid at a faster rate than it will breakdown. Some resins and formulations are prone to this problem, although the addition of a small quantity of an anti-foam compound will normally overcome the foam.

Drying problems

Solvent retention of prints is of great concern with food packaging. Given the low ink film weights deposited by flexography and the good drying equipment available, solvent retention is not normally a problem but occasionally on some substrates retention is encountered. A good solution is to increase the strength of the ink, as this will allow a thinner film of ink to be carried which is more easily dried. Choice of binder and solvents used in the ink will also determine the drying characteristics and with film inks the type of substrate will influence the solvent choice and it is best to avoid those which soften films or coatings.

Pick off is a possible fault when overprinting colour on colour and is caused by the first down ink not being sufficiently dry and remaining tacky when overprinted. Increasing the drying speed of the diluent mix on the first colour should help to overcome the problem or alternatively applying a thinner film of a stronger ink.

Screening which is the appearance of a screen pattern on the print can be caused by ink which dries too rapidly giving evaporation during transfer from the anilox to the stereo, showing the engraved pattern on the print. Retardation of the ink is the obvious solution but the ink may be formulated with solvents too fast for the press. The use of dual purpose flexo/gravure inks can give this problem since the choice of solvents is governed by the fast solvents required for gravure.

Problems in the reel

Blocking of print in the reel is mostly a problem when printing on non-absorbent substrates and appears as sticking or transfer of the image to the unprinted side of the web. Solvent retention can be a cause since this will make the ink soft and liable to transfer under pressure. Ink formulations which contain too much plasticiser or low molecular weight binders can also be soft and sticky after drying and these will cause blocking. Blocking tests are a standard procedure during ink formulation and these problems should be avoided at source.

Solvent retention may cause 'ghosting' in reels of printed coated films and this phenomenon appears as a faint cloudy image on the reverse side of the web due to a partial solution of the coating. The presence of traces of slow evaporating powerful solvents in the ink or diluent is the most common cause.

Odour in the print may be caused by the substrate or by retained solvent but certain resins used in ink have some odour which may be worse in the presence of retained solvents. All materials used in ink formulations require careful selection and screening although batches of materials can vary and mistakes can occur.

8.6 RECENT AND FUTURE TRENDS

The flexographic process has seen rapid change during its brief history particularly in the last 25 years and this is likely to continue. Ink technology is influenced and its development is stimulated by these and other changes from associated industries. Raw material developments have continuously progressed ink development and the suppliers now have a better understanding of what is needed, which assists the creation of new products. More easily dispersed pigments and pigment/resin dispersions are making stronger inks possible without the penalty of poor flow. Resin development is rapid with the availability of acrylic resins with increased adhesion, and polyamides with better properties of solubility and compatibility. The increased use of additives in inks is likely to give enhanced performance and special properties to existing formulation in the future. Raw materials are, however, subject to more scrutiny with respect to their toxicological properties and it may be expected that some materials now in common use will be proscribed as information of any harmful effects becomes available.

Flexographic process developments which are likely to become more widely used and will affect inks are the use of reverse angle doctor blades, ceramic anilox rollers and more halftone printing. The improved metering with reverse blades and lower ink weights deposited from ceramics will require stronger inks with better flow in the future and ink-makers will need to respond. Halftone printing is increasing and when used on high resistance specification work the ink will need better press solubility and stability as well as end use resistance. This will tax the ingenuity of the formulator. Water-based ink systems which have desirable slow drying properties on stereos may offer one possible solution or alternatively radiation curing inks.

Changes in substrate with new coatings, mixed coatings and laminated complexes have already become a feature of flexible packaging and this is likely to continue. Ink performance on these materials can alter in unpredictable ways and new problems of adhesion will have to be solved. Bio-degradable plastics may be used which will alter ink requirements depending on their surface characteristics. The economics of petrochemicals may change for the worse which could lead to a revival of cellulose-based substrates both film and paper.

The market for flexographic printed materials represents perhaps the largest stimulus for change. Packaging continues to run at ever increasing speeds and thus higher slip printed substrates are needed. Specification of slip level is becoming more common with narrow band ranges. Inks must conform but the degree of slip achieved is dependent also on drying, printing conditions and the substrate properties. All parties need to co-operate to achieve the results required. End users are also endeavouring to reduce costs and an example of how this trend can affect ink demand is the attempt to reduce the need for PVdC overlacquering when this is used only to give heat sealability. Inks which are both heat resistance and heat sealable under modified sealing conditions are thus being developed to meet these goals although a compromise in specifications may have to be accepted.

Finally environmental considerations are likely to be of heightened importance to flexographic printers with tighter restriction on the discharge of solvents into the atmosphere. Ink development will concentrate on improved and more widely usable water based inks. Single-solvent inks make solvent reclamation possible which is already being done in Europe. High solid inks and radiation cured systems are the subject of intense development at the present time and are likely to be used in specialist areas.

CHAPTER 9
Screen Inks

Screen printing is a stencilling process sometimes described as silk-screen printing or serigraphy. The process has grown considerably in importance during the last few years, particularly in the industrial area. Specific characteristics of the process have made it attractive outside the more established display and poster industries.

Screen printing on materials other than textiles began to attract attention in the 1920s, especially as an economic method of producing short-run posters. The process developed slowly, finding applications mainly in the display and point-of-sale market. The simplicity of the process combined with the thick ink film deposited governed its applications. These characteristics were particularly advantageous in the printing of fluorescent posters, water-slide transfers, display showcards, traffic signs and PVC window stickers.

During the 1950s screen printing was found to be a useful printing process for the decoration of polythene bottles and for the production of printed circuits. The growth of the plastics and electronic industries provided new substrates and techniques ideally suited to screen printing.

New applications continue to be found for this versatile printing process. Examples of this have been the incredible growth of the membrane touch switch industry and the printing of rub-removable metallic inks for lottery tickets.

As a result, screen printing must be placed alongside the other main processes of litho, letterpress, flexo and gravure in its importance.

9.1 IMPORTANT CHARACTERISTICS OF SCREEN INKS

The characteristics of screen inks become clear when one considers the basic characteristics of the process.

The stencil-like image is supported on a fabric mesh stretched across a rectangular frame. The ink is forced through the openings in the stencil on to the substrate by drawing it across the mesh with a rubber bladed squeegee (Fig. 9.1). Whether the printing machine is simple and hand

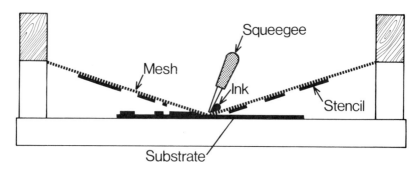

Fig. 9.1 The principle of screen printing

operated (Fig. 9.2) or large and automatically controlled (Fig. 9.3), the principles remain the same.

Firstly, during printing, the ink is held in a printing frame which is open to the atmosphere. The solvents must not be so volatile that there is excessive solvent loss during printing.

The ink solvents must not cause the squeegee rubber to swell or crack, nor remove the stencil film. An ink must have sufficiently low viscosity to pass easily through the mesh, but not so low as to give slurred printing.

It is undesirable that an ink has excessive flow; shortness is an advantage up to a point since it minimises the tendency for the ink to string when the screen is pulled off the printed surface.

Once the ink has been applied to the substrate, it must dry, within a reasonable period; depending on the application of the ink and the drying equipment available.

Screen and stencils

Although screen printing is often described as silk-screen printing, other mesh materials are employed, e.g. nylon, polyester and metal mesh. Monofilament polyester and nylon have many technical advantages and have largely replaced silk. Stainless steel mesh is used in the glass and printed circuit industries. Modern meshes are available in a wide range of mesh counts and qualities depending on the nature of the ink, printing application and end requirements.

There are two main classes of stencil, hand-cut and photographic. Photographic stencils can be further divided into direct emulsion, indirect film, indirect/direct and capillary film. Hand-cut stencils have limited application as they are less solvent resistant and less hard wearing, and of course, restricted to hand-cut designs.

The squeegee is produced from either rubber or polyurethane. The latter material maintains its sharp edge profile on long runs and is therefore often used on automatic printing press. In practice, there is often some swelling of the rubber in contact with screen inks but this is not normally a serious problem.

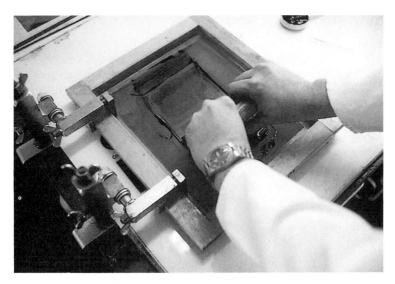

Fig. 9.2 Simple, hand operated screen printing
(courtesy of Coates Brothers Inks Ltd)

Fig. 9.3 A fully automatic screen printing press fitted with a jet dryer
(courtesy of Coates Brothers Inks Ltd)

Film thickness

Screen printing has been traditionally characterised by thick, opaque ink films. However, various developments in mesh materials, stencils and ink technology have made it possible to achieve a film thickness approaching that obtained with gravure. On the other hand, solvent free inks such as plastisols, are printed at a remarkably high film thickness.

Typical film thicknesses are as follows:

Printing process	Thickness (μm)
Offset litho	2
Letterpress	3
Flexographic	3
Gravure	7
Ultra thin film screen	8
Conventional screen	12
Thick film screen	30

The low film weight achieved with ultra-thin film screen inks is desirable to produce rapid drying and high mileage. Typical ethyl cellulose inks printed through a fine mesh (100 strands/cm) can give a total print area of 50 m^2/kg of ink and give drying times in the order of 10 seconds through a jet drier.

In other applications, high film thicknesses which cannot be achieved with any other printing process, are desirable, e.g. when printing fluorescent inks.

The pigment content of screen inks will vary considerably due to the wide range of film thicknesses that are possible. The coloured pigment content is still considerably lower than for letterpress or litho inks, although some screen inks contain a very high proportion of extenders.

Drying methods

Screen printing is almost unique in that it is possible to employ all available methods of drying. The very nature of the process allows the use of a wide range of solvents and a variety of catalytic curing resin systems. However, the most important drying method is evaporation. Most screen inks contain volatile solvents, sometimes representing 70% of the formulation.

In the past, white spirit was the most generally used solvent, but this is no longer true. Printing speeds have increased to such an extent that this solvent is considered too slow evaporating. In addition, the applications of screen printing are so varied that a wider choice of resins is necessary than is possible with white spirit.

An interesting point about evaporation drying screen inks is the wide range of driers that have been developed in an effort to remove the solvent. These include warm jet-air driers, wicket driers, simple drying racks, flame-driers, hot air blowers, oven driers and microwave driers (Fig. 9.4).

Oxidation drying inks are normally based on alkyd resins; linseed stand oil based inks are rarely used today. Inks for use on decals and metal signs

Fig. 9.4 A simple rack for air drying of screen inks

are normally oxidation drying. Catalytic curing systems are often employed for the more specialised applications of screen printing, e.g. polythene bottle printing and printed circuit production.

   Penetration drying is never used as the sole drying method; due to the
heavy ink film, this method of drying is impractical. When printing an
absorbent MG poster paper, penetration occurs but it is still essential to
rely on solvent evaporation. Certain inks, especially those for printing
PVC, incorporate solvents which penetrate the plastic sheet and promote
chemical adhesion.

   Ultraviolet curing screen inks are now firmly established for all aspects
of screen printing. The advantages of rapid setting and space-saving
curing units are utilised in the printing of plastic containers, circuit boards,
flat plastic and general purpose display printing.

Viscosity/flow

The consistency of screen inks is often described as short and buttery. As discussed earlier, this is advantageous in giving clean, sharp prints. However, it is not always possible to meet this requirement because some inks rely on good flow to achieve a specific effect, e.g. gloss inks for showcards.

If a screen ink has excessive flow there will be a tendency towards bubbling on printing and slurred prints. There will also be a tendency for the paper to stick to the underside of the printing frame although the use of a vacuum printing base minimises this effect. Although lack of flow is desirable, an ink with very poor flow will not flow out once printed to remove the pattern of the mesh from the printed ink film.

There is no one specific viscosity range which is suitable for all printing conditions. It is normal practice to supply screen inks at a higher viscosity than required; the printer adds solvent according to the nature of the design. Average printing viscosity of a screen ink would be 15–20 poise but may be much higher.

Drying and screen stability

The correct choice of solvent is the prime factor governing the balance between fast drying and screen stability. There has been an increasing tendency towards the use of faster evaporating solvents to meet the needs of fully automatic printing lines.

Solvents such as propylene glycol methyl ether or 160–180 °C aromatic hydrocarbons typify the basic solvents of thin film inks.

As a screen printing frame is exposed to the atmosphere, there will always be a certain amount of solvent loss during the printing operation. Therefore, solvents such as xylene are generally unsatisfactory, due to poor screen stability, and attempts to enclose the printing frame have been unsuccessful in promoting better screen stability.

Adhesion

The screen process is frequently employed to print on widely differing surfaces; therefore poor adhesion can occur if the correct ink is not selected or the nature of the substrates proves particularly difficult. One of the problems is that the thick ink film is often less flexible than the surface on which it is printed. Poor adhesion will often be the result of this situation. A good example of this is that a good quality polythene bottle ink may be completely unsatisfactory when printed on polythene film. It is not recommended that films of less than 2–3 thou (50–75 μ) thickness be screen printed.

PVC is widely used as a display material and is the most common flat plastic to be screen printed. Migration of plasticiser from flexible PVC sheet will impair ink adhesion immediately after printing and on storage of the print. Therefore, printing of flexible PVC normally requires inks with a higher proportion of ketone solvent and, due to the variation in PVC sheets, it is advisable to test the ink adhesion before commencing a printing operation.

Due to the inert surface of polythene, even when pre-treated, good ink adhesion is often difficult. Also, certain additives in the polythene migrate to the surface and destroy the adhesion. Unfortunately this effect can take several weeks to manifest itself. Even correctly formulated polythene bottle inks will not adhere to poorly pre-treated bottles. This is probably the most common problem with the decoration of polythene bottles.

When printing screen inks based on curing systems, excessive stoving must be avoided to obviate poor intercoat adhesion. This applies, of course, only when a colour is being overprinted. A curing screen ink must not cure so rapidly that it is too hard to accept later colours.

End-use requirements

The end-use requirement has the greatest influence on the choice of pigments.

When printing paper, the selection of pigment depends on whether it is for outdoor or indoor use. A pigment would also need to be lightfast if it is to be used indoors but subjected to daylight in, for example, a shop window. For this type of application, pigments with blue wool rating of 5 would be acceptable. A distinction must be made between pigments with good lightfastness and weather resistance. For the printing of road signs, a weather resistant pigment with lightfastness of blue wool 8 would be required, whether the road signs were made of plastic or enamelled metal.

The printing of polythene bottles does not require pigments of exceptionally good lightfastness; the most important factor is that the pigment is product resistant to the contents of the bottle, for example, detergent. This would also apply to the resin system.

Screen printing inks applied to printed circuits act as resists, therefore both resin and pigment must be functional. In the case of an acid resist, the pigment and resin must be selected to withstand the etchants. In addition, some resists are required to protect the circuit for many years.

9.2 REQUIREMENTS OF RAW MATERIALS

In common with most inks, screen inks are composed of resins, solvents, pigments and additives. However, many of the requirements are specific to screen inks.

Resins

The majority of screen inks that are for printing on paper dry by solvent evaporation. The first requirement is that resins should form a stable solution in acceptable screen solvents.

These resins should release the solvents readily to form a tack free film and give the film adequate flexibility and prevent set-off in the stack. It is advantageous if the resin forms solutions of fairly high viscosity at relatively low solids contents.

Inks based on cellulose derivatives can be used without further resin modification but it is possible to add maleic or modified phenolic resins to improve the finish. These resins must be selected to provide good solvent release.

Screen inks for the decoration of flat plastics are almost exclusively based on acrylic or vinyl resins. Acrylic resins have excellent adhesion to a wide range of plastics and have good outdoor durability.

Catalytic curing systems are frequently used for polythene bottle inks. These resins are selected mainly on account of their good chemical resistance.

Resins used in printed circuit inks perform a number of functions. The main requirements are that the resin should adhere to clean copper. It should act as a resist to various chemicals used in the involved processing technique of circuit production and, where necessary, be removed by a solvent or aqueous alkali.

Solvents

It is possible to use a wide range of solvents in screen inks, virtually all chemical types are employed. The limiting factor is the evaporation rate, as the printing frame is exposed to the atmosphere. Nevertheless, an adequate solvent range is available which, in turn, allows the use of most resin types.

The basic requirement of a screen ink solvent is that it provides adequate screen stability and will volatilise with the minimum of expenditure of energy. This energy can be applied from a warm jet-air drier; therefore, the vapour pressure at the drying temperature is of significance. It is unlikely that a single solvent will provide the exact balance of screen stability, fast-drying print and cost. Two or three solvents blends are normal. Propylene glycol ethers are frequently used in mixtures with aromatic or aliphatic hydrocarbons and this has led to the generally accepted term 'co-solvent ink'.

Attention should be paid to the effect of solvents on both photographic stencils and polyurethane squeegees. Prolonged contact with specific solvents can destroy both stencil and squeegee.

Screen printing shops are often not well ventilated therefore health and safety considerations are of paramount importance.

Pigments

Three points are of importance when considering the pigmentation of screen inks, namely:
   (1) the short soft desirable consistency;
   (2) the hue of the ink which is mainly determined by its mass tone; and
   (3) end-use requirements.
The consistency of a screen ink can be controlled by the type and amount of extender; calcium carbonate or china clay are frequently used. In addition to controlling the consistency of the ink, the use of extender will lower the cost, which is particularly important for poster inks. The extender will also prevent penetration of the ink into absorbent stocks such as MG poster paper.

Screen inks with adequate colour strength can be formulated with as little as 5% of organic pigment. The mass tone of the ink largely indicates its hue but this is not strictly true with transparent and ultra thin film inks.

It is very difficult to obtain a smooth solid print with completely transparent screen inks. Although these are available for special purposes, such as process inks and car indicator panels, they are not recommended for general purpose printing. Transparent inks tend to highlight any imperfections in the mesh or squeegee and accentuate any variation in film thickness. Transparent pigments such as phthalocyanine blue and green should be used in conjunction with a small amount of titanium dioxide.

The end-use requirement is very important in selecting a suitable pigment. Due to the numerous applications of screen printing, the properties required of a coloured pigment can vary considerably. Much screen printing is for display work and therefore requires good lightfastness and weather resistance. Few organic reds are sufficiently lightfast; much use has been made of Quinacridone red pigments. Phthalocyanine pigments are also suitable.

Poster inks for outdoor use do not require the same degree of lightfastness but fairly good lightfastness is still required. As a guide pigments with blue wool rating of 5–6 are acceptable.

When formulating screen inks containing solvents such as glycol ethers, ketones and esters, special regard must be paid to the solvent resistance of the pigment. PMTA pigments, for example, would be unsuitable for use with these solvents.

Where the printed ink film is subjected to certain processes or specific tests, this must be borne in mind when making pigment selection. Pigments with good resistance to bleaches will be required when printing polythene bottles to contain bleach. Naphthol red F2R and diarylide yellow are commonly used pigments in polythene bottle inks.

Additives

Screen inks require some form of control to avoid the tendency to bubble on printing. This can be achieved by the use of extender but where the use of extender is limited or impossible, then silicone fluids are frequently used.

Silicones of the polymethylsiloxane type are normally used in screen inks, usually in quantities less than 1%. However, inks containing silicone are not without problems. In certain circumstances, particularly curing systems, they lack intercoat adhesion or cannot be overprinted with an ink that does not contain silicone. If traces of silicone contaminate certain inks, then pinholing can occur on overprinting; this is most noticeable with metal printing inks. Non-silicone defoamers are available but these have proved of limited value in screen inks.

While extenders are frequently used to control the consistency of screen inks, this can be achieved with structuring additives such as Bentone. These additives are relatively expensive and can reduce gloss, and care should be taken to select the correct grade according to type of solvent being employed. The viscosity of screen inks is sufficiently high that pigment settling rarely occurs but anti-settle additives may be required where high specific gravity extenders such as blanc fixe are used in the formulation.

Wax pastes, such as micronised polyethylene wax, are added to formulations to promote good rub resistance. These are normally not required in the harder catalytic systems but are used in evaporation drying inks, particularly when the printed matter concerned is likely to be scuffed after printing.

Driers and anti-skinning agents are, of course, used with oxidation drying inks. As most oxidation drying screen inks are based on long oil alkyds, the most commonly used driers are calcium, manganese and cobalt naphthenates. Due to the heavy deposit achieved with alkyd inks, it is important to ensure good through-drying.

9.3 INKS FOR PAPER AND BOARD

The early screen inks for printing on paper were oil based. They usually contained a high proportion of coarse extender, such as whiting, and the vehicle was based on highly bodied linseed stand oil in white spirit.

Drying was by evaporation of the solvent followed by oxidation of the oil. As might be expected, such inks suffered from a number of disadvantages; they were slow drying, had poor scuff resistance, low mileage and were very coarse. Rather surprisingly this type of formulation was used for many years for the printing of posters and showcards but has now ceased to be of commercial interest.

The first type of screen ink to dry entirely by evaporation was the early 'cellulose ink'. This description refers to nitrocellulose-based inks although the formulations were different from modern nitrocellulose inks in that they were similar to cellulose paint. In fact, they were developed by industrial paint manufacturing companies. The disadvantages of such inks were objectionable odour, fire hazards associated with nitrocellulose and high cost of the solvents.

Thin film screen inks

The most significant development during this early period occurred with the introduction of EHEC (ethyl hydroxy ethyl cellulose). This allowed the formulation of the first thin film screen inks. Inks based on EHEC were a significant advance in that they overcame the problems associated with 'cellulose inks', with the added advantage of being white spirit reducible. The significance of this latter point is that all types of stencils are suitable and the cleaning solvent and thinners are relatively inexpensive.

A typical formulation for a thin film screen ink for posters would be:

Toluidine Red (CI Pigment Red 3)	5.0
China clay (CI Pigment White 19)	36.0
Low viscosity EHEC	6.0
Penta ester gum	10.0

White spirit	30.0
Aromatic hydrocarbon (160–180°C)	10.0
Propylene glycol ether	3.0
	100.0

The function of the ester gum resin is to increase the finish and assist pigment dispersion.

This type of ink would be suitable for printing on most paper stocks. Typical applications would be poster printing both for indoor and outdoor use, showcards and point of sale displays (Fig. 9.5). In comparison to oil based inks, EHEC inks offer better rub resistance, higher mileage, more finish and faster drying. Air drying time of the formulation given above would be approximately 20 minutes.

Ultra thin film screen inks

The increasing use of automatic screen printing machinery brought a demand for even quicker drying inks. This demand led to the development of the ultra thin film inks. The technique here is to use a cellulose derivative to give low solids and high viscosity vehicles. The introduction of higher speed printing machines permitted the use of more volatile solvents without problems of drying in the screen. The combination of

Fig. 9.5 Examples of screen printing on paper and board

lower solids vehicle and fast evaporating solvents formed the basis of fast drying, ultra thin film inks.

Typical formulation for ultra thin film screen ink is as follows:

Primrose chrome (CI Pigment Yellow 34)	20.0
Ethyl cellulose	15.0
Dioctyl phthalate	5.0
Propylene glycol methyl ether	25.0
Dipropylene glycol methyl ether	5.0
Aromatic hydrocarbon (160–180°C)	30.0
	100.0

A low cost aromatic hydrocarbon solvent would be a suitable reducer for this ink.

Apart from the obvious advantages of being fast drying, the ultra thin film ink has superior rub resistance, increased mileage and more finish. A higher standard of pigment dispersion is essential with this type of ink; this point and the lack of extender are the reasons for the increased finish.

The trend towards the ultra thin film screen ink has, to some extent, departed from the accepted image of screen printing. These types of inks no longer produce the characteristic strong, opaque prints. In addition, there is a greater tendency for the ink to penetrate absorbent stocks.

It is possible to formulate an ultra thin film ink using nitrocellulose in place of ethyl cellulose. The high cost of ethyl cellulose has encouraged ink-makers to formulate screen inks based on nitrocellulose. This now forms the basis of many high quality screen inks which are capable of jet drying and adhering to a very wide range of substrates.

A typical formulation for a nitrocellulose-based ultra thin film ink for printing paper, board, foils, hardboard and stove enamelled metal signs would be:

Phthalocyanine Blue (CI Pigment Blue 15.3)	5.0
Titanium dioxide (CI Pigment White 6)	3.0
Nitrocellulose	14.0
Dioctyl phthalate	7.0
High acid value maleic resin	3.0
Propylene glycol ethyl ether	38.0
Ethylene glycol isopropyl ether acetate	20.0
Aromatic hycrocarbon (160–180°C)	10.0
	100.0

However, a more expensive solvent reducer is required. The solvent balance given in the ink formulation would be a suitable reducer.

Oxidation drying gloss inks

Oil-based gloss inks are used to achieve a thick glossy film not unlike the effect achieved with diestamping inks. This type of finish has only limited applications, for example special effect showcards and labels.

The use of oil-based gloss inks has decreased in popularity for two reasons. Firstly, the inherent slow drying of the very thick film and secondly the improved finish of evaporation drying inks. The higher standard of pigment dispersion achieved with ultra thin film inks has resulted in improved finish. On coated paper, a semi-gloss finish can be obtained with a well-dispersed ethyl cellulose ink.

The slow drying of oil-based gloss inks, usually in the order of 6−8 hours, limits the production to the printing of one colour in a normal working day. A multi-colour run would take several days to complete.

A typical formulation for an oil-based gloss ink is as follows:

Diarylide Yellow (CI Pigment Yellow 14)	4.0
Blanc fixe (CI Pigment White 21)	40.0
Stand oil	10.8
Modified phenolic stand oil varnish	40.0
White spirit	4.0
Naphthenate driers	1.0
Methyl ethyl ketoxime	0.2
	100.0

Blanc fixe is chosen for its good gloss and levelling properties.

9.4 INKS FOR IMPERVIOUS SURFACES

Metal signs

Metal signs which are screen printed with non-ceramic inks are intended to have a life of 3−5 years. This is considerably more permanent than the relatively short lived paper poster but not as durable as a vitreous enamelled sign. Typical applications of metal signs would be traffic signs, semi-permanent window displays, fire exit notices (Fig. 9.6). In view of the long life of metal signs, the main requirement of a screen ink is durability. This refers to lightfastness, weathering and adhesion.

It is normal practice to use long oil alkyds as the basis of suitable screen inks. These alkyds are ideally suited in that thick deposits are possible, which is essential for outdoor durability. Alkyds are also capable of being air dried or stoved and have a well-proven history of reliable weathering properties.

The inks are normally applied over a stoved enamelled background; they may be stoved or air dried. Once the design has been printed it is then protected with a spraying or roller coating varnish. The completed sign is then subjected to a final stoving.

It is important that the screen inks are sufficiently dry to withstand attack from the solvents present in the spraying varnish. The base enamel, screen ink and protective varnish must be tested as a complete system. There should be good intercoat adhesion between the three systems. Formulation for a lightfast, durable, metal printing red ink is given below:

Fig. 9.6 Screen printing on metal used for a road sign

Quinacridone Red (CI Pigment Red 122)	6.0
Calcium carbonate (CI Pigment White 18)	24.0
Long oil linseed alkyd	50.0
Mixed napthenate drier	4.0
White spirit	16.0
	100.0

Drier content should be calculated on alkyd resin solids. The amounts and type of drier to be used is dependent on the specific alkyd used; however, it is normal to use calcium, manganese and cobalt naphthenate with long oil alkyds.

Inks similar to those described above are also used for the printing of nameplates. There are a large number of dials, clockfaces, instruction panels for household appliances which are screen printed. No general formulation can be given for these inks as there is such a variety of applications. Finish can vary between full gloss and dead matt. Resins used are those which give tough, chemically resistant finishes when stoved, such as melamine-alkyd, polyurethane alkyds, thermosetting acrylics and epoxy-polyamide.

Metal containers

It is often convenient to add identification or other marking to drums. This can, of course, be done by normal stencilling methods but a neater result can be achieved with screen printing. If a large number of drums are to be printed, then screen printing is more convenient. The inks are not usually applied to bare metal but drums enamelled with a background colour. The screen print is usually air dried, particularly if the drums have a large capacity. The requirements of such inks are that they exhibit good adhesion and are not removed by the chemicals packed inside the container.

Since the printing process is liable to be slow and intermittent, it is desirable to use high boiling solvents in the inks to avoid drying in the screen. It is also necessary to restrict the flow of the ink; this is particularly important with screens that are physically carried from drum to drum.

A suitable ink would be:

Carbon Black (CI Pigment Black 7)	6.0
China clay (CI Pigment White 19)	30.0
Medium long oil alkyd	45.0
Naphthenate driers	3.0
Anti-oxidant	0.1
Structuring additive	3.0
Dipentene	12.9
	100.0

Inks for sheet plastic

The decoration of sheet plastics is now a very important part of the screen industry. Many traditional materials have, for economic and technical reasons, been replaced by sheet plastics. A good example of this is the increasing number of traffic signs produced in acrylic sheet, to replace metal signs.

The variety of plastic substrates that exists presents problems to the ink-maker in that each substrate may require special consideration. An ideal situation is where one screen ink is suitable for all substrates; in practice this is not possible but an ink formulation should be as multi-purpose as possible. Possibly, the nearest approach to this is a screen ink based on acrylic/vinyl resins.

Polyvinyl chloride (PVC) sheet is probably the most frequently printed sheet plastic. This is available in rigid unplasticised sheet, self-adhesive sheet and heavily plasticised 'self-cling' PVC. Other important sheet materials are, acrylic, polycarbonate, polyester, polystyrene, cellulose acetate, corrugated polypropylene and spun polyolefin.

Relatively thick sheets of acrylic or PVC are frequently used for display signs. These displays can be in the form of an illuminated light box which functions day and night. Inks for illuminated signs must be translucent: if the inks are too opaque, the sign will not be visible at night. Translucent screen inks contain approximately half the normal pigment content, this provides an ink which is neither too transparent nor too opaque.

An ink for a road sign must be very weather resistant and lightfast; pigments with Blue Wool ratings not less than 7−8 should be used. In order to achieve a high degree of weathering, the ink must have excellent adhesion to the substrate. Some printed acrylic sheets are subjected to thermoforming, and the ink must be sufficiently flexible to withstand this process. However, if an ink has low softening point the prints will show mould marking if contact is made with the mould.

A lightfast red ink suitable for outdoor use on acrylic or PVC sheet would be:

Quinacridone pigment (CI Pigment Red 122)	10.0
Ethyl methacrylate copolymer	25.0
Vinyl resin	7.0
Methyl propoxol acetate	20.0
Aromatic hydrocarbon (186–214 °C)	20.0
Cyclohexanone	10.0
Diacetone alcohol	7.0
Silicone anti-foam	1.0
	100.0

Although 1.5–3.0 mm PVC sheet is used for roadsigns, the vast majority of PVC sheet screen printed is considerably thinner, i.e. 0.13–0.24 mm. The main application of this thinner grade of PVC is in the form of self-adhesive labels (Fig. 9.7).

There are two main methods of producing a self-adhesive printed label. The more conventional method is to screenprint directly on the surface of self-adhesive PVC with a gloss ink. The second method is to use transparent PVC and screenprint the legend in reverse (viewed through the plastic), apply a backing white ink over the whole print area and finally screen print a pressure-sensitive adhesive. The adhesive can be permanent or removable type but as it is tacky it must be protected with a release paper.

The main point of interest to the ink formulator is that the first method requires gloss inks and the second method, matt inks. Either system may be used to produce a double-sided window sticker, i.e. where the image is viewable from both sides. With such stickers 10 superimposed ink layers are not uncommon and this presents a problem to the ink formulator. A

Fig. 9.7 Vehicle marking on self-adhesive vinyl

high degree of flexibility is required which cannot simply be achieved by plasticiser modification. In the case of gloss inks this is normally achieved by careful selection of methacrylate resin. A wide range of resins are available which can produce screen inks with either excellent flexibility or hard drying properties.

The most common problem with printing PVC is obtaining consistently good ink adhesion and it has been the practice to use strong ketonic solvent to promote adequate adhesion. This presents some problems; the inks have an unpleasant odour and cockling can occur on some thinner grades of plastic. The basic resins used are normally polyvinyl choride/polyvinyl acetate copolymers and recent developments with these resins have led to inks with milder solvent balances with a high proportion of aromatic hydrocarbon solvent.

Pigment selection is important, ketone resistant pigments must be selected. Even the slightest contamination with a non-resistant pigment can cause staining of subsequent overprinting inks. Pigments must also be resistant to bleed into plasticisers because the high proportion of liquid plasticiser in flexible PVC can cause certain pigments to migrate into the plastic.

Typical formulations for a matt and gloss ink for either plasticised or rigid PVC are:

Matt ink for PVC sheet

Permanent Yellow H10G (CI Pigment Yellow 3)	5.0
Titanium dioxide (CI Pigment White 6)	7.0
China clay (CI Pigment White 19)	30.0
PVC/PVA copolymer	15.0
Aromatic hydrocarbon (160–180°C)	20.0
Pentoxone	16.0
Cyclohexanone	7.0
	100.0

Gloss ink for PVC sheet

Rutile titanium dioxide (CI Pigment White 6)	27.0
PVC/PVA copolymer	19.0
Methyl/butyl methacrylate	7.0
Cyclohexanone	20.0
Aromatic hydrocarbon (160–180°C)	18.0
2-butoxyethyl acetate	8.5
Silicone anti-foam	0.5
	100.0

Good lightfastness and weathering properties are not essential for all vinyl inks; certain PVC articles, e.g. sachets, require only minimum lightfastness. On the other hand, permanent labels must conform to a higher standard. Basic requirements for the more permanent labels are detailed in BS 4781 : Part 1 : 1973, specification for self-adhesive plastic labels for permanent use.

Inks for glass

Inks for screen printing on glass fall into two main categories:
  (1) conventional screen inks; and
  (2) ceramic inks.
The latter are normally produced by specialised pigment manufacturers and are outside the scope of the conventional ink manufacturer.

The application of conventional screen printing inks is mainly in the decoration of mirrors. Conventional inks have also been used on glass bottles, laboratory glassware, pinball machines and glass objects such as ashtrays and ornaments.

Printed mirrors are often used as advertising displays in shops. The screen ink can be applied to the front of the glass or printed in reverse on the back. In the case of reverse printing, the glass is screenprinted before the silvering operations. The ink must therefore withstand the mirroring process of degreasing with abrasive whiting, immersion in solutions of stannous chloride and hydrochloric acid and the final deposition of silver from a solution of silver nitrate.

For surface printing, an ink based on hard drying alkyd, e.g. vinyl toluene modified alkyd, of medium oil length would be suitable. However, when reverse printed, this type of an ink would not always withstand the mirroring process.

A two-pack epoxy/polyamide, screen ink would be suitable for this application; a typical starting-point formulation would be:

Paste

Titanium dioxide (CI Pigment White 6)	45.0
Epoxy resin	35.0
2-butoxyethanol	20.0
	100.0

Catalyst

Reactive polyamide	70.0
2-butoxyethanol	10.0
Aromatic hydrocarbon (186–214°C)	20.0
	100.0

Mixing ratio 100 parts of paste to 15 parts of catalyst.

Mirror pictures, i.e. where the silvered area covers only part of the glass, usually in the form of a simple picture or pattern, are produced with the aid of screen printing. An acid resisting ink is printed directly on to the silver. The mirror is subjected to an acid bath which removes the un-protected silver. This leaves a mirror pattern against a clear glass background. A suitable acid resisting ink can be formulated using isomerised rubber resins in aliphatic hydrocarbons.

Ceramic screen inks are used where a permanent mark is required on the glass container, e.g. reusable drink bottles, glassware, pottery and laboratory glassware.

Ceramic screen inks can take a variety of forms and may be applied by direct printing, decalcomania or hot melt screen. Regardless of the type used, the principle is the same; the ink is fired into the glass and becomes an integral part.

A ceramic ink consists of pigment, frit or flux and a varnish. The function of the varnish is simply as a temporary carrier of the other components. After the ink is applied to a bottle, it is fired at temperatures in the order of 800°C: the pigment and frit, known as the enamel, fuse into the glass and the varnish completely volatises at these temperatures.

Formulation for a ceramic screen ink for direct printing is as follows:

Ceramic enamel pigment	52.45
Thixotropic alkyd	26.23
Hydroquinone	0.07
Napthenate drier	0.13
White spirit	21.12
	100.00

9.5   INKS FOR PLASTIC CONTAINERS

Polythene containers

There are a number of alternative methods of decorating polythene bottles but, in the UK, the most popular process is screen printing (Fig. 9.8). The alternative processes are, letterset printing, paper labelling, hot-melt transfer printing, hot foil stamping, in-mould labelling and shrink labelling.

Before discussing suitable ink formulations, it must be stressed that correct treatment level of the surface of the polythene is absolutely essential. If the pre-treatment is incorrect, this can result in poor ink adhesion, poor flexibility and lack of chemical resistance. A further problem is that certain additives included in the polythene can migrate to the surface and destroy ink adhesion. Inks must be tested on the specific polythene being used and the adhesion tested over an extended period.

One of the main requirements of a screen ink is that it is resistant to the products that are being packed inside the containers. The most common products are detergents, bleaches, lubricating oil, household cleaners and cosmetics (Fig. 9.9). With such a variety of products it is unlikely that one ink system will be suitable for all bottles.

Good flexibility is required as some bottles, particularly those containing

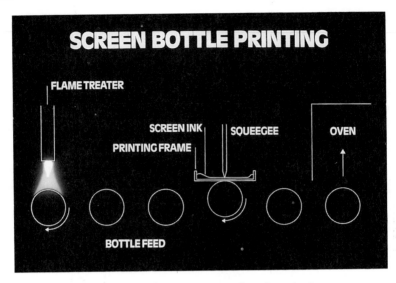

Fig. 9.8 Diagram of screen bottle printing
(courtesy of Coates Brothers Inks Ltd)

Fig. 9.9 Examples of screen printing on polythene containers

detergents, are subjected to frequent flexing when in use in the household.

The early screen ink formulations were oxidation drying, these were normally dried in an oven at 80 °C for 8–10 minutes. In many ways, oxidation drying inks have not been improved upon: they are simple to use, relatively inexpensive and give thick glossy films with fairly good chemical resistance. Their main drawback is the necessity to stove each colour for several minutes. Although oxidation drying inks cannot meet the demands of a high-speed bottle-producing plant, these are still used in certain circumstances.

The original polythene bottle inks were based on long oil linseed alkyds but in modern times better chemical resistance is required. An oxidation drying ink would be formulated by using epoxy ester resins as follows:

Phthalocyanine Green (CI Pigment Green 7)	6.0
Titanium dioxide (CI Pigment White 6)	10.0
Long oil epoxy ester	82.2
4% calcium naphthenate	1.0
6% cobalt naphthenate	0.6
Silicone anti-foam	0.2
	100.0

With the introduction of high-speed printing equipment it was essential to provide faster-drying inks, i.e. evaporation drying types. Traditionally polyamide resins have formed the basis of gravure polythene inks and these resins have also been found suitable for screen inks. Polyamides provide good adhesion to polythene but require some modification with a film former to improve solvent release and chemical resistance.

Polyamide inks can be developed which have fairly good resistance to oils, detergents and bleaches but are not resistant to cosmetics or aggressive products containing solvents. In addition, polyamide screen inks tend to lack gloss and are not very scuff resistant.

Where good chemical resistance is required, it is possible to use epoxy-polyamide based inks. These are two-pack systems with good chemical resistance but they are not particularly fast drying. There is a need, therefore, to provide an ink that combines the speed of drying of evaporation inks, the gloss of oxidation inks and the chemical resistance of epoxy–polyamide inks.

The inks which have most closely approached these requirements have been formulated using acid-catalysed amino resins. Unfortunately, these inks suffer from the disadvantage of being two-pack systems with limited pot life.

Considerable development work has taken place to produce a formulation that is one pack and with an extended shelf life. Very sophisticated one-pack catalytic curing inks are available which rely on chemically blocked catalysts. The action of stoving the printed bottle at 80°C unblocks the catalyst and allows a high degree of polymerisation. Excellent chemical resistance and fast production is achieved with the added advantage of ease of use. The one-pack catalytic ink is widely accepted as the most superior system for polythene bottle decoration.

PVC containers

Containers made from PVC have, to a limited extent, replaced glass bottles. Typical products that are packed in PVC bottles are soft drinks, disinfectants, shampoos, and sun tan oil. The vast majority of PVC bottles are, in fact, paper labelled, not screen printed.

As PVC is being considered as a replacement for glass, most bottles are made from a transparent grade of PVC. This presents some problem to the ink formulator because it is essential to formulate inks with good opacity. If a dark-coloured liquid is introduced into a printed bottle, the legend or design must remain clear. Where possible, inks should contain titanium dioxide or contain a higher proportion of pigment necessary to achieve good opacity.

Similar requirements to the polythene bottle inks are necessary, i.e. chemical resistance, adhesion and fast drying. A problem in the decoration of PVC bottles is that of stress cracking. Printed bottles can split very easily or have an unacceptable low impact strength. The source of the problem is in poor blow moulding technique. PVC bottles moulded in a highly strained condition will stress crack in the presence of ink solvents.

Although the problem can be avoided by moulding technique or the use of impact modifiers the ink formulator needs to minimise the effect by careful selection of resins and solvents. Formulations given under PVC sheet inks would be suitable but these would have a tendency to scuff; PVC bottle inks should be formulated using harder resins. A typical mar-resistant PVC black bottle ink would be:

Carbon Black (CI Pigment Black 7)	5.0
Hard acrylic copolymer	30.0
Dioctyl phthalate	6.0
PVC/PVA copolymer	8.0
Methyl propoxol acetate	20.0
Aromatic hydrocarbon (186–214°C)	20.0
Diacetone alcohol	5.0
Cyclohexanone	5.0
Polyethylene wax	0.5
Silicone antifoam	0.5
	100.0

9.6 TEXTILE INKS

It was noted in the introduction to this chapter that screen printing was first used in textile printing, primarily for the application of dyes and dye resists for subsequent chemical and physical treatment. Large textile producers normally blend these inks for themselves from dyes and bases.

However, there is considerable interest in the screen printing of pigmented inks, e.g. printed clothing, pennants, bags, etc. This work is normally carried out by the established screen printer rather than the textile manufacturer, although many specialist firms exist who produce printed sportswear such as sweatshirts and T-shirts (Fig. 9.10).

An ink should be capable of being applied to a variety of synthetic and natural fabrics and be washfast and dry clean resistant. Once applied the colour should not fade or impart a 'feel' or 'hand' to the material.

There are two main types of ink used, pigmented emulsions and plastisol inks.

The emulsion inks are used for direct printing of material; they are

**Fig. 9.10 Screen printing on textiles
(courtesy of Coates Brothers Inks Ltd)**

suitable for all types of fabric. They may be allowed to dry by evaporation at room temperature but to achieve resistance to washing and dry cleaning, they must be cured at 160°C for 2–3 minutes.

Suitable inks are based on aqueous dispersion of an auto-crosslinking acrylate copolymer.

A typical formulation would be:

Water	10.0
Emulsifier	1.0
Thickener	4.0
White spirit	62.0
Catalyst soln	3.0
Binder	15.0
Pigment dispersion	5.0
	100.0

This type of ink is an oil-in-water emulsion where the oil is, in fact, white spirit. The high proportion of solvent, in a basically water-based system, gives an ink with easy printing properties and a low filmweight. As a consequence the print does not impart any feel to the fabric.

It is sometimes desirable to produce an ink that is entirely water-based. This can be achieved by replacing the white spirit in the above formulation with water. The resultant ink will give less bright colours and a higher hand.

The basic medium, described above as a binder, could be an acrylate copolymer which is capable of cross-linking in the presence of a catalyst and under heat. Catalysts used are ammonium salts of inorganic acids, e.g. diammonium phosphate or ammonium thiocyanate.

The addition of proprietary chemicals such as definition improvers, lubricants, accelerators, thickeners, humectants and softeners is possible to achieve specific properties or variation in performance. However, an improvement in one area is usually at the expense of another property. For example, the addition of 1 or 2% accelerator will reduce the stoving schedule but have a dramatic effect on the pot life of the ink.

Pigmentation is achieved by adding pre-dispersed pigment pastes. These are normally organic pigment dispersions specifically formulated for textile inks. The lightfastness and washfastness are usually very good.

The second type of screen textile ink is based on plastisol, i.e. vinyl resin dispersed in plasticiser. This type of ink consists of virtually 100% non-volatiles; no solvent is present. As a consequence it applies an extremely thick opaque ink film and therefore is frequently printed on black or dark-coloured garments. Plastisols may be applied directly to the fabric or used as a transfer ink.

When used as a transfer the ink is screen printed on to a release paper, e.g. vegetable parchment. The ink is subsequently cured to a dry film and the print is stored. The plastisol, although partially cured, remains thermoplastic and can be transferred to a fabric simply by ironing or another heat transfer process.

Plastisol inks are basically formulated from PVC homopolymer dispersed in phthalate plasticiser. The vinyl resin should be high molecular weight emulsion polymer with fine particle size, i.e. a paste-making polymer. The liquid plasticisers frequently used are dialkyl phthalate and di-iso-octyl phthalate.

The PVC resin requires stabilisation against degradation by heat and light. The liquid barium/cadmium/zinc stabilisers combined with epoxy plasticisers are particularly useful.

A high proportion of extender is desirable to reduce the raw material cost and to improve the wet-on-wet printing properties. Printing on textiles is probably the only area of screen printing where wet-on-wet printing is common practice.

9.7 TRANSFER INKS

Transfers are sometimes described as decals, decalcomania or water-slide transfers. They are used where screen printing an object is impracticable, e.g. it is far easier to apply a transfer to a large object like a compressed air cylinder than to print directly on the cylinder. The screen printing is carried out on a flat sheet of transfer paper which has the ability to release the ink film completely.

A transfer can be defined as a printed image being removed from a substrate (transfer paper) and adhered to another surface (usually a large object).

The simplest type of transfer involves the use of a transfer paper which is an absorbent paper coated with a water release agent and adhesive. The gum serves as a release agent and adhesive for the transfer (Fig. 9.11). There are variations in this process but the ink requirements remain the same.

The main point of interest to the ink formulator is that an ink film must be sufficiently strong and flexible to completely support itself at the time of transfer. This flexibility must be retained over many years, some transfers are stored for long periods before being used.

In order to achieve a tough, flexible transfer it is necessary to apply a very heavy deposit of ink. Coarse meshes are commonly used, particularly for printing the background colour. A typical transfer could consist of a background white, two or three colours and a clear overprint varnish.

Transfer inks are usually based on long oil modified alkyds and are thinned with white spirit. The choice of alkyd and the manner in which it is modified is critical in achieving the desired flexibility. A resin should be selected with high solids content to achieve a thick ink film. Choice of pigment is dependent on the end use of the transfer.

Common criticism of transfer inks is the long drying times; evaporation drying inks can be formulated and are used for less critical work. Highly plasticised nitrocellulose usually forms the basis of these inks but it is not possible to achieve the flexibility and opacity of alkyd inks.

9.8 OVERPRINT VARNISHES

Overprint varnishes applied by screen printing often have to compete in terms of technical performance, cost effectiveness and ease of application with roller coating varnishes and film lamination.

The simplest application of a screen varnish would be a point-of-sale showcard requiring an overall gloss finish. Traditionally this type of varnish would be based on resin modified nitrocellulose, or for higher gloss, a non-yellowing medium-long oil alkyd. To a large extent these decorative overprint varnishes have moved towards UV curing formulations. These offer extremely high gloss finishes with the advantage of almost instantaneous drying.

It must not be assumed that the role of all overprint varnishes is to provide a gloss decorative finish. Many speciality screen varnishes exist

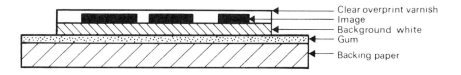

Fig. 9.11 A typical transfer print on simplex paper

which have other functions and indeed are not always full gloss. A few examples are given below with some indication of the chemical type.

(1) Alcohol-resistant varnish for labels on spirit bottles — chlorinated rubber.
(2) Sealing varnish for rub-removable lottery tickets — UV or modified nitrocellulose.
(3) Thermoplastic varnishes for blister packs — various.
(4) Barrier coatings for reducing gas permeability of plastics — PVdC.
(5) Alcohol and petrol resistant varnish for vehicle marking — vinyl/acrylic.
(6) Texturing varnishes for membrane touch switches — mainly UV but acrylic possible.
(7) Protective varnishes for PVC self-adhesive exterior stickers — vinyl/acrylic with UV absorber.
(8) Laminating varnish for PVC credit cards — thermoplastic vinyl/acrylic.

9.9 DAYLIGHT FLUORESCENT INKS

Fluorescence is a process of photo-luminescence by which light of short wavelengths, either in the UV or visible regions of the electromagnetic spectrum, is absorbed and reradiated at longer wavelengths. This re-emission occurs within the visible region of the spectrum and is manifested as colour. Fluorescent pigments are transparent organic resin particles containing dyes in solid solution.

Although fluorescent inks can be printed by most processes, the screen process gives incomparable results. The thick film of screen printing is the obvious method of printing fluorescent inks as fluorescence increases with film thickness.

Two main types of inks are available, evaporation drying or oxidation drying. The latter type gives thicker films and more brilliant results but is slower drying. In fact, oxidation drying fluorescent inks, due to the pigment loading, are touch-dry relatively quickly but then take several hours to become hard.

Two formulations for a fluorescent ink suitable for posters are given below:

Evaporation dry	
Fluorescent pigment	46.0
EHEC	4.0
Penta ester gum	16.7
White spirit	29.0
Dipentene	3.0
2-Butoxyethanol	1.3
	100.0

Oxidation drying

Fluorescent pigment	48.00
Long oil soya alkyd	46.00
6% cobalt octoate	0.25
Zirconium naphthenate	0.75
White spirit	5.00
	100.00

It is also possible to formulate screen fluorescent inks for printing PVC sheet. This type of printing is not very popular as the fluorescent pigment has limited outdoor life, while a PVC sticker may be regarded as semi-permanent.

A formulation for a screen ink for printing PVC sheet is as follows:

Fluorescent pigment (solvent resistant grade)	27.0
PVC/PVA copolymer	19.0
Methyl/butyl methacrylate	7.0
Cyclohexanone	10.5
Aromatic hydrocarbon (160–180°C)	32.0
Aromatic hydrocarbon (186–214°C)	4.0
Dioctyl phthalate	0.5
	100.0

When using fluorescent pigment in conjunction with ketone solvents, attention must be paid to the grade of pigment. Solvent-resistant grades are available which give complete stability with vinyl/ketone solutions.

It is essential to print all fluorescent inks over a white background because of the transparent nature of the pigment. It is possible to include some titanium dioxide in fluorescent inks but there is an inevitable loss of brilliance.

9.10 PROCESS INKS

The quality of screen process printing lines has improved dramatically, over recent years. This may be attributed to developments in inks, meshes, stencils, printing machinery, separation techniques and printing skills.

Printing of large posters would normally use rulings of 30 line screen but the average separation may be 65–85 line screen. It is not uncommon to use 120 line screen which produces superb quality that is sometimes difficult to distinguish from litho printing (Fig. 9.12). Achromatic reproduction techniques are applicable to screen printing and can further enhance the quality of print.

Unfortunately, there is no British Standard that is truly applicable to screen printing. It is therefore normal practice to quote screen process

Fig. 9.12 A screen printed poster using four colour process inks

inks as being matched to BS 4666 or DIN 16539 but strictly speaking this is incorrect.

As with other printing processes, suitable screen inks must be transparent and give good dot reproduction. Capability to jet dry quickly is nearly always required to allow all colours to be superimposed rapidly.

Screen process inks should be formulated on low solids varnishes to avoid a rough appearance on build up. They should be structured to give dot reproduction and contain a proportion of transparent extender to avoid penetration into the stock and also lower cost. Pigmentation is lower than normal screen inks but must be maintained around 2–3% to produce good colour strength through fine meshes. A considerable amount of process work is produced on self-adhesive vinyl. In this situation pigment selection is more critical, only lightfast, solvent resistant pigments may be used. As four colour superimposition is inevitable in process printing a degree of flexibility is required, particularly on self-adhesive vinyl which is subject to embrittlement with heavy ink deposits.

A good starting-point for formulating an ultra thin film process ink for poster printing would be:

Organic pigment	3.0
Alumina hydrate	8.0
Ethyl cellulose	15.0
Dioctyl phthalate	6.0
2-butoxyethanol	33.0
Structuring additive	2.0
Aromatic hydrocarbon	33.0
	————
	100.0
	————

Note that higher boiling point solvents are preferred to improve screen stability. For economic reasons and where appropriate it is possible to formulate either EHEC process inks or nitrocellulose-based inks.

9.11 METALLICS

The use of metallic inks in screen printing covers a very wide range of applications. Metallic effects are possible in the printing of T-shirts, point-of-sale showcards, self-adhesive vinyls, plastic containers, but possibly the largest application is in the printing of rub-removable lottery tickets and promotions.

Rub-removable tickets have gained worldwide popularity both as a lottery or as a promotional tool (Fig. 9.13). In situations where large prizes in cash or kind are at stake, security is essential. Screen printing offers the means of applying heavy ink deposits; a film thickness of $12-15$ µm is required for complete obliteration.

More than one printing process is used in the production of lottery tickets. The graphics are litho or gravure printed and the metallic ink is screen printed. The ticket is a four- or five-part construction, base board, printed design including symbols, sealant, obliterating ink and an optional litho print to add to the personalisation and security of the ticket.

The purpose of the sealant, which normally is litho printed UV curing varnish, is to prevent screen ink penetration and assist removability. It also serves to combat fraudulent abuse of the ticket.

The rub-removable screen ink is the most important component of the system. It must give total obliteration and give controlled rub-removal characteristics and must remain removable for many months after printing.

The selection of the aluminium powder is critical, both leafing and non-leafing grades can be used but they must be of fine particle size to give adequate opacity. If they are too fine this can result in finger staining or board staining. Coloured rub-removable inks can be achieved by dispersing 2% of organic pigment into an aluminium-based ink. The rub-removable characteristics are achieved with the use of pliable resins. The inks are often described as latex inks but modern inks are based on synthetic 'plastic' resins. No precise formulations can be given but the importance of thorough print testing particular on ageing must be emphasised.

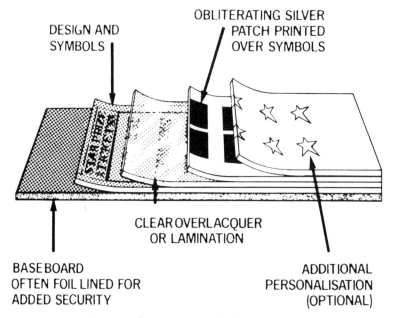

DESIGN AND SYMBOLS

OBLITERATING SILVER PATCH PRINTED OVER SYMBOLS

CLEAR OVERLACQUER OR LAMINATION

BASEBOARD OFTEN FOIL LINED FOR ADDED SECURITY

ADDITIONAL PERSONALISATION (OPTIONAL)

Fig. 9.13 The structure of a lottery ticket

Metallic inks are used frequently in the printing of double-side self-adhesive labels. The aluminium screen ink provides complete obliteration between both sides of the printed legend. The metallic content should be limited to avoid poor intercoat adhesion with overprinting colours.

Very attractive glitter effects can also be achieved with screen printing. Glitter powders are actually particles of highly polished aluminium foil which have been coated with a transparent pigmented lacquer. Various grades of glitter are available differing in colour, particle size and solvent resistance.

The coating lacquer is usually nitrocellulose, vinyl or epoxy and the solvent balance of the ink must be considered before selecting the grade of glitter. An average particle size of 200 μm is common and therefore may only be printed through coarse meshes.

Glitter particles may be dispersed in almost any medium providing the solvents are taken into consideration. Plastisol glitters are popular for the printing of T-shirts.

9.12   INK-RELATED PRINTING PROBLEMS

Most problems that occur in screen printing are often not directly attributable to the ink formulation. Problems occur in misuse or incorrect application of the ink or the final print. Listed below are common problems and suggested solutions based on manipulation of ink formulations.

Adhesion

As screen printing is such a versatile process it is possible to screen print on a very wide range of surfaces and thus each substrate can present its own ink adhesion problems. When screen printing on some plastics it is possible to base formulations on resins that bear similarity to the substrate, e.g. acrylic resins for sheet acrylic, amino resins for melamine formaldehyde mouldings, vinyl resins for self-adhesive PVC sheet. Resin selection is the most important factor in achieving good adhesion. Indifferent adhesion can sometimes be improved by solvent selection but this is not always reliable. A technique frequently used when printing on PVC sheet is to include a higher proportion of cyclohexanone. This has the effect of etching and improving chemical adhesion. This technique is never successful if the basic ink formulation is unsuitable and has poor initial adhesion.

Inadequate flexibility can result in poor ink adhesion, in this situation modifications with plasticiser, or flexibilising resins will improve matters. In thermosetting systems attention should be paid to the reactive component or catalyst, e.g. increasing polyamide content in an epoxy/polyamide screen ink will improve flexibility and adhesion.

Bubbling

This is a very common problem in screen printing. It is usually attributable to long flow or the inability of an ink film to break cleanly between mesh and substrate and may be overcome by the use of silicone defoamers, non-silicone defoamers, increasing extender content, structuring agents or increasing solvency.

Crazing

This has the appearance of hairline cracks in the print. Normally it only occurs on plastic and is believed to be attributed to shrinkage of the ink film. Crazing can arise due to the ink solvent being withdrawn rapidly from the film. It is most commonly seen when printing on solvent-sensitive plastics such as polystyrene or overprinting heavily pigmented inks. Crazing can be eliminated by reducing solvency, lowering pigmentation, correct resin selection or increasing plasticiser content.

Drying

Inks formulated for display application, i.e. printing of paper, board and self-adhesive plastics, are expected to jet dry rapidly in 20–30 seconds. It is obvious that in these conditions the evaporation rate of the solvent is critical. However, this alone does not govern the speed of drying. Most screen inks require a balance of solvent and diluent to achieve optimum drying. Attention should be paid to solubility parameters and solvent release of various resin systems. Poor solvent release can result in blocking and rewetting in the stack.

Intercoat adhesion

Poor adhesion between ink layers often appears in thermosetting inks, e.g. epoxys, amino and UV curing inks. Overcuring of the first ink layers is the obvious cause but this sometimes can be overcome by reducing the reactivity of the system, e.g. lower catalyst content or in the case of epoxy systems using a resin with high epoxide equivalent. If possible it is best to avoid the use of silicone defoamers in thermosetting inks.

Feathering

This appears as an unsightly spray pattern around the print. Although very often caused by static electricity charges, it can be attributed to poor ink formulation, Feathering can be overcome by increasing extender content, improving solvency, retarding solvent evaporation, inclusion of anti-static agents.

Mesh marking

A pattern of the mesh appears in the dried ink film. Always due to poor flow or overpigmentation. May be overcome by increasing resin content.

Pinholing

This is sometimes described as cissing or reticulation but there is a subtle difference. Pinholing is nearly always due to high surface tension. It is often caused by silicone contamination but can be eliminated by increasing silicone content. Restricting the ink flow will also assist.

Pick up

When overprinting screen inks it is possible to redissolve the first colour down with the overprinting colour. The fault shows itself as staining of the mesh, loss of finish and mesh marking. The problem is caused by using a high solvency blend and therefore may be overcome by including diluent or weaker solvents. Also selecting a less soluble grade of resin will assist.

Rub resistance

Unfortunately it is true to say that many screen inks give some degree of rubbing, this is not unexpected from the heavy ink deposit. It will be found that pigment selection and pigment/binder ratio is important. The use of polythene and fluorocarbon waxes will assist with this problem.

Stress cracking

This occurs when printing on plastics moulded under strain. Contact with ink solvents will release the strains and cause embrittlement and low impact strength. Apart from moulding considerations the problem may be minimised by weakening the solvent mixture or increasing the flexibility of the ink. In the case of printing on PVC it is necessary to reduce the ketone and ester solvent content and with polycarbonate and polystyrene reducing the aromatic hydrocarbon content.

9.13   RECENT AND FUTURE TRENDS

Health and safety

It has generally been agreed in the industry that four glycol ether solvents will be phased out of screen ink formulations, i.e. 2-methoxyethanol, 2-methoxyethanol acetate, 2-ethoxyethanol and 2-ethoxyethanol acetate. The trend is to replace these solvents with propylene glycol ethers and their acetates.

Lead chrome pigments continue to be used in screen inks but will be phased out gradually.

Paper and board inks

In the area of evaporation drying inks the high price of ethyl cellulose has now shifted the emphasis to nitrocellulose. This resin formed the basis of many early screen inks and is now being reintroduced but with a much higher level of performance. A proportion of the market has moved towards UV curing screen inks and this will continue to increase. Almost all of the screen overprint varnish market has now been converted to UV curing.

The increasing popularity of combination driers capable of drying UV curing inks and solvent-based inks offers the screen printer greater scope and flexibility.

Water-based inks

Restrictions on solvent emissions has increased the interest in water-based screen inks. A sector of the market may use water-based inks but the complete answer to reducing solvent emission is to use UV curing inks. Water thinnable UV curings have some technical advantages but will need to be developed further.

Flat plastics

The demand should increase for faster drying inks capable of running on high-speed screen cylinder machines. Formulations will need to be based on hard acrylic resins without losing flexibility and offering good solvent release.

Ultraviolet curing inks are increasing in use on flat plastics. They offer the facility of multi-colour printing without the fear of blocking.

Plastic containers

There is a strong movement towards UV curing inks. These inks offer exceptional gloss and the ability to print up to five colours on a bottle without large inter-stage drying units. Ink formulations will need to be fast curing to match the high-speed production rates and maintain good chemical resistance.

Textile inks

Water thinnable inks should be capable of curing and developing good washfastness without the need for long stoving schedules.

Plastisol inks should improve in cure speed to allow the printer greater safety margin in production. There is an increasing requirement for plastisols with soft hand and therefore attention should be paid to plasticiser content, selection of plasticiser and grade of vinyl resin.

New markets

The screen printing process continues to find new markets and of particular interest is the membrane touch switch market. This is a relatively new multi-million pound industry developing around the world. New ink formulations are demanded by this industry, particularly for the printing of polycarbonate and polyester.

Formulations will need to be developed along the following lines:
(1) Graphic inks with the ability to print on polycarbonate without causing embrittlement. The inks must maintain good flexibility when subjected to several million switching operations.
(2) Transparent filter inks for printing LED windows. These inks are based on dyes and need a very high degree of transparency.
(3) Ultraviolet curing inks and overprint varnishes for graphic work and producing matt textured finishes on glossy plastics. These texturing varnishes must be capable of curing under nitrogen-blanketed lamps.
(4) Conductive screen inks containing metals such as silver, graphite and nickel. These are used to screen print the two separated conductive tracks that perform the switching operation.

CHAPTER 10
Radiation curable systems

The fundamental concepts underlying printing inks as introduced in Chapter 1 have been developed and expanded in the succeeding chapters. It will be apparent, therefore, that a vital feature in the functioning of any ink or varnish is its ability to change rapidly from a fluid to a solid phase.

In the fluid state the rheological performance must be such as to give a satisfactory result with regard to particular printing processes and specific types of press. After drying, or curing, the solid ink film has to be capable of withstanding any additional processing that may be needed as well as meeting the physical and chemical demands imposed by the end user.

As the profitability of the printer depends, at least in part, on a quick turn-around of jobs it is essential that the ink drying takes place in as short a time as possible. The challenge for the formulating chemist is to devise a system which, while remaining stable during the printing phase, is nonetheless capable of undergoing very rapid chemical or physical changes.

It is convenient to divide the events that occur during the drying of an ink into those which involve simple physical changes and those in which new chemical bonds are formed. The term 'cure' is generally reserved for the latter situation for which definite chemical reactions are taking place.

Usually ink drying is brought about by a combination of factors, although it is often possible to assign particular significance to either a chemical or a physical process. With most flexographic and gravure inks, for example, the primary drying is dominated by solvent evaporation, together with penetration if the substrate is absorbent, high levels of solvent being essential to the printing process in this instance. On the other hand, for a paste ink the roles of penetration and evaporation are usually subordinate to chemical polymerisation reactions.

With conventional letterpress and lithographic sheetfed formulations in particular, once an initial tack free condition has been achieved by reducing the solvent level in the vehicle, slow oxidative coupling reactions can take place over a period of hours or days to yield the ultimate properties for the printed film of ink. The chemistry of these ink curing reactions has been discussed elsewhere.

Another approach to formulation for chemical cure is to use resins which contain a high density of sites at which are located very reactive chemical entities. Typical chemical species which are employed for this purpose include activated alkenic double bonds and species possessing steric strain such as the epoxide group.

A feature of such resins is that if activated at only a limited number of sites rapid chain reaction may be initiated. Reactive centres 'travel' through the vehicle raising the molecular weight by combination and quickly converting the fluid into a dense, high cross-linked solid. A cured ink, possessing most of its ultimate performance characteristics can thus be obtained in fractions of a second.

For liquid ink a wide range of reactive resins are available which can be readily dissolved in volatile solvents. Chemical cure is usually initiated thermally and occurs simultaneously with the solvent evaporation.

For paste inks there are fewer high functionality, high reactivity, resins with good ink-making properties to choose from. However, good formulations are possible, often as 100% solids systems.

As these resins contain centres of high reactivity it is a typical feature of their chemistry that once initiated their reactions are self-sustaining, that is they require no further input of energy.

Such vehicles are particularly suited to the so-called 'radiation curing' technology.

Radiation curing processes

Whether effected by chemical means, physical means or by a combination of both, ink drying requires at least an initial input of energy. There are essentially three ways by which the necessary energy can be applied to a wet printed substrate. The printed matter may be passed over a heated surface and the thermal energy conducted through the substrate into the ink layer. Alternatively, warm air may be blown across the surface of the print; a method particularly effective for drying solvent-based ink via forced convection currents. Finally the energy may be transmitted across space from a suitable emitter and absorbed directly by the molecules that comprise the ink vehicle.

Radiation is the term used to describe this passage of energy from a transmitting source to an absorbing body without interaction with any intervening matter. Four categories of radiation have found prominence in the field of ink curing, these are listed in Table 10.1.

Table 10.1   Radiation in ink curing

Radiation type	Radiative particle
Microwave and radio frequency	Photon
Infra-red	Photon
Ultraviolet	Photon
Electron beam	Electron

For technical reasons, that we will consider later, microwave drying is almost exclusively confined to water-based systems. The predominant areas of application are with adhesives and coatings. Uses involving printing inks are relatively minor.

Infra-red (IR), which is simply heat radiation, is relatively cheap and easy to generate and is extensively used by the printing industry. Any hot surface serves as an emitter of IR radiation. The appeal of IR lies in its ability to accelerate the drying of ink systems without the need for highly specialised resin formulations. It is not easy to date the introduction of true infra red drying into the printing industry. However, the recent development of good ceramic IR emitters has ensured its prominence as a radiation curing method.

Although the ability of UV radiation to initiate chemical reactions had been established for a long time it was not until the late 1950s that speciality resins became available which enabled UV curing to be exploited by the printing industry.

The technical and commercial advantages for printing with UV inks, together with the development of relatively simple and inexpensive lamp installations, have ensured a steady growth in its use. Currently ultra violet curing commands the largest volume of the specialised radiation curable ink and varnish market. Spectacular future advances in UV curing technology are not projected, but when considering new installations or when re-equipping, litho and letterpress printers will continue seriously to examine its introduction.

Electron beam (EB) curing is a comparatively new mode of drying printing inks and coatings. Vehicle systems for EB curable formulations are essentially the same as those used for UV systems, but do not require the addition of special initiating compounds.

The complexity, expense and bulk of the beam-generating equipment has limited its introduction into the printing industry. Nevertheless where commercial and technical factors are favourable there are many positive features which make an EB installation an attractive consideration.

10.1 ELECTROMAGNETIC RADIATION AND ELECTRON BEAMS

Electromagnetic radiation

Radiation has been defined as the process by which energy is transmitted across space. Electromagnetic radiation is a particular type of radiation for which a fundamental entity called a photon is the vehicle for energy transfer. Photons are massless particles of sub atomic matter which transverse space at a velocity of c. 3.13×10^8 m/s.

A characteristic of photons is that while certain aspects of their behaviour can be explained by considering them to be small particles of matter, other features can only be interpreted in terms of a travelling spacial wave possessing an electrical and a magnetic component. Hence the term electromagnetic radiation.

The peculiar behaviour of matter and energy at atomic and sub-atomic levels is the province of quantum mechanical theory. Although of mainly indirect interest to the ink technologist some appreciation of the concepts involved are useful in gaining an understanding of the principles which underlie radiation curing. Much of the terminology, used in radiation curing technology has its origin in nuclear physics and spectroscopy.

The standard form for wavelength is to give values in units of nano-metres (nm). However, spectroscopists, by tradition, also use the micron (μm) and the ångström Å as units.

$$1 \text{ Å} = 10^{-10}\text{m}$$
$$1 \text{ nm} = 10^{-9}\text{m}$$
$$1 \text{ μm} = 10^{-6}\text{m}$$

A further unit in popular use is the wavenumber. This is the reciprical of wavelength expressed in centimetres.

For electromagnetic radiations a fundamental relationship exists be-tween the wavelength, λ, the frequency, υ, and the velocity of propagation in vacuum, C:

$$\upsilon = C/\lambda$$

Thus regions in the electromagnetic spectrum can be designated in terms of frequency. The standard frequency unit is the hertz; 1 Hz = 1 cycle/second.

The minute quantity of energy, ΔE, carried by an individual photon is proportional to its frequency:

$$\Delta E = h\upsilon$$

where the proportionality constant, h, is Planck's constant. Hence energy units are acceptable in defining particular bands in the spectrum of elect-romagnetic radiation. It is customary to define the energy of radiation in terms of a mole of photons, called an Einstein. Thus units for energies may be expressed as kJ/mol or kJ/Einstein. Note that from Avogadro's number there are 6×10^{23} photon to the Einstein.

For convenience the spectrum of electromagnetic radiation can be divided into regions. Although there are no clear boundaries, each region defines a frequency range for which distinctive physical and chemical effects are observable when the radiation is absorbed by matter.

These bands of radiation and their inherent effects on matter are sum-marised in Table 10.2.

While all electromagnetic radiation involves the emission of photons, the most efficient type of emitter varies with each particular frequency band. Thus a hot ceramic surface is an effective source of IR radiation but radiates insignificant amounts of UV, while collisions between excited vaporised mercury atoms can emit substantial amounts of UV. This topic is further developed in later sections.

The physiological effects of a particular band of radiation are important as they dictate the degree of precaution that is necessary in order to operate a radiation curing system safely.

Table 10.2 Significant regions in the electromagnetic spectrum and their principal effects

Region	Radio		Microwave		IR		Visible and UV		X-ray
Frequency (Hz)		3×10^{10}		3×10^{12}		3×10^{14}		3×10^{16}	
Wavelength		1 cm		100 μm		1 μm		10 nm	
Energy (kJ/E)		7.1×10^{-4}		7.1×10^{-2}		7.1		7.1×10^{2}	
Effect on matter.	Subtle effects on the electrons and nuclei of atoms.		Changes in rotation and spatial translation of polar molecules.		Changes in the extent of vibrations between atoms in molecules.		Changes in the distribution of electrons bonding atoms in a molecule.		Electronic effects as for UV but involving electrons more closely bound to the atom.
Observable consequence.	Requires specialised equipment.		Increased evaporation of polar molecules.		Increased evaporation. Also chemical reaction, if intra-atomic vibrations lead to bond disruption.		Disruption of chemical bonds with consequential chemical reactions.		Ionisation with subsequent molecular disruption and chemical reactions.
Physiological effects.	None at normal intensities.		Deep tissue burns.		Experience as heat. Surface skin burns.		400–700 nm radiation experience as visible light. UV can cause damage to skin tissue.		Tissue damage due to ionising radiation.

Electron beams

For EB radiation the entity conveying energy across space is a free electron.

Electrons are particles of atomic matter which like the photon exhibit a particle-wave duality. In contrast to photons, however, electrons have a significant, albeit small, mass (9×10^{-28} g at rest) and carry a negative electric charge.

Usually in matter electrons are bound, by the positive electric force from the nuclear protons, in regions of space close to the nucleus.

In a metal there are certain electrons which are not bound to any specific nucleus, but are generally associated with all the atomic nuclei that comprise the metal structure. These so-called conducting band electrons are responsible for the electrical conductivity of the metal.

If energy is supplied, by say heating the metal, some of the conducting band electrons may become sufficiently energetic that they temporarily escape through the metal surface to form a cloud of effectively 'free' electrons. Thus around a hot metal a dynamic equilibrium develops with electrons continuously leaving and drifting back.

An electric circuit can be made under conditions of high vacuum, such that the hot metal forms the cathode, and a positive electrode, the anode, placed a short distance from it, attracts the 'free' electrons away from the cathode.

Taking advantage of either the electric charge or the magnetic moment possessed by the electrons, suitably positioned electrostatic or magnetic fields can be used to manipulate this current of free electrons.

In an EB generator such fields are used to concentrate the electrons into a narrow band, accelerate them to the desired potential energy and if necessary deflect the beam to allow scanning across a substrate.

Electron beam radiations are usually characterised qualitatively by the average kinetic energy possessed by the electrons on leaving the region of acceleration.

For convenience, the energies are often expressed in units of electron volts (eV), 1 eV being the kinetic energy increase acquired by electrons when accelerated across a potential of 1 V:

$$1 \text{ eV} = 1.602 \times 10^{-16} \text{ J}$$

Electron beam systems used in the printing industry provide electrons with energies of between 150 and 300 KeV.

In contrast with electromagnetic radiation where only photons with wavelengths shorter than c. 100 nm have sufficient energy to ionize molecules, electron beams are essentially an ionising radiation. When an electron with high kinetic energy enters matter, collision-induced disruption of the atoms and molecules occurs, resulting in the generation of ionic species. The subsequent formation of radical species which are responsible for the initiation of cure in inks and varnishes is discussed in Section 10.4.

10.2 MICROWAVE AND RADIO FREQUENCY DRYING

Electromagnetic radiation was introduced in an earlier section by describing it in terms of a flux of energy-conveying fundamental particles, called photons, which have the characteristics of a transverse wave. Each photon wave is composed of an electrical and a magnetic component. For photons with frequencies lying in the band between c. 1 and 3000 MHz there is an interaction between their rapidly alternating electric field and the electrical dipoles of polar molecules.

In molecules such as water or ethanol an electrical dipole moment exists by virtue of polarization, introduced via atoms of differing electronegativity, which is not balanced by molecular symetry. Such materials have a bulk, measurable property, their dielectric constant 'ε' the magnitude of which is the function of their internal molecular polarization.

When irradiated by radio frequency (RF) or microwave radiation, molecular motion can be induced in a polar material as the dipoles attempt to align with the oscillating electric field. However, mechanical inertia and interaction with surrounding molecules causes the molecular rotations to lag behind the oscillation of the photon wave. It is this internal friction that is responsible for the temperature rise in an irradiated material. The quantity of heat generated, ΔH, is proportional to the lag and can be quantified in terms of another measurable property of the material; its dielectric loss tangent. Usually written as tan δ.

A relationship exists thus,

$$\Delta H \propto \tan \delta f E^2 \varepsilon$$

where E is the applied field gradient (V/unit length) and f is the frequency.

The field gradient, E, is limited by the voltage that can be applied before either dielectric breakdown occurs and the material becomes conducting or a short circuit 'flash over' takes place between the transmitting array and the earthed press structure. From the equation it can be seen that the efficiency of energy transfer, and hence the heating effects, increases not only with the field strength but also with frequency.

Radiations with frequencies below 400 MHz correspond to the VHF radio region, hence the term RF drying. The more energetic frequencies, up to 3000 MHz, lie in the so-called microwave band.

Applications and formulation

The primary virtue of RF or microwave radiation is similar to that for short-wave IR, namely its ability to provide heat energy instantaneously and evenly throughout the ink film. However, the number of installations in use in the ink industry remains low. In allied industries significant numbers of units are to be found drying, for example, water-based adhesives and similar coatings.

A serious restriction with this type of drying is having to formulate inks using a substantial proportion of either water, very polar solvents, or water blended with polar co-solvent. This tends to bias the technology towards flexo-gravure printing.

Unfortunately water and other polar species tend to have high latent heats of evaporation. Furthermore, despite the cheapness of water, the resins required tend to be expensive and their overall performance is generally inferior to their non-aqueous counterparts.

There are restrictions on the types of pigments that can be used. Metallics can cause overloads and short circuiting with a possible risk of fire. Furthermore some substrates can give rise to problems. Foils cannot be used for the same reasons as metallic pigments. Microwave frequency, in particular, can cause shrinkage and embrittlement of substrates, such as paper, which have high dielectric loss factors.

10.3   INFRA-RED CURING SYSTEMS

Infra-red has been introduced in an earlier section as a band of radiation in the electromagnetic spectrum which, on absorption by matter, produces heating effects. It is of importance to the printing industry as a means of applying energy directly and rapidly into the mass of a wet ink or varnish. This may be contrasted with forced convection drying where the heat energy from hot air is applied to the top surface of the print, and also with conduction drying for which energy, applied at the surface, is transferred slowly through the ink layer by molecular collisions.

The emission of radiation from a hot body is governed by two important physical laws. Firstly, Stefan–Boltzmann's law quantifies the amount of energy emitted with respect to temperature. This law can be expressed mathematically as $E \propto T^4$, where E is the energy radiated per unit area and T is the absolute temperature. Secondly, Wien's law states that the maximum wavelength emitted is inversely proportional to the absolute temperature of the emitter, i.e.

$$\lambda_{max} \propto \frac{1}{T}$$

In summary these laws imply that the higher the temperature of the source, the greater will be the flux of radiant energy and the shorter will be its wavelength.

Application of Wien's fundamental relationship has allowed the design of IR equipments with very specific emission peaks. However, the actual emission profiles for most IR sources are quite broad; especially for those peaking in the longer wavelengths.

For convenience the IR region of the electromagnetic spectrum can be divided into three bands:

Short-wave IR	0.7–2 µm
Medium-wave IR	2–4 µm
Long-wave IR	4–100 µm

The longer wavelength radiation suffers from three characteristic properties which are detrimental to its use in the curing of printing inks:

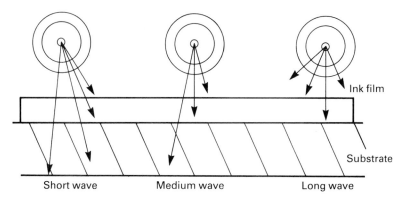

Fig. 10.1 Infra-red penetrations for the three significant wave bands

(1) It is readily scattered and is thus difficult to focus and direct towards the substrate using simple reflectors.
(2) It is extensively absorbed by air molecules, thus lowering the level of effective radiation that reaches the substance and also giving rise to stray convectional heating.
(3) Its penetration into an ink film is limited.

Infra-red radiations in the medium band are potentially more attractive for use by the industry. Focusing is comparatively efficient and absorption by air molecules, although still a problem, is much less than occurs at the longer wavelengths. These radiations can penetrate to a considerable depth into an ink film, but without excessive amounts of energy passing through and overheating the substrate (Fig. 10.1).

Theoretically this wave band should be ideally suited to enhancing the rate of cure of printing ink. Chemical bonds in the molecules from which the ink vehicle is comprised absorb radiation in this region with subsequent stretching or bending deformations. Table 10.3 lists a few typical examples of molecular functional group absorptions.

Deformation is a stage in the process that can lead to complete disruption of the molecular bond and result in a chemical reaction. It should be

Table 10.3 Characteristic IR absorptions for some common organic bonds

Group	Wavelength (μm)	Group	Wavelength (μm)
$- OH$	2.7	$>C = O$	5.7–6.25
$- NH_3$	2.9	$>C - C<$	6.0
$- CH_3$	3.3 / 3.4	$>C - N<$	
		$>C - C<$	8.3–10
$- CH_2 -$	3.41 / 3.5	$>C - O -$	

noted that many important chemical groups absorb just into the region that is arbitrarily defined as long-wave IR.

Short-wave IR radiation is virtually unattenuated by air. It is readily reflected and focused by simple polished reflectors. From Wien's law short-wave emittors need to operate at very high temperatures, 1000–2500 °C, and consequently, by Stefan–Boltzmann's law, the radiated intensities are very large.

Infra-red photons at these short wavelengths readily penetrate the 2–3 μm film thickness that is typical for lithographic printing. On absorption by the substrate its temperature is raised and it is this increase in heat that is responsible for the enhanced rate of cure.

It is debatable as to which of the two effective IR bands is the most suitable to use for curing inks. On the one hand medium-wave IR supplies energy directly into the ink film and specifically into certain chemical bonds. On the other, short-wave IR provides the heat energy more efficiently and stray radiation is less of a problem. In both cases, however, the desired rise in temperature is attained more rapidly and more evenly than can be achieved by convection or conduction and this factor permits faster production rates.

An unexpected drop in gloss level is sometimes reported with medium-wave IR drying. This has been attributed to the general lowering of viscosity, as the ink film warms up, which may result in excessive amounts of vehicle penetrating into the substrate prior to cure. Short-wave IR which produces a more even heating of the ink and substrate is less prone to this problem. However, short wave band IR shows a greater discrimination between colour and surface texture than is evident at the longer wavelengths. Consequently dark and matt areas of print are prone to overheating while, conversely, insufficient absorption may occur in lighter, glossy, regions.

Applications

In general IR is not attractive to flexo and gravure technologies. Here forced convection currents of warm air not only provide thermal energy for chemical cure but simultaneously carry away evaporated solvents.

Application, however, is extensive in the field of letterpress and litho printing of board. A difficulty with paper, and other thin stocks, can be a risk of overdrying which may damage dimensional and mechanical stability.

Many metal decorating installations are fitted with IR in preference to using large oil or gas fuel ovens. Plastic container printing lines often use medium-wave IR as its limited penetration prevents the sensitive substrates softening and deforming.

Formulation for infra-red

One of the attributes given to IR curing is that special formulations are unnecessary. Any of the formulations for oxidation, heatset or penetration inks will give enhanced drying rates under IR radiation.

Formulations which cure by thermally induced chemical polymerisation, polyester, alkyd-amino and epoxy-based inks for example, are particularly suited to IR drying. Addition of organic peroxides which decompose thermally to provide free radical initiation has been suggested as a means of increasing the effectiveness of certain vehicles towards IR curing. However, such a practice will carry an increased risk of instability and should not be used if a long shelf life is required.

A most important use of IR curing is with blocked-acid or blocked base catalyst systems. Penetrating IR radiation ensure that, for such ink, activation sites are generated evenly throughout the whole film thickness.

10.4   ULTRAVIOLET AND ELECTRON BEAM CURABLE INKS AND VARNISHES

Introduction to ultraviolet and electron beam formulation

A generalised formulation for a UV curable ink might have the following composition:

Pigment	15–20%
Prepolymers	35–25%
Monomers (including oligomers)	10–25%
Photoinitiator package	5–10%
Additives, e.g. stabilisers, waxes	1– 5%
	100%

Detailed examination of each component is given in later sections. At this point the formulation may be contrasted with that for a standard oleoresinous ink.

The low viscosity monomers, sometimes termed diluents, play the same rheological adjustment and pigment-wetting role as do the solvents in conventional inks. Usually, however, these monomers are capable of chemical reactions which result in their becoming fully incorporated into the ultimate polymer matrix. Selective use of non-curing solvent is possible in a UV vehicle formulation but this practice is limited.

In the main the very low viscosity reactive monomers are precluded by their volatility and, more importantly, their toxicology. Those which can be used in the printing industry have, if compared with conventional solvents, relatively high viscosities. This factor has restricted the development of UV or EB curable inks for flexographic and gravure printing.

The prepolymer provides the 'hard resin' portion of the formulation. Typically these are derived from conventional synthetic resins, urethanes, epoxides and polyesters are examples, in which residual hydroxyl groups have been reacted with acrylic acid to yield products possessing reactive ethylenic unsaturation.

The radiation sensitive part of the formulation is provided by the photo-

Fig. 10.2 The UV curing unit of a Steinemann varnishing machine at Glossifilm Ltd

initiator package. Only small amounts of the total photoinitiator present are used up in the curing reaction. Residual initiator and photodegradation products are in the main not bound into the cured polymer. Indeed their presence may be detrimental to the overall properties of the film; a fact that has encouraged the investigation of EB curing.

An EB formulation is generally the same as that for UV but without the photoinitiator components.

Ultraviolet and EB curable varnish formulations essentially consist of the vehicle as used for the pigmented ink systems. Appropriate adjustments are made in the selection of the prepolymers and monomers used in order to achieve the required viscosities.

Advantages and limitations for ultraviolet and electron beam inks and varnishes

The specialised nature of the raw materials used in the technology tends to make the formulations significantly more expensive than their conventional counterparts. Equipment to generate the radiation is sophisticated and particularly in the case of EB initially involves a considerable outlay of capital. Nevertheless UV printing has become well established. Table 10.4 summarises the principal advantages which can offset the additional expense incurred by a printer using a UV curable ink.

Advantages which are specific to particular printing processes and those

Table 10.4 Summary of advantages for printing with UV curable formulations

---

1. Dry prints, off the press, allow in-line post-print processing
2. Ink system remains open almost indefinitely during press stoppages
3. Elimination of spray powder leads to smooth prints and a clean press environment
4. No solvent emissions
5. Elimination of stack airing reduces 'holding' space requirements
6. Consistent, low odour level: no odorous species generated by post-cure chemistry
7. Compact lamp systems contrast with large IR or hot air drying ovens. Consequently considerable space savings are achieved, particularly for decoration of non-porous substrates such as plastics and metals

---

to be gained by the use of EB curing will be discussed in later sections. Disadvantages inherent in the use of UV technology are also noted.

Chemistry of ultraviolet initiation and cure

Rupture of chemical bonds

A covalent chemical bond supplied with sufficient energy may dissociate to give either ionic or free radical intermediates.

$$A - B + Energy \begin{cases} \rightarrow A^+ \ B^- & \text{ionic pathway} \\ \rightarrow A^{\cdot} \ B^{\cdot} & \text{radical pathway} \end{cases}$$

The ionic pathway is favoured if A and B are atoms of significantly different electronegativity and if the environment is conducive to stabilising charged ions. Generally speaking the raw materials in EB or UV inks and varnishes are comprised of molecules with limited polar functionality which make them comparatively poor ion solvating species. Consequently reaction mechanisms involving free radical pathways are favoured.

Overall the process of cure may be divided into three stages:

(1) initiation;
(2) polymer chain propagation;
(3) termination.

Initiation

Initiation is a term covering the early part of the curing chemistry and includes the generation of free radicals and their subsequent reaction with chemically active sites on the vehicle molecules. In most instances these sites will be alkenic double bonds, usually activated towards polymerisation by the chemistry of their neighbouring groups.

If irradiated with short-wave UV, alkenes will undergo polymerisation. This reaction is, however, usually too inefficient to be of use in practical ink and varnish systems. Additives, commonly termed photoinitiators, have to be employed to provide an effective source of photochemically generated radicals to start the polymerisation.

Although there are many chemical species that are potentially capable of generating radicals photochemically, UV curing technology has almost exclusively employed the chemistry of the carbonyl function.

The carbonyl group may absorb a photon of electromagnetic radiation whose energy lies in the UV region with the consequent elevation of one of its electrons from a non-bonding orbital into the lowest anti-bonding π-orbital. A so-called $n \rightarrow \pi^*$ transition.

Although the carbon to oxygen bond is weakened by this increased anti-bonding character it remains intact. The spin of the electron does not change during the transition and the resulting species is said to exist as a singlet excited state, $S_0 \rightarrow S_1$. The lifetime of the singlet state is very short and the chances of the uncoupled electron reacting via a collision encounter with nearby molecule is very low. A variety of decay modes from this excited state are possible. The electron may return to its ground state with re-emission of the absorbed radiation. Some or all of the excited state energy may be dissipated into the adjacent bonding system with a consequential increase in the vibrations between atoms. If sufficient vibrational energy becomes concentrated in a particular bond, homolytic dissociation into a free radical pair may result.

Another possible fate for the singlet excited state is for the electron to change its spin, via a quantum mechanical process called 'inter-state crossing'. This is symbolically written as: $S_1^* \rightarrow T_1^*$.

The resulting, so-called, 'triplet' excited state molecule may decay back to the ground state via either the re-emission of radiation or by increasing the inter-atom vibrational energy, with possible molecular fragmentation. However, the lifetime of the triplet state is long enough to allow reactions between what is essentially a diradical and surrounding molecules, during collison encounters. For excited state carbonyl, generated from a photoinitiator in an organic resin system, a hydrogen abstraction is the most likely subsequent reaction.

These theoretical ideas become clearer on examination of the proposed mechanisms for some of the initiators currently in use.

Benzil dimethyl ketal provides a good example of an initator which undergoes a so-called 'Norrish I' molecular fragmentation on irradiation:

$$
\underset{\underset{\displaystyle \text{OMe}}{|}}{\overset{\overset{\displaystyle \text{O} \quad \text{OMe}}{|| \quad |}}{\text{Ph} - \text{C} - \text{C} - \text{Ph}}} \xrightarrow{h\nu} \text{Ph}\dot{\text{C}}\text{O} + \underset{\underset{\displaystyle \text{OMe}}{|}}{\overset{\overset{\displaystyle \text{OMe}}{|}}{\cdot\text{C} - \text{Ph}}}
$$

Benzophenone and its derivatives, on the other hand, owe their effectiveness as photoinitiators to the, Norrish II, hydrogen abstraction mechanism:

$$
\overset{\overset{\displaystyle \text{O}}{||}}{\text{Ph} - \text{C} - \text{Ph}} \xrightarrow{h\nu} \underset{S_1}{(\text{Ph}_2\text{CO})^*} \rightarrow \underset{T_1}{(\text{Ph}_2\text{CO})^*}
$$

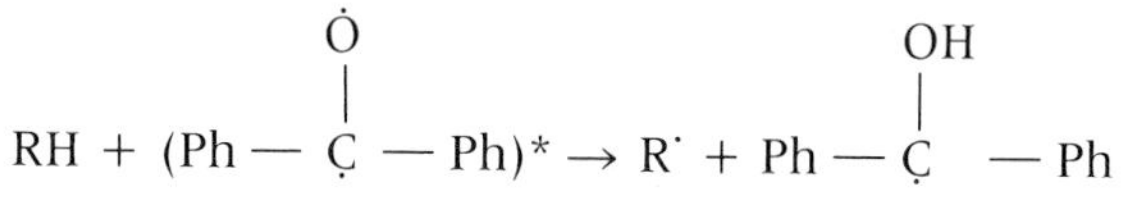

$$RH + (Ph - \overset{\overset{\displaystyle \dot{O}}{|}}{C} - Ph)^* \rightarrow R^{\cdot} + Ph - \overset{\overset{\displaystyle OH}{|}}{C} - Ph$$

Both mechanisms may be operative simultaneously, as for example with diethoxyacetophenone (DEAP):

(a) Fragmentation

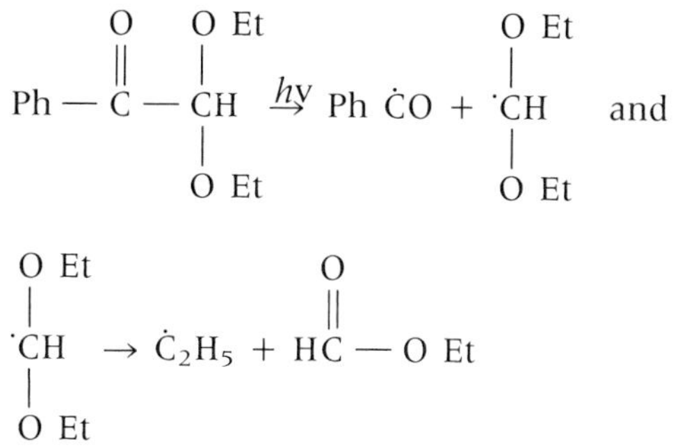

$$Ph - \overset{\overset{\displaystyle O}{||}}{C} - \underset{\underset{\displaystyle O\ Et}{|}}{\overset{\overset{\displaystyle O\ Et}{|}}{CH}} \xrightarrow{h\nu} Ph\ \dot{C}O + \underset{\underset{\displaystyle O\ Et}{|}}{\overset{\overset{\displaystyle O\ Et}{|}}{\cdot CH}} \quad \text{and}$$

$$\underset{\underset{\displaystyle O\ Et}{|}}{\overset{\overset{\displaystyle O\ Et}{|}}{\cdot CH}} \rightarrow \dot{C}_2H_5 + \overset{\overset{\displaystyle O}{||}}{HC} - O\ Et$$

(b) Intramolecular hydrogen abstraction

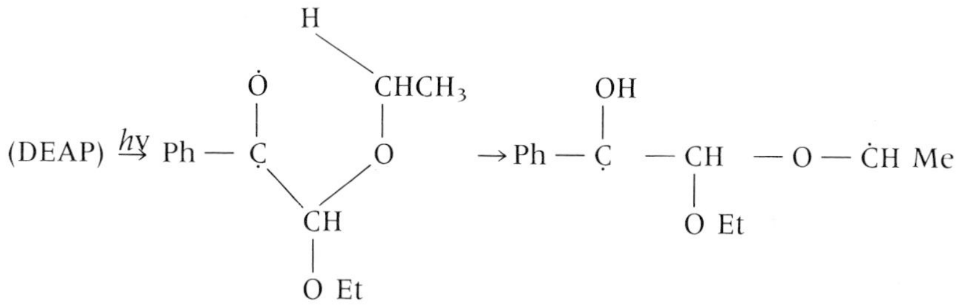

$$(DEAP) \xrightarrow{h\nu} Ph - \overset{\dot{O}}{C} \cdots \overset{H}{\underset{CH}{\overset{CHCH_3}{\diagup}}} O \rightarrow Ph - \overset{\overset{\displaystyle OH}{|}}{C} - \underset{\underset{\displaystyle O\ Et}{|}}{CH} - O - \dot{C}H\ Me$$

Table 10.5(a) lists some of the popular initiators that are available.

Photosynergists

For a Norrish II initiation mechanism to operate effectively a H-donor is necessary. Abstractable hydrogen is usually readily available from suitable sites on the vehicle molecules. Methylene groups adjacent to oxygen in epoxides and polyethers provide examples. It is thus usually unnecessary to add specific hydrogen donors to the formulation.

On their own, the curing performance of the ketone initiators is usually too slow to be of practical use for printing applications. Addition of amine containing molecules has been found essential if ink formulations are to have acceptable cure speeds.

Teritary and secondary amines are found to be more effective than primary. Table 10.5(b) lists a number of the amines which have proved popular in the industry. These amine containing molecules have sites with easily abstractable hydrogen, but this alone is insufficient to explain their enhancing effect on the photoformation of free radicals.

Table 10.5(a)   Some photoinitiators in common use

	Structure	Comments
Benzophenone	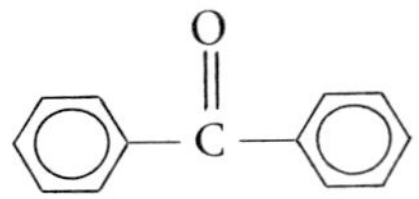	Readily soluble water white solid. Functioning via a H-abstraction mechanism. A cheap moderately effective initiator. Also provides plasticiser and pigment-wetting effects.
Benzophenone derivatives (a) *p*-chloro benzophenone (b) *p*-phenyl benzophenone	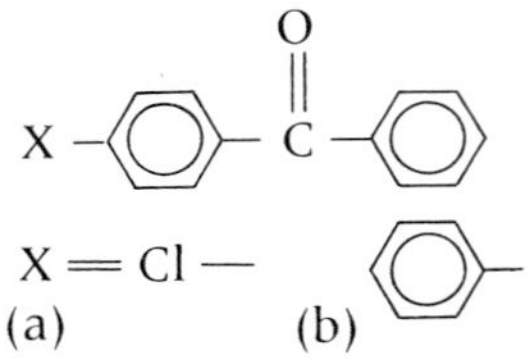 X = Cl — (a) X = (b)	Generally more effective than benzophenone. Functioning via a H-abstraction mechanism. Derivatives are more expensive than benzophenone and are often coloured.
Acetophenone derivatives (a) 1-Benzoyl cyclohexan-1-ol (b) 2-Hydroxy-2, 2-dimethyl acetophenone		Water white liquids or readily soluble solids. Initiating mechanism may be via H-abstraction or fragmentation or a combination of both. Highly effective initiators but are often expensive. Photo fragmentation products may impart an odour to the cured ink or varnish.

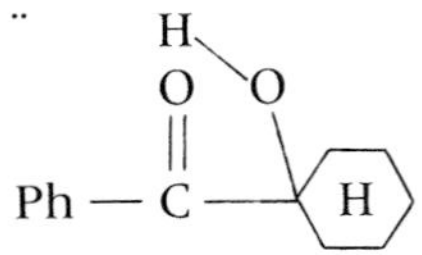

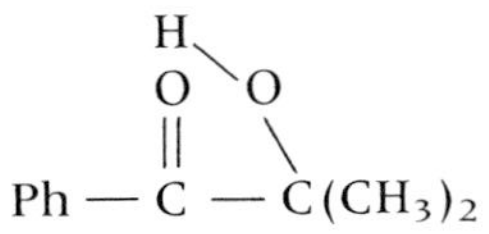

(c) 2,2-Dimethoxy-2 phenyl acetophenone

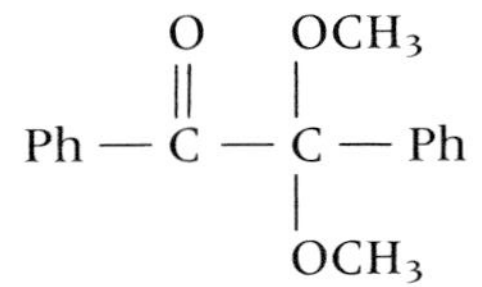

$$Ph—\overset{\displaystyle O}{\underset{\displaystyle \|}{C}}—\overset{\displaystyle OCH_3}{\underset{\displaystyle OCH_3}{C}}—Ph$$

White, reasonably soluble solids. Functioning via a fragmentation mechanism. Initiation is not sufficiently fast for sole use in ink and varnishes, but this class can provide good through cure if used in conjunction with other initiators.

Benzoin ether derivatives
(a) Methyl benzoin ether
(b) *n*-butyl benzoin ether

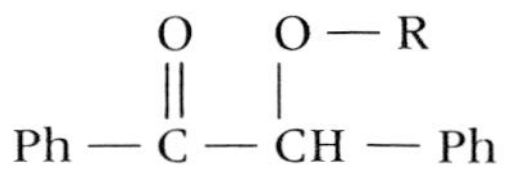

$$Ph—\overset{\displaystyle O}{\underset{\displaystyle \|}{C}}—\overset{\displaystyle O—R}{\underset{}{CH}}—Ph$$

(a) R = CH_3,
(b) R = $CH_3—CH_2—CH_2—CH_2—$

Thioxanthones
(a) 2-Chloro thioxanthone
(b) 2-Isopropyl thioxanthone

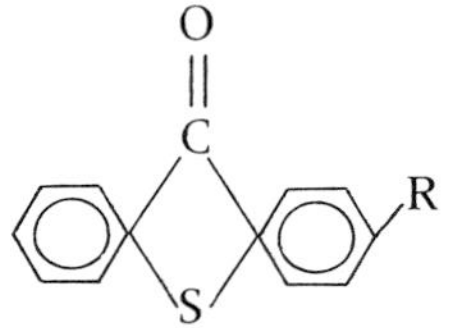

(a) R = Cl —, (b) R = $(CH_3)_2—CH—$

Very effective initiators. Sparingly soluble but forming effective dispersions. Functioning via a H-abstraction mechanism.

These initiators tend to be expensive. They are free from odour but their distinctive yellow coloration precludes their use in whites, varnishes and some pale tints.

A mechanism involving an encounter complex between the amine and excited state ketone has been proposed. Benzophenone and triethanolamine, a commonly employed initiator package, provides an example:

$$Ph_2CO \xrightarrow{h\nu} [PhCO_2]^*$$
$$[Ph_2CO]^* + (HOCH_2CH_2)_3N \rightarrow [Ph_2CON(CH_2CH_2OH)_3]^* \equiv A$$

It is probable that charge transfer takes place. The resulting excited state ion pair complex is called an exciplex.

$$[Ph_2\dot{C}O^- {}^+\dot{N}(CH_2CH_2OH)_3] \equiv B$$

In either case the excited state species readily decomposes into free radicals:

$$A \text{ or } B \rightarrow Ph_2\dot{C}OH + HO\dot{C}H\,CH_2N\,(CH_2CH_2OH)_2$$

An initiator combination which is extremely effective and has been extensively used in UV curable formulation is that of benzophenone and Michler's Ketone. Containing the aromatic carbonyl group Michler's Ketone is a photoinitiator in its own right. Its amino groups are also available to form complexes with excited state species from other Michler's Ketone molecules and also with benzophenone.

Unfortunately this ketone is toxic and is no longer used. The diethyl derivative is less hazardous and, despite its pronounced yellow colour, is widely used by the coatings industry.

Free radical generation by electron beams

Electrons from the beam accelerator bombarding an ink or varnish film have initially very high kinetic energies. The primary process is collision-interaction with electrons from the vehicle molecules. This results in the ejection of secondary electrons with reduced kinetic energies and the formation of molecular ions.

$$e + MX \rightarrow MX^+ + 2e$$

Further ionisation reactions occur as the ejected electrons collide with other vehicle molecules. Successive collisions result in a cascade of electrons with progressively decreasing energy. Ultimately the electrons have insufficient energy to effect ionisation and subsequent collisions with molecules lead only to the formation of excited states.

$$e + MX \rightarrow MX^* + e$$

Under favourable conditions some excited state molecules dissociate into free radical pairs which may initiate a polymerisation process.

$$MX^* \rightarrow M^{\cdot}\,X^{\cdot}$$

The fate of the ions generated by EB radiation is not well understood. Certainly some molecular fragmentation and rearrangement occurs. This accounts for the greater degree of branching found in EB cured systems when compared with the same vehicle after cure by UV initiation using

Table 10.5(b)   Photo synergists

	Structure	Comments
Aliphatic amines		Very effective liquid coinitiators. Water solubility precludes use in lithographic inks. Odour can be a problem. Many cause yellowing of prints on ageing.
(a) N-methyl diethanolamine	$(HO-CH_2-CH_2)_2 N-CH_3$	
(b) Triethanolamine	$(HO-CH_2-CH_2)_3 N$	
Aromatic amines		Effective liquid coinitiators. Insoluble in water. Low odour. More expensive than the aliphatic species.
(a) Ethyl-4-dimethyl aminobenzoate	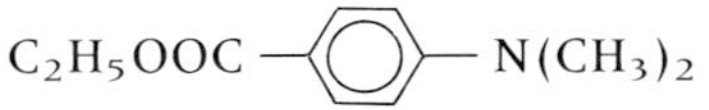 $C_2H_5OOC-\langle\text{ring}\rangle-N(CH_3)_2$	
(b) n-butoxy ethyl-4-dimethyl aminobenzoate	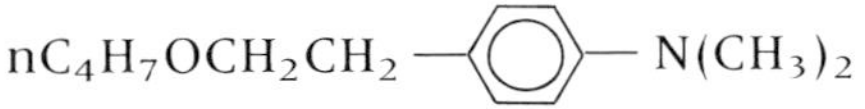 $nC_4H_7OCH_2CH_2-\langle\text{ring}\rangle-N(CH_3)_2$	
4,4'-Diethyl-amino benzophenone	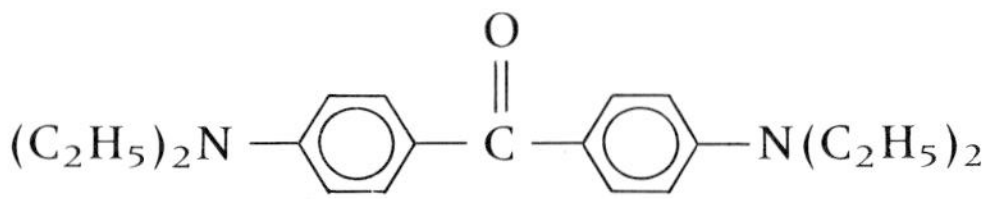 $(C_2H_5)_2 N-\langle\text{ring}\rangle-\overset{\displaystyle O}{\underset{\displaystyle \parallel}{C}}-\langle\text{ring}\rangle-N(C_2H_5)_2$	Very effective coinitiator for use with benzophenone. Highly coloured solid. Very effective with dark or highly pigmented inks.

photoinitiators. It is probable that some electron capture occurs and this provides a further source of free radicals:

$$e + MX^+ \rightarrow MX^* \rightarrow M^{\cdot} + X^{\cdot}$$

Propagation

Once generated, a number of reaction pathways are open to the free radical. Of primary importance is the process which results in chain polymerisation of the prepolymers.

Where $X^{\cdot}$ is any free radical species, generated via an initiation reaction polymerisation may be summarised as follows:

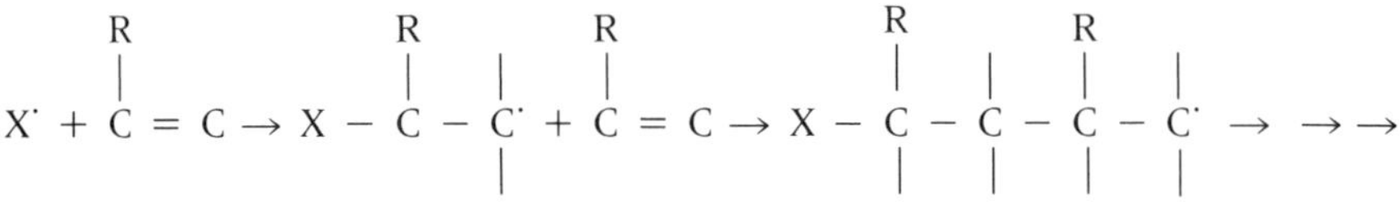

If the vehicle contains species with two or more ethylenic bonds per molecule the possibility exists for the growing chains to crosslink. As the chain lengths and the degree of cross-linking increases the viscosity of the ink or varnish film rises until a densely interlinked solid matrix is attained.

Termination

The rate at which an ink film is cured is dependent upon such factors as the number of initiating radicals generated, the number and reactivity of the unsaturated sites on the molecules of vehicle and the mobility of the growing polymer. Competing with the propagation reaction are any number of so-called termination processes which remove free radicals from the systems.

Ultimate termination occurs as the rate of polymerisation is reduced to zero by a depletion of the unsaturated reaction centres and also by a decrease in molecular mobility as the viscosity of the vehicle increases. Termination can result from dimerisation reactions; e.g.:

$$P^{\cdot} + P^{\cdot} \rightarrow P - P$$
$$P^{\cdot} + In^{\cdot} \rightarrow P - In$$
$$In^{\cdot} + In^{\cdot} \rightarrow In_2$$

P = growing polymer; In = initiating species.

Analysis of a cured coating containing benzophenone, for example, shows that a significant quantity of benzopinacol is formed:

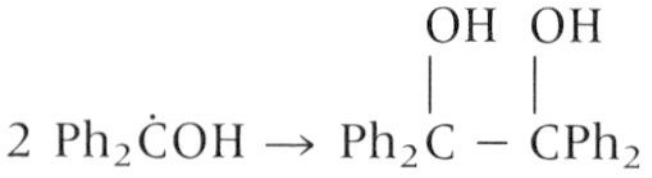

$$2\ Ph_2\dot{C}OH \rightarrow Ph_2\overset{\displaystyle OH}{\underset{\displaystyle |}{C}} - \overset{\displaystyle OH}{\underset{\displaystyle |}{C}}Ph_2$$

A serious termination process from the viewpoint of curing thin coating films arises from reactions with atmospheric oxygen. A free radical centre from either initiator or the growing polymer may react with oxygen to form a relatively stable peroxide:

$$X^{\cdot} + O_2 \rightarrow XOO^{\cdot}$$
$$XOO^{\cdot} + RH \rightarrow R^{\cdot} + XOOH$$
$$XOO^{\cdot} + R^{\cdot} \rightarrow XOOR$$

In addition to retarding the rate of cure peroxides can be detrimental to the chemical and physical properties of the cured film.

Ultraviolet and electron beam cure in thin films

With the exception of silk screen application printed ink and varnish films for UV or EB curing are essentially 'thin'. Being under 10 μm in thickness, surface effects become significant in relation to the bulk properties. The formation of peroxides in the upper surface layer has already been discussed.

Oxygen is an effective deactivator or 'quencher' of the triplet state and this process will further inhibit surface cure. With UV curable formulations two factors are active which counteract to some extent the oxygen inhibition. Firstly, the preparations usually contain an amine as a photosynergist. Although the mechanism has not been clearly established there is evidence to suggest that the effect of oxygen inhibition is substantially reduced by the presence of amine.

A second and more significant feature is the high density of free radical formation near to the surface on UV irradiation. The rate of formation of free radicals, and consequently the rate of cure, is proportional to both the intensity of the radiation and the concentration of initiator.

For an unpigmented, relatively thick film the Lambert–Beer equation provides an approximate description of the change in intensity with depth an concentration.

$$I = I_0 \mathrm{e}^{-kcl}$$

where I_0 = the intensity radiation at the surface;
  I = the intensity at a depth l from the surface;
  k = the absorption coefficient of the radiation absorbing species;
  c = the concentration of the absorbing species.

This simplified picture shows that the level of radiation falls off rapidly away from the surface of the film. Adjustment must be made to I_0 for partial reflection at the surface. For a thin film the situation is further complicated by multiple internal reflections from the substrate and the underside of the top surface. For a pigmented ink layer, scatter as well as absorption by the pigment particles, has to be taken into account.

The photoinitiator itself is an absorbing species. Thus a high level of initiator while enhancing the formation of radicals near to the surface, reduces penetration of the radiation and can be detrimental to through cure. Indeed over use of initiators typically results in soft, 'cheesy' films.

Electron beam radiation is more penetrating and results in the generation of proportionally more initiating radicals deep in the ink layer.

Through cure, even with thick highly pigmented inks is seldom a problem for EB systems. However, a comparatively low density of free radicals is created in the top surface where oxygen inhibition effects are dominant. Consequently inert gas blanketing is essential for effective EB surface cure.

Photoionic initiation

Free radical polymerisation of prepolymer resins containing activated carbon-carbon double bonds, has provided the dominant chemical process for UV curable formulations. There are, however, molecules that can be photo-decomposed to yield reactive species capable of effecting ionic polymerisation. Prepolymer resins containing epoxide (oxazirane) rings are used extensively with this technology.

The photo-decomposition of certain diazonium salts has been exploited in some areas of the surface coating industry, e.g.

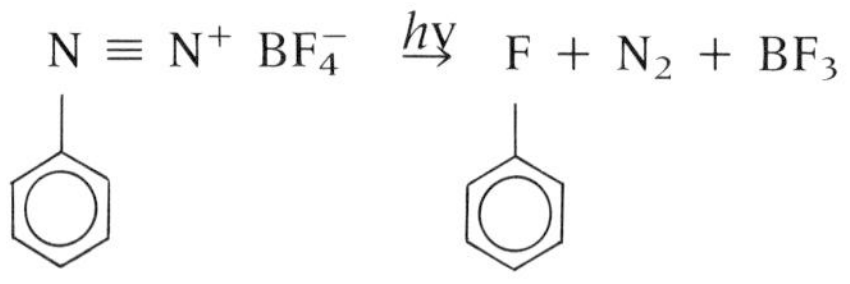

$$N \equiv N^+ \ BF_4^- \ \xrightarrow{h\nu} \ F + N_2 + BF_3$$

The electron-deficient boron trifloride can initiate polymerisation by acting as a Lewis acid:

$$BF_3 + R - \overset{\displaystyle O}{\overset{\displaystyle \diagup \ \diagdown}{CH - CH_2}} \rightarrow R - \overset{\displaystyle \bar{O}BF_3}{\overset{\displaystyle |}{CH}} - CH_2^+$$

However, the evolution of nitrogen, which forms disruptive bubbles in the cured film, together with poor thermal stability has restricted the use of this system in printing technology.

Recently, considerable development effort has been directed towards photoinitiator having a general formula $Ar_n X^+ Y^-$. Typically X is an atom of sulphur, iodine or phosphorus and Y is an anion of the type BF_4^-, PF_6^- or AsF_6^-. Examples of such molecules are

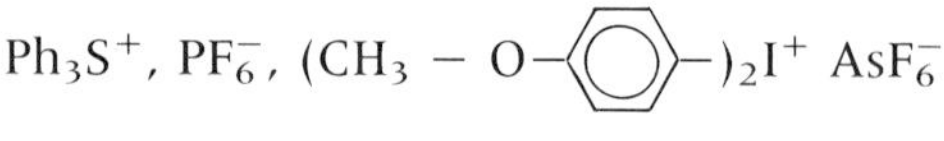

$$Ph_3 S^+, \ PF_6^-, \ (CH_3 - O\!\!-\!\!\bigcirc\!\!-)_2 I^+ \ AsF_6^-$$

These initiators are used to effect cationic polymerisation of epoxide functionalised resins and diluents by a mechanism which is still the subject of speculation.

Formulation for this means of cationic cure is not, however, without difficulty. Many potentially useful low viscosity epoxide monomers are sensitisers and skin irritants. The effectiveness of the initiators is reduced by trace amounts of basic impurities and they do not function at all in the presence of many pigment types.

None the less end-use properties of epoxide resins are sufficiently advantageous to ensure photoionic curing has a place in the future.

Prepolymer resins for election beam and ultraviolet formulations

Activated carbon–carbon double bonds can be introduced into a resin structure via a number of readily available unsaturated compounds. However, the rate of reactivity for an alkene group in a free radical polymerisation reaction is very sensitive to the nature of its neighbouring groups. Broadly, reactivity follows the order: vinyl < styrenyl < allyl < methacrylate < acrylate.

As far as the printing industry is concerned, it has been chiefly resins with acrylate functionality that have provided the necessary reactivity to bring about adequate cure at the throughput rates demanded by modern high-speed presses.

Table 10.6 lists the three main resin types that have provided the pre-polymer for EB and UV ink formulations.

Table 10.6   Some acrylated prepolymers

Resin type	Typical properties
Epoxy acrylate	Inexpensive, fast curing
Urethane acrylate	Toughness, chemical resistance
Polyester acrylate	Low viscosity, good pigment wetting

Epoxy acrylates

Both aromatic and aliphatic epoxy acrylates are in use. These are usually medium to high viscosity fluids that combine fast cure with generally good ink-making properties. The higher molecular weight, solid epoxy acrylates tend to give poor film properties. Possibly that is because their low mobility reduces the degree of cross-linking that can occur during cure. Manufacture of the resin is typically carried out by reacting acrylic acid with hydroxyl groups from a standard epoxy resin:

$$R(OH)_n + n(CH_2 = CH - COOH) \rightarrow (CH_2 = CH - COO)_n R + nH_2O$$

A disadvantage of epoxy acrylate resins is their comparatively poor pigment-wetting ability. Consequently, UV inks based entirely on simple epoxy acrylate resin tend to have low gloss and poor lithographic performance.

Improved pigment wetting, and at the same time enhanced substrate adhesion, may be obtained from resins with residual hydroxyl functions. Such functionality is achieved by opening an epoxide ring directly with acrylic acid:

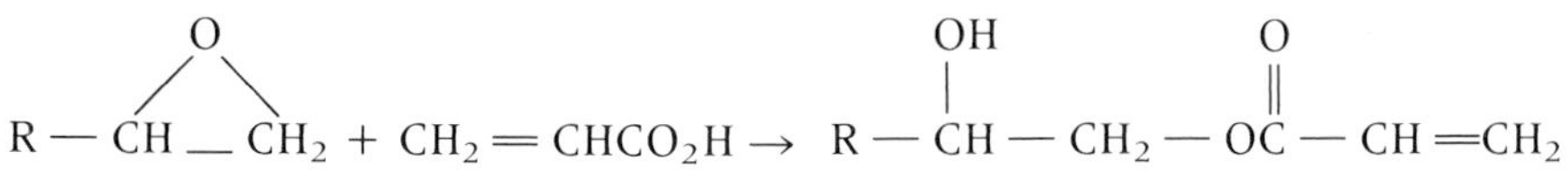

$$R - \overset{O}{\overset{\diagup \diagdown}{CH - CH_2}} + CH_2 = CHCO_2H \rightarrow R - \overset{OH}{\underset{|}{CH}} - CH_2 - O\overset{O}{\overset{\parallel}{C}} - CH = CH_2$$

where R is the resin backbone.

Typical aromatic resin backbones are those derived from bisphenol A epichlorohydrin condensates:

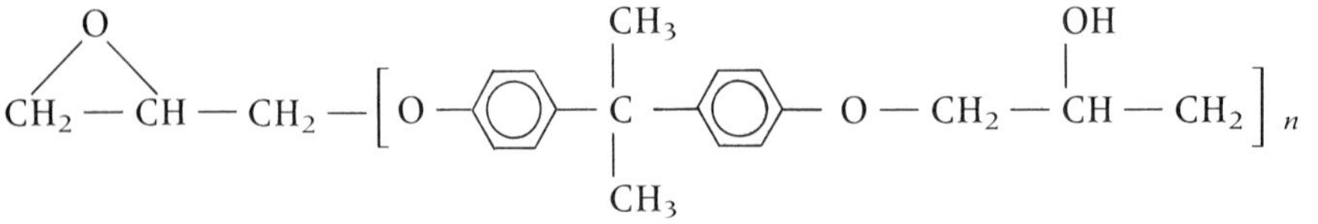

$$CH_2 - CH - CH_2 - \left[O - \bigcirc - \overset{\overset{CH_3}{|}}{\underset{\underset{CH_3}{|}}{C}} - \bigcirc - O - CH_2 - \overset{\overset{OH}{|}}{CH} - CH_2 \right]_n$$

Notable features of such a structure are the rotational flexibility imparted by the bisphenol A group, the chemical inertness of ether linkages and the hydroxyl sites which enhance adhesion and wetting while providing centres for further reaction if required. For example, to improve lithographic performance further, these residual hydroxyl functions are sometimes esterified with long chain fatty acids.

Unsaturated natural oils such as soya bean and linseed can be readily epoxidised. Subsequent reaction with acrylic acid yields radiation curable resins with good ink-making characteristics. Unfortunately their reactivity is low and their use in formulation can lead to a retardation in cure speed.

Polyurethane acrylates

Inks and varnishes formulated on urethane resins are noteworthy for their toughness and chemical resistance. Urethane acrylates provide a class of resins for UV curable products which more or less retain these desirable characteristics.

Products derived from resins based on aliphatic isocyanates have good colour retention on exposure to light or on stoving. These resins, however, are rather expensive. The somewhat cheaper aromatic urethane acrylates suffer the drawback that they tend to yellow significantly on exposure to light or heat.

A urethane acrylate prepolymer can be prepared by reacting isocyanate groups from a urethane resin with, for example, hydroxyl ethyl acrylate:

$$OCN - R - NCO + 2(HO - CH_2 - CH_2 - O\overset{\overset{O}{\|}}{C} - CH = CH_2) \rightarrow$$

$$CH_2 = CH - \overset{\overset{O}{\|}}{C} - O - CH_2 - CH_2 - O - \overset{\overset{O}{\|}}{C} - NH - R - NH - \overset{\overset{O}{\|}}{C} -$$

$$O - CH_2 - CH_2 - O - \overset{\overset{O}{\|}}{C} - CH = CH_2$$

where R is the urethane backbone.

For the simplest case where R is an aromatic di-isocyanate, such as toluene di-isocyanate (TDI), a lack of molecular flexibility leads to high-viscosity resins. Generally, therefore, prior to acrylation the isocyanate is chain extended by reaction with a polyol or alternatively with a polyester or a polyether which itself has residual hydroxylic functionality. In this way complex mixed prepolymers often, with extensive branching, can be constructed.

Polyester acrylates

The wide range of comparatively low cost polyesters that are available as precursors make these an attractive group of radiation curable materials.

Chemical and physical properties are very varied. General polyester prepolymers provide good pigment wetting and consequently make good inks, with if required, excellent lithographic performance. Their adhesion, particularly to non-porous substrates such as tinplate and plastic, is especially noteworthy.

The intrinsic ester linkages are more labile than are ethers thus when contrasted with epoxy acrylate the polyester acrylates tend to give products with inferior chemical resistance, particularly towards alkali. Furthermore, especially for the lower molecular weight polyester acrylates, reactivity can be poor and surface oxygen inhibition effects can be a serious problem.

On the positive side an advantage with many polyester acrylate prepolymers is that they have a low or relatively low viscosity. They may thus be used to replace part or all of a diluent monomer in a formulation.

Preparation is either by reaction of acrylic acid with residual hydroxyl groups from a polyester precursor or by reacting residual acid groups from a polyester resin with a hydroxyl acrylate such as 2-hydroxy ethyl acrylate. For example:

$$ROH + CH_2 = CH - \overset{\displaystyle O}{\overset{\|}{C}} - OH \rightarrow CH_2 = CH - \overset{\displaystyle O}{\overset{\|}{C}} - OR + H_2O$$

or

$$RCO_2H + HO - CH_2 - CH_2 - O - \overset{\displaystyle O}{\overset{\|}{C}} - CH = CH_2 \rightarrow$$

$$R - CO_2 - CH_2 - CH_2 - O - \overset{\displaystyle O}{\overset{\|}{C}} - CH = CH_2 + H_2O$$

where R is the polyester condensation product from a polyol and a poly basic acid and contains one or more residual hydroxyl or acid functions. However, in practice the esterification process is by no means simple.

The volatility and instability towards spontaneous polymerisation inherent with acrylic acid and its derivatives limit permissible reaction temperatures to under 100 °C. To drive the esterification towards completion strong acids, for example methane sulphonic acid, are required. These have to be removed from the final product together with any excess acrylic acid.

To remove the water formed, azeotroping solvents such as benzene have to be introduced and subsequently removed. Also inhibitors such as nitrobenzene and phenothiazine are added to prevent free radical polymerisation.

In addition to all these potential contaminates, the reaction conditions are such that saponification of the ester linkages in the resin backbone are possible. Acrylation of the original polyols can then occur to yield low molecular weight products with possible toxicity hazards.

Acrylated polyols and polyether acrylates

Resins which contain ether linkages that derive from polymerisation of epoxides are customarily called epoxide resins by reference to their origins. Of considerable importance are the polyether acrylates derived by chain extension of simple polyols, prior to acrylation. Usually extension is limited to only a few ether units and these products are thus often termed oligomers.

These polyether acrylates and the products derived by direct acrylation of polyols are often comparatively low viscosity liquids. As such their use is considered in more detail in the section covering diluents. It is possible, particularly in a varnish formulation, for the vehicle system to be comprised almost entirely from such molecules.

Water-reducible resins

The wood coatings and furnishing industry has made considerable use of UV curable water-based systems. Usually these have been formulated around emulsions of polyester acrylates. Such resins have so far attracted little interest in the printing field, being unsuitable for pigmentation and also lacking the gloss necessary for varnish formulations.

A few experimental water-thinnable resins have been developed. Despite the potential advantages of producing low viscosity inks cheaply these have so far met with very limited success. Possibly this is because their film-forming properties have not been adequate.

Other resins

In the coatings industries early UV curable formulations were based on blends of styrene and unsaturated polyesters. The unsaturation being introduced into the polyester from such precursors as itaconic acid and maleic anhydride. These systems generally had too slow a rate of cure to be of use for printing applications. Such resins can, however, be used in conjunction with the more reactive acrylated resins and diluents. Indeed it is posible to formulate a UV curable ink containing portions of fully saturated polyester, triazine or epoxy resins. These become 'occluded' in the polymer matrix but do not necessarily co-polymerise with the acrylated species.

Diluents for electron beam and ultraviolet and formulation

Diluents are materials used to reduce the viscosity of formulation to levels suitable for the required method of printing. They may also be needed to solubilise a solid prepolymer and they certainly contribute to pigment wetting.

A diluent can be a non-reactive solvent which is either ultimately lost from the formulation by penetration and evaporation or else remains in the cured film as a plasticiser. In UV and EB technology, however, diluents are more frequently species which have acrylate functionality and are thus capable of co-polymerising with the main resin system to provide a 100% solids formulation. Potentially there are a large number of molecular species which could function as reactive diluents. While these may vary in the number of acrylate groups they possess and can be complex branched structures, they are usually distinct from resins in that they are discrete molecular units. Hence the term 'monomers' is often applied to them.

In practice, the number of monomers that are acceptable for use by the printing industry are comparatively few. Many good viscosity-reducing acrylates are eliminated simply because they present a toxic hazard, but odour and volatility can also give problems.

Monofunctional monomers

Monofunctional monomers lack the ability to cross-link and, although good viscosity reducers, their excessive use may lead to poor film properties. An important use can be to impart flexibility to the print film. In this respect they may be thought of as reactive plasticisers. Two good examples in this respect are isodecyl acrylate (IDA) and phenoxy ethyl acrylate (PEEA).

In the coatings industry generally, 2-hydroxy ethyl acrylate (2EHA) is very widely used but like many of the acrylated low molecular weight alcohols, toxicity, volatility and odour restrict its use in printing applications.

Isobornyl acrylate and N-vinyl pyrolidone are two other examples of monofunctional monomers with excellent viscosity-reducing proporties, and which have good compatibility with multifunctional resins. However, neither are accepted by the printer as both have very pungent odours.

Difunctional acrylates

Most of the popular difunctional monomers are derived from simple diols. In this class may be included the low molecular weight species derived by acrylating the product of bisphenol A condensation with ethylene oxide:

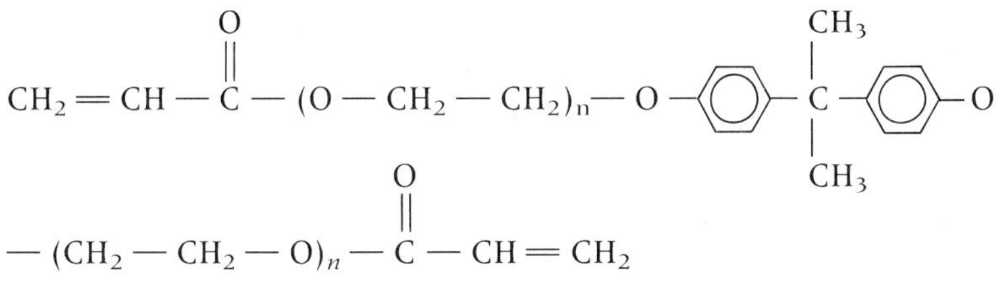

$$CH_2 = CH - \overset{\overset{\textstyle O}{\parallel}}{C} - (O - CH_2 - CH_2)_n - O - \bigcirc - \overset{\overset{\textstyle CH_3}{|}}{\underset{\underset{\textstyle CH_3}{|}}{C}} - \bigcirc - O$$

$$- (CH_2 - CH_2 - O)_n - \overset{\overset{\textstyle O}{\parallel}}{C} - CH = CH_2$$

where $n = 1$ or 2.

Table 10.7   Difunctional acrylate diluents

		Viscosity at 25°C (cp)
BDDA	1,4-Butane diol diacrylate	6
HDDA	1,6-Hexane diol diacrylate	7
NPGDA	Neopentyl glycol diacrylate	7
DEGDA	Diethylene glycol diacrylate	8
TEGDA	Triethylene glycol diacrylate	25
PEGDA(n)	Polyethylene glycol diacrylate	$10 \sim 30$
TPGDA	Tripropylene glycol diacrylate	16
DDA	2,2-Dionol diacrylate	$c.$ 1000

The viscosity of these molecules is not particularly low but this is compensated for by their good pigment-wetting and ink-making properties.

Table 10.7 lists a selection of the more important of the available difunctional acrylate monomers. Manufacture of this type of product is usually by reaction of a diol with acrylic acid.

Toxicity factors, in particular skin irritation and the potential sensitisation, limit the use of many difunctional diluents in ink and varnish formulation. Many of these chemically aggressive materials are, however, employed in the coatings industries and also for silk screen ink production. In both these areas, adoption of safe handling procedures by the user permits the more liberal approach to formulation.

Of the species listed in Table 10.7 all except TPGDA and DDA are classified as severe irritants. TPGDA remains as the best difunctional viscosity reducer that is not excluded by the SBPIM list of unacceptable materials. PEGDA for molecular weights of about 500 is classified as only a moderate irritant, but usefulness is restricted by the comparatively high viscosities prevalent at these molecular weights.

Trifunctional acrylates

The three most important polyols that have provided the industry with useful trifunctional monomers after acrylation are pentaerythritol, trimethylol propane and glycerol.

Pentaerythritol triacrylate (PETA)

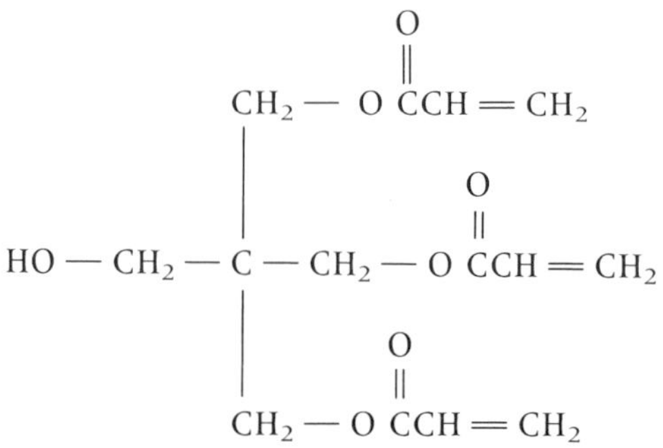

$$
\begin{array}{c}
\quad\quad\quad\quad\quad \overset{\displaystyle O}{\overset{\displaystyle \|}{}} \\
CH_2 - O\ CCH = CH_2 \\
| \\
\quad\quad\quad\quad \overset{\displaystyle O}{\overset{\displaystyle \|}{}} \\
HO - CH_2 - C - CH_2 - O\ CCH = CH_2 \\
| \\
\quad\quad \overset{\displaystyle O}{\overset{\displaystyle \|}{}} \\
CH_2 - O\ CCH = CH_2
\end{array}
$$

and trimethylol propane triacrylate (TMPTA)

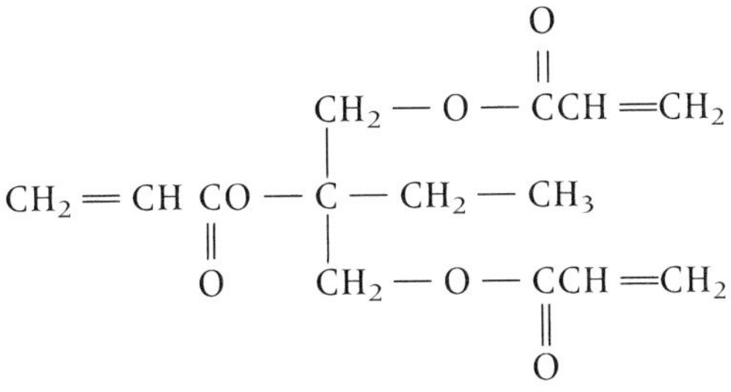

$$
\begin{array}{c}
\qquad\qquad\qquad \text{O} \\
\qquad\qquad\qquad \parallel \\
\qquad\qquad CH_2 - O - CCH{=}CH_2 \\
\qquad\qquad\qquad | \\
CH_2 {=} CH\ CO - C - CH_2 - CH_3 \\
\qquad\ \parallel \qquad | \\
\qquad\ \text{O} \qquad CH_2 - O - CCH{=}CH_2 \\
\qquad\qquad\qquad\qquad \parallel \\
\qquad\qquad\qquad\qquad \text{O}
\end{array}
$$

combine good viscosity reduction with fast cure and the ability for extensive cross-linking.

Unfortunately, PETA is a severe irritant and TMPTA, while classified as a moderate irritant, is a potential skin sensitiser and its use is declining.

To overcome these defects a new generation of trifunctional acrylates have been developed for which the hydroxyl functions have been ethoxylated or propoxylated prior to acrylation. This procedure has provided a number of proprietary diluents, the precise composition of which is often undisclosed by the supplier.

Propoxylation of glycerol with subsequent acrylation provides an example:

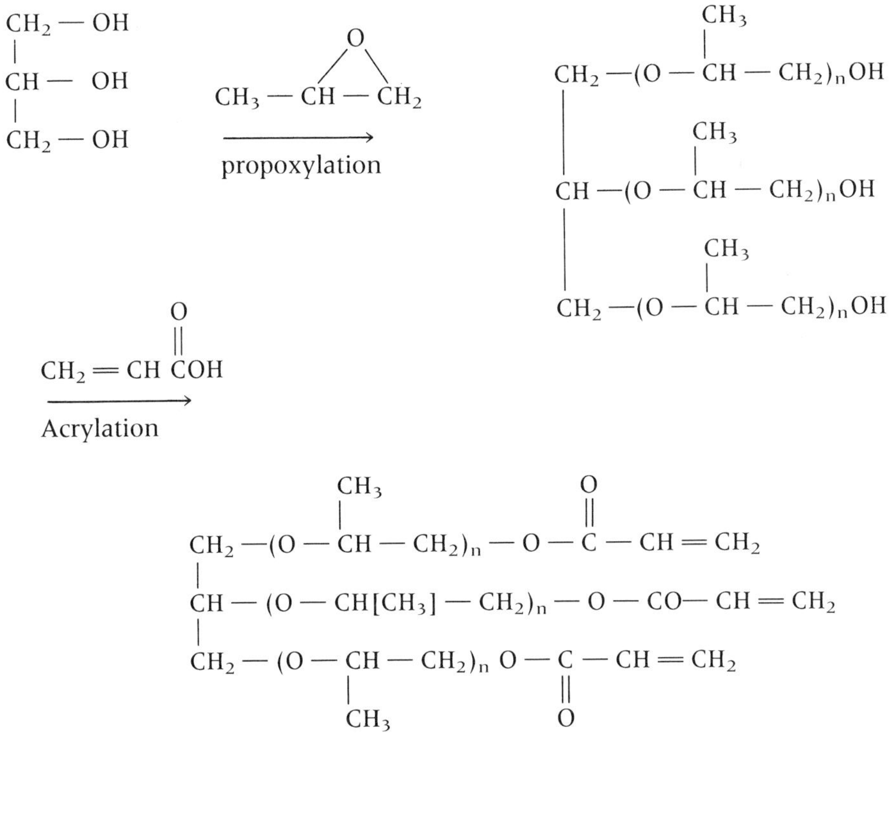

$$
\begin{array}{c}
CH_2 - OH \\
| \\
CH - OH \\
| \\
CH_2 - OH
\end{array}
\qquad
\begin{array}{c}
\qquad\quad O \\
\qquad\ \diagup \diagdown \\
CH_3 - CH - CH_2 \\
\xrightarrow{\quad\quad\quad\quad} \\
\text{propoxylation}
\end{array}
\qquad
\begin{array}{c}
CH_3 \\
| \\
CH_2 - (O - CH - CH_2)_n OH \\
| \qquad\qquad CH_3 \\
| \qquad\qquad | \\
CH - (O - CH - CH_2)_n OH \\
| \qquad\qquad CH_3 \\
| \qquad\qquad | \\
CH_2 - (O - CH - CH_2)_n OH
\end{array}
$$

$$
\begin{array}{c}
\qquad\quad O \\
\qquad\quad \parallel \\
CH_2 {=} CH\ COH \\
\xrightarrow{\qquad\qquad}
\end{array}
$$

Acrylation

$$
\begin{array}{c}
\qquad\qquad CH_3 \qquad\qquad\qquad O \\
\qquad\qquad | \qquad\qquad\qquad\quad \parallel \\
CH_2 - (O - CH - CH_2)_n - O - C - CH{=}CH_2 \\
| \\
CH - (O - CH[CH_3] - CH_2)_n - O - CO - CH{=}CH_2 \\
| \\
CH_2 - (O - CH - CH_2)_n\ O - C - CH{=}CH_2 \\
\qquad\qquad | \qquad\qquad\qquad \parallel \\
\qquad\qquad CH_3 \qquad\qquad\qquad O
\end{array}
$$

Typically $n = 1$ although the chemistry involved obviously leaves scope for variation in the molecular structure.

These products have viscosities lying in a range from about 1 to 10 poise. In combination with high molecular weight prepolymers they exhibit reasonable viscosity cutting power. The potential for high cross-link density and good cure speeds, characteristics of the non chain extended forms, are retained.

Of special importance are the very low skin irritation indices and the negative results in sensitisation studies that have been found for this class of materials.

High functionality monomers

Very few acrylate monomers with functionality of four or more have found favour in the ink industry. Pentaerythritol tetra-acrylate is a crystalline solid. Dipentaerythritol (mono hydroxy) penta-acrylate is liquid with acceptable toxicity, but in common with many multi-functional acrylates, the monomer is expensive. Although high functionality provides for a fast curing and high cross-link density, excessive brittleness may be imparted to the films.

Non-reactive plasticising diluents

Any of the typical plasticising molecules can be used to improve flow out and impart flexibility to the cured film. High boiling esters, highly branched alcohols and alkyl phosphates have been used. Benzophenone is interesting in that in addition to its photoinitiation properties it can reduce viscosity, aid pigment wetting and is a powerful solvent.

Formulation principles for ultraviolet curable inks

The formulation of any printing ink has to take account of:
 (1) the method of application;
 (2) the substrate;
 (3) the end-use requirements.
In these respects UV curable inks are no exception.

However, as the major portion of the vehicle system has to be comprised of acrylated species the range and number of combinations of raw materials available is more restricted than would be the case with conventional oleoresinous inks.

The required body and tack of a UV curing ink or varnish is primarily achieved by balancing carefully selected prepolymers and monomers. Addition of diluent to a formulation usually results in a lowering of tack, yield value and plastic viscosity. This is in contrast to experience with conventional formulation where it is often possible to adjust the rheological parameters individually by choosing the right type of solvent. Further, the viscosity reducing power of a reactive monomer is invariably less than that of a typical petroleum distillate.

Most pigment types can be used in UV ink formulation, but some caution in selection is needed for three reasons. Firstly, as a class, acrylated resins do not possess particularly good pigment-wetting characteristics. Consequently some pigments that might produce reasonable ink in a conventional vehicle are not suitable for UV or EB formulation. Secondly, some pigments are prone to initiating 'dark' polymerisation reactions in the inks during storage. Such pigments must either be avoided or else a suitable stabiliser must be incorporated in the formulation. As these reactions may only manifest themselves after several months storage, evaluation of individual pigments and stabilisers can be a problem. Acceleration oven tests may be useful but are not always reliable. Finally some pigments inhibit cure. This may be the result of chemical reaction with the initiating system, or it may arise because the pigment has significant UV opacity in critical special regions. Particular care is needed in formulating black inks.

All in all, formulating effective UV and EB ink probably calls for more skill and a greater understanding of the physics and chemistry involved than is necessary to achieve satisfactory results for corresponding inks based on conventional resins.

Formulation for ultraviolet curing lithographic inks

By far the largest application area for UV curable inks is sheet-fed lithographic printing of paper and board. Formulations of inks for this type of printing will thus be considered in some detail. Many of the principles introduced, however, apply generally to UV ink formulations regardless of the method of printing.

A typical UV curable dark blue litho ink might have the formula:

Phthalocyanine blue pigment (CI Pigment Blue 15.3)	16.0
Carbon black pigment (CI Pigment Black 7)	4.0
Epoxy acrylate resin	30.0
Fatty acid modified epoxy acrylate	25.0
Monomer viscosity modifier	8.0
Benzophenone initiator	8.0
Coinitiator	3.0
Aromatic amine photosynergist	4.0
Stabiliser, waxes, etc.	2.0
	100.0

The choice of resin blend reflects the need for achieving good lithographic properties while maintaining an effective rate of cure. If the ink is to be applied to a substrate lacking in absorbency, metal, plastic or polyethylene coated board for example, it may be necessary to use a portion of a polyester or a polyurethane acrylate to provide improved adhesion; albeit at increased expense and possibly a sacrifice of cure speed.

A first choice for the photoinitiator package would be benzophenone combined with an aromatic amine. However, the example chosen is a comparatively dark colour and a coinitiator such as 2-chlorothioxanone would probably be required in order to attain a satisfactory curing speed.

Alternatively the use of a cleavage initiator such as dimethoxyphenyl acetophenone might be considered. This latter choice would be important in the case of a white ink where the yellow tint from a thioxanone would be unacceptable. However, the ultimate end use must always be kept in mind. A side reaction of the above initiator leads to the formation of methylbenzoate:

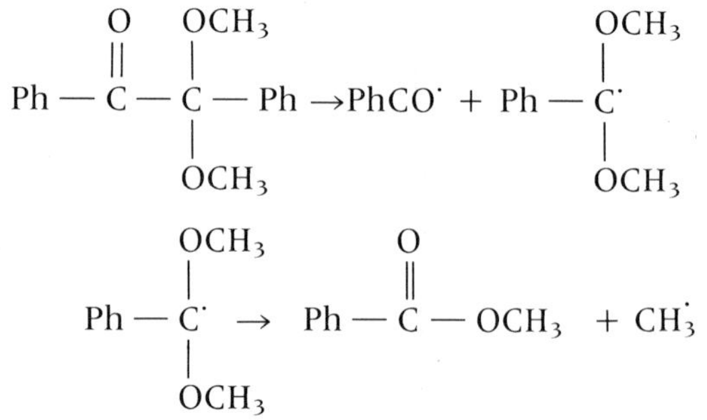

$$Ph-\overset{\overset{\displaystyle O}{\|}}{C}-\overset{\overset{\displaystyle OCH_3}{|}}{\underset{\underset{\displaystyle OCH_3}{|}}{C}}-Ph \rightarrow PhCO^{\cdot} + Ph-\overset{\overset{\displaystyle OCH_3}{|}}{\underset{\underset{\displaystyle OCH_3}{|}}{C^{\cdot}}}$$

$$Ph-\overset{\overset{\displaystyle OCH_3}{|}}{\underset{\underset{\displaystyle OCH_3}{|}}{C^{\cdot}}} \rightarrow Ph-\overset{\overset{\displaystyle O}{\|}}{C}-OCH_3 + CH_3^{\cdot}$$

The odour and tainting risk introduced into a cured ink by this compound might be unacceptable if the end use involved food packaging.

Even when one of the more exotic initiators is selected, it is customary to leave some or all of the benzophenone in the formulation to retain the advantage of its pigment-wetting and plasticising properties.

As the lithographic process is involved, use of water-miscible amines should be avoided. Ignoring the insignificant effect that these compounds may have on the emulsification of the inks, there would still be a risk of them leaching into the fount solution with a subsequent reduction in cure efficiency. Thus the more expensive, but water insoluble, N, N-dialkyl aniline derivatives are preferred as photosynergists for litho inks.

The body of a UV ink is controlled by the rheology of the prepolymer, the level of pigmentation, and the pigment type, adjustments to the tack and viscosity being made by the addition of monomer until a suitable printing rheology is attained. The selection and concentration of the monomer can best be made on the basis of experiments.

For process colour printing, tack grading of the inks for each different hue may be required to ensure satisfactory trapping. However, when process printing with UV inks it is popular, and relatively easy, to fit a lamp between print units, thus partially curing each colour before the next is applied and reducing the risk of backtrapping.

Ultraviolet curable letterpress inks
A UV curable letterpress ink might have the following generalised formula:

Pigment	23.0
Diluent monomer	20.0
Resin prepolymers	40.0
Benzophenone	8.0
Triethanolamine	3.0
Waxes	3.0
Additives	3.0
	100.0

The relatively short bodied inks required by the letterpress process are usually readily achieved from the available prepolymer and diluents. As the letterpress process does not require the sophisticated rheology demanded of litho inks the formulator may economise and select the relatively cheap epoxy acrylates as the total prepolymer component, provided that substrate and end-use requirements do not dictate otherwise.

The considerations for the selection of an initiator system for a letterpress ink are similar to those that apply to litho inks. However, as there is no involvement of water, cheap and highly effective alkanolamines, such as triethanolamine, can be used. While obtaining an adequate cure for blacks and very dark colours is a general problem with UV formulations, the existence of abnormally thick ink films in the regions around the image perimeter where squash may occur, aggravate this difficulty for letterpress formulations. In such situations the use of a combination of initiators, to take full advantage of all the available radiation, may be the only means of attaining a satisfactory cure.

Ultraviolet labelling inks

A rapidly growing market area for UV letterpress inks is in reel-to-reel printing of a narrow width web for the production of labels. Where paper substrates are employed a standard formulation based on, for example, epoxy acrylate prepolymer can be used. Increasingly, however, this type of label printing is turning to PVC substrates. To formulate inks which give adequate cure and adhesion on this type of substrate the use of more specialised resins, for example the polyester and polyurethane acrylates, are needed. The selection of monomer can contribute greatly to the success of this type of formulation. Unfortunately, most of the monomers which would provide a bite into the substrates and hence provide good adhesion, are not permissible in formulations on the grounds of their irritancy.

Inks for dry offset application on plastics and metal

Closely allied to the formulations for letterpess printing on impervious flexible substrates are inks for the decoration of rigid preformed-plastic and metal containers. Inks are required that will offset from a relief zinc or polymer plate to a blanket prior to transfering the image to the substrate.

A typical formula for such an ink would be:

Lithol rubine pigment (CI Pigment Red 57.2)	10.0
Talc (CI Pigment White 8)	2.0
Epoxy acrylate	14.0
Polyester acrylate	35.0
Polyurethane acrylate	20.0
Monomer, e.g. TPGDA	8.0
Initiator package	8.0
Waxes, surfactants, etc.	3.0
	100.0

Such an ink can be formulated using resins that are similar to those suitable for a letterpress ink. Rheologically, however, the inks require a body and flow characteristics, more aligned for a lithographic application. The inks have to transfer well and have adequate flow on the, sometimes, comparatively small printing units.

Of prime consideration is adequate adhesion and abrasion resistance. Curing is often carried out 'on mandrel', where despite the limited space for lamps and the short dwell times, a sufficient cure has to be achieved to withstand almost immediate handling and stacking. Both the resin and initiating system have to be chosen with this in mind. Often a blend of a number of different initiators is used to optimise the rate of cure.

Selection of the monomer can be critical to the success of these formulations, particularly with respect to adhesion. Ideally the diluent should slightly attack the substrate, so providing points of anchorage. Unfortunately, materials which behave in this manner are largely precluded from use in the ink industry because of their associated toxicity. The mechanical resilience of these inks can be enhanced by the careful use of selected surfactants and waxes and this, at least in part, can counteract their relatively poor adhesion.

Ultraviolet curable web offset inks

Despite the advantages of their being free from solvent emission UV web inks have made very little impact in the industry. Successful formulations have been run using inks based on those raw materials that are available for use in litho and letterpress inks. A few prepolymers are marketed which are specifically recommended by the manufacturers as suitable for use in web inks. The primary requirement is to make inks possessing good lithographic performance at high running speeds. Most currently available resin systems fail fully to meet this specification. Furthermore, because of the relatively poor pigment-wetting properties inherent with acrylated resins, the gloss levels fall somewhat short of the high standards set by conventional web offset ink.

Also critical is the fact that this type of printing involves immediate rereeling with high internal reel pressures. This allows no time for post curing and any inadequate cure will cause severe set-off and possibly blocking problems. Web UV remains an area of future challenge for the industry.

Ultraviolet curable silk screen

Silk screen ink films are typically between 20 and 60 μm thick. Although they can be much thinner, there is still a substantial depth for the radiation to penetrate when compared with the 3–5 μm thickness of a letterpress ink. It might be expected that this would present insurmountable problems for formulating an efficiently curing ink.

However, silk screen technology has two characteristics which have favoured the development of successful UV curable products.

(1) The pigmentation levels are low and while high proportions of extender are used this is invariably transparent to UV radiation.
(2) Printing speeds are relatively slow, e.g. 130 m/min.

The wide spectrum of applications for silk screen printing, ranging from printed circuitry to glass bottle decoration, has provided many areas for the successful introduction of UV inks.

A typical formula for a UV curable silk screen ink suitable for decorating a plastic container would be:

Pigment	8.0
Calcium carbonate extender	20.0
Epoxy acrylate prepolymer	25.0
Monomer, e.g. TMPTA	27.0
Polyester acrylate	5.0
Initiator (1) 2-chlorothioxanone	4.0
Initiator (2) Benzophenone	5.0
Amine, e.g. TEA	3.0
Waxes, surfactant, etc.	3.0
	100.0

Ultraviolet inks for flexography and gravure applications

Although a number of experimental formulations for flexo or gravure application have been developed, there has been little commercial interest for UV curing ink within these technologies. The primary problem has been to formulate inks which have the necessary low viscosities while using only toxicologically safe monomers. However, even if this difficulty can be resolved by, for example, the advent of water-reducible systems, there are other considerations to be taken into account.

Film weights and press speeds are high when compared with those of oil ink printing and in the event of insufficient cure there is greater risk of set-off or blocking in the reel.

Furthermore, the geometry of a flexo or a gravure press leave little room for interdeck lamp units. For multicolour work this would necessitate wet on wet printing; a difficult procedure at liquid ink viscosities, and especially so on the impervious flexible packaging substrates such as film or foil.

Ultraviolet curable varnish and coatings

Varnishes and coatings comprise a major portion of the market of UV curing products. Varnishing of printed material is carried out to provide protection and promote gloss. The ultimate target in both respects is the quality set by film lamination. Ultraviolet roller coat varnishes almost achieve this criteria for gloss and lay, while the UV printing viscosity products although an order of magnitude lower in gloss are still superior to conventional oleoresinous or emulsion varnishes.

Table 10.8 lists the various popular methods of application together with an indication of the viscosity range most suitable for each technique.

A UV varnish may be required to coat any of the multitude of substrates in general use in the industry. Problems with soak-in can arise if low viscosity varnishes with very little body are applied to porous material.

Table 10.8 Application methods and categories of UV curable coatings and Varnishes

Method of application	Approx. range of viscosities
Out of line: roller coating	Under 5 poise
In-line coating Head	8–12 poise
Dry offset via the damping system	20–30 poise
Dry offset via a printing plate	40–150 poise
Lithography and letterpress	100 poise to paste

Further complications are imposed by the need to overprint inks. A spectrum of possibilities exist some of which are summarised in Table 10.9.

Obtainable gloss levels usually decrease as the varnish viscosity increases. Indeed litho and letterpress printing varnishes can give disappointing results so far as appearance and gloss are concerned. Such products are mainly used to give a seal over the inks in circumstances where it is essential to leave certain image areas uncoated, for instance where clean flaps for subsequent gluing are necessary.

The majority of UV varnishing is carried out by a dry offset method. Application via the damping system can give excellent results, but control of film weight is not easy and incorrect press conditions may result in severe 'orange-peel' effects. In the UK application from the duct via the inking system is the more popular process. Film weight is readily controllable and, if desired, glue flaps can be left unvarnished by means of cutting out appropriate areas of blanket.

The above techniques suffer the disadvantage of utilising one of the printing units. A recent trend has been the introduction of in-line end-of-

Table 10.9 Substrates for overprinting with UV varnishes

Substrate	Comment
Unprinted substrate	Soak-in on porous substrates can occur.
Part printed: part unprinted	Disparity in gloss levels may be a problem.
Wet conventional inks	Gloss levels often poor and variable. Adhesion may be poor. Special ink formulations advised.
Dry conventional inks	Interfacial adhesion may be poor. Reticulation can be a problem. Low wax inks advisable.
Wet UV inks	Results usually good. Gloss levels may vary over different shades.
Dry UV inks	Results usually good.

the-press coating heads. Such an operation, of course, only allows an all-over varnish application. Despite the commercial advantages of an in-line single pass varnish procedure, a substantial amount of UV varnishing is still carried out off-line with trade varnishes via traditional single or double head coating machines.

Turning to end-use properties, the frequent need for adhesive receptive regions has already been mentioned. for an all-over varnish coating the question of its potential gluability thus arises. Experience suggests that this is very dependent on the type of adhesive that is used and also on the conditions of temperature, pressure and dwell time employed in its application. Such factors are usually beyond the control of the formulation. Often the best that can be said is that if formulated with minimal levels of surfactants and waxes then the varnish is potentially gluable. Similar considerations apply to the question of foil blockability; another end-use feature often demanded from a UV varnish.

Scuff and rub resistance for UV curable varnishes can be excellent but to optimise these properties requires a correct balance of the prepolymers and the photoinitiators. Waxes will improve rub resistance, but gloss is sacrificed. Similarly silicones will provide desired levels of surface slip, but incorrect use may lead to reticulation and poor interfilm adhesion.

Because each varnish has to be matched to its proposed mode of application as well as its end-use requirements, it is far from easy to give a generalised formulation. For application at around 10 poise the following would provide a starting-point for an acceptable product:

Bisphenol A epoxy acrylate monomer	13.0
Trifunctional polyether acrylate monomer	60.0
Tripropylene glycol diacrylate	5.0
Amine acrylate	12.0
Benzophenone	6.0
2,2-Dimethoxy-2-phenyl acetophenone	3.0
Silicone slip additive	1.0
	100.0

For a higher viscosity product a quantity of prepolymer resin would be needed. For a lithographic or letterpress varnish calcium carbonate would be required to provide body and other modification might be necessary. For example, 2–3% of talc would probably be added to such a formulation to reduce misting; albeit with a sacrifice in the gloss levels.

The choice of initiator is usually restricted by the need for water-whiteness. Thus use of the thioxanones with their yellow coloration is limited.

As with inks any end-use constraints on the allowable levels of odour and taint must be borne in mind when selecting the raw materials and in particular the photoinitiators.

The use of an acrylated amine is popular when formulating varnish. Free amines have a tendency to give rise to a surface bloom, which may not manifest itself until several hours after curing.

Electron beam curable inks and varnishes

The essential difference between a UV curable system and one designed for cure by EB radiation is that the latter does not require any photo-initiators. Thus the 8% or so of the UV formulation which comprises the initiator package is available for replacement by prepolymer or monomer.

The removal of this non-curing portion from the ink or varnish formula offers two significant advantages. Firstly, the 'purer' vehicle would be expected to provide enhanced mechanical and chemical resistance properties after curing. Secondly there is no risk of any odour, taint or toxicological contamination that might arise from either unreacted photo-initiator or its photo-degradation products.

In all other respects an EB formula can be identical to one designed for UV curing. Some manufacturers of acrylated propolymers and monomers have made specific recommendations concerning which of their products are best suited to EB curing. However, current practical experience has not shown any outstanding distinctions between EB and UV performance for most of the popular resins in use and selection is probably best based on their potential printability and end-use performance. An exception to this, however, is the possibility that some resins are less prone to surface oxygen inhibition and could thus give a greater guarantee of satisfactory cure in the event of a partial failure in the inert gas blanketing.

Electron beam radiation is sensitive to the density of the formulation but not to its optical properties. Thus the formulation is free from the constraints that exist which strongly absorb dark colour in UV ink technology. This leads to consideration of a further advantage for EB curing, namely that the technique gives greater assurance of in-depth cure of the print film. Again risk of uncured, migratory, odorous or tainting material in the cured film is minimised. These factors are important as, although the present number of installations are very limited, the current trend suggests that the immediate future for EB lies in printing of reeled food packaging materials. For such operations the customary long print runs make EB viable commercially and the end-use requirement in the case of food packaging materials will demand maintenance of negligibly low levels of contamination from the print.

10.5   RADIATION CURING EQUIPMENT

Equipment for microwave and radio frequency drying

Restrictions on freqencies and usage
Specific frequency bands are allocated by governments for industrial use in order to avoid interference with communication channels. In the UK these are centred on 13.6, 27.1, 84 and 168 MHz, RF, and 896 and 2450 MHz in the microwave region. Even at the permitted frequency there may be restrictions on the allowed power levels and further regulations regarding shielding.

Sources and press mountings

At radio frequencies, radiation can be generated using standard oscillating circuitry employing a conventional triode valve. In the microwave region radiation is produced by subjecting electrons to strong alternating magnetic fields in special 'magnetron' valves.

Radio frequency radiation can be conveyed from the power source along coaxial cables to rod or plate electrodes extending across the web and mounted a few centimetres above the substrate. The field gradient can, to some extent, be controlled by the geometry of the transmitting array. Microwave radiation is transferred from the generator along specially dimensioned tubes, called wave guides, which terminate above or around the substrates.

In either case the radiation source and its associated power supply are usually too bulky to be sited close to the substrate pathway on the press.

Equipment for infra-red drying

Sources

Long-wave emitters are constructed with a heating element embedded in ceramic or surrounded by a mineral insulator, e.g. magnesium oxide, which in turn is encased in metal. These sources have operating temperatures of 400–800°C.

Medium-wave IR is generated from a filament source electrically heated to between 800 and 2000°C. Usually the filament is suspended in an unsealed quartz tube which can be up to 10 feet (3 m) in length. Power ratings are of the order of 7–8 kW. The life span of such a source is measured in years.

Short-wave IR is produced from a tungsten filament operating at up to 2200°C. Lamps of various shapes have been produced, usually these are sealed envelopes made from borosilicate glass. Typical power ratings are about 300 W with about 80% of the radiation emitted as short-wave IR. High powered halogen lamps, which can take a load of up to 1 kW, are available, but these have a comparatively short life span.

Reflectors

There is debate over the relative merits of using focused or unfocused radiation. In either case reflectors are needed to utilise fully emission

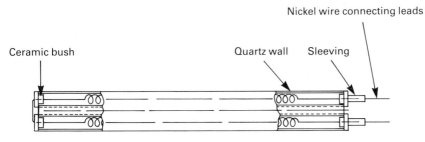

Nickel wire connecting leads

Ceramic bush   Quartz wall   Sleeving

Fig. 10.3 A medium-wave IR radiator

from the rear of the lamp. Simply anodised aluminium reflectors with air or water cooling are available, but for high efficiency gold plating is necessary.

For medium and, especially, short-wave sources it is common to coat parts of the interior of the lamp envelope with a reflective material. This improves efficiency and can help prevent overheating.

Press mounting

Infra-red lamps are invariably fitted at the delivery end of the press. Usually a number of lamps are required. With short length, say 10 cm lamps, a herringbone configuration with a degree of overlap across the delivery is popular.

Mountings usually position the lamps about 5 cm above the substrate. An additional reflector plate can be placed opposite the lamps and beneath the substrate to further increase efficiency and prevent unwanted heating of the press structure.

Maintenance and facilities

Lamps are comparatively cheap, have very long lives, are mechanically robust and do not require a special power supply. Full output is obtained virtually instantly and lamps may be switched on and off as necessary during temporary stops, thus avoiding the need for mechanical shuttering.

Air or water is needed to cool the lamp mountings and reflector housing. For short-wave IR good reflection, minimal air losses and the close proximity to the substrate means that stray radiation is not a problem. With medium-wave systems screening panels and additional cooling may be required to prevent unacceptable heating of the press superstructure.

Ultraviolet curing equipment

Sources

The most widely used, practical, source of UV radiation is the mercury vapour lamp. Commercially available lamps can be divided into three categories (1 Torr $\equiv$ 1mm mercury):

(1) low pressure ($10^{-3} \rightarrow 10^{-2}$ Torr);
(2) medium pressure ($c.$ 10^2 Torr);
(3) high pressure (10^3 Torr).

Typically the standard design is a linear tubular envelope of quartz glass with electrodes at either end. The envelope retains a charge of mercury and an inert gas, usually argon.

Energy is applied, which vaporises and subsequently ionises the mercury. Collision interaction between mercury atoms, mercury ions and free electrons possessing high kinetic energies, results in the population of excited states in some of the mercury atoms and ions. Decay transitions, back to the ground state, occur with the emission of radiation. The dominant transitions, and hence the spectral emission profile, depend upon the temperature and pressure prevailing in the tube.

Low-pressure lamps emit UV predominantly at two wavelengths, 185 and 234 nm. Penetration into ink or varnish films is relatively poor for radiation of these wavelengths. This fact, together with the low radiative intensity, only $10-15$ W/cm, has limited the use of low-pressure lamps in the printing industry.

High-pressure mercury lamps emit a continuum throughout the UV and visible into the IR wavelengths. These lamps have relatively short lifetimes, are necessarily small in dimesion and are little used in UV curing technology.

Linear medium pressure lamps with lifetimes in excess of 5000 hours can be constructed in lengths exceeding 1 m. They have efficient spectral output with two dominant UV bands at 365 and 366 nm, and are thus ideal as a press-mounted UV source.

Two types of medium-pressure mercury lamps are used by the industry. The most popular, the electrode lamp, utilises a high voltage discharge to excite the mercury atoms. A warm-up period of $10-15$ minutes is required with this type of lamp, over which time the mercury vaporises and full spectral output is developed. If switched off, condensing mercury prevents restrike for several minutes.

The effective life of this lamp is limited by the slow erosion of the hot electrodes. Also vaporised tungsten from the electrode is deposited on cool inner wall of the quartz envelope gradually reducing transparency. These problems can be overcome by eliminating the internal electrode completely.

In the so-called electrodeless lamp, microwave radiation is transmitted through the quartz envelope to excite the mercury atoms. Such lamps claim to offer a longer life, rapid attainment of their full spectral output, and almost instant restarts. They suffer the disadvantage of being short, 250 cm is a typical maximum length. Also the necessary microwave source is expensive.

Heat output and shuttering

The medium-pressure mercury source has a significant output in the IR. This can be an advantage to curing as the polymerisation reactions proceed faster at elevated temperatures. However, the lamp envelope requires some means of cooling, which is usually achieved by mounting the tube in the air stream of an extraction system. Air or water cooling is also needed for the reflector and any other parts of the press exposed to incidental radiation.

In the event of temporary line stoppage electrode lamps are normally left on to avoid incurring a restrike delay. To handle such events without the risk of fire, from the over-irradiated substrate, interlocked shutter assemblies are generally fitted.

Ozone generation

Ozone is generated from atmospheric oxygen on exposure to UV radiation in the region of 185 nm. In addition to being an irritating gas, ozone may damage polymeric parts of the press and can reduce cure efficiency by absorbing the longer wavelength UV. The use of an effective extraction, venting into the outside atmosphere, is therefore essential.

Lamps are available for which the quartz has been doped to filter out short-wave UV and hence eliminate ozone production. However, under practical conditions such lamps are found to show a significant loss of cure efficiency.

Press assembly and reflectors

Lamps are generally fitted a few centimetres above the substrate at some convenient position on the press. Typically two or three 80 W/cm full length electrode type lamps might be fitted at the delivery end of a four-colour sheet-fed press. However, on a multicolour press it is not unusual to find interdeck lamps sited between each printing unit. The high voltage transformer and controls for electrode lamps can be situated away from the press if desired.

Microwave excitation systems often have the lamp and power source housed in a single module. Multiple units are mounted across the press with overlap as required, to provide a continuous field of radiation.

Reflectors are obviously needed to utilise all the radiated output from a tubular lamp. Three basic configurations are in common use:

(1) elliptical;
(2) parabolic;
(3) planar (non-focusing).

Elliptical reflectors are the most popular design in use, particularly on web or sheet-fed presses where the substrate is flat. The lamp may be positioned so that a partially focused beam is obtained and hence a compact band of UV radiation is applied across the substrate.

With the lamp positioned at the focus of a parabolic reflector, a partially parallel beam is achieved. This configuration is often recommended for the curing of printed preformed containers and other non-flat substrates.

Non-focusing reflectors are less popular. There is however, some evidence that greater cure efficiency is achieved with the longer exposure at a lower flux density which occurs with planar reflectors as compared with an equal dose of high flux exposure from a focused system.

Cool ultraviolet lamp systems

The need arises for partial, or total, removal of the IR component from a UV lamp to protect sensitive substrates such as plastic films from over-heating and deforming. Additionally some authorities suggest that lower substrate temperature leads to improved handling and stacking immediately off the press because the cured cold ink film carried less risk of set-off.

Filtration of the output by way of doping the quartz used to construct the lamp envelope has been mentioned. Dichroic reflectors, which absorb IR while reflecting the short UV, provide another possible means of partially removing the unwanted radiation. Alternatively a number of designs provide so-called cool UV, by circulating water in a quartz jacket either around the lamp or immediately above the substrate.

In all cases a significant loss of UV radiation also occurs.

Electron beam curing equipment

Introduction
In contrast to those for IR, UV and to a lesser extent microwave, systems for EB curing are bulky and expensive to install and run. Three factors contribute to these disadvantages. Firstly, the necessary high vacuum requires ancillary pumping and control equipment. Secondly, wherever the high-energy electrons impinge on a surface the risk of generating X-ray occurs. Unlike the electrons whose penetration in air is limited to a few centimetres, the X-ray radiation may present a health hazard several metres from the source. Thus extensive lead shielding in conjunction with a failsafe radiation detection system is necessary. Finally, for the reasons discussed earlier, inert gas blanketing is essential for efficient EB cure.

Installation is made at the delivery end of press with the beam window sited a few millimetres above the substrate.

Source design for electron beams
Two configurations of beam generator have been developed to the point of commercialisation:
  (1) the scanned beam type;
  (2) the linear cathode type.
In both cases the source of electrons is a hot metal cathode situated in a high vacuum.

For the scanned beam system the electrons originate from a point cathode and are accelerated towards a positively charged grid. Passing through the grid they are foccused into a narrow beam by a magnetic lens. This accelerated pencil beam of high-energy electrons is deflected on its way to the emission window by a varying magnetic field which is applied via electromagnet coils positioned close to the beam pathway. In this manner the beam is scanned across the width of the window.

With the linear cathode design a filament cathode, which runs the effective width of the press, is suspended in a vacuum tube. Electrons leave the cathode and are accelerated towards a positive grid placed between the cathode and the emission window.

A curtain of unfocused electrons pass through the window to bombard the ink or varnish film which travels a few millimetres below. Although the linear cathode assembly is unable to deliver the very high energies at high flux densities that can be achieved with the scanner type, they have a compact design and are more easily shielded against X-ray emission. They are thus the more popular model for fitting to a printing press or coating machine.

A variation on the basic linear cathode design gives enhanced output by having an array of parallel cathode emitters within a single vacuum tube.

The dose
The operational characteristics of an EB curing system are defined by consideration of two features:

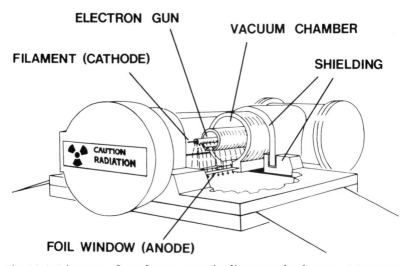

ELECTRON GUN

VACUUM CHAMBER

FILAMENT (CATHODE)

SHIELDING

CAUTION
RADIATION

FOIL WINDOW (ANODE)

Fig. 10.4 Diagram of an electro-curtain, linear cathode type, EB system

(1) the dose delivered;

(2) the beam penetration.

The dose is measured in terms of the quantity of energy absorbed per unit mass of irradiated material. Units in general use are the megarad (Mrad) and the Grey (Gy). The relationship between these units is:

$$1 \text{ Mrad} = 10 \text{ kGy} = 10 \text{ kJ/kg}$$

Dose is directly proportional to the number of electrons in the beam flux as measured by the current, I, flowing in the cathode circuit. It is inversely proportional to the throughput rate, S, of the substrate undergoing cure. Dose may thus be defined by the relationship:

$$\text{Dose} = kI/S$$

The proportionality constant k is determined experimentally and is dependent upon such factors as the machine geometry and the efficiency of electron transmission through the foil window.

Penetration

The depth of material that can be penetrated by an electron beam radiation is directly proportional to the energy processed by the accelerated electrons, E, and is inversely proportional to density, ϱ, of the irradiated material, i.e.

$$P = K E/\varrho$$

Penetration is essentially independent of the chemical nature of the material.

For a given accelerating voltage the penetration can be expressed in terms of the weight of material irradiated per unit area. Usually this is given in units of grams per square metre. The energy of an electron entering the material to be cured is the final accelerating voltage less potential losses which occur at the window and across the gas filled gap between the window and the substrate.

For a given system it is possible to plot profiles displaying the fraction of surface dose delivered against the penetration for a given accelerating voltage and substrate density.

Shielding

Penetration distances for free electrons are relatively short. For example, a 200 keV electron will penetrate only about 40 cm of air. However, when arrested by collision with matter, X-ray radiation can be generated.

With the scanner type quite extensive overall shielding is required, which may include several feet of external concrete. For the linear cathode machines, which operate at voltages below 300 keV, stray electrons and the associated X-rays are less penetrating and can be contained by only a few millimetres of lead shielding. This is usually incorporated as a part of the unit housing.

Press installations include extensive electro-mechanical safety interlocks coupled to radiation monitors to ensure that an operator cannot enter an irradiated zone while the beam is on. With EB equipment there is no radioactive source and thus no risk from residual radiation when the beam is not operating.

Inert gassing

The generation of ozone from atmospheric oxygen on interaction with an ionizing radiation and the effects of oxygen on cure efficiency have been discussed earlier. The gap between the window and the substrate has to be purged free of oxygen. This is usually carried out by maintaining a positive pressure of nitrogen gas in the region of the window. After a period of purging the curing region becomes virtually free of oxygen. Maintenance of a slow stream of nitrogen prevents ingress of more air from the slipstream above the moving substrate.

Maintenance

While a 'hard' vacuum has to be achieved and maintained during running it is customary to retain a rough vacuum throughout any stand-by period.

Over a period of time the cathode filament will evaporate and have to be replaced. However, the more usual limitation on a filament's life is imposed by mechanical shock arising from heating and cooling on start-up and shut-down.

The window which is constructed from an extremely thin foil of metal, typically aluminium or titanium, is also subject to long-term thermal damage and thus requires periodic replacement.

A supply of water is usually required to remove excess heat, particularly from the window and the filament mountings. Additionally a source of inerting gas, usually nitrogen, has to be supplied whenever the system is in operation.

10.6   STATE OF THE ART AND FUTURE TRENDS

The ultraviolet scene

The dominance of UV products in the field of radiation curing technology seems set to continue into the foreseeable future. Furthermore the major application area for UV curable inks will remain the lithographic process. Indeed the strong commercial pressure to include a UV system, at least as an option, when installing a new or replacement press will ensure its steady continued growth.

It is unlikely that the price of UV curable resins and monomers will ever match those of the, comparatively cheap, oleoresinous resins and solvents used in conventional oil ink formualtion. It is anticipated that the manufacturers will strive to develop more speciality UV materials for use in application where novelty can offset the extra cost. Additionally improvements in the performance of existing products can be expected. In such areas as gloss attainment and lithographic behaviour currently available UV resins fall short of their conventional counterparts.

Monomers with extremely low viscosity, satisfactory toxicology, and a competitive price are not likely to become available in the near future. Thus the 100% curable liquid ink formulations which, might make inroads into the solvent-based flexographic and gravure markets, are improbable.

Already available, however, are a number of experimental water-miscible or water-reducible UV curing resins, together with a number of water-soluble photoinitiating species. These may provide interesting varnish formulations and possibly allow the production of low-viscosity UV curing inks.

Electron beam: an expanding technology

Along with the growth in UV, an increase in the number of EB installations can be expected. However, the extremely high capital cost and the complexity of these systems should restrict this means of curing to very large commercial printers who have long print runs on speciality substrates and where packaging of sensitive products demands minimal contamination from the printed surface. If the commercial outlets become greater it is probable that suppliers will place more emphasis on the development of resins that are specifically designed for efficient EB initiation.

Microwave and radio frequency drying inks: a low growth technology

Apart from the drying of water-based systems, these methods offer little advantage over other less complex heating processes. Unless unforeseen developments in water-soluble resins occur or unexpected legislative or commercial pressures threaten the use of standard ink solvents, the comparatively lowly positions occupied by RF and microwave drying in the ink industry are unlikely to change. Despite the low cost of water the resin component of aqueous inks and varnishes remains expensive and this is usually not offset by any outstanding performance feature.

Infra-red

Displacement of indirect method of applying thermal energy to wet print by the more direct IR approach will probably in the future, as now, depend to a greater or lesser extent on the relative prices and availability of the different fuels. Infra-red systems should continue to be favoured in circumstances where there is minimal evaporated solvent to be removed or where substrates demand a very careful control of the temperature.

The industry will always be looking for novel heat-setting materials and processes. These being either resins which undergo direct thermal polymerisation or thermally activated catalysts which bring about a controlled initiation reaction. Development of ink with high solids content may be of even greater importance in the future if environmental pressure impose restrictions on the emission of solvents. Ink of these types are often particularly suited to curing by the mass-heating effects induced by IR radiations.

FURTHER READING

Holman, R. J. (ed.) (1986). *UB and EB Formulations for Printing Inks and Coatings*. SITA.
Roffey, C. G. (1982). *Photopolymerisation of Surface Coatings*. J. Wiley and Sons.
Phillips, R. (1983). *Sources and Applications of Ultra Violet Radiation*. Academic Press.

CHAPTER 11
Inks for Special Purposes

In the context of this manual, the term 'special purposes' is used to describe ink applications either totally outside the mainstream of the five major printing processes — lithography and letterpress, flexography, gravure and screen — or which have their roots in one of these technologies, but for which ink development has been unique to suit a highly specific end use.

In the first category come the non-impact print processes — ink-jet and electrostatic printing. Here are two relatively new processes that are making a very strong case for being classified as 'major' and which have attracted the specialist skills of the ink-maker, despite the fact that the 'inks' they use bear no resemblance to any other inks available.

Under the second category, the versatility of screen printing has, not suprisingly, given rise to a wide variety of novel and unique applications. Examples of this include thermochromic and encapsulated perfume inks, but screen has also had a powerful influence on the electronics industry on the printed circuit and membrane switch front. Other examples of the influence of screen printing will be discussed.

Flexo and gravure also feature under this heading, as they have been adapted to the specialist markets of wallcoverings, textile transfer and sterilisation inks. Metal decorating relies on the combined principles of letterpress and offset, as does the decoration of plastic tubs and tubes. Lastly, the security printing market takes ideas from a number of different disciplines and has developed a very specialised, and of necessity secretive, ink technology all of its own.

All these ink applications will be discussed and will demonstrate how the science of ink technology has been extended to fulfil some highly specialised applications.

11.1 NON-IMPACT PRINTING

Two types of non-impact printing are of interest to the ink industry — electrophotography and ink-jet. The former is the principle used in photocopying machines which has developed over the years to fill a demand for short-run duplication work that would be slow, uneconomic or practically impossible by conventional means. The ink-jet process uses a stream of minute ink droplets projected from a very small nozzle and is widely used for batch and date coding of packaging and in the generation of variable data.

Electrostatic imaging

The foundation of modern electrostatic imaging lies in the invention of Xerography in 1938 and does not rely on any chemical reaction process to form an image, but utilises a purely physical process based on the interaction of light and electricity. The process relies on the principle that two objects of unlike charge will attract each other, while those of like charge repel each other. This principle has been adapted in electrophotography into a technique that produces a latent, or invisible, image of electrostatic charge on a photoconductive material, coated on to a special paper or an intermediate drum. Such coatings are good insulators in the dark and able to retain electrical charges on their surface. When exposed to light, their resistance reduces considerably and any retained surface charge is rapidly dissipated.

Latent image formation
The first stage in forming a latent electrostatic image is uniformly to sensitise the photoconductor in the dark using a corona unit. An illuminated original is projected on to the surface of the charged photoconductor and in those areas where light strikes, the material undergoes a change which substantially increases its electrical conductivity. The surface charge in the areas exposed to light can now leak away, whereas the non-exposed areas, corresponding to the image pattern of the original, retain their surface charge. It is this remaining charge pattern on the photoconductor which is called the latent image.

Developing the image
The image can now be developed by a toner consisting of particles which have an electric charge opposite to that of the image. Areas not possessing a charge will not attract toner and no development will take place.

If the image has been developed on to a photoconductive drum, it must now be transferred to the paper by applying a charge which is opposite in polarity to that of the toner, thus stripping the image from the drum on to the paper. The image is fixed on to the paper by heat or pressure, producing a faithful reproduction of the original.

Development methods
One of three basic image development techniques may be used — two-component, single-component and liquid development. The first system uses a powder developer comprising a carrier material and a coloured toner, which become electrically charged when they are agitated together. The bond they form for each other, because of their equal and opposite charges, is only broken when the developer comes into contact with the latent image on the photoconductor. The toner transfers to the

photoconductor in the pattern of the latent image, leaving the carrier behind to be returned to the hopper. The toner comprises colouring matter and a 'fusible' resin which enables a permanent image to be generated on the paper (Fig. 11.1).

The carrier is a magnetic material, such as iron powder, which generates the charge on the toner and carries the fine toner particles from the hopper to the image. Because the carrier is being stripped of toner particles and returned to the hopper, the material in the hopper will gradually become less active unless there is a technique for replenishing the toner in the correct proportion. A disadvantage of the two-component system is the relatively large size of the toner particles, which means that fine lines and halftones are difficult to reproduce. A two-component toner (Fig. 11.2) must have the following key properties:

(1) It must be capable of generating a charge of the right polarity when rubbed against a selected carrier material.
(2) The charge must be of the right magnitude.

TWO-COMPONENT
DEVELOPMENT

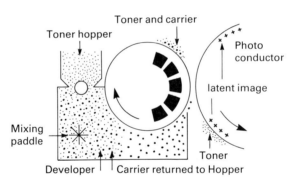

MONO-COMPONENT
DEVELOPMENT

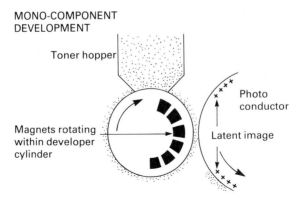

Fig. 11.1 Development principles of electrophotography

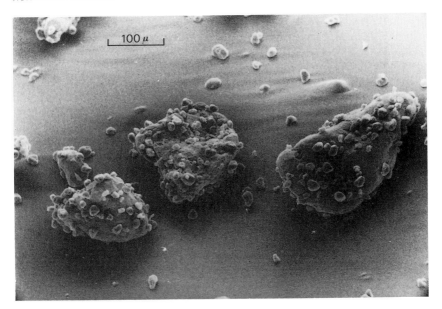

Fig. 11.2 Photomicrograph of powder toner

(3) It must be capable of being fused on to the paper, either thermally or by pressure.

A single-component toner overcomes the problem of replenishment, but still has the same disadvantage of particle size of a two-component toner. A single-component toner is a magnetic material, such as iron oxide or ferrite, which acts as both carrier and colouring matter, dispersed in a 'fusible' resin that will hold an electrical charge. The charge of the latent image on the photoconductor induces an equal and opposite charge on to the toner. As opposite charges attract, the complete toner transfers to the photoconductor in the pattern of the latent image.

A single component developer must have three basic properties:

(1) It must be magnetic in order to adhere to the developer roller.
(2) It must be capable of having a charge induced on its surface.
(3) It must be capable of being fused on to the paper, either thermally or by pressure.

There are literally thousands of possible raw material combinations for single and two component developers and it is the skill of the technician that establishes the right formulation to suit a particular machine.

Liquid toners

Liquid toners are a variation on the two-component toners previously discussed. They are based on carbon black suspended in a low dielectric constant (insulating) hydrocarbon solvent. The carbon black particles acquire a surface charge and will migrate towards an area of opposite charge. The latent image on a sheet of photoconductive paper will pick up pigment if the sheet is passed through a bath of liquid toner. Excess solvent is removed by squeegee rollers and heat.

Liquid toners are based on a number of different ingredients and are formulated to have the following properties:

(1) a suitable resin which has the right melting point and maintains a charge of right polarity;

(2) pigment, e.g. carbon black, which is easy to disperse and gives the right flow characteristics and conductivity charge;

(3) charge control agents to maintain the charge of the toner on the carrier;

(4) dyes to enhance the density;

(5) dispersion aids.

The principal advantage of liquid development is the high resolution that can be achieved because of the much smaller size of the toner particles. A film thickness more comparable to a printing ink can be achieved. Because of this improved print definition and the relative ease with which liquid toners can be converted to colour, the technique is increasingly being used in 'high tech' applications such as microfilm reader/printers and computer aided design and manufacture (CAD/CAM).

Inks for jet printing

Traditional printing processes work because of contact of an inked image against the stock. Ink-jet printing is a non-impact process that uses a stream of minute ink droplets projected onto the stock from a very small nozzle. There are two main systems.

The first uses a continuous stream of droplets which are electrostatically charged as they leave the nozzle and are deflected in flight by the application of a high voltage to a set of deflector plates. The charge on each individual droplet can be varied so that as the stream passes through the deflector plates, which are maintained at a constant electromagnetic potential, the trajectory of each drop (and hence its point of contact on the substrate) can be accurately controlled (Fig. 11.3).

The other technique is known as 'drop-on-demand' or 'impulse' printing, because droplets are only released from the nozzle as they are required. The drops are not normally charged or deflected in their travel to the substrate. Thus a matrix consisting of a bank of nozzles is required to create the image.

Ink-jet is an extremely versatile, high speed, relatively low cost process and can be used, unlike any other process, to print delicate, uneven or recessed surfaces. Very high speeds can be achieved (60 000–100 000 drops/second in a continuous jet printer), partly because the printers contain no moving parts but chiefly because the charging or release of the individual droplets is controlled by electronic data stored on computer tape or disc. Unlike conventional printing where the input, i.e. flexo stereo, lithoplate, etc. is fixed, the input to an ink-jet printer can be constantly changing. The process has the unique ability to print variable information from one 'impression' to the next.

The inks used must have physical properties suitable for jet formation

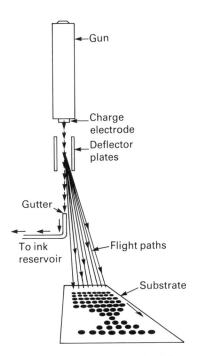

Fig. 11.3 Diagram of continuous jet printing — in this set-up the droplets are deflected to print

and streaming while being capable of producing sharp, dense and permanent images. In addition, they should be stable on storage and must present no long-term health, or short-term chemical, hazard.

For accurate and consistent drop formation, the inks must possess a careful balance of rheology, surface tension and (for continuous jet machines) conductivity. Flow should ideally be Newtonian, but viscosity may vary from machine to machine. Impulse jets place more critical requirements on ink viscosity than continuous jets because of their mode of operation which demands that they maintain a stable viscosity during long periods of use. Unfortunately, viscosity varies with temperature and for this reason a number of impulse jet machines incorporate thermostatically controlled nozzles.

Surface tension is also a critical factor in the formation and maintenance of discrete drops. Water, at 72 dynes/cm, would provide the optimum surface tension, but because it is blended with dyes, resins and additives, this figure is not achieved. A more approachable figure is 50–60 dynes/cm. Solvent-based inks for continuous drop printers have even lower surface tensions.

Conductivity, the reciprocal of resistance, is an important property in inks for use in printers which deflect drops. Polar solvents are used to obtain conductivity and water achieves the highest conductivity figures. Where inks are to be printed on non-absorbent materials and/or need to be faster drying, organic solvents have to be used. Early continuous jet printing inks for this type of application contained methanol, since this produced good conductivity. Methanol is, however, highly toxic, and fast drying inks now use less toxic solvents such as ethanol and methyl ethyl ketone (MEK). These solvents produce less conductive inks than water and methanol and it is common practice to add soluble salts to improve their electrical properties.

Dyes, particle size and stability

Most ink systems contain soluble dyes which give satisfactory colour properties for most work. Colour strength is limited by the solubility of the dye, but with careful selection, ranges of bright strong colours can be produced. Pigments are rarely used, because they present flow, clogging, nozzle wear and stability problems, although it is now possible to make pigmented white inks for added opacity on dark backgrounds.

Dye solubility in water tends to be dependent on pH and ink-jet systems are normally adjusted to be neutral or slightly alkaline. To prevent pH variation due to absorption of carbon dioxide from the atmosphere, such systems may also be buffered with additives such as sodium carbonate. pH can also have an effect on the corrosiveness of the ink system in contact with certain metals and this must be borne in mind when formulating for specific machines.

Most ink jet systems incorporate filters to remove particles which would otherwise clog the fine printing nozzles. These filters are themselves subject to clogging by particles of a size roughly equal to their pore size. A typical filter would be 3 μm and inks must therefore be prefiltered to at least this quality. They should also be extremely stable, so that particles do not seed out or grow by chemical or biological action during storage and use.

The solubility of the dye is a critical factor in the stability of the ink on storage, and the possibility of crystallisation is reduced the higher the solubility of the dye. To aid stability, dyes with low salt content are normally chosen and water inks use deionised or distilled water. Water-based systems are sometimes prone to attack by biological organisms such as spores and bacteria and have to be protected by a biocide. All the additives used to buffer pH, minimise biological contamination, etc. must also be highly soluble if they are not to present storage problems themselves.

Finished print requirements

In many cases prints must dry quickly. This does not present too much of a problem with continuous jet printers as fast drying solvents such as ethanol and MEK can be used. An ink based on such fast drying solvents would dry in the nozzles of an impulse jet printer and it is therefore difficult to formulate a true 'fast-drying' ink. Most impulse jet inks are based on slow evaporating glycols to minimise evaporation.

Glycol and water-based jet inks dry by absorption into the paper. On more absorbent papers, drying will be quicker but there will be a greater tendency for the ink to spread through the paper fibres giving poor print definition and strike through. The degree to which this occurs depends largely on the surface tension and viscosity of the ink, properties that are dictated by the printing machine specifications. By formulating to the limits of the viscosity and surface tension specifications, it is possible to minimise this problem, but a greater improvement can generally be achieved by changing to a less absorbent paper.

Ink-jet systems are often used for batch and date coding and may be used on beer cans, wires, plastic bottles and electrical components (Fig. 11.4). Solvent-based inks are used for this type of work and they must adhere well and produce scuff resistant, hard images. Adhesion to a wide range of plastics can be achieved by using strong solvents, such as MEK, which are absorbed and carry the dye and binder into the surface.

In the case of non-absorbent surfaces, binder choice is critical. Almost all solvent-based jet printing inks contain resinous binders to give key and hardness. Binder resins are selected for their general adhesion and solution properties and include acrylics, polyamides and maleics.

Typical ink formulations
A. Solvent-based continuous jet ink

Aniline Blue	3.0	Solvent-soluble dye
Phenol-formaldehyde polymer	6.0	Film-forming polymer to give water resistance and adhesion to substrate
Alcohol	49.5	Solvent
Dimethyl formamide	41.0	Solvent
Soluble electrolytes	0.5	Conductivity aid
	100.0	

Final characteristics
Viscosity 2.1 cp at 20 °C
Conductivity 1200 micromhos/cm
Surface tension 25 dynes/cm at 20 °C

B. Water-based continuous jet ink

Direct black dye	4.25	
Distilled water	83.15	
Polyethylene glycol	5.00	Crusting inhibitor
N-methyl pyrollidone	4.00	Dye solvent
Ethylene glycol monobutylether	3.00	Paper penetrant
Sequestering agent	0.20	Heavy metal suppressor
Buffering agent	0.30	pH control
Biocide	0.10	Anti-mould
	100.00	

Fig. 11.4 Ink-jet is widely used for 'sell-by' dates

Final characteristics
Viscosity 2.28 cp at 20 °C
Surface tension 43.5 dynes/cm
Conductivity 11 000 micromhos/cm
pH 10.3

C. *Drop-on-demand (impulse) system*

Direct dyestuff	3.0	Soluble dyestuff
Polyethylene glycol	14.0	Anti-clogging solvent
Diethylene glycol	12.0	Humectant
N-methyl pyrollidone	15.0	Dye solubiliser
Biocide	0.1	Anti-fungal
Buffering agents	0.3	pH control
Polyvinyl alcohol	3.0	Viscosity controller
Triethanolamine	1.0	Surface tension control
Distilled water	51.6	Solvent
	100.0	

Final characteristics
Viscosity 9.0 cp at 20°C
Surface tension 45 dynes/cm
pH 10.5

11.2   SPECIALITY SCREEN INKS

Screen printing is a versatile technique which allows all sorts of jobs to be undertaken that are impossible by any other means. In recent years, a number of speciality ink systems have been developed to extend this versatility even further.

Encapsulated perfume inks

The technique of encapsulation is not new — it has been used for many years for 'no carbon required' multi-part stationery. The most recent application of this technique is in the encapsulation of fragrances which are then screen printed on to a paper substrate. They have the advantage of offering a stable fragrance in the form of a dry print, which is only released when pressure is applied. The fragrance remains locked in until the capsule is broken by scratching, unlike a conventional perfumed ink which will lose its fragrance rapidly.

The capsules are incorporated into a water-based medium which can be screen printed in the normal manner and jet or air dried. It is important that the capsules are not broken during printing and, as they are relatively large, a coarse mesh (77–90T) must be used. It is normal to print solid blocks in selected areas of pre-printed material. The potential range of fragrances is enormous, although the most popular tend to be flowers and fruits.

Thermochromic inks

Thermochromic inks are based on liquid crystals that change colour as the temperature varies and have a number of novelty and practical applications. The inks may be used to indicate very specific temperatures or produce a vivid change of colour over a selected temperature range.

Specific temperature inks are used in the production of thermometers or fever strips whereas inks with a colour play of several degrees are used for novelty products and promotional material. Thermochromic inks can be formulated on liquid crystals that react to any specified temperature or temperature range between -20 and $50\,°C$. They can be printed on a wide range of plastics, including PVC and polyester, and can find application on some papers and boards. For best visual effects, the inks must be viewed against a black background.

The use of the term 'liquid crystal' is somewhat curious since the molecules of a 'liquid' generally exist in a haphazard configuration, whereas a 'crystalline' structure implies an ordered arrangement of molecules. Such regular geometry normally means that crystals are solid and their molecules are unable to move. Liquid crystals exhibit the same ordered geometry but, because they are fluid, their molecules are able to twist and move in relation to each other. This can be brought about by gentle heat and as the disturbance takes place, a change in the wavelength of reflected light occurs — the crystal changes colour. On cooling, the crystal reverts to its original colour. Thermochromic inks are almost colourless but as temperature increases, the sequence of colour play is red/brown, yellow/green, blue and violet. Apply more heat and the ink becomes colourless again. The colour play is continuously reversible.

Phosphorescent inks

There is sometimes confusion between luminosity, fluorescence and phosphorescence. A phosphorescent material is defined as one that has the ability to emit light in total darkness, following excitation by various light sources.

	Activated by	*After-glow*
Phosphorescence	Any light source	2 hours
Fluorescence	UV light	None
Luminosity	Radioactive chemicals	Long

It is now possible for ink-makers to incorporate phosphorescent pigments into screen inks for both promotional and industrial applications. Currently available is a phosphorescent green which in normal light has the appearance of very pale green, but appears yellow-green in the dark after excitation by daylight, UV or artificial light. The resultant light emission gradually fades but can be perceptible to adapted eyes for 1–2 hours. Re-exposure to light will reactivate the print.

Phosphorescent pigments are activated zinc sulphide and as they contain large particles can only be printed through very coarse meshes. As the duration of the after-glow is dependent on the filmweight of ink deposited, it is normal to carry out two or three workings. Suitable substrates are acrylic, unplasticised PVC, polystyrene and polycarbonate films, paper and some enamels.

Phosphorescent inks have a number of applications. Warning signs can be produced in the case of light failure; light switches and keyholes can be identified in the dark; other applications include protective clothing, accident prevention, dial illumination, toys and novelties.

11.3 INKS FOR THE ELECTRONICS INDUSTRY

Printed circuit products

Despite the fact that the term 'ink' is rarely used nowadays to describe products used in the manufacture of printed circuit boards, it is not surprising to find that a few ink manufacturers, with their specialist knowledge of coating technology, have taken up the challenge offered by the electronics industry. The term used for such coatings is 'resists' and the three types most commonly used are etch resists, plating resists and solder resists (see Fig. 11.5).

Strictly speaking, we could not use the description 'printed circuit', as this implies that the circuit is applied by a printing process. This is not the case and the pattern of the electrical circuit is achieved in two ways:

(1) Screen printing an etch resist on a copper clad base board in the image of the circuit required. The unprotected copper is removed by treating the board with an etchant solution. The etch resist is then removed either by spraying with a solvent such as trichloroethylene or a mild alkali. In the latter case, the acid resists are based on high acid value maleic resins.

Alkali-soluble etch resist

Phthalocyanine Blue	2.0
Extender	46.0
High acid value resin	30.0
Glycol ether	20.0
Levelling agent	2.0
	100.0

The pattern of the circuit is inspected after printing and, to assist visual inspection, the resist should be matt and contrast with the background copper. As the requirements for higher printing definition increase, the flow characteristics of the resist become more important and excessive flow must be controlled. This can be achieved by using extender or structuring additives.

(2) Screen printing a plating resist to a copper clad board in the areas where the circuit is not required, leaving the circuit pattern exposed. After cleaning, the board is placed in a plating bath containing a metal salt which may be gold, nickel, silver, tin or lead (or a combination of the last two). By electrolysis of the solution, a thick layer of the relevant metal is deposited in the exposed, circuit pattern areas. After plating, the resist is removed by solvent or alkali, depending on the type used, and the copper is selectively removed using an etchant that does not attack the plated metal.

Such boards are very expensive to produce and are therefore only used for the highest quality circuits, for example in military equipment. A plating resist must have a high standard of print definition, must resist the plating solution and must be easily removed in solvent or alkali. As with acid resist, finish is matt for

Fig. 11.5 Preparation of a printed circuit board

inspection purposes, but the proportion of extender to control flow must be minimised to avoid excessive sludge formation in the removal solutions.

Once the circuit pattern has been established by one of the above methods, the electrical components can be soldered into place. Solder resists are applied to the board to protect those areas where solder is not required. They remain on the board as a protective coating and a high degree of adhesion is therefore required. They must be heat and solder resistant and must not be attacked by the solvents used to remove excess solder flux. They must also be electrical insulators. They are lightly pigmented, usually green, and transparent to allow inspection of the underlying circuit. The two most popular forms of screen-printed solder resists are conventional two-pack curing products based on epoxy/polyamide and UV curing resists based on epoxy acrylates.

Two-pack epoxy solder resist

Phthalocyanine Green	2.0
Extender	30.0
Epoxy resin	50.0
Glycol ether	16.0
Levelling agent	2.0
	100.0

Aromatic adduct amine hardener mixed 2 : 1 resist : hardener. Stove at 120−140°C for 30−45 minutes, convection oven.

UV solder resist

Phthalocyanine Green	1.0
Extender	25.0
Thickening agent	3.0
Epoxy acrylate	30.0
Reactive monomer	37.0
Levelling agent	2.0
Photoinitiator	2.0
	100.0

The print definition of solder resists has become extremely critical as circuit tracks become narrower and narrower, particularly for professional boards. As the screen process is being pushed to the limit of its definition capabilities, a new technique using a photo-imaging system has been developed. Here, screen printing applies a solid all-over coating of photo-definable resist which is then exposed through a photographic negative. The exposed area hardens to give the resist design and the unexposed areas are washed off. Definition is only limited to the accuracy of the negative, not the printing process.

The final printing operation in printed circuit board manufacture, before the components are added and the solvent is applied, is a board marking ink used to identify component serial numbers, etc. These are truly inks, as they are fully pigmented and the only element in PCB printing that is used for a decorative function. They must, however, resist solder at high temperature and the flux removal solvents. Conventional two-pack curing inks based on epoxy resins are used as well as UV curing products.

Thermal-curing notation ink

Titanium dioxide	20.0
Extender	20.0
Epoxy resin	40.0
Glycol ether	18.0
Levelling agent	2.0
	100.0

Amine hardener mixed 6 : 1 ink : hardener. Stove at 120°C for 30 minutes.

Much of the technology of resists and board-marking inks is now based on UV curing because of the following advantages:
  (1) Dimensional stability of the boards is maintained. The temperature of the board does not rise above 50°C.
  (2) Resists and inks are solvent-free and thus have excellent screen stability. The flammability hazard is removed.
  (3) Productivity is increased. The boards can be processed very rapidly.

Membrane switches

Membrane switches are a relatively new technology and are finding widespread use in place of push button and other mechanical switches. As there are no moving parts, a membrane switch is extremely reliable and could easily outlast a domestic appliance. They are easy to clean, are attractive to look at and are relatively cheap to produce.

A membrane switch uses a number of different types of screen printing inks and is typically constructed with a plastic graphic overlay, top circuit, spacer, bottom circuit and a mounting panel. Finger-tip pressure on the top circuit brings it into contact with the bottom circuit completing the electrical connection (Fig. 11.6).

Each of the circuits is screen printed with a conductive ink based on silver, gold, nickel, copper or graphite, depending on the degree of electrical conductivity required.

Silver gives the best conductivity (lowest resistance) but at a premium price. Low-cost graphite is often used for less critical applications and it is quite common to blend silver and graphite to provide a cost/resistance compromise.

The graphic overlay can be either polyester or polycarbonate; the latter is popular because it is relatively easy to produce a textured, non-reflective surface without loss of clarity. Polyester is cheaper and has better chemical resistance for applications such as laboratory equipment, where it might come into contact with industrial cleaning compounds. The clarity of polyester suffers when producing a textured finish grade and special matt coatings are used to give a non-reflective finish and maintain clarity.

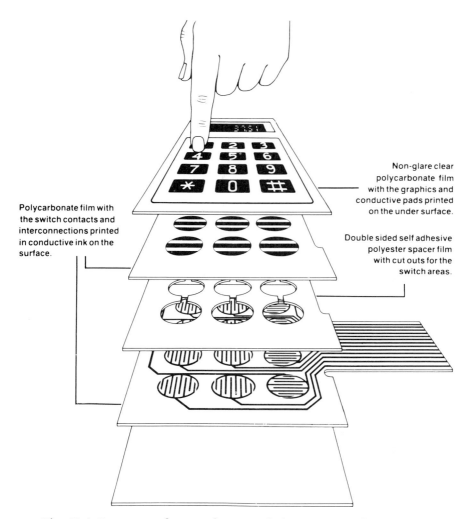

Non-glare clear
polycarbonate film
with the graphics and
conductive pads printed
on the under surface.

Polycarbonate film with
the switch contacts and
interconnections printed
in conductive ink on the
surface.

Double sided self adhesive
polyester spacer film
with cut outs for the
switch areas.

Fig. 11.6 Structure of a membrane switch (courtesy of Autotype)

overlay materials, but they can, for example, give poor lay and pinholing on polycarbonate. Careful solvent selection can overcome these problems. The overlay may also incorporate transparent window areas through which an LED (light emitting diode) can be viewed. For maximum clarity a transparent ink is essential and a high gloss overlay is desirable. This has the disadvantage of creating a highly reflective surface on the overlay, making it difficult to read.

To counter this, a screen-printed anti-glare varnish is applied over the surface, but leaving the LED windows exposed. The switch then has a matt, non-reflective finish with patches of glossy transparent windows. These varnishes are available in a variety of finishes and in UV curing grades. Nitrogen-blanketed UV equipment is available that can produce a variety of finishes from just one varnish by altering the conditions of cure. These are described as texturing varnishes and produce a grain finish.

The windows are coloured using highly transparent, dye-based inks which have good lightfastness in order to survive the working life of the switch.

11.4 INKS FOR LAMINATED PLASTICS

The most common application for laminated plastics is for hard-wearing, highly durable work surfaces. Decoration is achieved by including printed matter in the plastic structure during the lamination process which is described in general terms as follows:

(1) Special paper that has been impregnated with thermosetting resin is printed and dried thoroughly to avoid all traces of solvent. Choice of print process is governed by the length of run, colour strength and type of image required; screen and gravure are most commonly used. The thermosetting resin is normally of the urea or melamine formaldehyde type.

(2) Further sheets of impregnated paper are stacked to a height which is just greater than the thickness of the laminate required. The dry, printed sheet of impregnated paper is added to the top of the stack and this is overlayed with a sheet of pure thermosetting resin.

(3) The stack is sandwiched between highly polished metal plates subjected to high temperature and pressure, typically 150–200°C for 45–90 minutes at 600–1000 lb square inch. The stack is compressed and the thermosetting resin flows and cures to produce a solid plastic.

It is quite common for laminate manufacturers to produce their own inks and exact formulations are not readily available. Obviously, the inks must be resistant to the heat, pressure and chemical reactions of thermosetting. They must also be lightfast and highly resistant pigments, such as chrome yellows, phthalocyanine blues and greens and naphthol reds, must be used. Care should be exercised in the use of additives, such as waxes and silicones, as these may cause blistering and delamination. Both evaporation drying and two-pack catalytic systems are available, although the latter are not popular because they require mixing and have limited pot life.

11.5 INKS FOR WALLCOVERINGS

The development of inks for wallcoverings has become a very specialised technology because of the unique end use of the product. Flexo and gravure are both used to print wallcoverings and ink systems are now available to deal with the demands for washability on paper substrates and the specific needs of the popular vinyl materials (Fig. 11.7).

Water-based flexo inks are now widely used, based on acrylic emulsions which give the necessary toughness and resistance properties.

Fig. 11.7 Wallpaper flexo press (courtesy of Coates Brothers Inks Ltd)

Water-based flexo wallcovering ink

Pigment Yellow 83	10.0
Acrylic emulsion	74.0
Anti-foam	1.0
Water	5.0
Isopropanol	5.0
Wax paste	5.0
	100.0

Vinyl-coated substrates have exceptional durability and inks must be formulated to meet the highest specifications. This is achieved by using vinyl resins and with the introduction of resistant stereos, there has been a trend towards flexo using hydrocarbon-free inks.

H/C free flexo/gravure vinyl wallcovering ink

Pigment Yellow 83	10.0
Vinyl/acrylic varnish	55.0
Ester	20.0
Propylene glycol ether	10.0
Anti-blocking agent	1.0
MEK	4.0
	100.0

Pigment selection is important for wallcovering inks and a minimum lightfastness reading of 5 on the Blue Wool scale is essential. Also, pigments for vinyls should have sufficient heat resistance to resist heat embossing and should not bleed in PVC and plasticiser. While the traditional method of achieving these properties was to use inorganic pigments, the industry in general has moved away from their use and more expensive, resistant grades of organic pigments are utilised.

Embossing is a technique for improving the appearance of wallcoverings and can be achieved either by physical or chemical means. In the second case, a specially formulated vinyl coating is used which contains a 'blowing' compound and an activator which will cause the coating to expand when heated. In those areas of the design where expansion is not required, a special ink containing an inhibiting agent is applied. At high temperature, the inhibitor reacts preferentially with the activator in the coating and prevents the blowing action. Designs can be varied, register is good and deep emboss can be achieved. More recently, an even deeper soft-touch emboss can be achieved by pattern printing the vinyl coating by rotary screen, thus eliminating the need for inhibitor.

11.6   TEXTILE TRANSFER INKS

Sublimation transfer printing involves the dry sublimation of dye from a printed paper when the paper and synthetic fabric are brought together at around 200°C. Only the dye is transferred, leaving the ink resin on the paper. The transfer can be regarded as taking place in two stages. Firstly, the dye is sublimed on to the surface of the textile fibres; secondly, the dye diffuses into the fibres to give the necessary standard of fastness. Thus, adequate dwell time must be allowed for the full transfer process to be achieved (Fig. 11.8).

The essential simplicity of sublimation transfer printing has been a major factor in its growth alongside the following advantages:

(1) High quality reproduction of halftones and four-colour work which cannot be achieved on textiles by any other method.
(2) Strong, bold colours.
(3) Good fastness to washing and dry cleaning.
(4) Non-polluting.
(5) Simplicity of transfer and low cost of local transfer print machinery.
(6) Reduced stock of expensive, printed fabric. Only pre-printed transfer paper needs to be stocked and transfer to fabric carried out as demand occurs.

Transfer paper can be printed by most printing techniques. Gravure was the earliest method to be used and remains the most convenient method of producing high quality continuous print despite the lower cost of common impression flexographic machines specially developed for the purpose. Sheet-fed printing can be by lithography or screen.

Dyes used for transfer printing are 'disperse' dyes which are non-polar sublimable dyes. Fairly low molecular weights of 240–340 are suitable

Fig. 11.8 Examining a textile transfer

but in any ink or system it is important to have dyes with similar molecular weight and sublimation performance. Although they are classed as dyes, from the ink-maker's point of view they must be regarded as pigments and dispersed rather than dissolved. Normal ink-making techniques are used but a very high quality of dispersion is necessary to avoid uneven transfers.

Ink performance and properties are similar to conventional inks, but ink vehicles must be carefully selected to allow good dye release and sufficient heat stability to resist the temperatures in the transfer press. Resins similar to polyester, nylon, etc. are to be avoided as they may hold the dye and give a weak transfer. Cellulose derivatives such as ethyl cellulose are ideal for gravure and flexo. Flexography also uses water-based systems for which polyvinyl acetate, fumaric and acrylic resins can all be used. Thickeners such as bean gum or alginate are also recommended.

Modern high performance quicksetting vehicles form a good basis for lithographic transfer inks, but careful selection of driers and additives is necessary. Dyes can be destroyed during drying and inks are usually slower drying than normal to avoid this problem. There can be problems of ink performance or colour stability with these inks.

Many printing papers are suitable for transfer printing but some general principles can be applied. Papers for gravure and flexographic printing are typically $55-65$ g/m^2 with low air porosity and a smooth surface. For gravure, the paper should be blade coated or supercalendered, but for flexo a more absorbent MF paper is preferable. For offset printing, 80 g/m^2 machine-glazed paper is often chosen, but other smooth papers such as cartridge are also satisfactory. Papers are selected for their good ink setting and good dye release.

Sublimation transfer printing is principally for use with synthetic fibres, notably polyester, but also acrylics, nylon 66 and triacetate. Considerable advances have been made in the use of transfer printing in combination with natural fibres such as cotton, but there still remain some problems to be solved, e.g. fastness properties. Techniques have also been developed for deep pile rugs and carpet tiles where vacuum transfer is adopted to aid dye mobility and penetration.

Flexo water-based heat transfer ink

Dispersed dyestuff	15.0
Alkali solubilised aqueous acrylic	65.0
Anti-foam	2.0
Wetting agent	0.5
Water	15.5
Amine	2.0
	100.0

11.7 STERILISATION INKS

Specialised printing inks that undergo an irreversible colour change during sterilisation are widely used on the packaging of surgical or medical products. Such inks are used to identify if the contents of the pack have been properly sterilised before they are used. There are several ways that sterilisation can be carried out — dry heat, wet heat (auto-claving) and by using gas or radiation.

Autoclaving, or steam sterilisation, involves the exposure of the pack to high temperatures ($> 100\,°C$) under pressure in a high humidity atmos-phere. This method is more effective than dry heat, since the tempera-tures are lower, dwell times are shorter and because bacteria are less resistant to wet heat. The inks must show a significant colour change at low filmweight, but this must only occur under sterilisation conditions and not in dry heat or in the wet ink. The reactive components must also be dispersible in the relevant ink media and give a stable material.

Raw material choice is severely restricted. Most commercial systems make use of the white to black colour change that occurs when some colourless lead compounds are converted to lead sulphide. Various systems are available which differ in their reaction time, but the more reactive types are less stable and more prone to change under dry heat. Some metallic carbonates also exhibit a satisfactory colour change.

Gravure and flexo are the preferred application techniques. The reactive materials are not suitable for use in litho inks and dry offset puts down such a low filmweight that colour changes are difficult to detect. Screen printing would overcome the filmweight problem, but is too slow for the long runs needed. Ink jet can be used on high-speed packaging lines using a special ink containing a material that bleeds out under autoclave condi-tions.

One-pack steam sterilisation ink

Lead thiosulphate ⎱	27.5
Zinc sulphide ⎰	
Extender	10.0
Butyrate resin	13.0
Plasticiser	2.5
Wax	2.0
Methylated spirit	45.0
	100.0

Gas sterilisation normally uses ethylene oxide, so any indicator must be based on a raw material which gives a colour change response on reaction with the gas. One such is hydrated magnesium chloride which is converted to magnesium hydroxide which in turn causes a colour change in the presence of an alkali/acid indicator such as bromophenol blue or methyl red.

Other colour change and security inks

Sterilisation is not the only circumstance under which an ink is formulated to change colour. Colour change reactions are used in the production of toys and games and they have many security applications.

Invisible inks are designed to be imperceptible on a properly chosen substrate, but become visible under specific conditions. They rely on a reaction between two chemicals. One of the reagents is despersed in a suitable medium and printed in the desired pattern. The other can be conveniently incorporated into a felt-tip pen or the image might be developed by gentle heating, lightly damping with water or some other reagent, exposure to a specific vapour, or by viewing under specified light conditions. Children's magic painting books use a letterpress black containing an edible dye printed in a halftone design. When the page is painted with water, the dyes bleed into the area around the dots creating the image. Another type of invisible ink contains a mildly abrasive material and is only revealed when rubbed with a coin or soft pencil.

Security inks are used in the production of items of immediate cash value, such as banknotes, cheques and stamps. They are designed to make the forgery of valuable documents more difficult or to prevent fraudulent alteration. For obvious reasons, the formulation and use of the majority of these systems are closely guarded secrets.

The use of security inks represents only a part of the many varied processes employed by security printers. Security paper and printing technologies are equally important. For example, although most people are aware that special papers are used in the production of banknotes, they may not realise that the paper/ink combination may have been chosen because it is difficult to photograph. Also, most banknotes have inks standing out in relief which have been printed by the intaglio process and which it is extremely difficult to reproduce.

Intaglio has many factors in common with gravure printing. In both cases the design is etched below the surface of the plate which is inked, so that the recesses are filled and excess ink is wiped off. However, in gravure the design areas are divided into a series of discrete cells separated from each other by cells walls which prevent the doctor blade from touching the bottom of the etchings. In intaglio, there are no such walls and the process is restricted to line printing up to a maximum depth of about 2 mm. However, the great advantages of the process are that it can print very fine lines, can deposit a thicker filmweight of ink than any other print process and by using a mixture of deep and shallow engravings can develop a continuous graduation from heavy to fine work.

Unlike gravure inks which have a very low viscosity, intaglio inks are extremely stiff and are applied from a warm duct (30–35°C) on to a warmed plate (45–50°C). Surplus ink is removed first by an ink-saving unit which returns ink to the duct and any ink still left on the non-printing areas of the plate is then removed by a further wiping system consisting of paper or cloth which passes over a pad and is pressed against the plate. High pressure is used to transfer the ink from the plate to the paper. As a relatively thick film of ink is applied, the ink is prone to set-off. Where fast drying is not critical, heavily pigmented inks based on drying oils are used and the sheets are interleaved to avoid this problem. Non-interleaving intaglio inks are also available which use the quicksetting principle and are based on synthetic resins and high boiling solvents.

Apart from the 'reproduction' of genuine documents, there is the important aspect of forgery referred to as 'adulteration'. This involves the modification of a genuine document to enhance its face value. The security printer is constantly involved in a battle to devise new methods to avoid the fraudulent adulteration of documents and only a few of the more well-documented techniques can be discussed here. One method is to print a background with an ink which changes when treated with aqueous ink eradicators. Most cheques, passports, airline tickets, etc. are letterpress printed with fugitive inks based on glycol and containing soluble dyestuffs which bleed or change colour when treated with a large number of organic and inorganic solvents and solutions.

Double-tone inks are black or dark coloured inks that produce a background tint of a contrasting colour which can only be seen if the primary colour is erased. They are used as numbering inks and if alteration is suspected, the parent colour can be erased to reveal an image of the second colour which has penetrated into the paper fibres. As it is impossible to remove this second colour from the paper, any alteration of the numbers can be detected.

Long-wave UV fluorescent inks can be printed just like any other litho or letterpress ink and show good non-metameric properties under a number of different light sources. When exposed to long-wave UV light, these inks give a bright visible response. Large variations in colour and design are possible and document design may be changed regularly, although it is difficult to formulate suitable blues and violets.

These are just a few examples of security printing inks. Many other techniques are regularly employed, although these are rarely described and are often formulated and made by the printers concerned.

11.8 METAL DECORATING

Metal has tremendous scope because of its strength and the ease with which it can be formed into shapes. Thus it has become widely used for cans for foods, household products, beers and beverages, aerosol cans, tins for sweets and biscuits, caps and closures for bottles and jars, collapsible tubes, drums, trays, ashtrays and advertising signs. Its decoration and protection can often involve the use of a complex system of enamels, varnishes, lacquers and inks. The technology of enamels, varnishes and lacquers is a study in itself and falls outside the scope of this book. This section will concentrate on the role of the printing inks used in the decoration of tinplate, tin-free steel and aluminium.

There are two ways of decorating metal:
(1) print a flat sheet which is subsequently formed into a can, closure, tray, etc.; or
(2) print a pre-formed container.

Decoration of sheets

For sheet-fed printing, decoration can take place directly on to the metal surface or over a transparent size or pigmented enamel previously applied by roller coating. Lithography and dry offset are both used for applying metal printing inks. Unlike paper, metal is non-absorbent and the ink must be dried by the application of heat or some other form of energy (Fig. 11.9).

Metal printing ink — sheet fed

Diarylide Yellow	30
Long oil alkyd	58
Antioxidant	1
Driers	2
Reducer	9
	100

The traditional technique of using a long, travelling oven is still widely used. The printed sheets are picked up on an endless chain of racks (known as 'wickets') which carry them through in a vertical position. Hot air is blasted at the sheets as they pass through the oven and the total stoving time is in the region of 10−15 minutes. With some exceptions, the inks are protected by a suitable finishing varnish.

The curing process is the slowest stage of the printing operation and the development of inks places strong emphasis on faster curing rates which will allow shorter stoving times and higher line speeds. Ultraviolet curing has secured a niche in this market and several printers have set stack machines which no longer require thermal curing ovens. However, very few metal printers are totally dependent on UV inks and the majority have opted for a combination, i.e. a UV set stack machine followed by a conventional press which can apply an overvarnish wet-on-wet and where one thermal stoving cures both inks and varnishes.

By the very nature of the process involved, metal printing inks have

Fig. 11.9 Examples of sheet-fed metal printing

properties over and above those expected of inks for other processes. For example, they must have the ability to react to high temperature stoving schedules without suffering any detrimental effects. Where deformation of the sheets is to take place during the formation of cans, closures, etc. the inks must possess excellent adhesion and flexibility characteristics. Decorative coatings must also be hard and abrasion resistant to withstand mechanical handling.

Heatfastness

The heat resistance requirements go beyond the initial stoving temperature of the ink itself, which is in the region of 130–170°C. Additional heat resistance is required to allow for the stoving of an overprint varnish which could be at 180°C for 10–12 minutes. Some packs also require the application of a protective lacquer on the reverse of the sheet, applied as a last coat down after the external decorative system. Typical of these is an epoxy-phenolic meat release lacquer which needs to be stoved at 200°C for 10 minutes at peak temperature. Coating systems for lug cap and crown closures must be resistant to the heat of the plastisol gasket forming operation. Certain foodstuffs, vegetables, meats and fruits are cooked in the can or jar and the decoration must withstand a retorting process in steam and water, sometimes lasting for several hours. Some cans have an externally soldered side seam where temperatures as high as 370°C can be reached for short periods.

In all these cases, there must be no tendency for the ink to fade, discolour, bleed, blister or scorch.

Flexibility and adhesion

Some of the most severe tests of the suitability of an ink system occur when the sheets are being stamped and formed into shape.

The manufacture of crown closures presents a severe test of coating materials. The stamping process requires toughness and good adhesion to avoid damage and the finish must be very hard and abrasion resistant to withstand packing, transport and hoppering in the filling plant.

Roll-on closures, the threaded caps used on bottles, present their own particular coating requirements. The inks must be able to withstand the severe distortion as the metal is drawn into shape from a flat sheet. This is followed by the thread forming and knurling operations and the ink must not suffer from lack of adhesion or surface damage.

Drawn and redrawn, or DRD, cans are finding increasing favour for food packaging, particularly for patés and pastes. Flat sheets are precoated and sometimes distortion printed, then formed into can bodies by drawing in several stages to the full height. The ink performance must be of a high order to withstand the difficult tooling requirements.

Printing a pre-formed container

One of the most radical advances in can making in the last few years has been the development of the two-piece drawn and wall-ironed (DWI) container (Fig. 11.10) It consists of a body and base made in a single unit from a flat strip or coil of aluminium or tinplate.

The bodies are first made by drawing out a cup from the flat base stock and then ironing the walls to reduce their thickness and increase the height of the can. The body is then washed to remove lubricating oils, coated and decorated externally and given an internal protective spray. The finished body, ready flanged to take the can end, is palletised for transport to the filling plant.

These DWI cans are economical in use of materials and energy and can

Fig. 11.10 Two-piece DWI cans

be made at very high production rate (800–1000 cans/minute). The nature and speed of the operation, which is essentially continuous, calls for some very special properties from the coatings and printing inks. The decorative inks are applied to the pre-formed container by rotary offset and must be capable of curing very quickly at high temperatures. Curing times of 6–7 seconds at 315–370 °C are now commonplace.

Inks for DWI cans are rarely varnished and must therefore have as high a gloss level as possible and must be resistant to damage by high-speed mechanical handling.

Printing in the round at such high speeds requires an ink rheology that must be controlled to minimise ink misting over a wide range of temperatures, but still allows good flow on the can to produce a glossy result. The absence of a finishing varnish means that the inks must have high surface slip, or mobility, to resist damage by high-speed mechanical handling.

Ink for DWI cans

Calcium 2B toner	25.0
Polyester	44.0
Melamine	15.0
Sulphonic acid catalyst	2.0
Anti-fly paste	5.0
Wax	6.0
Solvent	3.0
	100.0

Raw materials for metal decorating

The selection of pigments for use in metal decorating is more restricted than that for paper and board. Resistance to heat is obviously important and only minimal colour change on stoving is acceptable. Colour permanence, can stability and bleed in solvents are also important.

The vehicles used are based on blends of alkyds, vinyl toluene modified alkyds, phenolic or maleic resins. Good colour retention is important, particularly for white or pastel shades, bearing in mind the heat treatment that the inks must undergo. The choice of alkyd is governed by a number of factors. It must possess good pigment-wetting properties, to assist in the high level of pigment dispersion necessary for this type of ink. It must have good lithographic properties where relevant, plus good press stability and transference.

11.9 LETTERSET PRINTING

Letterset (or offset letterpress) has developed over the years to fit very specific needs of the printing and packaging industry. It has adapted particularly well to the decoration of plastic tubs for yoghurt, margarine and other dairy products, and to the production of plastic and metal tubes.

Letterset printing uses a raised printing surface fed with ink from a short chain of oscillating rollers to give an even distribution. The ink is offset on to a blanket, then transferred to the object to be printed, which, if it is a tub or a tube, is held on a mandrel by vacuum (Fig. 11.11). The tub or tube is dropped on to a conveyor and dried. In a traditional tub printer, the tub is carried past a bank of medium-wave IR heaters, is constantly turned to ensure even heating and will travel for at least 15 seconds before being nested. The temperature is controlled to ensure that it is below the softening point of the plastic. Because of this long drying sequence, efforts have been made to adapt UV curing to tub printing and it is now being used commercially in the UK and other countries.

One of the main problems in formulating a conventional tub printing ink has been the need for an adequate compromise between stability of the ink on the rollers and fast drying on the tub at a temperature below the softening point of the resin. Two types of ink are available. The first is based on solvent and can be used on all types of tub, including non-absorbent varieties such as polypropylene. They lack, however, a certain amount of roller stability associated with the second type, which contain a percentage of non-volatile plasticiser.

Solvent-based tub printing ink

Pigment	20.0
Hydrocarbon resin	30.0
Ethyl cellulose	10.0
Solvent	39.0
Slip aids	1.0
	100.0

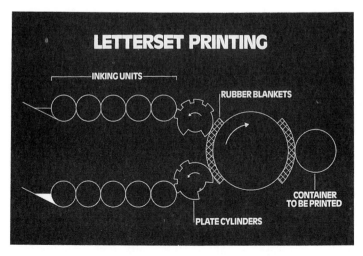

Fig. 11.11 Principle of letterset printing

Plasticiser-based tub printing ink

Pigment	20.0
Nitrocellulose	8.0
Synthetic resin	25.0
Plasticiser	34.0
Solvent	12.0
Slip aids	1.0
	100.0

Inks containing plasticiser are unable to dry on non-absorbent stocks, as they rely on penetration of the plasticiser into the plastic to give very rapid drying. The importance of rapid drying cannot be over-emphasised, as it is vital that set-off should be avoided when the tubs are stacked. As such tubs are widely used for food packaging, any ink transferred to the inside of the tub is a potential contaminant. Although a properly formulated and dried ink will not set off, there may be times during make-ready or after a machine stop when the ink is not 100% dry when it reaches the delivery pile. Particular care should be taken to ensure that the inks are non-toxic by following the recommendations issued by the Society of British Printing Ink Manufacturers regarding materials that should not be included in ink for immediate food wrappers.

Both types of ink must be resistant to the final contents of the tub. In most cases, this is some form of dairy product, although many other materials are packed in this way. The inks must also resist the effect of low temperatures and condensation for products that are likely to be stored in the refrigerator or deep freeze. Where boiling water is used for snack foods in polypropylene, wet rub resistance is important.

Many different plastics are used. Polystyrene is particularly used for a wide range of foodstuffs, including yoghurt, cream, cottage cheese and salad products, as it is lightweight, chemically inert, gives good rigidity in low gauges and is relatively cheap. It does not, however, provide a very effective barrier to UV light and has partly been replaced by PVC or ABS. Polypropylene has emerged as an important new material, one reason being its cost benefit. Mastic cartridge tubes, formerly made of card, are now made from polypropylene to improve storage, handling and appearance. Paint containers, for emulsion paints at least, are already available in polypropylene and work on barrier coatings is in hand to enable gloss paints to be similarly packed. Polypropylene also has greater heat stability and this has opened up new markets for powdered, instant snack foods which require boiling water to be added (see Fig. 11.12).

Collapsible tubes were originally made from lead, but are now made from aluminium or plastic. Many products are packed in tubes, including toothpaste, cosmetics, creams, pharmaceuticals, foods and adhesives. Aluminium tubes are formed by extrusion and an internal lacquer is applied when necessary. They are then externally coated and printed, the top receives its cap or seal and they are packed ready for despatch to the filler.

Fig. 11.12 Examples of printed tubs

External decoration usually consists of white enamel and inks, with coloured enamels being used occasionally. The system must have good adhesion and excellent flexibility to tolerate the crushing and distortion of the tube in use and, particularly in the case of toothpaste and some cosmetics, flexibility must be maintained after long periods of immersion in soapy water. Product and sterilisation resistance may also be needed.

CHAPTER 12
Manufacture of inks and varnishes

The manufacture of printing inks is not a complicated process. Most printing inks consist of pigment dispersed into varnishes and manufacture is mainly concerned with the mechanical means of producing these dispersions efficiently, economically, reproducibly and, often, very quickly. Usually two techniques are employed, mixing and milling, which are briefly defined in respect of inks as follows.

Mixing is the mechanical agitation of pigment and vehicle/varnish together and is usually continued until no dry pigment is discernible and an agreed level of dispersion has been achieved. This is followed by milling, the process by which the pigment/vehicle mixture is refined and the pigment particles are reduced to a size suitable for the intended printing process. In some cases these two processes are combined in the same item of plant.

The emphasis of the last few years and for the foreseeable future is on speed of production and the ability to reproduce a consistent quality, meeting the delivery time and the print requirements of the customer. To achieve these objectives, the operation of an ink manufacturing plant demands an efficient and flexible organization since the number and size of orders to be processed can vary considerably. The equipment in common use is relatively simple.

12.1 GENERAL REQUIREMENTS

Printing ink production units are today becoming more specialised, organised on mass-production lines making a single product or a limited number of products of similar type. The manufacture of rotary news inks, heatset web offset and sheet-fed process inks, flexo white, gravure process colours and the bulk manufacture of blending scheme bases or concentrates are examples of this change of emphasis. However the ink-maker must cater for a wide variety of requirements, which explains why an ink factory, although with specific lines of bulk manufacture, has to retain the ability to be flexible. This has led to the blending of bulk lines for small quantities, a process which is costly and can be time consuming.

The real cost of small batch production should always be borne in mind. In general, it is far easier to select equipment and to design and operate a plant for large-scale production of a few similar products than to operate a plant manufacturing many products in varying batch quantities.

Apart from the manufacturing capabilities of the equipment, there are other considerations to be borne in mind when designing an ink plant. Among these are the services which may be required, including electrics, health and safety, flame-proofing, water supply, drainage and cooling, steam supply and, compressed air. Space requirements and access to the machinery, ease of cleaning, wear rate and maintenance (see later for possible control methods), cost of replacement parts, ease of feeding the equipment and transferring the finished products, are all items to be considered before the equipment is purchased. While the rate of production is of great importance, the nature and quality of the material produced and the likely profitability of the unit concerned, have also to be considered.

12.2   THE MANUFACTURING PROCESSES

A study of the individual chapters on the inks for different printing processes will have shown that the majority of printing inks fall into two catagories; basically these can be defined as:
  (1) oleo-resinous systems for letterpress and lithographic printing, and
  (2) volatile solvent/resin systems for flexographic, gravure and most
      screen printing. The former are formulated on non-flammable
      solids dispersed into vehicles containing mineral or vegetable oils
      or high boiling point solvents.
        The latter types of ink can be subdivided into two classes: (a)
      those based on water-reducible vehicles, and (b) those based on
      highly volatile and flammable solvents, which require special
      buildings and plant with flameproof electrics for their manufacture.
The manufacture of inks can be broken down into the following operations:
  (1) Varnish manufacture, i.e. the vehicle;
  (2) The manufacture of additives;
  (3) Dispersion of pigments into vehicles using mixing and milling
      techniques.
Providing that the formulation has been carefully considered and appropriate raw materials have been selected, the process of dispersion is relatively simple. To understand the process it is necessary to review the materials and the routes by which dispersion can be achieved.

Primary dispersions for ink production demand that the pigment particles are thoroughly wetted by the liquid phase and that the particles in the finished ink for most processes should be in a range between 5 and 1 μm. If this level of dispersion is not achieved, printing problems will arise which are considered in other relevant chapters.

Pigments are generally precipitated from aqueous solutions. At the time of the chemical reaction causing the precipitate, these particles are molecular in size. However, the molecules of the pigment are strongly attracted to each other by physical forces and immediately form larger pigment particles, the size of which will depend on the nature of the material and the physical conditions existing at the time of reaction. These primary particles cluster together during processing to form agglomerates. The particles in an agglomerate are strongly bound together and high energy is required in the dispersion process to break them down. As the pigment is filtered from the liquid phase, the agglomerates tend to bind together more loosely but in larger groups known as aggregates. If the subsequent press cake is dried, the resulting solid will contain a mixture of primary particles, aggregates and agglomerates.

The pigment is finally crushed to reduce the size of the largest agglomerates, giving what appears to be a relatively fine powder. However, it is an essential part of the dispersion process to break down these large particles exposing a greater surface area of pigment to the wetting medium in order to produce a finished, evenly wetted product.

There are several routes by which ink dispersions can be produced. In some cases, pigments can be purchased or manufactured in a form which makes the ink-making process, rather than the dispersion process, much easier.

While there are similarities in the way the individual classes of inks are made and, since the introduction of bead mills, much of the machinery involved is also similar in concept, there are distinct differences between them and they will be considered separately.

The manufacture of oleo-resinous systems

Varnish manufacture

Varnishes have three primary functions. Firstly, to wet the pigment particles and thereby to assist the process of dispersing the pigment in the vehicle system; the better the wetting property of the varnish, the easier it is to disperse the pigment thoroughly. The second function is to incorporate printability into the system. The system must be stable enough to perform on the press and must ensure faithful reproduction of the printing surface on the printed material. Thirdly, the varnish must bind the pigment on to the printed surface so that the final print dries to an acceptable film which is suitable for the use for which the finished print is designed. Thus, varnishes impart such characteristics as gloss, rub resistance, product resistance, flexibility and adhesion to the finished print.

In the case of oleo-resinous inks, these functions will be carried out by the use of varnishes based on drying oil–resin combinations with, or without, the addition of high boiling solvents. Vehicles based on hard resins dispersed in these solvents will also be used, and if the latter have a low aromatic content and are refined to give a narrow boiling point range material, the varnish may require the addition of 'bridging' solvent to give the required stability and to avoid the resin separating out.

The correct drying oil, resin and distillate combination in the finished ink, may be achieved by combining the varnish raw materials at the manufacturing stage but in practice is more often carried out by the ink-maker, by combining one or more varnishes in the formulation. Where practical, the varnishes may be pre-blended to create a composite vehicle system, thereby simplifying the weighing or metering stage operation.

Oleo-resinous varnishes are produced in closed kettles, of, mainly, up to 5 tonnes capacity, and these are heated by oil heat exchangers or, on older plants, by gas-fired furnaces. Cooling of the larger sized vessels is also carried out by heat exchanging oil from a primary source. Relatively high temperatures are involved in the processing, usually in the range 120–260°C, and complex chemical reactions take place during this cooking process. These conditions require strict control of the procedures used, especially the rate of temperature increase and decrease, the maximum temperature achieved and the length of time for which the varnish is held at this temperature. After the initial cooking of the oil–resin mixture has been carried out, the structure of the varnish can be further modified by the addition of gelling agents or 'chelates' which are based on metallic molecules held inactive in an organic complex. The method of addition of such compounds and the control of the temperature at which they are added are also critical if a stable varnish, and hence ink system, is to result.

Since many of these varnishes are formulated to dry by oxidation polymerisation, it will be clearly seen that, at all stages of their manufacture, every effort should be made to exclude atmospheric oxygen. For this reason, the varnish is usually cooked under a nitrogen 'spurge' or 'blanket'. The use of a condenser helps to prevent the loss of distillate and 'bridging' solvents.

In modern varnish plants, all functions can be monitored from a control panel (Fig. 12.1) and temperatures, times, heating and cooling rates, can all be recorded on flow charts.

A typical cooking cycle takes the following progression during the manufacture of a high gloss varnish used in a lithographic ink:
— Distillate metered from tank farm to clean vessel under nitrogen flow.
— Preheated alkyd charged by back weighing.
— Charge heated to required temperature.
— Hard resins added slowly maintaining temperature.
— Resins cooked $\frac{1}{4}$–2 hours between 160 and 240°C to achieve solubility/modification.
— Batch cooled to required temperature for addition of gelling agent. This is specific to the agent used.
— Gelling agent addition made by spraying a solution in distillate.
— Batch reheated to the reaction temperature to achieve optimum structure/modification. Again, temperature and time specific to the gelling agent used.
— Batch tested during cooling stage with adjustment of distillate, as necessary.
— When completed, batch filtered (or strained) through fine filter into tared 45-gallon (205-litre) drums.
— Vessel distillate washed for next batch.
— Final sample of varnish product tested prior to delivery.

Fig. 12.1 Varnish manufacture control panel (courtesy of Glasurit Beck Ltd)

Figure 12.2 shows a typical 'cook cycle' achieved during the production of a batch of oleo-resinous varnish which has been 'structured' by chelate addition, while Fig. 12.3 shows the arrangement of a varnish kettle with the ancillary equipment required for the manufacture of this type of vehicle. The careful recording of previous experience is necessary to scale up the production of a batch of varnish, from a laboratory pilot plant size to that of full-scale works production, if the most efficient manufacturing process is to be devised. For consistent results, the raw materials used should have been checked carefully to ensure that they were up to the specification agreed with their supplier. From previous records a working specification can be set up, to be used during manufacture, which may be modified after the initial batches, so that a final vehicle formulation is evolved where any adjustments are kept to a minimum. These adjustments will usually be small additions of the high-boiling petroleum distillate used to cut the viscosity of the varnish. Besides the final viscosity measurements, earlier checks for clarity and resin dispersion will have been carried out during the cooking cycle, while checks for tack, colour and grind, will usually be carried out prior to filtering and running off to storage.

Development work on varnishes requires a combined knowledge of resin and ink technologies, and expertise, gained over many years of varnish manufacture. This has led many users today to choose not to make their own vehicles, but to purchase proprietary varnishes from the specialist manufacturer. There are arguments for and against this policy but, whichever route is taken, an essential part of ink manufacture is a

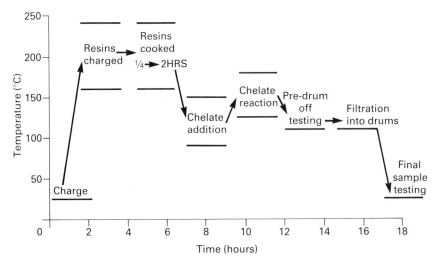

Fig. 12.2 Temperature/time related cooking cycle (courtesy of Glasurit Beck Ltd)

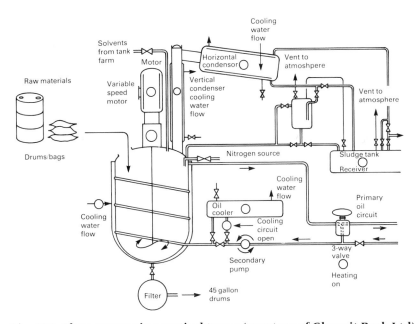

Fig. 12.3 Flow pattern in practical terms (courtesy of Glasurit Beck Ltd)

knowledge of the vehicle and what it will do, when used in a letterpress or lithographic ink formulation.

Where smaller quantities of specialised varnishes are required, a 'cold cut' technique is sometimes employed. The term 'cold cut', in this context, describes the process of dissolving soluble resins in oils or high-boiling-point solvents at temperatures between 160 and 180 °C which are not far above the melting points of these resins. Modern techniques for cold cutting use high-shear, high-speed mixers, which can be used to achieve the temperatures required to get the resins used into solution. External heating bands can be used to assist the speed of temperature increase and hence the solubilising process.

The resulting oleo-resinous varnishes can be considered as ultra-fine particles of resin which are dissolved in solvents which have just sufficient strength to hold them in solution. Although clear and stable varnishes are produced in this way, their resultant properties do show differences from a varnish of similar formulation produced by the cooking method, because the higher temperatures employed (230–265 °C) by the latter allow ester interchange reactions and then chemical complexing of the resin and oil. The latter method also permits the use of less soluble resins.

The 'cold cutting' method of varnish production is much more relevant to liquid ink vehicle production and will be discussed further in that section.

Summary
Producing paste ink vehicles by cooking at elevated temperatures under carefully controlled conditions gives the following benefits to the formulator and manufacturer:
— Good batch consistency.
— Economy in large batch sizes.
— Low oxidation (pale) colour when nitrogen 'blanket' used.
— Good lithographic performance due to high temperature solubility process.
— Good gloss.
— Uses gelling agent to maximum effect giving consistent structure.
— Controlled solubility leading to crystal clear filtered product.
Although the oils and resins used are classified as non-hazardous, at these elevated temperatures great care must be taken to minimise the potential fire risk and operators should be well trained in the safe handling of the hot varnishes to prevent any skin contact with the product.

Bulk varnish storage and handling
How this is achieved depends on certain criteria, and although the list may seem long, all of the following aspects have a bearing on the type of storage and handling employed:
(1) Is the varnish made on the same premises as it is to be used?
(2) What is the temperature at completion of the manufacture and can the residual heat be used to aid transmission of it to the point of use?
(3) Are there any special requirements for special materials to be used for the tanks and pipework construction? Is it oil based, or glycol based for water-reducible inks?
(4) Does it need further processing?
(5) What is the frequency of use or turnaround of the varnish? Long-term storage at elevated temperatures can lead to changes in the characteristic of the product.

(6) If the products are bought or made at a distance from the point of use, do they need special transport facilities and storage in a warm area to enable them to be pumped or used effectively?

In principle, liquid materials, whether at 10 or 1000 poise, flow more efficiently if at an elevated temperature. The head of storage is also most significant, so that gravity feed can be used wherever possible, and surprisingly heavy materials flow with the aid of small amounts of pressure or warmth. To retain the heat, lagging of pipework and in some cases oil heating of valves to encourage this flow is well worth while and has a short payback period. The pressure can be achieved easily by compressed air or hydraulic pressure. The latter has a history of being more accurate across the wide range of viscosities which may be encountered.

'Fail-safe' devices need to be incorporated into the design of any system to avoid discharging into vessels with valves in the open position or filling beyond the capacity of the container.

Manufacture of wax pastes and other additives

The waxes used in the construction of printing inks are as various in their type as in their physical appearance, and have been described in Chapter 4.

The latest mode is to incorporate micronised versions but these do not always give the desired or expected results. There are several reasons for this; for example, the particle size may not be large enough to provide the abrasion resistance required, or the material may not be uniformly dispersed. The wax content of a finished ink is small, and must be thoroughly mixed to give the necessary protection to the ink film. As a result, waxes are, more often than not, added as a compound containing between 30 and 40% wax. The word 'compound' in this context needs to be further defined as the waxes are normally neither soluble nor compatible with the vehicle or solvents into which they are dispersed, which will be chosen so that the final paste is compatible with as wide a range of inks as possible. The only purpose of making a compound is to reduce the particle size of the wax to a suitable level for incorporation into the ink, and reduce the chances of localised additions of large agglomerates of wax formed by the attraction of smaller particles due to static forces.

Dispersion methods are crude and are normally by use of a star or rotor high-speed mixer, or alternatively the wax is heated as a mixture to a temperature around softening point and refined through a heated three-roll mill until an acceptable reduction in particle size is achieved.

Although waxes are the largest group of materials which need to be added to oleo-resinous systems in small amounts, similar techniques can be used for other minor additions of solid raw materials which an ink technician may wish to include in his formulation.

With the use of highly accurate electronic balances, the weighing of liquids as ink additives becomes a simple operation; however, dilution with some of the vehicle, or the solvent used in the formulation, aids transfer of the material and helps ensure a homogeneous mix. For example, see the addition of gelling agent, during oleo-resinous varnish manufacture. If the reduction is done in bulk, a simple paddle mixer should be sufficient, but care needs to be taken to ensure that the reduced additive does not form layers of separated material and it should be well stirred before use.

Manufacture of oleo-resinous inks

Until the early 1960s the production of paste inks had followed the traditional path of mixing the items just sufficiently to wet out the dry pigment and then passing this rough mixture over a three-roll mill until sufficiently dispersed, i.e. most of the work was done at the milling stage. With the introduction of faster running presses the required levels of dispersion meant that the ink had to be passed over the three-roll mill several times and this was time consuming and very expensive.

Combined with this, the increasing use of heatset inks meant that a totally enclosed system was needed to overcome evaporation of the lower boiling point solvents being used in increasing quantities. Both these factors led to the introduction of bead mills, which were not only enclosed but also gave improved rates of production and permitted the easier handling of larger batch sizes. This method of manufacture is efficient for dispersing most pigment/oleo-resinous varnish mixtures, although it would be premature to suggest that all the problems such as machine downtime, temperature control and maintenance of gloss in the final letterpress or lithographic inks have been overcome. In fact, some companies use bead mills to disperse the premix and follow this by passing the ink through a three-roll mill to improve gloss and remove air.

Whatever the equipment being used, the following processes will be carried out when making a paste ink from dry colour and varnish:

(1) Weigh out pigment and vehicle and pre-mix.
(2) Disperse using a three-roll mill or a bead mill.
(3) Incorporate additives and solvents.
(4) Transfer to bulk or pot off into tins, buckets or drums, once passed by quality control.

Using his experience of the pigments being used, the ink formulator should have broken down his formulation so that the initial stage consists of weighing out, or metering, sufficient vehicle to 'wet out' the pigment. There is a vast range of equipment which can be used for this purpose, with a gradual move from mechanical scales to the load cell principle. The readings are either digital or make use of an optical print-out. Electronics in this type of equipment can be complex and if it is computer linked, boosters along the line may be necessary. Since these are costly, the location of the weighing and weighing-related activities is very significant. The addition of the dry colour needs to be done with care to avoid airborne contamination of the working place. To ensure thorough wetting out of the individual particles, the pigment may be added in stages and thoroughly dispersed between each addition. It may be necessary to adjust the viscosity of the pre-mix, at this stage, to achieve the maximum working pressure and hence the most efficient condition for shearing and dispersion action in a particular mixing. This is especially important in the early stages of mixing, using a high-speed disperser, where under the desirable dilatent conditions, brought about by pigment agglomerates locking, the variable impeller speed permits extremely high forces to be applied to the mixture. When the resistance is overcome, the agglomerates are pulled apart far more effectively than by very high speed alone in a semi-paste or fluid state.

This type of machine could very well switch the emphasis of ink manufacture away from roller milling to high-intensity change pan mixing, at a much reduced capital and operational cost. Short predispersion cycles would be followed by a light milling, or in the case of some base stock items, no milling at all (see Fig. 12.4 and section on hyperconcentrates).

Harsh pigments, which are difficult to disperse, will require extra work done on them to break down the agglomerates and this will be carried out on the roll mill. However, the wide use of 'easily dispersible' (ED) pigments in modern oleo-resinous formulations has speeded up production rates and enabled premixes to achieve 'off the gauge' readings when drawn down on a Hegman gauge. These pigments are treated, usually in aqueous solutions, with a wide range of surface-active material and resins, so that the pigment surface is compatible with the medium into which the pigment is to be dispersed. The offset ink formulator needs to ensure that the surface-active wetting agents do not react with the fount solution and lead to printing problems.

Following the mixing stage, the viscosity of the ink formulation may be further adjusted depending on the ink type and the nature of the mill to be used for the next processing stage. For three-roll milling, the mixture

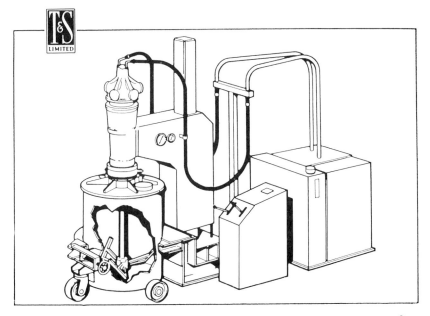

Fig. 12.4 Heavy duty paste mixer (courtesy of Torrance & Sons Ltd)

should have sufficient structure to ensure shearing takes place, but should also have sufficient vehicle present to complete the wetting out of the pigment particles. For bead milling, there is an additional requirement to keep the temperature in the chamber low enough to avoid changes in shade and opacity with certain yellow and red pigments.

Once the ink is fully dispersed, final additions can be made and the ink passed by quality control. It may then be potted off manually by passing through a three-roll mill with the roller gap set to give a high rate of throughput without aerating the ink. Alternatively, a modern ink-filling line may be used (see section on Ink Packaging).

An alternative method of manufacture may be used for small to medium sized batches (say up to 100 kg size) or when using flushed pigments or special concentrates:

(1) Blend the predispersed base (made from dry colour and varnish, as above) or the flushed pigment or the concentrate, with the extra varnishes and additives required.

(2) Refine the mixture over a three-roll mill.

(3) Transfer to bulk or pot off as above.

The range of bases available to the ink formulator will depend on the type of inks being supplied, but ideally should contain a high pigment concentration and should be as versatile as possible. A simple mixing process using a paddle or 'butterfly' type mixer is normally used.

Flushed colour

Pigments had been flushed into various vehicles from the 1860s, but it was not until 1914 that pigment flushing, as it is known today, was invented by Robert Hochstedder. The heart of the process was the use of vacuum to remove all traces of water that could not be poured off. The double arm, jacketed mixer used, was equipped with a cover so that the system could be put under vacuum (see Fig. 12.5).

The original flushing technique was developed to produce finished inks in a more efficient, less expensive manner than by three-roll milling, since the pigment is transferred directly from the water phase to the vehicle without the necessity of drying, crushing, blending and milling. Since the water paste avoids the drying process, agglomerates and aggregates are not formed. Much of the early flushing was done by ink-makers themselves as most at that time made pigments 'in house'.

Nowadays, the organic pigment manufacturer has greatly extended the process to produce highly concentrated bases which are in turn sold to the ink-maker for formulating into finished inks.

Whereas a few years ago the quality of flushed organic pigment was recognised to be superior to that of dry colour, nowadays, with the improvements in dry pigment technology, this difference has been narrowed significantly. Why then use flushed colour?

By far the most important property of flushed colour is that it contains fully dispersed pigment, free from any agglomerates which require grinding equipment. Moreover, in spite of the significant improvements in dry colour, especially in terms of dispersion efficiency, the basic need to

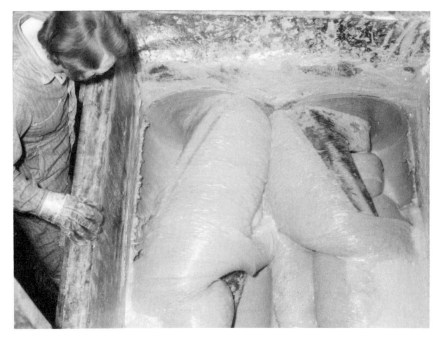

Fig. 12.5 Z-blade mix of flushing (courtesy of Sun Chemical Ltd)

grind out pigment agglomerates formed during the dry process is still the greatest limiting factor and the cause of high cost ink production.

The justification for flushed colour as an alternative method of manufacture becomes more apparent from the line diagram for the production of a theoretical 500 kg of ink (see Fig. 12.6).

Clearly, with existing three-roll mill equipment, using flushed colour in place of dry, production can be somewhat more than doubled, resulting in output significantly greater than that achieved by the introduction and increased capital cost of introducing shot mills. Inevitably, the economic viability of using flush depends very much on the price relationship between pigment powder price and flush, the savings in capital outlay and the increase in plant production, all of which are best assessed by the individual ink-maker.

Flushed pigments do offer significantly cleaner working possibilities to the ink manufacturer who does not wish to install sophisticated dust extraction systems. But if the ink-maker has a large capital commitment to shot and/or three-roll mills, it is as well to recognise the fact that flush is unlikely to figure in the immediate future as a major process; change to this philosophy has to be planned and may be introduced in part as a result of plant breakdowns or a sudden surge in demand.

One method for the effective handling of flushed pigment pastes is to have air lines with free air, to connect to the base of the containers, thus ensuring easy and economical emptying. Alternatively, under the action

Powdered pigment 3-roll mill 1 grinding pass/ 1 processing pass	Powdered pigment shot mill/3-roll mill	Flushed colour

1. Weighing — 15 mins
2. Mixing — 30 mins
3. Mill 1 pass — 4 hours
4. Control adjustment —
   15 mins
5. Mill pass processing —
   2.5 hours to can
6. Control during processing
   run — 15 mins
	1. Weighing — 15 mins
	2. Mixing — 30 mins
	3. Shot mill — 2 hours
	4. Control adjustment —
	   15 mins
	5. Mill pass processing and to
	   can — 2.5 hours
		1. Weighing — 30 mins
		2. Mixing — 30 mins
		3. Control adjustment —
		   15 mins
		4. Processing to can —
		   ....
		
		3-roll mill Filter
		2.5 hours 1 hour

7.75 hours 6 stages	5.5 hours 5 stages	3.75 hours 2.25 hour 4 stages

**Fig. 12.6 Comparison chart of manufacture of 500kg of finished ink
(courtesy of Sun Chemical Ltd)**

of high shear and the temperature induced, the flushing can be made sufficiently mobile to transfer to the blending plant. Whichever method is used, some care needs to be taken to avoid affecting the dispersion when added to a large volume of cold solvent and varnish.

Hyperconcentrated pigment dispersions

From time to time new approaches are made to an old problem and it is worth mentioning such a development in the field of dispersion. Hyperconcentrated pigment dispersion is a new production technology for printing inks which has been growing in popularity during the last few years.

The reasoning behind this is that since the dispersion process is the most time-consuming part of the ink production process, this part should be carried out with only the pigment concerned and without the film-forming polymers being present. This will give the maximum degree of flexibility, enabling the ink manufacturer to achieve a more rational and cost-effective way to counter the ever increasing demand for shorter delivery times. The traditional answers have been installation of more machinery and keeping larger stocks of finished inks and intermediates, but this type of investment can tie up large amounts of capital. Based on the new technology, an optimised set of highly concentrated, readily dispersed pigment pastes are stored in tanks and subsequently pumped to mixers and mixed with specific individual or composite let-down varnishes to produce the large variety of cold or heatest web offset, screen process, rotary and sheet-fed inks which the market demands. By changing the solvent, the same technology can be applied to publication gravure inks.

The main condition to be fulfilled in order to achieve this new level of technology is super-wetting and stabilisation of the dispersed pigment phase. This is attainable by using the so-called hyperdispersants which are specially activated, non-film-forming polymers. The properly chosen pigment having sufficient activity, and the combination of this with the correct hyperdispersants with the opposite electronic activity, will lead to the desired state of super-wetting, resulting in pumpable, readily dispersed, pigment pastes, with a pigment content from 50 to 65% (see Figs. 12.7 and 12.8).

This new technology effectively enables the ink-maker both to improve strength development per unit of input energy, reducing consumption, sometimes avoiding peak load situations, with less problems, and increased productivity and flexibility in manufacture, as well as reducing storage of finished inks.

These hyperconcentrated dispersions may be produced either on a three-roll mill or a shot mill and before any milling takes place, a pre-mixing operation should be introduced.

At this point, it is important to realise that the physico-chemical wetting process should be given sufficient time for the molecules to move into position before any kind of mechanical high shear force is applied to the system. A butterfly mixer at low to medium speed is very suitable for the purpose. when all pigment has been fully wetted (15–20 minutes) and the mill base has become fluid, the shaft speed is increased to maximum. If a twin-shaft mixer is used the dissolver shaft now may be started and the speed gradually increased to cavitation conditions and for a short time, approximately 5 minutes, held in that position.

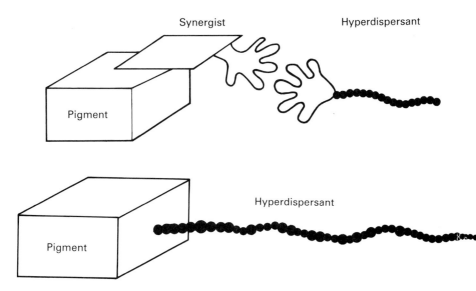

**Figs 12.7 and 12.8 Pigment/hyperdispersant relationship
(courtesy of Tennant-KVK Ltd)**

This sequence is recommended to provide sufficient time for the hyper-dispersants to be absorbed on to the pigment surface ensuring that it is the only active polymer coming in contact with the pigment during the initial premix. For this reason, it is recommended that no traditional film former or resin is added during the dispersion process. By this method, it is possible to obtain a constant and reproducible wetting for each batch.

If a three-roll mill is used for the final dispersion, it should be emphasised that some concentrate mill bases can have a low viscosity, and as a result, low roll adhesion.

The three-roll mill should therefore be operated at lower pressure and thus the rolls should have a corresponding low camber to obtain the same nip clearance along their entire length. The three-roll mill should be equipped with the most efficient 'across roll' cooling which can be used to aid the production rate, along with a reduction and accentuation in the camber, as required.

For the manufacture of these concentrates by bead milling, it is most important to ensure that the mill base does not become too thixotropic during grinding at elevated temperatures such as 60–80 °C. If the mill base is very fluid without any thioxtropic structure at this stage good bead milling will be achieved. If not, variation in the amount and type of hyperdispersant, synergist and pigment, should be explored.

From these dispersed pigment pastes, a large variety of inks can be made by stirring. Thorough studies and subsequent production tests tend to confirm that the so-called butterfly mixer is the most suitable equipment for the final stirring process (see Fig. 12.9).

Other types of mixers can also be used of course. It should be stressed, however, that the mixture should not be subjected again to dispersion conditions, namely high impeller speed and high temperature, as this might affect stability already achieved.

Deaeration and potting (see Fig. 12.10)
When concentrates are used, less air becomes trapped in the ink during the mixing process, nevertheless it remains advisable for inks which rely on oxidation for drying to go through a deaeration procedure. This can be done either by the traditional light potting pass or alternatively by filtration and vacuum deaeration techniques. This would naturally be aided if the initial mixing was carried out under vacuum also.

Summary
Figure 12.11 illustrates a few of the possibilities from the multiplicity of methods and individual machines all with their own conceptual design style and benefits. Routes (1) and (2) could only be used for the lower viscosity types of paste ink and (3) to (8) are by far those in most common use.

The individual pieces of equipment can be put together to form an on-line unit, but before purchasing, any machinery under consideration should be evaluated for throughput, quality and general handling conditions. This is particularly important with envisaged automatic controls, either remote or through a computer, as the program can only be written when the aspects just mentioned are known. A ridiculously simple

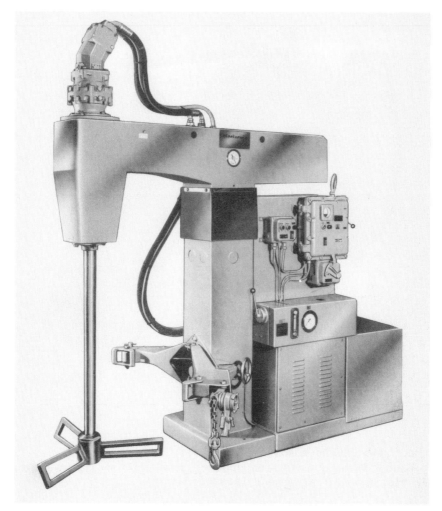

Fig. 12.9 High torque mixer (courtesy of Mastermix Ltd)

example of this would be to have a mixing capacity of 1000 kg per hour, when the product can only be refined at 500 kg per hour, which would mean the purchase of another refiner or using the mixer for only half the time. It is foolish to dismiss such simple examples because they do occur, and all too frequently.

Although a fairly rigid set of parameters is needed to program in-line production, the constituent parts need to be flexible to accommodate changes in the products' physical characteristics. For example, if, due to temperature the viscosity of a heatset colour is liable to change, it is important that the three-roll mill being used has some facility to adjust the rolls to enable a similar throughput even though, due to the viscosity drop, the cohesiveness of the product has reduced. With the exception of

Fig. 12.10 De-aerator (courtesy of Mastermix Ltd)

the major bulk lines, some ability to vary certain dwell times in the
production line is useful, remembering at all times to reprogram the
microprocessor, if one is being used.

The manufacture of polymer/solvent systems

This section covers the manufacture of flexographic, gravure and most
screen inks. As noted previously, there are two main types of formulation
— water based and solvent based. The volatile nature of the liquid phase
of both solvent- and water-based inks requires similar manufacturing
techniques but they are often produced in separate units to avoid contamination. The very inflammable nature of a large number of the solvents
used requires special precautions to be taken to prevent the risk of fire
and explosion. Solvent extraction and fresh air ventilation are both used
to keep the level of solvents in the atmosphere below the threshhold limit
value (TLV).

The area where inflammable and highly volatile products are manufactured is normally classified as a Zone 1 flameproof area and every effort
must be made to eliminate the risk of fire; manufacturing equipment and
containers must be earthed to eliminate the generation of static electricity;

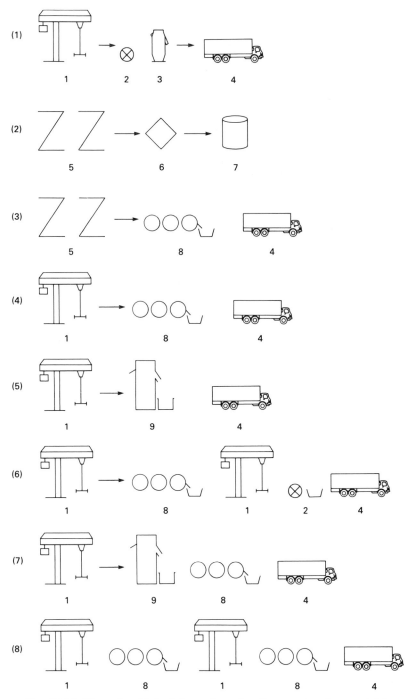

Fig. 12.11 Thumbnail sketches of production routes: 1 = variable speed mixer; 2 = pump; 3 = filter; 4 = distribution; 5 = z-blade mixer; 6 = de-aeration; 7 = container; 8 = 3-roll mill; 9 = bead mill (courtesy of Lorilleux & Bolton Ltd)

all motors, lights and switches must be to Class 1 flameproof standards and any hand tools used must be non-sparking. Further, the clothes worn by operatives should be so manufactured that overalls do not cause static and protective footwear does not cause sparks.

A number of the materials used in liquid ink formulation could cause a health risk if adequate precautions are not taken. The extraction systems should keep the overall levels of solvent in the atmosphere below the 8-hour weighted maximum allowed concentration (MAC), but care must be taken not to allow pockets of solvent fumes to build up.

Face filter masks must be worn if there is a risk of powder in the atmosphere and gloves are essential to reduce the risk of skin contact related health problems.

Varnish manufacture

It is possible when producing liquid inks on a ball mill to load pigment, resin, solvent and other ingredients to produce a finished ink from basic raw materials. However, it is more common to produce varnishes using the resins and solvents before the pigment dispersion process.

Liquid ink varnishes are formulated using a very wide range of resins such as nitrocellulose, ethyl cellulose, shellac, maleic resins, metallic resinates acrylics, polyamides, etc. The solvents used range from aliphatic and aromatic hydrocarbons, to esters, alcohols, ketones and water. Varnishes are normally produced on high-speed stirrers. In view of the volatility of the solvents used and the heat generated, it is essential that the stirrers are totally enclosed and, where the design allows, water cooled.

Typically, the following types of mixer are employed.

Cavitation mixer

The cavitation mixer is a driven shaft fitted with a saw-tooth disc impeller which, combined with a variable speed drive, provides for flexibility of tank size and volume of charge. The impeller acts to keep the whole mass of the change in motion and its size and velocity is determined in direct relation to the tank diameter.

This type of mixer is better suited to products in the higher viscosity range and in this respect, process time can be considerably reduced if a proportion of the total liquid for the batch is omitted in the initial stages of the mix, to be added once the resin solids have been dissolved. This 'addition' can be used to advantage as a coolant to the batch where heat-sensitive materials such as wax compounds are part of the formulation.

Rotor/stator mixer

This type of mixer has a fixed rotor/stator size and is normally operated at only one or two fixed velocities. The charge size and vessel size is there-fore limited. The overall agitation is comparatively less vigorous than that produced by the cavitation mixer and viscosities must be kept corre-spondingly lower. The reduction of solids particles is achieved by their subjection to shear forces as they continuously pass through the narrow gap between the rotor and stator.

The application of heat by using a jacketed vessel will accelerate both processes although this is not always advisable when volatile flammable solvents are used unless strict temperature controls are used.

If multi-solvent systems are used, it is prudent to dissolve the resin in true solvents before adding the diluents. The normal procedure used is to agitate the liquid while adding the resin gradually to it. After the addition is complete, a further period of agitation dissolves the resin. For efficient dissolution the agitation must be of sufficient velocity to maintain resin particle suspension and separation. The liquid will thus act upon the whole surface of the particle, giving maximum opportunity for the solubilising process to take place.

Storage and handling
In addition to the questions raised about storing oleo-resinous varnishes, the following important points must be considered.

The low viscosity, possible flammability, and solubility characteristics of liquid ink varnishes requires pipework and storage containers to be constructed of materials adequately resistant to attack by them. A study of relevant legislation covering the storage of inflammable products is essential.

Manufacture of additives
When a formulation requires very small amounts of material, i.e. anti-foaming agents or ingredients that need pretreatment to highly controlled levels, i.e. wax pastes, a separate preparation is normally manufactured for incorporation into the formulation.

When the additive is a dilution of the active ingredient, it only requires a simple mixing of the ingredients but where predispersed waxes are required, these must be separately manufactured on normal liquid ink equipment.

Liquid ink manufacture
Referring back to Fig. 12.11 which gave optional manufacturing paths for oleo-resinous inks, routes (1) — variable speed mixer followed by filtration and (5) — variable speed mixing followed by bead milling and filtration — are also applicable to polymer/solvent systems. In addition, liquid inks can be made by charging a ball mill with either the whole mixing or a pigment vehicle base which can be thinned *in situ* once the initial dispersion has taken place.

Ball mills
The traditional method of manufacturing low viscosity inks during the last 50 years was in ball mills. A ball mill consists essentially of a metal or ceramic cylinder rotating on a horizontal axis. This cylinder is partly filled with steel or porcelain balls or pebbles. During the rotation of the cylinder at the correct speed, the grinding balls rise and cascade over each other subjecting the ink components to crushing and shearing stresses.

The voids between the balls in a normal charge are estimated to be 40% of the charge volume, and this is the amount of ink material charge recommended for general work. In practice, good results are obtained when the ink charge is slightly above the ball charge level. If, as recommended this is 45%, it should make up the additional volume to a total ball and ink charge of 50%, thus a half-full mill will contain 23% of ink by volume. If the pigments are very easily dispersed, then figures of up to 65% total volume of ball and ink charge are suitable. In this case, the mill is acting mainly as a mixer.

The choice of pigment/vehicle ratio has a considerable influence on the performance of the mill and it may need experimental work carried out to determine the charge for maximum efficiency. It is claimed that the best result is obtained from a high viscosity mix, allowing for the limitation that the cascading effect of the balls must not be impeded.

Higher production speeds may be obtainable by milling the pigment in solvent or in a low viscosity first and then adding the higher viscosity medium later, but more wear takes place. A rough guide is that the volume of the liquid should be 40% of the apparent bulk of the pigment; however, the viscosity of the system will change during mixing. Thixotropy should not cause any problems while the mill is rotating but discharging the mixture may be a problem.

Bead Mills

While the ball mill may be considered as the original totally enclosed system for manufacturing polymer/solvent type inks, the introduction of bead mills gave opportunities for introducing faster and continuous production methods. Bulk amounts of concentrates and inks can be manufactured from dry pigment and varnish using a combination of mixers and bead mills by the following processes:

(1) Weigh out pigment and vehicle and premix on a stirrer.
(2) Disperse on a bead mill or other continuous dispersion mill.
(3) Add additives and solvents.
(4) Filter and pump to the storage tanks or to containers for delivery, once passed by quality control.

These processes are very similar to those used for making an oleoresinous ink from pigment and varnish. Decisions about the amount of vehicle required to 'wet out' the pigment at stage (1) and the correct pigment/vehicle ratio for optimum dispersion at stage (2) have to be taken to give the conditions for maximum shear. The increasing use of ED pigment powders makes the grinding stage a much simpler operation in polymer/solvent types of vehicle and gives the level of dispersion and high colour strength required for quality inks with less energy.

Stage (3) consists of the addition of the remainder of any vehicle not previously used, also solvents to achieve the correct viscosity, together with other preparations not suitable for inclusion in the original grinding phase.

To prevent any chance of any of the bead charge or other adventitious material getting into the final product, the ink is sieved through a fine mesh filter. This will often be a multi-stage process, especially if the ink is held in stock for any length of time.

Chips

An alternative method of producing liquid inks with exceptional colour strength, transparency and high gloss properties, is by the use of the pigment 'chip'. Chips are produced by dispersing pigments into solid resins which may also include some plasticiser. Various wetting agents and antioxidants can be added to improve the manufacturing time and prevent instability on storage.

Basically, a 'chip' is a solid ink which needs the addition of liquid to make it usable. Care needs to be taken to enable the 'chips' to be dispersed in the solvents used in the formulation without flocculation. Various methods may be used:

(1) Soak the chip in diluent until soft and then use a high speed mixer to dissolve in the true solvent. Varnish and additives are then added.
(2) Disperse the 'chip' in high resin content varnish and add solvent slowly, avoiding shock 'cooling'.
(3) Add the chip slowly to the varnish/solvent component of the formulation. When dispersed, add the other ingredients.
(4) It is also possible to use a ball mill although care has to be taken to ensure that the formulation of the initial charge does not cause the chip to form a conglomerate with the ball charge.

Figure 12.12 gives a sketch form of the manufacture and use of the pigment 'chip' and more detail regarding their production will be found later. However, the use of chip has diminished considerably over the last 5

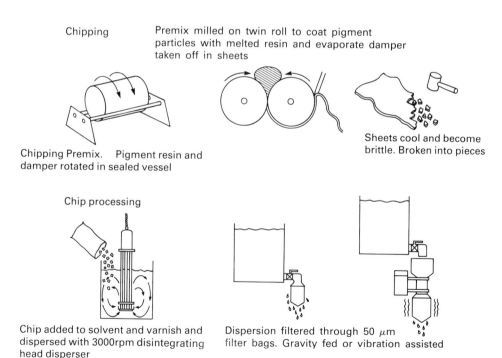

Chipping

Chipping Premix. Pigment resin and damper rotated in sealed vessel

Premix milled on twin roll to coat pigment particles with melted resin and evaporate damper taken off in sheets

Sheets cool and become brittle. Broken into pieces

Chip processing

Chip added to solvent and varnish and dispersed with 3000rpm disintegrating head disperser

Dispersion filtered through 50 μm filter bags. Gravity fed or vibration assisted

Fig. 12.12 Liquid ink manufacture from chips (courtesy of Lorilleux & Bolton Ltd)

years, mainly due to the cost of manufacture, while improvements in pigment manufacture and dispersion equipment have offered alternative production methods giving inks of nearly equivalent properties. For top quality results and when using high cost pigments the process has cost-effective advantages.

Other types of pigment dispersion can be handled in the same way as 'chips', particularly in the preparation of water-based flexographic inks. These simply require mixing with the rest of the formulation, again avoiding flocculation, and refining in the normal way. The new generation of hyperdispersants may be used for manufacturing publication gravure inks (see p. 604).

Pigment 'chip' manufacture

The hazards associated with the use of these materials are encountered mainly during their manufacture, especially where nitrocellulose, chlorinated rubber and oxidising pigments are employed. After manufacture, although nitrocellulose and the oxidising pigments remain flammable, they are generally less hazardous and more stable than a liquid ink being solids. This gives the product a greater shelf-life, improved environmental tolerance and allows easier storage. The oxidising action of specific pigments can be overcome by the use of inhibitors in the formulation.

The cycle of manufacture usually involves four separate stages, namely pre-mixing, dispersion, dry/cooling, final size reduction.

The initial pre-mixing of pigment and resin can be carried out using a Z-blade mixer, solvent and plasticisers being added to give cohesion, producing a viscous paste. This method is commonly used for nitrocellulose systems as it contains any potentially dangerous nitrocellulose dust, at an early stage. Alternatively a comparatively dry, although homogeneous mix, can be prepared either by tumbling or with a planetary type mixer, using a minimum of solvents/wetting agents. In both cases, the levels of solvent, plasticiser or other wetting agents must be strictly controlled as excessive use will prolong eventual milling operation and could lead to loss of dispersion. Heavy duty Z-arm mixers are available from several manufacturers, some featuring bottom unloading by an extrusion worm-screw fitted to the base of the mixing vessel. These machines should also incorporate a water jacket for cooling around the mixing pan.

After pre-mixing, the material is transferred to a two-roll mill and placed in the nip. There it forms a plastic band around the rolls which is repeatedly cut off, turned and replaced into the nip until dispersion is complete. When using resins which have low softening points, for example polyamide, close control of roll surface temperature is desirable, as the mass can, if allowed, become very plastic, and therefore difficult to work, which in turn reduces shear efficiency. It is also advantageous to maintain a differential in roll surface temperatures as the mass (during its final stages) can be confined to one roll allowing easier removal. The nip setting varies with materials used, although it is usually between 1 and 1.6 mm. Some machines have a mechanism for changing this setting while the process is carried out. Shear is produced in the mass by the roll speed differential. The pressure applied during the milling operation is around 50 tonnes per square inch; power input is usually between 5.9 and 8.9 kW (8–12 hp) per kilogram of product. The mills are the types used by the plastics and rubber industries and are generally between 44.7 and 111.8 kW (60–150 hp). The rolls must have a speed differential, the ratio being around 1.1 : 1 and will be either cored or drilled peripherally to allow for cooling by water. The rolls have an adjustable nip and should incorporate a removable blade across one or both rolls to facilitate removal of the finished product.

When milling is complete a plastic sheet is produced (see Fig. 12.13) which is allowed to cool during which any residual solvent evaporates off. The time taken for this varies with the resin system employed; for example, polyamide or proprionate require little drying time, while nitrocellulose systems can take several hours. Some nitrocellulose dispersions are made with a water-dampened resin which can present difficulties when drying. The use of a de-humidifying system can be contemplated if large volumes of this chip are required.

The sheeted dispersion becomes brittle when dry and is ready to be reduced to a convenient particle size. This is often carried out mechanically through a high-output cutting mill. Such machines are supplied with a selection of graded screens which determine the particle size; care should be taken to select a screen which produces a minimum of dust as this is detrimental to the quality of the finished ink.

Extreme caution is required when reducing dry nitrocellulose dispersions as any dust created during the process can be highly explosive,

Fig. 12.13 Chip manufacture (courtesy of Lorilleux & Bolton Ltd)

requiring purpose-built plant to be carried out safely, An alternative dispersion technique is to blend pigment and resin in a Banbury mixer. This machine produces shear under conditions of high pressure; the ram applies and maintains the pressure while the rotating arms provide the differential moving surfaces. Within the chamber the temperature rises rapidly and control is necessary to prevent the subsequent rise in viscosity and eventual loss of shear.

Manufacture of dye-based inks

Dyestuffs are soluble in the solvents and diluents used in the formulation and the method of manufacture is normally to dissolve the dye in the solvent part of the formulation with constant stirring.

The application of heat will increase the solubility of the dye and therefore lessen the time taken for the process to be completed. When the dye is dissolved, varnish and other additives are added to complete the formulation. Once again, care must be taken not to 'shock cool' the system, causing the dye to become insoluble.

12.3 MIXING EQUIPMENT

The dispersion of the dry materials, pigments, fillers, etc. is normally done in a mixing vessel. Whether they are dropped in from the top manually, or by means of a screw feed or, alternatively, drawn in from the bottom using vacuum in the mixer, is academic. The function of the mixer at this stage, is to to achieve an accepted level of dispersion before it is further refined or filtered, also to facilitate the addition of liquid reducers and other additives to already dispersed material. *Paddle mixers* find a wide use for mixing all types of ink. They are also used to agitate or keep products in solution or suspension.

The types of mixers available are numerous but, in principle, they fall into fairly simple groups, mainly determined by the type, viscosity and volatility of the ink to be made. High-viscosity oleo-resinous formulations are best dispersed using a low shear pre-mixer, but as the ink viscosity falls, higher-shear machines give better results. Low-shear equipment employs slow speed mechanical movement, coupled with a high-viscosity product to maximise the shearing effect.

Vacuum pre-mixing has become general practice in the last few years, especially where oil inks are concerned. The main reasons for this are twofold. Firstly, it is known that air dispersed in inks can be as high as 40% of the mix and this reduces the efficiency of the pre-mixer considerably. Secondly, dispersed air can lead to problems in subsequent bead milling. Heat transfer within the mill is reduced considerably, leading to high processing temperatures which would affect the milling of some yellows and other heat-sensitive pigments.

Z-arm mixers have been used in the production of viscous pastes for many years and their use in manufacture of flushed pastes has already been mentioned. Modifications can be made to the basic mixer, such as the provision of heating/cooling jackets, while vacuum or pressure can be applied to the mix. In general, they are of heavy construction with relatively powerful motors and a low operating speed.

The mixer contains two gear-driven Z-shaped blades rotating, usually at different speeds, the blades almost scraping the bottom and sides of the mixer. The blades rotate in opposite directions, both towards the centre of the mixer, and the whole unit can be rotated through 90° to empty the contents for further processing. Usually, part or all of the varnish is added to the mixer and then the pigment and/or filler, added in stages, allowing each addition to be fully wetted out before the next one is added.

As the blades rotate, the mix is kneaded and thus strong internal shear forces result in a tearing action within the mix. Further shearing occurs between the blades and the sides and bottom of the mixer.

Butterfly mixers (change pan) (see Fig. 12.9)

In some cases, combined with a fixed speed shearing head, these mixers are best suited for middle to high viscosity materials, such as paste inks. They produce a 'folding' movement, where flow is restricted, but a kneading effect is achieved although the shear effect is limited. It is in fact a compromise between the planetary-type mixer and dissolver and in its field is a widely used machine.

High-speed impeller mixers (see Fig. 12.14)

For fast, efficient dispersion of middle to high viscosity materials, this type of mixer is gaining in popularity today because of its speed, reproducibility, ease of cleaning and general flexible efficiency.

High speed impeller-type mixers originally used a serrated blade (Cowles type) which creates high turbulence and intense shearing action. Generally known as dissolvers, they can be of fixed type, sited on large tanks or pedestal type which rise on a column allowing portable vessels to be used. The speed is generally variable allowing for a range of viscosities and loadings.

If the serrated blade is aligned on the central axis of the vessel, dispersion is achieved by the adhesion of the material to the disc surface combined with the configuration of the teeth. The teeth force the material radially outwards to the side of the vessel where they are deflected upwards causing a vortex effect. Additionally, the teeth act in breaking down clusters or lumps of solids. Peripheral speeds of such discs are between 20 and 25 m/s.

An infinitely variable, hydraulically driven disperser is an accepted processing tool in the surface coatings industry and has been supplemented with other similar units such as the high torque mixer and the twin shaft disperser (see Fig. 12.15).

For the manufacture of liquid inks, the mixers described below find particular use.

Rotor and stator high speed mixers

In machines like the Silverson, a stator is introduced between the rotor and vessel wall, in order to increase the shear. The disadvantage of this

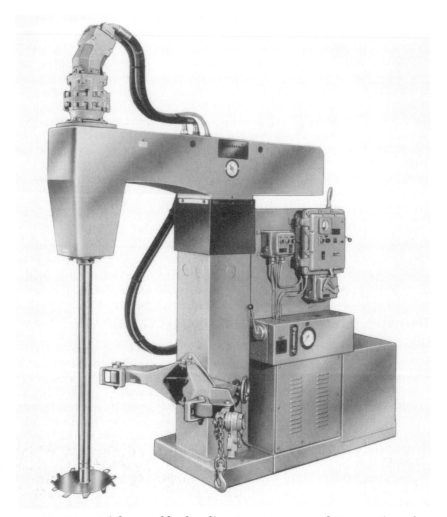

Fig. 12.14 High-speed hydraulic mixer (courtesy of Mastermix Ltd)

type, however, is that the radial velocity beyond the stator is reduced and it is therefore limited to low viscosity materials and chip dispersion. (See also previous section on Varnish Manufacture).

The 'star' impeller type (see Fig. 12.16)

This represents the latest development where contrary to the Silverson type head, the rotor is the outer element and the stator the inner. This combines the advantages of the toothed blade and the combined rotor/stator system. A comparison of the disc impeller with the star head shows that, for a Newtonian fluid, the reduction of flow to the side of the vessel is linear, in the case of the disc. The star head, however, based on similar

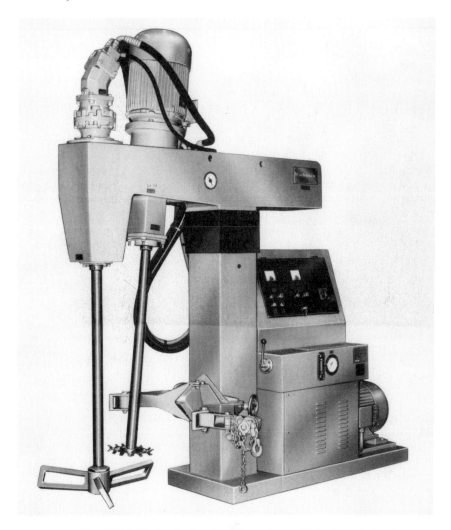

Fig. 12.15 Twin shaft mixer (courtesy of Mastermix Ltd)

speeds, shows similar velocity parameters as the disc from rotor to vessel wall, but exhibits very high shear rates between rotor and stator, without loss of vortex.

The high-speed disperser

The field of use for these mixers is in the premixing of low to medium viscosity inks, such as the range of the various liquid ink formulations and web offset inks for fast presses.

The apparatus used in high-speed dispersion usually comprises a cylindrical tank in which is operated a dispersion disc or head, driven from a central shaft, normally capable of variable speed operation from

Fig. 12.16 Star mixer under vacuum (courtesy of KWR Chemicals Ltd/Draiswerke GmbH)

zero to approx. 1440 rpm (different speed ranges are usually necessary for high torque mixers).

There are numerous variations with respect to both dispersion disc design and position of the central shaft, but the commonly accepted formula for operation is as follows:

Diameter of turbine	D
Diameter of vessel	$2-3D$
Batch level	$2-3D$
Height of turbine from bottom	$0.6D$
Peripheral speed of turbine	$1100-1500$ m/min.

It should be emphasised that these operating conditions are merely those which are most commonly used. Obviously, it is necessary to vary, for example, dispersion disc diameter, batch depth, etc. in accordance with the requirements of the particular process under consideration.

The ratio of mixing tank diameter to disc diameter is an important factor, in that, as scale-up occurs to larger batch sizes, the actual distance between the dispersion disc and the inner wall of the mixing vessel increases and problems may sometimes occur with lack of circulation of material, particularly in those systems of unusual rheological properties, of thixotropy or dilatancy. It should always be remembered that the objective is to create a suitable condition of flow for the batch as a whole, when the intense shear of the high speed dispersion disc can be imparted to create acceptable dispersion of the solid phase.

If processing difficulties causing loss of material flow are encountered, then, should it not be possible to adjust the formulation, attempts may be made to operate the mixer at a higher peripheral speed. This may be achieved either by increased rotational speed or the fitting of a dispersion disc of greater diameter. Regrettably, on occasions this can only lead to a further work input into the batch and can result in excessive temperature levels and the loss of volatile solvents, and indeed, colour change.

It should be mentioned at this stage that the raising or lowering of the dispersion head during the initial wetting out and mixing phase can be a most useful processing technique, and while this is often carried out manually in the high-speed disperser, there are other machines available with an automatic process control function, which will raise and lower the dispersion head in accordance with the degree of turbulence created within the batch and hence will greatly influence mixing and dispersion efficiency.

Another principle using the technique of asymmetric agitation is also widely used in the liquid ink field. This system relies on the intensive flow patterns achieved by placing the agitator shaft, off-centre which initiates high shear forces between adjacent streamlines having different speed patterns. Particle collisions are enhanced by such mixing and providing this can be maintained, this is a highly energy efficient pre-mixing system.

In summary, while the classical dispersion technique of using a high-speed disperser does offer certain advantages, some outstanding problems remain. The most important of these is with materials having unusual flow characteristics, particularly in larger batch sizes, when adequate flow and circulation of material are not maintained at the periphery of the vessel. Hence, efficient dispersion is not obtained or alternatively, time cycles can be extremely lengthy. In such cases, a slow scraper on the periphery of the container can be attached to rotate at a speed of 8−12 rpm wiping clear the inner surface and pushing the dead areas back into the vortex.

The fixed or on-line mixer

The Mastermix PMD disperser comprises a cylindrical vessel with smooth internal surface and reversed coned bottom, the vessel being suitable for jacketing for heating or cooling. The slow-speed agitator driven from the bottom of the vessel is fitted with scraper blades to both sides and bottom and the agitator is supported in a large bearing and seal assembly. High-speed dispersion is achieved by an angled fixed disperser unit driven by a fixed speed motor which is welded on a telescopic mast and which permits the shaft to be raised and lowered, as required.

A closed cover includes a charging port for solids and various liquid addition points, together with provision for an autoraise/autoload facility and a cleaning mechanism. Alternatively, the cover can be modified to provide for automatic solids loading. Vacuum operation is possible with this type of machine.

Probably the most important single factor in relation to the PMD type of mixer (see Fig. 12.17) is that distinction is made between the requirements of mixing, i.e. adequate circulation of materials at all times including thorough wetting out, and the technique of dispersion. This is

Fig. 12.17 Enclosed PMD mixer (courtesy of Mastermix Ltd)

carried out in an area of high shear, provided by a high-speed disperser disc, at the same time, ensuring that all material passes through this intense dispersion area very often. This is critically important as batch sizes increase. It is, for example, difficult to imagine adequate dispersion using the classical technique of a single high-speed disperser, above say, a batch size of 2500 litres, without consuming vast quantities of power. The PMD mixer is able to carry out this technique commonly using working capacities of 7000 litres. However, machines having recommended working capacities of up to 20 000 litres are available.

The ideal dispersion system should have the following qualities:

— Be enclosed and with safe operation at low noise levels.
— Maintain circulation of the batch at all times with no possibility of undispersed solids being deposited on the walls or bottom of the vessel.
— Have maximum adaptability for full automation and solids handling, e.g. process control functions including automatic raising/lowering of dispersion head, automatic pigment/solids loading, fitment of cleaning installations and provision for solids loading systems.
— Have the ability to handle materials of unusual flow characteristics, particularly thixotropic materials.
— Have minimum power consumption.
— Have easy discharge facility, even for products of poor flow properties

High-speed mixing

Over recent years, high-speed mixing techniques have shown little change as far as low viscosity mixture are concerned. In the highly viscous paste field, however, significant technical advances have been made, and more regard is being given to the importance of good wetting out and better pre-mixing. Few product manufacturers when purchasing or using a mixer, such as a Trifoil type, realise how much efficiency and money may be lost between the motor drive and the blade. The blade, after all is the only working part in contact with the materials being mixed, yet its inefficiency is obscured by the mixer horse power; even so, between 30 and 50% power losses have been recorded.

To obtain the optimum results, it is essential to achieve the correct balance of torque, shear and vortex. The use of a hydraulic drive has shown great promise of improving these requirements but mixer manufacturers generally have been reluctant to develop this. However, at least one particular range of equipment has a full programme of development and is currently proving its new technical advantages within the industry. To maintain torque and shear, the hydraulic system has been specially designed, balancing the pump, transmission lines and motors, so ensuring a minimum loss of power at the point of mixing.

To impart shear with particle reduction, the relation of the Trifoil blade to the bottom and side surfaces of the mixing containers is crucial, and the vortex should develop to ensure that the batch passes through the high energy zones.

As an example, the conventional high-speed mixer fitted with a 30.5 cm diameter rotor and revolving at 1440 rpm, should obtain a peripheral speed of 1379 m/min when used to disperse a low-viscosity ink. A 70 cm diameter Trifoil rotor at 600 rpm will have a peripheral speed of 1320 m/min. It can be seen that they are almost equal, but if the Trifoil blade is being forced through a high viscosity oleo-resinous ink, cooler conditions prevail because the work is done much nearer the container sides and bottom. This means that the work is far more effective in the earlier stages at a reduced speed, the force being very much higher with a shorter distance being moved, resulting in less heat. A higher speed is desirable as the viscosity drops and the resistance falls as the state of dilatency changes to pseudo-plastic. With high-speed open rotor systems the movement near the container sides is slow with most of the work being done in the centre when heat exchange is made less efficient due to the lagging effect of the slow-moving outer mass.

Having achieved a uniform and acceptable quality it now needs refining by filtration.

Filtration has its own parameters; its standards are normally measured in microns and its restrictions are viscosity and pressure. To filter a liquid requires some pressure, even if it is only gravity as in the case of many of the thinner liquids and varnishes. However, in the paste ink and liquid ink toner areas, cartridge filtration has made very considerable advances. Caution is needed here to avoid filter bag and cartridge blocking and the choice of material and pressure resistance is paramount because if either causes deterioration the batch is spoilt, but just as important is the danger if undue pressure builds up in the line. A pressure gauge is essential in such lines.

12.4 MILLING EQUIPMENT

Three-roll mills

These have been the mainstay of the paste ink industry for many years, but the machines today bear little resemblance to the early mills, being much more sophisticated.

Modern hydraulic three-roll mills, such as the 1100−1300 mm range, while being considerable improvements over their predecessors, are costly and, relative to output, very expensive to run.

In order to optimise output, the major mill manufacturers have developed sophisticated systems of gearing ratios, specific roll cambers and automatic ink feed, all of which have improved the specific results but have done nothing to assist the flexibility of these machines, which was once one of their greatest assets.

The mill is in fact a three (or in some rare cases, a five) roll disperser made to very exacting engineering standards. Roll widths are between

150 and 1300 mm with almost any gear ratio and roll camber to ac-
comodate the vast range of products which are refined in these machines.
A word of caution concerning the operation of these machines. They
have two in-running nips which in the UK have to be electronically
guarded with a fail-safe device to avoid the entrapment of fingers or any
other appendages. They also have an extremely sharp 'doctor' take-off
blade which must be handled with respect. Should the machine run dry,
the roll surfaces will be damaged and this will necessitate regrinding. To
avoid this, an automatic feeding device such as the SAT control system
which both regulates the feeding of the mill and protects their rolls from
running dry can be used. This does not predetermine the method of filling
the hopper as the system can accommodate tub tilting, pressurised pan
emptying or pumping, allowing the system to work automatically without the need for manual control. As can be seen in Fig. 12.18, a float is
scanning the product level in the hopper of the roller mill. As soon as the
product arrives at level I, II or III preselected at the transmitter, this level
will be kept constant between the limits 2 and 3 by switching the feeding
mechanism on or off. If material supply fails and the hopper empties, both
the roller mill and the feeding device will be automatically stopped at
point 1. At the same time, this state will be announced by an optical signal.
For cleaning purposes or when grinding small batches, the automatic
device may be put out of operation and the float removed. A monitoring
light indicates the state of supervision of the mill.

As has been said, this type of roller mill has been with us for some years
but the degree to which it has been refined is most impressive, as also has
the feeding of the hopper. (See Fig. 12.19 outlining these features whether
the ink is pumped, pressed or gravity fed.) The SDV-1300-E (see Fig.
12.20) is now available in a programmable format offering clear advantages in factories where optimum output with uniform quality is required
even with wide and frequently changing production programmes. The
advantages are that the preselected values are electronically controlled

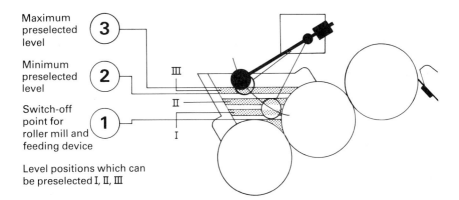

Fig. 12.18 Hopper with float (courtesy of Buhler Bros Ltd)

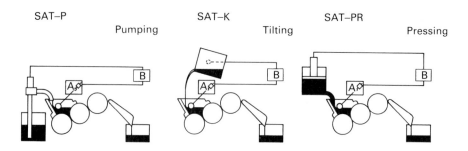

Technical data

- Preselection of the level in 3 positions
- Optical signal when mill is automatically stopped
- Flame-proof design of transmitter A, class ex G4

- Control circuit B in closed box, not flame-proof
- Standard output for SAT-P 1, 1kW, other rates available on request

Fig. 12.19 Feeding options (courtesy of Buhler Bros Ltd)

and thus kept constant avoiding continuous attendance by the operator or influence from unauthorised persons with the knowledge that the machine's safety is assured. By recording and evaluating the production data the results can be used to optimise the efficient operation of the mill and enable the product to be made consistently from batch to batch. This is a 'fail-safe' piece of equipment where roll temperatures and hydraulic pressures are automatically adjusted and there is electronic locking of selector switches, to avoid error, even in the limited operator related aspects of the process.

These machines and their performance figures are not comparable with the machines built 10 years ago or earlier. A modern roller mill is capable of meeting the following requirements with regards to a rational manufacture of quality products:

— High throughput.
— Optimum level of grinding with good wetting.
— Optimum and at all times uniform dispersion.
— Simple and safe operation.
— Continuous operating process without the supervision of an operator.
— High space utilisation factor.
— Rapid and simple colour change.
— Robust and maintenance free design.

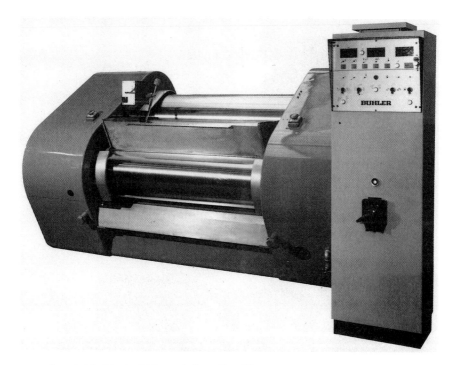

Fig. 12.20 SDV-1300-E triple roll mill (courtesy of Buhler Bros Ltd)

Three-roll mill in operation

It is often incorrectly assumed that during the dispersion process on the three-roll mill, a reduction of solid particles occurs. It is more correct to say that the passage through the mill breaks down the pigment agglomerates formed during production, storage and transport of the dry colour into primary particles which may be smaller than the required terminal fineness of the dispersion as well as being smaller than the roll gap (see Fig. 12.21).

Figure 12.22 shows how the levels of dispersion are improved by the increasing amount of energy used. All of these micrographs were taken using a Zeiss Universal Light microscope with bright field illumination. The overall magnificantion is 200X and the practical limit of resolution is estimated to be around 4 μm.

The pigment particles are coated with a layer of binder replacing the skin of moisture and air previously present — this is equivalent to a microflushing process, otherwise designated as 'wetting'. The breakdown of these agglomerates, achieved by the high shear forces present in the nip areas between the rollers, causes an increase in the pigment surface area. This, in turn, leads to the absorption of more binder, during the wetting process, producing an increase in viscosity for a given level

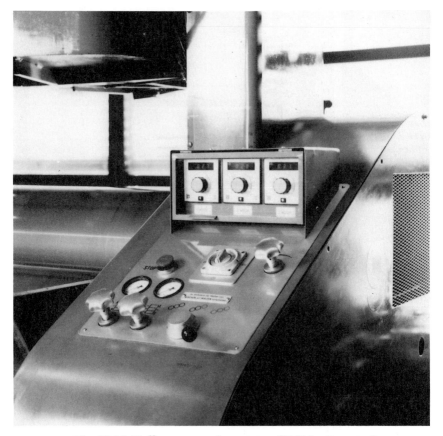

Fig. 12.21 Roll pressures (courtesy of Bühler Bros Ltd)

of vehicle. The improvement in the degree of dispersion is normally accompanied by an improvement in the level of colouring power.

An important factor is to obtain, by the whole working process, a uniform and at all times reproducible depth of colour. This means that all physical parameters concerned in the dispersion process, namely roll pressure, speed ratios and dispersion temperature, should be held automatically constant and should be as reproducible as possible.

Today it is possible by means of the centrifugal casting process to cast a 10−15 mm thick wear-resistant surface layer on the rolls; the hardness and elasticity of which is determined by the composition of the metal and can be adapted to meet requirements. With statically cast rolls, on the other hand, the surface hardness is obtained by quenching, namely by the prevention of graphite formation on the surface so that it falls rapidly with a decrease in the diameter.

The centrifugal casting process causes diffusion between the inner bearing cast-iron layer and the surface. Shrinkage cavities and hardness stresses are virtually excluded. All rolls must, nevertheless, undergo

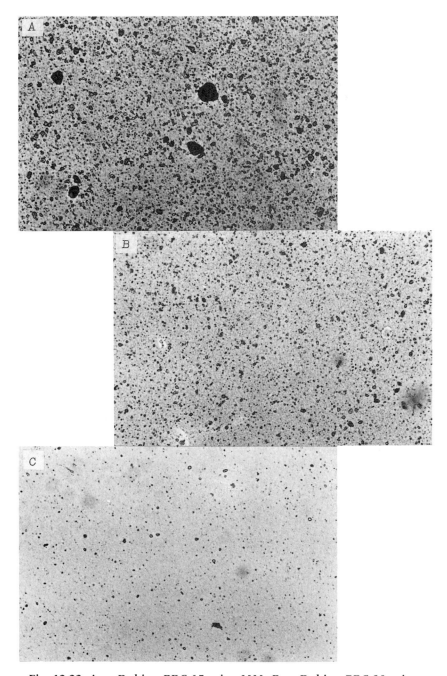

Fig. 12.22 A — Rubine PBC 15 mins MM; B — Rubine PBC 30 mins
MM + 1LP; C — Rubine PBC 30 mins MM + 1LP + 1HP *Note:*
MM = Laboratory Mastermix; LP = Light pass over three-roll mill;
HP = Heavy pass over three-roll mill

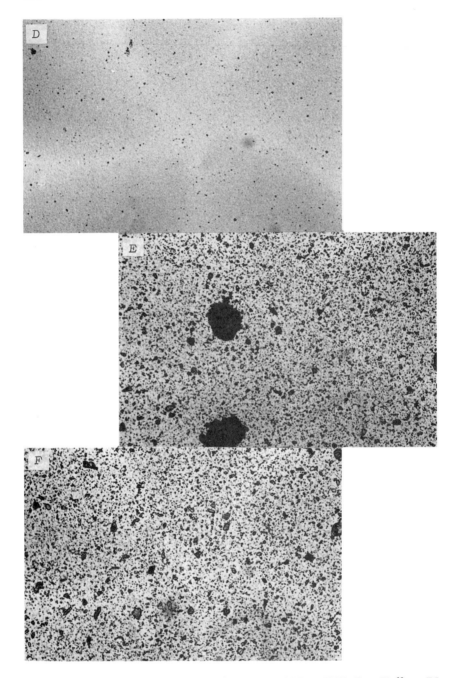

Fig. 12.22 D — Rubine PBC 30 mins MM + 1LP + 2HP; E — Yellow BL
Premix 15 mins; F — Yellow BL 2 mins Eiger. *Note:* MM = Laboratory
Mastermix; LP = Light pass over three-roll mill; HP = Heavy pass over
three-roll mill

Fig. 12.22 G — Yellow BL 5 mins Eiger; H — Yellow BL 10 mins Eiger;
I — Yellow FG4 Flocculation
(courtesy of Ciba-Geigy Pigments)

ultrasonic testing. A well-designed structure and maximum care during machining, influence the good and uniform temperability of the rolls and guarantee vibration-free running even at high rpm.

Automatic roll cooling

The importance of the correct roll temperature during the processing of pigmented systems is often overlooked.

Since the heat ouptut is not identical on all rolls, each roll should be fitted with its own thermostat system. The temperature gradient for each roll is produced as a result of friction heat. If the actual temperature, measured in the water bath in the rolls, exceeds the preselected required figure, automatic regulation of the cooling water feed is effected. This ensures the uniformity and reproducibility of all the required working temperatures. It also provides an estimated 30% saving in water consumption.

The same automatic system includes a magnetic valve which permits the inflow of water only while the main motor is running — this again saves water and obviates the formation of condensed water on the rolls, when the machine is not running. Should the cooling water supply fail, for any reason, it stops the machine.

Hydrodynamic roll pressure

Dynamic roll pressure involves a system consisting of a pump, maintaining an incompressible or compressible liquid or gas under a constant pressure under all circumstances, irrespective of external factors such as temperature, time, etc.

The mechanical and static hydraulic systems are still widely used; in such systems the transfer of force is taken over by an incompressible liquid or paste. Any desired transfer of force is possible by selection of the actuating and working cylinder system.

The dynamic system guarantees a constant pressure while the working distance resulting from the movement of the hydraulic cylinder (the quantity of oil in the cylinder and thus the roll gap) can be varied independently of the pressure.

The position is exactly the reverse with the static system — any variation in the roll gap changes the contact pressure.

This means that in relation to the three-roll mill, only the dynamic pressure system is capable of taking into account variations due to temperature differences or changes in the rheology of the product, without corresponding changes in the pressure. This is, however, critical for the reproducibility and uniformity of the operating process and the performance of a roller mill will be decisively affected thereby. Although the dynamic system involves a slightly greater technical and financial outlay, this will be rapidly recouped by the resulting decisive advantages outlined as follows:

— Reproducibility of all operating processes.
— Increased output.
— Constant pressure.
— Uniformity over the whole operating process and thus a uniformly high quality of the end product.

— Extremely simple operation, the pressure being preselected and subsequently controlled merely by the movement of a lever.
— Ancillary assemblies, such as hopper cheeks, stripping blades and tub-tilting devices, operate at constant oil pressure and can be very simply regulated and automated.

Floating-rolling system

In the floating roll system, which also has considerable advantages from the point of view of operation, the third roll (delivery roll or cutter roll) is mounted in a fixed position. The high rpm of this roll (in excess of 400 rpm) is thus achieved without vibration. The angle of the blade to this roll also remains constant without subsequent adjustment. Pressure is applied to the first roll (feed roll) with the middle roll floating between the outer rolls. Correct adjustment of the milling gaps between the roll is thus automatic. Any spraying is avoided, since the first milling gap only releases as great a quantity as the second milling gap is capable of processing.

Development of single cast rollers

The principle in manufacture of rollers over the last 100 years or more in the foundry is to pour molten cast-iron plus additions around the rope type core, and within the cast. The molten metal chills forming a wall of a given diameter. The addition or reduction of alloys to the material relates to the depth and hardness of the chill to the roll surface. The original roll mills ran slowly, some with closer ratios and a number of material passes were acceptable. Today, the roller mill runs at high speeds, produces more shear and uses more horsepower. All relate to increased temperatures of operation and product. The need for efficient water cooling became essential. Roller wall thickness had to be reduced. Metallurgy development to maintain strength was necessary. It was also established that the chilled areas reacted more slowly to efficient water cooling than other areas. The chilling process has retained its hardness and its material penetration but has a calculated depth related to wall thickness giving greater cooling efficiency to roller surface and subsequently, the product. Research has established the volume of cooling water to be in a roll at any point of time in operation. The development of the single cast roll without losing its established long life has been significant in the last 30 years. This, combined with improvements of process materials and formulations, is producing much higher quality products more efficiently.

Bead mills

Bead mills or stirred ball mills have been known for almost 60 years although their use in the ink industry has been limited to the last 20 years, the breakthrough in liquid inks being the introduction of glass media rather than 'Ottawa sand' which, until the early 1960s, had been the only medium available. Oil inks, due to their viscosity, required further technical developments in mill design and the use of bead mills for their manufacture came considerably later.

History

The first stirred ball mills were basically, fixed vessels with an internal rotor, unlike the standard ball mill which rotated about its own axis. The grinding in the tumbling ball mill is that of impact and compression where one can only affect the efficiency by mill diameter and speed — its limit being that where centrifuging takes place. Grinding in the case of bead mills is by compression and shear where accelerations of more than 50 times that of gravity are normal.

The original open top 'attritor' type mills used slow speeds and large beads. The restriction of this type of mill was that there was a limit to energy input and hence grinding efficiency. These mills were followed by the open top Dupont sand mills using 'Ottawa sand' as the medium, which became a standard production unit in the paint industry until the early 1960s, when they were replaced firstly by mills using glass beads and then by a multiplicity of closed mills which are today's standard. These operate at high speeds using very small grinding media — as small as 200 μm as against a minimum size of 13 000 μm ($1/2$ inch) in the tumbling ball mill.

There are many ways of grinding/dispersing solids in liquids. A particularly efficient means is by using the combination of the dissolver (impeller mixer) and bead mill, the first part of the process being batchwise, the bead mill continuous.

A dispersion produced from a mixture of liquids and solids needs a certain given energy input to produce the required quality. This total energy input can be reached by sharing the work between the dissolver and bead mill. However, the more efficient the dissolver, less is required by the bead milling process and vice versa. It is therefore important to achieve a high degree of pre-dispersion both to maximise the bead mill and increase product quality. It also avoids the production of agglomerates within the pre-mixing process itself.

Recent developments have now made it possible to avoid pre-dispersion by using a new direct dispersion process where the dry and liquid phase are fed directly to a bead mill and this process will be described later.

The following are types of bead mill, in chronological order, which have contributed towards its development.

Open sieve mill (Dupont and Sussmeyer)

This was the original sand mill and was basically a vertical narrow cylinder with an inner shaft to which were attached agitating discs shaped as a ring supported by bars to the centre. Subsequent shapes of disc have been: flat cam type, a plate with inner holes, etc. Separation of the medium from the product was by a wedge wire screen at the top of the grinding pot, the medium being Ottawa sand. Peripheral speeds were in the range of 9–11 m/s. The major development in this type of mill was the introduction in 1962 of the use of glass media by Draisewerke GmbH in what became known as the Perl Mill.

The various types of media now in use will be discussed later.

The open sieve type of machine was mainly used in the pigment, dyestuff and paint industries and was manufactured in sizes of 1 litre to 1000 litres. It still remains in use today, mainly in the production of TiO_2

and $CaCO_3$ based coatings. However, its drawbacks are that the viscosities of feedstock are limited due to bead lift and overflow in the sieve area. Thixotropy can also be a limiting factor as the lack of movement around the discharge area can result in a build-up of ground material. This could be moved by vibrators but they were not particularly successful.

Closed sieve mill (Dupont and Draiswerke)
The design is basically as the open sieve type but with an enclosed housing around the sieve area and the addition of a scraper to clean the sieve discharge. This machine enabled products such as water-based pigment pastes, which tend to foam, to be handled as once the air had been purged, the mill operated in a totally closed condition. Viscosity limitations, however, appeared to hamper further development of this mill.

Gap separation mill (Draiswerke GmbH)
Developed by Herr Durr and Herr Engels of Draiswerke GmbH, this major development enabled higher viscosities to be handled without the discharge problems experienced with the sieve designs.

The gap separation consists of a rotating disc fixed to the agitator shaft and an annular ring fixed to the grinding pot. The device is adjusted so that there is only a few tenths of a millimetre between the ring and rotor which retains the medium but allows the product to be discharged through what basically is a self-cleaning gap.

With this development, and its ability to handle pastes, a change of medium was required due to the low specific weight of sand/glass relative to the paste viscosity. Steel beads (ball bearing quality) were introduced, mainly 2 mm size being preferred. These easily destroyed the available agitator discs due to the turbulence they produced and a 'finger' design of disc was introduced. These discs were stellite coated to resist wear and created a 'rise and fall' effect rather than turbulence. Such mills were first used for the production of rotary news inks but then moved quickly into the cold and heatset inks field.

Limitations of this mill were mainly those of temperature. High energy inputs were necessary and the heat transfer through the only cooling surface, the pot wall, was limited. Blacks are not affected by the high temperature; in fact, dispersion is improved at temperatures of 80–90°C, but pigments such as yellow and some reds cause problems through changes in appearance.

John mill (Netzsch GmbH)
Developed by Willi John of Netzsch in the mid-1960s, for the manufacture of high quality paste inks, it came to prominence in the early 1970s because of its superior cooling system. Using an annular shaft gap consisting of an outer sleeve over the shaft, the gap between creating the separation zone, the John mill was able to cool the central shaft, increasing the cooling surface area by more than 30%. A further difference was the use of agitator pins in place of discs and counter pins in the pot wall creating increased shear. The relative increase in energy input required was countered by the ability to extract the heat more efficiently. This development enabled the heat sensitive pigments to be milled.

Tex mill (Draiswerke GmbH) (see Fig. 12.23)

This was the first totally closed grinding chamber, and the first step towards the development of what we now know as the high duty agitation mill. By virtue of the closed chamber, the medium is forced to keep a 'mean distance' apart irrespective of other influential factors. The separation of the medium from the product is by means of two sieve type cartridges which fit from the pot wall into the centre of the chamber and are therefore in the actual medium agitation zone, discharge being through the wedge wire cartridge into the outlet pipe. It is also necessary to use a double-acting mechanical seal on the shaft to avoid wear by the medium and also counter the inner chamber pressure which is developed. The benefits of the closed system are considerable:

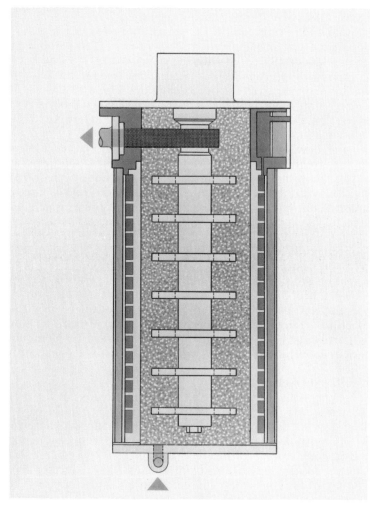

Fig. 12.23 Tex mill (courtesy of KWR Chemicals Ltd/Draiswerke GmbH)

— Higher throughput;
— Ability to handle higher viscosities;
— Elimination of evaporation losses;
— Better environmental control;
— Full use of the grinding chamber volume;
— Foaming avoided;
— Possibility of vacuum operation.

Dyno mill (Willy A Bachofen Ag) (see Fig. 12.24)

This machine changed the parameters of existing milling technology by incorporating a number of already known features into a totally new package. Milling to this time had been based on vertical mills with narrow grinding chambers and relatively slow peripheral speeds. The Dyno mill is a horizontal mill with a large diameter to length ratio and peripheral speeds ranging from 13 to 18 m/s. Separation is either by wall-mounted sieves or gap separation depending on viscosities and product need.

STS mill (Draiswerke GmbH) (see Figs 12.25–12.27)

The STS is a supercooled mill similar to the John system previously described but with considerably increased cooling area. In this machine

Fig. 12.24 Dyno mill (courtesy of Glen Creston Ltd)

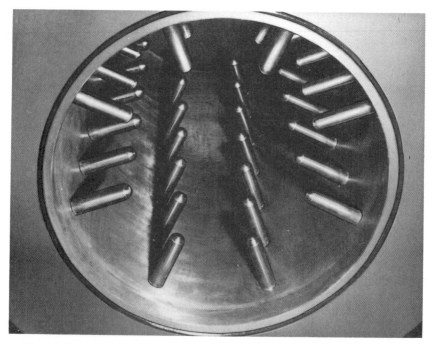

Fig. 12.25 STS chamber (courtesy of KWR Chemicals Ltd/Draiswerke GmbH)

not only the shaft is cooled but also the discs are cooled enabling even more useful energy input for grinding. Although mainly used for paste inks, the STS can also be used for liquid ink production, particularly solvent based, but high input energies lead to high medium wear.

Electronically controlled Cobra mill (Buhler Brothers Ltd)
A recent addition to the bead mill scene is the Cobra mill, using a known system of variable bead charge which can be controlled during milling. This enables an easily ground product such as a white to be milled at a low bead loading and a subsequently more difficult pigment at the full loading without the need for medium addition.

Boa 500 mill (Buhler Brothers Ltd) (see Fig. 12.28)
This is a mill with a horizontal ring chamber which can handle a wide range of viscosities, with simplified cleaning facilities. For product re-production the settings can be preset to obtain batch to batch comparity.

Co-ball mill (Fryma Ag)
A novel approach to dispersion is the Co-ball mill which uses the ink plus grinding medium without any moving agitator. Though still in early development, it is finding acceptance in liquid dispersion. It still, how-ever, has a number of problems to solve regarding flow, heat transfer, viscosity changes, etc.

There are many other types of mill on the market incorporating features

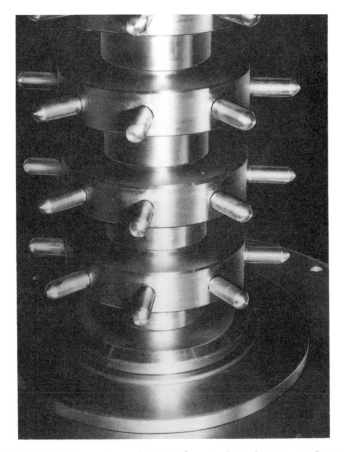

Fig. 12.26 STS shaft (courtesy of KWR Chemicals Ltd/Draiswerke GmbH)

first developed in the previously discussed mills but none have particularly contributed to the further enhancement of the dispersion technique, although they play a large part in ink production.

Grinding media
The original medium 'Ottawa sand', used in the Dupont mill is still in use today but to a very small extent. It has been replaced by glass and oxides for liquid inks and steel for paste inks (see Fig. 12.29 for a list of the various media available and their densities).

Horizontal versus vertical mode (see Fig. 12.30)
In the past few years, there has been considerable discussion as to the relative merits of using mills of horizontal or vertical design. Before any direct comparison can be made, it must be based on like parameters and it is not possible to compare make X horizontal with make Y vertical where both have differing features, speeds, etc.

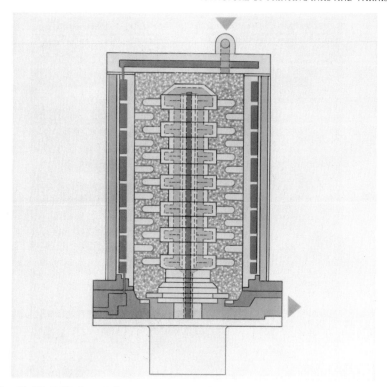

Fig. 12.27 STS sketchform (courtesy of KWR Chemicals Ltd/Draiswerke GmbH)

It is known that, to achieve the optimum conditions for grinding/ dispersion, it is necessary to achieve a maximum of energy input while obtaining a state of 'bead flotation'. This varies from product to product and also with temperature, surface area, behaviour under shear, etc. Some inks are particularly difficult to disperse and throughputs are of necessity very low. In the vertical configuration it is not possible to lift the beads with such low throughputs and so the horizontal mode is superior. The converse is true with higher throughputs. However, there is no hard and fast rule which can be applied. When one refers to oil inks or higher viscosity liquid inks, then the horizontal mode is not recommended due to compaction of the medium and product at the outlet. This leads to flow reduction resulting in high temperatures and excessive wear. The decision, therefore, as to whether a machine should be horizontal or vertical is not just a matter of design or fashion but it should be related to achieving correct bead flotation in the mill, while processing a particular grade of ink.

A mill exists which is able to swivel through 90° during production in order to attain the most satisfactory mode relative to the product being processed. It has yet to realise its potential.

Fig. 12.28 BOA 500 mill (courtesy of Buhler Bros Ltd)

Direct dispersion

All the machines discussed so far are based on the standard two-stage process of pre-dispersion and subsequent bead milling. This process has drawbacks in that it is labour intensive, and requires high capital outlay. Research has therefore been conducted by a number of companies to realize a totally enclosed process where the solid and liquid phases can be fed to a single machine without the need for pre-dispersion or pre-mixing, and the required fineness of product achieved in a single pass.

The Direct Perl mill from Draiswerke GmbH was the first to reach this requirement and is now in use in the ink industry, mainly for the production of rotary and offset news inks. This in due course may well spread to liquid inks with the introduction of coloured granular pigments.

A standard bead mill is used but with the addition of an integral screw feeder which ensures, by means of a positive sweeping device, that the unmixed materials are transferred from the feed hopper to the bead mill.

Materials used as grinding medium in stirred ball mills

Material		Density (g/cm^3)	Diameter (mm)
Siliquarzite*	Ottawa sand	2.8	0.4–0.8
	Hard glass, unleaded	2.5	0.3–3.5
	Hard glass, leaded	2.9	0.3–3.5
Draisal*	Aluminium oxide	3.4	0.5–3.5
	Zirconium silicate	3.8	0.3–1.5
	Zirconium oxide	5.4	0.5–3.5
	Steel shot	7.85	1.0–12.0

* Registered trade mark for DRAISWERKE.

Fig. 12.29 Materials used as grinding media in stirred ball mills (courtesy of KWR Chemicals Ltd/Draiswerke GmbH)

The feed screw itself does not undertake any dispersion, although a certain amount of pre-mixing and wetting occurs. A major problem was the avoidance of agglomerate formation. This was solved by the use of an annular nozzle system which sprays the liquid phase to the sides of the hopper while the solids are fed through the centre.

It is claimed that, as no pre-mixing is made, agglomerates which tend to be produced in standard dissolvers are not evident and therefore even higher production rates can be obtained than by conventional methods of production, namely dissolver/bead mill.

Feeding of the two streams is made through a loss in weight feeding system controlled by computer giving accuracies of ±0.5%.

Bead mill/roller mill — paste inks
Bead milling is the random disentegration of primary particles by the use of a retained grinding charge the important word being 'random'. It cannot be guaranteed that all particles will be subjected to shear within the mill and therefore the process is liable to produce a 'tail' of stray large particles. It is also not a self-sieving device so that these particles will be discharged with the majority of product which is already ground to the required standard. Any air which is also in the feed will also be finely dispersed within the product.

Bead milling also produces strong boundary line attraction between the ground particles which in itself restricts flow of the ground product. In the early days of bead milling, much work was necessary to formulate inks which would retain flow after passing through the bead mill. Even so, it is not always possible to achieve the required rheology in spite of formulation changes.

The three-roll machine is still widely used after beadmilling to 'finish'

Fig. 12.30 Horizontal versus vertical (courtesy of Lorilleux & Bolton Ltd)

or 'polish' the ink. Basically, it breaks down the molecule attraction boundaries by rolling shear and not acting simply as a deaerator. In fact, ink passed over a three-roll mill may have more entrained air than after bead milling. Developments to date have still not been able to eliminate the use of the three-roller.

Conclusion

Modern high-speed bead mills are very versatile devices. The number of operating parameters and the variations possible allow for optimised grinding processes over a wide range.

The requirement for finer grinding and higher quality inks will demand even more from the bead mill which will inevitably lead to further developments of the process. The full potential of the bead mill, however, has still not been fully developed.

Microflow mill

The microflow, still in limited use for small batch manufacture to avoid losses, is a large dispersion chamber.

Developed by Torrance for liquid ink manufacture, in particular, the microflow works on an attrition principle. By the use of a jacket round the dispersion chamber it can be heated or cooled depending on the dispersion conditions required. This type of mill can process recipes containing conventional pigments to give good colour development and gloss and the totally enclosed grinding chamber minimises solvent loss.

The microflow dispersion chamber is a stationary horizontal cylinder. The rotor inside this chamber carries a series of twelve toughened steel blades, called 'ploughs' around its periphery and lies flush with the length of the cylinder. A number of special steel rods, held loosely in the rotor frame, act as the dispersion medium. As the rotor turns, centrifugal force moves the rods to the outside of the rotor where they form a single layer against the ploughs and the inside of the cylindrical dispersion chamber. The grinding rod layer rotates in this position at a lower speed than the rotor and ploughs. Centrifugal force moves the product into a thin layer in the area where the ploughs and rods are rotating.

The rotational velocity differential between the ploughs and the rods and the change of direction of rotation of the rods as they pass over the ploughs, create intense shear in the thin layer of product in which they are turning. The ploughs continually remove the product from the chamber into the dispersion area thus ensuring product uniformity. By varying the throughput rate, 'dwell time' can be adjusted to assure quality, degree of dispersion and fineness required.

Ball milling

This has diminished considerably in the last few years, but for reference and history, as they are still common in other parts of the world, we recommend the section on page 327 of the third edition of the *Printing Ink Manual* published in 1979, which is précised here.

Advantages of ball mills
— After installation cheap to run as they run through the night on low tariff electricity.
— They act as mixer as well as mill.
— Low maintenance costs.

Disadvantages of ball mills
— Cumbersome and bulky.
— Noisy.
— Only 50–60% of the volume can be used for the batch or the cascading is influenced detrimentally.
— Cannot be speeded up as demand increases as there is a set production time cycle.
— Emptying can be tedious particularly if the product has any thixotropic tendencies.
— Can only handle low viscosities.

12.5 HANDLING, STORAGE AND MANUFACTURE OF UV INKS

It should be understood that the manufacture of UV inks and lacquers is an entirely different proposition from making the equivalent product using conventional materials. In fact this type of manufacture is probably more closely related to the pharmaceutical industry where cleanliness and handling care are of paramount importance. These comments apply not only to the way the components are put together to make the final product, but also to the way the ingredients are stored at controlled temperatures. Skin contact is to be avoided as certain of these materials are complex chemicals which, although very carefully selected, have been known to cause irritation to some people. The materials used are also considerably more expensive than conventional materials and as a result justify the extra care required.

Since these materials are generally poor wetters of the dry phase of the ink, the mixing procedures are critical for satisfactory dispersion. They do not differ in principle from conventional operations but enforced cooling is a prerequisite to avoid the temperature of the mix exceeding 55–60°C. Certain of the ingredients can be crystalline so these have to be added at a crucial temperature during manufacture or the product seeds.

Milling is best achieved on modern three-roll mills with comprehensive temperature control on the individual rolls which means that they can be warmed or cooled individually or *en bloc*, as required. This is an essential requirement as certain UV products are not enhanced by the shock cooling which can occur during the start up period of milling. Handling of the finished product is less sensitive but it must be stored in opaque containers and not exposed to extremes of temperature change or sunlight.

12.6   MANUFACTURE OF NEWSPAPER INKS

Newspaper inks lend themselves to automated and semi-automated bulk manufacture primarily because large volumes of standard products are required and secondly because a relatively few basic materials are involved in their formulation.

Nevertheless, because of the critical nature of newspaper manufacture, these inks must be of high and consistent quality to perform on long runs on high-speed presses.

The inks fall into three main groups: rotary letterpress, di-litho and coldset web offset. Reference to Chapters 5 and 6 will show that these are mainly produced from mineral oils modified by small additions of pitch varnish with carbon as the pigment. Other additives include wetting agents and anti-misting additives such as clays and extenders, while oleo-resinous varnishes are added to coldset litho inks to improve rub and lithographic properties.

The bulk of furnace black currently used in news ink manufacture is fluffy carbon because of its relative ease of dispersion, but great interest is being taken in both oil and water wet beaded forms as a primary raw material. The advantage of the beaded grades over the fluffy material is that they are cleaner to handle and feed more easily in bulk plants. The beaded form is also more compact and because the specific gravity is almost twice that of the fluffy grade (hence less storage space), is easier to handle in bulk. Beaded carbon is supplied in bulk tote bins, big (0.5 tonne) bags and tanker vehicles, whereas fluffy is only supplied in 12.5 kg sacks. The disadvantage of beaded carbon is that the bead is more difficult to disperse and small seeds of cokey material form a residue which is unnacceptable even in low concentrations (below 0.1%) in the finished ink.

The handling of carbon in any quantity can give rise to significant dust levels in the workplace, and manual handling of bags should be carried out under effective extraction; even under these conditions it is almost impossible to prevent some contamination, particularly with fluffy grades. Most modern bulk plants (see Fig. 12.31) incorporate some sort of mechnical bag slitter, these machines not only open the bags efficiently but also compact the empty bags and automatically wrap the scrap in polythene film. In many plants without this facility the most significant carbon contamination arises from the handling and disposal of the empty paper sacks. Once the sacks have been opened the carbon is processed in a totally enclosed system, thus further contamination is reduced significantly.

The main problem in bulk handling fluffy carbon is the difficulty of getting the product to flow from bulk containers, and it is primarily for this reason that it is supplied in 12.5 kg sacks. Even when the sacks have been emptied into a suitable hopper it is difficult to transport the powder into the mixer or mill. Without some form of mechanical assistance fluffy carbon will block pipes and valves and will feed irregularly in the system. Several mechanical techniques have, however, proved quite successful. The best of these is fluidization, which is achieved by blowing low-pressure air into the carbon. The air separates the light carbon particles and the resulting fluidized carbon will flow through the system satis-

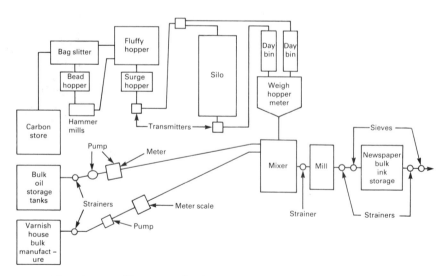

Fig. 12.31 Bulk news ink plant (courtesy of Usher-Walker plc)

factorily. The air is diffused into the system through sintered pads, usually placed at the base of all hoppers and cones in the plant where blockages are likely to occur.

The use of 'live bottom' hoppers is also a satisfactory technique in keeping fluffy carbon mobile. This process involves vibrating the base of hoppers or cones at high frequency, preventing a build-up of carbon at these points, by agitating and thus separating the carbon particles.

Other mechanical means of agitating the carbon include mechanically driven rotating paddles of various designs located at the base of the hoppers.

Screw feeding has been used for fluffy carbon but has two major disadvantages. Firstly, because there is a gap between the screw and the feeder body, the carbon tends to be compacted against the vessel wall. This compacted carbon will flake off in the screw feeder and the resulting material is more difficult to disperse. Secondly, the screw itself tends to aerate the fluffy carbon which, if the feed is needed to lift the carbon through the system, limits the angle of elevation of the screw feed by the tendency of the carbon to flow back through the screw.

The most satisfactory method of tranporting fluffy carbon through the system is by pneumatic transport. The carbon is blown through smooth-bore PVC pipes using pulsed air supplied from a Roots blower. The carbon is carried to elevated day bins by the air which then escapes from the system through reverse air jet self cleaning filters mounted on top of the day bins and storage silos. A transfer rate of over 5 tonnes/hour can be achieved easily.

In modern plants the flow of carbon is controlled by micro processors. The day bins and storage silos are filled automatically and the level of carbon in these vessels is monitored by level probes.

The above methods of handling fluffy carbon are equally applicable for handling beaded product. However, because of its higher density and because of the bead form, this product flows perfectly well through the system without fluidisation. The screw feed method is quite effective for this grade of material.

Oil handling

For economy, oils are usually purchased in bulk from the refiners and delivered in tankers of up to 20 tonnes capacity. The oils are supplied in a range of viscosities but those in the range up to 10 poise are the most widely used. The heavy news ink oils are pumped into large holding tanks. These tanks should be sited as near to the production area as possible to facilitate ease of pumping. The holding tanks for the heavy oils need to be heated by steam coils or immersion heaters in order to keep them within a practical viscosity range for pumping and metering.

Production method

Bulk production can be undertaken by either batch process or by continuous process.

In the batch system the process oils and varnishes are metered into a large mixing vessel. While these are being metered the carbon is weighed in a weigh hopper usually situated above the mixers. In practice half the oil is metered into the mixer together with the varnishes, wetting agents and extenders, then the pre-weighed carbon is dosed into the mixer at a controlled rate, together with the remainder of the oil. Controlling the rate of carbon addition at a speed which is compatible with the ability of the mixer to incorporate the pigment into the oil is important as it gives more efficient pre-dispersion and, particularly with fluffy carbon, takes advantage of the fact that the carbon particles have already been separated by the fluidising air.

The mixers used are usually high shear mixers or of the cavitator type, fitted with additional slow speed stirrers to increase mixing efficiency. When the pre-dispersion is complete the pre-mix is processed by high dispersion bead mills, using 2 mm steel beads as the grinding medium. The resulting ink is filtered both mechanically and magnetically before being pumped into bulk ink storage tanks. The magnetic filters remove any ferrous particles which may be generated in the bead mill, while the mechanical filters remove any cokey undispersed carbon particles. This attention to filtering is necessary because even low concentrations of undispersed matter can cause press problems due to the high volume of ink which is used and stored in press installations over long periods in enclosed systems. Web offset inks are generally stored in heated and lagged tanks and are kept stirred to breakdown the natural thixotrophy of the system.

Continuous production is more easy to achieve with beaded carbon due to the relative ease by which the carbon feed can be controlled. The granular material can be fed by a screw-feed mechanism directly into the mill. Oil is fed by means of a loss-in-weight system giving a controlled and consistent flow of product together with the dry pigment. The loss-in-weight system is a microchip system constantly monitoring the weight of the hopper, in which the oil is held, together with the control of the carbon flow.

12.7 HANDLING AND STORAGE OF INKS

The adoption of bulk handling or tanking makes it possible to ensure better use of available space and appropriate controls on ink stocks. If the tanks are linked with an in-house blending scheme then the whole process is subject to a degree of simplification, in that it is possible to operate from a reduced number of materials, in greater bulk and with less frequent deliveries. The ink-maker has the ability to control or to make up special colours when they are needed.

What has to be considered when deciding whether tanking is an appropriate method to adopt? Experience has shown that with liquid inks it frequently pays for an assessment to be made where there is a monthly consumption of a material of between 300 and 400 kg. However, in storage, there can be a gradual decline in quality as pigments flocculate, settle out, or the binder systems can separate. It is therefore recommended that the tank size be related to approximately 2 months' consumption. This ensures that stocks are turned over in a tank at least six times a year (in the case of newspapers, etc. this is measured in days). Furthermore, agitators need to be installed which are time controlled to give periodic stirring. With regard to the movement of liquid materials, experience has shown that the rectangular tanks do promote better agitation in bulk, without undue turbulence.

The tanks, having been equipped with automated agitation, have to be provided with a means of charging and discharging, level control, venting and sometimes temperature controlling. Changes in legislation have led to the simple venting valve on the top of each tank being scrapped and fixed piping conducted to external atmosphere so that the factory is free of solvent vapours (see Fig. 12.32).

A well-designed tanking system will make maximum utilisation of the available space. It will also provide a very much safer and cleaner working environment. Finished inks are blended by mixing the single colour bases held in the tanks. In the case of lower viscosity bases dispensing is by means of gravity. As the viscosity increases it becomes increasingly necessary to provide some other force. An air driven, low geared pump, features as a fairly ideal method of providing that extra pressure behind the fluid to ensure smooth and adequate dispensing rates (see Fig. 12.33).

With a driven delivery system, it is possible to incorporate the dispens-

Fig. 12.32 Typical liquid ink storage (courtesy of Fulbrook Systems Ltd)

**Fig. 12.33 Small style tank farm indicating pumps (courtesy of Fulbrook
Systems Ltd)**

ing valves from individual tanks into automated dispensing stations. The
valves are mounted on a frame which is powered backwards and for-
wards to bring each individual nozzle, as required, into the appropriate
dispensing position over the centre of the receiving tank. Normally, these
heads would be arranged over dormant mounted scales (see Figs 12.34
and 12.35). The sophistication of the system can vary from punch card
control to computerised controls.

The agitators which would normally be used for mixing the blend are
powered by electric motors. Air-operated agitators are occasionally used,
but the electric agitators are very much more practical, not only for
periodic timings, switching on and off automatically, but also for overcoming the varying forces met by the impellers as the level of the material
in the tank drops. A constant speed is maintained with an electric motor,
whereas the 'revs' of air-driven versions can vary appreciably.

Holding the materials in bulk necessitates efficient stock control. Using
steel tanks, an effective form of level indicator is required. This is one area
in which equipment has gone through quite dramatic changes, recently.
The development has been a continual process. The preferred method
now is the ball float linked to an external indicator needle and scale. This
mechanism has been subjected to very rigorous testing on the part of
certain of the EEC countries, and has been given safety certificates for its
ability to resist ingress of solvent and hazardous or corrosive materials. It
is now shown to be a very safe application. By using this system of con-

**Fig. 12.34 Flexible dispersing end for filling varying container heights
(courtesy of Fulbrook Systems Ltd)**

tent indicator, it is possible to link it with end contact switches. These come into play when filling the tanks, prevent accidents caused by over-filling and, at the opposite end of the scale, they will switch off the agitators when the level of the contents is too low. The latest thinking on this matter is the Weka gauge which is electronically operated by pressure. The selection of the pipework, the fittings, the glands, is a specialist operation (see Fig. 12.36).

Charging the tanks can be done in one of two ways, either with an external barrel pump which will be linked to the coupling pipes at the base of the tank or with slow-moving geared ink transport pumps. With appropriate valving these units do double duty both as dispensing and filling pumps, thus simplifying the system. A recent development is the introduction of pouring valves with a two-stage flow-rate. These flow-rates can be controlled to such accuracy as to permit the delivery of as little as 75 g per charge. These valves are PTFE seated right at the exit point of the nozzle. Conical seating obviates the problems that would otherwise occur with ink drying into the valve thus causing clogging, jamming and leakages. The unit has been developed precisely to over-come those problems. The valves are spring loaded so that it is a fail-safe system, whether it is a hand or powered operation. The valves, hand or electronically operated, can be coupled to the transport pumps giving maximum safety and avoiding blow-backs.

The application of automated weighing does minimise waste in that the

Fig. 12.35 Console and dosing head (courtesy of Fulbrook Systems Ltd)

accuracy is guaranteed if one is using an electronic data processor or a punch card controlled system. The accuracy of the dispensing is such that shading is normally unncessary.

What are the economies available from bulk tanking? The obvious one is the limitation of stocks to a number of singly pigmented toners for blending, or finished inks sold in bulk. This is a very real asset to the ink manufacturer who can plan his production so much more effectively. The fact that there is no need to fill shelves with special shades surplus to requirement, thus tying up vast quantities of money, has to be another effective way of making economies.

Much of the above is also true for paste inks, but here one is dealing with materials of a much higher viscosity, which may very well flow quite adequately just after manufacture with the internal heat processing has produced, but when stored at ambient temperature in the UK between say $-5\,°C$ and $+30\,°C$ the rate of flow is slow. The methods to move such products at 250 poises/$20\,°C$ are gravity, or gear pumps which are accurate and instant, or air-driven displacement pumps ideally suited for 200 litre

Fig. 12.36 Tanking system filling end (courtesy of Fulbrook Systems Ltd)

straight-sided drums, and finally and more sophisticatedly, hydraulic pressure, for example Schwerdtel (see Fig. 12.37). The air or hydraulically driven following plate method is now commonplace. In the higher viscosity ranges if distances are to be covered using pipework the bore and jointing are of great significance and must be calculated very accurately, with the same care being given to the driving (moving) force to cause the product to be mobile through small pipes and valves. Thorough heating of these pipes is essential to ensure mobility particularly if any of this pipework is outside.

Ink packaging

During the last 20 years, considerable strides have been made in packaging and ink supply to the printer.

For letterpress and lithographic inks, slip-lid tins of various sizes, particularly those capable of holding 1 and 3 kg of ink, are still widely used. However, with the advent of larger web presses, 10 or 15 kg capacity polythene buckets are now widely used. For still easier handling, 200 kg straight-sided drums can be used, where air pumps with a follower plate, such as the Graco pump, are installed on press lines. In cases where the ink consumption justifies it, such as on the larger heatset web offset presses, supply by use of 800 kg or 1 tonne drop tanks of each colour is becoming more widespread.

The main change in oil ink packaging was the appearance of the vacuum tin developed by Gustav Gruss of Düsseldorf. Introduced at the 1964

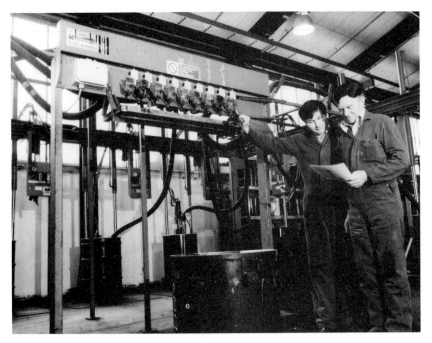

Fig. 12.37 Dosing and weighing of high-viscosity materials (courtesy of Ludwig Schwerdtel (Munich))

Interpack Exhibition as a new food can, it was a three-part construction. Michael Huber of Munich carried out trials using the original can and although these were accepted by printers, the leakage rate of 0.5% made the system unreliable. It was evident, however, that the can had many advantages over the standard slip lid can such as saving, storage space, easy emptying, no necessity for the inner paper seal, etc. Investment for a single deep-drawn can, however, was too high for a single ink outlet and so Huber were joined by Kast & Ehinger and the can we now know was produced by Gruss in 1970.

Paste ink filling lines were already being made by Schwerdtel GmbH of Munich (see Fig. 12.38), but no machine was available to seal the vacuum can automatically: this Gruss built themselves. It became evident, as the use of the vacuum can grew, that a specialist machine company was needed to take the development of this vacuum closing unit further and in 1973/4 Schwerdtel made the first such unit, the S10-ZV.

Since this time, the vacuum can has become a standard in Europe, but only recently in the UK and Japan. It still has to be accepted in the USA.

Recent developments in oil ink packaging have been in the use of 100 cc, 1 kg and 2 kg flexible sachets. The ink is filled into an aluminium foil/polythene sachet under an inert blanket of nitrogen which gives a totally impervious and air-free sachet. Obviously, much cheaper than the can, it is now on extended trials in Germany and could well be the next phase in oil ink packaging.

Fig. 12.38 Canning line (courtesy of KWR Chemicals Ltd/Ludwig Schwerdtel)

There is a growing trend to supply the larger flexographic and gravure printers with a limited range of concentrates, supplied in 45-gallon (205-litre) drums, which are pumped into individual tanks for use in blends as required. For smaller quantities, polythene containers, of various sizes, can be used for water-based inks but the generation of static electricity makes them totally unsuitable for solvent-based inks and sundry items. Steel cans and drums with an internal coating of lacquer are most frequently used for the latter type of ink.

Bulk rotary letterpress and coldset web offset blacks are often delivered in custom-built tanker vehicles. Rotary blacks are fairly easy to transport, but tankers for offset inks need to be heated with isotapes or steam coils and lagged to facilitate efficient pumping at the customer's premises, owing to the higher viscosity involved. Attention should be paid to the type of pumps and the design of the tank bottom or pumping problems may occur, particularly in winter or on longer journeys.

12.8 MODERN PRODUCTION TRENDS

Computerisation

The computer has now become an established tool in ink manufacture. The routine procedures of preparing manufacturing instructions, progressing of orders, preparation of labels, stock control, etc. are carried out at high speed and high efficiency. Furthermore, the recent advances in electronic, computer and microchip technology are changing the face of ink manufacture. It is now theoretically possible to design a completely automated ink-making factory in any sector, from gravure through to screen process (whether this concept would, in the short term, be cost effective remains to be studied, but in the long term it looks very attractive). This has been made possible not just because of the increasingly greater scope of computers and their on-line electronic attachments, but because these previously cumbersome installations have been streamlined, to the point where the most complex electronics are pocket-sized. Such units can be used to actuate valves, monitor the weighing (whether mechanical or via load-cells) and flow-rates of liquid products.

Costs of production and related subjects

In manufacturing operations it is essential to know the costs of labour, be it direct or indirect, power and utilities, rent, rates, maintenance costs, depreciation costs of plant and machinery, etc. because the product being made has to bear an apportionment of all these costs in relation to its real cost of manufacture so that they can be recovered from the customer with a realistic selling price being constructed. One of the major problems facing the ink industry is the pricing of small batches of ink which rarely bear any relation to their real costs and can be more expensive in a production sense than batches many times their size.

Maintenance strategy in the printing ink industry

A maintenance strategy, where this exists, usually embraces one of a number of separate philosophies. There are those who consider maintenance to be the repair or replacement of items which have failed, a necessary evil and an undesirable overhead, and consequently such maintenance is frequently given a low priority within the management strata. In some organizations, sufficient spare capacity is available and a failure may not adversely affect production. In this case the 'on failure' approach may appear to be adequate. This is not often the case, however, as most companies have to operate as near to maximum plant utilization as possible and a prolonged failure normally has serious consequences for production and profitability.

Within the printing ink manufacturing industry, safety traditionally has a high priority, especially where hydrocarbons and other potentially hazardous substances are employed. The environment in which the processes are carried out and the materials employed require a high level of maintenance effort to meet the appropriate standards. Modern equipment is also becoming increasingly complex, demanding skills not usually found in the traditional maintenance department. This has resulted in a higher level of contractor involvement in maintenance, where these skills are only employed when necessary and can be put to competitive tender.

The high capital cost of modern process equipment, require standards of safety and the cost consequence of excessive downtime, make the introduction of a preventative maintenance system an attractive proposition. Industrial equipment has a life-cycle which can be considered as a number of stages, the first being design, the last replacement. The maintenance objective within this cycle must be to maximize the reliability and availability of the plant for optimum cost.

The primary objective in most cases is to move from the 'on failure' strategy to either a 'fixed time' or 'condition based' maintenance system.

'On failure' maintenance

To replace or repair after failure is the default strategy sometimes known as corrective maintenance. It requires no pre-care of the plant and all management effort goes into organising manpower and stores. This technique encourages the 'fire-fighting syndrome' where the fastest fitter is the 'superman' of the team.

The disadvantages of this method are:
— There is no warning of failure and therefore safety risk;
— There is often secondary damage leading to longer and more extensive repairs;
— A need to purchase and maintain standby or back-up plant;
— The need to carry large stocks of spares.

'Fixed time' maintenance

This is strategy most commonly applied to reduce 'on failure' maintenance. It requires only a simple planning structure and is often known as 'planned preventative maintenance'. Fixed time maintenance can reduce failure and uses the workforce effectively with a planned work schedule. It allows work to be planned well in advance.

The disadvantages of this method are:
— The maintenance activity and therefore costs increase because the maintenance interval will be shorter than the mean time to failure (MTTF).
— Maintenance carried out on a time-related basis only can induce failure of other parts through disturbances.
— It can only be applied effectively where the deterioration is age related, and is predictable.

These three factors severely restrict the effective use of a fixed-time strategy, and as industrial plant rarely fails in a predictable way this method has to be used in conjunction with other systems.

'Condition based' maintenance (see Figs 12.39 and 12.40)

This is an attractive concept because corrective maintenance will be carried out before failure, but not before it becomes necessary. It is therefore akin to 'on failure' maintenance, being corrective in nature. If condition monitoring is adopted, failure prediction is enhanced and the effect of failure minimized.

This strategy embraces the advantages of the others and also allows a more flexible response to any potential failure situation. It is sometimes known as the 'multi-role maintenance strategy'.

The advantages of this system are:

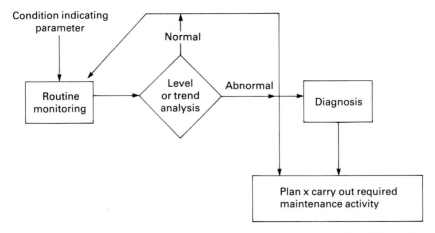

Fig. 12.39 Condition-based monitoring system (courtesy of Lorilleux & Bolton)

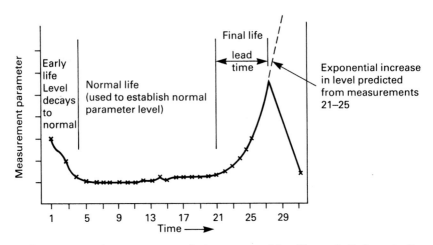

Fig. 12.40 Maintenance graph (courtesy of Lorilleux & Bolton Ltd)

— The opportunity to take remedial action before a stoppage or severe damage occurs;
— Allows plant to be on-line for longer under controlled and known operational parameters;
— Allows maintenance work to be planned in advance, although the planning time may be shorter than with time-based;
— Spares need only be purchased when required, reducing expensive spare stock levels;
— The cause of the failure is immediately apparent and can be related to operating conditions as data are compiled during operations;
However, the management effort required to organize and operate this

strategy is considerable, as is the time-lag between implementation and the realization of benefits, which are not always tangible.

The objective of condition monitoring is to assess the condition of the moving parts of a machine, usually while it is in operation, and from the data collected to determine the correct moment to carry out a repair/ replacement before failure occurs. These inspections often involve drawing on many fields of unrelated expertise. The parameters to be measured include vibration monitoring or thermal monitoring, current flow, pressure, oil debris analysis, etc.. Various instruments, which can be fixed permanently to the plant or can be portable, are available for their implementation. Vibration monitoring is of great advantage in determining the condition of rolling elements and can detect a wide variety of faults.

It is, however, imperative that a data-recording system is established as condition monitoring relies on either trend or deviation in performance to predict faults.

Computers and maintenance

It is often believed that a computer is the solution to many maintenance problems when in fact it is merely a tool to be used by maintenance management to assist in the implementation of a previously established maintenance strategy. Before a computer can be utilized there must be a detailed specification of requirement describing the maintenance functions which are to be computerized, the methods with which the data are to be collected, analysed and used, by whom and for what purpose.

The range of hardware and software aimed specifically at potential users is increasing dramatically for both the corporate mainframe and the larger 'mini' and 'micro' user market. These systems usually offer the same broad functional classifications:

— Plant inventory
— Spares and stock facility
— Job scheduling
— Plant history
— Data analysis.

The system considered should be one which either mimics or resembles as closely as possible the manual system currently in operation.

Choosing a system is a potential minefield for the unwary and all options should be thoroughly vetted before a committment is made.

Some software manufacturers are offering energy management systems combined with a maintenance function, which is tagged on as an added extra. These are not usually worthy of serious consideration as the maintenance function is complex in operation and must be dealt with separately.

Companies who already operate a large corporate mainframe with sufficient capacity are often inclined to have an in-house maintenance program written. There are a number of pitfalls to be aware of if this course of action is considered.

Large mainframe computer manufacturers serve a comparatively small sector of the total potential computer user market, and therefore to maintain their hold in this sector they jealously guard their system operating instructions. If the use of 'condition monitoring' instrument interfaces is envisaged these would prove difficult to operate on a mainframe system. The manufacturers of these instruments cater for the wider 'mini' and 'micro' market where operating systems allow the necessary 'auto entry' interface.

In-house systems can also be more expensive than comparative 'mini' or 'micro' packages. The buyer has to submit to a software house a specification of requirement which is unique. If, after a lengthy programming session, the system does not run or does not fulfil the needs of the men then the system has to be rewritten or corrected at the buyer's expense.

Again, because the operating instructions are usually secret, making the eventual program in effect locked (inaccessible by others), can be a costly exercise.

'Mini' and 'micro' systems are usually supplied as a package which can contain all the broad functional areas required for maintenance application and can usually be tailored to fit a specific requirement.

After advocating a 'mini' or 'micro' package system it is wise, however, to state that a similar degree of caution is required before any commitment of resources is made.

Health & safety at work

Although ink making is not complicated there are considerable areas for caution and as a result considerable time and effort is spent on training. It is said that, since the Health and Safety at Work Act 1974, which put the responsibility for safety in the workplace jointly in the hands of the employer and employees, training in the workplace has had more significance. Whether or not this Act has been the incentive, continual training within the factory area along with a good knowledge of the materials used in ink manufacture has always made sense, and never more so than today. For further information, see Chapter 15.

12.9   THE FUTURE

A scheme for a modern ink plant designed for the automatic production of publication gravure inks gives an example of current thinking about liquid ink production and indicates the trend towards automation and computerised control of ink manufacture.

Plant control system

A modern plant control system ensures smooth operation of the production plant. The control system is designed so that each production line is allocated its own field on the panel.

The formulae are stored on punched cards. Each line is equipped with a punched card reader and a tape printer for recording the actual weights, batch sizes, production sequence, bead mill data, and alarms. Digital displays continuously show information on the formula number, the gravimetric proportions of ingredients and the major data concerning the bead mills.

Further plant features

Great importance has been attached to safety and environmental compatibility:

— The entire electrical system is of explosion-proof design.
— Clean pigment intake system with the application of fabric dust collection filters.
— Entirely enclosed system, hence no loss of solvent vapours.
— Gas displacement device with flame arresters on all production vessels; exchange of air with atmosphere through activated carbon filters.

Manufacturing plant

The entire plant consists of several production lines. The capacity depends on the ink formulations and the size and number of bead mills. Figure 12.41 shows the process schematically.

The production plant consists of the following systems.

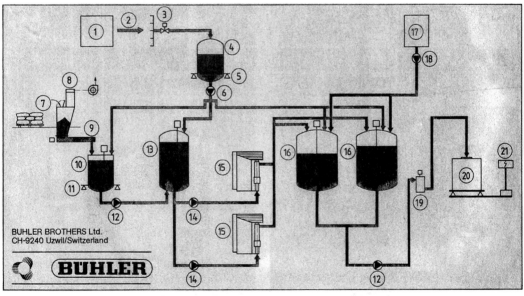

Fig. 12.41 Schematic plan of rotogravure plant (1) Tank storage for bonding agents and solvents; (2) conveying lines for bonding agents and solvents; (3) metering valve; (4) Weigh tank for bonding agents and solvents; (5) Load-cell scale; (6) pump; (7) pigment hand tip; (8) dust collection filter; (9) screw feeder for pigments; (10) mixing vessel with agitator; (11) load-cell scale; (12) pump; (13) buffer tank with agitator; (14) regulated feed pump; (15) COBRA bead mill; (16) finished ink vessel with agitator; (17) additives; (18) pump for additives; (19) cleaning system; (20) container; (21) electronic potting scale (courtesy of Buhler Bros Ltd)

Metering scale for bonding agents and solvents (see Fig. 12.42)
The required bonding agents and solvents are weighed exactly according
to the formulation by means of a hopper scale and delivered to the
relevant tanks in the form of liquids. The actual weights of all liquid
ingredients are printed out after weighing.

Manual pigment addition and predispersion
Each production line is equipped with a manual pigment intake hopper
which is kept dust-free by means of a textile envelope (pocket) filter. The
pigments are added manually according to the formulation, and the addition is acknowledged when completed, after which the pigment is fed to
the liquid by a screw feeder. The bonding agents and solvents propor-
tioned according to the formulation plus the pigments are predispersed in
a disperser. The batch weight is registered and printed out. The product is
then delivered via a buffer tank to the bead mills.

Fig. 12.42 Weigh tank (courtesy of Buhler Bros Ltd)

Bead mills

Here, the product is dispersed until it has an acceptable fineness. An explosion-proof control console is located next to the bead mills. The cabinets of the control and regulation systems are installed separately in a less hazardous area. The COBRA 251V bead mills are characterised by the following special features (see also section on Bead Mills, see Fig. 12.43):

— Ring chamber pin agitator for application in high viscosity ranges, equipped with high-precision separating gap.

— Corrosion-resistant cooling system with stator/rotor cooling system.

— Pressure unit for changing the bead charge in the grinding chamber.

— Electronic regulation of all major production parameters using several interfacing electronic control loops; application of controllers of high accuracy; dependable reproducibility of parameters; high specific capacities and optimal utilisation of electric power.

Fig. 12.43 COBRA 251V bead mills (courtesy of Buhler Bros Ltd)

Fig. 12.44 Finished ink tanks (courtesy of Buhler Bros Ltd)

Finished ink tanks (see Fig. 12.44)
In the finished ink tanks, the concentrate is blended with a liquid consisting of a bonding agent and solvent to become the finished ink. It is also possible to add additives by means of pumps. From the finished ink tanks, the ink is pumped via a cleaning unit to the container for feeding to the potting scale.

Other developments

It is difficult to see any further major developments in the basic bead mill other than continued refinement. The next steps could well be towards the direction of the Fryma Co-ball mills where only the medium and product are in movement but many problems have yet to be solved to produce a universal machine, suitable for wide viscosity ranges. Ultrasonics have been tried and tested, as yet without success, and technology such as the use of laser beams is in the very early stages.

In the next few years, the more efficient handling and automatic control of bead milling will be to the fore. It will no longer be possible or viable to produce small batches by hand and automatic processes will become more the norm.

CHAPTER 13
Rheology of Printing Inks

Rheology is the study of the flow and deformation of matter. The aim of this science is to define and evaluate such words as consistency, tack length of flow, stiffness and body in terms of the physical properties of the materials. These words are used in everyday language and with special significance in many sections of industry. At first sight such terms may appear to be self-explanatory and simple to define, in fact the converse is true and the greater the need for precision the greater is the complexity of separating the factors involved.

In the printing industry there are two main areas of interest for the rheologist:

(1) Lithographic and letterpress inks where the ink is required to be ready for the press without addition or modification.

(2) Gravure and flexographic inks where the ink-maker supplies ink at a viscosity greater than that required for use on the press and the printer accepts the responsibility for adjustments in viscosity to satisfy his printing conditions.

In lithographic and letterpress inks the need for precise evaluation on numerical scales arises because of two factors which are diametrically opposed. For good printing a lithographic and letterpress ink should be as stiff as possible but when a printing machine is run at high speed the stiffness of the ink must be sufficiently low to avoid the paper surface being torn away from the body of the paper (i.e. it will not pluck) and to avoid the generation of too much heat on the machine.

With flexographic and gravure inks there are two reasons for the ink-maker to supply ink at a viscosity much higher than required by press conditions. Firstly it overcomes the tendency of the pigment to settle to the bottom of the container used for transport. Secondly it allows the printer to thin the inks to a viscosity suitable to the press and running speed and to enable him to adjust the balance of solvents to achieve the desired solvent release and drying rate.

Most laboratory workers concerned with printing ink do not require a specialist knowledge of rheology. They will be better served if they acquire a more general appreciation of this subject and refer problems to the specialist for explanations when the need arises. The depth of knowledge required may be stated as follows:

(1) In branch operations where the laboratory worker is called upon to mix inks to pattern, an understanding of 'tack' and an elementary appreciation of flow should suffice.

(2) The chemist responsible for ink development should understand tack, Newtonian and non-Newtonian aspects of flow and gain experience of how these change in relation to the choice of pigment, pigment concentration, the ink media and their composition particularly with regard to molecular size and shape and pigment media interaction. A consideration of these parameters will lead to a greater appreciation of the visco-elastic nature of most ink systems under the high shear stress and shear rates encountered on printing press.

(3) The specialist rheologist will have an appreciation of the flow of liquids and pastes which will extend far beyond the information recorded here. Such knowledge may be gained by study of the extensive literature. Membership of the British Society for Rheology would be useful for their literature surveys and conferences on a wide range of rheological topics.

13.1   FLOW IN IDEAL SYSTEMS

In commencing this study it is convenient to begin with an examination of flow in an ideal or simple system. Fluids which show ideal viscous behaviour are known as Newtonian named after Isaac Newton who was responsible for the original theoretical work on the flow of liquids [1].

Consider two parallel planes, abcd and efgh of area A, in a fluid. (Fig. 13.1). Holding efgh stationary and subjecting abcd to a force F resulting in a movement of velocity V, the intervening fluid may be looked upon as having many planes sliding over one another. The nearer the planes are to plane abcd the greater is their velocity. Consequently there is a velocity gradient between the planes abcd and efgh:

$$\text{Velocity gradient} = \frac{dv}{dx} = D \text{ (rate of shear) } (s^{-1})$$

The shearing stress is given by the force per unit area

$$\tau = \frac{F}{A}$$

For a Newtonian liquid τ is proportional to D, i.e.

$$\tau = \eta D$$

where η is a constant for the liquid under examination and is known as the coefficient of viscosity. The unit of viscosity is the pascal second and is derived as follows:

$$\text{Shearing stress} = \text{force per unit area}$$
$$= \text{newton/m}^2$$
$$= \text{pascal}$$
$$\text{Shear rate } D = \text{velocity gradient}$$

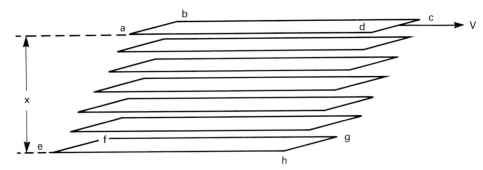

Fig. 13.1 Newtonian flow

$$D = \frac{dv}{dx} = \frac{ms^{-1}}{m}$$
$$= S^{-1}$$

$$\text{Coefficient of viscosity} = \frac{\tau}{D}$$

$$= \frac{newton.s}{m^2}$$

$$= pascal.s$$

$$1 \ pascal.s = 10 \ poise$$
$$1 \ millipascal.s = 1 \ centipoise$$

European literature is now predominantly using the pascal.s. However, the poise is still in wide use in the ink industry.

Many fluids, such as the solvents used in inks, e.g. light mineral oils, ethanol and water are Newtonian liquids so also are lightly bodied stand oils, low viscosity alkyds and hydrocarbon resin solutions. Changes from Newtonian behaviour occur with pigmentation of the ink medium and by chemical modification of the medium [2, 3].

Flow curves or rheograms for two Newtonian liquids are shown in Fig. 13.2, the rheograms are straight lines passing through the origin, the steeper the line the higher the viscosity so in the diagram liquid A is more viscous than liquid B. With a Newtonian liquid measurement of shear stress and shear rate at a single point on the curve is sufficient to define the viscosity at all shear rates and shear stresses.

13.2 DEVIATIONS FROM NEWTONIAN BEHAVIOUR

The introduction of pigments into Newtonian liquids produces considerable deviations from Newtonian behaviour. This is due to particle association, by chemical bonds and physical interaction during flow. The following rheograms illustrate this non-Newtonian or anomalous behaviour.

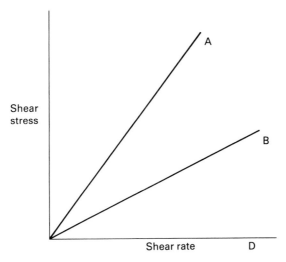

Fig. 13.2 Rheogram for Newtonian liquids

The ideal plastic substances

The rheogram Fig. 13.3 illustrates the ideal plastic substances, here flow commences only when the shear stress reaches a specific value in this case τ_0.

Once flow starts the substance shows a simple relation between τ and D resulting in a straight line plot. The plastic viscosity U is the slope of the curve and is given by

$$U = \tan \theta$$

$$= \frac{\tau - \tau_0}{D}$$

This value is a constant. However, if measurements are obtained at only one point on the curve the resulting calculation for viscosity

$$\eta = \frac{\tau_1}{D_1}$$

will give a different value. This is called the apparent viscosity and will be different for different shear stresses and shear rates. At high shear rates and shear stresses the apparent and plastic viscosity values will approach one another.

This type of rheogram is typical of highly pigmented low viscosity media an example of which is toothpaste. When the toothpaste tube is squeezed the shear stress at the edge of the hole exceeds the minimum stress required for flow to occur and thus the cylinder of toothpaste is extruded. This minimum value τ_0 for flow to occur is known as the yield value.

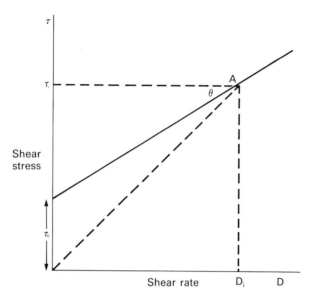

Fig. 13.3 Ideal plastic substance

Pseudoplastic substances

When the rate of shear increases more in proportion to the shear stress the substance is described as a pseudoplastic or shear thinning material. At the higher shear rates the curve becomes approximately linear and it has become a practice in the ink industry to extrapolate this portion of the curve to the stress axis, as shown in Fig. 13.4 and record this value of shear stress as a yield value. This is not sound practice as all pseudoplastic materials will give apparent yield values using this method. Pseudoplastic materials with yield values have rheograms as shown in Fig. 13.5.

Dilatancy

When the rate of shear decreases more in proportion to the shear stress the material is known as a dilatant or shear thickening substance and gives a rheogram of the type shown in Fig. 13.6. Dilatancy occurs when particles in the fluid tend to 'pack' to form rigid structures. Consequently dilatancy is associated with inks having a high proportion of pigment or extender.

Substances which are dilatant would be useless as lithographic or letterpress printing inks because at the high shearing stresses encountered with fast running presses, the ink would fail to distribute properly. Dilatancy may occasionally be found in stiff pastes of very high pigment concentration designed to function as the base of inks to which varnishes are subsequently added in preparing an ink for printing. Dilatancy is also sometimes specifically formulated into intaglio inks for copper and steel engraving to assist in clean wiping of the plate.

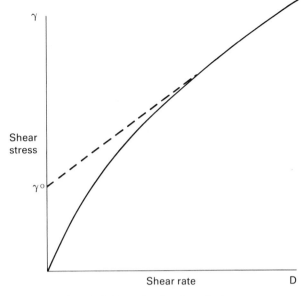

Fig. 13.4 Pseudoplastic substance

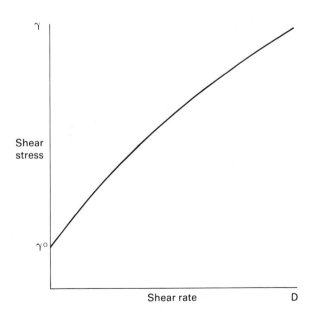

Fig. 13.5 Pseudoplastic substance with yield value

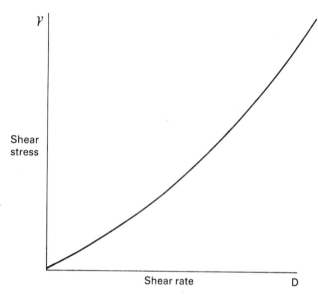

Fig. 13.6 Dilatant substance

Thixotropy

Most printing inks thicken on standing and the gel structure can be broken down by stirring, a phenomenon generally described as thixotropy.

When a thixotropic substance is stirred at a constant shear rate the apparent viscosity falls with time and eventually attains a constant value. At this point the breakdown and build-up of gel structure are balanced for that particular shear rate. Graphs of viscosity against time at three different shear rates are illustrated in Fig. 13.7. Thixotropy is a time-dependent phenomenon. It is assumed in flow curves such as in Figs 13.2–13.6 that the breakdown of structure and its re-establishment, when the shearing stress is removed are instantaneous. This is true within the limits of measurement for some materials, and curves such as these could be followed upwards and downwards an indefinite number of times without any variation. For many other materials, however, including printing inks the processes of breakdown and especially set-up take time and this leads to further complications.

In the flow curve for a thixotropic material readings taken at advancing shear stress differ from those where the shear stress is being reduced (see Fig. 13.8) As the stress is increased the structure in the ink is reduced and the curve AB results. If after B is reached, the stress is progressively reduced curve BO results because structure recovery takes much longer.

The area in the hysteresis loop may be taken as an indication of the degree of thixotropy existing in the substance under test. Care should be taken to establish that this thixotropic loop is in fact real because fluid systems (even if Newtonian) which are subjected to stress will become

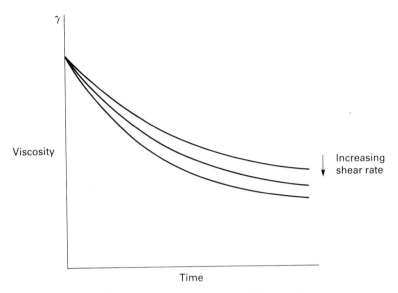

Fig. 13.7 Thixotropic substance at different shear rates

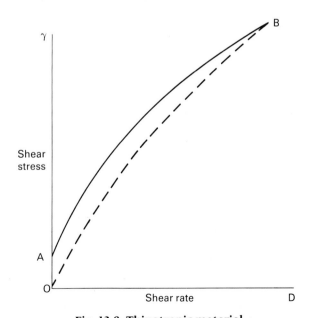

Fig. 13.8 Thixotropic material

heated and yield similar hysteresis loops if the thermostatic control system is inadequate.

Influence of temperature

Flow behaviour is strongly dependent on temperature which means that during measurement great care must be exercised to ensure that the temperature is kept constant. The viscosity of liquids is always reduced by increasing the temperature, for example a rise of a few degrees centigrade transforms sugar syrup from an immobile liquid almost to the fluidity of water.

Alternative methods of expression of ink flow

The use of the rheogram is the classical approach to representation of the flow of a material and most who are familiar with this from a graphical expression of rheological behaviour are happy with it.

However, for students who seldom refer to a rheogram, this method of expression may be confusing particularly as the important interest of the ink-maker is to understand the apparent viscosity on application of differing shearing stresses (i.e. the structure when in the process of being broken down). On the rheogram viscosity is equal to tan α which is inconvenient in making a quick appreciation. (α is the angle between the tangent to the curve and the shear rate axis).

An approach similar to that of Goodeve and Whitfield in expressing the apparent viscosity against reciprocal shear rate may perhaps be more readily understood [4].

Figure 13.9 shows the apparent viscosity of a printing ink plotted against reciprocal rate of shear. This means that the rate of shear at the origin is infinite and the residual viscosity of the ink when the structure is completely broken down may be obtained by extrapolating to infinite rate of shear at A. In the graph, the point A is found in printing ink to be slightly in excess of the viscosity of the medium. In addition this value at A may be found to bear some relationship to values ascribed to the tack of an ink.

It is also found that this value at A is in general conformity with the Einstein equation for a system of particles suspended in a liquid.

$$\eta_\infty = \eta_\theta \ (1 + \phi C)$$

where η_∞ = the residual viscosity;
η_θ = the viscosity of the medium;
C = a constant;
ϕ = the volume of the particles in the system per unit volume.

Visco-elasticity

A purely viscous liquid will move on the application of a minimal stress, the deformation remaining on removal of the stress. An elastic solid, on the other hand, will recover its original form immediately on removal of the stress. No substance is either totally elastic or viscous and so most

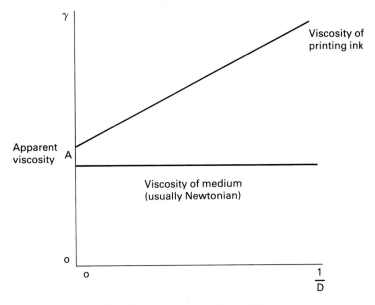

Fig. 13.9 Apparent viscosities

materials contain both characteristics the extent of each depending on the molecular structure, the presence of particles, any particle to medium or interparticulate interactions.

These materials are called visco-elastic and most inks come within this classification. For example, an ink which has a yield value will behave as an elastic substance at shear stresses below the yield value, i.e. the stress will cause a deformation and on release of the stress the original form will be regained. On exceeding the yield value the ink behaves as a liquid.

These considerations are important in the conditions which occur on printing presses where the response of the visco-elastic material will depend on the rate of application and release of the shear stress on distribution rollers and in the separation of paper and blanket.

The visco-elastic nature of inks becomes more significant with higher pigmentation. The choice of medium with respect to its molecular nature can have considerable influence, rosin-based derivatives will behave differently to long polymer chains based for example on linseed oil.

13.3 APPARATUS FOR THE MEASUREMENT OF THE VISCOSITY OF NEWTONIAN LIQUIDS

The range of liquids suitable for this equipment includes mineral oils, solvents, drying oils and varnishes (other than those with a gel structure). The rheology of Newtonian fluids does not require the establishment of a rheogram because the curve is simply a straight line passing

through the origin. The slope of the line is therefore the only matter of
interest, this being the viscosity.

Some viscometers, e.g. Ostwald U-tube type, are essentially one point
instrument. That is to say they have the capability of making a measurement at one set of conditions of shear stress or shear rate. These viscometers cannot be used to obtain pairs of values or τ and D necessary to
plot a rheogram. However, when the liquid has been confirmed as being
of Newtonian behaviour a single point apparatus can be used and several
are available.

Reference should be made to the determination of the viscosity of
liquids in CGS units BS 188:1957 issued by the British Standards Institution which provides information on capillary viscometers and falling
sphere viscometers. Details are given therein of the design, use and
calculation of results of these two pieces of apparatus which in the
printing industry may be used for solvents, mineral oils, etc. and
varnishes other than gel varnishes.

Capillary viscometers

These are used for liquids of low viscosity. Since the stress involved in
this apparatus is that due to the difference in liquid level between the two
arms of the 'U', the density of the liquids enters into calculation and the
result is expressed as a kinematic viscosity the unit of which is the stoke.

The relationship between the stoke and the poise is as follows:

$$\eta = \delta\varepsilon$$

where δ = the viscosity (stokes);
     η = the viscosity (poise);
     ε = the density of the liquid.

The working formula is as follows for capillary viscometers:

$$\delta = Ct - \frac{B}{t}$$

where t = time of efflux;
     C = constant;
     B = constant.

The second term can be neglected if t is greater than 100 s (see BS
188:1957).

Falling sphere viscometers

This method also is applicable to Newtonian liquids although some
concept of non-Newtonian behaviour may be established by dropping
spheres of varying size through the liquid. It is essential however that the
Faxen corrections are applied in any attempt to establish non-Newtonian
behaviour.

Again this method gives the results as a kinematic viscosity centistokes.
Dependent on convenience, either formula (A) or (B) may be used:

$$v = Mg \frac{\delta - \varrho}{0.03\pi \, vd\varrho\delta} \times F \qquad (A)$$

or

$$v = \frac{d^2 g \, (\delta - \varrho)}{0.18 v\varrho} \times F \qquad (B)$$

where M = the mass of the sphere (g);
   d = the diameter of the sphere (cm);
   δ = the density of the sphere (g/cm^3);
   ϱ = the density of the liquid (g/cm^3);
   v = the velocity of fall (cm/s);
   g = the local acceleration due to gravity, (cm/s^2); and
   F is known as the Faxen term and is a correction for the effect of
   the wall of the fall-tube on the motion of the sphere:
   $F = -2.104d/D + 2.09d^3/D^3 - 0.9d^5/D^5$
where D is the diameter of the fall-tube in centimetres and d is the diameter of the sphere.

This Faxen correction should be used unless D is so large that d/D can be neglected for the largest diameter of sphere employed.

BS 188:1957, gives the range of viscosity measurements employing the capillary viscometer as 0.01–1500 stokes for clear liquids, and 0.02–4500 stokes for opaque liquids. When using falling sphere viscometers for clear liquids it is 20 stokes and upwards.

Use of falling sphere viscometer for the measurement of the viscosity of opaque liquids

For opaque mineral oils one can fill the tube with water or glycerol (dependent on the specific gravity requirements) to the lower mark and between the marks, with oil. The time of fall of the sphere is taken between its disappearing at the surface and its emergence into the water layer. The equation normally applied may be used in these circumstances but with the obvious error associated with the sphere not having achieved uniform velocity. The method is convenient and the error is small.

Flow cups

Flow cups should be used only for the measurement of the viscosity of Newtonian fluids and can be calibrated by means of an oil of known viscosity. In practice where the fluid is quite thin, the deviation from Newtonian behaviour is ignored for practical purposes and so flow cups may be used for the control of varnishes and gravure and flexographic inks. However, it is essential to be alert to those circumstances where the structure in the fluid is such that the flow cup will produce any result dependent on the method of operation. BS 1733:1955 gives a complete description of flow cups and their use and lays emphasis on the care and checking of the cup if erroneous figures are to be avoided.

The results are expressed as the flow time from commencement to the instant at which the stream first breaks into droplets. They are not recommended for use where the time of efflux is very short. Normally the results are expressed as the flow time, although conversion to stokes is possible by reference to conversion factors.

Flexographic and gravure printers frequently use Zahn flow cups for routine control. The cups have a handle and can be conveniently dipped into tanks. The ink-maker usually supplies ink at a viscosity which the printer can reduce by solvent addition to achieve the optimum printing conditions for the job in hand. Some presses have automatic control which adds solvent to maintain a predetermined viscosity. The formulator should anticipate such additions to ensure that the flow properties remain satisfactory.

Flow cups have been reviewed by Stockhauser [5].

13.4 PRACTICAL MEASUREMENTS FOR NON-NEWTONIAN SYSTEMS

Most printing inks are non-Newtonian because the element of structure is involved. In the case of lithographic and letterpress inks, the pigmentation and extender particles may occupy 25% of the volume. The particles of pigments and extender are therefore sufficiently close together to exert an attractive influence by the van der Waal's forces so that in time some orientation may be produced, giving an appearance of solidity in the ink. Such systems are normally said to be thixotropic because they will break down with agitation, but when the agitation is removed the structure will set-up again. The time of setting-up of structure may be protracted and an increase in rigidity may continue for many days. In rheological measurements this means that depending on the shearing stress or the degree to which the structure is broken down, the apparent viscosity will vary. In other words, the greater the shearing stress, the greater becomes the fluidity of the system, i.e. the viscosity falls.

The rheogram for the typical ink system is frequently that of an ideal plastic substance (Fig. 13.3) or pseudoplastic (Fig. 13.4). It will be seen that the ideal plastic substance does not move until the shearing stress reaches a critical value. This value is described as the yield value.

Since non-Newtonian substances, in which many printing inks are classed, vary in apparent viscosity with increase in rate of shear or shear stress, it is obvious that a picture of the rheological behaviour must include measurements taken at varying rates of shear to enable a graph to be produced relating shearing stress to the shear rate.

Rotational viscometers

The most generally used of these is the coaxial cylinder type, shown diagrammatically in Fig. 13.10. The material under test is sheared in an annular gap between a driven outer cup and a cylinder bob, suspended from a wire or spring. The viscous drag tends to rotate the bob, until it is

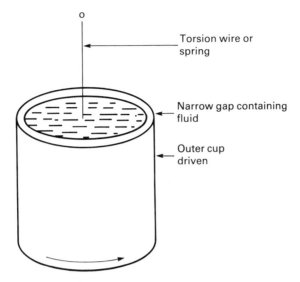

o

Torsion wire or spring

Narrow gap containing fluid

Outer cup driven

Fig. 13.10 Schematic diagram of a rotational viscometer

balanced by the torsion in the wire or spring. This type of viscometer was first employed by Couette, and numerous modifications of it have been devised, including some in which the bob is rotated and the viscous drag measured at the cup. Rotational viscometers have the particular advantage that a sample can be subjected to a known rate of shear for any chosen period, and a graph of the shearing stress against the velocity gradient may be obtained (i.e. a rheogram).

The rate of shear will obviously be proportional to the speed of rotation of the cup. Since it is desirable to measure printing inks at high rates of shear comparable to those on a printing machine, this leads to the choice of instruments with narrow gaps, to avoid the use of a very high speed of rotation of the cup.

It should be noted that the rate of shear in rotational viscometers varies over the distance of the gap. However, if the gap is sufficiently narrow, the mean value may be taken. With large gaps care must be exercised in comparing viscosities obtained with instruments of differing geometries.

A further reason to favour a small gap is that it helps temperature control. At high rates of shear, considerable heat is developed which must be conducted away if the temperature is to be held constant. In practice, this is difficult to achieve. Printing inks are generally poor conductors of heat, and the best results are obtained by shearing thin films between metal surfaces with thermostatic cooling. The practical disadvantage of using narrow gaps may be the difficulty, particularly with still materials, of forcing the cylinder into the cup.

The temperature effect is not the only limitation in the use of co-axial cylinder viscometers at high rates of shear. Above certain values some materials including printing inks tend to 'climb' out of the annular gap producing anomalous results. Unfortunately these limitations occur at or before rates of shear (of some thousands of reciprocal seconds) known to occur in the manufacture of the ink at the milling stage and also during the distribution on the printing press. This phenomenon is connected with the visco-elastic properties of the materials and is known as the Weissenberg effect. 'Visco-elastic' describes materials which have some of the properties of both elastic solids and viscous liquids, and such properties can be investigated with more complex instruments called rheogonio-meters which measures forces at an angle to the plane of shear. (6.7)

A modified version of the rotational viscometer provides convenience and speed of operation for the evaluation of gravure and flexographic inks. It is particularly useful when high pigmentation gives rise to struc-ture which makes the use of flow cups inappropriate. The outer cup is a standard can which thereby avoids transference of the ink to a specific cup for the instrument. The shearing force is applied to the inner cylinder which by means of holes placed within it produces a vigorous stirring action on the liquid. The force produced on the outer cylinder (i.e. the can in which the ink is placed) via the liquid in the can is opposed by a spring and the apparent viscosity may be read in poises on the dial.

Cone and plate viscometers

A variation of the rotary principle is found in the cone and plate visco-meter, the Ferranti–Shirley instrument being one of the first to be used in the ink industry. The principle is shown in Fig. 13.11. The torque to drive the cone is measured and by using different sized cones and a variety of speeds a range of measurements are possible. The angle AOB between the cone and plate is normally so acute, approximately 1°, that the rate of shear is constant throughout the gap. Cone and plate viscometers are relatively easy to fill and clean and therefore very convenient to the ink-maker's requirement.

The introduction of inexpensive powerful computors has given rheo-logists the opportunity to control the variables of shear stress and sheer

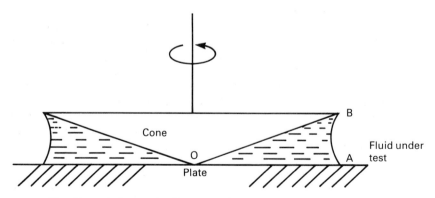

Fig. 13.11 Schematic diagram of a cone and plate viscometer

rate to a much greater extent than ever before. Complex shear stress–shear rate programs can be applied to the test substance under strictly controlled and repeatable conditions.

The resulting rheograms can be analysed to see how closely the data fit to Newtonian, pseudoplastic, Bingham, Casson (see section on falling bar viscometers) dilatant or thixotropic behaviour and the best fitting curves found. An estimate of the thixotropic nature can be found by measuring the area in the hysteresis loop on the rheogram. All these curves can be displayed on monitors or x–y chart recorders and compared with established standards. For routine quality control methods, batch detail can be stored on discs together with all the relevant rheological data.

A programmable calculator can be used to find the best least squares fit for the straight line extrapolation on the falling bar viscometer rheogram.

Several companies are now producing a wide range of instruments for investigational work on ink flow. These include both concentric cylinder and cone and plate systems, which are most convenient for measurement on offset inks.

Haake, Contraves and Carrimed produce a range of computor controlled viscometers suitable for making viscous and visco-elastic measurements. The Haake and Contraves systems control the speed of rotation of the system, as already described, whereas the Carrimed viscometer controls the shear stress applied to the sample.

Although expensive these viscometers are being used for quality control work in the ink industry where their simplicity of use, combined with the advantage of computer control should produce benefits in accuracy and time saving.

Simple versions of the cone and plate system are available for routine work for paints and similar substances with apparent viscosities of between 1 and 10 poise. Their employment in the ink industry for flexographic and gravure ink is limited by the fact that the solvent evaporation rate is too rapid. However, for inks with solvents evaporating at the rate of white spirit, this instrument is most useful.

The principles of this apparatus are contained in ref. 8.

Falling bar viscometers

For routine control in a printing ink laboratory the falling rod viscometer is now the most widely used and convenient method of determining the rheology of lithographic and letterpress inks. In common with cone and plate and rotational viscometers it is possible to evaluate the plastic viscosity in poises, and the yield value in dynes/cm^2. The numerical values obtained depend upon the precise method of use, but once that is established repeatable results can be achieved. This makes the instrument suitable for control of batches, and also for the characterisation of inks which have been found particularly satisfactory on the printing machine for some specific process.

The apparatus consists of a vertical rod which passes through an annulus. A sample of ink is placed between them, after which the rod is forced through the aperture by applying weights to it (see Fig. 13.12). The

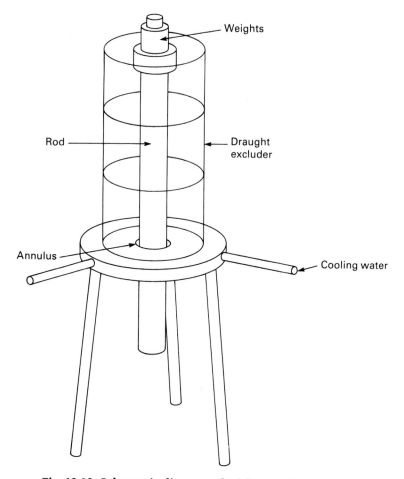

Fig. 13.12 Schematic diagram of a falling rod viscometer

time of fall is measured at each load. The readings can then be converted into shearing force (load) and shearing rate (time) by applying appropriate viscometer constants.

A series of pairs of values may be plotted to obtain a rheogram. The apparatus is particularly useful in the viscosity range of letterpress and offset inks and the concentrated pastes from which they are often made (approximately 5 pascals to 100 pascals). Materials of this kind are usually pseudoplastic, and many are very thixotropic.

The yield value of an ink is defined as the force applied to produce some significant flow and it is of interest in predicting, for instance, whether a given ink will feed readily from the duct where shearing forces may be small. However, it is difficult to make reliable measurements at low rates of shear to see if a yield value is apparent. The difficulty is commonly overcome by the use of Casson's equation, and charts based upon it may be supplied by the equipment makers. It is summarised as:

$$\sqrt{\lambda} = K_0 + K_1 \sqrt{D}$$

where λ = shearing stress (N.m^{-2}) (pascal);
  D = shear rate (s^{-1});
  K_0 = square root of yield value;
  K_1 = square root of plastic viscosity.

The application of the Casson equation produces a straight line which can be extrapolated to zero rate of shear to determine the square root of the yield value. The slope of the line represents the square root of the plastic viscosity.

Manufacturers' charts have been prepared so that the loads and their respective times of fall can be plotted, and the viscosity and yield value directly read off (Fig. 13.13). It should be noted that these charts are prepared on the basis of the original dimensions of the instrument, which changes with wear.

Frictional heat is generated when the rod passes through the annulus, and cooling is provided by circulating water at constant temperature through a jacket surrounding the aperture.

The recommendation about the performance of measurements vary both in terms of the choice of loads, and whether they are applied in ascending or descending order. Generally determinations using high loads produce relatively high figures for yield values and low figures for plastic viscosity [9]. This effect was found to be reinforced if the loads were applied in descending order. An article prepared by the Experts Committee in Rheology of the European Association [10] recommends a maximum load corresponding to a minimum falling time of 5 s (rate of shear 500s^{-1}), or if an electrical timer is in use, this can be reduced to about 1 s (rate of shear 2500 s^{-1}) It is advisable not to use loads above 2000 g, because excessive heating will occur. The committee also recommends that the highest load is first applied, then the lowest, then the second highest and so on. In this way the influence of thixotropy is reduced.

Special attention is needed to the cleaning of instruments, because small amounts of dried ink on the rod or in the annulus will upset the results. It is also important to ensure that all solvent is removed before a new measurement is made.

Flow at low shear stresses

The shear stresses involved in producing ink flow in the duct of a press are due entirely to gravitation and are thus quite small. In these circumstances the rate of build up of structure is of considerable importance. If too rapid then the ink will not feed properly into the nip. A simple test to assess the build up of structure under low shear stress is to run the ink down a vertical or near vertical glass plate. The sample of about 1 ml is agitated and collected on a palette knife and the sample drawn across the top of the glass plate. Timing of the rate of flow down the plate and the ultimate length of flow can give useful information on the low shear stress characteristics of the ink. Reproducibility will depend on the care taken to standardise the method of sample preparation and the adequacy of the temperature control.

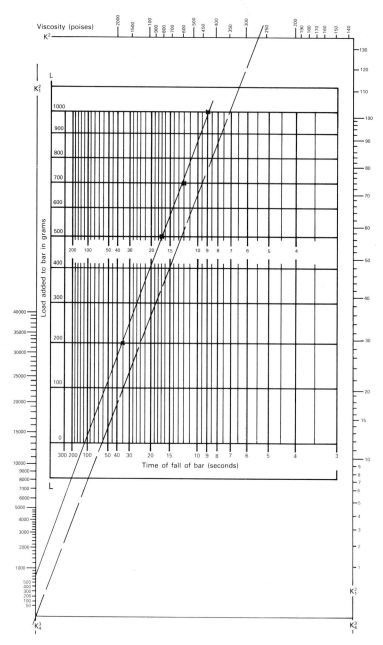

Fig. 13.13 Chart to determine apparent viscosity from a plot of the weights and times on a falling rod viscometer

Ranges of rates of shear

The rates of shear with which the ink-maker is concerned in lithography and letterpress range from zero to very great in the order perhaps of 10^6 s^{-1}.

This arises in the several situations encountered:

(1) Stiffness of ink in the can. Inks kept in store can produce a degree of rigidity which gives inconvenience to the printer in removal from the can.

(2) The behaviour of ink in the duct of the printing machine under very light shearing stress. It is essential that the ink responds to the movement. If the shearing stress is less than the yield value, the ink will fail to feed unless it is constantly agitated.

(3) The rollers of a fast-running press subject the ink to exceedingly high rates of shear.

(4) The capacity of ink to be pumped from large containers. The supply of inks by container is common in web offset news printing and heatset web offset. These containers are delivered and located in the printing plant and ink is pumped from them direct to the printing machine. The rheology of the ink requires that such containers are designed so that the pumping operation is performed easily and efficiently. Recently container supply has commenced in the field of offset four-colour sheet-fed printing which requires stiffer inks than those previously supplied on a commercial basis for pumping systems.

13.5   TACK

Essentially 'tack' may be defined as the 'stickiness' of a substance. Alternatively it may be looked upon as the force required to separate two planes of unit area at a set speed when they are held together by a liquid.

'Tack' would seem to be the factor governing the process of applying ink to paper. With fast-running presses, particularly when printing coated paper, the tack of an ink must be below the critical point at which the ink will pick the paper at the required printing speed. Otherwise, tack should be as great as possible (allowing a reasonable safety margin) to produce the sharpest printing by the letterpress process and the cleanest working in offset lithography. With increase in printing speed it has become a matter of great importance both to the ink-maker and the printer that inks should be supplied ready for the machine at a given tack (plus or minus a small tolerance). This implies the need for an acceptable standard scale of tack to satisfy these requirements.

Much has been written about this, notably by Voet [11], by Strasburger [12], by Myers, Miller and Zettlemoyer [13], by Mill [14], by Whitfield [15] and in European Study of Tack Measurement — interim report [16]. It is clear, however, that the theoretical side is much less well developed than that of viscometry. It is probable that 'tack' involves not only the viscous behaviour of fluids but also the elastic rupture properties of solids.

Nevertheless, it is a property of great practical importance, and instruments have been developed to measure it, necessarily empirically.

The measurement of tack is usually carried out by evaluating the force exerted by a rotating roller on another via an ink film joining them but such a system is not amenable to a rigorous mathematical treatment. Tackmeters available to the printing industry therefore rely on empirical methods for evaluation and standardisation.

The relationship between tack and viscosity

Whitfield found in a series of lithographic and letterpress inks that tack was related directly to the residual viscosity [15].

Mill [17] noted that tack can be closely associated in some cases with viscosity and Turner [18] also found a very rough general relationship between tack and plastic viscosity. Thus when extremes are considered high and low plastic viscosities are normally associated with high and low tack values respectively.

13.6   TACK MEASUREMENT

Originally, the evaluation of tack in printing ink was made by tapping out an ink with the finger on a glass plate. Particularly with the introduction of fast-running machinery, the need arose for a numerical scale and attempts were made to estimate tack by applying a force to separate two flat plates with an ink film between them. The result of such experiments did not lead to the design of any instrument of commercial acceptability.

Those tackmeters now commercially available measure tack under dynamic conditions at known ink film thickness, temperature and speed. The basic principle of such tackmeters is that two rollers, one of which is driven, rotate with a film of ink of known thickness on them. The second roller may be supported in a manner which permits the measurement of the force required to hold it in a stationary position relative to the driven roller (Fig. 13.14).

An alternative system identifies the movement of the axis of the second roller to a position of equilibrium against a strain gauge. In all cases, a third oscillating roller is employed but its function is solely to help in the distribution of the ink over the roller surface and to eliminate ribs which otherwise would form (Fig. 13.15).

At least five instruments designed to measure the tack of printing ink are available. While there are similarities in design, there are also important differences particularly in the nature of the rollers. In the case of the LTF Inkometer and the Churchill, these instruments involve a metal roller operating on a roller of some resilient substance employed in the printing industry for the manufacture of distributing rollers [19–21].

There has been discussion concerning the desirability of having metal on metal rollers or metal on resilient rollers. Metal on metal has the advantage of not complicating the issue arising from the absorption of ink solvents into the roller. On the other hand, some observers believe that

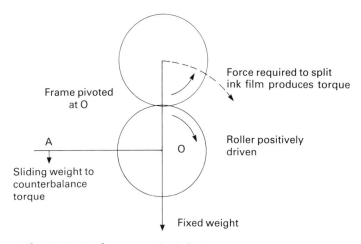

Fig. 13.14 Tackmeter principle (torque measurement)

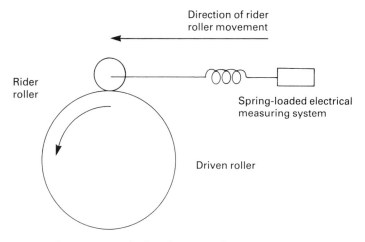

Fig. 13.15 Typical tackmeter (drag movement)

with well-seasoned rollers this difficulty is avoidable. The resilient roller
has the advantage that uniform contact is more readily achieved throughout the length of the roller systems, although some workers in this field
would not agree that this is so. A resilient roller will, however, make a
contribution to the value obtained in a tack measurement.

It has been stated that the two roller system with an ink film between
them is not susceptible of a rigorous mathematical analysis. Consequently
tackmeters need to be calibrated with a standard substance. The Society
of British Printing Ink Manufacturers and PIRA were members of a
European co-operative committee which studied tack measurement, and
a report was issued [16] discussing the several devices available at that

time. A recommendation was made that tack measurements should be carried out with room temperature 23 °C and circulating water 25 °C.

The instruments under study by the European Committee involved the following:

(1) LTF inkometer (five instruments);
(2) PIRA (PATRA) tackmeter (five instruments);
(3) Churchill tackmeter (five instruments);
(4) Tack-O-Scope (four instruments);
(5) DIAF tackmeter (five instruments).

(See also ref 22.)

In all, six samples of printing inks were examined. From the figures obtained, it was apparent that in order to compare instruments of the same design, it is essential first to calibrate all instruments. This calibration should be carried out in accordance with the manufacturers' instructions. However, such calibrations may not take into account all the factors which affect the tack reading. Consequently, any final correlation must be made with suitable inks, and the use of available tack standards is to be recommended for this purpose. This procedure has the additional advantage of revealing drift arising from any gradual deterioration of an instrument.

Instruments of differing design cannot be correlated without recourse to a series of ink samples covering a range of tack values.

Tack standards

Tack standards are manufactured from well-matured printing ink pastes by adjustment to some specific figure on a research laboratory tack-measuring device. Unfortunately such pastes are not absolute standards, but they can be regarded as convenient reference samples.

Further work on the use of polyisobutene fluids as surface strength test liquids, the rheological properties of black letterpress and litho inks and tack graded inks has been carried out by Aspler et al. [23] using a Rheometrics mechanical spectrometer.

Tack testing

The equipment is widely used for routine quality control by comparison with a master sample. In combination with a printability tester such as the IGT, it helps to avoid picking, and provides some indication of the safe machine speed for any paper and ink combination. In high speed wet-on-wet printing, tack grading of letterpress inks in particular has proved beneficial to the attainment of good trap. It should be remembered however, that the figure obtained will most closely relate to the condition of fresh ink in the duct, which may be different from that at the printing head.

Press stability can be predicted by running an ink for a set period and noting the change of ink tack with time. A graph having one of the general forms illustrated in Fig. 13.16 will be obtained. Inks having relatively steep slopes and high peaks when tested in this way, will often have poor stability and may cause piling on the blanket due to incipient picking of the paper surface.

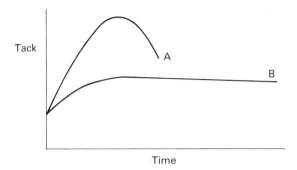

Fig. 13.16 Press stability prediction from inkometer readings. Curve B typifies an ink with better stability than one giving curve A

The equipment is also useful for assessing the tendency for an ink to mist, and the speed at which its onset is likely. The general method is to place a piece of graph paper adjacent to the rollers for a standard time, and then to count the number of ink dots over a unit area, using a magnifying glass. It is important to compare the ink under test with some reference ink whose practical misting tendency on a printing machine is already known, because small changes in atmosphereric conditions may modify the results.

Testing of tack without recourse to an instrument

The experienced operator is able to produce results for comparative tack measurement by tapping the ink with a finger on a glass surface and appreciating by 'feel' the comparative stickiness. It has been suggested in the past that such methods of measurement are unscientific and therefore must be inaccurate. In practice, the fully experienced operator will be found to have the accuracy of many approved tack measuring devices.

Such evaluations are only relative to known samples and unless a series of tack standards is available for which values have been obtained on a suitable tackmeter, it is not possible to ascribe numerical values to unknown samples.

13.7   INK DISTRIBUTION AND RELATED MATTERS

The rheological characteristics of an ink appear to determine its performance on the press, and many attempts have been made to relate the rheological parameters of viscosity, yield value, thixotropy and tack both to press performance and to printing quality. So far, most of the relationships have defied a rigorous quantitative treatment, but rheological measurements can be useful for a semi-quantitative assessment of probable machine performance. Additionally, for quality control, they provide a helpful way of ascribing numbers to an ink whose working properties are known to be commercially satisfactory on the press, so that later batches can be supplied to a similar quality.

The qualitative properties needed in inks can be evaluated by some consideration of the inking system. In letterpress and litho for example, the ink is applied by a roller system from a duct. In the duct, the ink requires a consistency high enough to prevent leakage by flow between the duct blade and roller, but not so high that it will not feed, i.e. 'hang-back' on the rollers, the structure must be broken down, and the ink must spread and transfer from roller to roller; it must therefore have an adhesion to the rollers greater than its own cohesion. However, the ultimate fluidity at the point of transfer must not be too great, otherwise the ink will flow laterally at impression and so not print the image accurately. However, if the ink is insufficiently fluid, the forces developed between the forme and the paper may damage the print.

The ideal rheological description for an ink therefore differs from stage to stage throughout the printing process, and clearly, practical formulations are a compromise.

'Hanging-back'

The shear stress conditions in the duct are comparatively low, so good feeding will be consequent upon a low yield value and low viscosity. Inks having high yield values and high viscosities also feed because they have good cohesion and so act as one mass in the duct. It is an ink with a high yield value and a low viscosity which feeds badly, because its lack of cohesion means that action of the duct roller liquifies only a portion of it. Once this has been fed, the remainder of the ink in the duct becomes isolated from the duct roller, and must be agitated to maintain printing (Fig. 3.17).

The phenomenon is likely to occur when the formulation of a low to medium viscosity ink of high pigmentation is needed, such as might be desirable for adequate results on strong solids for instance.

Distribution

It is evident that the ink must be fluid to distribute, however, deviation from Newtonian flow should not be considerable. Highly pigmented

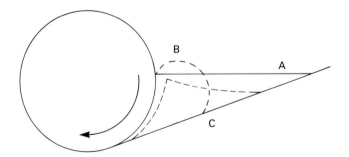

**Fig. 13.17 Ink profiles in a duct: (A) low yield value, low viscosity —
feeds; (B) high yield value, high viscosity — feeds; (C) high yield value,
low viscosity — hangs back**

systems may exhibit visco-elastic or dilatant properties at the high shear rates encountered. This will reduce the ability of the ink to feed through the roller train and transfer to the printing surface.

Consideration should then be given to the reasons for poor distribution in terms of the effect of pigment concentration, pigment–medium interaction and medium type. The structure of inks can be controlled by the use of gelled varnishes which reduce in structure on agitation, i.e. they are thixotropic. This means that the structure will be reduced as the ink passes through the ink train. Thus an ink can have high structure in the duct, at low tack levels, and low structure and tack at the printing surface.

Ink stability

Printing inks are a paradox in the sense that they are required to dry as quickly as possible after printing, and yet remain fluid on the machine. This becomes a matter of great compromise in letterpress and litho inks, which must retain stability when spread into thin films for prolonged periods over the large surface areas of a roller system running at 30–40 °C. The viscosity may increase by oxidation polymerisation, but solvent losses by evaporation are usually particularly critical. On print, it is a great advantage to arrange for small losses of diluent into the substrate to produce a correspondingly large increase in viscosity; the ink left on the surface of the print is thereby rapidly stiffened or 'set' and this is one basis for quick-setting inks. However, premature loss by evaporation from the rollers must be suppressed by the choice of diluents having a limited volatility, otherwise an increase in tack may produce a vicious circle of heat generation and drying on the rollers, and a tendency for the picking of the paper. Piling on the blanket sometimes arises because the ink acquires sufficient tack in this way to pull off a little more paper coating with each successive impression.

Occasionally ink starts to pile as a different manifestation of the basic problem of solvent loss. In this instance dilatency may be induced. Some quicksetting varnishes consist of hard resins barely dispersed in diluent mixtures, so that very rapid phase separation occurs when the ink is printed on to paper. Should the separation happen through evaporative loss on the rollers, the resin seems to become part of the disperse phase instead of the continuous phase and the tack falls. If as is the case in litho, water droplets also augment the disperse phase, the ink may become dilatant, lose its proper capacity to distribute and transfer, and so pile up on the rollers and perhaps the plate.

Stability can be assessed by testing on a tack meter and observing the increase in tack with time of running. Curves of the general shape shown in Fig. 13.16 may be obtained.

Transfer characteristics

The rheological characteristics ideal for printing solid areas are rather different from those ideal for printing halftones. Inks with high plastic viscosities usually do not lay well, particularly if the yield value is also high. This may be connected with the difficulty of 'bottoming' the low surface contours of the stock, particularly in letterpress; lower viscosities and yield values often achieve better lay of the ink by allowing it to flow into surface irregularities.

On the other hand, sharp dots and small dot gain is generally favoured by high yield values and high plastic viscosities, because lateral flow of the ink at impression is thereby minimised.

It must be noted, however, that factors other than ink greatly affect dot gain on image sharpness, including impression, press speed, temperature and the substrate. Consequently, it is not always possible by adjusting the rheology of the ink alone to overcome all difficulties which may need to be investigated in a logical manner.

Transfer of publication gravure inks is a low sheer process and investigations by Schubert *et al.* [24] have shown that, in this area of $0.01-300$ s^{-1} pseudoplastic flow can greatly improve transfer and give better reproduction. Demonstrations of these results are given in photomicrographs and viscosity studies on the Haake PV100/CV100 viscometer.

Mottle

When ink is the cause of mottle (sometimes it might be due to an inconsistency of ink acceptance by the paper) it is thought to be connected with the ink being drawn out into strings as the film splits (Fig. 13.18). The strings then collapse on to the print, producing a characteristic mottled appearance due to slight differences in film thickness all over the surface. It often appears to be aggravated by some combination of a heavy impression, a paper of high surface absorption (which may prevent the strings of ink flowing out when they relax back to the print surface) and inks of low viscosity and low yield value. Mottle may be somewhat reduced by adding an extender whose oil absorption will reduce the flow of the ink, or a paste containing starch and wax which has a similar effect.

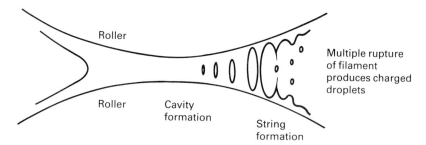

Fig. 13.18 Representation of an ink film splitting

Squash

When printing on a non-absorbent surface in particular, the impression must be kept as low as possible to reduce the lateral flow of ink which shows up as squash beyond the true image boundary. The extent will be determined by printing pressure and viscosity; to avoid excessive flow, it is desirable to formulate for the highest possible viscosity. In lithography, physical squash is not the only factor concerned, because the ink and water balance keeps the image sharp. Although the rheological properties may appear to be less critical in litho, inks generally need a high viscosity in order to minimise the scumming and water marking associated with excessive take-up of water.

Misting

The tendency for inks to mist or fly follows from the formation of strings as the ink film splits on emerging from the nip. High-speed photography has shown the formation of cavities in the ink as a prelude to splitting, the walls of which eventually draw into strings; sometimes multiple rupture seems to occur and the free droplets produced in this way can form a stable mist which disperses into the machine room (Fig. 13.18). It is an objectionable factor at a time when attention is being increasingly focused on evironmental issues.

The phenomenon appears to start at some critical speed for a given ink film thickness, and it is a particular nuisance in high-speed rotary newspaper printing. Empirically, it can be suppressed by the addition of an extender having a high oil absorption which reduces the capacity for the ink to flow into long strings. Bad misting sometimes occurs even at moderate speeds on sheet-fed machines, and the usual cause is the use of resins containing long linear molecules similar to those found in rubber. These materials are extremely soluble in the usual ink solvents and there are indications that very soluble resins and/or the use of strong solvents may increase the tendency to mist. Increasing the concentration of extender also increases the elastic nature of the ink. This property ensures that the ink snaps on elongating of the ink filament on leaving the nip of the rollers, thus preventing formation of ink droplets.

A tendency for misting can be assessed by running an ink through a series of increasing film weights and speeds on a tack machine, and assessing the mist drops which fall on to a piece of graph paper sited adjacent to the rollers.

Wet-on-wet trapping

In multi-coloured printing, the degree of acceptance of an ink film on one previously printed but still wet, will depend on the relative tacks of the inks. Theoretically it is desirable for the first down colour to have a tack as great as possible so that subsequent colours of lower tacks are pulled from their printing heads. If the reverse happens, that is successive units partly scavenge ink already on the print, 'back-trapping' is said to occur, with a consequent loss of colour balance and printing control.

Tack grading is more desirable in high speed wet on wet letterpress printing than in litho, where the resilience of the blanket and the lower ink film thicknesses seem to make trapping less critical. The nature of the paper also plays a very important role in the degree of acceptance of quick-setting inks. The first colour down, having lost solvent to the paper, presents to the oncoming colour an ink film with a tack greatly increased over its original value.

It must be remembered too that the tack of a set of printing inks measured on a tack machine are numbers derived under uniform and standard conditions. In the case of practical printing, the film weights of inks may not be uniform among the printing units; the losses of solvent on passage through the distribution system from duct to printing surface may not be consistent; and in the case of litho, the take-up of water may be different in each ink. Each of these factors would have made some considerable difference if they could have been allowed for during the original measurements on the tack machine, and together with the effect of paper absorbency, they suggest why the theoretical requirements may sometimes be of reduced significance. Nevertheless, the tack machine remains a convenient and most useful empirical method of sorting inks into their correct order for printing in succession, provided other factors of importance are taken into account.

In the case of printing non-absorbent substrates, it is essential to observe all the factors which influence the printing of a sequence of tacks in the correct order.

On contact with the paper there will be a rapid loss of low viscosity diluent into a porous substrate. This will mean a considerable increase in the pigment concentration in the ink with a subsequent increase in the elastic component of a visco-elastic ink. This change should ensure that the first ink down is not pulled from the paper.

Ink uptake can be related to the following parameters, the thickness of the previous layer, the speed with which this first layer sinks in, the degree of tack and viscosity, structure and chemical composition of this paper, printing speed, excessive damping temperature, roughness of the plate, hardness and porosity of the blanket.

13.8 RHEOLOGICAL MEASUREMENTS AND MACHINE DESIGN

The mechanism of feeding ink on to the roller trains places demands on the ink-maker in providing inks with appropriate rheological properties to satisfy this aspect of machine design [25]. It is suggested in that article that the design of the ink duct should not call for inks of defined consistency because the necessity of satisfying the requirements of the ink duct prevents the ink-maker from formulating an ink which would give the best results in the printing process.

Many letterpress and lithographic presses continue to operate with the ink duct roller and ink duct blade system. The degree of agitation with ducts of this type is frequently very small which means that structure in the ink must be reduced to a low level. More recently such ducts have been fitted with duct agitators which permit the use of structured inks.

In the field of web offset news printing, ducts of new design have been introduced. Many machines are now fitted with a device operating on two rollers with serrated faces not in contact which meter the quantity of ink to the press. Such ink feeding devices call for a reasonable degree of structure in the ink to avoid the inconvenience caused when the ink dribbles through the rollers on to the floor.

In some web offset machines of very sophisticated design the ink is fed on to the machine through a series of jets producing an ink jet feed on to the duct roller. With these systems the quantity of ink is metered for each point on the rail, and control is by remote press button. This system calls for detailed study on the part of the ink-maker if the best results are to be obtained.

Measurement of flow properties

Descriptions of a device for measuring ink properties on the press are given by Ritz and Rech [26, 27]. The instruments consists of two rollers, the top one of which oscillates from side to side to imitiate the movement of a distribution roller. The flow behaviour of inks in the presence of water, the viscosity of the ink when oil is added and the influence of water in the press on the distribution capacity of the rollers were tested. The drop in shear between rollers is highest in the transverse direction.

The relationship between tack and viscosity has been drawn up by Bauer and Decker [28]. The errors in rheological values are discussed and the determination of these values on the press with the help of a newly designed viscometer.

Tasker [29] has devised three new laboratory tests for determining the water pick-up of lithographic inks. These involve emulsification of the ink in a Litho Break tester; the Delta plastic viscosity test compares the viscosity of emulsified and non-emulsified ink, the a−k fit test analyses the emulsion for water and the emulsified water particle test measures this parameter.

Surland [30] has investigated the relationship of ink emulsification curves to rate of change of ink flow properties and ink tack increase, also the ink transference, misting and sling.

The effect of ink viscosity in relation to water absorption has been examined by Braun [31]. Other parameters using simulated offset conditions on a Lithomat tester from Prufbau include the effect of pigments and the calcium solubility of lithol rubine pigment in acidified medium.

The rheological characteristics of inks have been described by Bassemir [32] and the relationship between flow properties, emulsification rate and printability. Particular attention has been given to dot resolution; however, much work remains to be done in order to understand the phenomenon fully.

The need for instruments which will measure viscosities at both high and low shear in liquid inks is discussed by Anderson and Jensen [33]. At high shear rates the visco-elastic nature of the binder solution plays a major role, while at the low shear rates the flow properties are controlled by the dispersion characteristics of the pigments. For measurements of the flow at high shear rates which are most interesting from the printing point of view the widely used flow cup will not give any definite impression of the flow properties. The viscometer must be able to achieve high shear rates and measure visco-elastic effects.

Babula [34, 35] has described the measurement of the rheological properties of liquid inks in relation to printing properties, e.g. ink split, dirty printing, pigment dispersion and rewettability.

The relationship between the flow properties of screen inks and their performance has been well described by Frecska [36, 37] and Messerschmitt [38–42]. The effect of fillers on the flow properties of ink and their thixotropic nature is reviewed. The range of viscometers is discussed and their place in viscosity control is considered. A suitable viscometer for assessing performance of screen inks at a practical price is not yet available.

Ink control on the press and in tanks

The development of fully automatic ink pumping systems has been discussed by Kirby [43, 44] and Quilliam [45]. The pump equipment and its operation to pump ink from the bulk container meter the quantity and maintain levels in the ducts are described.

The viscosity of the ink will rise rapidly in the winter months and a knowledge of the ink flow properties at low temperatures will be required for efficient delivery. Storage tanks should be narrow and deep with a tapered bottom. Pipe diameter is important and there should be no tight bends which would reduce flow.

REFERENCES

1. Newton, Isaac. *Philosophiae Naturalis Principia Mathematica*, Book 2. (1687).
2. Vincent, J. E. Chemistry of aluminium — organics in the preparation of ink vehicles and gel varnishes. *Am. Ink Maker* **62**(10), 25–6, 28–30, 32, 34, 34B, 104 (Oct. 1984).
3. Love, D. J. The modification of lithographic inks with Manchem aluminium compounds. *Polym. Paint Col. J.* **173**(4109), 837–8 (14/28 Dec. 1983).
4. Goodeve, and Whitfield, *Trans. Faraday Soc.* **44**, 652 (1940).
5. Stockhauser, N. Cup viscometers: a testing method with limits. Paper presented at Fogra — Symposium *Drucktechnik Forschungsergebnisse für die Praxis*, Fogra, Munich (4–5 April 1984) 17pp. (PM10263G.)
6. Van Wazer, Lyon, Kim, and Colwell, *Viscosity and Flow Measurement*. Chapter 6.
7. Douglas, A. F., Lewis, G. A., and Spaull, A. J. B. The investigation of the dynamic visco-elastic functions of printing inks. *Rheol. Acta* **10**, 382–6 (1971).
8. Routine measures of the viscosity of paint samples. *JOCCA* **49**(7) (1966).
9. Turner, *British Ink Maker* **15**(2) (1973).
10. Experts Committee in Rheology. *British Ink Maker* **17**(4) (1975).
11. Voet, *Ink and Paper in the Printing Process*. Interscience Publishers (1952).
12. Strasburger, F. *Coll. Sci.* **13**, 218 (1958).
13. Myers, Miller, and Zettlemoyer, F, *Coll. Sci.* **14**, 287 (1959).
14. Mill, *British Ink Maker* **6**(2) (1964).
15. Whitfield, *British Ink Maker* **7**(2) (1965).
16. European study of tack measurement — interim report. *British Ink Maker* **15**(2) (1973).

17. Mill, *Chemistry and Industry.* (1952), p. 159.
18. Turner, *British Ink Maker* **18**(2) (1976).
19. PATRA tackmeter. *British Ink Maker* **5**(3), (1963).
20. Cartwright and Lott, The Metal Box/Churchill tackmeter. *British Ink Maker* **4**(4), (1962).
21. Van Hessen, The Tack-O-Scope. *Am. Ink Maker* **41** (Jan. 1963).
22. Souty, G. R. A comparison of commercial tackmeters. *British Ink Maker* **10**(3) (1968).
23. Aspler, J. S. *et al.* Rheological properties of new inks and surface strength test liquids. *Advances in Printing Science and Technology,* Vol. 16, Proceedings of the 16th International Conference of Printing Research Institutes, Key Biscayne, Florida, June 1981, pp. 235–51. (Pentech Press for IARIGAI, London, 1982, 467pp. (1310/BK1859).)
24. Schubert, F. G. *et al.* Low shear viscosity of publication gravure inks and its implications. *Taga Proc.* 219–29 (1984).
25. Whitfield, *Lithographic Technology.* PIRA/IARIGA (1970).
26. Ritz, A. and Rech, H. *Determination of the Rheological Magnitudes of Inks in the Inking Unit of an Offset Machine.* Forschungsbericht S. 204, Deutsche Forschungsgesellschaft für Druck-und Reproducktions-technik e v (Fogra), Munich, (1978) 292pp. (PMS 847.)
27. Ritz, A. Determination of the rheological magnitudes of inks in the inking mechanism of an offset litho press. *FORGA Mitt.* **27**(96), 10–11 (Aug. 1978). (In German.)
28. Bauer, M. Decker, P. and Bosse, R. *Relationship between Ink Viscosity and Tack; Theoretical Analysis and Laboratory Experiments Carried Out on an Offset Press.* Forschungsbericht 5.00G, Deutsche Forschungsgesellschaft für Druck-und Reproduktions-technik e v (Fogra), Munich (1977) 35pp. (In German.) (PM 4119.)
29. Tasker W. *et al.* Water pick-up test for lithographic inks. *Taga Proc.* 176–90 (1983).
30. Surland, A. Factors determining the efficiency of lithographic inks. *Taga Proc.* 191–236 (1983).
31. Braun, F. Studies of offset inks — the effect of pigments on the formation of emulsions. *Am. Ink Maker* **63**(2), 26–48 (Feb. 1985).
32. Bassemir, R. W. The physical chemistry of lithographic inks. *Am. Ink Maker* **59**(2), 33, 36, 41–2, 44, 46, 98 (Feb. 1981).
33. Andersen, J. H. and Jensen, H. *Rheological Properties of Liquid Inks.* T2-82T Horsholm Denmark: Scandinavian Paint and Printing Ink Research Institute (1982) 69pp. (In Danish.)
34. Babula, S. Viscosity is only one aspect of rheology that determines ink distribution. *Flexo* **9**(3) 21–3 (March 1984).
35. Babula, S. Viscosity is not enough — a study of the behaviour of ink in the distribution system of a flexographic press. In *Report of the Proceedings of the Colour Conference,* 6–7 Dec. 1983, Atlanta, Georgia, pp. 61–4. (New York Flexographic Technical Association, 163pp.)
36. Frecska, T. Viscosity measurements. A challenge for screen printers. *Screen Print* 56–9, 124 (July 1984).
37. Frecska, T. The operating characteristics of screen printing inks. *Screen Print.* **72** (6), 62–5 (June 1982).
38. Messerschmitt, E. The continuing story on viscosity. *Screen Print.* 50–7, 60 (Aug. 1984).
39. Messerschmitt, E. Rheological considerations for screen printing inks. *Screen Print.* **72**(10), 62–5 (Sept. 1982).

40. Messerschmitt, E. Rheological considerations for screen printing inks. Part II. *Screen Print.* **72**(11) 136–9 (Oct. 1982).
41. Messerschmitt, E. Rheological considerations for screen printing inks. Part III. *Screen Print.* **72**(12) 80–3 (Nov 1982).
42. Messerschmitt, E. The release of printing ink from the screen. *Siebruck* **26**(11), 702, 704, 706, 708–10 (Nov. 1980). (In German.)
43. Kirby, S. From tankers to tank. *Ink and Print.* **3**(1) 16–17 (March 1985).
44. Kirby, S. Saving multicolour costs with automatic ink pumping systems. *Lithoweek* **6**(35), 42 (29 Aug. 1984).
45. Quilliam, B. L. Development of a system of computerized automatic ink-flow control. *Taga Proc.* 89–103 (1983).

CHAPTER 14
Testing, Control and Analysis

The performance of an ink relies largely upon the optimum selection of raw materials, and upon correct fabrication. Once this selection has been carried out, the continuing operational performance of that product will then depend upon the maintenance of standards, both in raw material quality and manufacturing. The extreme value of testing and controlling raw materials and finished inks is thus apparent, and it is proposed to examine this area in the detail that is necessitated by its importance.

Methods of testing belong essentially to two distinct types:
(1) Numerical determinations that can be expressed in identifiable quantities such as tack values, acid values, etc.
(2) Comparative tests based on simple non-quantifiable tests of factors such as shortness, flow characteristics.

As many of the properties of printing inks are either ill-defined, or not well understood, it is not surprising that many tests fall into the second category.

In some cases, it is necessary to check the product in the form of a laboratory print. This invariably involves visual assessment, and highlights the still dominant reliance upon the human observer, who is more efficient in many ways. As long as the ink-maker is not less critical than the printer, printer's customers or other assessors, it is likely that his decisions will be acceptable, particularly as his method of assessment will be the same as the others.

Any planned testing programme should first involve an estimate of the testing required. In estimating, certain questions need answering, such as:
(1) Is the test realistic?
(2) Can the length of time required for the test be justified?
(3) Is the test necessary?
(4) Is the testing adequate?
(5) What are the limits of any test?

Decisions must be made in all testing departments on the tolerance limits required: If these limits are too narrow, then ink production can be impeded and late deliveries could ensue, whereas too wide limits can, of course, have a detrimental effect on both quality and reproducibility. For commercial purposes, a working compromise must be reached. Some testing will be regarded as essential, some as desirable but not essential, and further testing, while desirable, may be uneconomic. The ink formulator must bear a great responsibility in setting these limits as he alone is fully aware of both the final ink specification and of the raw materials which are essential to the success of his particular formulation.

Comprehensive testing of raw materials has the advantage of reducing the range of tests needed at the finished product stage, and hence a faster service to the consumer. However, this advantage is partially offset by the necessity to test deliveries of some of the more versatile raw materials in a wider range of formulations than those in which the batch may eventually be used. The individual ink companies must, therefore, balance their testing resources to give the most advantageous results.

In the remainder of this chapter, a relatively comprehensive guide to the testing of both raw materials and finished inks has been attempted, but it is impossible in the scope of such a section to include all the highly specific tests which could be carried out. Where standard test methods are included, only the salient points have been given and the reader should refer to the published test procedure for the practical details.

14.1   STANDARD TESTS

It is obviously desirable that some degree of standardisation of test procedures should exist, so that results obtained in one laboratory can be directly compared with those of another. This is true both with raw materials (that is between the raw material supplier and the ink-maker) and also with the printed product (where an ink supplied to a printer may have to comply with a specification imposed by the printer's customer).

Some tests have been widely agreed and standardised by such bodies as:
The British Standards Institution (BS Standards).
The European Committee of Paint and Printing Ink Manufacturers' Associations (CEI Standards).
The Technical Association of the Pulp and Paper Institute (TAPPI).
The American Society of Testing Materials (ASTM Standards).
The Institute of Peteroleum (IP Standards).
The Research Association of the Paper and Boards, Printing and Packaging Industries (PIRA).
Some standard test methods have also been put forward by the National Printing Ink Research Institute (NPIRI) in the USA, and several of these tests have been published in the *American Ink Maker*. Alternatively, members of the Society of British Printing Ink Manufacturers can obtain information on these test methods direct from the society.

14.2   SAMPLING TECHNIQUE

In order for raw material testing to have any meaning, the sample taken from a delivery of material must be representative of the bulk in all respects. This is a simple and perhaps obvious statement, but if the sampling is not carried out efficiently, then a batch may be incorrectly assessed and this could possibly result in costly repercussions at some future date.

A single delivery of material may contain several drums with the same batch reference number, as well as others with a different reference. Ideally, samples should be taken from all containers and each tested accordingly, but where this is not considered practicable, one may take a calculated risk and examine samples taken at random from containers of nominally the same batch.

The following is intended as a guide to sampling different kinds of material.

Resins

Resins which are delivered in lump form, may well also contain some smaller pieces and fine powder. A composite sample should then be taken which is representative of the proportions of each grade present. It must also be remembered that there is a specific reason for resins being delivered in lump form (usually to prevent oxidation) and so they should only be crushed, where necessary, immediately prior to the actual raw material testing.

Pigments

The vibration in transit of a cask of pigment may well have caused some segregation of particles of different size. Hence the cask should either be tumbled before opening, or several samples taken at varying levels in the cask with a special auger-like tool.

Liquids

Liquids should be thoroughly stirred or the barrels tumbled before opening, to ensure that no separation into distinct phases has occurred.

14.3   PIGMENT TESTING

Quality control of pigments is particularly difficult as a single pigment can be used in a wide variety of vehicle systems to give inks for many different end uses. Hence, the testing procedure must be selective and it can logically be divided into three sections. Firstly, limited testing can be carried out on the dry pigment itself to ensure the batch meets the specification for shade, particle size, oil absorption, moisture and volatile content. The second and possibly more important consideration is to compare the ink making characteristics of the batch against the master standard in the ink maker's own vehicle systems. Finally, provided the batch meets the specification to date, prints from the sample and standard inks are checked for their resistance properties.

Tests on the dry pigment

Particle size

Since the pigment particle size will affect such fundamental properties as strength, hue, rheology and stability of the finished ink, it would seem desirable to check particle size or at least the average size distribution, as part of the quality control procedure. However, in the size range of most pigments (less than 1 μm) the majority of the standard techniques for sizing (e.g. gravitational sedimentation, optical microscopy) are slow and inaccurate. Only two methods are suitable for particle size measurement in this range — electron microscopy and centrifugal sedimentation. Electron microscopy is an expensive method and requires an experienced operator, but is capable of producing very accurate readings (Figs 14.1 and 14.2). The disc centrifuge (Fig. 14.3) is perhaps the preferred technique and results compare favourably with those obtained using electron microscopy. However, neither method is simple and quick enough to be used as a standard quality control procedure and the ink-maker therefore places greater emphasis on the dispersion of the pigment particles in the finished ink (see later). For further reading on particle size determination by centrifugal sedimentation the reader should consult refs 1 and 2.

Oil absorption

The oil absorption of a pigment will depend on:
(1) The chemical nature of the pigment.
(2) Particle size and size distribution.
(3) Particle shape.

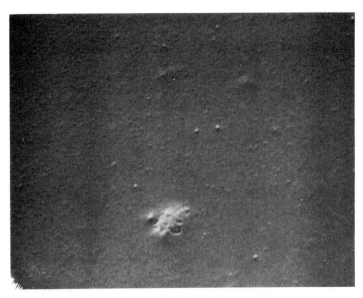

Fig. 14.1 Electron micrograph (x 5000) of a well-dispersed pigment

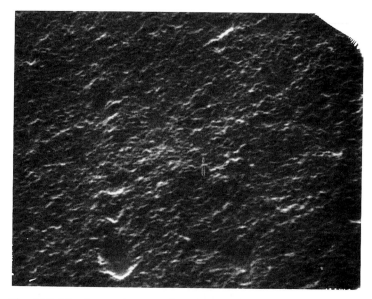

Fig. 14.2 Electron micrograph (x 5000) of a flocculated pigment

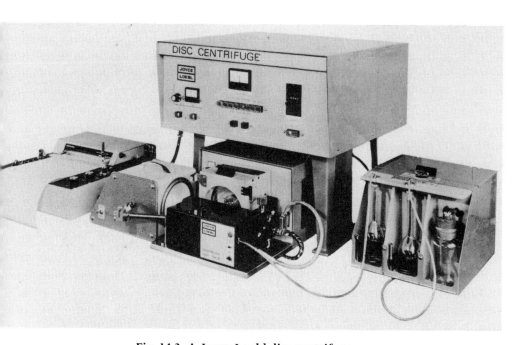

Fig. 14.3 A Joyce Loebl disc centrifuge

(4) Surface characteristics of the sample, such as those obtained by treatment to improve wettability or even simply due to absorbed moisture.

If the oil absorption of the pigment varies from batch to batch, then formulation changes will have to be made to the inks, and clearly this is undesirable. Since the factors affecting oil absorption cannot easily be measured, the test is reduced to a purely empirical comparison between the batch and master standard. The actual test method is given in BS 3483 : Part B7 : 1974, the result being quoted as the mass (or volume) of oil required to wet 100 g of pigment. Alternatively, the sample and standard can be dispersed in a measured amount of oil and the relative consistencies of the two samples can then be compared.

Moisture and volatile content

Absorbed moisture may affect the ease of dispersion of some pigments, and this should be determined on each batch of pigment according to the method in BS 3483 : Part B6 : 1982. A temperature of 105 °C is chosen for this test, as at higher temperatures volatile matter may be lost which is an essential part of the pigment: This is especially true of carbon black and pigments on an aluminium hydrate base.

Ink-making characteristics

All pigments should be checked wherever possible in the vehicle systems in which they will be used in production, but in certain cases where the pigment is used in several different types of inks, a lengthy testing procedure will result and a selective range of formulations must then be chosen. This should include oil and liquid ink formulations (where appropriate), a vehicle system which is used in bulk with the test pigment and one which is a more difficult medium for ease of dispersion. The method of dispersion in the laboratory should:

(1) Simulate production milling methods.
(2) Be rapid and comparative.
(3) Utilise small quantities of pigment, so as to retain a single master standard for testing as many batches as possible.

The first test must be to ensure that the batch of pigment disperses at least as readily as the standard and to an equivalent degree, so that costly production milling time will not be lost. Dispersion is measured on a fineness of grind gauge (Fig. 14.4). This is a hardened steel block containing a shallow groove which is graduated in depth from zero to, for example, 25 or 50 μm. A sample of ink is placed at the deep end, drawn down with a steel blade towards the shallower part, and the dispersion is taken as the point at which the continuous ink film breaks down across the groove. However, care must be taken in interpreting the result if discontinuity in the ink film occurs higher up on the gauge due to the odd oversize particle. It is then important to determine whether these scratches are caused by undispersed pigment particles, or by, for example, wax or dried ink, and this can be confirmed by examination of a smear of the ink under a microscope.

Fig. 14.4 A fineness of grind gauge

Once the correct dispersion level has been attained, the trial ink can be compared with the standard for strength, hue, gloss and flow (or consistency). Firstly, the flow is checked as it is impossible to obtain meaningful comparative prints unless the batch and standard have very similar consistency. Liquid inks having near Newtonian flow properties can be compared by viscosity measurements using any of the standard techniques (e.g. Ford, Sheen and Zahn cups). Oil inks, however, are more difficult to assess by instrumental measurement and for the present purpose are probably best compared by checking their consistency with the aid of a palette knife, and by a comparison of their relative flow properties on an inclined glass plate.

The strength of the trial pigment is then assessed on a drawdown at full strength and at reduced strength as a 'bleach'. The drawdown is carried out by placing small quantities of the standard and test inks about one inch apart on the top of a securely held pad of paper, and then firmly scraping them towards the operator with, for example, a 3-inch (75-mm) wide blade. Initially, the blade should be held nearly vertical to reproduce a film thickness comparable to a printed film weight and then finally the blade should be brought almost parallel to the pad to give a thick film weight — the masstone of the ink. The drawdown should be carried out on a semi-absorbent paper on which a broad black band (19 mm wide) has been previously printed, so the opacity (or transparency) of the batch and standard pigments can also be compared.

The bleach technique involves dilution by mixing (not grinding) a coloured ink with a much larger quantity of opaque white, so as to eliminate any minor differences of gloss and hue. In the case of yellow inks, the eye cannot differentiate small strength differences, and so sufficient blue is added to the bleach to bring the hue into the blue-green region.

Hue can be measured by instrumental techniques (see Chapter 3), but normally a visual comparison of a drawdown and print is sufficiently accurate. The undertone and top tone of the inks can be assessed using the thin film of the drawdown and the thicker masstone respectively. Gloss can also be checked on the drawdown, but more accurately by examination of a print of the batch versus standard on an appropriate stock.

Resistance tests

Provided the pigment matches the standard on the tests conducted to date, the appropriate resistance checks are carried out. Many of the characteristic properties of a pigment are a function of its chemical nature, and hence are unlikely to vary from batch to batch, unless an impurity is present. Pigments are, however, coated to provide easier dispersion in different vehicle systems and this can give rise to batch variation, especially when the pigment is obtained from an alternative supplier.

All the resistance tests are carried out on prints of the batch and standard inks, rather than dry pigments themselves, but it is important that an appropriate vehicle system be chosen; for example, it is pointless to try and check the heat resistance of a pigment when a low softening point resin is present.

Lightfastness

As the lightfastness of a pigment is a function of its chemical nature, it is only necessary to carry out random spot checks, but attention should be paid here to pigments used in inks where lightfastness is an important criterion, e.g. wallpaper inks.

Pigments will show a higher lightfastness at full strength and therefore it is important that a realistic strength print be submitted for test. For example, a PMTA pigment should rarely be tested at full strength, since its main application will be as a toner pigment. As pigments for use on wallpapers could often be used in tints, it is advisable to test these at both full and tint strengths. Prints should be prepared on a good quality paper which does not yellow on exposure to light or contain artificial brightening aids which are destroyed by UV radiation.

The exposure of prints to daylight is too lengthy a procedure to be suitable and the use of an artifical source is therefore adopted. The only light source to have an emission spectrum close to that of daylight is the xenon arc and provided a filter is used to reduce the amount of UV radiation, a result comparable to the daylight exposure is obtained in a much shorter time (Fig. 14.5). Other light sources such as the carbon arc and mercury vapour lamp are not recommended as the results bear little comparison to those obtained under daylight conditions.

**Fig. 14.5 A xenotester for checking lightfastness
(courtesy of Heraeus GmbH)**

Another factor which greatly affects the lightfastness figure is the effective humidity which is dependent on the relative humidity and temperature of the surrounding air and the surface temperature of the pattern. The average effective humidity value for daylight exposure in Britain is 45% and therefore the test apparatus should be capable of approximating to this figure.

In order to numerically classify the lightfastness results obtained, the prints are compared with a set of eight pieces of wool, each dyed with different blue dye which has been carefully graded to fade after a set exposure to light. This is known as the Blue Wool Scale, and the scale ranges from one (poor lightfastness) to eight (excellent) according to the number of the wool dyes which have faded under the same exposure as the test method. The procedure for mounting and assessing samples is given in BS 4321 : 1969, but it is worthy of note that the term 'fading' includes not only a reduction in strength but also changes in hue, brightness, darkening or any combination of these factors.

Heat resistance

Heat resistance of pigments is a critical factor in the areas of tinprinting inks (and other metal coatings) and wallpaper inks, which will be subjected to hot embossing. The pigments are simply tested by placing prints in a suitably heated oven for a set period of time, the actual conditions depending on the end usage of the print. The prints are then compared with an unheated sample for changes in strength, hue and brightness. Again, prints should be at a realistic strength, as heatfastness, like lightfastness, will vary from full to tint strength.

Soap and detergent resistance

Pigments to be used on soap and detergent wraps should be checked to ensure no impurities are present which are liable to cause bleed problems. Although test methods for both soap and detergent resistance are given (BS 4321 Test methods 6 and 7 respectively) the manufacturers of these products have their own test methods which are generally more severe than the British Standard tests.

A typical soap resistance test as provided by a leading manufacturer, is as follows. A 25% soap gel is made up and allowed to set. The gel is moistened and a print of the pigment to be tested is first soaked in water to prevent curling (1 minute for paper and 5 minutes for board). The print is then placed face down on the gel and left for 15 minutes, after which time it is examined. There should be no sign of bleeding at this stage. The print is then replaced and left for three hours before being further examined. After this period there should be preferably no bleed, although a slight bleed could be considered acceptable.

Testing for detergent resistance is becoming more complex with the advent of enzyme and 'solvent' powders. Provided a pigment passes the above soap test, it is unlikely to fail with the standard soap powders, and therefore detergent resistance should be assessed with an enzymic powder. (*Note:* To date no standard test has been published for resistance to the 'solvent detergents'.)

A typical detergent resistance test is as follows: A 5% solution is made up using the appropriate enzymic detergent in water. Two filter papers are soaked in the solution, one is placed on a glass slab and the print to be tested is placed face down on filter paper. The second filter paper is then laid on top and a further glass slab provides a sandwich effect. The print is then examined after three hours and although ideally there should be no bleed, a slight bleed could be considered acceptable.

The disadvantage in interpretation of these test methods is that the degree of staining is only expressed in vague terms such as 'slight bleed', whereas both the British Standard tests (numbers 6 and 7) quote a numerical means of assessment of staining by use of the grey scale (BS 2663C).

Alkali resistance

This test can be conducted either by streaking dilute sodium hydroxide solutions on to a print of the pigment (e.g. 2, 3 and 5%) and examining the dried streaks for discoloration, or as in the British Standard test (BS 4321 Test method 5) by sandwiching the print between filter papers previously soaked in a dilute alkaline solution. In this case, the filter papers are dried after a set contact period, and then the degree of staining is assessed with the aid of a grey scale.

Acid resistance

Acid resistance can be carried out in a manner similar to BS 4321 Test method 5 above, but replacing the alkaline solution with either dilute hydrochloric acid or dilute acetic acid. The latter reagent is normally used where the print could come into contact with vinegar or pickles.

Wax resistance

Wax resistance is essential for inks on waxed paper wrappers, ice cream cartons, certain vending machine cups, and similar applications where the substrate is made more water resistant by immersion in molten wax (usually paraffin wax). The standard test (BS 4321 Test method 10) involves immersion of a print of the pigment in the molten wax for five minutes, after which the wax is allowed to run over the margin surrounding the print. The margin and the wax in the basin are then examined for discoloration.

Plasticiser bleed

This test is extremely important for pigments that will be used in inks to be printed on plasticised PVC such as for vinyl wallpaper and shrink wrap applications. Prints of an ink made from the pigment and a standard ink on plasticised PVC are sandwiched between plain sheets of the unprinted material and placed between glass plates in an oven at 45 °C for 3 days. A small weight is often placed on top of the sandwich to increase contact pressure. On examination of the unprinted substrate adjacent to the print there should be no evidence of colour bleed at the end of the test period.

Deep freeze resistance

This test is becoming increasingly important as more and more products are being stored under deep freeze conditions. On carton board, the vehicle system (or overprint varnish) will normally provide adequate deep freeze resistance and testing of the pigment is not necessary. However, for polyethylene printing the deep freeze resistance of the finished pack is a function not only of the vehicle (usually polyamide), but also of the pigment as well. The deep freeze resistance of the pigment in these cases is not so much a function of its chemical nature, but of the pigment coating and therefore quality control of each batch of pigment is essential for this application. The tests are carried out on prints made from the pigment under test.

The severest condition met by the print is not when the pack is actually under the deep freeze conditions, but during the stage of thawing on removal from the refrigerator, and hence the test normally conducted is one of water immersion.

A typical test is as follows: The print is immersed in water for a period of at least 1 hour and a maximum of 5 hours. The print is then flexed between the knuckles ten times and should show no more than five percent removal. A second portion of the print is then blotted dry and should show no more than 5% removal when checked by the Scotch tape test.

14.4 CHIPS AND PRE-DISPERSIONS

In chips and pre-dispersions the physical process of dispersing the pigment in the resin system has already taken place and in theory an ink can be made simply by solubilisation of the resin component in the solvent system. However, in production dissolution usually takes place on a high shear mixer and therefore the laboratory technique should simulate this method.

A useful comparative method of testing such pre-dispersions is to solubilise the pigment 'chip' in the solvent by shaking in a sealed container on a Red Devil paint mixer for a set period. Finally, if the dispersion is less than adequate both batch and standard can be further processed for a short period on a high shear laboratory mixer such as a Silverson blender. A comparison between batch and standard inks can then be made according to the test methods given in the section on pigment dispersion (14.3). The method of first shaking in a sealed container and then mixing under shear enables the tester to ensure that dissolution takes place without excessive solvent loss as well as providing a technique for assessment of relative ease of dispersion of the standard and batch.

14.5 DYE TESTING

Solvent dyes

Solvent dyes are finding increased use because of their brightness and transparency on foil and metallised substrates. They are tested by ensuring that the dye is soluble to the required concentration in the appropriate solvents. On standing for a further period (usually up to 48 hours) no precipitation should occur. The strength, hue and brightness of the resulting solution may be determined on a colorimeter or by making up a suitable ink and comparing with a master standard in the normal manner. Lightfastness and bleed tests can then be carried out on a print as described earlier in this chapter.

Basic dyes

These dyes, have poor fastness properties, high tinctorial strength, and find their greatest use as toners in flexographic inks, after laking with, for example, tannic acid. They should be checked for solubility and resistance properties against a master standard in an appropriate formulation.

Acid, reactive and direct dyes

These are little used in the printing ink industry, but do find specific end uses, such as in security inks and inks for printing tissues and paper kitchen towels. Again these are tested by comparison with a master standard in a typical formulation.

When these dyes are to be used for printing tissues, etc. particular attention must be paid to the properties of the dried print (e.g. soap bleed resistance, ease of bleachability) since the dyes play an important part in determining whether or not the inks will meet the customer's specification.

14.6 RESINS

There are very few classes of ink which do not contain a resin of one form or another, thus before discussing the importance of testing and control procedures an appreciation of the many forms a resin may take is essential. Encompassed within the name 'resin' is a wide range of differing polymeric chemical entities ranging from low molecular weight oligomers to high molecular weight complex polymers and from low viscosity mobile liquids to plastic solids. Beyond these boundaries, resins may also be chemically reactive or comparatively inert, thermally stable or sensitive to light and heat.

The properties of resins selected by the ink-maker are governed by their chemical and physical nature, which in turn are controlled by the manufacturing procedure and the purity of the raw materials. Both of these factors may be influenced at any one time for any number of reasons, the economic climate and changes in health and safety legislation to name only two. To ensure that a consistent end product is manufactured, suitable control procedures must be applied to the materials used in manufacture. The simplest and most useful form of resin testing would be to incorporate it into an ink and then monitor the desired properties of that ink. This would be very costly in terms of manpower and time if every resin used by an ink-maker had to undergo the same testing. It is therefore desirable to apply empirical tests which are rapid and will convey sufficient information to determine whether a resin is acceptable or not.

The following methods are typical of those employed by an ink-maker to determine whether incoming materials are acceptable against a proven standard.

Acid value

The stability, printability, flow, gloss and pigment wetting are just some of the properties which are affected by the acid value of a resin. Changes in acid value may not always be detrimental, a higher acid value resin in an oleoresinous lithographic system may improve the water take-up and thus the press performance of the ink. However, a reproducible product is essential and therefore this property must be controlled within set parameters.

The acid value of a resin may be determined by the direct titration of a known weight of the resin dissolved in alcohol or a toluene/alcohol mixture. This may be titrated against a standard deci-normal alkali solution to a colorimetric end point using a suitable indicator. Typically phenolphthalein is selected but if the resin solution is strongly coloured more appropriate indicators may be used. Conventionally acid value is expressed in terms of the number of milligrams of KOH required to neutralise 1 g of resin.

Hydroxyl value

Residual hydroxyl groups in a resin may influence different factors according to the end use of that resin. Hydroxyl groups may affect resin and varnish viscosity via hydrogen bonding, solubility and solvent release characteristics may be influenced for the same reason. The water resistance of a system may be impaired if the value is too high, similarly reactive systems may not 'cure' properly if the value is too low. Pigment wetting and thus gloss, transparency and flow may be adversely affected and 'can' stability may suffer if the hydroxyl value is outside of specification.

The hydroxyl value is determined by refluxing a weighed sample of resin with 10% acetic anhydride in pyridine, when sufficient time has been allowed for the hydroxyl groups to be converted to acetate the residual anhydride is hydrolysed by the addition of water and titrated with standard alkali. The hydroxyl value is expressed as the number of milligrams of KOH required to neutralise the acetic acid capable of combining with 1 g of the resin. Other less volatile anhydrides have been successfully used (stearic anhydride) however, acetic anyhdride despite its volatility is the accepted one.

Solubility

Resins are often formed into a varnish prior to incorporation in an ink and this may itself be used as a quality control test on the resin. This is not always so and in the case of plastisols and some aqueous systems the resin is not in solution in the finished ink at all. In plastisols the resin forms a plastic solution in plasticiser only when the final print is heated. The complex interactions which occur within ink systems are not as simple as this and many different solubility factors come into play, solvent on solvent, solvent on resin, resin on solvent, plasticiser on resin to mention a few of the many possibilities. If the resin is required to be compatible and stable within a defined system it is relatively simple to test this characteristic. Solubility is a good means of testing a resin's performance within a system and changes in the solubility of a resin are an excellent indication that there is some significant batch to batch variation. Differences in the solubility of resins manifest themselves mainly in a change in the solution viscosity. Significant variations in the viscosity of a resin solution indicate that either the chemical or physical nature of the resin is different. As the molecular weight of a resin increases so does its solution viscosity. Similarly, subtle changes in the solvent being used may have significant effects upon the solution characteristics being measured. It is therefore essential to either keep a stock of standard test solution or compare each new batch of resin with a retained sample of standard material under identical conditions.

Melting range

The melting point of a resin is also a characteristic of its molecular weight, thus batch to batch changes in melting point indicate a variation in molecular weight distribution which will directly influence solution viscosity, film hardness and gloss. The melting range of a resin is also a very useful test to apply when resins are being considered for certain applications. For example a resin of too high a melting range may have no application in heatsealing inks, similarly if the melt range is too low, marking due to resoftening in reverse printing may prove a problem.

There are two common methods for determining melting ranges. The standard method is the ball and ring method in which a steel ball is placed upon a cast resin film supported in a ring cradle which in turn is then immersed and heated in a liquid bath (Fig. 14.6). As the temperature increases, the resin softens and the ball eventually passes through the film. The ball and ring softening point is the temperature at which the ball is clear of the film according to BS 2782 : 1970. The other common method is the capillary melting point whereby the melting point of powdered resin contained in a glass capillary tube is determined. The tube is heated at a constant rate in a metal heating block, monitored by thermometer and viewed through a lens. Reproducibility of this method for resins is not good and results tend to be lower than for the ball and ring method.

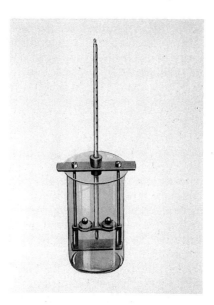

Fig. 14.6 Ball and ring apparatus for determination of softening point

Colour

Gross colour variations although immediately obvious may not always have a significant effect upon the finished product. Visual colour measurements of a resin are far more realistic when carried out in solution and instrumental techniques such as a colorimeter or standard colour guides are normally employed.

One of the most important aspects of colour testing is the indication of either age or overheating that may have caused the darkening of a resin. If the colour of a resin is acceptable but outside normal standards, it would be advisable to test thoroughly the other resin properties discussed above.

Infra-red spectroscopy

This is an instrumental technique which is employed to identify the chemical nature of a compound. Further details of its use are to be found later in the section on analysis; suffice to say that two identical resins should give identical IR spectra.

The IR specturm obtained when compared with a standard can be used to obtain information on the following parameters, acid value, hydroxyl value, amine content, hydrocarbon content, aromaticity and many other fundamental properties of the resin. An experienced operator can obtain this type of information in as little as 15 minutes which makes it a very good quality control procedure.

Gel permeation chromatography (GPC)

It is worth briefly mentioning this instrumental technique at this stage although more detailed information may again be found in the section on analysis. It is used to generate numerical data relating to the molecular weight of a resin. The molecular weight distribution of a resin is fundamental to the majority of its physical properties. Two batches of the same resin with identical GPC curves are certain to have virtually identical physical properties. Once again GPC is a rapid technique requiring in some instances as little as 15 minutes.

14.7 VARNISHES AND OILS

Varnishes in this category may be simple solutions of a single polymer or they may be complex mixtures such as the 'cooked' varnishes encountered in paste ink technology. Oils may be pure or partly polymerised vegetable oils or animal and mineral derived products.

Often the most successful test method for these materials is to fabricate them into a finished ink. This, however, is time consuming and therefore would not generally be a realistic quality control measure. Good quality control practice should make use of a sufficient number of reproducible, rapid tests to ensure that the product is the correct material of a suitable grade. Invariably economic pressures force the ink-maker to test when necessary and thus to rely upon his supplier to submit raw materials of an acceptable quality. Improved manufacturing technology and a competitive supply market has made this possible in many cases. This is never a wholly acceptable situation, the ideal procedure would involve comparing an incoming raw material with a retained sample of standard material.

General tests

These would include the following:
— Infrared spectroscopy
— Refractive index
— Colour
— Acid value
— Iodine value
— Diene value (i.e. degree of unsaturation)
— Viscosity
— Saponification value
— Unsaponifiable matter
— Hydroxyl value
— Water content

Infrared spectroscopy

Once again this useful instrumental technique may be used to establish that the correct chemical class of compound has been supplied and that there are no significant variations from the standard. Useful information concerning the level of unsaturates in, for example, varnishes containing tung oil, can be gleaned from the spectrum by examining the absorption at 985–995 cm^{-1} due to the ethylenic unsaturation in eleostearic acid.

Refractive index

At the interface between two media i.e. glass/liquid, a beam of light is both reflected and refracted in a direction which differs from both the incident and the reflected beams. The relationship between the refracted and the incident beam is known as the refractive index.

The refractive index (RI) of an oil should be a reproducible value. Any significant variation in RI would indicate that the product is not up to standard (Fig. 14.7).

Colour

The colour of an oil or varnish may influence the final colour of an ink or print but generally the colour is used as an indication of consistency of manufacture. For comparative purposes one can use standard colour tubes or colour comparators such as the Tintometer or Gardner tubes. The colour of an oil or varnish, if it has darkened, may indicate that the sample has deteriorated and on oxidation drying oils and varnishes may no longer have the same drying characteristics.

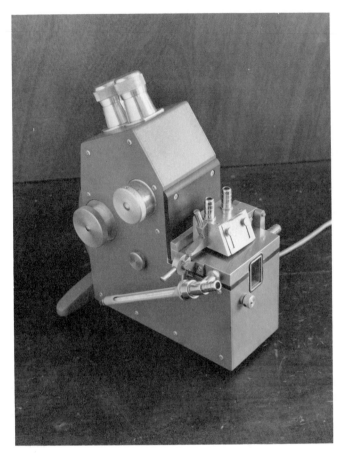

Fig. 14.7 Abbé refractometer

Acid value

This has been previously discussed (see p. 711); suffice it to say that the method applied to resins is equally applicable to oils and varnishes.

Iodine value

The drying potential of a vegetable oil is related to the level of ethylenic unsaturation present in that oil. This may be determined by reacting the material with free Iodine. The reaction is not always directly proportional but is reproducible for one batch of oil to another and thus may be used to directly compare an incoming batch with a standard sample. A typical example of this is the presence of conjugated unsaturation in tung oil; all the ethylenic groups not all react with iodine and thus an experimental value of 160 is obtained which is significantly less than the theoretical value. A typical procedure for the determination of iodine values is the Wijs method which may be found in many organic chemistry textbooks.

Diene values

This once again is a method for examining the level of unsaturation of an oil and involves reacting a known quantity of oil with an excess of maleic anhydride. The excess anhydride is hydrolysed to maleic acid and back titrated with alkali.

Viscosity

Viscosity according to Newton is a ratio of shear stress applied to a liquid versus the rate of shear and in an ideal system should be constant. Newton's interpretation of ideal is rarely the same as that of the ink-maker who tailors the viscosity/time characteristics of his ink to the varying shear forces encountered in different printing applications. Figure 14.8 shows the relationship of shear stress and shear rate in a liquid which is 'Newtonian'. Deviations from this encountered in thixotropic systems would exhibit a curve as shown in Fig. 14.9. Figures 14.10 and 14.11 are typical viscosity/time plots for a thixotropic system. A suitable measurement technique must be chosen which will reflect the characteristics of the oil or varnish.

As the viscosities of oils and varnishes are temperature dependent it is important that this testing is carried out under controlled conditions.

There are many types of viscosity measurement equipment available and care must be taken when selecting which to use. Efflux or orifice viscometers such as the Ford, Shell, Zahn and Sheen cups are useful when 'Newtonian' flow behaviour is expected. Extrusion rheometers and more commonly rotational viscometers such as the Brookfield or Stormer viscometers may be used to monitor 'non-Newtonian' behaviour.

Other methods of viscosity measurement include: cone and plate viscometers, falling rod viscometers (Fig. 14.12), falling sphere visco-

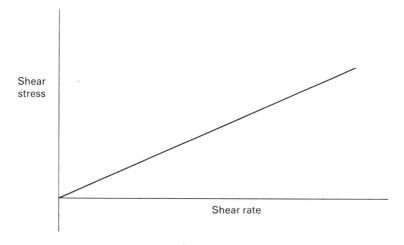

Fig. 14.8

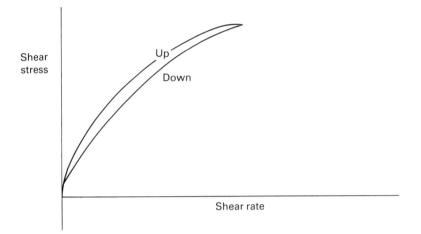

Fig. 14.9

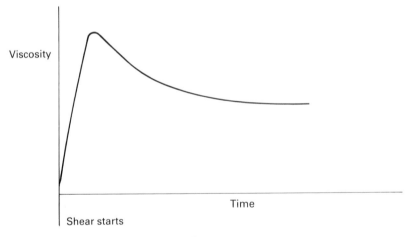

Fig. 14.10

meters and bubble tubes. The latter are commonly employed in varnish manufacture.

Saponification value

Vegetable oils consist of triglycerides and certain waxes and have high saponification values which vary in accordance with the molecular weights of the acids and alcohols involved. Mineral oils and materials which are not esters have a value of zero. The usefulness of this test is in assessing the purity of a sample.

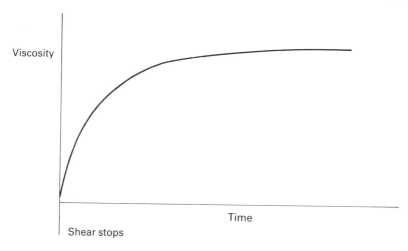

Fig. 14.11

The test is carried out by refluxing the sample with an excess of alcoholic KOH and then titrating the excess alkali with standard acid. The result is expressed as milligrams of KOH per gram of oil.

Unsaponifiable matter

The sample is saponified with caustic alkali and the alkaline solution extracted with diethyl ether. The ethereal solution is evaporated and the residue dried at 80 °C. The residue is the unsaponifiable matter.

This value may be useful in determining whether a vegetable oil has been contaminated with or contains an unusual excess of materials other than triglycerides.

Hydroxyl value

It is sometimes useful to be able to measure the hydroxyl content of a substance, e.g. to measure the residual castor oil in a batch of dehydrated castor oil. Similarly a varnish where the presence of hydroxyl groups take place in a reaction in a 'curing' system should be tested. The procedure for determining the hydroxyl value of an oil or varnish is the same as previously described on p. 712.

Water content

The presence of water in oils and varnishes may lead to incompatibility, poor pigment dispersion, loss of reactivity or 'bumping' and 'spitting' during the cooking process.

The level of water may be determined quantitatively by azeotropic distillation with xylene in a Dean and Stark apparatus. Alternatively a more accurate titrimetric method according to the Karl Fischer procedure may be utilised. It should be noted that the presence of aldehydes and ketones will give erroneous results with the Karl Fischer method.

Fig. 14.12 Falling rod viscometer with temperature controlled water-bath (courtesy of Betta-Tech Equipment Ltd)

14.8 SOLVENTS

Solvents are all too often considered to be discrete chemical compounds yet even highly refined single compound solvents contain low levels of impurities. Solvents commonly used in the printing ink industry are invariably a technical grade and therefore contain residual impurities resulting from their method of manufacture. Most solvents have a characteristic odour which may be used as a first indication of the nature of the solvent. Quality control of solvents involves a blend of modern instrumental techniques such as IR spectroscopy and gas liquid chromatography (GLC) with traditional testing procedures.

Infrared spectroscopy may be used to give a chemical 'fingerprint' of the solvent under test. No two different compounds will give exactly the same spectrum. The IR spectrum of the test sample should be exactly superimposable over that of the standard. Variations from this would indicate the presence of impurites. It should be noted that solvents of large extinction coefficient (absorbance in the IR region) may well conceal significant levels of low extinction coefficient impurities.

Gas liquid chromatography, a vapour phase separation technique, may be successfully used to identify the test material, its boiling range and its purity. Spurious chromatographic peaks obtained from a single compound solvent may automatically be labelled as impurities. It is not often possible to do this for solvents derived from petroleum as the chemical composition will vary with the raw material supply.

Petroleum solvents may be usefully 'fingerprinted' using GLC and significant changes in composition or aromaticity may be detected.

High performance liquid chromatography (HPLC) and column chromatography are also finding application in solvent control. A particular application of these techniques is the determination of aromatic content.

Boiling range

The boiling range of a solvent may be cited as the literature value provided that the level and nature of impurities is within the acceptable level. The boiling range may be readily determined by distilling a known volume of the solvent, recording the initial boiling point, dry point and the temperatures at which specific volumes of the solvent have distilled over. This method is covered by BS 658 : 1962.

Simulated distillation by GLC may also be used especially when utilising modern very high performance capillary columns with computerised data handling.

Relative density

The relative density (or specific gravity (SG)) of a liquid is the ratio of the density of the liquid and the density of air free distilled water at the same temperature. The density of a solvent is commonly determined using a hydrometer or by weighing using a calibrated density bottle. Useful information about the purity of a solvent can be obtained from density information. Aliphatic and aromatic solvents of the same boiling range may be distinguished as the aliphatic hydrocarbons normally have a lower density.

Flash point

The flash point is defined as the lowest temperature at which the vapour above a liquid will ignite when brought into contact with a flame or spark. The flash points of the more volatile solvents encountered in flexographic and gravure ink technology are usually very low. Ethyl acetate has a flash point (closed cup) of $-4°C$. The hazards involved in handling, transporting and storage of flammable solvents have led to definitive legislation covering all aspects of exposure, handling, transporting, packaging and labelling. There are three approved methods for flash point determination using the Abel apparatus (Fig. 14.13), Pensky-Martens

Fig. 14.13 Abel apparatus for measurement of flash point

apparatus and the Cleveland open cup. The open cup method is preferred for solvents of flash point greater than 80 °C. The details of these methods may be found in BS 2000 : Part 4 : 1983, IP 34, 35 and 36.

The flash point of a mixture of flammable solvents can usually be taken as that of the lowest. If any doubt exists the flash point of a mixture should be determined according to one of the approved methods.

There has been a review of current legislation to allow for mixtures which have a flash point when measured using one of the approved methods, but which do not support combustion. Any solvent with a flash point below 21 °C is deemed **highly flammable** whether it supports combustion or not. Any solvent with a flash point between 21 and 55 °C and which supports combustion at 55 °C is deemed to be **flammable**.

Typically an aqueous emulsion containing a low percentage of isopropanol may have a flash point between 21 and 55 °C but would not support combustion. Such a mixture does not require a 'flammable' label.

Aromatic content

Commercial grades of petroleum hydrocarbons contain varying proportions of aliphatic, naphthenic and aromatic compounds. The levels of these vary according to the source of the raw materials and the methods of refinement. The presence of aromatic compounds leads to greater solvency, a tendency to swell the compounds used in producing litho blankets, increased odour and toxicity. The aromatic content is normally measured by the Aniline point or Kauri-butanol value.

Aniline point

The aniline point of a solvent is the lowest temperature at which a clear solution is obtained with an equal volume of aniline. High aromatic solvents are miscible with aniline at lower temperatures than more aliphatic solvents. Equal volumes of the solvent and pure aniline are mixed in a test tube. The mixture is warmed until it clears and is then allowed to cool. The temperature at which it becomes cloudy is the aniline point. This method requires standardisation with solvents of known aromatic content.

Kauri-butanol value

This test consists of adding the solvent from a burette to a standard solution of kauri gum in *n*-butanol until the mixture become turbid.

Other methods for determining the aromaticity of solvents include reduction of the volume of the solvent when reacted with concentrated sulphuric acid, bromine values, column chromatography and high performance liquid chromatography. Some of these are covered by standard methods IP 128 and IP 156.

Water content

The presence of water in solvents invariably leads to problems such as poor adhesion, lack of gloss and flow in inks. The water content may be determined by the Karl Fischer method. This technique will readily detect less than 0.1% of water. The Karl Fischer titration employs the following reaction:

$$I_2 + SO_2 + CH_3OH + H_2O \rightarrow 2HI + CH_3H.SO_4$$

The end point is determined electrometrically.

Water in solvents may also be determined chromatographically using GLC equipped with a katharometer detector.

Refractive index

When a ray of light is incident at an interface between two media, i.e. glass/liquid, air/liquid, observation reveals that some of the light is reflected from the surface while the rest of the light travels in a new direction. On account of the change in direction the light is said to be 'refracted'.

For two given media, the relationship between the incident light and the refracted light known as the 'refractive index' is a constant. The magnitude of this constant is dependent upon the colour of the light and is commonly specified as that obtained for yellow light from sodium vapour.

The RI for a liquid as measured by a defined method may thus be used as an indication of its purity. A typical device for measuring RI is the Abbe refractometer. The liquid sample is squeezed into a thin film between two prisms arranged in such a manner that the RI may be read directly from a scale.

It is important that measurements made with this instrument are at the same temperature each time and that the instrument is properly calibrated. The following example shows how useful RI can be in assessing the purity of an incoming raw material.

	RI
Isopropanol	1.380
Ethanol	1.361
25% Ethanol in isopropanol	1.378

Using RI is a very rapid, reproducible technique ideally suited to quality control.

Sulphur content

Petroleum products may be contaminated with sulphur compounds from two routes. Sulphur compounds occur naturally in crude oils and may not be removed by distillation or purification, and may also be introduced when aromatics are reduced by treatment with concentrated sulphuric acid. Sulphur compounds lead to tarnishing and odour problems.

The usual test for sulphur consists of immersing a polished copper strip in a known volume of the solvent at elevated temperature. A quantitative approach is described in IP 155.

Compounds of this nature may also be detected by GLC using a flame photometric detector which is very sensitive to sulphur compounds.

Odour

Odour, although a subjective parameter, is becoming of increasing concern to the ink-maker, particularly in the field of food packaging.

Very simple odour tests such as allowing solvent to evaporate slowly from a filter paper and then examining for residual odour may be employed.

Odour tests are normally carried out using a panel of testers and comparisons are made against proven standard materials using a simple numerical scoring system.

14.9   RADIATION CURING PRODUCTS

Products which are polymerisable by the action of UV light or by the bombardment of electrons in the case of EB curing are by their very nature reactive and unstable. Any quality control procedure or testing method must be concerned with the unique properties of these compounds. The compounds used in radiation curing inks fall into four categories:

(1) Diluents — these consist of reactive monomers generally based on the reaction product of a polyol, or alkoxylated polyol with an ethylenically unsaturated carboxylic acid.

(2) Resins — this category consists of reactive polymers of the acrylate type varying in molecular weight from viscous liquids to solid resins.

(3) Photoinitiators — compounds which form free radicals when exposed to UV radiation.

(4) Synergists — a class of compounds, generally amines, which promote the action of photoinitiators.

Diluents and resins

The role of the diluent and resin is to produce a curable film via a printing process, therefore all the usual tests for colour, viscosity, curing speed and odour may be applied.

Some reactive monomers are a blend of isomers whose reactivity may vary. It may therefore be of use to examine this distribution using gas chromatography or HPLC.

Residual acrylic acid in either the monomer or polymer can affect the curing rate, odour, shelf-life and skin irritancy of the product and the acid value should therefore be determined.

The unstable nature of these reactive products require that the shelf-life should be examined. Stabilisers such as methoxyphenol are incorporated into diluent monomers to prevent premature reaction leading to viscosity increases, gelation and darkening and it may be advantageous to monitor the levels of these. Inhibitors can be readily determined using HPLC. If instrumental techniques are not available the conventional method for examining shelf-life by accelerated testing may be applied. Accelerated testing for stability may be done by placing samples of a standard material and the test material in an oven at temperatures ranging from 30 to 60°C and then comparing viscosities and colour after suitable time periods.

Photoinitiators and synergists

These classes are, with the exception of amine acrylates, invariably pure compounds and as such may be tested using analytical methods. Infra-red spectroscopy can be used to generate a spectrum of the sample which may then be compared with that of the standard. The majority of these compounds may also be examined using GLC or HPLC. All of these techniques can be used to identify the compound and its purity. The only other applicable test is to fabricate the initiator into a standard vehicle and assess its performance under fixed curing conditions.

14.10 MISCELLANEOUS MATERIALS

Driers

Driers are oil-soluble metallic salts or soaps in which the metal is rendered soluble by reaction with a long chain acid (e.g. octoates) or complex mixture of acids (e.g. naphthenates). Variation from batch to batch can therefore be associated with the organic portion or with the type and quantity of metal ion.

Comparison of the batch against a known standard using IR spectroscopy will characterise the organic portion of the drier. The type and quantity of metal can be accurately determined by either atomic absorption (AA) or HPLC. All these techniques will be discussed in the analysis section of this chapter (14.16).

Plasticisers and pure chemicals

These materials can be simply tested by IR spectroscopy. The spectrum of the test material should be exactly superimposable on that of a standard.

Waxes

Waxes fall into two main categories for raw material testing purposes. The simple chemical types such as fatty acid amines can be treated as pure chemicals and analysed by IR spectroscopy as indicated in the previous section. The more complex polymeric waxes (e.g. paraffin and polyethylene waxes) cannot be identified by this analytical method and more traditional methods have to be employed as shown below:

(1) Melting point determination similar to that used for resins.
(2) Solubility characteristics at room and elevated temperatures.
(3) Empirical end-product testing.

Others

Many miscellaneous items are often included in small amounts in printing inks for specific uses, e.g. pigment-wetting aids, flow additives, anti-foams, and these can only usually be tested by assessment against a standard in a typical formulation.

14.11 INK QUALITY CONTROL

Provided the raw materials have been adequately tested, the prime objective of quality control of finished inks is to ensure that no deviations from the formula have occurred during manufacture and that the batch is therefore identical to the master standard in all respects. In both oil and liquid ink manufacture there are fundamentally two types of grinding apparatus, i.e. continuous and batch mills. The latter type of operation is relatively simple to quality control, a sample, representative of the total batch, being removed from the mill at various time intervals until it compares with the master standard. Any alterations which may be necessary can be made as soon as an adequate level of dispersion has been obtained. The continuous mills present more of a problem to the quality control laboratory, since while the sample is being tested the mill, if left running, could be producing ink inferior to the standard, or alternatively if stopped will result in lost production. Hence a series of quick tests must initially be carried out, followed by a more rigorous testing procedure when the milling operation is complete. Where mills of the continuous type are used, the ink should be stirred before a final sample is taken for testing to ensure that it is representative of the whole batch rather than being just the last portion off the mill.

In any ink quality control area, it is important to have good lighting which should be as near 'white' as possible. The more complex forms of lighting used for colour matching purposes are not necessary as all quality control procedures should be on a comparative basis of batch versus master standard, and any distortions of hue due to an off white light source will be equal for both inks.

For convenience, the quality control of inks will be divided into the following four sections:
(1) Short-term ink testing.
(2) Long-term ink testing.
(3) Press performance tests.
(4) Dry print performance tests.
Under each heading, tests common to both oil and liquid inks will first be considered and then specific tests on each of the two basic types of inks will be given.

14.12   SHORT-TERM INK TESTING

There are tests which can be carried out rapidly while the ink is still on the mill and includes checks for dispersion, flow, strength, hue, opacity (or transparency) and gloss.

Dispersion

A satisfactory level of dispersion (i.e. equal to the standard) is the fundamental requirement for any printing ink and until this has been achieved, any further testing is unnecessary. Dispersion can be adequately checked on a fineness of grind gauge as described earlier in this chapter.

Flow

Before comparative prints can be obtained, to compare the strength and hue, the test ink must be of similar rheology to the standard. Provided the ink is not potted directly from the mill, it is not necessary at this stage to adjust the ink accurately for any small flow variations as this can be carried out after the ink has been taken off the mill.

Tests on the rheology of oil and liquid inks are quite different and hence will be considered separately.

Oil inks

Rheological parameters are extremely important in controlling the quality of oil inks. While many of these parameters can be assigned limits, others involve an assessment based purely on a comparison with an agreed standard. This subject has been discussed fully in Chapter 13 and only points pertinent to quality control will be further considered.

Feel on the slab

Observation of the body of an ink, the length of flow and 'stringiness' resulting from the rapid lifting of a flexible palette knife from the body of an ink on a slab can provide valuable information in a short space of time concerning certain vital properties of the finished product. Comparison against the master standard can be effected in a similar manner and substandard products can be readily identified even by the novice ink technician.

Ink viscosity

Ink viscosity can be measured in a variety of ways, giving values for both the plastic viscosity of the sample and its yield value. These two values will in many instances give information concerning the anticipated performance of a product. A lower ink viscosity will be required for web offset ink, for example, than for a similar product for sheet-fed lithographic applications.

Certain applications will also demand a high yield value particularly in metal decorating applications where varnishing on wet inks is desired. A low yield value, conversely, may be paramount in overprint varnishes where good flow and subsequently high gloss may be required.

Ink tack

The 'tack' of an ink can be measured on several different instruments; none really giving a true tack or stickiness factor, but a related property (Fig. 14.14). The tackmeter which is commonly used in many laboratories can give an indication of the so-called tack of the ink on an arbitrary scale.

Tack, in this instance, is a measure of the resistance of the ink film to splitting, such as occurs during the distribution of ink over the rollers of a printing machine and during the actual printing operation. A knowledge of the tack values of inks is of help to the ink formulator in obtaining the best results, particularly in multicolour wet-on-wet printing operations where ink trapping is important.

Ink flow

Flow characteristics can often best be measured comparatively by way of reference to a standard. It is important to remember in measuring flow that freshly manufactured products may change considerably when allowed to stand overnight. In carrying out immediate checks, it is important to work the standard vigorously on the slab before making comparisons, taking care not to aerate the material during the working-up. Smoothing vigorously with a palette knife is the usual method.

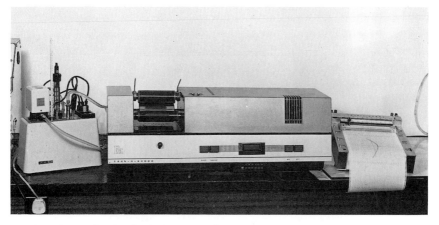

Fig. 14.14 The Tack-O-Scope tackmeter (courtesy of Testprint BV)

Flow can be compared very simply by observing the rate of movement of two inks or varnishes down a glass plate placed vertically on a flat surface.

Liquid inks

Many liquid inks when first sampled off a mill will be above room temperature and may also be quite heavily aerated. All efforts should, therefore, be made to equalise as far as possible the batch sample to the standard. The viscosity can then be checked with a Ford or Zahn cup, while the relative flow properties can be assessed by stirring the inks with a palette knife and then comparing the flow off the knife as it is withdrawn from each of the inks.

Strength, hue and opacity

The 'drawdown technique' as described earlier in this chapter is sufficient as a quick test for comparison of the relative strength, hue and opacity of the batch against the standard.

Gloss

Relative gloss is more difficult to assess on a drawdown and therefore prints of the test ink against the standard should be prepared. For liquid inks, a drawdown with a wire wound applicator is suitable (Fig. 14.15). This technique gives an excellent comparison between two inks provided they have similar flow. The applicators are available for manual drawdowns. Automatic applicator models with a built-in applicator bar can be obtained so that a consistent pressure can be applied on consecutive prints.

Gloss, in the opinion of the writer, is still best assessed by visual means as this method will generally be used in practice by the customer. Reflect-

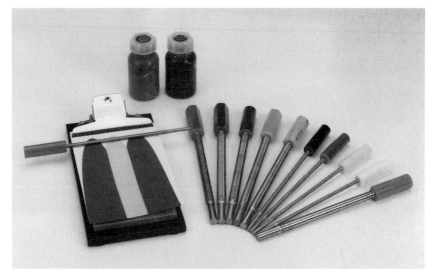

a

b

Fig. 14.15 A wire-wound applicator (a) K bars (b) Applicator (courtesy of RK Print-Coat Instruments Ltd)

ance spectrophotometers are available, however, if a numerical assessment is preferred. These consist of a beam of light which is arranged to fall on the print at a set angle and the reflected ray then falls on a photo-electrical cell. All readings are compared against that obtained from a standard surface. The results obtained do not necessarily agree with those obtained visually, since the eye will judge the relative gloss levels at a variety of angles and will also take into account other factors such as print imperfections and colour variation. Visual assessment is therefore preferred, especially since the printer is normally only concerned with the visual appeal of his finished print. Discrepancies in the gloss level of the batch sample could be caused by a misweigh of one or more of the ink components or on the continuous type of mills it could be due to an incorrect mill setting or condition. Hence, by checking the gloss level at this early stage, any mechanical adjustments necessary can be easily be carried out.

Additional tests on inks cured by radiation

Inks 'curing' under radiation (UV or electron beam) vary in their rate of polymerisation. The end point when the ink is 'dry' on the substrate is somewhat more speculative.

For quality control purposes, it is necessary to ascertain whether an ink batch is curing at the same rate as the standard. This is normally carried out qualitatively using a finger rub test at a given exposure time under the source of radiation (e.g. 2 lamps at 500 fpm at a film weight of 2 mμ). There are certain situations, however, when it is difficult to distinguish between cure and adhesion. An ink may be dry to the touch, but easily removed under the finger rub test. This can best be judged by experience.

Very slow rates of polymerisation can still result in 'chalking' as in the case of rapid setting conventional inks on certain absorbent stock. Cure rate can also be measured by employing a rub test on the substrate after a given radiation exposure. Indeed, the toughness of a cured film, which may reflect the degree of cure, can be checked, for example, using the PIRA rotary rub tester. A cured ink can thus be described in terms of its resistance to rub under given conditions.

14.13 LONG-TERM INK TESTING

The tests already outlined under Section 14.12 should first be confirmed on a representative sample of the final ink. The strength of the ink can additionally be assessed by the bleach technique and the rheology can be adjusted, where necessary, by the addition of varnish or diluent. The strength of the ink must of course be rechecked if further additions have to be made. In general terms, an ink with short flow compared to the standard indicates a low resin solids content and a further addition of varnish is then required. Alternatively, a stout batch sample with good flow points to a diluent correction. For liquid inks, before any diluent (i.e. solvent) is added the drying rate of the test sample against the standard must be assessed.

Additionally, both oil and liquid inks must be tested for stability on ageing. This test is usually accelerated by placing the inks in a well-sealed container in an oven pre-set at between 35 and 50 °C. After 24 hours, the sample is allowed to cool to room temperature and checked for any separation and/or loss in flow characteristics compared to the bulk. A comparative drawdown of the heated and unheated sample is then prepared and examined for changes in strength and hue. Caution should be exercised with this test, however, as some pigment–vehicle interactions can cause a shade change to occur at these temperatures over a period of time.

Specific tests on oil inks

Drying time

The drying time of an oleoresinous ink is important to know for reasons of both press stability and print drying. Optimum drying rates are a compromise between these two extremes and will obviously depend upon the nature of the particular application. Before discussing methods available to measure ink drying times, it is important to note that this term refers to the time taken to dry a print of a particular ink film thickness on a certain substrate (e.g. paper/board/glass).

Factors affecting drying time of an oleoresinous system, include:

(1) Temperature.
(2) Humidity.
(3) Availability of atmospheric oxygen. A single laboratory print left on a workbench will have a greater access to the atmosphere than a production situation of printed stacks of varying pile heights.

In carrying out a manual assessment care should be taken to ensure all fingers are clean, as a build-up of deposit on the fingers can severely affect the test. In cases where powdering of the ink may be mistaken for drying failure, the print should be warmed in an oven for a few minutes. This will aid drying, but have little or no effect on powdering. For quality control purposes, one rarely needs to know the exact time of drying, but merely whether or not the print is dry within a certain time. However, since atmospheric conditions can vary quite widely, it is always imperative to carry out the test against a standard ink.

Automatic drying time testers

Exact measurements of drying times on prints is obviously both time consuming and laborious, and one to which automation has been applied. Unfortunately, no equipment exactly reproduces a real situation. Automatic test equipment can only be used to give relative drying times.

PIRA ink drying time tester

This equipment was described in a paper by Coupe entitled 'Ink and paper testing' in the Proceedings of the PATRA Letterpress Conference, 1955. Modifications to this equipment are currently available in other similar forms.

The equipment utilises a glass plate upon which is printed the ink in question (or the standard) and which is subsequently rotated. A track is scored in the wet film using a metal stylus and as the ink dries, the smooth track pattern changes to a tearing pattern. At this point where the track changes its pattern, the ink is designated dry, and the time required to reach this point is classed as the drying time.

The main drawbacks to methods based on glass plate prints are the reproducibility (which can be poor) and the remoteness of the system from behaviour on absorbent substrates. It is, however, a rapid method suitable for quality control purposes and gives information concerning the ink drying time on non-absorbent surfaces such as rollers.

IGT drying time recorder
The IGT drying time recorder measures the drying time of an ink or the suitability of a particular paper for a given ink in two ways. The first is the set-off at preselected time intervals over a period of up to 24 hours and the second is related inasmuch as one examines the ink removed during the dwell time of a pressing-bar, over the same period. Various ink film weights can be examined on any paper. Whenever the moving bar is in contact with the wet ink film either disturbance of the test strip is observed or set-off on a backing strip is noticed.

Setting time
Ink setting time is governed by the choice of paper as well as the ink film weight used. A control should thus be run under similar conditions. An approximate qualitative test for setting time can be carried out by careful finger assessment checking surface tackiness. It is also noteworthy that setting is often accompanied by a change in surface gloss and the wet glossy appearance after initial printing is reduced as the ink sets.

A more quantitative test utilises a 'back-up' technique on a laboratory proofing press or printing unit. 'Back-up' in this context refers to the process whereby an extra sheet of paper is applied to the surface of the printed substrate and run through the press again. Wet ink will be transferred to the second sheet of paper. The time interval between printing and backing up is kept constant for a series of prints (e.g. 1, 3, and 5 minutes for quick-setting inks) and the amount of ink transferred to the backing paper is inversely related to the setting time. The time required for no back-transfer of ink is thus a function of the setting rate and gives an indication of the 'back-up' time when printing the reverse of the sheet.

Ink fly

The problem of ink 'misting' or 'flying' in the vicinity of a press is a serious nuisance which can largely be overcome by formulation changes. This phenomenon is often a function of the nature of the vehicle, but will obviously also depend upon a multiude of other parameters, e.g. pigment, extender type and level.

It is comparatively easy to monitor ink fly in the laboratory on a given batch of ink, given a two roller system. It is preferable to have one roller oscillating to reduce the tendency of the ink to form ridges. A useful piece of equipment in the laboratory for this purpose is the tackmeter or inkometer.

Comparisons between different inks can be undertaken quantitatively by either:

(1) Weighing the weight increase per unit area on the collection device.

(2) Visually assessing the droplet distribution per unit area.

Any test for ink fly ideally should try to reproduce the most severe conditions likely to occur in the field on a given press. The parameters for consideration should thus be:

(1) Ink film thickness on the roller surface — Generally ink fly will increase as the ink film thickness increases.

(2) The temperature of the roller surfaces — Certain inks will dramatically change their rheological properties as the temperature changes and will generally decrease in viscosity as the temperature rises. As modern roller temperatures may be appreciably above the surrounding air temperature, this factor should be remembered when assessing ink fly.

(3) Roller speed — It is important to remember that ink misting primarily occurs in the ink-roller chain. It is worth noting that speed of separation is higher as the roller diameter decreases, hence a greater tendency for ink fly will arise on small diameter high speed rollers such as those in an inking-roller chain.

(4) Ink consistency or rheology — A low bodied ink will show a predictably greater tendency to mist than a high bodied ink.

(5) Siting of the target — In measuring misting levels of a variety of inks, it is important to site the target at a point which reflects this level. On a tackmeter for instance, a convenient point is at the same height as the nip between the two rollers, on the rear wall of the measuring chamber. It is perhaps surprising that the position of maximum fly should be on the opposite side to the parting nip of the rollers, where the filaments are formed and broken: the explanation appears to be that particles are carried around in a cloud which follows the roller surface until it collides with the cloud following the other roller which will be at this very position. It is also important to note that a percentage of heavy droplets, which are often termed 'sling' will be formed on the base of the measuring chamber. Although this may or may not be a nuisance in the same way as ink mist is (depending upon the design of the press), it may result in the necessity for constant press cleaning, and does not lead to a clean operation. This should also be monitored carefully.

(6) The complete enclosure of the test roller system is important to minimise operator exposure to ink mists, by restricting the spread of particles, and to exclude draughts which could affect reproducibility.

Specific tests on liquid inks

The drying rate of liquid inks is a critical factor since an ink which is too fast drying will cause printing problems, whereas a slow drying ink could give rise to excessive solvent retention. Also, an incorrect solvent balance, especially in a system where two or more resins are present, could lead to premature precipitation of one of the resins leading to poor printability and a drop in the overall gloss level. Tests on the drying rate can be carried out by drawing down side by side the test and standard inks on a grinding gauge and comparing the relative evaporation rates at different film thicknesses.

Pigment settling cannot easily be tested except by examination of a batch sample over a long period of time which is of course completely impracticable. However, the batch sample should be allowed to remain undisturbed for 24 or 48 hours before being examined for any signs of either a supernatant layer or pigment sediment on the bottom of the can.

Water-based inks should additionally be checked for foaming and pH variation. The batch and standard inks are either stirred or shaken for a set period of time and compared for the degree of foaming. Both inks are then allowed to stand for a few hours to check on the ease of dispersability of any foam formed on the above test. Excessive use of antifoams should be avoided as they can lead to printability problems and stereo swelling.

The pH of a conventional water-based ink (i.e. one in which an acidic resin is solubilised with ammonia or an amine) is an important factor since if it is too high the ink would be too slow drying due to the excess of amine present. Alternatively, if the pH is too low, then large viscosity fluctuations can occur and the resin can even be precipitated out of solution. A check on the pH of the finished ink can simply be carried out by using a standard pH meter.

14.14   PRESS PERFORMANCE TESTS

Oil inks

While the ultimate test for examining the printing performance of an ink is to run it on a press, it is obvious that this is impracticable with every batch of ink. The number of press types required and the variations in printing conditions prohibit this kind of test in all but a few cases.

It is important to distinguish between a development product where extensive testing may be important and batches of an existing product where preliminary press trials should have established the performance of the product. It is thus necessary only to emulate the successful system by laboratory means. In routine testing, one makes the assumption that the standard ink runs satisfactorily, and laboratory tests are designed not so much to show that the particular batch works well, but that it resembles the standard.

Laboratory experiments can go some way to examine certain printing properties, but no test is in itself a substitute for examination of the printing properties of the ink on the press itself.

The IGT printability tester (Fig. 14.16) exhibits certain features of the printing operation and is well used for both ink and paper testing. It consists of an ink distribution system and a print unit (Fig. 14.17). This printing unit is comprised of a 150° sector of impression cylinder suitably covered, on which a 1-inch (25-mm) wide strip of paper is placed. Ink is applied from the distributor to the rim of the black-centred wheels, which act as the rotary printing plates. Power is furnished by a pendulum or spring and in later models by electrical means. One has thus the choice of either having a constant speed or varying speed impression. In the case of the latter, the paper strip can be graduated along its length in terms of speed at given pressures, which themselves can be varied.

Under controlled conditions of humidity and temperature, this instrument can be used to examine a whole variety of printing properties together with comparisons of ink and paper. The manufacturer's information is recommended for a full list of these applications.

One property missing from the IGT list involves the lithographic properties associated with water/ink interfacial phenomena. Although the IGT is not designed specifically for examining water/ink relationships, there are pieces of equipment available to do this. The Pope & Gray Litho-

**Fig. 14.16 Distribution unit of the IGT printability tester
(courtesy of Testprint BV)**

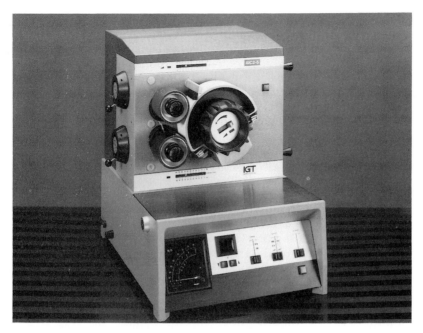

**Fig. 14.17 The print unit of an IGT printability tester
(courtesy of Testprint BV)**

Break Tester is perhaps the most simple and commonly used equipment, whereas the Prufbau Lithomat can also be extended to measure changes in tack, film thickness and water take-up.

The Lithomat consists of a steel roller that revolves in the fountain solution. This roller rotates a rubber roller which presses against another steel roller that lies above it. Both steel rollers oscillate to ensure an even ink distribution. The effect of the fount solution on tack and thickness of the ink on the rollers can be assessed by combination with a graphemetronic instrument which measures the tack as the force exerted on a roller driven by the inked roller. The change in tack and film thickness is recorded with time. The final water content take-up by the ink is measured by removing a sample from the centre rubber roller and analysing by the Karl-Fisher method.

As one moves away from conventional oleoresinous systems to, for example, ester chemistry as in UV curing inks, it is important to ascertain the effect the ink may have on the rollers, blankets and plates of the press. While this is of paramount importance in introducing a new product, it is not necessary for routine quality control testing. The assumptions made earlier should hold.

Liquid inks

The majority of solvent-based inks are printed by either the flexo or gravure processes, although screen printing is also important in certain specific areas. Whatever the process, the performance of the ink on the press is obviously a critical factor and is one which should be carefully examined in the quality control procedure.

Gravure printing can be carried out in the laboratory by using a scaled down version of the production model, the rollers being either mechanically driven or even simply turned by hand. Alternatively, small gravure proofing presses are available (Fig. 14.18) and these, although ideal for quality control work, suffer from the disadvantage that only a single print can be obtained before cleaning up the rollers. All inks should be thinned where possible with the recommended reducers, although the use of slower evaporating solvents may be necessary to obtain clean prints at the relatively low printing speeds obtained in the laboratory.

Before any printing is undertaken, the proposed end use of the ink should be carefully considered and the following questions raised:

(1) Is the ink expected to be used for printing solids, tone areas, or a combination of both?
(2) Is the ink part of a process set?
(3) Is the ink to be used for surface or reverse printing?
(4) For coloured inks: Is the ink to be used in conjunction with a white? Conversely, for whites: Is there a typical coloured ink also used when printing this white?
(5) What substrate(s) should the ink be printed on?

The answers to these queries will provide the quality control tester with a guide in assessing the press performance of the test ink. It is important in this context that any inks to be used in conjunction with the test ink (e.g. for a process ink, the remaining colours of the process set) should also be

Fig. 14.18 A gravure proofer
(courtesy of RK Print-Coat Instruments Ltd)

printed so that the effect of superimposition of the inks can be carefully examined.

Other problems associated with the gravure process, such as examination for cylinder wear and scumming, are more difficult to reproduce as a standard procedure and a great reliance must be placed here on having a good ink formulation. Several tests have been designed for measuring cylinder wear, but perhaps the most common uses a small trough containing the test ink in which is immersed a glass slide coated with a thin layer of chromium. A weighted arm with a brass weight on the end is then passed back and forth in the ink over the chrome plated slide for a set time period. The test is then repeated using a control ink on a second chrome slide. A comparison between the degree of chrome abrasion on the trial and control ink slides is an indication of the wear potential of the ink.

Scumming, i.e. the formation of an ink film in non-image areas, can result either from an incorrect press setting or from an ink fault. Assuming correct mechanical adjustments on the press scumming can still be evident, in certain extreme cases from the beginning of a print run, and this is usually due to oversize particles in the ink causing the doctor blade to lift fractionally away from the cylinder surface. This fault should have been eliminated, of course, in the initial quality control, but in other cases where scumming is obtained only after a longer press run the cause is normally more difficult to ascertain and hence to rectify as a standard procedure.

In flexography, the film weight applied is a much less readily defined factor than in the gravure process. This is because the film weight carried can be varied simply by altering, for example, the pressure between the rollers, the speed of running, the ink viscosity, as well as varying the screen size of the anilox and the Shore hardness of the rubber rollers. Hence, flexographic proofing presses have a limited value and a greater reliance is usually placed on the more versatile anilox hand roller (Fig. 14.19).

This consists of an engraved roller (termed the anilox) and a rubber roller. A print is obtained by placing the ink in the nip of the rollers and rolling the whole unit across the substrate. The film weight applied by this method should roughly parallel a typical commercial flexographic print and can be varied by:

(1) Altering the ink viscosity.
(2) Applying more or less pressure on the roll out.
(3) Using a different screen size anilox roller.
(4) Using a different Shore hardness rubber roller.

Some raw materials, especially some solvents and antifoams, can cause swelling of the stereos leading to poor printability and in certain cases it may be necessary to specify the type of stereo material (i.e. natural or a synthetic rubber or a photopolymer plate) to be used with a particular ink. Tests can easily be carried out by immersing a sample of the stereo material in the ink for a short time and measuring the thickness of the sample before and after the test using a micrometer gauge and also the hardness before and after immersion.

Fig. 14.19 An anilox roller (courtesy of Weller Patents Development)

Screen inks

Before the press behaviour of a screen ink can be examined, the formulation of the ink should be checked since inks for this process vary considerably in composition depending, for example, on the substrate to be printed and the drying facilities available. Generally, screen inks can be dried by evaporation, absorption, oxidation, polymerisation or more often by a combination of these methods. Inks drying mainly by oxidation (e.g. by the incorporation of a drier) resemble a letterpress ink in many ways and would normally be manufactured in a similar manner (e.g. triple roll mill). Alternatively, an ink drying mainly by evaporation (e.g. using a cellulose ester in solvent) is more akin to a high viscosity liquid ink and would normally be manufactured on conventional liquid ink milling equipment. Hence, in assessing the press performance of screen inks, many of the difficulties encountered may be a direct result of the formulation type and the reader should then refer to the appropriate section of this chapter.

In the actual printing of screen inks, the best results are normally obtained when the inks are soft and buttery to allow easy passage through the screen and short and thixotropic to give smooth lay and sharp definition. Testing can be carried out in the laboratory simply by drawing a squeegee over a hand screen, the film weight applied being varied by the use of different mesh screens (Fig. 14.20).

14.15 DRY PRINT PERFORMANCE TESTS

Paper and board

Although inks for paper and board printing may have different properties, many tests carried out on the final print have similar performance ratings.

Properties such as lightfastness, scuff resistance, soap and alkali resistance, etc. on the printed substrate are generally similarly tested in both paper and board printing and for convenience are treated together.

Fig. 14.20 A screen proofer

Lightfastness
It is not normally necessary to check the lightfastness of each individual ink except for cases where contamination may cause a specific problem. Details of the test have already been described earlier in this chapter and are also covered in BS 4321 : 1969.

Chemical resistance tests
Several chemical resistance tests on finished prints have already been covered in the testing of pigments (Section 14.3). These are:
(1) Soap and detergent resistance.
(2) Alkali resistance.
(3) Acid resistance.
(4) Wax resistance.

Specific test methods
The following specific resistance tests are all fully described in BS 4321 : 1969 and are all self-explanatory.

Determination of resistance of printing inks and prints to cheese: Test Method 8, p. 27. Determination of resistance of printing inks and prints to edible oils and fats: Test Method 9, p. 29. Determination of resistance of printing inks and prints to spices: Test Method 11, p. 34. Determination of resistance of printing inks and prints to water: Test Method 3, p. 17. Determination of resistance of printing inks and prints to detergent: Test Method 7, p. 25. Determination of resistance of printing inks and prints to solvents: Test Method 4, p. 19.

Abrasion resistance
The resistance to abrasion is an important aspect of many printed articles, particularly on paper and board. Demands for improved abrasion resis-

tance continue while the speed of packaging rises, giving greater possibility of severe scuffing from other printing surfaces or from contact with the production machinery itself. In certain cases (e.g. oleoresinous inks where excessive spray powder may give an abrasive surface) the package in itself can have a very rough surface and should be examined prior to any test. Failure on laboratory testing may result from incorrect or over use of anti set-off powders. It is also worth noting that problems may arise in certain packaging areas where spillage of the contents can lead to an abrasive material being trapped between two areas of print. Any laboratory test should be designed to cover the severest conditions likely to be encountered.

Several instruments are available for checking rub resistance of prints and three such instruments are described in the British Standard *Methods for Measuring the Rub Resistance of Print*, BS 3110 : 1959.

While the simplest method described in BS 3110 is the Procter & Gamble tester, it is not used much outside the soap and detergent pack area for which it is specified. This rub tester is composed of a frame acting as a clamp to a print sample which is held against the base of the test instrument. The frame also acts as a guide to a block on which is attached a second piece of print in surface contact with the first sample. Hand movement of the block for a number of times at a steady speed (around 1 ft/sec or 30 cm/sec) gives direct comparisons between prints.

The Sutherland instrument and the PIRA instrument (Fig. 14.21) are discussed in BS 3110 and the reader is recommended to consult this for further details.

It is important that the age of prints being tested is known. With conventional oleoresinous prints a week is the normal time allowed to elapse before rub testing. Even with radiation cured prints 24–48 hours normally elapse before testing as inks dry to the touch can change with respect to their rub resistance properties within these times.

Crease resistance

In board printing the finished print is often required to be formed into a carton (e.g. a cigarette packet) and the ink must therefore have sufficient flexibility to prevent cracking when the board is folded. This test is particularly important where higher film weights of ink (and/or overlacquer) are carried as in gravure printing. Ideally, the test should be conducted using an apparatus such as the PIRA carton board crease tester (Fig. 14.22) which reproduces quite accurately production conditions. A cruder test can be carried out by either folding the board back and forth through an angle of 180° until the first signs of ink split is apparent or by folding the board once through an angle of 180° and lightly scuffing the surface of the fold with the finger nail.

Film printing

Immediate adhesion of the ink to the substrate should normally be obtained in film printing to prevent set-off occurring in the reel. All film

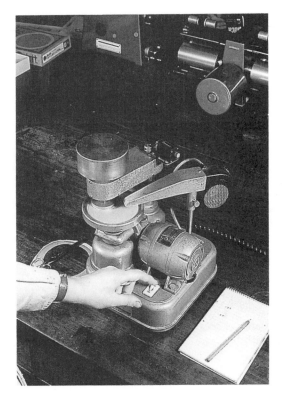

Fig. 14.21 A PIRA rub tester (courtesy of PIRA)

prints should be tested for adhesion, flexibility, scuff, slip and set off as
outlined below.

Scotch tape adhesion

A length of standard Scotch tape is placed over the printed areas, smoothed
down with the ball of the finger and then pulled back slowly for half of its
length and snatched back rapidly for the remainder. The print is examined
for ink removal and the tape for any ink transfer. Care should be exercised
with coated films so as not to remove the base coating and so obtain a
completely false result.

Flexibility

The printed film is held firmly between the thumb and forefingers of each
hand with about half an inch (13 mm) in between the hands and flexed
10 times. The print is then examined for any ink removal.

Scuff

The print is repeatedly scuffed with the back of the fingernail, then
examined for ink abrasion.

Fig. 14.22 A PIRA carton board crease tester (courtesy of PIRA)

Slip

The friction of film to film and film to metal are important factors in processing packaging materials. If the coefficient of friction is too high, this can result in:

(1) Movement of the packaging material over fixed elements becoming difficult.
(2) Problems in controlling web tension.
(3) Print which is stacked becomes 'adhered' and this can give rise to blocking.
(4) Generation of a high static electrical charge.
(5) Excessive creasing during packaging.

Alternatively, if the coefficient of friction is too low, then:

(1) Guiding rollers on the packaging line will not rotate and this results in irregular flow.
(2) Web tension is difficult to adjust.
(3) Finished packages tend to slip during transportation.

Two types of friction can be measured, i.e. static and dynamic friction. In static friction there is no relative velocity between the boundaries of the materials and the static coefficient of friction can be described as the energy required to break down the localised 'adhesion' between the substrates in order to allow them to move freely. The static coefficient of friction can be measured using the tilting plane where a print is fastened face upwards to the plane and a metal sled is then placed on the film. The plane is slowly raised until the sled slowly and evenly moves down the plane and the tangent of the angle of inclination is equal to the static coefficient of friction. Alternatively, the static coefficient of friction between two printed surfaces can be measured by attaching print to the metal sled and repeating the test as above.

Dynamic friction measures the friction between two materials which have a relative velocity to each other and is caused by molecules on the surface of the two materials colliding. These collisions are imperfectly elastic and hence energy is consumed. The dynamic friction can be measured on a flat bed by drawing a sled across a print at an even velocity (Fig. 14.23) or alternatively by the pendulum method where the print is held in a fixed position and the pendulum is rotated.

Blocking

The print immediately after drying is placed in contact with the back of the same film and subjected to a pressure of between 750 and 3000 1b psi for 16 hours in a hydraulic ram to simulate conditions produced in a reel of printed film (Fig. 14.24). The pieces of film are then slowly pulled apart and the test ink examined for degree of cling relative to the standard and any signs of ink transfer to the back of the plain film. Additionally, the blocking tests can be repeated at elevated temperatures if the facility exists for heating the platens of the hydraulic ram. This test can be used to simulate the effect of entering the re-reel stage while still warm or alternatively as a guide when film is used for packaging hot goods. In carrying out these tests, care must be taken in selecting a realistic temperature, time and pressures.

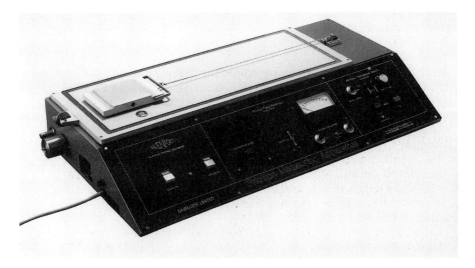

Fig. 14.23 Daventest dynamic friction tester (courtesy of Daventest Ltd)

Fig. 14.24 Hydraulic ram for blocking tests (courtesy of Specac Ltd)

Additional tests for film

The following tests are dependent on the type of ink and its end application and need only be checked where stated in the ink specification.

Heat resistance

Heat-resistant inks are necessary where the film is printed in the heat seal area so as to prevent ink pick-off on to the heat seal jaws. An area of print in contact with virgin foil is placed between the heat seal jaws (pre-set at the heat seal temperature of the film under test) for a short time dwell (usually 1 second). After cooling, the print and foil are pulled apart and examined for ink pick-off and ink transfer on to the foil. Ideally, the print and foil should show no cling and there should be no ink transfer from the print to the foil. A further and more stringent test involves folding the print inside the foil so that the two ink surfaces are in contact. The heat resistance test is then repeated and when the ink surfaces are separated there should be no cling (Fig. 14.25).

Adhesive Lamination

If necessary the adhesive is mixed according to the manufacture's instructions and reduced to a suitable printing viscosity with solvent. The print

Fig. 14.25 A BCL heat sealer (courtesy of British Cellophane Ltd)

under test is set aside for 1–2 hours to allow for any waxes to migrate to the surface and then coated with the adhesive. After evaporation of the solvent, the tacky print is smoothed on to the laminating film (or foil) taking care that all air bubbles are removed. In order to allow for easy access into the laminate after the adhesive has cured, a strip of paper is wedged between the substrates down one edge of the adhesive layer. The bond strength of the laminate is tested after 24 hours and ideally also after 1 week, on a spring balance. Laminates for boil-in-the-bag work also require an additional bond strength test after being immersed in boiling water for 30 minutes.

PVdC coating
In certain types of packaging (particularly biscuits) the print is coated with a PVdC emulsion to allow for better seal integrity on A-B and A-A seals. To simulate this process, the PVdC emulsion is simply applied by a wire wound applicator over the print and the adhesion checked with Scotch tape. An additional problem can often occur after heat sealing, when interlayer adhesion between the PVdC and the ink film or the ink film and the base film breaks down. To ensure good interlayer adhesion is maintained the coated print is heat sealed (PVdC to PVdC) at 135°C/0.5sec/0.5 psi and the strength of seal tested on a spring balance as for a laminate described in the preceding section. If poor interlayer adhesion occurs a poor seal strength will result.

Cold seal packaging
Cold seal packaging is being increasingly adopted to prevent heat contamination of sensitive products and for faster packaging line speeds. In this process a cold seal adhesive is applied in line to the reverse side of the

printed film so that on re-reeling the adhesive is in direct contact with the inks or release lacquer. It is therefore imperative that:

(1) No set off of inks on to the adhesive occurs or blocking will result.
(2) Pigments, dyes, waxes do not migrate into the cold seal and cause discoloration and/or loss of seal.
(3) No chemical interaction takes place between components in the inks or lacquer and the adhesive which might poison the adhesive.

Testing of inks and lacquers must be such as to show if any of the above situations can occur in practice. The normal test involves both short- and longer-term blocking tests with ink/lacquer in contact with the cold seal adhesive and with a maximum pressure of 750 psi in the hydraulic ram. In addition to no cling being obtained on separation of the inks/lacquer from the cold seal adhesive, the cold seal adhesive must be sealed to itself to ensure no drop in seal strength is given compared to a print of adhesive which has not been in contact with the ink/lacquer.

Deep freeze

This test was fully described earlier under pigment quality control (see p. 709).

Odour and taint

Since many liquid inks are used on films for food packaging, it is necessary that they must have low odour and do not taint the packaged product. Samples of the test print, standard print and plain film (as a reference) are placed in separate tightly sealed containers, and heated in a 50 °C oven for 1 hour. On removal the samples are tested for residual odour whilst still hot, then rapidly resealed and tested again after cooling to room temperature. A more elaborate odour test and taint test procedure is given in BS 3755 : 1974.

Non-curl

A few films are especially prone to curl when printed with certain types of ink, causing great problems on the packaging lines. In cases where curl could be a problem, a sample of film, printed all over with the test ink, should be allowed to lie undisturbed on a bench. After a period of time (e.g. 1–2 hours), there should be no signs of the film curling at the edges.

Grease resistance

Printed film used for packaging certain foods, e.g. potato crisps, must be resistant to the oils used in the manufacture of these goods. A test can be carried out by rubbing the printed film a standard number of times with a piece of cotton wool saturated in the appropriate oil, followed by a flexibility test on the greased print. The print and cotton wool are then examined for print breakdown and stain respectively.

Alternatively, the grease rub test can be carried out using the SATRA rub tester by impregnating the felt pad with a measured amount of the oil and then rubbing the test print for a set number of revolutions with an agreed pressure. This test, although apparently more scientific, is subject to wide variations in results due to variability in the felt pads and an average result of three readings should be taken and compared to those of tests on a master standard print carried out at the same time.

Foil

The testing of foil prints closely resembles the tests already outlined for plastic films. Care should be exercised to ensure the correct type of foil is used whether it be, for example, the simple virgin foil (such as used on many chocolate wraps) or the paper/board backed foil which has a surface wash (normally either nitrocellulose, shellac or vinyl) to prevent tarnishing and improve print receptivity.

The procedure for film printing should be adopted, where appropriate, with the exceptions of the flexibility test. For virgin foil, flexibility is tested by folding the printed foil through a 180° bend and then with the thumb and forefinger at either side of the crease the foil is rolled back and forth through the fold 10 times. The ink should then show no signs of break-down at the fold. Paper and board backed foil is examined for crease resistance using test methods already outlined for paper and board (see p. 742).

Metal

While some of the tests already described for paper and board prints will apply in the case of metal decorating inks, others are extensions of these (e.g. abrasion resistance), to take into consideration the tougher film involved, or the subsequent processing requirements of the metal container (e.g. high temperature process resistance), that is not possible with other substrates.

Abrasion resistance
The tougher film formed on metal surfaces, necessitated by the abrasion resistance requirements of the ultimate package, demands a different approach to testing the scratch susceptibility or hardness of that film. Several instruments are available for these tests:

Scratch testers
The apparatus consists of a weight loaded needle arm which is placed in contact with the panel under test. Operation of the lever gently lowers the needle on to the test surface and slowly withdraws the panel so that the needle makes a scratch on the surface. As the load on the needle increases, a point is reached where penetration of the ink film is achieved and the needle strikes the metal substrate. This contact completes an electrical circuit as signified by the illumination of a lamp. Results are thus expressed as the maximum weight at which the film passes the test.

Hardness tester:
Two types of hardness tester are used in the main, these are the Sward type and the pendulum type (DN 53–157). The Sward type utilises a rocker action and the hardness is described as the percentage of oscillations made by the wheel type rockers on the test plate as opposed to the reference glass plate.

The pendulum type hardness tester is primarily used on the continent. The apparatus consists of a triangular shaped pendulum, free to swing on two ball points which rest on the sample. An electronic counting device records the number of swings made by the pendulum between the 6 and 3 degree marks on the scale. Comparison is then made between the number of swings on the glass reference plate and the number on the test surface.

Heat resistance

This test depends primarily upon the properties and concentration of the pigment, but may be affected by the medium.

While knowledge of the pigment will provide the basis for a selection, to be decisive, a test should always be carried out on the final print. Routine checking on ink batches need not be necessary for reasons outlined earlier. In the case of new products, however, this test must always be carried out where heat resistance is required.

The test itself using a print of known film thickness on the metal to be used employs a stoving operation in an oven at a suitable temperature for a realistic time. An average schedule is 150°C, for 5 minutes, but this can vary depending upon the type of container under test, and higher temperatures may be encountered for example on DWI can lines. There are specific applications (such as solder seaming) where much higher temperatures (e.g. 230°C) are encountered albeit for much shorter times.

The stoved panel is then compared for hue, strength and brightness with the control panel, making due allowance for the wet state of the latter.

Flexibility or elongation properties

Simple flexibility tests can be carried out by bending the printed metal substrate and examining for failure, but the method can show poor reproducibility and often has little relationship to the stresses that the printed substrate is subjected to in commerical situations.

Reproduction of the actual stress situation on a tinprinting assembly line is not easy in the laboratory, but some equipment is available that can give information on the flexibility of a printed or coated film. One such piece of equipment is the Conical Bent Test apparatus designed to measure the elongation potential of coated metal panels. The panels are clamped in position and formed round the conical mandrel by rotation of the roller frame. The 200 mm long mandrel tapers from 37 to 3 mm diameter along its length and after the test the panel is examined to determine the minimum diameter at which failure by stretching occurred.

Adhesion

Adhesion to the metal and good interfilm adhesion on overprinted areas are both important requirements for printing inks in metal decorating applications. Failure of abrasion resistance tests may indicate that insufficient adhesion to the base metal or the underlying film is the cause. The principal test for poor adhesion is probably a simple scuffing test with a hard, blunt surface. Soft films or brittle films can be disturbed relatively easily. Poor interfilm adhesion can also be noticed in the same test as indicated by removal of surface layers only. The use of Scotch tape can also prove helpful, particularly for radiation-cured inks where adhesion has not been assisted by thermal treatment. It is particularly useful to predict the behaviour of cured ink films on interdeck drying as poor adhesion or interfilm adhesion can give rise to stripping of the ink from the sheet on to subsequent blankets.

Pasteurisation and processing

Film cracking, softening or adhesion failure can also occur during the processing or pasteurising of the printed metal. Where the ink is to be used in this application, the necessary laboratory tests should be performed.

Processing tests are normally carried out under steam pressure (15 psi) for various times, depending upon the particular application. The print is then checked for failure.

Pasteurisation varies somewhat from laboratory to laboratory but an average test uses partial immersion of the print in a water batch at 150 °F (66 °C) for 30 minutes. Print failure tests are performed as for processing. It is important when carrying out these tests to ensure that the print has been treated as would the finished article. Where for instance bending or stretching occurs this should be carried out under laboratory conditions. Also where a base coating or lacquer is present, it is important that the correct material is used in conjunction with the ink under test.

Plastics

The tests required for prints on polymeric surfaces (plastic containers, etc.) tend to lie between the paper/board and metal areas.

As many of these containers hold food, the main tests are contact tests with the material in question. These are similar to those described for paper and board, and are carried out by immersing the printed substrate in the material for a certain period and testing for film weakness or pigment bleed.

Ink adhesion and scuff resistance

Ink adhesion is tested using Scotch tape in a similar manner to that described for film inks (see p. 743). While complete resistance to Scotch tape may not be required, the inks must possess good scuff resistance. A fingernail test as performed for film inks generally suffices.

14.16   ANALYSIS OF PRINTING INKS

Analysis within the printing ink industry is not confined solely to the investigation of inks. Substrates, raw materials, printing sundries, finished prints, environmental sampling and personal monitoring are now integral parts of the analytical function. The last 20 years has seen a significant change in the way analysis is carried out which has been influential in the ever increasing role the ink analyst plays in the industry.

The conventional essentially wet chemical techniques are being replaced by rapid, precise instrumental analytical techniques. This is due almost totally to the microprocessor revolution which has swept through industry. Modern high performance instruments with sophisticated data handling are now available at prices well within the reach of most laboratories. Despite these innovations analytical results must always be viewed in context, a formulation from the analytical laboratory detailing what chemical entities are present in an ink often cannot be converted directly into an ink as vital information regarding the mixing, blending, processing and final combinations of those entities is not available.

Therefore an experienced technician is required to translate the information into a workable form. Similarly the analyst, in order to gain the most useful information from a sample, needs a good grounding in all aspects of ink technology and the raw materials used in order to select a starting point and suitable analytical method.

The analysis of single component or simple mixtures as a raw material quality control step is often required and relatively straightforward; the most difficult area is the analysis of an ink. The diverse range of printing inks in use today means there are a great many varying routes of analysis, some of the simpler liquid ink systems lend themselves to analysis whereas oleoresinous systems which may contain as many as 30 different raw materials, invariably subsequently reacted together, are a real challenge. The use of modern instrumental techniques now allows a great deal of information to be gained on complex systems and the problem is deciding whether it can be put to any advantageous use. Too much information can be more confusing than too little.

There is no universal answer to the analytical approach to a specific problem but it can be simplified into three distinct phases, separation, identification and quantification.

If an ink is considered, then a manual solvent fractionation scheme is often the simplest means of partial semi-quantitative separation of the ink components. A weighed sample of ink, usually 1 g, is sequentially washed with at least three 10 ml aliquots of solvents of increasing solvent power starting with a very weak solvent such as petroleum ether. At each extraction stage centrifugation is used to separate the solvent from the insoluble matter. The extracting solvent can then be removed from each fraction by heating and the residue weighed. Each of the fractions thus obtained are identified using an instrumental identification technique such as IR spectroscopy. The simple extraction scheme shown diagrammatically (Fig. 14.26) will give sufficient information on any ink system for an experienced analyst to select the optimum method of analysis.

The instrumental techniques available have enabled the analyst to advance and in many cases to combine the separation and identification procedures.

Gas liquid chromatography is used to separate and identify quantitatively volatile materials, i.e. solvents, plasticisers, antioxidants.

Techniques such as HPLC and GPC allow us to separate soluble materials according to their chemical nature or molecular size respectively.

These techniques and others will be discussed in depth.

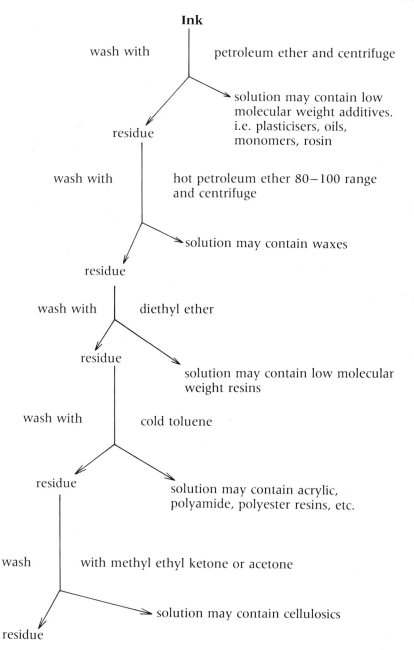

Fig. 14.26 Ink extraction scheme

Such a scheme can be readily adapted to suit the type of ink under investigation.

Chemical tests

In spite of the invasion of instrumental techniques specific wet chemical tests still play an important role in the analytical laboratory. These are the tests which lend themselves to on-the-spot analysis by giving definitive reactions with distinct chemical entities. For example, the simplest way of determining if an amine has been used to produce an aqueous acrylic solution is by the distinctive odour obtained when the sample is heated gently after the addition of sodium hydroxide.

The following colorimetric tests are very useful when only small samples are available or if there is insufficient time to carry out a purification process prior to identification.

Diphenylamine test for nitrocellulose (NC)

This is a rapid sensitive test which may be used to identify whether NC contamination is present in an acrylic ink or if a print contains NC. Diphenylamine of 1–4% may be made by dissolving 1–4 g of pure diphenylamine in 70 g of concentrated sulphuric acid, this solution is then diluted by the *very careful* addition of 30 g of deionised water. When spotted onto the test sample a strong blue colour develops if NC is present.

This reagent also gives a colour change when ketones are present; a weaker red coloration results. This may be very useful in distinguishing styrenated ketone resin from styrenated acrylic and styrenated allyl alcohol. It is also used as a confirmatory test for the presence of a ketone resin. Every precaution has to be taken to ensure no ketone solvent is present.

Liebermann–Storch test for rosin

A small amount of the sample is boiled with acetic anhydride. A few drops of this are spotted on to a porcelain spot test plate and concentrated sulphuric acid added. A violet colour indicates rosin. Other resins such as maleics and fumarics also react under these conditions.

Beilstein test for halogens

A halogen containing material when ignited in a bunsen flame in the presence of copper gives rise to a strong green-turquoise coloured flame.

This is very useful for instances where the presence of vinyl chloride is suspected. The presence of chlorinated photoinitiators or amine hydrochlorides may also be tested in this manner.

Amine-formaldehyde resins

An amine formaldehyde resin when warmed in a test tube with 70% sulphuric acid and one crystal of chromotropic acid will give rise to an intense violet coloration.

Urea-formaldehyde resins

A small portion of the resin should be boiled gently with a mixture of 20 parts acetone, 30 parts water, 10 parts of 30% HCl and 1 part of furfuryl alcohol. A red coloration indicates urea.

Epoxy resins

Either the resin or a print is dissolved in concentrated sulphuric acid with minimal charring. By means of a glass rod some of this solution is streaked on to a filter paper. A purple coloration indicates the presence of epoxy.

This is a very useful test for identifying whether a print is conventional litho or UV, assuming of course that the UV ink contains a bisphenol-A-epoxy resin. At this moment in time the assumption is usually valid.

Polyurethane resins

A simple test may be applied to polyurethane resins to determine whether they are TDI or MDI based. A filter paper previously impregnated with 4-nitrobenzene diazonium tetra fluoroborate is subjected to the fumes given off from a polyurethane resin when contacted with a red-hot glass rod. MDI gives a yellow coloration, TDI gives a red colour.

This test is specific for urethanes based on aromatic isocyanates and will also give a colour reaction with the less common 1,5-naphthalene diisocyanate. A violet colour is obtained. No distinct colour reaction occurs with aliphatic polyurethanes.

Phenolic resins

The most sensitive test for phenolic resins is the Gibbs–Indophenol test. There are two ways of performing this test and although the former is quick and simple the latter has been shown to be considerably more sensitive.

A little resin dissolved in alcohol is treated with a drop of Gibbs reagent (1% solution of 2,6-dibromoquinone-4-chloroimine in alcohol). The solution is made alkaline with dilute sodium hydroxide. Phenolic resins generate a strong blue colour and a modification of this test can also be used to detect the presence of phenolic stabilisers in UV diluent monomers.

The alternative is to impregnate a filter paper with some Gibbs reagent and subject it to the fumes given off when a sample of resin is pyrolysed in a test tube over a bunsen flame. The filter paper may then be developed by the alkaline fumes from a bottle of 0.880 ammonia. Once again a strong blue colour is obtained.

Tests for PVdC

The presence of polyvinylidene chloride (PVdC) on the surface of flexible packaging film may be readily detected using IR spectroscopy but some simple colorimetric tests are often valuable.

Application of the Beilstein test will determine whether any chlorine is present. A hot piece of copper is briefly scraped across the surface of the film and then placed in a bunsen flame. A bright blue-green coloration confirms the presence of chlorine. A distinguishing test for PVdC may then be applied. A spot of morpholine dropped on to the surface of the film will darken through brown to black if PVdC is present. This reagent does not react to the other forms of chlorine which may be encountered in packaging films.

Physical techniques

Physical separation techniques play an important role in the analysis of surface coatings. Manual solvent fractionation has already been discussed and instrumental techniques will be covered in the following pages. However, there are several other simple techniques of note. These are thin layer chromatography, paper chromatography and column chromatography.

Thin layer chromatography (TLC)

The mechanism employed in (TLC) is such that a mixture of compounds is separated as it is moved through a thin layer of sorbent material by a migrating solvent. The most common sorbents are silica gel and aluminium oxide, the thin layer of sorbent may be mounted on either glass plates, aluminium or plastic sheet. Using these materials the separations are achieved by a process of adsorption, which is very useful for the separation of organic compounds of medium or low polarity. The sample being separated proportions itself between the developing solvent and the active sites of the sorbent. The polarity of the components determines what this proportion will be. Polar solvents cause the greatest movement of the sample. The following solvents commonly used in TLC are listed in order of increasing polarity: hexane, benzene, toluene, trichlorethylene, diethyl ether, chloroform, ethyl acetate, butanol, propanol, acetone, ethanol, methanol, water.

Paper chromatography is a form of partition TLC which finds application in separating water soluble inorganic compounds and polar organic compounds. Other forms of partition TLC utilise cellulose or silica containing absorbed water as the stationary phase. Reverse-phase partition TLC utilises a non-polar stationary phase and a polar mobile phase. The non-polar stationary phase may be a coated silica. Identification of components in TLC is achieved by comparing the R_F value of the sample with a standard. R_F values are computed by dividing the distance the component travels by the distance the solvent front travels.

Some typical applications of TLC are:
(1) Separation of the acids from 2B, 4B, Lake Red C and other pigments.
(2) Separation of plasticisers.
(3) Separation of dyestuffs.

Visualisation of the different components may be achieved by a variety of methods including: exposure to iodine vapour, charring with concentrated sulphuric acid, colour reactions with specific reagents. More commonly TLC plates are manufactured with a fluorescent agent incorporated in them. Under a UV lamp the components appear dark against a light background.

Column chromatography is an extension of TLC often used as a large-scale preparative technique to isolate components of a mixture for further investigation. Once a separation has been achieved using TLC then a similar stationary phase packed into a hollow glass column and eluted using the same solvent via gravity feed may be used to separate a larger quantity of the same mixture.

Instrumental techniques

Whatever the reason for having an ink or related substance analysed be it a substandard product from the factory, a customers print requiring matching or, as happens on occasions, a competitors product, the analytical results are always required within a very short space of time. Modern electronics have put sophisticated instrumentation within the reach of every ink manufacturer. A skilled analyst using such equipment to its limit can gain valuable knowledge concerning the sample under investigation very rapidly.

Most of the common analytical instruments discussed in the following pages are available to the ink-maker.

Infra-red spectroscopy

This technique is undoubtedly the most important in a modern analytical laboratory. Infrared spectroscopy is an identification technique used to characterise the chemical nature of the sample under investigation. Nearly all molecules absorb radiation in the IR region of the electromagnetic spectrum. The instrument (Fig. 14.27) is used to obtain spectra characteristic of the chemical nature of the sample under investigation. Any type of sample may be studied using appropriate accessories, solids, liquids, gases, pastes, powders, prints, etc. No two different pure chemical compounds will give exactly the same spectrum and even physical differences in the same compound, solid or liquid for example, will give rise to differences in the IR spectrum. Variations in the spectra of polymers of the same type may not be so obvious as polymers are complex mixtures and not pure compounds.

At temperatures above absolute zero molecules are in a constant state of motion, the vibrational energy of the molecule causing a change in its dipole moment which gives rise to an alternating dipolar electric field. It

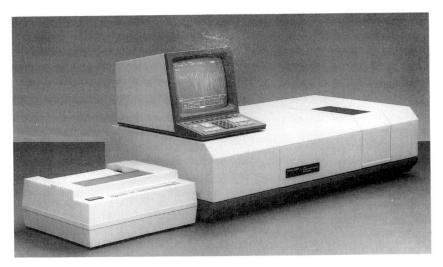

**Fig. 14.27 A Perkin-Elmer infra-red spectrometer
(courtesy of Perkin-Elmer Ltd)**

is this which interacts with the electrical components of electromagnetic radiation giving rise to an absorption of energy. The vibrational transitions are classified as stretching modes and deformation modes generally occurring in the wavelength range 2.5–15 μm.

Spectroscopists commonly use the term 'wavenumber' which is the reciprocal of wavelength and is expressed in units of cm^{-1}. The usual range of an IR spectrum is therefore between 4000 and 667 cm^{-1}. A complex molecule has a large number of vibrational modes which involve the whole molecule; however, most of these molecular vibrations are associated with the vibrations of individual bonds or functional groups. These vibrations are either stretching, bending, rocking, twisting or wagging. The fundamental stretching and deformation vibrations for the methylene group are:

Asymmetric stretch	2926 cm^{-1}
Symmetric stretch	2853 cm^{-1}
Scissoring	1468 cm^{-1}
Wagging	1305 cm^{-1}
Twisting	1305 cm^{-1}
Rocking	720 cm^{-1}

Infrared quantitative analysis is made possible as the absorption of IR radiation takes place in accordance with the Beer–Lambert law $A = Ecl$, where A is absorbance, E is a constant for the sample under investigation, c is the concentration of the sample and l is the pathlength of the IR radiation.

The most common use of this relationship is in 'Ratio' quantitative analysis. In this approach the strength of a peak in compound A compared to the strength of a peak in compound B is used to calculate the ratio of the two compounds in a mixture by comparison with a calibration table generated from known mixtures of A and B.

Quantitative and qualitative IR analysis are only successful when the sample is properly presented to the instrument. To present the correct quantity of sample to the instrument it is necessary to support the sample in the IR beam. Typical supports are constructed from the following: NaCl, KBr, CsI, KRS-5 (thallium bromide and iodide), AgCl or polyethylene. Liquids may be supported in cells of fixed or variable path length, the windows of which are usually NaCl or KBr.

Solids when they are film formers may be cast from solution as a thin layer on a salt plate. The most common technique for sampling solids is to prepare a paste (mull) of the solid in a viscous liquid.

Films and prints may be examined by placing them directly in the IR beam. However, a more common technique widely used in the surface coatings industry is that of surface reflectance.

There are two useful variations of reflection IR spectroscopy. The first involves specular reflectance of the IR beam from a metal or metal oxide surface. An organic layer on this surface will selectively absorb IR radiation as the beam travels into and returns through the layer. The strength of the absorption is then dependent upon the material and the layer thickness.

The second variation is multiple internal reflection (MIR) spectroscopy. The surface of the sample to be investigated is clamped against a crystal of TlBr–TlI (KRS-5) or Germanium. Infrared radiation enters this reflection element and is totally internally reflected at the interface between the crystal and the sample. Some penetration of the sample takes place thus leading to selective absorption.

The following equation derived by Harrick gives the depth of penetration of the IR beam:

$$d_p = \frac{\lambda_\theta}{2\pi n_1 [\sin^2 \theta - (n_2/n_1)^2]^{1/2}}$$

where θ is the angle of incidence between the IR beam and the surface normal, n_1 and n_2 are the refractive indices of the crystal and the sample respectively, λ_θ is the wavelength of the radiation.

Utilising reflection techniques information regarding the surface or coating on a sample as small as 1 cm^2 may be obtained.

In the analysis of printing inks interpretation of IR spectra is rarely done by identifying every peak. Analysis is normally done by comparing the unknown spectrum with known references. Most of the compounds used in the surface coatings industry are well documented. Typical examples are shown in Figs 14.28–14.30.

The use of microcomputers to enhance the data obtained from IR spectrophotometers has improved every aspect of IR analysis. Modern IR data stations can manipulate the raw data (spectrum) in a variety of ways: baseline smoothing, expansion of weak spectra, spectral overlay, abscissa

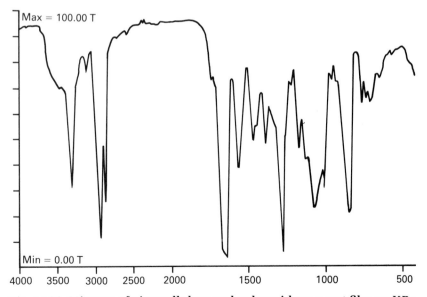

Fig. 14.28 Mixture of nitrocellulose and polyamide as a cast film on KBr

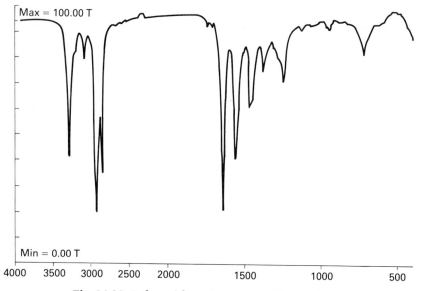

Fig. 14.29 Polyamide resin as a cast film on KBr

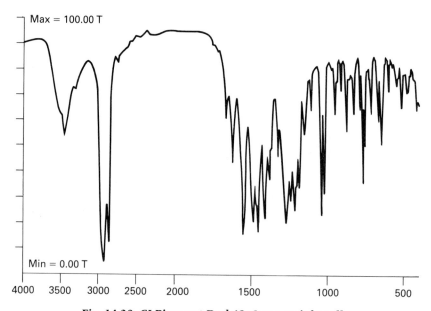

Fig. 14.30 CI Pigment Red 48 : 1 as a nujol mull

expansion, baseline flattening, conversion from transmittance to absorbance and vice versa, spectral subtraction, etc.

One of the standard routines, that of spectral substraction, may be used to illustrate the fact that this type of data handling brings its own unique problems. Firstly it should be noted that the absorption of IR radiation by a compound in a mixture is subject to matrix effects. These effects invariably lead to band broadening or even a slight shift in the frequency at which the absorption takes place. The result of subtracting the spectrum of a pure compound, e.g. plasticiser, from the spectrum of a mixture may result in the appearance of 'peaks' which do not truly exist.

Once the spectral data are in a satisfactory form the data station may be further used to interpret spectra, search software libraries to find the closest match and carry out automated quantitative analysis.

Once again computerisation has brought yet another technique, Fourier-transform infra-red spectroscopy (FTIR), from the world of pure research to the industrial laboratory. This is a very high speed scanning technique which can generate spectra of very small samples in seconds. This technique finds application in rapid quality control, process control and as a detector for other techniques such as gas chromatography.

High performance liquid chromatography

This is a relatively new laboratory technique (Fig. 14.31). It is once again a separation technique that when properly utilised may be used to both identify and quantify substances. A separation is achieved because the samples undergo a competitive distribution between a mobile liquid

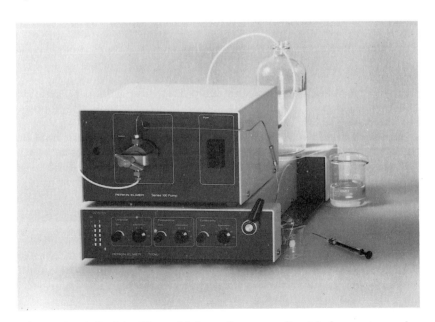

**Fig. 14.31 A Perkin-Elmer high performance liquid chromatograph
(courtesy of Perkin-Elmer Ltd)**

phase and a stationary phase. The stationary phase is contained within either a glass or steel column, the liquid phase is pumped through the column under pressure at precisely controlled flow rates. There are three different modes of separation employed in HPLC:

(1) Adsorption
(2) Liquid–liquid partition
(3) Gel permeation or size exclusion.

The latter of these will be dealt with in detail in the next section.

Adsorption chromatography

This is also known as liquid–solid chromatography. Typical adsorbents are silica and alumina. Silica has a slightly acidic surface and preferentially retains basic compounds, alumina has a basic surface and as would be expected retains acidic compounds. For the purpose of determining the order of elution in liquid chromatography it is necessary to classify compounds according to their polarity. The following list depicts the polarity of compounds with respect to polar adsorbents starting with the least polar. Fluorocarbons < saturated hydrocarbons < olefins < aromatics < halogenated compounds < ethers < nitro compounds < esters < ketones < aldehydes < alcohols < amines < carboxylic acids.

As HPLC is dependent upon a competitive distribution of sample between stationary and mobile phase, the polarity of the solvent used has a marked effect upon the separation achieved. In order to achieve optimum separation careful selection of solvents or even a programmed solvent gradient method may be required.

Liquid–liquid partition

The separation in liquid–liquid chromatography (LLC) is achieved by the distribution of the sample between two immiscible liquid phases. The stationary liquid phase is chemically bonded to the finely divided support particles. There are two distinct types of LLC: (a) normal — where the stationary phase is polar and the solvents used are non-polar; and (b) reverse phase — where the stationary phase is non-polar and the solvents polar. Solute elution in reverse phase is usually the opposite of that in normal LLC, i.e. polar compounds are eluted first.

In order to achieve a separation in reverse phase HPLC the components of the mixture submitted for analysis are encouraged to have a greater affinity for the stationary phase than the solvent. This is done by progressively increasing the polarity of the mobile liquid phase until the required separation is achieved. To ensure that all the compounds introduced on to the column are subsequently removed, gradient programming to a less polar solvent may be required.

One of the significant advantages of HPLC is that the detectors are mostly non-destructive and therefore specific detectors may be connected in series. Of the detectors available the following are the most common:

— Refractive index detector
— UV detector — fixed or variable wavelength
— Fluorescence.

Other available types are:
— Electrochemical detector

— Conductivity detector
— Infra-red detector
— Mass specific detector.

Of these the RI detector is the most universal although it may not be used for gradient solvent systems and lacks sensitivity when compared to other detectors. It is widely used as the detector in GPC as gradient elution is not employed and it responds to any compound present in the solvent.

The possible applications of HPLC to the analysis of surface coatings are continually growing. Some typical applications which have already been investigated are:
— Phthalate plasticisers
— Surfactants
— Polyols
— Amines
— Antioxidants
— Polynuclear aromatics
— Dyes
— Fatty acids
— Triglycerides
— Solvents
— Acids
— Inorganic anions and cations (using ion exchange columns and buffered aqueous mobile phases)
— Stabilisers
— UV monomers
— UV initiators.

Any separation which may be achieved by TLC may be achieved far better using similar HPLC conditions. Thus TLC is often used as a guide for the selection of a starting point in HPLC. Typically a system for examining polynuclear aromatic compounds would require a reverse phase liquid–liquid partition separation using acetonitrile and water as the mobile phase. Fluorescence detection in this case would give the greatest sensitivity.

Gel permeation chromatography

Gel permeation chromatography (GPC) is essentially a process for fractionating polymers according to their molecular size in solution resulting from permeation of the molecules through a porous matrix. Polymer solution is injected into a continuously flowing stream of solvent and passes through a bed of highly porous particles packed together in a column. If the pore size is so large that all the molecules can enter the pores or so small that none can, then no efficient separation can occur. The pore size range of the column packing must match the size of the polymer molecules under investigation. If this condition has been met, as the polymer solution flows through the column the smaller molecules will penetrate more pores than the larger and thus will travel a greater distance from injection to elution. As a result of this the large molecules elute first. GPC may be used as an analytical or a preparative technique.

One distinct advantage this has over HPLC is that everything put on the column, assuming it is dissolved in the solvent and not of such large particle size that it physically blocks the column, will be eluted.

There are two detectors used in GPC which are considered to be virtually universal in their response, differential refractive index detectors and low angle laser light scattering detectors. The data from gel permeation chromatography may be presented in the form of a distribution curve on a chart recorder or numerically. The numerical data, $\bar{M}_w$ (weight average), $\bar{M}_n$ (number average), $\bar{M}_v$ (viscosity average), usually encountered are very useful for comparing bulk properties of resin samples. These data, however, often fail to account for the low and high molecular weight tails of the polymer distribution and thus a direct comparison of two polymer distribution curves may prove more useful. Overlaying curves in GPC is a widely used technique (Fig. 14.32) but molecular weight distributions are often expressed in terms of the numerical data, which can be obtained from the molecular weight distribution curve. These values are all different for a given polymer and were originally obtained using classical methods. $\bar{M}_n$ may be derived from osmotic pressure measurements, vapour pressure measurements or other colligative properties such as elevation of boiling point or depression of freezing point.

$\bar{M}_w$ is obtained from light scattering experiments and is a measure of the size and shape of the molecules present.

$\bar{M}_v$ is measured by examining the viscosities of dilute solutions of polymer. The actual viscosity obtained is dependent upon temperature, the nature of the solvent and polymer, polymer concentration and the size of the polymer molecules.

These same numbers may be obtained with reasonably good agreement by using a calibrated GPC column set. Molecular weight standards are readily available for this. The elution time at any point on the GPC curve may be related to a molecular weight and the height of the curve when using a RI detector may be taken as proportional to the number of molecules of that molecular weight. By taking data from several points

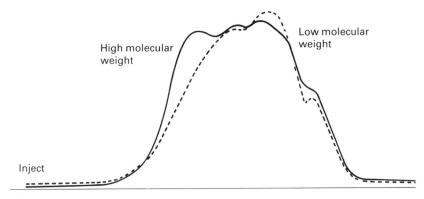

Fig. 14.32 Gel permeation chromatograms: (------) 'bad resin'; (———)
'good resin'

on the curve $\bar{M}_w$ and $\bar{M}_n$ may be obtained from the following equations:

$$\bar{M}_w = \Sigma h_i \, . \, M_i / \Sigma h_i$$

$$\bar{M}_n = \Sigma h_i / \Sigma (h_i / M_i)$$

h_i = height of GPC curve from baseline at elution time i

M_i = molecular weight of species eluting at time i

As previously mentioned, GPC can be used as an aid to the classification of a polymer. The molecular weight distribution affects many of the characteristic physical properties of a polymer. These may include:
— Tensile strength
— Brittleness
— Melt viscosity
— Softening temperature
— Cure time
— Coefficient of friction
— Hardness
— Gloss.

One very underrated application of GPC is as a separation technique. Even analytical grade columns may be used to fractionate a mixture of compounds although several injection/collection cycles may prove necessary in order to obtain sufficient sample to examine using other techniques such as IR spectroscopy.

Gas chromatography

Gas chromatography (GC) since its inception in the early 1950s has become a very powerful analytical technique. The instrument may be used to separate, identify and quantify almost any compound volatile below 450°C. The technique may be used either in a microanalytical mode or on a preparative scale.

The main principle of GC is that of a competitive distribution of a compound between a moving gas phase and a stationary phase.

There are two modes of GC, packed column and capillary column chromatography.

Until recently packed column chromatography was the method exclusively used in ink analysis and this may be subdivided into two categories:
(1) Partition — where the stationary phase is a non-volatile liquid.
(2) Adsorption — the stationary phase is an active solid.

Partition or gas liquid chromatography (GLC)

In conventional GLC the stationary liquid phase is coated on to a fine, particulate, 'inert' support, which is packed into a tubular column, normally made of glass. The moving gas phase or 'carrier' gas is then passed through the packed column. The mixture to be separated is then introduced into the carrier gas and swept through the column.

If no partition of sample between the carrier gas and the liquid phase takes place then the column behaves like a distillation column and compounds are eluted from the column in boiling point order. To separate two different compounds of the same boiling point then a stationary phase must be selected which will have a greater affinity for one of the compounds. In order to separate a polar from a non-polar compound, a polar liquid phase would be chosen. The degree of partition for the polar compound would be greater than the non-polar and thus the polar compound would be eluted from the column last.

The degree of separation in GLC is influenced by the following factors:

(1) Column temperature. As temperature increases the partition equilibrium is shifted to the gas phase and the compounds elute more rapidly.

(2) Stationary phase. The greater the affinity for the stationary liquid phase the longer a compound will take to elute.

(3) Carrier gas flow rate. Increasing this decreases the elution time from a compound, however separation efficiency of the column also depends upon this factor.

(4) Particle size of support. This controls the area of stationary phase in contact with gas phase.

(5) Column dimensions. Length and internal diameter may be adjusted to increase or decrease analysis time.

(6) The percentage loading of stationary phase on the support. This greatly affects the amount of sample that may be analysed and the retention time.

Careful control of these parameters eventually leads to the optimum separation of a mixture of compounds.

Adsorption or gas solid chromatography (GSC)

The process of GSC is very similar to that of GLC; only the mechanism by which compounds are preferentially retained by the column is different. These are two types of stationary phases used in GSC, active adsorbents and molecular sieves. Active adsorbents, carbon, alumina, retain compounds according to the chemical/physical nature of the active sites. Molecular sieves retain small molecules longer and therefore separate on the basis of molecular size. Gas solid chromatography is particularly applicable to the study of small molecules such as gases, typically H_2, He, Ne, Ar, Kr, O_2, CO, CO_2, N_2, NH_3, CH_4 to name a few.

Capillary gas chromatography

Capillary columns have been known for a long time but have only recently emerged as a routine laboratory technique. Typical capillary columns are made of fused silica which when coated with a polyamide resin become very flexible. Dimensions of these columns are typically 0.2–0.3 mm ID by 10–100 m in length. The incredible length of these compared with a typical packed column of 2 m is possible because capillary columns have a hollow core and therefore generate very low back pressures.

The liquid phase is bonded to the inside wall of the capillary column in thickness of 0.1–5 µm. Capillary columns compared to packed columns (see Figs. 14.33 and 14.34), give very efficient separations but have a very low sample capacity, limited normally to nanogram quantities per component. Column efficiency is measured in theoretical plates, a packed columns are made of fused silica which when coated with a polyamide capillary column will have 1000–4000 plates per metre.

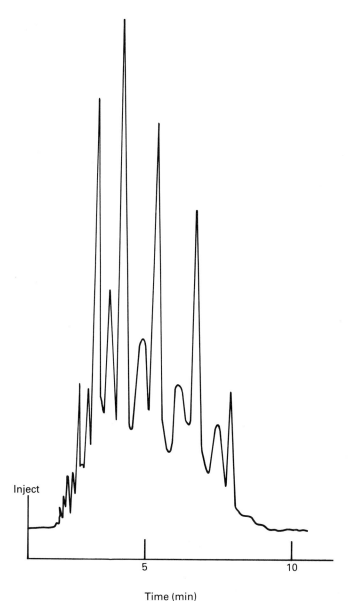

Time (min)

Fig. 14.33 Gas chromatogram of a distillate sample on a conventional packed column

The very low sample capacity of capillary columns means that special injection techniques must be used to prevent overloading and any resulting loss of resolution.

Perhaps the most useful of the various types of capillary injector for the examination of inks and related products is the split/splitless system. This

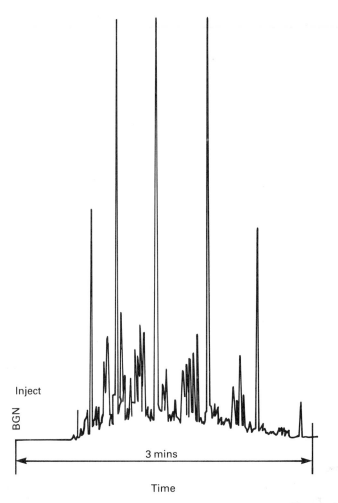

Fig. 14.34 Gas chromatogram of a distillate on a 25 m capillary column

technique allows the reproducible introduction of very small samples into the capillary column.

Other factors influencing the performance of a capillary column are the carrier gas flow rate and temperature.

Of the carrier gases available, H_2 offers the fastest analyses with maximum resolution, the greatest number of theoretical plates being available when H_2 is used at a velocity of 40 cm/s. The flow rates through a capillary are of the order of 1–2 ml/min and thus difficult to control; flow rate is normally controlled therefore by pressure.

Capillary columns have very thin walls and being of low thermal capacity are very sensitive to temperature fluctuations.

A modern gas chromatograph has three main component parts (Fig. 14.35): An injection block, a column oven and a detector. The injection block is where the sample is introduced into the column without leakage.

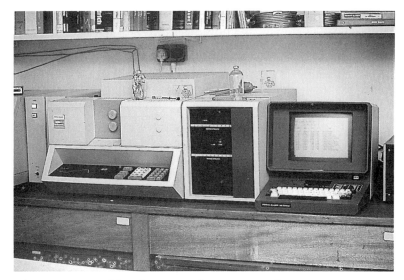

**Fig. 14.35 A Perkin-Elmer gas chromatograph with data station
(courtesy of Perkin-Elmer Ltd)**

The block is heated in order to vaporise the sample. A septum injection technique is normally employed using a precision sampling syringe.

Column ovens are necessary for maintaining reproducible temperature conditions. Modern chromatographs allow almost infinite variation of oven temperature with time and this is known as temperature programming.

Various detectors are available, flame ionisation detector (FID), katharometer or thermal conductivity detector (TCD), being the most common. Other more specific detectors are available, electron capture, nitrogen/phosphorus and flame photometric to mention a few.

None of the detectors listed are able to identify the compounds eluting from the chromatographic column. Identification is made by comparison of retention times with those of known materials.

There is, however, one detector which can be used to identify the eluted components very efficiently, i.e. a mass spectrometer.

Gas chromatography/mass spectroscopy, or GC/MS as it is known, combines the separating power of a gas chromatograph with the identification power of a mass spectrometer to produce one of the most powerful analytical tools available.

The components separated by the GC are eluted in turn into the MS. The component is fragmented into ions by an electron beam. The ions are then accelerated through a varying electromagnetic field formed by a radio frequency generator and a magnet in the form of a hollow centred quadrupole. The varying electromagnetic field serves to elute the ions in order of charge to mass ratio. Thus a mass spectrum is obtained for the component eluted. Every compound has its own unique mass spectrum, and therefore these instruments are equipped with computers and extensive data banks for the purposes of identifying the compound.

Whatever type of detector is employed, the output is proportional to the amount of component, and thus the technique may be used quantitatively. Two types of quantitative analysis are normally encountered, external standard and internal standard. The external standard technique involves calibrating the instrument by injecting a known quantity of reference material and relating this to a known quantity of the sample. This method is reliant upon the operator's ability reproducibly to inject accurately known quantities.

The most common quantitative analysis method involves the use of an internal standard. With this technique a known quantity of a standard compound is added to a known weight of both the sample and calibration mixture. This amount then figures in the final calculation and eliminates injection errors and changes in detector sensitivity.

Applications

Gas chromatography may be applied to the analysis of the following compounds met in printing inks:

Solvents	Alcohols, esters, etc. (Fig. 14.36)
Distillates	Petroleum based or synthetic
Plasticisers	Phthalates, adipates, citrates
Amines	Diethylamine, alkanolamines
Fatty acids	From oils or alkyd resins
Photoinitiators	Benzophenone, acetophenone
Antioxidants	BHT
Polyols	By prior derivatization
Resins	By prior pyrolysis or chemical cleavage followed by derivatization
Solvents retained in prints	By headspace techniques (see BS 6455: 1984) or discontinuous gas extraction

Sampling techniques

Samples are normally introduced into the GC by the use of precision sampling syringes. Syringes are available for injecting liquids, solids or gases. Liquid samples lend themselves to precise repeatable injections more so than solids or gases. It is preferable therefore to have samples in the form of non-viscous liquids. This may be done by diluting the sample with a known solvent or extracting the desired components from the rest.

Resins may be studied directly by GC using a pyrolysis technique; this has been shown to be of particular use for acrylic resins.

Headspace gas chromatography is a technique which has found increasing popularity in the packaging industry as it enables the quantitative analysis of retained solvent levels in prints to be carried out.

The technique involves enclosing a known area of print in a fixed volume sample jar fitted with a septum cap. The jar is heated such that any solvent retained in the print is driven into the 'headspace'. A fixed volume of this is then sampled using a gas tight syringe and the quantities of each solvent determined chromatographically. The quantity of solvent

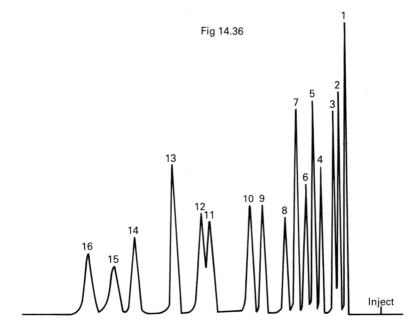

Fig. 14.36 Gas chromatogram of liquid ink solvents on a packed column:
(1) acetone; (2) ethanol; (3) isopropanol; (4) ethyl acetate; (5) MEK; (6)
isopropyl acetate; (7) *n*-propanol; (8) sec. butanol; (9) *n*-propyl acetate;
(10) isobutanol; (11) propylene glycol ether; (12) *n*-butanol; (13) toluene;
(14) *n*-butyl acetate; (15) ethylene glycol ether; (16) MIBK

retained in a print is normally expressed in terms of milligrams per square metre, typical values for an acceptable print are 10–30 mg/m^2, depending upon the final use of the print.

The headspace technique is not the best quantitative method available but is the most appropriate for giving rapid, reproducible results and is a technique which has been standardised throughout the industry in BS 6455.

Gas chromatography calculations

The two most common calculation methods encountered are external standard and internal standard. With these methods the area under a given peak is related to the concentration of the component by comparison with a chromatogram obtained from a standard mixture.

The calculation used in external standard methods is: area of component A in sample/area of component A in standard times amount of component A in standard = amount of A in unknown.

Obviously with this method the amount of sample injected must be accurately known.

The calculation used in the internal standard method is quite different:

$$\frac{A_s}{A_c} \times \frac{AI_c}{AI_s} \times \frac{QI_s}{QI_c} \times Q_A = \text{Amount of component A in sample}$$

where

A_s = area of component A in sample;
AI_s = area of internal std in sample;
QI_s = quantity of internal std in sample;
A_c = area of component A in calibration mix;
AI_c = area of internal std in calibration mix;
QI_c = quantity of internal std in calibration mix;
Q_A = quantity of component A in calibration mix.

This calculation is far more precise as the quantities of the various components can be accurately weighed and by relating the calculations via the internal standard component the amount of either the sample or the calibration mixture injected no longer matters as long as the quantities of components are within the linear working range of the detector and do not exceed the capacity of the column used.

Atomic absorption spectroscopy

Since its introduction in the 1950s, atomic absorption spectroscopy (AAS) has proved itself to be among the most powerful instrumental techniques for the determination of metals (Fig. 14.37). Between 60 and 70 metals may be determined invariably without prior separation. The technique is not restricted to aqueous solutions as many organic and mixed phases are suitable.

Atomic absorption spectroscopy is the measurement of the absorption of photons by an atomic vapour. As an atom absorbs a photon one of the outer electrons is transferred to an excited energy level. The wavelength

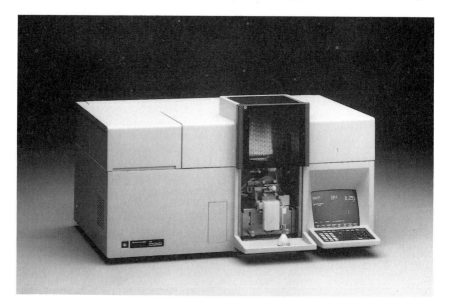

**Fig. 14.37 An atomic absorption spectrophotometer
(courtesy of Perkin-Elmer Ltd)**

at which this occurs is characteristic of the element and the degree of absorption is directly proportional to the concentration of the solution.

There are two popular techniques employed to create the atomic vapour, flame and electrothermal. The most widely used technique is flame atomisation, whereby a solution of the sample is sprayed via a nebuliser into an air/acetylene or nitrous oxide/acetylene flame. The light source for AA work is specific for the element being analysed and is normally a hollow cathode lamp. The lamp filament is coated with a layer of the element to be investigated, thus the light given out is at the appropriate wavelengths for absorption by the sample. The light given out is of several wavelengths and only one is used for the measurement, this is selected after passing through the aspirated sample by a monochromator. A photomultiplier detector amplifies the amount of unabsorbed light. The technique is applicable to solutions containing ppm levels of the element. Typical sensitivities are shown in Table 14.1.

Disadvantages of the technique are that the sample must be in the form of a solution. Only 5–15% of the aspirated solution reaches the flame and is then further diluted by the flame volume. The minimum sample size is 1 ml.

The majority of new equipment is available with flameless atomisation. This technique employs the use of an enclosed furnace and the sample is atomised by a heated filament which may be carbon or tungsten. The furnace is flushed with an inert gas, i.e. argon, to prevent oxidation of the heating source. As the sample chamber is enclosed during the measurement stage this technique is far more sensitive. Solid samples may also be used, although solutions are often preferable. Typical sample size would be 50 μl.

Atomic absorption spectoscopy may be used for the following determinations typical of the type of analysis found in the surface coatings industry:

Table 14.1 Detection limit ($\mu g/cm^3$)

Element	Flame atomisation	Electrothermal
Aluminium	0.09	0.002
Barium	0.05	0.0045
Cadmium	0.003	0.00002
Chromium	0.006	0.0016
Copper	0.005	0.001
Iron	0.008	0.002
Manganese	0.003	0.0003
Molybdenum	0.15	0.004
Nickel	0.012	0.002
Lead	0.02	0.001
Silicon	0.3	0.02
Titanium	0.2	0.012
Vanadium	0.15	0.009

— Determination of soluble metals from pigments.
— Determination of metals from driers.
— Metals in resinates.
— Metals from leafing agents.
— Metals from chelate type adhesion promoters.

The sample solutions may be prepared by either dry ashing followed by acid digestion of the metal oxides or salts, or by a wet ashing technique, depending on the volatility of the raw metal or its compound.

Dry ashing is carried out by heating a weighed sample until completely oxidised and the ash is then acid digested and diluted to a known volume.

Wet ashing involves completely charring a sample with concentrated sulphuric acid and then oxidising with nitric acid. The resultant solution is again diluted to known volume.

One of the many uses of AAS is determining whether an ink meets the specifications listed in the *Toys (Safety) Regulations 1974*, Statutory Instruments Number 1367. This legislation is under review but currently requires that inks for use on toys do not exceed the following amounts (ppm of the dry film):

Total arsenic	250
Total lead	2500
Soluble arsenic	100
Soluble cadmium	100
Soluble barium	500
Soluble antimony	250
Soluble chromium	250
Soluble mercury	100

Ultraviolet–visible spectroscopy

Ultraviolet–visible spectroscopy is concerned fundamentally with the electronic transitions between energy levels within a molecule. The transitions are generally between a bonding or lone-pair orbital and an unfilled non-bonding or anti-bonding orbital. The wavelength at which this transition takes place is then proportional to the separation of the energy levels of the orbitals involved. Excitation of electrons from p, d and π-orbitals and π-conjugated systems, gives rise to very useful spectra. But UV–visible spectroscopy is not confined solely to inorganic salts and conjugated organic systems, for even simple ketones will give rise to a weak spectrum; however, the former are the most widely investigated using this technique.

One of the most useful aspects of this technique is that the absorption is directly proportional to the concentration of the sample in dilute solutions. This relationship tends to break down in concentrated solutions. The absorption of energy in the UV–visible range obeys Beer–Lambert's law such that the absorption is directly proportional to the concentration of sample in solution. The relationship may be expressed by the following equation:

$$A = Ecl$$

where A = absorbance;
$\quad\quad\ E$ = extinction coefficient (constant for the compound being investigated);
$\quad\quad\ c$ = concentration, and
$\quad\quad\ l$ = pathlength.

The UV–visible spectrum is usually taken on a solution of typically 0.001% concentration. A portion of this is transferred to a silica cell of pathlength 1 cm. An identical cell containing the pure solvent is also prepared and the two cells placed in the appropriate part of the spectrophotomer (Fig. 14.38). Two equal beams are then passed through these cells and the intensities compared over the wavelength range of the instrument. The spectrum of absorbance versus wavelength is plotted automatically. The excitation of electrons is accompanied by changes in both the vibrational and rotational quantum numbers giving rise to a broad peak. Further interactions with the solvent molecules blur this out and ultimately a smooth curve is obtained.

The most common solvent is 95% ethanol as the commercial absolute ethanol contains benzene. Fine structure due to vibrational and rotational absorbance may be revealed by using non-polar solvents where possible such as cyclohexane. Dipole–dipole interactions with the solvent molecules lower the energy of the excited state and therefore ethanol solutions give longer wavelength maxima than hexane solutions. This is known as a red-shift.

One important factor to note is that in UV–visible spectrophotometers the beam is split into discrete wavelengths by a diffraction grating before passing through the sample. This is because UV light is at a frequency and energy likely to cause decomposition of a molecule, therefore the minimum necessary is passed through the sample at any one time.

Despite having many applications in the ink industry UV–visible

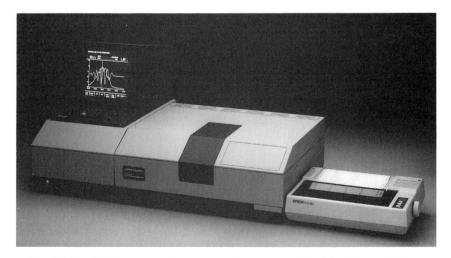

Fig. 14.38 A UV spectrophotometer (courtesy of Perkin-Elmer Ltd)

spectrophotometry is not a widely used technique due to the problems of sample preparation. The sample under investigation needs to be free from interfering compounds, thus the obvious application of UV curing inks is probably one of the least exploited.

Typical applications of UV–visible spectrophotometry are:
(1) Identifying and quantifying dye mixtures.
(2) Identifying and quantifying optical brighteners.
(3) Determination of phenolic antioxidants.
(4) Determination of aromatics in solvents.

Nuclear magnetic resonance spectroscopy (NMR)
This analytical technique depends on the fact that nuclei of certain atoms act like spinning bar magnets when placed in a magnetic field. The most commonly studied nucleus is that of the hydrogen atom, i.e. the proton, and in any organic molecule the various hydrogen atoms react differently to a magnetic field, depending on their position in the chemical structure.

This response can readily be measured and interpreted thus helping the analyst to determine the chemical structure of the material under investigation.

Figure 14.39 shows the spectrum obtained with ethanol using tetramethyl silane (TMS) as the internal standard and clearly identifies the different H atoms in the molecule.

For more detailed information on NMR spectroscopy the reader is recommended to read one of the specialist books on this subject.

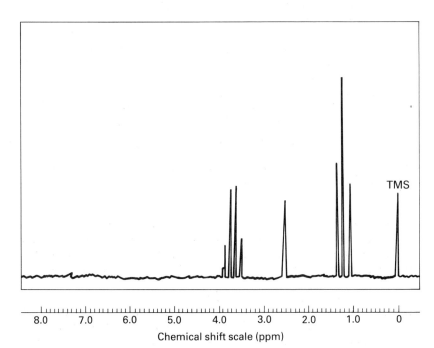

Fig. 14.39 NMR spectrum of ethanol

Surface analysis techniques

Earlier in this chapter we had discussed testing procedures which may be used to identify a 'good' or 'bad' ink or an acceptable or unacceptable raw material. An ink, in order to perform properly on a press, must have particular rheological properties, flow, thixotropy, tack, etc. In order to perform well the final printed product must have good adhesion, slip, gloss, colour strength, flexibility, product resistance, and an assortment of other properties. Of all of these adhesion is possibly the least understood. Flame treatment of polyethylene bottles prior to screen printing for example modifies the surface to give improved adhesion. Exactly what happens to the surface is only just beginning to be understood. In order to understand the phenomenon of adhesion or failure to adhere we need to study the chemical and physical nature of the surfaces at the interfaces between ink to substrate and ink to ink layers. There are several important new techniques which will be discussed. The instruments are prohibitively expensive to purchase but access to such techniques is available through research institutes and universities.

X-ray photoelectron spectroscopy (XPS) or electron spectroscopy for chemical analysis (ESCA)

In this technique the sample is bombarded with X-rays of typical energies 1253.6 and 1486.6 eV. Photoelectrons are emitted from the surface as a result of this bombardment and are analysed in terms of kinetic energy. Photoelectrons emitted from the atomic core, i.e. the carbon 1s electrons, are highly characteristic and allow the identification of all elements except hydrogen. The peak intensities at the emitted energy levels are proportional to the numbers of atoms present. But XPS is a surface technique despite the fact that X-rays may penetrate several microns into the sample the emitted electrons still have to 'escape' from the sample matrix. The maximum escape depth for this technique is typically 3 nm.

Most XPS equipment permits the variation of the angle at which electrons leave the surface and thus changes the escape depth of the electrons. The actual depth sampled is given by the equation

$$d = 3\lambda \sin \theta \quad (\theta = \text{'take-off angle'})$$

A direct comparison of peak intensities at high and low angles enables a form of 'depth profiling' to be carried out.

Ion scattering spectroscopy (ISS)

This instrument bombards the sample with low energy (1−2 keV) noble gas ions. A proportion of these are scattered by the surface atoms and their energies are analysed. This technique is not applicable to organic compounds because the ions are not scattered by hydrogen atoms.

Transmission electron microscopy (TEM)

This technique makes use of the fact that electrons are similar to light having both a wave function and a particle nature. The electrons having passed through a very thin sample (10−100 nm width) are focused using electromagnetic lenses. Any investigation of an organic compound requires that the sample is coated with a thin layer of metal or carbon to prevent charging. Damage to the sample by the electron beam is common.

Scanning electron microscopy (SEM)

This technique is perhaps the most useful tool for gaining information regarding the topography of a surface. A finely tuned electron beam is scanned over the sample surface. The beam undergoes a variety of inter-actions leading to emission of electrons of varying energies, X-rays and light. The image in an SEM is produced by the detection of low energy secondary electrons. The scanning electron microscope has a depth of field some 10–100 times greater than an optical microscope would have at the same magnification, the resolution of an SEM is also of the order of 10 better. The only major problem in SEM is the charging of insulating materials, and this may be overcome by applying a 30–50 nm conductive coating by vapour deposition.

The use of energy dispersive analysis of X-rays (EDX) is a technique which may be incorporated in an SEM. This utilises a high energy electron beam which induces the emission of X-rays. The frequency of the X-rays is very specific to the emitting atom and proportional to the number of similar atoms. Therefore, EDX may be used to identify the chemical nature of the surface under investigation. The X-rays generated are not confined purely to the surface layer but may originate from a depth of several microns.

Secondary ion mass spectrometry (SIMS)

The sample is bombarded with a beam of ions e.g. Ar^+, O_2^+, of 2–20 keV. This bombardment gives rise to a few charged particles being emitted from the surface. These are then examined using a mass spectrometer from which secondary ion mass spectra are obtained.

This technique may be used in either one of two modes of analysis. Low energy or 'static SIMS' is used for gaining the maximum information from the surface being investigated. High energy 'dynamic SIMS' utilises a high current density ion beam which may be used to give a 'depth profile' of the sample.

Environmental monitoring

Historically the printing ink industry has used whatever materials are necessary to achieve the performance required. Solvents, catalysts and other additives were chosen purely for their characteristics. Ignorance of potentially unhealthy side effects led to a great deal of misuse. With the increasing awareness of the possible damage to both the environment and the health of workers, legislation has followed to ensure that standards of cleanliness are met. Factory effluents are monitored as is the working atmosphere. Atmospheric monitoring is that which greatly concerns this industry due to the very volatile nature of the solvents used.

The measurement of fundamental importance is that of the average level of airborne contamination to which any one person may be exposed.

The recommended limits for occupational exposure levels are available from the Health and Safety Executive. However, every manufacturer should attempt to eliminate the presence of contaminants from the working atmosphere.

There are several ways in which the atmosphere may be tested. 'Gas sampling' involves taking an atmospheric sample using a gas tight syringe or sampling cylinder and then determining the contents of the device by gas chromatography. This technique as with sampling tubes such as the Draeger tubes gives a result only for the atmospheric conditions at the time of sampling. In order to obtain a time-weighted average, passive or active sampling devices must be worn by a worker for a significant time period. These devices contain absorbents which may consist of activated charcoal or porous polymers. Passive samplers absorb solvents from the atmosphere by diffusion. Active samplers require a pump to draw air over the absorbent. The absorbed contaminants may be desorbed by a suitable solvent or heat for subsequent analysis using gas chromatography.

Environmental monitoring is an ever-increasing, important aspect of an analyst's duties.

REFERENCES

1. *Paint Tech.* **29**(5), 30 (1965).
2. Beresford, J. *J. Oil Col. Chem. Assoc.* **50**, 594–614 (1967).

FURTHER READING

Feigl, I. F. (1939). *Qualitative Analysis by Spot Tests.* Elsevier.
Hachenberg, H. (1973). *Industrial Gas Chromatographic Trace Analysis.* Heyden and Son Ltd.
Hachenberg, H. and Schmidt, A. P. (1977). *Gas Chromatographic Headspace Analysis.* Heyden and Son Ltd.
Haslam, J., Willis, H. A. and Squirrell, D. C. M. (1972). *Identification and Analysis of Plastics.* Butterworth and Co. Ltd.
Mark, H. F., Othmer, D. F., Overberger, C. G. and Seaborg, G. T. (1984). *Kirk-Othmer Encyclopedia of Chemical Technology.* John Wiley and Sons.
Miller, R. G. J. and Stace, B. C. (1972). *Laboratory Methods in Infrared Spectroscopy.* Heyden and Son Ltd.
Vogel, A. I. (1968). *Elementary Practical Organic Chemistry.* Part 3, *Quantitative Organic Analysis.* Longman.
Williams, D. H. and Fleming, I. (1969). *Spectroscopic Methods in Organic Chemistry.* McGraw-Hill.

CHAPTER 15
Health, Safety and the Environment

> '*Although many substances may present hazards at work, they can still be used without risk to health and safety provided those hazards are understood and the appropriate precautions are taken.*'
> *(Health and Safety Executive,* Substances for Use at Work, *1986)*

General introduction to UK legislation

This chapter is aimed at a review of UK legislation and other constraints affecting health safety and environmental aspects associated with the manufacture of printing inks. It is not intended to set out in fine detail all the provisions of the regulations discussed but simply to highlight the existence of relevant legislation and note the provisions that impinge on the ink-maker's art.

The first section presents an overview of three principal areas of legislation:
— Health and Safety at Work, etc. Act 1974
— Control of Pollution Act 1974
— Product liability legislation.

Brief mention is also made of a number of isolated statutory instruments made under other legislation.

The second section covers the handling of chemicals and mixtures (preparations) under the following headings:
— Occupational exposure limits
— ECOIN/EINECS and notification of new substances
— Lead
— Fire hazards
— Data Communication: Labelling
                                  Health and safety data sheets
                                  Tremcards
— Transport
— Control of substances hazardous to health
— Control of industrial major accidents.

The third section deals with mechanical and operational aspects of health and safety in the manufacture of printing inks.

The fourth section covers four particular applications for printing ink which require more than general attention to health and safety matters.

Health and Safety at Work, etc. Act 1974 (HSW Act)

The quotation at the beginning of this chapter embodies an unexpressed principle of current UK health and safety legislation. With few exceptions, UK legislation does not seek to impose an absolute ban on the use of chemical substances; what it does seek to do is to ensure that those who decide to use chemical substances at work do so in the knowledge that they are required to make adequate provision for the health and safety of those expected to come into contact with chosen substances.

Prior to 1974, health and safety legislation was dispersed across a wide range of statutes. As a result of the deliberations of the Robens Committee of Inquiry [1], the Health and Safety at Work, etc. Act [2] set out to consolidate earlier legislation and established general duties on employers and employees. It also gave ministers the power to issue specific regulations to deal with specific problems. Every employer needs to consider, within the context of his own organisation, what action is necessary to fulfil the duties imposed by the Act [3, 4].

Section 1 of the Act made specific provision to replace a range of earlier statutes dating back to 1875. Existing duties, spelt out by earlier statutes, remain in force until the latter are replaced by specific statutory instruments under the HSW umbrella. The Act had immediate effect on the method of enforcement. The gradual replacement of those earlier provisions continues.

A two-component governmental agency was charged with overseeing actions to promote health and safety at work and their inspectors were given the powers to issue improvement notices or prohibition notices. Improvement notices set a specified time within which remedial action must be taken; prohibition notices require immediate suspension of an operation until deficiencies are rectified.

The Act aims to ensure that employers look after the health, safety and welfare not only of their employees but also of those outside their employment who may be affected by their work activities. Employers are required to control the handling and storage of dangerous chemicals and the emission of noxious or offensive substances into the atmosphere (not just those which are dangerous to health).

The primary responsibility for accident prevention and combating occupational ill health, lies with those who create the risk. However, the Act aims to involve everybody at the work place in achieving high standards of health and safety.

The employer's general duties will be carried out so far as is reasonably practicable. Specifically he must provide safe plant and systems of work, safe arrangements for use, handling and transport of raw materials and finished products, necessary information and training and a safe working environment. He must also provide a written statement of general policy with respect to health and safety at work. Recognised trade unions may appoint safety representatives who may, under prescribed conditions, require the establishment of a safety committee.

Manufacturers of substances will ensure that their products are so formulated and manufactured as to be safe and without risk when properly used, carrying out such tests and examinations as are necessary to ensure this, but it is not necessary to repeat tests that have already been completed. In this context 'properly used' will be taken to to imply having regard to any relevant information or advice which has been made available.

Adequate information about the safe use of the substances will be made available by suppliers. Research will be conducted to identify and eliminate (or minimise) any risk to health or safety to which the substance may give rise. An employee is required to take reasonable care for the health and safety of himself and others who might be affected by his actions at work, and is also required to co-operate with his employer to enable the employer's duties to be discharged.

In general, the coatings industry is not picked out for special treatment and ink-makers are required to comply with the general provisions imposed on processors of chemicals, including the provision of adequate advice to downstream customers. For the time being, special provisions for classification and labelling are included for the coatings industry, but these provisions are likely to be generalised in the foreseeable future.

Control of Pollution

The Control of Pollution Act [5] aims to consolidate earlier legislation and supplement common law constraints with statutory provisions. Again the coatings industry is not picked out for special treatment. The Act has four parts covering land pollution, water pollution (inland and coastal), noise and atmospheric pollution.

Under the land provisions there are two classes of waste. Controlled waste is defined to be any household, commercial or industrial waste. Disposal authorities are responsible for the provision of a written disposal plan for their areas which will be reviewed from time to time. It is illegal to deposit controlled waste on land or use equipment for disposal without a disposal licence.

The second class of waste, designated 'special waste' [6], contains listed substances that are either dangerous to life (defined) or have a flash point below 21 °C. The latter criterion is more significant to ink manufacturers as many ink solvents, particularly gravure and flexo, have flash points below this temperature. Such materials are subject to incineration disposal and are not for land-fill disposal. Producers of special waste must complete a consignment note which will be further completed by carriers and the disposer. Producers and carriers must maintain a register containing copies of all consignments of special waste which they handle, and their records will be kept for at least 2 years.

Part II of the Act covers pollution of water and continues the provisions of the earlier Acts. Under the Drainage of Trade Premises Act 1937, trade effluents may not be discharged into a public sewer otherwise than in accordance with a trade effluent notice served on the water authority at least 2 months before the discharge is made. The notice is to be regarded as an application for consent and the authority can consent with or without conditions. Separate agreements between the owners of premises and a water authority are permitted. By-laws may be enacted by the water authority to apply local constraints to prevent water pollution and the Minister is empowered to make (local) orders.

All coating materials are potential contaminants and it is even inappropriate to consider disposing of water-based coatings into public sewers. Factory layout must also take into account the possibility of accidental disposal via surface water.

Part III of the Act covers noise and does not rely on earlier statues (in contrast to Part II for example). The Noise Abatement Act 1960 has been repealed. The topic is covered in the section on noise (p. 812).

Local authorities are empowered to take summary proceedings in respect of noise amounting to nuisance. Noise abatement zones may be designated. These may cover the whole or part of an area and the order will specify the classes of premises covered. European discussions on noise at work have been going on for several years and these discussions have now culminated in a final directive to be implemented by member states by January 1990.

Part IV of the Act covers atmospheric pollution, is relatively short and relies on earlier statutes (Alkali and Works Regulations 1906 and the Clean Air Acts of 1956 and 1968) which have now been added to, empowering local authorities to collect and publish information on atmospheric pollution in their area.

The 1906 Act set up the Industrial Air Pollution Inspectorate, at one time under the control of the Health and Safety Executive but now transferred to the Department of the Evironment with effect from 1 April 1987 when it was incorporated into the unified HM Inspectorate of Pollution. It is an offence to fail to use best practicable means to prevent the escape of noxious or offensive gases from regulated works. The provisions may now be enforced under the Health and Safety at Work, etc Act by prohibition and/or enforcement notices.

The 1956 and 1968 Acts distinguish between 'scheduled' and 'non-scheduled' works according to the potential risks of the process involved. The former, where the risks are considered to be greater, are the responsibility of the National Inspectorate while the latter are administered by local authorities through their environmental health officers.

They control the emission of smoke from chimneys (industrial and domestic) covering visible vapours, but not invisibles such as sulphur dioxide.

The Department of the Environment is reviewing the working of present pollution control legislation and has proposed the introduction of a two-tier list of 'scheduled' works. Under the revised scheme 'Part A' works would be controlled by the National Inspectorate and a new list of middle-tier processes ('Part B' works) would be controlled by local authorities but using new powers with more explicit specification of the standards that must be achieved (in partial contrast to the existing concept of best practical means). Current drafts of the 'Part B' list specifically includes, under 'Other Categories', the entry 'd. Manufacture of printing inks and textile dyes.' The proposed changes are intended to take into account an EEC directive (84/360/EEC) and would set statutory emission limits (for emissions yet to be agreed).

Product liability

Product liability is that branch of the law requiring retailers, distributors, manufacturers, designers and installers to provide goods which are fit for purpose and are not inherently dangerous. Unfitness and danger are distinct but overlapping concepts. In general, the fitness for purpose is left to agreement between contracting parties in the UK. Only in the case of dispute is it necessary for the law to intervene in determining the meaning of a contract. However, safety is an area where the law sets itself a preventative rather than an arbitration role. Increasingly, safety has become the subject of criminal (as opposed to civil) sanction as under the Health and Safety at Work Act 1974.

In the UK, under the Sale of Goods Act [7], seller and buyer enter into a contractual agreement which guarantees the reasonable suitability and safety of the goods received by the buyer. Should the goods prove defective in either sense, the contract has not been fulfilled and the buyer can claim against the seller. During the course of any such claim, the law does not require that the buyer proves 'negligence' on the part of the seller. The seller remains liable to the buyer whether the defect is known or unknown. Also, the contract is between these two parties only; the buyer has no contractual arrangement with the manufacturer (if he is not the seller). The primary responsibility rests with the seller not the manufacturer. In legal terms, this is the concept of strict liability.

However, in the absence of contractual rights (for example, where the injured party was not the buyer) the injured party must prove negligence. In the USA and some European countries, this unreasonable situation where a third party has to prove negligence has been removed and there is considerable pressure on the European Community (EC) to adopt this broader application of strict liability. In the UK, there has been what might be termed isolated recognition in various Acts (e.g. in both the Toys Regulations [8] and the Pencils and Graphic Instruments Regulations [9] made under the Consumer Safety Act 1978) that this wider application of strict liability is appropriate and desirable.

As a result of this pressure, a European directive [10] has been published and adopted. It is likely to become UK law [11] in 1987 or 1988 following the publication of a composite Consumer Protection Bill [12] in November 1986. It must be emphasised, however, that for the purposes of the new directive, defectiveness is with respect to safety only and not by its fitness for purpose in a technological sense. Thus, the directive is not concerned if an electric fire, for example, fails to heat a room adequately, but is concerned if, in doing so, the electric fire electrocutes the operator. Also, the seller is not under any obligation to provide an absolutely safe product as some products (one extreme example might be dynamite) are inherently dangerous. What he must do is provide adequate information regarding the safe handling and use of such products. The legislation does not affect contractual obligations to provide products which meet the performance criteria agreed between buyer and seller.

The legislation will organise compensation, without the need to prove negligence, for death or personal injury resulting from defective products. Compensation will extend to damage to property but is restricted to property ordinarily intended for private use and so used by the injured party. Manufacturers will need to check that their insurance policies cover these liabilities.

Clause 36 of the Bill [12], by reference to Schedule 3, will amend the general duty under the Health and Safety at Work, etc. Act to ensure that manufacturers exercise reasonable foresight in product design and construction.

Other UK Acts

In addition there are a number of isolated statutory instruments issued under other Acts which potentially impinge on ink manufacture. These will be dealt with in greater detail in Sections 15.3 and 15.4. They are:
— Consumer Protection Act 1978: Toys Regulations
Pencil and Graphic Instruments Regulations
— Factories Act
— Fire Precautions Act
— Food and Drug Act.

15.1 HANDLING OF DANGEROUS SUBSTANCES IN THE MANUFACTURE OF PRINTING INKS

Occupation exposure limits

On an annual basis, the Health and Safety Executive publishes guidance [13] on exposure limits for some 750 airborne substances, including many solvent vapours and dusts which are of direct interest to the ink-maker. In the intervening period, quarterly amendments are noted in their periodical *Toxic Substances Bulletin*. The guidance is issued especially to assist manufacturers of substances to discharge their responsibilities under Section 6 of HSW. Overall the intention is that exposure to such substances should be eliminated or minimised.

Within the present regime, it is a manufacturer's responsibility to ensure that his processes are so controlled as to ensure that the concentrations of solvent vapours and dusts in the factory environment do not exceed any of the published limits. Measurement of solvent vapours can be either for single point, single substance using for example a

Draeger tube, or for monitoring mixed atmospheres over an extended period of time using for example an ORSA gas badge.

The Draeger tube is similar to a breathalyser in operation as it samples a known volume of air and gives an indication of the atmospheric concentration of a single substance by means of a colour change. It is quick and simple to use and gives a reasonably accurate assessment although it is unable to take account of the possible cumulative effect caused by mixtures of vapours of a similar chemical type.

The gas badge, while much slower in operation, is capable of measuring the atmospheric concentration of mixed solvents by diffusion into an absorption tube under controlled conditions followed by measuring of the absolute quantity of each component by gas/liquid chromatography.

The frequency of measurement in each case is currently left to the judgement of the manufacturer which is in contrast to the expected legislation to be introduced under COSHH — see p. 807.

The published exposure limits take the form of a two-part listing, the first entitled 'control limits' covering some 30 substances or groups of substances which the authorities believe require stricter control than the further 700 or so substances in a second listing entitled 'recommended limits'. Collectively 'control limits' and 'recommended limits' constitute occupational exposure limits (OEL).

Both control limits and recommended limits are further subdivided into 'long-term' and 'short-term' exposure limits. The long-term exposure limit, or 8-hour time weighted average (TWA) is that level of exposure to which it is believed that nearly all workers may be repeatedly exposed for 8 hours per day without adverse effect. The short-term exposure limit (10-minute TWA) is set for those substances for which there is evidence of a risk of acute effects occurring as a result of brief exposure.

Control limits

The control limits derive from Regulations, Approved Codes of Practice, European Community Directives, or are by adoption of the Health and Safety Commission (HSC). They are deemed to be reasonably practicable on the basis of scientific and medical evidence. Failure to comply with control limits, or to reduce exposure still further where this is reasonably practicable, may result in enforcement action.

The glycol ethers have been the subject of extensive review by ECETOC [14] and more recently by ACTS [15]. The OELs for all but 2-Methoxyethanol and 2-Methoxyethyl acetate in the 'Control limits' listed in Table 15.1 (for which the data represent the current figures) are to be reduced from January 1988 to the figures indicated.

Recommended limits

The recommended limits are simply recommendations of the Health and Safety Executive (HSE) on advice from HSC's Advisory Committee on Toxic Substances (ACTS). These limits are considered to represent good practice and realistic criteria for the control of exposure to the relevant substances. Inspectors will use the recommended limits as part of their criteria for assessing compliance with HSW Act. Table 15.2 illustrates

Table 15.1

	Long-term exposure		Short-term exposure	
	ppm	mg m^{-3}	ppm	mg m^{-3}
2-Butoxyethanol	25	120	—	— sk[a]
2-Ethoxyethanol	10	37	—	— sk
2-Ethoxyethyl acetate	10	54	—	— sk
2-Methoxyethanol	5	16	—	— sk
2-Methoxyethyl acetate	5	24	—	— sk
1-Methoxypropan-2-ol	100	360	—	— sk

[a] sk = materials where absorption through the skin represents a significant health risk.

typical examples from the 'recommended limits' list of ink raw materials. However, it is important to note that the Guidance Note from which they are taken is reviewed and reissued annually. **The most recent Guidance Note should always be consulted for the current figures.**

The HSE is at pains to point out that exposure to all substances, including those listed, should be kept as low as is reasonably practicable at all times. In addition, for listed substances, the exposure limits specified should not normally be exceeded.

The exposure limits should not be used as an index of relative hazard

Table 15.2

	Long-term exposure		Short-term exposure	
	ppm	mg/m^{-3}	ppm	mg/m^{-3}
Ethylene glycol (vapour)	—	60	—	125
Isophorone	5	25	5	25
n-Butanol	50	150	50	150 sk[a]
Toluene	100	375	150	560 sk
White spirit	100	575	125	720
n-Butyl acetate	150	710	200	950
Butan-2-one (MEK)	200	590	300	885
Methanol	200	260	250	310 sk
n-Propanol	200	500	250	625 sk
Ethanol	1000	1900	—	—

[a] For those materials where absorption through the skin represents a significant health risk, the entry is marked 'sk'.

or toxicity. Also it should be noted that they apply to concentrations of single substances. Exposure to additional substances, either simultaneously or sequentially, could give rise to greater hazards to health. There is no universally applicable method for the derivation of exposure limits for mixtures. The suitability of the wide range of formulae, devised in recent years, for a particular mixture should be assessed by toxicologists or other competent persons.

While occupational exposure limits apply to solvent vapours and dusts, dusts in general are perceived to represent an additional hazard by virtue of their physical form irrespective of their chemical nature.

Dusts in general should therefore be controlled [16] to the minimum that is reasonably practicable, although not all dusts have been assigned occupational exposure limits. The lack of such limits should not, however, be taken to imply an absence of hazard. In the absence of exposure limits for a particular dust, exposure should be controlled to the minimum that is reasonably practicable (a point which is emphasised in the proposed COSHH regulations). In addition, where there is no indication of the need for a lower value, personal exposure should not exceed 10 mg/m^3 8-hr TWA total dust and 5 mg/m^3 8-hr TWA respirable dust.

An official method for the measurement of total and respirable dust has been published by the HSE in the Methods for Determination of Hazardous Substances series [17].

Similar guidance is published by many other countries. The ACGIH in the USA publish an annual listing of threshold limit values and biological exposure indices [18] together with an updated publication [19] which presents reviews of the documentary basis for the values assigned. Corresponding data, assigned in Germany, are referred to as maximum allowable concentrations (MAK).

In the particular case of asbestos dust [20, 21] specific regulations have

Table 15.3   Gives specific dust limits which are set [13] for some powders.

	Long-term exposure		Short-term exposure	
	ppm	mg/m^{-3}	ppm	mg/m^{-3}
Calcium carbonate (total)	—	10	—	—
Calcium carbonate (respirable)	—	5	—	—
Carbon Black	—	3.5	—	7
Mica (total)	—	10	—	—
Mica (respirable)	—	1	—	—
Silica, amorphous (total)	—	6	—	—
Silica, amorphous (respirable)	—	3	—	—
Talc (total)	—	10	—	—
Talc (respirable)	—	1	—	—

been issued. Under these regulations any work undertaken which is likely to involve asbestos dust must only be done by licensed contractors.

ECOIN, EINECS and new substances

Several years ago, the EEC set itself the task of establishing toxicity parameters for all known chemical substances in order to control the information made available to users through, for example, labels and health and safety data sheets. Relatively early on in this exercise, it was realised that this was a virtually impossible task.

The EEC sought to overcome this problem by the simple device of establishing a definitive list of 'known' substances for which it is presumed that their manufacturers are sufficiently knowledgeable to provide relevant information on the basis of experience in use.

Thus, all chemicals demonstrated to have been on the market within the period 1 January 1971 to 18 September 1981 fall into this category of 'known chemicals' and were published initially as the European Core Inventory (ECOIN [22]) drafted by EEC itself. Subsequently, industry was requested to submit suggested additions to this list and the updated list, to be known as the European Inventory of Existing Chemical Substances (EINECS), will be published in due course. It must be remembered that EINECS is simply a listing of chemicals and makes no attempt to provide information relating to the toxicity of the listed materials.

Any chemical substance which appears in EINECS will be defined to be a KNOWN SUBSTANCE [23]; those which do not will, by definition, be NEW SUBSTANCES for the purpose of notification legislation and as such are subject to pre-marketing notification requirements. 'Substances' is not taken to include polymers which are deemed to be dealt with by virtue of the inclusion of their component monomers.

Any 'new' substance must undergo extensive physical and toxicological testing if a total quantity of 1 tonne or more is expected to be put on the market in a 12-month period. This is covered by the Notification of New Substances Regulations [24] which came into effect at the end of November 1982 and the requirements are amplified by a guidance booklet [25].

Recommended test methods for toxicity [26], physico-chemical properties [27] and ecotoxicity [28] are given in three associated Approved Codes of Practice. Early in 1987 the HSE expects to replace these three codes of practice by the adoption of the annex to EEC Directive 84/449/ EEC which itself amends two earlier directives on classification, packaging and labelling.

The dossier of information included in the notification to the competent authority will include a proposed classification, and labelling recommendations.

At the time when registration for EINECS closed, all significant ink raw materials had already appeared on ECOIN or were the subject of notification by their manufacturers.

Much speculation has been voiced about the curtailment of commercialistion of new chemical compounds but it remains to be seen what effect the 'Notification of New Substances Regulations' will have on the availability of new materials. Suppliers will need to assure themselves that likely financial returns on investment will justify the considerable expense involved in the toxicological testing required by these regulations.

Control of lead at work

It is generally recognised that lead is a significant cumulative poison and the HSC has issued specific regulations to control exposure to lead at work. The regulations are important for the ink industry to the extent that the screen ink sector continues to use lead pigments in limited quantities. SBPIM members outside the screen ink sector have voluntarily withdrawn all lead-based raw materials.

Requirements needed to control exposure to lead (including lead compounds) are spelt out in regulations [29] under the Health and Safety at Work Act, and a Code of Practice [30] details acceptable methods for meeting those requirements. At this stage it is worth noting that failure to comply with the provisions given in a Code of Practice is not in itself an offence, though such a failure may be used in evidence in criminal proceedings. It is open to the defendant to satisfy the court that he complied with the relevant regulation in some way other than that spelt out in a Code of Practice. In contrast, failure to comply with regulations can result, directly, in criminal proceedings.

Central to the control is the assessment by employers of the nature and extent of exposure. Under no circumstances should the lead-in-air standard (currently 0.15 mg/m^3 of air as an 8-hour TWA) be exceeded. To ensure this a further lower 'trigger' level has been set, above which exposure is defined to be 'significant' and regular air monitoring and medical surveillance are required. This second, trigger level is set at half the lead-in-air standard (i.e. 0.075 mg/m^3). At levels below that defined to be significant, only some of the regulations apply.

The underlying philosophy is one of control by prevention of lead releases; only as a last resort should protective clothing be relied upon.

The Code of Practice identifies some of the types of lead work where significant exposure will occur unless adequate controls are provided; these include work which gives rise to lead dust as exemplified by the manufacture of lead paints. Works in which compounds of leads are used in processes which give rise to dust have a statutory duty [31] to use the best practicable means for preventing lead dust emissions to the atmosphere. Air sampling techniques have been published in the Guidance Note series [32].

Fire

Legislation
The Fire Precautions Act [33] aims to ensure that adequate fire precautions to protect people from fire risks are established in certain classes of buildings, now including 'places of work'.

All factories in which more than 20 people are employed, must possess a fire certificate. The certificate imposes requirements relating to the maintenance of fire exits and means of escape, fire warning systems and signs, fire-fighting equipment and the training of employees on a case-by-case basis.

Because of split responsibility for the enforcement of separate controls, especially in the area of flammable solvents (fire certificates and the Petroleum Consolidation Act 1928) there exists potential for conflict between enforcing authorities. Any overlap will be a matter for discussion between the licensing authority and the HSE.

Separate regulations define 'special premises' [34] where fire risk is deemed to required special attention but this is most unlikely to apply to printing ink factories.

The Home Office has recently reviewed [35] the workings of the Fire Precautions Act and has provisionally concluded that it is necessary to replace the existing certificates with a more efficient procedure, placing greater emphasis on identification of greatest risk areas and introducing the use of Improvement Notices and Prohibition Notices (as in HSW). This review has been taken into consideration in the drafting of Part 1 of the Fire Safety and Safety of Places of Sport Bill [35]. Places already carrying a certificate under the 1971 Statute would not be subject to Improvement Notices.

The 'Highly Flammable Liquids Regulations' (HFLR) [36, 37] are of more direct relevance to the printing ink industry. A highly flammable liquid is defined to be a liquid giving off a flammable vapour at a temperature of less than 32°C and which supports combustion (according to the prescribed test). (This definition should not be confused with the EEC labelling criterion of 'below 21°C'.)

The regulations apply to all factories where any highly flammable liquid is present in connection with a business purpose, with the exception of those subject to a licence under other related Acts.

The regulations control the storage, marking of storage facilities, precautions against spillage, source of ignition, prevention of escapes of vapours and their dispersal, explosive pressure relief, means of escape, solid residues, smoking, taking of samples, fire fighting and duties of employees.

The storage provisions [38, 39] are intended to ensure that highly flammable liquids, when not in use, are stored in a safe manner. The main purpose is to separate storage from process areas. Small quantities (total amount not to exceed 50 litres) may be kept in a work room in a suitably placed cupboard or bin of fire-resisting material.

Marking requirements specify the wording on store-rooms and containers, (excluding small containers below 500 cc capacity) but the container-marking provision may now be discharged by labelling according to the Classification, Packaging and Labelling Regulations (CPL; see pp. 798, 802). However, the HSE still recommend that some indication be included in the labelling which shows the applicability of the HFLR or Petroleum Consolidation Act (whose flash point criteria are different from those of CPL).

Ideally, spills will be prevented by transporting liquids in totally enclosed systems incorporating pipelines. If this is not possible, conveyance vessels will be so designed as to avoid the risk of spills. Open buckets and tins are not acceptable.

Sources of ignition must be eliminated. Smoking must not be permitted and 'no-smoking' areas clearly defined. Escape of flammable liquids into the workplace must be prevented and, where this is not possible, safe dispersal arrangements must be made.

A general guide to safe practice in the storage and use of highly flammable liquids for the manufacture of paint and other highly flammable surface finishes has been published [40]. Two further booklets in the same series [41] give basic guidance on the types of electrical apparatus available for use in flammable atmospheres.

In passing it should be noted the HFLR covers those nitrocellulose solutions which are dissolved in solvent systems having a flash point below 32°C which were previously covered by the Cellulose Solutions Regulations.

Fire and fire prevention

Fire theory

The tragedy of fire will fortunately have been experienced only by a minority of people, but with modern communications the horrific scenes at a football stadium fire were televised almost immediately illustrating how a match or lighted cigarette carelessly discarded into waste material caused a flaming carnage within a matter of minutes.

The theory of fire is very simple and can best be illustrated by the fire triangle in Fig. 15.1. Fire is the result of the interaction of oxygen, heat and fuel. If any one these factors is removed the other two are unable to support combustion and the fire will go out. Taking each of the factors in turn, oxygen is present in the atmosphere at approximately 21% v/v and this concentration is sufficient to aid combustion. Heat may be in the form of a naked flame, mechanical metallic spark, electric spark, spontaneous combustion, static electricity, or even generated by sunlight. The fuel can be paper, wood shavings, waste material, solvents or any other combustible material.

Solvents used in flexo and gravure inks give cause for concern because of their highly flammable nature. It is important to remember there are two flammable limits, the lower flammable limit and the upper flamm-

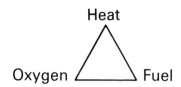

Fig. 15.1 Fire triangle

able limit. A flammable gas or vapour in air can only be ignited and sustain continuous combustion when its concentration is neither too low nor too high. The lowest concentration to support combustion is known as the lower flammable limit and the highest concentration to support combustion is termed the upper flammable limit; these are the limits of flammability (see Fig. 15.2). All intermediate concentrations are said to be within the flammability range. Flammability limits are usually quoted as volume percentages but may be given by mass concentrations. The lower flammability limit can best be illustrated when a car will not start because of insufficient petrol getting through, and the upper flammability limit when the carburettor is flooded. (Flammability limits are sometimes referred to as explosion limits).

Flammable limits usually relate to limits determined in air under a total pressure of 1 bar at 20°C. The theory of fire and flammable limits should be thoroughly understood.

As can be observed, from Table 15.4, most solvents used in the printing ink industry have a flammable range between 1 and about 12 vol. %, IMS having an upper value of 19%.

Sources of heat

Static electricity
The principles of static electricity were taught to us at school and we remember such experiments as those of rubbing ebony and glass rods with various materials and the effect the charges had on a gold leaf electroscope. Today these simple experiments are illustrated in our everday life. Shocks from static electricity often occur when walking over a synthetic fibre carpet and the body is earthed by the hand coming in

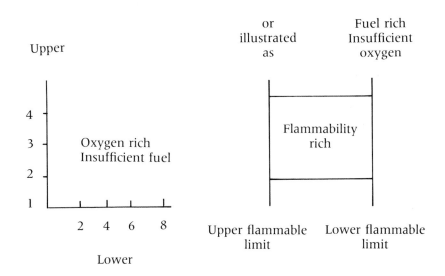

Fig. 15.2 Flammability limits

Table 15.4   Flammable limits

	Lower vol. %	Upper vol. %
IMS (Industrial methylated spirit)	4.3	19.3
Propyl alcohol	2.1	13.5
Isopropyl alcohol	2.0	12.0
Butyl aclochol	1.4	11.2
Isobutyl alcohol	1.6	10.9
Sec. butyl alcohol	1.7	9.8
Ethyl acetate	2.2	11.0
Propyl acetate	1.8	8.0
Toluene	1.2	7.1
Xylene	1.0	7.0
Acetone	2.6	13.0
MEK	1.8	10.0
White spirit	0.9	8.0
SBPs	0.9	8.0
HBPFs	0.6	6.0

contact with a filing cabinet or other fixture in an office or room. When a synthetic fibre jumper is taken off the body in the dark it is even possible to see the static electrical charges jump across from the garment. Clearly static electricity can be generated by very simple means and is not only restricted to man-made fibres. Static electricity can be generated by the flow of liquids and powders and therefore can be a potential danger in promoting a fire risk in an atmosphere of highly flammable vapours.

To prevent build-up of static electricity it is essential to earth all tanks, vehicles, drums, etc. before starting any operation. Such earthing systems must be kept clean and in good condition and regularly inspected. The faster solvent is pumped through pipes the more readily static is generated, so flow rates should be kept to a minimum, 1 m/s is the recommended flow rate. Static can build up in a system and special static footwear may be purchased to give good conductivity and allow a charge to dissipate. Ordinary leather shoes act as a good conductor and are much safer than many synthetic materials used in shoe manufacture.

Internal combustion
If a vegetable drying oil such as linseed oil is spilled and cleaned up with rags and wipers which are then left in a bundle in a restricted space, oxygen and heat could be generated possibly leading to spontaneous combustion causing a fire. For this reason all soiled and dirty wipers should be stored outside a building in a closed-top metallic container and should not be left in a building overnight. It has been known that even dirty wipers in a closed top metallic container can cause an exothermic reaction, the heat generated may be detected by touch or vision. Should this occur the container must not be opened, as the flow of fresh air could cause immediate combustion. The closed container should be doused

with water until the temperature is reduced to a safe level and then carefully investigated to ensure complete safety.

Fire extinguishers

While brief instructions for use are given on fire extinguishers, it is highly recommended that all personnel working in areas that present a higher than usual fire risk receive instruction on fire prevention and the use of extinguishers. It is common practice for the Fire Brigade and insurance companies to state the types and number of fire extinguishers to be placed in any location. The size, type and number of extinguishers will be assessed by the fire hazard and the volume of the room, laboratory or workshop. As far as possible the type of extinguisher placed in any location would be the type necessary to deal with the fire hazard; for example, it would be hazardous to use foam against fires involving electrical risk, therefore foam extinguishers should not be located near electrical equipment. Extinguishers should be placed in an accessible position and not be blocked in, should be regularly serviced and be in good working order.

Fire alarm procedure notices should be exhibited in all departments giving clear and precise instruction what to do in case of a fire. It is essential that all personnel be aware and know what exactly to do in case of a fire. Should a fire break out, the fire alarm procedure and drill must be strictly followed, but the following points on fire fighting are important if the fire can be tackled by local personnel.

The largest hand fire extinguisher has an active life of less than 3 minutes; if a fire cannot be brought under control in this period of time it is unlikely that the fire will be extinguished with just one extinguisher. Before tackling a fire it is essential that there is a safe escape route and if possible never tackle a fire alone. When attempting to extinguish a fire, aim at the base of the fire, and if there is excessive smoke keep down near the floor where the air is cleaner and breathing and visibility are easier.

Spillage of printing and auxiliary products should be cleaned up at once using sand or some other absorbent material. Most good fire prevention is simple and based on common sense, and is usually called good housekeeping. Fires do not occur — they are caused. One cardinal rule to remember with regard to fire is that life is infinitely more valuable than property.

Types of fire extinguishers
(1) Water — painted red.
   Usually of 2-gallon (9-litre) size; discharge water under carbon dioxide pressure; although not suitable for chemical fires they are very effective against ordinary combustible materials such as wood, paper, textiles. Water must not be used where live electrical equipment is involved.
(2) Foam — painted cream.
   Usually 2 gallon (9-litre) size, foam extinquishers are suitable for use on small fires in flammable liquids, especially where these are in a container or are overheated. A 2 gallon (9-litre) extinguisher can extinguish 10 sq. feet (1 m^2) of burning liquid. Foam must not be used where live electrical equipment is involved.

(3) Carbon dioxide (CO_2) — painted black.

Carbon dioxide extinguishers are suitable for use on small fires of flammable liquids and on electrical equipment, especially live equipment, where it is necessary to avoid damage or contamination by foam by dry powder. When using a CO_2 extinguisher never hold it by the horn as this will give severe freezing to the hand.

(4) Bromochlorodifluoromethane (BCF) — painted green.

Ideal for electrical, laboratory and computer protection. BCF is slightly toxic. Personnel should not be subject to a concentration greater than 4.5% and then only at this level for about 1 minute. The agent should not be used in a confined space and rooms should be thoroughly ventilated after use.

(5) Dry powder — painted blue

These are useful all round extinguishers for fires of flammable liquids, especially where the liquid may be spilled over a large area. They are also effective on electrical fires. They have the disadvantage of leaving a very large deposit of dusty white powder. Ensure the gauge indicating the pressure of the extinguisher is in the charged position.

See Fig. 15.3 for illustration of colour-coded fire extinguishers.

Automatic detection systems

Sprinkler installations detect fires and operate automatically to extinguish them, at the same time operating the fire alarm. Where a sprinkler system is not installed an automatic fire detection and alarm system should be considered as a means of an early warning.

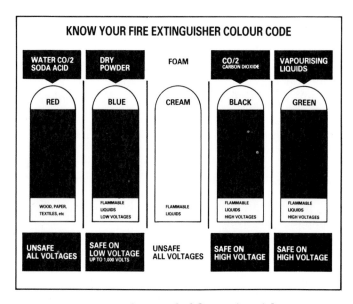

KNOW YOUR FIRE EXTINGUISHER COLOUR CODE

WATER CO/2 SODA ACID	DRY POWDER	FOAM	CO/2 CARBON DIOXIDE	VAPOURISING LIQUIDS
RED	BLUE	CREAM	BLACK	GREEN
WOOD, PAPER, TEXTILES, etc	FLAMMABLE LIQUIDS LOW VOLTAGES	FLAMMABLE LIQUIDS	FLAMMABLE LIQUIDS HIGH VOLTAGES	FLAMMABLE LIQUIDS HIGH VOLTAGES
UNSAFE ALL VOLTAGES	SAFE ON LOW VOLTAGE UP TO 1,000 VOLTS	UNSAFE ALL VOLTAGES	SAFE ON HIGH VOLTAGE	SAFE ON HIGH VOLTAGE

Fig. 15.3 Colour-coded fire extinguishers

Fire blanket
Non-asbestos fire blankets should be available to smother flames where clothing and other materials catch fire.

Sand buckets
Sand buckets are very frequently abused and used for waste, rubbish or cigarette ash trays. Dry, soft sand is a very effective means of extinguishing a small fire and can be used to prevent flammable solvent from flowing into public drains.

Zone classification and flame-proofing

The regulations which govern the use of highly flammable liquids set down a very strict code of practice for bulk use of such solvents to give the maximum protection against fire and explosion. Areas in which highly flammable solvents are used must be clearly marked 'No Smoking', and in order that highly flammable liquids are identified immediately by any persons using or coming into contact with them the sign 'Highly Flammable Liquid' should be clearly and boldly displayed. Highly flammable solvents, in this context, are those with a flash point below 32 °C.

In dealing with the risk of fire or explosion in the presence of flammable liquids or vapour three sets of conditions are recognised.

Div. 0
An area or enclosed space in which any flammable or explosive substance, whether gas, vapour or volatile liquid, is continuously present in concentration within the lower and upper limits of flammability. These conditions do not occur in printing ink manufacture.

Div. 1
An area within which any flammable or explosive substance, whether gas, vapour or volatile liquid, is processed, handled or stored and where during normal operations an explosive or ignitable concentration is likely to occur in sufficient quantity to produce a hazard. A risk of the nature described under Div. 1 can be combated by the use of flameproof or intrinsically safe electrical equipment.

Flexo and gravure ink plants will come into this classification and all electrical equipment will be flameproof. However, in a modern ink plant with good extraction ventilation, volatile liquid concentrations should not accumulate in the workplace for significant periods.

Div. 2
An area within which any flammable or explosive substance, whether gas, vapour or volatile liquid, although processed or stored, is so well under the conditions of control that the production (or release) of an explosive or ignitable concentration in sufficient quantity to constitute a hazard is only likely under abnormal conditions. An area falling into the category of Div. 2 is sometimes known as a 'Remotely Dangerous Area'.

Div. 3
Electrical equipment and switches need not necessarily be flame proofed but situated in an area placed well outside the risk area or in an adequately ventilated compartment.

Data communications to the user

One of the intentions of Section 6 of the Health and Safety at Work, etc. Act is to ensure that adequate information about products, with respect to their safe use, is available. Amendments to this general requirement have been under consideration since late 1984 [42], under which manufacturers would be required to provide (rather than make available) necessary health and safety information, and they will also be required to take into account foreseeable operator error when compiling their information documents.

Despite this direct attention to product safety data sheets, it has long been recognised that their provision is not always the most effective way of alerting those most at risk to the hazards associated with exposure to substances at work.

Supply labelling

In 1967, before the UK had joined the Common market, Directorate XI of the EEC had begun to draft what was to become a family of directives aimed at providing hazard labels for dangerous substances. The intention was to make available health and safety advice at the point of use in a harmonised way across the whole of the community.

The gestation period was very long; the UK enacted its first statute deriving from these directives in 1978 [43]. This first instrument was concerned with single substances (as opposed to preparations). This was followed by a number of amendments, until 1984 [44] when a major revision was published which brought UK legislation up to date with respect to substances and, for the first time, implemented the solvents and paints directives which date back to 1973 and 1977 respectively. A further UK update was published in 1986 [45].

For surface coatings manufacturers, detailed provisions are spelt out for the assessment (by calculation), of the hazards of individual coatings preparations and for the content of the associated hazard label. Manufacturers need to prepare lists of raw materials, noting those which fall into one of the statutory classes of danger and which type of calculation each will be subjected to (SOLVENTS, PAINTS or other SUBSTANCES).

The primary source of hazard data is the Authorised and Approved List [46] although, for materials not listed there, it will be necessary to seek advice from suppliers and the literature. A first amendment [47] to the Authorised and Approved List has been published and is primarily directed at conveyance data. Other amendments are envisaged, taking into account amendments required by the seventh adaptation directive 86/431/EEC Official Journal L247 (29) 1 September 1986, and it is expected that the Authorised and Approved List will be republished in loose-leaf form. A Guide [48] to the regulations and two Codes of Practice on classification [49, 50] have been published by the HSE.

Manufacturers need to subject their formulations to the specified calculations and to assess the flammability characteristics as determined by flash point measurements and, where necessary, the combustibility test. For those preparations which are classified as either TOXIC, HARMFUL,

CORROSIVE or IRRITANT, it is necessary to name components which are present above the labelling threshold set in the regulations [44] or the Authorised and Approved List [46].

The label will contain the symbol or symbols corresponding to the classification, the unique designation, including any nameable components, the name and address of the supplier and risk and safety phrases chosen, so as to ensure that the specific dangers identified by the classification procedure are properly expressed on the label.

Additionally, the regulations set out the labelling requirement for conveyance purposes, but to avoid the need for double labelling, for supply and for conveyance, regulation 10 permits the construction of a composite label for road journeys within the UK, and regulation 11 permits a construction for international journeys (see Fig. 15.4).

The SBPIM issued to members a short guide to the regulations, but it is emphasised that this was not intended as a substitute for reading the regulations themselves, the Guidance Note and the associated Approved Codes of Practice.

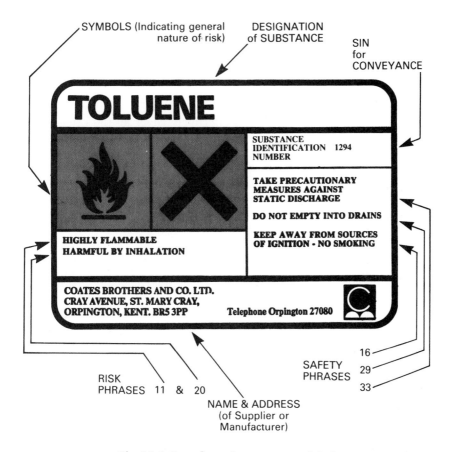

Fig. 15.4 Supply and conveyance label

A proposal for another labelling directive was published in August 1985 [51]. It seeks to extend the labelling of specific preparations (such as paints and inks) to cover all preparations using uniform criteria for classification. At the time of writing it seems likely that the paints and solvents directives will eventually be incorporated into a new General Preparations Directive (GPD) though the exact form that it will take is the subject of active discussion following the publication of the proposal.

A single classification calculation was proposed for assessment of very toxic, toxic and harmful preparations to replace the separate calculations option for SOLVENTS, PAINTS and other SUBSTANCES. At the time of writing the proposed classification procedure was the subject of vigorous debate.

Health and safety data sheets
Until recently, the HSE had preferred to leave the detailed form of the 'information made available' to the discretion of individual manufacturers and suppliers. Emphasis is being increasingly applied to the provision of information to those likely to be affected by exposure to products. The original provisions were contained in Sections 2 and 6 of the HSW Act, though the latter is the subject of amendment proposals [42] as a result of its application in practice.

The HSE has published interim guidance [52] covering the broad spectrum of information provisions. This has included illustrated examples of effective data sheets containing the necessary basic information and guidance to users. At the same time, it is pointed out that 'No single model format can adequately and effectively cover all the categories of information required under Section 6.... Indeed the converse may be true.'

This general guidance has been augmented by special advice [53] to the printing industry from the Printing Industry Advisory Committee (PIAC). A typical raw material safety data sheet as designed by SBPIM is illustrated (Fig. 15.5). A suggested check-list of essential items which need to be considered when preparing an information sheet is given in both the HSE and PIAC documents. The health and safety sheet must provide a means of communicating to the next employer/manufacturer in the supply chain, adequate information to allow the latter to formulate appropriate use and handling procedures for his own employees and customers. It may be necessary at each level of information assimilation to augment the material provided by the supplier by relevant information from other sources such as HSE (the regular Guidance Note series [54] for example or the occasional *Toxicity Reviews* [55]).

In contrast to the data sheet, a CPL label in accordance with SI 1984, No. 1244, communicates directly with the user of a product as supplied; the information cannot be regarded as exhaustive. The user is being given first-line alerting advice each and every time the container is handled. Although the label may be in compliance with CPL, it will usually be necessary to make further information available to comply with Section 6.

Coates Brothers PLC
Raw Material Safety Data Sheet [SBPIM MEMBER]

Raw Material Name _____ TOLUENE

Chemical Family ___ Aromatic Hydrocarbon

Synonyms _____ Toluol, Methyl Benzene

Appearance

Paste ☐   Liquid ☑   Solid ☐

Other _____ Clear Colourless

Health Effects

Harmful by inhalation. Irritating to skin and eyes, can cause dryness of skin due to defatting action.

Handling Precautions

Gloves ☑   Goggles ☑

Protective clothing ☐

Other _____ Exhaust ventilation

O.E.L.   100 ppm   L.D.50   3000 mg/kg

Emergency & First Aid

Inhalation

Remove victim to fresh air, obtain medical advice.

Ingestion

Do not induce vomiting. Seek medical advice.

Skin contact

Remove contaminated clothing, wash skin with mild soap and water. Apply skin conditioner.

Eye contact

Immediately flush eye with clean low pressure water. Seek medical advice.

Fire Hazard ___ Take precautionary measures against static discharges.

Flashpoint _____ 4°C _____   Flammability classification: UK ___ H.F. ___ EEC __ H.F. __

Autoignition Temperature ___ 480°C ___   Explosion Limits: Lower ___ 1.2% ___ Upper ___ 7.1% ___

Extinguishant Media: Waterspray ☑ Foam ☑ CO^2 ☑ Dry Chemical ☐ Other (state) _____

Physical Data

Characteristic Odour _____ Aromatic _____   Boiling Range _____ 111°C

Mpt _____ N.A. _____   Soluble in _____ Insoluble in water.

Specific Gravity _____ 0.87 _____   Other _____

Storage

Protect from Frost ☐   Protect from Direct Sunlight ☐   Flameproof Area ☐

Spillage & Disposal

Spillage should be covered with an absorbent material such as Vermiculite. Dispose by evaporation, if safe, or as local authority instructions.
This Product Safety Data Sheet is supplementary to, and should be read in conjunction with, the general advisory leaflet, COATES Health & Safety at Work.

Origination Date	86.06
Ref No.	NNN-30
P D No	12.1.5

Customers Signature _____ Date _____

Fig. 15.5 Raw material safety data sheet

In summary, labels and data sheets are complementary; clearly the separate messages must be consistent, but the one does not replace the need for the other.

Tremcards® (registered trade mark)

It was generally recognised that there is a need to provide different information to emergency services in the event of an accident during conveyance. Under the Tankers Regulations the driver of the vehicle is required to be given information in writing to ensure that he knows the identity of the substance(s) being conveyed, the associated dangers which might arise in an accident and the necessary emergency action. The CEFIC system of transport emergency cards is intended to meet the need for a uniform European system enabling the driver of a road vehicle carrying dangerous substances to have clear information in writing on immediate action that should be taken in the event of an emergency.

The European Agreement concerning the International Carriage of Dangerous Goods by Road (ADR) [56] at marginal 10385 of Annex B requires that, as a precaution against any accident or emergency the driver shall be given instructions in writing in the language of the originating state and those of the transit and destination states. The instructions will be kept in the drivers cab and will indicate the nature of the danger inherent in the substance(s) carried, the safety measures needed to avert the danger, the treatment for persons who come into contact with the goods, necessary fire-fighting equipment and counter-measures in the event of spillage.

Two volumes of recommended Tremcards have been published, one covering single substances [57], the other covering groups of substances and mixtures (Group Text) [58]. These documents will find additional application with the introduction of the Dangerous Goods in Packages Regulations. A Tremcards guide has also been published by the Chemical Industries Association [59]. (Figs. 15.6 and 15.7).

Unfortunately it is not possible, at present, to make firm recommendations for the selection of Tremcards for use during conveyance of printing inks. It is not clear whether early-published ink-related cards such as CEFIC TEC(R)595 and 596 are now adequate to meet CPL [44] and Carriage of Dangerous Substances in Packages [63] requirements.

The topic is the subject of active discussion with CEFIC and CEPE in Europe and SBPIM and PMA in the UK. Group Text cards currently under consideration include:

— 30G30   Highly flammable, non-nitrocellulose containing
— 30G34   High flammable, nitrocellulose containing
— 30G35   Flammable, non-nitrocellulose containing
— 30G36   Flammable, nitrocellulose containing
— 30G80   Flammable in consumer-sized packages
— 61G80   Toxic and flammable in consumer-sized packages
— 80G80   Corrosive and flammable in consumer-sized packages.

TRANSPORT EMERGENCY CARD (Road)

CEFIC TEC (R)-696
May 1972
Class 3 ADR
UN No. 1263

Cargo	**INFLAMMABLE PAINTS, VARNISHES, PRINTING INKS, ADHESIVES, THINNERS AND OTHER ANCILLARY PRODUCTS** Liquid or paste with perceptible odour Immiscible with water
Nature of Hazard	Highly inflammable Highly volatile The vapour is invisible, heavier than air and spreads along ground Can form explosive mixture with air particularly in empty uncleaned receptacles Heating will cause pressure rise with risk of bursting and subsequent explosion The vapour causes giddiness, may cause headache and in high concentrations induces unconsciousness
Protective Devices	Goggles giving complete protection to eyes Plastic or rubber gloves Eyewash bottle with clean water

EMERGENCY ACTION — Notify police and fire brigade immediately

- Stop the engine
- No naked lights. No smoking
- Mark roads and warn other road users
- Keep public away from danger area
- Use explosionproof electrical equipment
- Keep upwind

Spillage

- Shut off leaks if without risk
- Absorb in earth or sand and remove to safe place
- Sewers must be covered and basements and workpits evacuated
- If substance has entered a water course or sewer or contaminated soil or vegetation, advise police

Fire

- Extinguish preferably with dry chemical, foam or sand
- Do not use water jet

First aid

- If substance has got into the eyes, immediately wash out with plenty of water for several minutes
- Remove soaked clothing immediately and wash affected skin with soap and water
- Seek medical treatment when anyone has symptoms apparently due to inhalation, swallowing, contact with skin or eyes or inhalation of the fumes produced in a fire

Additional information provided by manufacturer or sender

TELEPHONE

Prepared by CEFIC (CONSEIL EUROPEEN DES FEDERATIONS DE L'INDUSTRIE CHIMIQUE, EUROPEAN COUNCIL OF CHEMICAL MANUFACTURERS' FEDERATIONS) Zurich, from the best knowledge available; no responsibility is accepted that the information is sufficient or correct in all cases
Obtainable from THE WHITEFRIARS PRESS LTD, MEDWAY WHARF ROAD, TONBRIDGE, KENT TN9 1QR.
Acknowledgment is made to the European Committee of Paints, Printing Inks & Artist's Colours Manufacturers' Associations for its help in the preparation of this card

Applies only during road transport　　　　　　　English

Fig 15.6

TRANSPORT EMERGENCY CARD (Road)

CEFIC TEC (R)-31
May 1971 Rev. 1
Class 3 ADR
Marg. 2301, 1°a
UN No. 1294

Cargo

TOLUENE

Colourless liquid with perceptible odour
Immiscible with water
Lighter than water

Nature of Hazard

Highly inflammable (flashpoint below 21°C)
Volatile
The vapour is invisible, heavier than air and spreads along ground
Can form explosive mixture with air particularly in empty uncleaned receptacles
Heating will cause pressure rise with risk of bursting and subsequent explosion
The substance poisons by absorption through skin and inhalation
The vapour has effects resembling alcoholic intoxication

Protective Devices

Suitable respiratory protective device
Goggles giving complete protection to eyes
Plastic or synthetic rubber gloves
Eyewash bottle with clean water

EMERGENCY ACTION — Notify police and fire brigade immediately

- Stop the engine
- No naked lights. No smoking
- Mark roads and warn other road users
- Keep public away from danger area
- Use explosionproof electrical equipment
- Keep upwind

Spillage

- Shut off leaks if without risk
- Contain leaking liquid with sand or earth
- Prevent liquid entering sewers, basements and workpits, vapour may create explosive atmosphere
- Warn inhabitants—explosion hazard
- If substance has entered a water course or sewer or contaminated soil or vegetation, advise police

Fire

- Keep containers cool by spraying with water if exposed to fire
- Extinguish with dry chemical, foam, halones or waterspray
- Do not use water jet

First aid

- If the substance has got into the eyes, immediately wash out with plenty of water for several minutes
- Remove contaminated clothing immediately and wash affected skin with soap and water
- Seek medical treatment when anyone has symptoms apparently due to inhalation

Additional information provided by manufacturer or sender

TELEPHONE

Prepared by CEFIC (CONSEIL EUROPEEN DES FEDERATIONS DE L'INDUSTRIE CHIMIQUE, EUROPEAN COUNCIL OF CHEMICAL MANUFACTURERS' FEDERATIONS) Zürich, from the best knowledge available; no responsibility is accepted that the information is sufficient or correct in all cases
Obtainable from THE WHITEFRIARS PRESS LTD, MEDWAY WHARF ROAD, TONBRIDGE, KENT TN9 1QR.
Acknowledgment is made to V.N.C.I. and E.V.O. of the Netherlands for their help in the preparation of this card

Applies only during road transport   English

Fig 15.7

Transport of dangerous goods

Historically the Tanker Regulations [60] were the first of a group of three sets of regulations dealing with the labelling and conveyance of dangerous goods. These were followed in 1984 [61] by the CPL Regulations (though these superseded the earlier 1978 [62] regulations, which only covered the supply of single substances). The third arm of the group, dealing with conveyance of dangerous goods in packages, was published in November 1986 [63]. The Packaged Goods Regulations appeared in draft form in mid-1984 [64], and were the subject of much consultation and redrafting. The Tanker Regulations cover the transport of all dangerous substances in tankers and impose duties on the operator to ensure that the vehicle is fit for the purpose and is regularly examined and tested.

The operator must obtain information from the consignor to enable him to be aware of the risks involved and the driver must have sufficient information to make him aware of the identity of the substance and its risks. A copy of such information will be kept in the cab of the vehicle. The driver must receive adequate instruction and training.

The subsequent Packaged Goods Regulations follow a similar pattern to that for the Tanker Regulations but with their own definition of dangerous goods. They apply to the conveyance of any quantity of a dangerous substance in bulk, or in receptacles with a capacity of 200 litres or more (together with a number of categories not likely to be of interest to ink manufacturers). They only apply to packaging group I and II materials (see below). Under Regulation 9 of the UK's Classification, Packaging and Labelling (CPL) Regulations (discussed in the section on Supply labelling, p. 798), substances dangerous for *conveyance* will be labelled to provide specified information (except for packages of capacity 1 litre or less).

The label will carry an indication of the contact point in the UK from where expert advice may be obtained, the designation of the substance, the substance identification number (SIN) and the appropriate hazard warning sign. For conveyance in receptacles greater than 25 litres, supplementary information (nature of the danger and emergency action) will be provided either on the label or as a separate sheet accompanying the package (such as a Tremcard for example). However, to avoid double-labelling (*supply* and *conveyance*), regulations permit the construction of composite labels for road journeys within the UK (regulation 10) and for international journeys (regulation 11).

Most international rules relating to the transport of dangerous goods are based on UN recommendations [65]. Each substance or article is assigned a UN number (the basis for UK SINS). The primary listing, in UN number order, records the official description together with recognised alternatives, the assigned UN (danger) class and where appropriate the division number, subsidiary risks and any special provisions. The listing also assigns substances or articles to a packing group and gives recommendations for the packing method. This information appears in the revised UK Authorised and Approved List [47] as Part 1A2 with supplementary information in Part 1B.

There are nine UN classes, the most important of which to the ink maker is Class 3 (Inflammable Liquids). Such liquids have a closed-cup flash point of not more than 60.5 °C or an open-cup flash point of not more than 65.6 °C. Having classified goods, subsequent chapters detail the recommendations for the transport of each UN class.

Substance identification numbers of importance to ink-makers, together with their associated packing group, are given in Table 15.5.

Based on the UN recommendations, the International Maritime Organisation (IMO) has issued and updates the IMDG Code [66] on a regular basis to control the transport of dangerous goods by sea and provide readily accessible advice to ships' masters as to requisite action in the event of an emergency. Regulations for transport of dangerous goods by air similarly follow the UN recommendations, as published in the IATA regulations [67]. Not surprisingly there are restrictions on what can be carried on passenger or cargo aircraft. ADR [56] is an agreement between nineteen contracting states whose purpose is to ensure that dangerous goods arriving at a frontier in or on a road vehicle have been suitably packed and are being carried safely. Under these circumstances the contracting states agree to allow the transport by road of dangerous goods in their territories.

Table 15.5

SIN	Description	Packing group
1210[a]	All 'flammable' — only inks with a flash point below 60.5 °C[b]	II or III (depending on flash point and viscosity)
1993	All 'flammable' — only thinners and diluents which are mixtures (single-substance solvents use their own SIN)	II or III (depending on flash point)
2059	Nitrocellulose solutions with flash point below 23 °C, less than 55% nitrocellulose of less than 12.6% nitrogen[b,c]	II
2060	Nitrocellulose solutions with flash point between 23 and 60.5 °C, less than 55% nitrocellulose of less than 12.6% nitrogen[b,c]	II
2557	Nitrocellulose with more than 18% plasticising substance and less than 12.6% nitrogen	II
1950	Aerosol-packaged materials irrespective of their contents	Not assigned

[a] It is fairly certain that 1210 includes inks containing NC.
[b] There are minor differences between authorities as to the exact flash point boundaries. The above are according to UN (transport) criteria.
[c] The original UK criterion was 12.3% nitrogen.

Control of Substances Hazardous to Health [COSHH]

The prospect of an increasing number of statutes, like those controlling exposure to lead [29] and asbestos [20], gave rise to a proposal in August 1984 [68] to replace existing piecemeal controls with an umbrella regulation controlling exposure to all substances hazardous to health at work. The consultative document gave rise to a strong response from many sides, and the comments received are still under consideration by the HSE. The proposal goes far beyond that outlined in an earlier EEC directive [69].

COSHH would apply to all substances defined as dangerous (toxic, harmful, corrosive or irritant) by the criteria of the CPL Regulations and, in addition, to substantial dust-in-air concentrations of any kind. In contrast to those CPL Regulations, COSHH places emphasis on the conditions of use while still bearing in mind the inherent properties (the latter being the specific concern of CPL).

Users of potentially hazardous materials will be required, on a formal basis, to examine their raw material inventories against the handling and safety data received from suppliers and other sources. An informed judgment of what monitoring and preventative action, if any, ought to be taken will be made by a competent person, noting where necessary the control limits imposed which must be met by adequate means. In the context of health and safety legislation, 'adequate' is intended to imply having regard only to the nature of the substance and the nature and degree of exposure. This concept should be contrasted with 'so far as is reasonably practicable' which has come to imply a need to comply until the cost of additional control measures becomes grossly disproportionate to further reduction in the risk under consideration.

Very much more formal procedures of record-keeping would be required for all staff exposed to dangerous substances as defined. In a sense, COSHH is reinforcing the Section 6 provisions of HSW and ensuring that the information is acted on by the user and reinforces the need to provide information to employees at risk from exposure in the workplace. Recently a further EEC proposal [70] has been issued amending their earlier occupational exposure directive.

The initial proposal had a relatively short list of materials picked out for special control (only lead compounds were of potential interest to ink-makers) though it did include a general requirement to minimise exposure of workers to all agents likely to be harmful to health. Minimising exposure by engineering controls was recommended. The use of individual protective measures was only envisaged when exposure could not reasonably be avoided by other means.

The amendment proposal increased the number of agents covered, by adding an appendix containing approximately 100 compounds together with assigned limit values. At the time of its publication these values were generally the same as those published in the UK's, then current, OEL list [13]. However, the permitted exclusion limits were in most cases higher than the UK's short-term exposure limits.

If, as a result of initial assessment, it is considered that values in excess of a fifth of an assigned limit value might occur, measurement of the concentration in the atmosphere would be required. A further annex lays down frequency of measurements.

The following materials used in the ink industry would be affected by the proposal:

Cyclohexanone	Talc (non asbestiform)
Ethyl acetate	Titanium dioxide
Ethylene glycol	Toluene
n-Hexane	Xylene
Silica (amorphous and amorphous respirable)	

The proposals are not wholly compatible with the UK's published COSHH proposal and further amendment of both the proposed directive itself and of COSHH can be expected. Final regulations are intended to be published in the first half of 1987 and to come into operation in 1988.

It should also be noted that the raw materials listing necessary for the labelling provision (discussed in the section on Supply labelling, p. 798) would form the basis for the review of raw materials inventories and the required assessment of health risks involved in working with those raw materials.

The Control of Industrial Major Accident Hazards

The Control of Industrial Major Accident Hazards Regulations (CIMAH) [71–73] implemented the so-called 'Seveso' directive and set out to impose controls on sites which use or store significant quantities of toxic, explosive, reactive or flammable substances and represent an extension of the controls earlier presented in the Notification of Installations Handling Hazardous Substances Regulations (NIHHS) [74].

The main aim of the new regulations is to ensure that adequate emergency plans are developed (by both local authorities and industrial organisations) for handling major accidents involving a number of specified dangerous substances. For each substance, there is a threshold quantity which must be present before the regulations apply. One class of materials covered by the regulations is cobalt compounds in general and this is of potential concern to the coatings industry. Their coverage was the subject of much discussion at the time that the regulations were published. The EEC will make a proposal to amend the scope of the definition of cobalt compounds, but in the meantime it is necessary to discuss cobalt driers, for example, with the local factory inspector.

With this exception, the ink industry is unlikely to find itself within the scope of the 1984 regulations. Nevertheless it is prudent for all sites to have a contingency plan to deal with emergency situations.

However, further controls are under discussion at the time of writing and proposals were published in 1985 [75] to require notification to the HSE and to require the marking of sites within which more than a total of 25 tonnes of all dangerous substances are, or are likely to be, on site at any one time. The definition of dangerous substance is taken from the CPL Regulations and includes materials which, although not listed in Part

1A1 of the Authorised and Approved List, have properties comparable to materials which are so listed.

Hazchem signs

Hazchem code signs [76] give information to the police and fire service about the action to be taken within the first 15 minutes in dealing with an incident. The sign gives an immediate indication of the hazard to the public, immediate information on the action required by the emergency services and is simple to use. It does not need an elaborate reference text for interpretation but can be understood from a simple card of the type illustrated. The information provided covers fire-fighting media, personal protection, risk of violent reaction, spillage and evacuation. The Hazchem sign can be attached either to vehicles such as tankers or to the entrances of factories and other premises (Fig. 15.8).

15.2 MECHANICAL AND OPERATIONAL ASPECTS

First aid at work

In the UK, the provision of adequate first aid facilities is one of the many aspects of the Health and Safety at Work Act. The First Aid Regulations [77] place a general duty on employers to make or ensure adequate first aid provisions for their employees if they are injured or become ill at work.

In determining the total number of first-aiders need in an establishment, account should be taken of all the relevant factors. These include not only the number of employees but also the nature of the work. As a guide, it is recommended that in establishments with relatively low hazards such as offices, etc. it would not be necessary to provide a first-aider unless 150 or more employees were at work; a ratio of 1 per 150.

In establishments with a greater degree of hazard such as factories, dockyards or warehouses there should be at least one first-aider present between 50 and 150 employees. When there are fewer than 50 employees at work in a factory then there may be no need for a first-aider, but an employer if he is not providing a first-aider, has to provide an 'appointed person' at all times when employees are at work.

Employers whose establishment present special or unusual hazard will need to provide more than one first-aider. Where establishments have a larger number of employees and are divided into a number of more or less self-contained working areas, the employer will need to provide both centralised facilities and supplementary equipment and personnel in other locations.

A first-aider is a person who has been trained and holds a current first aid certificate issued by a recognised authority. The certificate lasts for 3-year period and then is renewed by examination. An appointed person is a person provided by the employer to take charge of the situation (e.g. to call an ambulance) in the absence of a first-aider. Soap and water and disposable drying materials, or suitable equivalents, should also be avail-

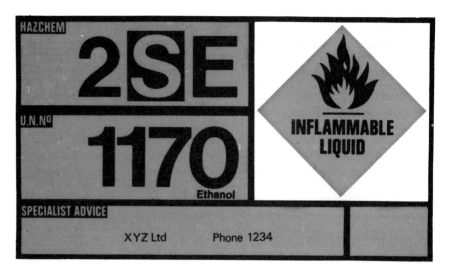

Hazchem card   *front*   *back*

Hazchem Scale				

1	JETS		
2	FOG		
3	FOAM		
4	DRY AGENT		

P	V	FULL	
R		FULL	
S	V	BA	DILUTE
S		BA for FIRE only	DILUTE
T		BA	
T		BA for FIRE only	
W	V	FULL	
X		FULL	
Y	V	BA	CONTAIN
Y		BA for FIRE only	CONTAIN
Z		BA	
Z		BA for FIRE only	
E		CONSIDER EVACUATION	

Notes for Guidance

FOG

In the absence of fog equipment a fine spray may be used.

DRY AGENT

Water **must not** be allowed to come into contact with the substance at risk.

V
Can be violently or even explosively reactive.

FULL
Full body protective clothing with BA.

BA
Breathing apparatus plus protective gloves.

DILUTE
May be washed to drain with large quantities of water.

CONTAIN
Prevent, by any means available, spillage from entering drains or water course.

Fig 15.8

able. Where tap water is not available, sterile water or sterile saline, in disposable containers each holding at least 330 ml, should be kept easily accessible, and near to the first-aid box, for eye irritation. First-aid boxes should be clearly marked and made of materials to protect the contents.

A first-aid box should contain a guidance card, individually wrapped sterile adhesive dressing, sterile eye pad, triangular bandages, sterile covering for serious wounds, safety pins, medium, large and extra large sterile unmedicated dressing.

Personal hygiene

Where dermatitis is a hazard, proven barrier creams and hand cleansers, preferably antiseptic, are necessary and a weekly hand inspection by a factory nurse or a supervisor is advisable. Barrier creams are formulated for various operations, some water repellent, others oil repellent, etc. The correct types must be chosen in the light of the work to be performed.

Reporting of injuries, diseases and dangerous occurrences

The 1985 [78] regulations cover three specific topics and their reporting procedures. The requirements and implications are extensively reviewed in an associated HSE Guide [79].

Death or injury at work

This topic covers fatalities and is intended to apply, in practice, to any death which occurs either at the time of the accident or subsequently. Specific injuries and conditions are illustrated, such as fracture of the skull, fracture of any bone in the arm (but does not include collar bone), fracture of the leg, amputations and injury from electric shock for example.

All accidents no matter how trivial or small must be recorded in the local accident book. A set procedure should be established to deal with all accidents and dangerous occurrences during and out of working hours.

Dangerous occurrences

Dangerous occurrences, having a high potential to cause death or serious injury must be reported to the HSE. This reporting gives the HSE the opportunity to accumulate information on the circumstances under which such accidents occur, adding to the pool of knowledge which can be called upon in accident prevention.

Locations of occurrences which require reporting include warehouses, factories, fairgrounds, conveyance by road, etc. and cover collapse of lifting machinery, building structures, scaffolding, release of harmful substances into the atmosphere, explosions and fires.

A free leaflet covering reporting of injuries and dangerous occurrences is available from the HSE [80].

Reportable diseases

The diseases are grouped into poisonings, skin diseases, lung diseases, infections and a group of miscellaneous diseases which are associated with exposure to particular chemicals or physical agents. A free leaflet on reporting diseases is also available from the HSE [81].

Training in health and safety

Accident prevention can most effectively be achieved by a well-trained and able workforce. The HASAWA 1974 clearly states that instruction, training and good supervision are essential. Therefore all personnel must be thoroughly trained in all aspects of their work. The training should be performed by a competent person and a record kept of all training on the individual's personal file. The training syllabus for health and safety could be formulated from the objectives listed in the safety audits. Individual training programmes can be formulated using the whole of this chapter and should be integrated with general training programes.

Lifting

Twenty-five per cent of all industrial injuries occur during the handling of materials. Many of these injuries are to the back causing much pain and the possible recurrence of the injury. The introduction of mechanical aids for lifting progresses at a steady rate but manual handling is employed at a very high level in industry and in the home. It is very difficult to legislate for which loads should be lifted by manual means, the size and shape of the object are never the same and the average worker does not have the capabilities of a trained weightlifter. A very simple guide for lifting is that the person concerned should judge for himself if he is capable of lifting an object. If the object is too heavy, mechanical assistance should be obtained, or the help of another person enlisted.

A scientific study has been made on lifting, known as the Kinetic method, and is best illustrated by a training demonstration and visual aids. Under the principle of kinetic lifting the correct stance is to have the body straight but not vertical, the feet placed apart (the width of the hips) and one foot ahead of the other in the direction of the movement. The knees should be bent and the arms close to the body.

Noise

The human ear does not judge sound powers in absolute terms but judges how many times greater one power is than another. A logarithmic scale is used to measure sound level. The decibel (dB) is the unit used for sound intensities. Noise can be divided into two different groups. Environmental, and occupational noise.

Environmental noise would be that of a nuisance noise audible outside a factory, and can be illustrated by the running of ball mills overnight causing disturbance to a neighbouring residential area. If the level of noise from a mill could not be reduced to a satisfactory level it could result in an order from the Public Health Department of the local authority forbidding the running of the ball mill overnight.

Occupational or work area noise has frequently been associated with industry and although recognised as a hazard it is only in recent years that work has been performed on deafness inflicted by industrial noise. In 1972 a Code of Practice [82] for reducing exposure levels of noise in industry was published. Basically the maximum level of noise was given as 90 dB for an 8-hour working day, which is now regarded as the legal upper limit.

It is the duty on the part of the manufacturer, suppliers and importers of machinery, etc. to ensure there is not a noise risk to health when

Table 15.6 A guide to comparative sound levels

	Sound level (dB)
Quiet room	30–35
Noisy office with typewriter	60–70
Saloon car at 40 mph	60–70
Paste ink manufacturing mill room	70–80
Liquid inks manufacturing plant	80–85
A shouting person	95–100
Riveting hammer on steel	110–125
Four airliner jet engines	160–170

installed and operating. If an operational machine generates more than 90 dB either it should be silenced or notices put up to give warning about its danger to hearing. Suitable ear protectors should also be provided and be readily available. However, the Code of Practice is not directly enforceable and is applied by factory inspectors as a description of what employers may reasonably be expected to do to ensure the health and safety of employees.

As already noted, the Noise Abatement Act 1960 was repealed when Part III of the Control of Pollution Act 1974 was brought into force. Nevertheless, it must be noted that consents issued by a local authority cannot be used as a defence in actions brought in a magistrate's court by the occupier of one premises against the occupier of other premises alleging a noise nuisance.

Neither the Factories Act 1961 nor the Health and Safety at Work, etc. Act 1974 specifically mention noise. Both lay general duties on employers to take reasonable care for health, safety and welfare of their employees. Four statutory instruments detail circumstances for specific noise reduction, but these do not relate to ink making. (All three carry 90 dB(A) as their noise threshold.) Forthcoming British regulations, based on the final EEC directive [83] will include a requirement that employers not only make available hearing protectors at noise levels above 85 dB, but also that such workers be informed about the associated risks of noise exposure, and the right to have hearing examinations.

Areas where worker exposure is likely to exceed 90 dB(A) will require the use of hearing protection and access to such areas must be restricted. Table 15.6 gives a guide to comparative sound levels.

Extraction

Many solvents, particularly the more volatile types, can be hazardous and this can be recognised by their documented occupational exposure limits. Continued exposure to solvent vapour is not recommended and adequate extraction ventilation is necessary to provide an acceptable working environment. Solvent vapours should be monitored in the workplace to ensure a safe level. The use of a fume cupboard is recommended for laboratory work [84]. Dust can form a source of irritation to the lungs and eyes, therefore addition extraction is recommended in areas used for weighing out fine powders and pigments. A specially constructed weighing booth with good extraction is advisable if large numbers of weighing are to be performed. The nuisance dust level is 10 mg/m^3.

Lighting

The intensity or brightness of light is expressed in lux and some idea what this term means can be gained by noting that an overcast summer day is estimated to be between 30 000 and 42 000 lx and a midwinter day 10 000 lx. BS 8206 Part 1 deals in general terms with the code of practice for artificial light. The following gives some general guidance for the light requirements in the workplace:

General offices, laboratories, kitchens medium bench work	500 lx
Drawing office	750 lx
Gauge and tool rooms, retouching paintwork	1000 lx
Inspection of graphic reproduction	1500 lx

In the laboratory artificial daylight is most important for colour matching. This subject is covered in BS 950, *Artificial Daylight for the Assessment of Colour*.

Machinery protection

It is estimated that over half the annual 16 000 UK accidents involving machinery could have been avoided by the adoption of well-established safeguarding standards. Legislation stemming from the Health and Safety at Work, etc. Act and the Factories Act state that plant and machinery must be safe and the employer must provide a safe system of work. To achieve this, dangerous parts of every machine shall be securely protected with properly constructed guards. The guards must be inspected, maintained and remain firmly in position while the machine is in motion and in use. Manufacturers of machinery have a responsibility to ensure that the products they supply comply with legislation regarding safety. If there is doubt on the safety protection of a machine there are BS and HSE publications covering most machinery. If doubt exists on the best method or if it is necessary to protect a piece of machinery the local factory inspector will offer advice.

Types of guard

Fixed guards
These should prevent contact of any part of the body with the dangerous part.

Interlocking guards
These should ensure that an opening is automatically closed before the machine can be operated, and will remain closed until the operation is completed.

Trip guards

These should ensure that, if the position of the operation is such that the operator is liable to be injured, the guard will automatically stop the machine.

Automatic guards

These are usually of the photoelectric type for power presses.

The types of mills, mixers and stirrers used for the manufacture of printing inks fall into the following categories: mixers, ball and bead mills and multi-roll mills.

Mixers

A large number of different types of vertical shaft mixers exist and may be placed into two separate groups, change pan and fixed pan. The main hazards associated with change pan mixers are caused by the rotating shaft and impeller blade. In some instances the rotating shaft can be protected by an outer casing, but when this is not possible the entire mixer should be protected by a metallic mesh safety cage. The safety cage should be fitted with an electro micro safety switch that will prevent the machine operating until all safety doors are securely closed and locked.

In recent years modern types of fixed pan mixers have been introduced which, because of their design and construction, allow more advanced safety techniques to be employed. The safety principle of the mixer is very simple, i.e. the mixing pan vessel protects the stirrer and mixing blade and cannot be turned on unless it is in the correct position and locked down. Similarly other rotating parts are suitably protected by micro switches and do not rotate or move unless the whole of the machine is secured in the correct working position.

Ball mills

Ball mills vary in size and capacity and must be positioned in such a way to allow safe rotation and be completely protected by a fixed safety cage. Access to the ball mill at ground floor level will be essential for maintenance and discharging. The entrance gate through the ball mill cage must be protected in such a way that the mill cannot rotate and is in a fixed position by a safe brake system. For the larger capacity ball mills it would be essential to have a mezzanine floor erected above the ball mill. Access to the platform would be by a stairway and the platform suitably protected by a safety barrier. A manhole in the ball mill is essential for loading, discharge is often performed using a separate valve system. Both these systems must be protected in such a way that they cannot be left in the open position when the mill is operated. To prevent build-up of internal pressure in the mill it should be fitted with a vent valve. The mill must be vented before either the manhole or discharge valve are open.

On occasions it will be necessary for maintenance engineers to enter the ball mill. A permit to work system must be employed for any workman entering into an enclosed space and the vapour concentration inside the ball mill must be measured and tested as safe by a competent person. The mill must be disconnected from the electrical supply at the main isolator, e.g. fuse removed and a notice placed on the switch stating 'Do not switch on — workman working on mill.'

Bead mills
Bead mills may have the dispersion cylinder in a vertical or horizontal position. Bead mills are moderately modern in design and the machine does not give a high hazard while in operation. Any exposed moving or rotating parts should be well protected.

Multi-roll mills
The most common multi-roll mill in common use is the three or triple roll mill. The hazards associated with a three-roll mill are the doctor blade and the nip of the middle and furthest roller from the blade. The nip between these two rollers must be protected by a safety bar which is interlocked with the mill. The doctor blade is extremely sharp and is mounted in the apron and fits onto the first roller. Great care must be taken when re-placing the doctor blade and a blade guard should be employed to prevent cuts. Should a foreign object be caught in the nip of the rollers, the mill should be switched off, the pressure released and the object withdrawn.

Permit to work
To ensure that maintenance staff and contractors are aware of the correct procedures that are necessary for safe working, a 'permit-to-work' should be issued. Such a document is illustrated in Fig. 15.9(a) and (b).

Safety audits

Safety audits or inspections provide an excellent means of ensuring the workplace, machinery, apparatus and raw materials are safe and of satis-factory standard or condition. The audit list could be prepared in con-junction with a local health and safety committee and different lists used in different departments or laboratories. The audit should have listed the manager/supervisor responsible for the location and the nominated safety officer or representative.

The audit team should submit a dated safety report to the manager of the department who will give instructions for any faults to be remedied. The audit team must not take on the job of repairing any fault.

The preparation of a safety audit list could be formulated using the following topics:
   — Structural damage, fittings which may cause a safety hazard, safe means of escape, noise level, extraction ventilation.
   — Electrics, all wiring plugs and connections earthing points and clips.
   — Fixed equipment and machinery. All guards securely in position and giving protection from all dangerous moving parts. Safety equipment such as mill bars and interlocks should received special attention.
   — Portable equipment. Pumps, motors, generators, etc. should be inspected for mechanical and electrical features.
   — Laboratory glassware, reagents.

— Ensure cracked or dangerous glassware is not in use and broken glassware is disposed by a safe method. Laboratory reagents inspected to ensure correct storage with special attention to flammability and toxicity.

Safety equipment

Fire extinguishers shall be regularly inspected. Ensure all fire and safety signs, fire precautions and notices are exhibited.

Personal protection

Ensure ear protectors, eye protectors, first aid box, washing and emergency showers come up to an acceptable standard of hygiene.

Stores and working areas

Ensure a good standard of housekeeping is achieved. Remember to inspect stores, dark rooms and instrument rooms. If ladders or steps are used to reach shelves, ensure they are in a safe condition.

Safety committee

Section 2 (7) of the Health and Safety at Work, etc. Act states that an employer must establish a safety committee [85, 86] if requested by at least two safety representatives appointed by a recognised trade union. The membership of the safety committee should be established to include representatives from all major departments of a factory or site, and this should be a matter for management and union to decide. The committee should meet at regular intervals, chaired by a senior manager and minutes should be recorded.

The objectives of a safety committee are to bring together employer and employee and promote all aspects of health and safety. It would be beneficial for a safety committee to draw up a statement of intent so that all members were aware of its aims and objectives.

The health and safety committee should include the following items for its meetings:

(1) Study accident reports and offer recommendations for accident prevention.
(2) Fire alarm procedure.
(3) Report on fire drills and fire training.
(4) Health and safety training.
(5) Study health and safety data sheets for raw materials and products.
(6) New health and safety legislation.

Other items relevant to health and safety should be included as necessary.

Safety representatives

A safety representative is appointed by a recognised trade union [85–87] and, as far as is reasonably practicable, should have 2 years' experience in the same or similar employment. An employer shall permit a safety

Notice to Contractors working at   Co Ltd

This is a Chemical Works and contains substances which must be handled with caution, some being highly flammable, and therefore it is imperative that the works rules for Health & Safety are adhered to.

The essential rules are listed below. If in doubt the Site Engineering Department or the Safety Officer of the working location will advise.

1.   A Permit-to-Work must be obtained from the Site Engineering Department before commencement of work in any location on the   site.

2.   No Smoking except in permitted areas.

3.   Any drilling, cutting, burning, welding must be authorised in writing by the Site Engineering Department.

4.   Do not enter buildings, interfere with machinery, electrical wire systems or pipe systems unless authorised by the Site Engineering Department.

5.   In the event of fire or any emergency, all contractors' employees should make their way to the main gate and report to the gatekeeper.

6.   Private cars are not allowed on site. If tools and materials are brought in cars, then after unloading the vehicle must be taken to the car park.

7.   The Speed Limit on this site is 10 m.p.h.

8.   Contractors must arrange for the removal of rubbish, and for ensuring that the work area is left in a clean condition.

Fig 15.9(a)

PD 9-2-1   SERIAL **N°** 5001

COATES BROTHERS PLC
SOUTH WALES FACTORY
PERMIT-TO-WORK CERTIFICATE

1 LOCATION	2 PERMIT VALID (MAXIMUM OF 8 HOURS)
	FROM............ TO............ DATE(S)........

3 WORK TO BE CARRIED OUT ....

4 (a) ISOLATION OF PLANT/WORK AREA (cross out items not applicable)
Precautions have been taken for the isolation of the following :— STEAM/WATER/GASES/FLUID HEAT MEDIUM/ MECHANICAL DRIVES/PIPE LINES. SLUDGES AND DEPOSITS HAVE BEEN REMOVED. APPROPRIATE SAFETY TICKETS HAVE BEEN ATTACHED.

SIGNED..DEPT. MANAGER.   DATE & TIME................

(b) ISOLATION OF ELECTRICAL POWER. I certify that all electrical power has been isolated to allow work to be safely carried out.

SIGNED..AUTHORISED PERSON   DATE & TIME................

5 ATMOSPHERIC TEST CERTIFICATE. I certify that the atmosphere in the work area/vessel has been examined for the presence of toxic, asphyxiating or flammable gases having regard to the previous contents and is considered to be safe for entry (or the commencement of work).
NOTE Where gases or fumes may be present, entry without breathing apparatus must not be permitted.

SIGNED..QUALIFIED TESTER.   DATE & TIME................

AN ADEQUATE SUPPLY OF FRESH AIR IS UNNECESSARY/NECESSARY ....DEPT. MANAGER

6 NAKED FLAME CERTIFICATE. Naked flame or other sources of ignitions ARE/ARE NOT permitted.
TYPE PERMITTED (i.e. Burning, welding, grinding, drilling)................

SIGNED..DEPT. MANAGER

7 PRECAUTIONS TO BE TAKEN. The following protection/precautions must be used during the work (State TYPE against each item required and State NONE against each item not required).

(1) Eye protection, type ....   (5) Respirators, type ....   (10) Vessel ....COLD/HOT

(2) Gloves, type ....   (6) Fire extinguishers, type ....   (11) Rescue equipment in good order and available............

(3) Protective clothing, type............   (7) Flameproof lamps............   (12) Surrounding areas are safe............

(4) Breathing apparatus, type............
(Duration of cylinders to be made known   (8) Safety belt and line ....
to user)   (9) Safety man ....

OTHER PRECAUTIONS............

SIGNED..DEPT. MANAGER

8 CERTIFICATION : I certify that the precautions in section 4 have been taken and the work may commence subject to the requirements of SECTIONS 5, 6, and 7.	**9 ACCEPTANCE.** I have read the permit and understand the nature of the work and the safety precautions to be taken.
SIGNED................DEPT MANAGER DATE	SIGNED (1)............ C/NO ....
If the certified conditions change, work must be suspended immediately and the certifier informed.	SIGNED (2)............ C/NO ....
	SIGNED (3)............ C/NO ....

10 WORK COMPLETED/SUSPENDED. All persons assisting in work, tools, and gear have been withdrawn	**11 ACCEPTED AND PERMIT CANCELLED**
SIGNED .... DATE....	SIGNED .... DATE & TIME ....

Fig 15.9(b)

representative to take time off with pay during working hours to receive necessary training at a recommended establishment centre.

The major functions of a safety representative are:

(1) To attend safety committee meetings.
(2) Submit reports to the safety committee following investigations and safety audits.
(3) Investigate potential hazards and dangerous occurrences.
(4) Investigate complaints relating to employees' health, safety and welfare.
(5) Conduct safety and fire preventions audits.

It is common practice for the factory inspector to send the safety representative a copy of all correspondence sent to the management.

Safety policies

A company employing five or more people must produce a written statement of its health and safety policy. The government has indicated its intention to raise the threshold number of employees to 20. The policy will spell out the implications of HSW and other health and safety constraints relevant to the work carried out. It is important to stress the need for co-operation from the workforce and good communication at all levels. The policy should be signed by a senior director of the company.

Employees should be able to determine from the policy statement how they fit into the health and safety structure, what their own duties are and where to go for advice, where to report an accident or potential hazard and how to obtain first-aid or other assistance. The written policy must detail the management structure responsible for implementing the policy.

The HSC has published a free leaflet providing advice to employers [88]. The HSE reviewed the experience of some of their staff on policy matters in an earlier publication [89]. More recently, the HSE has issued guidance with the small business in mind [90].

General guidance to printers

A general guide to health and safety issues for both employers and employees in small printing firms has been prepared by PIAC and published by the HSE [91]. PIAC are also working on a guide to noise reduction at web-fed printing presses.

15.4 SPECIFIC PRINTING INK APPLICATIONS

Food wrapper inks

Extensive use of wrapping and packaging materials for food continues to ensure good hygiene during handling and increases sales-appeal. Printing inks are applied to such packaging materials to provide directions for use, to identify the package contents and to improve sales appeal. However, printing inks cannot be made from truly edible materials, except in very special circumstances, as the range of suitable components is, in general, too limited. For this reason printing inks must not be used in such a way that they come into contact with food.

In the UK there are no specific regulations relating to the use of printing inks for the printing of food wrappers or packaging provided the printed material is not intended to come into contact with food. The SPBIM has promoted recommendations [92] that its members adopt to ensure that printing on food wrappers does not pose any health hazard and can be relied upon with confidence.

For a wrapper which will come into direct contact with food the ink must be applied to the outside of the wrapper. The wrapper itself must form a functional barrier between the printed surface and must prevent the ink from migrating or bleeding through the barrier and contaminating the foodstuff. The ink should be printed in such a way that set-off is avoided, ensuring that the wrapping surface in contact with the foodstuff is free of printing ink. An ink film is extremely thin (see section on screen inks covering typical film thicknesses). Consequently the total quantity of ink involved is extremely small.

To ensure maximum safety, inks for immediate food wrappers should be formulated on non-toxic raw materials and should not contain materials listed on the SBPIM exclusion list. The application of an over-print varnish or lacquer to provide a functional barrier between the ink and food stuff will not necessarily provide a suitable barrier to prevent migration and therefore cannot be accepted as a suitable material to ensure contamination does not occur. The suitability of an overprint varnish or lacquer as an impermeable barrier should be assessed before finally deciding on a particular formulation.

If ink films or their by-products were to come into contact with food, then the provisions of the 'Materials and Articles in Contact with Food Regulations' would apply to the printed composite. The printing process employed will be governed by the type of stock being printed but will generally be flexo, gravure or litho. Dependent upon the process, inks dry by evaporation or oxidation and, in the absence of strict quality assurance controls, there exists the possibility of adverse taint and odour effects on the packaged food.

The odorous volatile solvents coming from the inks are chemically identifiable and well known. Trained testing panels can readily identify both the taint and odour. However it is possible that, for example, a trace of fatty acid in a coating could react with an additive or intermediate and finally break down to odorous by-products such as aldehydes or ketones which in turn could introduce alien taint or odour to the food. The term 'organoleptic' refers to odour and taint characteristics and can only be perceived by nose and tongue. Threshold taste values for various substances have been reported as low as 5 ppb and the odour threshold for butyric acid (rancid butter odour) is 1 ppb.

There is a wide range of statutory instruments [93–96] under the various Food and Drugs Acts in the UK, which specify permitted food ingredients under such headings as emulsifiers, solvents, antioxidants, colouring matter and preservatives. In addition, one set of regulations

[97] referred to above specifically controls the marketing and use of materials and articles which are intended to come into (direct) contact with food.

The regulations require that materials which are in their finished state and are intended to come into contact with food will, *inter alia*, be labelled 'for food use'. Such materials will not endanger human health or bring about a deterioration in the organoleptic characteristics of the food. Clearly, wet inks are not 'in their finished state'. The Food Contaminants Branch of the Ministry of Agriculture Fisheries and Food have confirmed that, in their view, printing inks manufactured in accordance with SBPIM's policy would not conflict with the provisions of the Materials and Articles in Contact with Food Regulations 1978.

In the absence of direct UK legislative guidance, it is common in the UK for print customers to seek reassurance in respect of so-called 'FDA Approval', for example. It is now well established that, in the USA, the Food and Drug Administration (FDA) has no authority under the food additives provisions of the US Federal Food Drug and Cosmetic Act to 'approve' specific products, and that no food additive regulations on the general subject of printing inks exist. The FDA are on record [98] as believing that printing inks applied to the outer surface of food packaging materials are not food additives within the meaning of section 201 (3) of the Act, provided that the packaging material serves as a functional barrier between the printing ink and the food.

This said, it is common for UK converters to seek from their ink suppliers assurances as to the safety of the coatings to be applied, by reference to FDA regulations. This is understandable in the absence of direct guidance from our own legislators.

Under such circumstances, it is necessary for printing ink manufacturers to understand the relevant sections of the Federal Regulations [99], but at the same time to ensure that enquirers are aware of the real scope of those regulations. The sections that are relevant in this context are:

21CFR174   Indirect food additives: General
21CFR175   Indirect food additives: Adhesives and components of coatings
21CFR176   Indirect food additives: Paper and paperboard components
21CFR177   Indirect food additives: Polymers
21CFR182   Substances generally recognised as safe
21CFR186   Indirect food substances affirmed as generally recognised as safe

In addition, parts of two further sections are of some relevance:

21CFR178   Adjuvants, production aids and sensitizers
21CFR181   Prior sanctioned food ingredients

21CFR175.300, covering resinous and polymeric coatings for use as food contact surfaces, is the most useful guide in this context.

In broad terms, these regulations describe the components of coatings, for example, which may become added to food as a result of their use on articles that contact food (packaging material, for example).

Substances that, under conditions of good manufacturing practice, may
be safely used as components of articles that contact food, are:
  (a) Substances generally recognised as safe in, or on, food;
  (b) Substances generally recognised as safe for their intended use in
       food packaging;
  (c) Substances used in accordance with a prior sanction or approval;
  (d) Substances permitted for use by specific regulations (such as
       21CFR175.300, for example).

It should be noted, however, that coatings are described not only in terms
of their permitted ingredients but also in terms of the level of extractibles
present in the coating in the finished form, i.e. migration level into defined
simulants from the coated substrate.

In 21CFR175.300, for example, coatings intended for one-time use on
containers having a capacity in excess of 1 gallon, will not yield an extr-
actable content of greater than 1.8 milligrammes per square inch under
the extraction conditions (solvent type and time/temperature regime)
appropriate to the food type in question (for example, dairy products, dry
solids or acidic foods).

Appeal to EEC sources for guidance is less than reassuring at the
moment. For at least 10 years, discussions have been in progress with
a view to producing a positive list of monomeric precursors to plastics
intended to come into contact with food, and a separate listing covering
additives for such plastics. For the time being, this activity has deliberately
excluded consideration of surface coating compositions and such consideration was not expected to begin before 1988, though there is now
evidence that the timetable is likely to be hurried a little.

Thus, ink-makers must take direct responsibility for the formulation of
indirect food wrapper inks themselves, using such individual guidance as
is available to them. The responsibility is greatest in the selection of
pigments, since colouring materials listed in various states' positive lists
are very limiting. For members of the Society, this means relying on the
published recommendations.

Nevertheless it is the ultimate responsibility of the packer to ensure that
the final package composite does not adversely affect the contents and
this dictates a high level of co-operation between all those in the processing chain from ink-maker to food packer.

Coatings for toys, pencils and graphic instruments

The Toys Regulations [100], made under the Consumer Protection Act,
were drafted to limit the use of paints based on lead, other heavy metals
and cellulose nitrate. The definition of paint, to include lacquers, varnishes
and other similar substances, is taken to mean that the regulations apply
to printing inks. Regulations 4 limits the content of heavy metals to listed
proportions by weight with all but lead being determined as 'soluble'
element as measured by the test given in Schedule 1. However, the test
method is conducted on a detached dried film. In the case of printing inks,
it is not possible to completely detach the film from a paper or paperbroad
substrate, so the prescribed analysis cannot be adhered to strictly.

Table 15.7

	Pencils and graphic instruments (ppm)	Toys (ppm)
Antimony (soluble)	250	250
Arsenic (soluble)	100	100
Barium (soluble)	500	500
Cadmium (soluble)	100	100
Chromium (soluble)	Not specified	250
Chromium (hexavalent soluble)	100	Not specified
Lead (soluble)	250	Not specified
Lead (total)	Not specified	2500
Mercury (soluble)	100	100

The same range of heavy metals is controlled in the Pencils and Graphic Instruments Regulations [101] but with variation in the conditions and concentrations for lead and chromium.

The Society has issued a recommendation to members relating to the formulation of inks for toys (where the ink film cannot be detached from its substrate). This recommendation relies on the specific exclusion of heavy metal components (except barium sulphate) rather than relying on an analysis of the dried film. It is important to note that regulations apply to the dried film, and it is thus the responsibility of the printer or toy maker to satisfy himself that the dry coating meets the requirements of the regulations. No written guarantee can be given by the ink manufacturer that inks 'conform' to the Toy Regulations (see Table 15.7).

Printing inks for use on children's comics

The SPBIM considers that, as the print on children's comics could affect the health of children, the same limitations should be applied to printing inks for children's comics as are applied to printing inks for use on immediate food wrappers.

SBPIM general exclusion list

The SBPIM have consulted with their members and agreed on a list of banned materials which will not be used because of possible health hazards. This list, known as the general exclusion list, includes known carcinogens, cumulative poisons or other toxic substances that should not be used in printing ink manufacture. Additions are made to this list as new knowledge becomes available.

General exclusion list
The following materials must be excluded from any printing ink or varnish:

(1) Compounds of lead (except where specific exemption has been given by the SBPIM).
(2) Compounds of cadmium.
(3) Zinc chromate.
(4) Pigments and dyestuffs based on selenium, arsenic, antimony and mercury.
(5) CI Solvent Blue 7 50400 (Induline).
(6) Toluene di-isocyanate (TDI).
(7) 2-Nitropropane (2-NP).
(8) Polychlorinated biphenyls (PCBs).
(9) Tricresyl phosphate.
(10) 4,4-tetra methyl diamino benzophenone (Michlers Ketone).
(11) 2,4-dimethyl-6-*t*-butylphenol (Topanol A).
(12) Diamino stilbene and derivatives.
(13) Acrylates with a Draize figure greater than 3.4 on the 0 to 8 scale. The Draize figure for any acrylate to be determined by a recognised research institute. (Specific acrylates to be excluded are listed in a separate exclusion list — see next section.)

Ultraviolet and electron beam curable inks

Ultraviolet inks, varnishes and lacquers are formulated on polyfunctional acrylates and other materials which can be polymerised to a hard film when exposed to high intensity UV light. The Printing Industry Advisory Committee (PIAC), on which the SBPIM is represented, has made recommendations [102] and laid down guidelines on safety in the manufacture and use of inks, varnishes and lacquers cured by UV light. Some functional acrylate/methacrylate monomers can cause skin irritation and others in addition cause skin sensitisation. Sensitised persons must be removed from any further exposure to these materials and transferred to another job. Extraction ventilation must always be employed when handling UV products.

Electron-beam curable inks use similar raw materials to those employed in UV inks and similar handling and formulation techniques should be applied.

Primary Irritation Index (PII), or the Draize rating, is determined from a patch test performed on animals and measures the degree of irritation imparted by a substance. The scale ranges from 0 to 8; 0 is low and 8 high irritancy. The SBPIM has recommended to its members that only materials with a Draize rating of less than 3.5 and with no sensitising effect be used in UV cured inks. The Draize test has changed little since it was developed in 1944 and has aroused protests which can be justified on both scientific and humanitarian grounds. Work is in progress to develop alternative testing with the intention that animal testing can then be drastically reduced or better still, no longer carried out.

The following acrylates are currently recommended for exclusion from from UV-curable ink formulations and are taken from the PIAC booklet as updated by SBPIM in their exclusion list:

2-Hydroxyethyl acrylate	Phenoxy ethyl acrylate (phenyl cellosolve acrylate)
2-Hydroxypropyl acrylate	Dicyclopentenyl acrylate
Methyl carbamyl ethyl acrylate	2-Ethyl hexyl acrylate (2EHA)
1,3-Butylene glycol diacrylate	Di ethylene glycol diacrylate (DEGDA)
Butyl glycidyl ether acrylate	Tetra ethylene glycol diacrylate (TEGDA)
1,6-Hexanediol diacrylate	Mixtures of pentaerythritol tri and tetra-acrylates (PETA)
Phenyl glycidyl ether acrylate	Tri metholpropane triacrylate (TMPTA)
Acrylate of glycidyl ester of C_{9-11}-carboxylic acid	Butanediol diacrylate (BDDA)
Neopentylglycol diacrylate	

Under the Health and Safety (Emissions into the Atmosphere) Regulations, acrylates are controlled under some circumstances. Thus:

> Works in which acrylates are...stored in fixed tanks with aggregate capacity exceeding 20 tonnes have a statutory duty to use best practical means to both prevent acrylate emissions into the atmosphere and render harmless and inoffensive such emissions that do occur.

Normally we would expect acrylates to be stored in 200-litre barrels within ink factories.

15.5 SOME INTERNATIONAL CONSTRAINTS

Throughout this chapter we have deliberately concentrated on UK legislation. Nevertheless, since the UK joined the Common Market, UK legislation has increasingly reflected European practice (though in some cases we were rather slow to adopt Community directives). O'Neill has prepared [103], and has regularly updated [104, 105], an excellent review covering health and safety and environmental legislation of some 20–30 countries with special attention to the UK, USA, EEC and to some extent Japan.

Chemical inventories

US regulations [106] require the Environmental Protection Agency to 'identify, compile, keep current and publish' a list of chemical substances (as defined) which are manufactured, imported or processed for commercial purposes in the USA. The initial inventory was published in 1979; the current edition [107] contains 63 000 chemical substances. In contrast to EEC's EINECS, TSCA Inventory is a growing list of chemical substances ('keep current') and as such provides a rolling definition of 'new'. Neither lists *toxic* substances: they simply list commercially existing chemical substances of their respective territories.

The Australian Core Inventory [108] (ACOIN) serves the same role as both the EINECS and TSCA Inventory.

The UN has published [109] a list of recommendations and legislation of various countries and international organisations on a selected range of potentially toxic chemicals. Of particular interest are the various labelling requirements, suitability for food contact and occupational exposure limits. In the latter context the International Labour Office has published [110] a list of some 1200 substances with their occupational limits as assigned by 19 countries and includes a useful review of the countries covered. At the time of its publication, the UK followed ACGIH recommendations and so was not covered separately.

In contrast to the TSCA Inventory, the Registry of Toxic Effects of Chemical Substances (RTECS) [111] provides details of the known toxic data of some 80 000 chemical substances, though NIOSH estimate that the number of unique substances for which toxicity data may be available is approximately 250 000. The ultimate intention of the registry is to identify all known toxic substances and to provide pertinent data on the toxic effects from known doses entering an organism by any route described. It has long been the single most significant source of LD_{50} data in the world.

The data must not be considered as providing definitive values for describing 'safe' versus 'toxic' doses for human exposure. Rather it is one source of data for critical evaluation by experienced investigators. Indeed NIOSH themselves conduct critical reviews of occupational exposure hazards and the resulting criteria documents are used as the basis for recommendations to US Department of Labor, Occupational Safety and Health Administration, for US occupational exposure standards.

The last complete printed edition of RTECS was 1982 with a supplement in 1983. For non-US residents it is probably more convenient to use the microfiche edition which, in any event, is more up to date, being republished with amendments on a quarterly basis.

The Japanese Ministry of Labour has promulgated [112] lists of chemical substances which require both regular air monitoring and labelling. Associated enforcement regulations [113] classify solvents according to the level of ventilation required for their safe use (and also specifies the personal protection required):

— Class 2 local ventilation required, e.g. toluene, 2-ethoxyethanol
— Class 3 general ventilation required, e.g. petroleum distillates

New chemicals need to be notified [114] to the Ministry of International Trade and Industry (MITI) by reference to a definitive list of existing substances [115].

US data communication

Under the provisions of OSHA's Hazards Communications Standard [116] a greater emphasis is placed on the use of labels to communicate hazards to employees than is the case under analogous UK requirements.

Published in November 1985 and coming into effect in May 1986, the OSHA standard requires ink manufacturers to:

— review their material inventories (raw materials, intermediates and finished goods);
— establish a hazardous materials list;

— prepare materials safety data sheets and labels (for all materials hazardous or not);

— prepare a written hazard communications program;

— provide employees with information and training on all hazardous materials in their workplace.

The National Association of Printing Ink Manufacturers (NAPIM) have developed a range of documentation to enable their members to discharge their communications responsibilities. NAPIM first published their material safety data sheets (MSDS) in 1971. These have been revised and a guide published describing in detail the requirements of each section of the new MSDS.

They have also published [117] a guide to a Hazardous Materials Identification System (HMIS) based on a system developed by National Paint and Coatings Association (NPCA) and the National Fire Protection Association (NFPA) for example. HMIS is a visual system using colour coding and simple numerical ratings (on a scale from 0 to 4; 0, low hazard and 4, high hazard) and is designed to communicate risk information to workers with respect to health hazard (blue), flammability hazard (red) and reactivity (yellow). Recommendations for personal protection equipment is indicated by a two-letter code and by symbol (see Figs 15.10 and 15.11).

Health hazard rating are assessed from acute oral, dermal and inhalation toxicity data, generally with a tenfold difference between labels. On this basis 1H corresponds to the middle of the European 'Harmful' range and 2H embraces the bottom end of 'Toxic' and the top end of 'Harmful'. Flammability criteria bear some resemblance to European criteria; thus 4F corresponds, approximately, to extremely flammable, and 3F corresponds, approximately, to highly flammable. Thereafter the two sets of criteria diverge.

Food wrapper coatings

In 1982 the UN's Chemical Industry Committee of the Economic Commission for Europe decided to undertake a feasibility study in the area of food additives legislation. Its aim was to provide a condensed survey of, then, current legislation in various countries and publish a short description of the coverage of that legislation. Twenty-three European countries and the USA replied to the questionnaire and their responses were published in 1984 [118]. The level of detail varied considerably from country to country.

In 1985 CEPE [119] published a review of European national legislation and practice for coatings for metal substates in direct contact with food. The compilation was intended to assist producers and users of coatings compositions, particularly bearing in mind the needs of countries which have no national legislation of their own. The document goes some way to compensating for the continuing absence of a European directive in this area.

EEC directives and drafts on packaging in contact with food have so far concentrated on plastic substrates and, as noted in the section on food

HEALTH	1
FLAMMABILITY	3
REACTIVITY	0
Personal Protection	SF

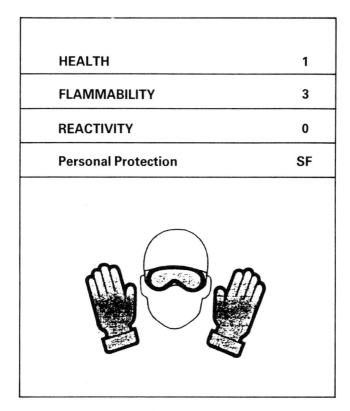

SF – Employee must wear
gloves and splash
goggles

Fig 15.10

wrapper ink, p. 820 have specifically excluded surface coatings such as inks and paints from their scope. Difficulties with overall migration limits have slowed progress on the directive but it now seems likely that the Single European Act may have the effect of overcoming delays (by introducing qualified majority agreement to replace unanimity in the decision procedures) and at the same time hasten progress on subsequent daughter-directives dealing with paper and metal substrates and associated coatings.

In the meantime various trade bodies, such as the Association of Plastics Manufacturers in Europe (APME) for example, have prepared 'temporary' guidance either for the use of their members or as discussion documents for EEC consideration. APME have prepared such guidance covering colourants for food contact plastics.

EEC drafts have drawn heavily on West German and Dutch sources. In the absence of more territorially relevant guidance these two sources can

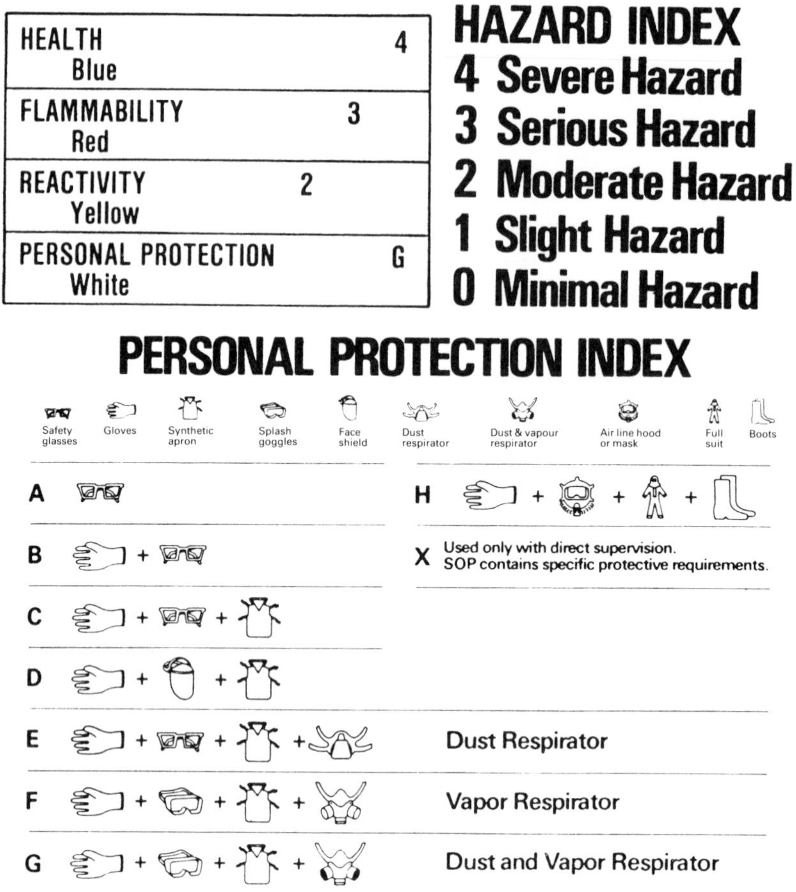

HEALTH Blue	4
FLAMMABILITY Red	3
REACTIVITY Yellow	2
PERSONAL PROTECTION White	G

HAZARD INDEX
4 Severe Hazard
3 Serious Hazard
2 Moderate Hazard
1 Slight Hazard
0 Minimal Hazard

PERSONAL PROTECTION INDEX

Safety glasses	Gloves	Synthetic apron	Splash goggles	Face shield	Dust respirator	Dust & vapour respirator	Air line hood or mask	Full suit	Boots

A (safety glasses)

B (gloves) + (safety glasses)

C (gloves) + (safety glasses) + (apron)

D (gloves) + (splash goggles) + (apron)

E (gloves) + (safety glasses) + (apron) + (dust respirator)   Dust Respirator

F (gloves) + (splash goggles) + (apron) + (vapor respirator)   Vapor Respirator

G (gloves) + (splash goggles) + (apron) + (dust and vapor respirator)   Dust and Vapor Respirator

H (gloves) + (air line hood or mask) + (full suit) + (boots)

X Used only with direct supervision.
   SOP contains specific protective requirements.

Fig. 15.11 Example of a type of hazard label (adapted from the American Industrial Hygiene Association)

go some way to providing positive guidance in the UK. In particular Dutch legislation has a growing section on global and specific test regimes for migration of coatings components. An English translation [120] of the fifth version of the draft packaging and food utensils regulation was prepared by the Netherlands Plastics Federation. West German legislation [121] presents a positive list of materials, in the main subdivided by chemical type, which may be used for food packaging applications. Chapter XL is entitled 'Lacquers and Varnishes for Food Containers and Packs'. Permitted ingredients include pigments among the auxiliary materials.

In the UK frequent appeal to the US Food, Drug and Cosmetic Act for guidance is encountered. For this reason, and despite their non-territorial relevance, FDA constraints have been dealt with in the section on food wrapper inks, p. 820.

Coatings for toys

Under a proposed new EEC directive [122], apart from physical, mechanical and flammability constraints, toys will be subject to control of their chemical composition by limiting the bioavailability of a range of heavy metals (on a daily intake basis). The numerical relationship between the acceptable daily intake and the corresponding concentration of the relevant components in the coating, is currently under discussion as a topic for a revision of European Standard EN 71. This will have a knock-on effect on European national legislation once it achieves an adoptable form.

Transport

There are separate international recommendations for the transport of dangerous goods multimodally, by sea, by air, by road and by rail. These continue to be developed but there is evidence of increasing harmonisation with UN recommendations [65]. Mostyn has compiled a series of guides [123], bringing together the various route requirements for the main UN classes of dangerous goods.

Light gauge metal packaging, made of steel or tin plate (thickness less than 0.5 mm) is extensively used in the conveyance of printing inks. From 1989 such packaging will be subject to regulations for sea, air, and international road/rail transport and will be required to pass performance tests and bear an official mark as evidence of compliance [124].

GLOSSARY OF ABBREVIATIONS

ACGIH	American Conference of Governmental Industrial Hygienists
ACOIN	Australian Core Inventory
ACTS	Advisory Committee on Toxic Substances (to HSE)
ADR	Accord européen relatif au transport international des merchandises Dangereuses par Route
CEFIC	European Council of Chemical Manufacturers' Federations
CEPE	European Committee of Paint, Printing Ink and Artist's Colour Manufacturers Associations
CIA	Chemical Industries Association
CIMAH	Control of Industrial Major Accident Hazards
COSHH	Control of Substances Hazardous to Health (proposals)
CPL	Classification, Packaging and Labelling of dangerous substances regulations (SI 1984, No. 1244, as amended)
ECETOC	European Chemical Industry Ecology and Toxicology Centre
ECOIN	European Core Inventory
EINECS	European Inventory of Existing Chemical Substances
EPA	(US) Environmental Protection Agency
FDA	(US) Federal Drug Administration

HASAWA	See HSW
HAZCHEM	(CIA voluntary scheme for vehicle marking — Dangerous goods)
HMIS	Hazardous Material Identification System
HSC	Health and Safety Commission
HSE	Health and Safety Executive
HSW	Health and Safety at Work, etc. Act 1974
IMDG	International Maritime Dangerous Goods (code)
IMO	International Maritime Organisation
IATA	International Air Transport Association
MSDS	Material Safety Data Sheet
NAPIM	(US) National Association of Printing Ink Manufacturers
NFPA	(US) National Fire Protection Association
NIOSH	(US) National Institute for Occupational Safety and Health (Dept of Health and Human Services)
NPCA	(US) National Paint and Coatings Association
OEL	Occupational exposure limits
OSHA	(US) Occupational Safety and Health Administration (Dept of Labour)
OTS	(US) Office of Toxic Substances (EPA)
PIAC	Printing Industry Advisory Committee (to HSE)
PMA	Paintmakers Association (of Great Britain)
PRA	Paint Research Association (Teddington)
RIDDOR	Reporting of Injuries, Diseases and Dangerous Occurrences Regulations
RTECS	Registry of Toxic Effects of Chemical Substances (NIOSH)
SBPIM	Society of British Printing Ink Manufacturers
SI	Statutory instrument (UK)
TLV	Threshold limit values (established by ACGIH)
TREMCARD	Transport Emergency Card (Trademark CEFIC)
TSCA	(US) Toxic Substances Control Act
TWA	Time-weighted average (applied to OELs)

REFERENCES

1. Lord Robens. *Safety and Health at Work: Report of the Committee 1970–1972.* Command Publication 5034. HMSO, (July 1972). (ISBN 010 150 340 7)
2. Health and Safety at Work, etc. Act 1874: Chapter 37. (ISBN 010 543 774 3)
3. Health and Safety Commission. *HSC 3: Health and Safety at Work, etc. Act 1974. Advice to employers.* Reprinted 1985. (Free)
4. Health and Safety Commission. *HSC 2: Health and Safety at Work, etc. Act 1974. The Act outlined.* Reprinted 1985. (Free)
5. Control of Pollution Act 1974: Chapter 40.
6. Statutory Instrument 1980, No. 1709. Control of Pollution (Special Waste) Regulations 1980. (ISBN 011 007 709 1)
7. Sale of Goods Act 1979: Chapter 54.
8. Statutory Instrument 1974, No. 1367. The Toys (Safety) Regulations 1974. ISBN 011 041 367 9.
9. Statutory Instrument 1974, No. 226. The Pencils and Graphic Instruments

(Safety) Regulations 1974. (ISBN 011 040 226 X)

10. European Community. Directive 85/374/EEC: On the approximation of the law, regulations and administrative provisions of the Member States concerning liability for defective products. *Official Journal of the European Communities* L210, 29–33. (07.08.85).

11. Department of Trade and Industry. Implementation of EC Directive on Product Liability. An Explanatory and Consultative Note. November 1985.

12. Consumer Protection Bill as amended in Committee (House of Lords 59). 1987. (ISBN 010 405 907 7)

13. Health and Safety Executive. *Guidance Note EH 40/87: Occupational Exposure Limits 1987.* (ISBN 0 11 883 940 3) (1987).

14. European Chemical Industry Ecology and Toxicology Centre. *Technical Report No. 4: The Toxicology of Ethylene Glycol Monoalkyl Ethers and its Relevance to Man.* (1982). *Report No. 17: The Toxicology of Glycol Ethers and its relevance to Man:* (an updating of ECETOC Technical Report No. 4). (1985).

15. Health and Safety Executive. *Toxicity Review No. 10: Glycol Ethers.* (ISBN 0 11 883 807 5) (1985).

16. Health and Safety Executive. *Guidance Note EH 44: Dust in the Work Place: General Principles of Protection.* (ISBN 0 11 883 598 X) (1984).

17. Health and Safety Executive. MDHS 14: General methods for the gravimetric determination of respirable and total dust. (ISBN 071 760 142 0)

18. American Conference of Governmental Industrial Hygienists. *Threshold Limit Values and Biological Exposure Indices for 1986–87.* (ISBN 0 936 712 69 4) (1986).

19. American Conference of Governmental Industrial Hygienists. *Documentation of the Threshold Limit Values,* 4th ed. (ISBN 0 936 712 55 4) (1986).

20. Statutory Instrument 1983 No. 1649. The Asbestos (Licensing) Regulations. (ISBN 0 11 037 649 8) (1983).

21. Health and Safety Executive. *Guidance Note EH 10: Asbestos — Control Limits, measurement of Airborne Dust Concentrations and the Assessment of Control Measures* (revised). (ISBN 0 11 883 596 3) (1984).

22. Commission of the European Communities. *Constructing EINECS: Basic documents: European Core Inventory (ECOIN),* vols I–IV. (ISBN 92 825 2454 X) (1981).

23. Commission of the European Communities. *Construction EINECS: Basic Documents: Compendium of Known Substances,* vols I–III. (ISBN 92 825 2455 8) (1981).

24. Statutory Instrument 1982, No. 1496. The Notification of New Substances Regulations 1982. (ISBN 0 11 027 496 2)

25. Health and Safety Executive. *Health and Safety Booklet HS(R)14. A Guide to the Notification of New Substances Regulations 1982.* (ISBN 0 11 883 660 9) (1982).

26. Health and Safety Commission. *Approved Code of Practice No. 10: Methods for the Determination of Toxicity.* (ISBN 0 11 883 657 9) (1982).

27. Health and Safety Commission. *Approved Code of Practice No. 9: Methods for the Determination of Physico-chemical Properties.* (ISBN 0 11 883 655 2) (1982).

28. Health and Safety Commission. *Approved Code of Practice No. 8: Methods for the Determination of Ecotoxicity.* (ISBN 0 11 883 656 0) (1982).

29. Statutory Instrument 1980, No. 1248. The Control of Lead at Work Regulations 1980. (ISBN 0 11 007 248 0)

30. Health and Safety Commission. *Approved Code of Practice: Control of Lead at Work,* 2nd edn. (ISBN 011 883 780 X) (1985).

31. Statutory Instrument 1983, No. 943. The Health and Safety (Emissions into the Atmosphere) Regulations 1983. (ISBN 0 11 036 943 2)

32. Health and Safety Executive. *Guidance Note EH28: Control of Lead: Air Sampling Techniques and Strategies.* (ISBN 0 11 883 393 6) (1981).

33. Fire Precautions Act 1971, Chapter 40. (ISBN 10 544 071 X) and Home Office: *Guide to the Fire Precautions Act 1971: 2. Factories.* (ISBN 011 340 44 1) 1977.

34. Statutory Instrument 1976, No. 2003. Fire Certificates (Special Premises) Regulations 1976. (ISBN 011 062 003 8); and Statutory Instrument 1976, No. 2008. The Fire Precautions (Application for Certificate) Regulations 1976. (ISBN 011 062 008 9)

35. Home Office: Consultative Document: A Review of the Fire Precautions Act 1971 (ISBN 0 86 252194 7) (July 1985); Fire Safety and Safety of Places of Sport Bill, as amended in Committee Bill No. 107 (ISBN 010 310 787 8) (1987).

36. Statutory Instrument 1972, No. 917. The Highly Flammable Liquids and Liquefied Petroleum Gases Regulations 1972. (ISBN 011 020917 6)

37. Department of Employment. *Highly Flammable Liquids and Liquefied Petroleum Gases: Guide to the Regulations 1973.* (ISBN 0 11 360904 3) (1973).

38. Health and Safety Executive. *Guidance Note CS 2: The Storage of Highly Flammable Liquids.* (ISBN 0 11 883027 9) (Jan. 1977).

39. Health and Safety Executive. *Guidance Note CS 17: Storage of Packaged Dangerous Substances.* (ISBN 0 11 883526 2) (Jan. 1986).

40. Health and Safety Executive. Health and Safety Booklet No. HS(G)4. *Highly Flammable Liquids in the Paint Industry.* (ISBN 0 11 883 219 0) (1978).

41. Health and Safety Executive. Health and Safety Booklet No. HS(G)22. *Electrical Apparatus for Use in Potentially Explosive Atmospheres.* (ISBN 0 11 883 746 X) (1984); Health and Safety Booklet No. HS(R)15 rev. *Administrative guidance on the European Community 'Explosive Atmospheres' Directive (76/117/EEC and 79/196/EEC) and Related Directives,* 2nd edn. (ISBN 0 11 883 880 6) (1987).

42. Health and Safety Commission. *Consultative Document: Proposed Changes to Section 6 of the Health and Safety at Work, etc. Act 1974.* (ISBN 0 7176 0225 7) (1984).

43. Statutory Instrument 1978, No. 209. Packaging and Labelling of Dangerous Substances Regulations 1978. (ISBN 0 11 083209 4)

44. Statutory Instrument 1984, No. 1244. Classification, Packaging and Labelling of Dangerous Substances Regulations 1984. (ISBN 011 047244 6)

45. Statutory Instrument 1986, No. 1922. Classification, Packaging and Labelling of Dangerous Substances (Amendment) Regulations 1986. (ISBN 0 11 067922 9)

46. Health and Safety Commission: *Authorised and Approved List: Information Approved for the Classification, Packaging and labelling of Dangerous Substances for Supply and Conveyance by Road.* (ISBN 0 11 883 712 5) (1984).

47. Health and Safety Commission. *Revision No. 1 to the Approved List (Information Approved for the Classification, Packaging and Labelling of Dangerous Substances).* (ISBN 0 11 883 888 1) (1986).

48. Health and Safety Executive. Health and Safety Series Booklet HS(R)22. *A Guide to the Classification, Packaging and Labelling of Dangerous Substances Regulations 1984.* (ISBN 0 11 883 794 X) (1985).

49. Health and Safety Commission. *Approved Code of Practice: Classification and Labelling of Substances Dangerous for Supply and/or Conveyance by Road.* (ISBN 0 11 883 773 7) (1984).

50. Health and Safety Commission. *Packaging of Dangerous Substances for Conveyace by Road.(rev. 1): Classification, Packaging and Labelling of Dangerous Substances Regulations 1984 Approved Code of Practice.* (ISBN 0 11 883 904 7) (1987).

51. European Community. Proposal for a Council Directive: 85/C211/03 Official Journal of the European Commuities, 22 August 1985, pp. 3–15.
52. Health and Safety Executive. *Health and Safety Booklet HS(G)27: Substances for Use at Work: The Provision of Information* (ISBN 0 11 883 844 X) (1985).
53. Health and Safety Commission, Printing Industry Advisory Committee. *Chemicals in the Printing Industry: The Provision of Health and Safety Information by Manufacturers, Importers and Suppliers of Chemical Products to the Printing Industry.* (ISBN 0 11 883852 0) (1986).
54. For example: Health and Safety Executive. *Guidance Note EH16: Isocyanates Toxic Hazards and Precautions.* (ISBN 011 883 581 5); *Guidance Note EH22: Ventilation of Buildings* (ISBN 011 883 190 G) (1979).; *Guidance Note EH26: Occupational Skin Diseases* (ISBN 011 833 374 X) (1981).
55. *For example:*
    Health and Safety Executive. *Toxicity Review No. 2: Formaldehyde.* (ISBN 0 11 883 452 5) (1981); *Toxicity Review No. 14: Review of the Toxicity of the Esters of o-phthalic Acid (Phthalate Esters).* (ISBN 0 11 883 859 8) (1986).
56. Department of Transport. *European agreement concerning the International Carriage of Dangerous Goods by Road.* ADR 1985 edn. (ISBN 0 11 550735 3) (1986).
57. Chemical Industries Association. CEFIC Tremcards, Reference edn (1979); CEFIC Tremcards, Reference edn Supplement (1982).
58. CEFIC. *CEFIC Transport Emergency Cards, Group Text.* 1st edn. (1980).
59. Chemical Industries Association. *Tremcard numbers and ADR Appendix B5,* A Guide to Selecting CEFIC Tremcards.
60. Statutory Instrument 1981. No. 1059. The Dangerous Substances (Conveyance by Road Tankers and Tank Containers) Regulations 1981. (ISBN 011 017059 8)
61. Statutory Instrument 1984, No. 1244. The Classification, Packaging and Labelling of Dangerous Substances Regs 1984. (ISBN 011 047244 6)
62. Statutory Instrument 1978, No. 209. The Packaging and Labelling of Dangerous Substances Regulations 1978. (ISBN 011 083209 4)
63. Statutory Instrument 1986, No. 1951. The Road Traffic (Carriage of Dangerous Substances in Packages etc) Regulations 1986. (ISBN 0 11 067 951 2); Health and Safety Commission. *Approved Code of Practice: Operational Provisions of the Road Traffic (Conveyance of Dangerous Substances in Packages, etc.) Regulations 1986.* (ISBN 0 11 883 898 9) (1987).
    Health and Safety Executive: *Health and Safety Booklet HS(R)24: A Guide to the Road Traffic (Dangerous Substances in Packages, etc.) Regulations 1986.* (ISBN 011 883 899 7). (1987).
64. Health and Safety Commission. *Consultative Document: Proposals for Dangerous Substances (Conveyance by Road in Packages etc) Regulations 198–.* (ISBN 071 760181 1) (1984).
65. United Nations. *Transport of Dangerous Goods: Recommendations of the Committee of Experts on the Transport of Dangerous Goods,* 4th rev. edn. (ISBN 921 139022 2) (1984).
66. International Maritime Organisation. *International Maritime Goods Code; Medical First Aid Guide for Use in Accidents Involving Dangerous Goods* (1985); *Emergency Procedures for Ships Carrying Dangerous Goods.* (1985).
67. International Air Transport Association. *Dangerous Goods Regulations,* 28th edn. (ISBN 92 9035 072 5) (1987).
68. Health and Safety Commission. *Consultative Document: Control of Substances Hazardous to Health: Draft Regulations and Draft Approved Codes of Practice.* (ISBN 0 7176 0215 X) (1984).
69. EEC Council Directive: 80/1107/EEC: On the Protection of Workers from

the Risk Related to Exposure to Chemical, Physical and Biological Agents at Work.

70. EEC Council Proposal: 86/C164/04: On the Protection of Workers from the Risks Related to Exposure to Chemical, Physical and Biological Agents at Work (COM(86) 296 final).
71. Statutory Instrument 1984, No. 1902. The Control of Industrial Major Accident Hazard Regulations 1984. (ISBN 0 11 047 902 5)
72. Health and Safety Executive. *Health and Safety Series Booklet No. HS(R) 21: A Guide to the Control of Industrial Major Accident Hazards Regulations 1984.* (ISBN 0 11 883 767 2) (1985).
73. Health and Safety Executive. *Health and Safety Booklet No. HS(G)25: The Control of Industrial Major Accident Hazards Regulations 1984. (CIMAH): Further Guidance on Emergency Plans.* (ISBN 0 11 883 831 8) (1985).
74. Statutory Instrument 1982, No. 1357. Notification of Installations Handling Hazardous Substances Regulations 1982. (ISBN 0 11 027 357 5)
75. Health and Safety Commission. *Consultative Document: Dangerous Substances (Notification and Marking of Sites) Regulations and Guidance Note.* (ISBN 0 11 883 486 X) (1985).
76. Chemical Industries Association. *Hazard Identification. A Voluntary Scheme for the Marking of Tank Vehicles Carrying Dangerous Substances by Road and Rail,* Rev. edn. (1976); Chemical Industries Association. *Hazchem Codings,* 2nd edn. (1978).
77. Statutory Instrument 1981, No. 917. The Health and Safety (First Aid) Regulations 1981. (ISBN 011 016917 4).
78. Statutory Instrument 1985, No. 2023. The Reporting of Injuries Diseases and Dangerous Occurrences Regulations 1985. (ISBN 0 11 058023 0)
79. Health and Safety Executive. *Health and Safety series booklet, HS(R)23: A Guide to the Reporting of Injuries and Dangerous Occurrences Regulations 1985 (RIDDOR).* (ISBN 011 883858 X) (1986).
80. Health and Safety Executive. *HSE 11 (Rev): Reporting an Injury or a Dangerous Occurrence.* (1986). (Free)
81. Health and Safety Executive. *HSE 17: Reporting a Case of Disease.* (1986). (Free)
82. Department of Employment. *Code of Practice: Reducing the Exposure of Employed Persons to Noise.* (ISBN 0 11 880340 9) (1972).
83. EEC Council Directive: 86/118/EEC: On the Protection of Workers from the Risks Related to Exposure to Noise at Work 1986.
84. British Standards Institution. *Draft for Development: DD 80. Laboratory Fume Cupboards: Part 1 1982 Safety Requirements and Performance Testing; Part 2 1982 Recommendations for Information to be Exchanged between Purchaser, Vendor and Installer and Recommendations for Installation; Part 3 1982 Recommendations for Selection, Use and Maintenance.*
85. Statutory Instrument 1977, No. 500. The Safety Representatives and Safety Committees Regulations. (ISBN 011 070500 9)
86. Health and Safety Commission. *Safety Representatives and Safety Committees.* (ISBN 0 11 880335 2) (1977)
87. Health and Safety Commission. *HSC 9: Time Off for the Training of Safety Representatives.* (1978). (Free)
88. Health and Safety Commission. *HSC6 (revised): Writing a Safety Policy Statement: Advice to employers.* (1985). (Free)
89. Health and Safety Executive. *Effective Policies for Health and Safety. A Review Drawn from Previous Work and Experience of the Accident Prevention Unit of HM Factory Inspectorate.* (ISBN 0 11 883254 9) (1980).
90. Health and Safety Executive. *Health and Safety Policy Statement.* (ISBN 0 11

883 882 2) (1986).

91. Health and Safety Commission: Printing Industry Advisory Committee. *Health and Safety for Small Firms in the Printing Industry.* (ISBN 0 11 883 851 2) (1986).

92. Society of British Printing Ink Manufacturers. *Guide to Materials and Substances for Exclusion from Printing Inks and Varnishes.* (February 1986).

93. Statutory Instrument 1980, No. 1838. The Emulsifiers and Stabilisers in Food Regulations 1980. (ISBN 011 007838 1)

94. Statutory Instrument 1980, No. 1831. The Antioxidants in Food (Amendment) Regulations 1980. (ISBN 011 007831 4)

95. Statutory Instrument 1973, No. 1340. The Colouring Matter in Food Regulations 1973. (ISBN 011 031340 2)

96. Statutory Instrument 1967, No. 1582. The Solvents in Food Regulations 1967.

97. Statutory Instrument 1978, No. 1927. The Materials and Articles in Contact with Food Regulations 1978. (ISBN 011 084927 2)

98. Editorial, *American Ink Maker* **58**(9), 21 (September 1980).

99. Code of Federal Regulations Title 21. Food and Drug Administration (published annually).

100. Statutory Instrument 1974, No. 1367. The Toys (Safety) Regulations 1974. (ISBN 0 11 041 367 9)

101. Statutory Instrument 1974, No. 226. The Pencils and Graphic Instruments (Safety) Regulations 1974. (ISBN 0 11 040 226 X)

102. Health and Safety Commission; Printing Industry Advisory Committee. *Safety in the Use of Inks, Varnishes and Lacquers Cured by Ultra-violet Light.* (ISBN 0 11 883678 1) (1983).

103. O'Neill, Leonard A. *Health and Safety, Environmental Pollution and the Paint Industry,* 2nd edn. Paint Research Association, (1981).

104. O'Neill, Leonard A. *Health and Safety, Environmental Pollution and the Paint Industry,* Supplement to the 2nd edn of 1981. Paint Research Association (1986).

105. O'Neill, Leonard A. *Survey of Hazards, Pollution and Legislation in the Coatings Field.* Paint Research Association (Quarterly Bulleton No. 1 May 1986, No. 2 August 1986, No. 3 November 1986, No. 4 February 1987).

106. Toxic Substance Control Act 1976.

107. US Environmental Protection Agency; Office of Toxic Substances; *Toxic Substances Control Act (TSCA) Chemical Substances Inventory,* vols I–V. (January 1986).

108. Department of Home Affairs and the Environment. *Australian Core Inventory of Chemical Substances (ACOIN).* Canberra (1984).

109. United Nations Environmental Program. *IRPTC Data Profile Series No. 4: Interional Register of Potentially Toxic Chemicals Legal File,* vols I & II. (1983).

110. International Labour Office. Occupational Exposure Limits for Airborne Toxic Substances, 2nd edn. (1980).

111. US Department of Health and Human Services, National Institute for Occupational Safety and Health. Registry of Toxic Effects of Chemical Substances.

112. (Japanese) Ministry of Labour. Industrial Safety and Hygiene Law, 8 June 1972; Regulation for Prevention of Injury by Specified Chemical Substances, 30 September 1972; Regulation for Prevention of Injury by Specified Powders and Dusts 25 April 1979.

113. (Japanese) Ministry of Labour. Regulation for Prevention of Poisoning by Organic Solvents, 30 September 1972.

114. (Japanese) Ministry of International Trade and Industry. Chemical Sub-

stances Inspection and Control Law, 16th October 1973.
115. *Handbook of Existing and New Chemical Substances*, 2nd edn. The Chemical Daily Co., Tokyo (March 1987).
116. (US) 29 Code of Federal Regulations (CFR) 1910.1200 November 1985.
117. National Association of Printing Ink Manufacturers. *HMIS: Guidelines for a Hazardous Materials Identification System for Raw Materials.* (December 1984).
118. United Nations: Economic Commission for Europe. *Regulations and Legislation on Food Additives and Chemicals for Food Packaging. ECE/CHEM/54.* (1984).
119. CEPE *Criteria for Evaluation of Coatings in Direct Contact with Food A Summary of Existing Legislation and Practice in Europe Coatings for Metal Substates.* (1985).
120. Nederlandse Vereniging–Federatie voor Kunststoffen. Translation of the fifth version of the draft packaging and food-utensils regulations (food laws). (1975).
121. *Kunststoffe im Lebensmittelverkehr.* Carl Heymans Verlag (1984).
122. European Community: Proposal for a Council Directive on the Approximation of the laws of the member states concerning the safety of Toys COM(86) 541 final. *Official Journal of the European Communities* C282 pp. 4–13 (8 November 1986).
123. H. Mostyn Packaging. *Dangerous Goods for Transport: General Guide; Guide to Class 3; Guide to Class 6.1; Guide to Class 6.2; Guide to Classes 8 and 9.* Aurigny Ltd (1986).
124. PIRA Packaging Division: *Light gauge metal packagings for the transport of dangerous goods.* Summary report of project 52/SP/28. PK/DE/D23. March 1987. (A full report was issued at the same time under the reference PK/DE/D24).

Index

Illustrations are indicated by reference to a Figure in brackets.

The word *bis* following a page reference indicates two separate references to the subject on that page. The word *passim* means that the references are scattered throughout the pages indicated.

Page numbers in **bold** indicate that more than a few lines are devoted to the subject in the text.

Proprietary products are indicated by the symbol * immediately following a name. The absence of such a symbol, however, does not mean that proprietary rights may not exist in a particular name.